Springer Handbook of Auditory Research

For further volumes:
http://www.springer.com/series/2506

Arthur N. Popper • Richard R. Fay
Editors

Perspectives on Auditory Research

With 82 Illustrations

Springer

Editors
Arthur N. Popper
Department of Biology
University of Maryland
College Park, MD, USA

Richard R. Fay
Marine Biological Laboratory
Woods Hole, MA, USA

ISSN 0947-2657 ISSN 2197-1897 (electronic)
ISBN 978-1-4614-9101-9 ISBN 978-1-4614-9102-6 (eBook)
DOI 10.1007/978-1-4614-9102-6
Springer New York Heidelberg Dordrecht London

Library of Congress Control Number: 2014932275

Printed on acid-free paper

Springer is part of Springer Science+Business Media (www.springer.com)

Series Preface

The following preface is the one that we published in Volume 1 of the Springer Handbook of Auditory Research back in 1992. As anyone reading the original preface, or the many users of the series, will note, we have far exceeded our original expectation of eight volumes. Indeed, with books published to date, and those in the pipeline, we are now set for more than 60 volumes in SHAR, and we are still open to new and exciting ideas for additional books.

We are very proud that there seems to be consensus, at least among our friends and colleagues, that SHAR has become an important and influential part of the auditory literature. While we have worked hard to develop and maintain the quality and value of SHAR, the real value of the books is very much because of the numerous authors who have given their time to write outstanding chapters and to our many coeditors who have provided the intellectual leadership to the individual volumes. We have worked with a remarkable and wonderful group of people, many of whom have become great personal friends of both of us. We also continue to work with a spectacular group of editors at Springer, currently Melissa Higgs and formerly Ann Avouris. Indeed, several of our past editors have moved on in the publishing world to become senior executives. To our delight, this includes the current president of Springer US, Dr. William Curtis.

But the truth is that the series would and could not be possible without the support of our families, and we dedicate all of the SHAR books, past and future, to them. Our wives, Catherine Fay and Helen Popper, and our children, Michelle Popper Levit, Melissa Popper Levinsohn, Christian Fay, and Amanda Fay, have been immensely patient as we developed and worked on this series. We thank them and state, without doubt, that this series could not have happened without them. We also dedicate the future of SHAR to our next generation of (potential) auditory researchers—our grandchildren—Ethan and Sophie Levinsohn; Emma Levit; and Nathaniel, Evan, and Stella Fay.

Preface 1992

The Springer Handbook of Auditory Research presents a series of comprehensive and synthetic reviews of the fundamental topics in modern auditory research. The volumes are aimed at all individuals with interests in hearing research including advanced graduate students, postdoctoral researchers, and clinical investigators. The volumes are intended to introduce new investigators to important aspects of hearing science and to help established investigators to better understand the fundamental theories and data in fields of hearing that they may not normally follow closely.

Each volume presents a particular topic comprehensively, and each serves as a synthetic overview and guide to the literature. As such, the chapters present neither exhaustive data reviews nor original research that has not yet appeared in peer-reviewed journals. The volumes focus on topics that have developed a solid data and conceptual foundation rather than on those for which a literature is only beginning to develop. New research areas will be covered on a timely basis in the series as they begin to mature.

Each volume in the series consists of a few substantial chapters on a particular topic. In some cases, the topics will be ones of traditional interest for which there is a substantial body of data and theory, such as auditory neuroanatomy (Vol. 1) and neurophysiology (Vol. 2). Other volumes in the series deal with topics that have begun to mature more recently, such as development, plasticity, and computational models of neural processing. In many cases, the series editors are joined by a coeditor having special expertise in the topic of the volume.

College Park, MD, USA — Arthur N. Popper
Woods Hole, MA, USA — Richard R. Fay

SHAR logo by Mark B. Weinberg, Bethesda, Maryland, used with permission.

Contents

Contributors

Leslie R. Bernstein Departments of Neuroscience and Surgery (Otolaryngology), University of Connecticut Health Center, Farmington, CT, USA

William E. Brownell Bobby R. Alford Department of Otolaryngology/Head and Neck Surgery, and the Departments of Neuroscience, Molecular Physiology and Biophysics, and the Structural and Computational Biology & Molecular Biophysics Program, Baylor College of Medicine, Houston, TX, USA

John F. Brugge Department of Neuroscience, University of Wisconsin-Madison, Madison, WI, USA; Department of Neurosurgery, University of Iowa, Iowa City, IA, USA

Nell Beatty Cant Department of Neurobiology, Duke University Medical School, Durham, NC, USA

Jeffrey T. Cheng Department of Otology and Laryngology, Harvard Medical School, Boston, MA, USA; Massachusetts Eye and Ear Infirmary, Boston, MA, USA

M.-Dominique Crapon de Caprona Department of Biology, College of Liberal Arts, University of Iowa, Iowa City, IA, USA

Ruth Anne Eatock Department of Neurobiology, University of Chicago, Chicago, IL, USA

Jos J. Eggermont Department of Physiology and Pharmacology, University of Calgary, Calgary, AB, Canada; Department of Psychology, University of Calgary, Calgary, AB, Canada

Richard R. Fay Marine Biological Laboratory, Woods Hole, MA, USA

Mary Florentine Bouvé College of Health Science and College of Engineering, Northeastern University, SPLA Department, Boston, MA, USA; Department of Electrical and Computer Engineering, Institute for Hearing, Speech and Language, Northeastern University, Boston, MA, USA

Thomas B. Friedman Porter Neuroscience Research Center, Bethesda, MD, USA

Bernd Fritzsch Department of Biology, College of Liberal Arts, University of Iowa, Iowa City, IA, USA

Jay M. Goldberg Department of Pharmacological & Physiological Sciences, University of Chicago, Chicago, IL, USA

Sandra Gordon-Salant Department of Hearing and Speech Sciences, University of Maryland, College Park, MD, USA

John J. Guinan Jr. Eaton Peabody Lab, Massachusetts Eye and Ear Infirmary, Boston, MA, USA; Department of Otology and Laryngology, Harvard Medical School, Boston, MA, USA

Anthony D. Hawkins Loughine, Aberdeen, UK

Henry E. Heffner Department of Psychology, University of Toledo, Toledo, OH, USA

Rickye S. Heffner Department of Psychology, University of Toledo, Toledo, OH, USA

Ronald R. Hoy Department of Neurobiology & Behavior, Cornell University, Ithaca, NY, USA

Toshio Irino Faculty of Systems Engineering, Wakayama University, Wakayama, Japan

Nina Kraus Auditory Neuroscience Lab, Departments of Communication Sciences, Neurobiology & Physiology, Otolaryngology, Northwestern University, Evanston, IL, USA

Ole Næsbye Larsen Department of Biology, University of Southern Denmark, Denmark

Paul B. Manis Departments of Otolaryngology/Head and Neck Surgery, Cell Biology and Physiology, University of North Carolina, Chapel Hill, NC, USA

Geoffrey A. Manley Fakultät V, Research Centre for Neurosensory Science and Institute for Biology and Environmental Sciences, Carl von Ossietzky Universität, Oldenburg, Germany

Axel Michelsen Department of Biology, University of Southern Denmark, Denmark

John C. Middlebrooks Departments of Otolaryngology, Neurobiology & Behavior, Cognitive Sciences, and Biomedical Engineering, University of California at Irvine, Irvine, CA, USA

Brian C.J. Moore Department of Experimental Psychology, University of Cambridge, Cambridge, UK

Hideko H. Nakajima Department of Otology and Laryngology, Harvard Medical School, Boston, MA, USA; Massachusetts Eye and Ear Infirmary, Boston, MA, USA

Trent Nicol Department of Communication Sciences, Auditory Neuroscience Lab, Northwestern University, Evanston, IL, USA

Alan R. Palmer Medical Research Council Institute of Hearing Research, Nottingham, UK

Roy D. Patterson Centre for the Neural Basis of Hearing, PDN, University of Cambridge, Cambridge, UK

George D. Pollak Section of Neurobiology and Center for Perceptual Systems, The University of Texas at Austin, Austin, TX, USA

Arthur N. Popper Department of Biology and Center for Comparative and Evolutionary Biology of Hearing, University of Maryland, College Park, MD, USA

Sheikh Riazuddin Centre of Excellence in Molecular Biology, University of the Punjab, Lahore, Pakistan; Allama Iqbal Medical Research Centre, Jinnah Hospital, Lahore, Pakistan

John J. Rosowski Department of Otology and Laryngology, Harvard Medical School, Boston, MA, USA; Massachusetts Eye and Ear Infirmary, Boston, MA, USA

David K. Ryugo Garvan Institute of Medical Research, Sydney, NSW, Australia

Jochen Schacht Kresge Hearing Research Institute, The University of Michigan, Ann Arbor, MI, USA

Robert V. Shannon Department of Otolaryngology, University of Southern California, Los Angeles, CA, USA

Karen P. Steel Wolfson Centre for Age-Related Diseases, King's College London, London, UK

Constantine Trahiotis Departments of Neuroscience and Surgery (Otolaryngology), University of Connecticut Health Center, Farmington, CT, USA

Neal F. Viemeister Department of Psychology, University of Minnesota, Minneapolis, MN, USA

William A. Yost Speech and Hearing Science, Arizona State University, Tempe, AZ, USA

Chapter 1
A Brief History of SHAR

Richard R. Fay and Arthur N. Popper

Richard R. Fay

Arthur N. Popper

R.R. Fay
Marine Biological Laboratory, Woods Hole, MA 02543, USA
e-mail: rfay44@aol.com

A.N. Popper (✉)
Department of Biology, Center for Comparative and Evolutionary,
Biology of Hearing, University of Maryland, College Park, MD 20742, USA
e-mail: apopper@umd.edu

A.N. Popper and R.R. Fay (eds.), *Perspectives on Auditory Research*, Springer Handbook of Auditory Research 50, DOI 10.1007/978-1-4614-9102-6_1,

1.1 SHAR Background

This volume is the 50th in the 22-year history of the Springer Handbook of Auditory Research (SHAR). The first volume, *The Mammalian Auditory Pathway: Neuroanatomy*, was published in 1992 and was edited by Douglas Webster and the two of us. Volumes 2 to 49 and the volumes that will follow number 50 are on specific topics in hearing and related areas. Appendix 1 to this volume provides the complete Tables of Contents for SHAR Vols. 1–49.

When we originally conceived of SHAR, we had the expectation of publishing perhaps eight volumes on the fundamental and more mature topics of auditory neuroscience. Our goal was to have volumes geared not toward experts in the given field but to beginning investigators, students, and younger faculty. We wanted the chapters and volumes to be synthetic, not exhaustive, reviews of the book topic.

In short, our vision in developing SHAR was to direct the series at persons like us (the founding editors). Both of us have had very strong interests in auditory science, but our backgrounds in fish bioacoustics place us at the periphery of the entire field and, in fact, in an ideal position to benefit from the SHAR subject matter presented in the way we originally envisioned. Thus, back in 1992 we asked authors, and continue to ask them today, to view us as their audience and help teach us, as well as new people entering the field, the major *concepts* of the authors' disciplines. In other words, our focus in 1992, today, and in the future is to have chapters that present the most expert conceptual overview of a particular field. Thus, SHAR aims to provide the overview that students need to enter a field and that others, like the two of us, could use to learn about and give a few lectures on a particular topic.

Moreover, our philosophy for SHAR is to have each volume focus on a particular topic related to auditory neuroscience, although we have gone somewhat far afield at times. For example, we have included titles such as *Electroreception* (Vol. 21, Bullock, Hopkins, Popper, & Fay, 2005), *The Lateral Line* (Vol. 48, Coombs, Bleckmann, Fay, & Popper, 2013), and *The Vestibular System* (Vol. 19, Highstein, Fay, & Popper, 2004).

In deciding on books, we generally prefer topics that are reasonably "mature" so that the chapters will provide a background into an enduring topic and have immediacy that is greater than in a regular review or research paper. The chapters for each volume arise from our request to senior editors to come up with the book topic and the chapters he or she wants to include before even thinking about authors. The reason is that we want the volumes to cover a particular topic and not choose chapters based on the interests of the authors the editors think should be in such a volume. Thus, the volumes are idea rather than author driven.

1.2 Volume 50

In contrast to our past 49 volumes, we have asked senior colleagues to provide essays for Volume 50 that focus on their contributions to auditory neuroscience in the past, on their views of the current state of the field, and on their thoughts on the future of their field, including the outstanding questions that are still unanswered. We also asked authors to write in the first person, and provide, if they wished, "autobiographical" information. Our goal was to have senior scholars think about their disciplines and even their careers and write whatever they wanted. Our intent was to have a highly diverse series of chapters, all of which might be different in style and approach and that would be interesting and "fun" for the reader. In fact, we suggest that readers browse the book for the "fun of it" rather than look for specific topics.

The various authors approached their writing opportunities rather differently. All authors were encouraged to include personal information on their careers in science but were not required to do so. We saw this as an opportunity for authors to go in whatever direction they wanted. Some essays are a mini memoir of research careers combined with personal experiences in research with scientific and theoretical content (e.g., Chapter 3 by John Brugge and Chapter 22 by Alan Palmer Dick Art), while others wrote essays on a rather specific scientific topic or research program without giving any personal information (e.g., Chapter 23 by Roy Patterson and Toshio Irino and Chapter 31 by Tino Trahoitis and Les Bernstein).

In most cases, the author's essay was on one to several areas of research that took place over his or her long career, and our impression is that they got a chance to explain the thread(s) in a way that would be impossible in a typical report appearing in a research journal. These essays are thus unique in scope and content and therefore in providing senior investigators the kind of writing opportunities that they normally would not have.

We should add that we learned something very interesting in asking people to write chapters. We proposed that authors model their essays on our mutually favorite American literary magazine, *The New Yorker*. Articles in *The New Yorker* are always very well written and always tell a story, although the style of writing differs for each author. What we learned is that writing essays comes very easily to some scientists and not for others.

Of course, people will ask why some people are included in the volume and others not. We did solicit chapters from perhaps twice as many authors as in this volume, with a focus on "senior" (we will not define that age, but in the United States, we consider it to be people who are reasonably close to, or who are already receiving, Social Security) and on people who have been SHAR authors in the past. Regretfully, some people who we really wanted to write chapters turned us down for various reasons (e.g., fully retired, too busy with other activities). But we are delighted that so many of our friends and colleagues agreed to be part of this book.

As a final note on the organization of this volume, we originally conceived of the book as having sections on different broad-topic areas. However, as we received the manuscripts, we realized that our original organizational scheme fell apart and that there was no one logical way to sequence the book. Indeed, we concluded that no matter how we organized chapters, there would always be another equally logical way to do it. Thus, we decided that the "most logical" approach would be to sequence the chapters alphabetically by the first author and let readers either browse (our preferred approach to this volume) or develop their own scheme for sequencing chapters. Indeed, these days, with most SHAR volumes being primarily downloaded from the Web, it would be easy for each reader to create his or her own personal Table of Contents.

1.3 A Brief Overview of SHAR

When we first came up with the idea for SHAR, although we considered the topics to cover in individual volumes, there was no overall vision for the scope of the series itself and the direction the volumes should take. However, as we moved forward and added volumes, it became clear that SHAR started to encompass several "themes."

For example, one clear theme is comparative auditory neuroscience, which in retrospect is not surprising because we are, as discussed later, "card-carrying" comparative biologists. A second theme is the basic science of auditory neuroscience and how the auditory system works. Indeed, our first several volumes clearly were on the basic science of the auditory system, and we keep coming back to this very important foundation.

However, particularly in the past few years, we have added a clear third theme, human-oriented translational auditory neuroscience. Thus, a growing number of volumes in SHAR focus on specific human-related issues such as development of hearing in young humans, perception of music, and hearing in aging humans (a particular favorite of ours as we get a bit older). Finally, we see a fourth theme of applied topics that focus primarily on developing clinical issues; for example, cochlear implants, the middle ear, and a forthcoming volume on hearing aids fit this niche, and we expect it to grow.

It also becomes apparent that the SHAR volumes support one another. Indeed, at the end of the each volume preface, we try to show how chapters in that volume are supported by chapters in earlier volumes. And we find this interaction between chapters to be growing. Indeed, at some point, it would be interesting to look at all SHAR chapters irrespective of volume and see how they can be reshaped into new volumes on different topics. Perhaps we will leave that as an exercise for our successors or our readers using the Appendix of this volume as a source of all chapter information.

1.4 Our History

We first met in Hawai'i (then spelled Hawaii) in 1971 as young investigators, Popper as an assistant professor in the Department of Zoology at the University of Hawaii and Fay as a postdoc with Georg von Békésy at the Laboratory of Sensory Sciences at the University of Hawaii. We met the day after Christmas when Fay and his family joined Popper and his wife for a BBQ. We "hit it off" instantly, and as a forecast of how we would work ever since then, we immediately started to come up with ideas for research and various other projects.

Indeed, our shared interest in research focusing on the hearing of fishes led to a lifelong collaboration in writing, editing, and research. We are quite convinced that our differing backgrounds, Popper in biology from the laboratory of William N. Tavolga and Fay in psychology from the laboratory of Ernst G. Wever, and yet with our very similar interests in hearing by fishes, were an ideal match for the field of auditory neuroscience that combines biology and psychoacoustics, as the contents of this volume demonstrate well.

We came to realize that as SHAR grew over the last 22 years that our backgrounds were perfect, in a way, to shepherd SHAR to many more volumes in auditory research because our interests were general and broad, but our specific research areas would not interfere with our vision for SHAR as a set of reviews designed for relative generalists like us. Indeed, we have often wondered whether other editors could have wound up with a series like SHAR. We view our broad interests and deep passion for comparative issues to have greatly enhanced the breadth of SHAR and its value to the auditory neuroscience community.

1.5 Some SHAR Statistics

It is very clear to us, based primarily on conversations with colleagues and perusal of journal and book articles, that SHAR has had a major impact on auditory neuroscience, and we are very pleased that we could have conceived and executed an idea with such an impact. Our colleagues tell us that they have multiple volumes of the "green books" (the color of SHAR covers) on their shelves and that SHAR is a focal part of their libraries. We also see that SHAR volumes are cited often in research papers and reviews, suggesting to us that they are used widely. What it would be nice to know is how often individual SHAR chapters are cited; doing an ISI or Google Scholar search is not productive because authors don't cite chapters as being from SHAR but cite by the title of the individual volume.

But now that SHAR (and most other books) are available for chapter download from the Web, we do have a few statistics that perhaps give a sense of the use of SHAR. For example, between 2007 and 2012, there were 78,227 downloads of SHAR chapters (and note that not all SHAR volumes are yet available digitally) (Fig. 1.1). And for those who are curious, the most downloaded volume is *Auditory Perception of Sound Sources* (Vol. 29, Yost, Popper, & Fay, 2008), with 6,398 chapter downloads from 2008 to 2012. A close second, with 5,476 downloads, is *Electroreception* (Vol. 21, Bullock, Hopkins, Fay, & Popper, 2005). Although we don't have really great records, our sense is that up to now the "best seller" in print has been *The Cochlea* (Vol. 8, Dallos, Popper, & Fay, 1996).

As indicated, this volume is number 50. Over the course of the 49 preceding volumes, we have worked with 54 coeditors. Of these, Geoffrey Manley and Fan-Gang Zeng have each edited three volumes and Edwin W Rubel and William A. Yost have edited two volumes. All five of these individuals have been particularly great to work with, and we keep encouraging them to suggest new volumes. (And, we are always open to new ideas from other past editors and potential new editors.)

Over the course of 49 volumes, we have had 473 chapters and contributions from about 675 authors. Geoff Manley has contributed to the most chapters (11). Fay also has 11 chapters and Popper has 9. We estimate that about 150 authors have contributed 2 or more chapters.

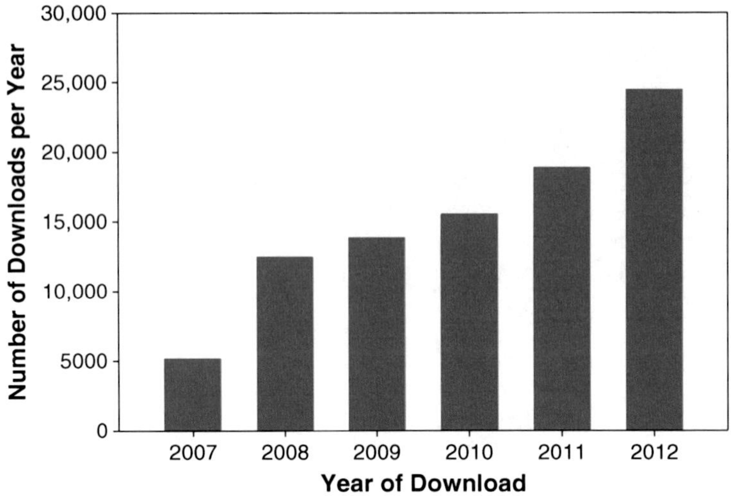

Fig. 1.1 SHAR downloads per year

1.6 Working with Springer

The two of us published our first non-SHAR volume with Springer in 1980 (*Comparative Studies of Hearing in Vertebrates*, Popper & Fay). This arose from a chance meeting at the annual Society for Neuroscience conference between Popper and Dr. Mark Licker, then a senior editor at Springer and now retired VP and publisher for science at McGraw-Hill. This volume was followed by several others devoted to particular meetings, including *Hearing and Sound Communication in Fishes* (Tavolga, Popper, & Fay, 1981), *Sensory Biology of Aquatic Animals* (Atema, Fay, Popper, & Tavolga, 1988), and *Evolutionary Biology of Hearing* (Webster, Fay, & Popper, 1992).

But then, around 1990, Popper was visiting Fay at his home in Falmouth, MA, and the idea for SHAR struck us while we were sitting around talking about some research project or other. The whole concept for SHAR came quickly, and we were able to interest Springer in publishing the series.

Since the inception of SHAR, we have had several Springer editors. Dr. Robert Garber was our editor for the first 15 volumes (he is now VP and publisher at John Wiley & Sons), followed by Janet Slobodien for about 15 volumes, Jasmine Ben-Zvi for 3 volumes, and our immediate past Ann Avouris for 17 volumes. We now look forward to a long and productive relationship with our new editor, Melissa Higgs. We also worked closely with Dr. William Curtis, first a senior editor at Springer and now its U.S. president. Bill has been highly supportive of SHAR throughout and we are most grateful for his encouragement and friendship.

We have been delighted with the people we work with and the way we and SHAR have been treated. Springer has gone out of its way to ensure that SHAR continues to be successful, and the quality of the organization and the people is exceptional. We also want to make note of a very special person, Terry Kornak. Terry started out as our production editor at Springer. She is now a freelance copy-editor and we are fortunate that she is "permanently assigned" to edit all SHAR volumes. Terry has been a joy to work with, and her caring about SHAR as her "baby" is deeply appreciated by both of us and by every author who has had her "eagle eye" reviewing their manuscripts.

1.7 The Future

But what of the future? Several years ago, we mentioned to Ann Avouris that we are at points in our careers where "retirement" is a distinct possibility. Ann instantly retorted that we were not allowed to retire and that SHAR had to go on. Of course, this support from our publisher was heartening for any number of reasons. But at the same time, we wanted to be realistic. So, when we met with Ann and Melissa Higgs in the fall of 2012, we proposed that we bring on two new coeditors who would ultimately take over SHAR. Springer was very supportive of this idea.

Thus, we invited Dr. Allison Coffin and Dr. Joseph Sisneros to be "editors apparent" for SHAR. Alli and Joe are exceptional young auditory neuroscientists who have stellar careers ahead of them. They join SHAR having interests similar to ours, with strong backgrounds and interests in comparative and evolutionary issues. At the same time, both are doing research in hearing that is cutting edge and so they bring fresh perspectives as to the future of the field.

But perhaps what is most important to the two of us is that we are ultimately passing the series to our academic "kids." Alli was Popper's doctoral student (co-mentored by Matt Kelley at NIH) and Joe has worked extensively with Fay. Like the two of us, Alli and Joe are good friends, have published together, and care deeply about broad issues in auditory neuroscience. We are convinced that Joe and Alli will not only maintain the quality and importance of SHAR but will also bring to it fresh approaches and a fresh understanding of our field. And they have promised that our children (none of whom are scientists) will get a copy of every future volume of SHAR.

1.8 Dedication

Starting several volumes ago, we decided that the whole series should be dedicated to our families and we want to reiterate that point here. SHAR started out in Fay's house in Falmouth, and every time we would get together, the idea for a new volume (or volumes) was hatched. Indeed, it got to the point where our older children,

Michelle Popper Levit and Christian Fay, would "conspire" to keep the two of us apart because they knew that "no good" would come from even the briefest time we were left alone. Still, despite the hassles of Michelle and Chris and with their support as well as that of their mothers (Helen Popper and Catherine Fay) and their siblings (Amanda Fay and Melissa Popper Levinsohn), we were able to continue SHAR. Our families have been immensely supportive of our work on SHAR and our more than 42 years of collaboration and friendship and that close friendship extends to our families.

So, this volume, and SHAR as a whole, are dedicated to our wives and children. And, because we are looking to the future, we also add six very special people in our world to this dedication: Ethan and Sophie Levinsohn, Emma Levit, and Nathaniel, Evan, and Stella Fay. Maybe, just maybe, one of our grandchildren will decide to become an auditory neuroscientist and take over the series from Alli and Joe when they are ready to retire. We hope so!

Chapter 2
Structures, Mechanisms, and Energetics in Temporal Processing

William E. Brownell and Paul B. Manis

Paul B. Manis and William E. Brownell

W.E. Brownell (✉)
Bobby R. Alford Department of Otolaryngology/Head and Neck Surgery,
and the Departments of Neuroscience, Molecular Physiology and Biophysics,
and the Structural and Computational Biology & Molecular Biophysics Program,
Baylor College of Medicine, Houston, TX 77030, USA
e-mail: brownell@bcm.edu

P.B. Manis
Departments of Otolaryngology/Head and Neck Surgery, Cell Biology and Physiology,
University of North Carolina, Chapel Hill, NC 27599, USA
e-mail: pmanis@med.unc.edu

A.N. Popper and R.R. Fay (eds.), *Perspectives on Auditory Research*, Springer Handbook of Auditory Research 50, DOI 10.1007/978-1-4614-9102-6_2,

2.1 Introduction

Rapid temporal processing of environmental events is crucial to the survival of animals, and the auditory system provides the fastest detection time in higher vertebrates. The ability of mammals to discriminate which ear is the first to receive acoustic stimuli with a resolution of approximately 10 μs is remarkable. It requires precise detection of the acoustic signal by the inner ear and equally precise communication and refinement of the information through at least three synapses before dichotic integration in the superior olive. In addition, mammals discriminate temporal events spanning nearly five orders in frequency. Three small, light ossicles define the mammalian middle ear and function to match the impedance of airborne sound vibrations with the fluid environment of the inner ear throughout this frequency range. Additional specializations are found in the cochlea and auditory brain stem that facilitate the faithful encoding of temporal information. High-speed performance is energetically costly, as is the case with automobiles, and provides the rationale for the high metabolic activity associated with peripheral and brain stem auditory structures. Short electrical time constants are required for rapid temporal processing. Hair cells and auditory brain stem neurons possess high- conductance ionic currents that contribute to short time constants and the resulting large currents contribute to high metabolic demands. These currents are regulated by membrane proteins that have a shallow voltage dependence, resulting in a broad dynamic range that further increases metabolic demand. The Jeffress place theory for sound localization was introduced in 1948 (Jeffress, 1948) and remains a touchstone for research in the area. At its heart are assumptions about the ability of the auditory system to detect, transmit, and process temporal information. Our perspective on progress in characterizing and understanding the mechanisms found in mammals is the basis of this chapter. We find that the rapid processing of acoustic information in hair cells and brainstem auditory structures is achieved with similar mechanisms having high metabolic demands.

2.2 The Cochlea

2.2.1 Pre-1992 Active Hearing and Its Battery

Cochlear processing of acoustic signals had been investigated for well over a half century before 1992. A great deal about the history of cochlear research can be found in *The Cochlea* (Vol. 8 of the Springer Handbook of Auditory Research; Dallos et al., 1996). Thomas Gold was the first to propose an active electromechanical process in the cochlea (Gold, 1948), arguing that the performance of the ear required a piezoelectric-like feedback to overcome viscous damping. Two decades later a physiologically vulnerable enhancement of basilar membrane tuning was described (Rhode, 1971). The discovery of otoacoustic emissions (Kemp, 1978) demonstrated the cochlea could generate acoustic energy and the concept of the cochlear amplifier (Davis, 1983) gained wide acceptance. Shortly thereafter we found outer hair cells (OHCs) possessed electromechanical properties (Brownell, 1983, 1984; Brownell et al., 1985), making them a strong candidate for Gold's active process. The OHC motor mechanism was found to be (1) independent of cellular stores of ATP (Brownell & Kachar, 1986; Kachar et al., 1986); (2) able to generate movements in the kilohertz range (Ashmore & Brownell, 1986; Ashmore, 1987); (3) a function of voltage and not current (Santos-Sacchi & Dilger, 1988); and (4) able to be reversibly blocked by salicylate (Shehata et al., 1991). These functional properties were matched by distinct structural features including the fact that the OHC is a hydrostat (Brownell, 1990) with a trilaminate lateral wall. The unique subsurface cisterna is the innermost of the three layers and it is disrupted by salicylate (Dieler et al., 1991) and acoustic overstimulation (Evans, 1990) resulting in reduced turgor and electromotility. Several reviews on OHC structure and function (Brownell, 2006; Ashmore, 2008; Hallworth & Jensen-Smith, 2008) provide additional pre- and post-1992 information on their role in temporal processing.

Cellular electrochemical processes generate potential energy that is used to perform work. Neuron, muscle, and sensory cells are familiar examples of cells that utilize the energy represented by potential differences to drive ionic currents. With the exception of vertebrate inner ears the generation of potential energy is typically done by the same cells that use it. Most of the energy required for OHC mechanoelectrical and electromechanical transduction is provided by the stria vascularis. The division of labor in the cochlea is unique because the stria vascularis is an organ that is spatially separated from the hair cells in the organ of Corti. In addition to elevated potassium concentrations in the scala media, the stria vascularis creates a positive potential within the endolymph relative to the perilymph and increases the electrochemical gradient that drives a constant flow of K^+ ions from the endolymph through the hair cells. Von Békésy was the first to describe a positive endocochlear potential and negative potentials in the organ of Corti (von Békésy, 1952). Davis and his colleagues appreciated the novel arrangement of potentials and fluid compositions in the cochlear compartments and proposed the stria vascularis acted as a "battery" providing the energy source for cochlear microphonics (Davis et al., 1952). Implicit in their model was the presence of a standing current generated by the stria vascularis that exits scala media through the mechanically dependent variable resistance associated with hair cell stereocilia and flows radially across scala tympani, through the spiral ligament and back to the stria vascularis, completing a local circuit.

We made the first direct measures of AC current in this circuit finding that approximately 500 pA passes through individual OHCs in response to acoustic stimulation (Brownell et al., 1983). The measures were made using a current density analysis initially developed to examine the synaptic organization in the dorsal cochlear nucleus (DCN; Manis & Brownell, 1983). Struck by the functional similarity between the retinal dark current (Hagins et al., 1970; Baylor et al., 1984; Yan & Matthews, 1992) and the cochlear standing current, we called it the silent current (Brownell, 1982). DC measures of the silent current were made after the location of the ground electrode was changed and the preparation stabilized (Brownell et al., 1986; Zidanic & Brownell, 1989; Zidanic & Brownell, 1990, 1992). Analysis of the silent current and its modulation in scala vestibuli and scala tympani demonstrated that the stria vascularis acts as a constant current source. Movement of the basilar membrane toward scala vestibuli increased the current through the OHCs about as much as movement toward scala tympani decreased the current, suggesting that mechanoelectrical transduction channels were half open in the absence of sound. The magnitude of the silent current places a large metabolic demand on the energy requirements of the stria vascularis, which is highly vascular, metabolically active, and rich in mitochondria in the marginal and intermediate cells where ion pumps are located. Experimental evidence showed that the magnitude of the endocochlear potential affected cochlear mechanics (Mountain, 1980) and eighth nerve discharge (Sewell, 1984).

2.2.2 A Twenty-Year Perspective as Viewed from the OHC

There have been numerous developments in understanding OHC motor mechanisms over the past 20 years. One of the more exciting functional discoveries was its ability to generate isometric force at constant amplitude and phase (re the applied electric field) at greater than 50 kHz (Frank et al., 1999). The membrane protein prestin was a key structural discovery (Zheng et al., 2000) because it is an essential component of the OHC motor. The demonstration of traveling waves in the tectorial membrane (Ghaffari et al., 2007) is important because the mechanical excitation of OHC stereocilia results from interaction between them and basilar membrane traveling waves. The role of OHC in hearing is to enhance and shape the inner hair cell (IHC) stimulus which releases neurotransmitter to generate action potentials in auditory nerve fibers. Successful recording of excitatory postsynaptic potentials (EPSPs) in the eighth nerve terminals on IHCs (Glowatzki & Fuchs, 2002) has revealed important clues about how temporal information preserved by the cochlear amplifier is relayed to the brain. Other important findings abound and some are discussed in the text that follows.

2.2.2.1 Cochlear Amplification for High-Frequency Hearing and Temporal Processing

Viscous damping is proportional to velocity, placing greater demands to counteract it as frequency increases. Diverse strategies to counteract viscous damping and increase the upper frequency limit of hearing are found in animal ears and involve the production of a "negative damping" force. It is likely that force production by the stereociliary bundle was the active process used for the relatively low-frequency hearing of early vertebrates [see SHAR Vol. 30, *Active Processes and Otoacoustic Emissions in Hearing* (Manley et al., 2008)]. Mammals appeared more than 200 million years ago and eventually adopted a mechanism associated with the lateral wall of their cylindrically shaped OHCs. The ability of OHCs to counteract viscous damping is referred to as the cochlear amplifier. Amplification is a function of vibration magnitude, operating with the greatest gain for low-intensity sounds at the center or best frequency of mechanical tuning. The amplifier suppresses movement for frequencies immediately surrounding the center frequency, and has no effect for lower frequencies where the response is proportional (linear) to stimulus amplitude. The dynamic nonlinearity at the center and surrounding frequencies improves the sensitivity and frequency selectivity of hearing.

In addition to improved sensitivity and frequency selectivity, the cochlear amplifier is thought to be responsible for the temporal invariance of basilar membrane vibrations with changes in sound intensity. Temporal shifts of basilar-membrane vibration zero-crossings and local peaks and troughs would occur in the absence of mechanical feedback and these shifts are not observed experimentally (Recio & Rhode, 2000). OHC mechanical feedback preserves the temporal fine structure of basilar-membrane vibrations throughout a broad range of intensities (Shera, 2001).

2.2.2.2 The Membrane-Based Lateral Wall Motor

The motor mechanism is piezoelectric-like in that mechanical deformation of the cell changes the transmembrane potential, comparable to the direct piezoelectric effect, while electromotility is comparable to the converse piezoelectric effect where a change in potential leads to mechanical deformation (Brownell, 2006). Cellular models that include the direct piezoelectric effect in their equivalent circuit push the roll off frequency to higher values (Spector et al., 2003). When the interplay of mechanical and electrical energy (direct and converse effects) is included, the roll off frequency is extended and resonance is introduced at a frequency higher than the roll off (Weitzel et al., 2003). Resonance in the electrical admittance of isolated OHCs has been reported (Rabbitt et al., 2005). Additional elevation of the roll off frequency may result from the presence of stretch activated channels in the lateral wall plasma membrane (Spector et al., 2005).

Membranes from a variety of cell types generate electromechanical force at acoustic frequencies [see (Brownell et al., 2010) for summary] and the mechanism is thought to be a piezoelectric-like property of membranes (including pure lipid bilayers), known as flexoelectricity. Biological membranes are thin (~5 nm) ensembles made of lipids, proteins, and other molecules. They are intrinsically polarized having a different surface charge on the two sides of the membrane. The charge asymmetry and the huge electric field across the membrane set the stage for electromechanical generation of force. These can manifest themselves as a bending force and the amount of force is greater for highly curved membranes. Transmission electron micrographs reveal highly curved membrane ripples in the OHC lateral wall plasma membrane (Dieler et al., 1991) and aspiration of excess membrane contained in the rippled surface provides support for their presence in the living cell (Morimoto et al., 2002). A flexoelectric model of electromotility based on the highly curved lateral wall plasma membrane generates sufficient force to deform the spectrin in the lateral wall cortical lattice (Raphael et al., 2000). The tonotopic structural relation between decreasing OHC length with increasing frequency is described in the results of a power efficiency analysis of lateral wall piezoelectricity (Rabbitt et al., 2009). A power efficiency analysis applied to a flexoelectric based model of stereocilia force production is likewise capable of describing the progression of long to short bundles found on hair cells in the low- and high-frequency cochlear locations both within the same cochlea and between species (Breneman et al., 2009). Optimizing power efficiency is likely to be as important to living cells as it is to society.

2.2.2.3 Full Expression of OHC Electromotility Requires Prestin and Small Intracellular Anions

The membrane protein prestin (Slc26A5) is an important component of the OHC motor. It was discovered by subtractive cloning to identify proteins that were expressed in the OHCs but not IHCs (Zheng et al., 2000). When prestin is expressed in a cultured cell line its presence increases the electromechanical force generated by the native membrane three to five times (Ludwig et al., 2001; Zhang et al., 2007). Prestin is a member of the Slc26A family of anion transporters, with up to 12 transmembrane helices. The family member with the closest sequence similarity is pendrin Slc26A4. Pendrin is expressed in the tissue forming the membranous labyrinth where it may be involved in the maintenance of endolymph. When prestin is in a membrane, a displacement charge movement into and out of, as opposed to through, the membrane can be measured. This prestin-associated charge movement is used by many investigators as the electrical signature of electromotility. Small intracellular anions such as chloride and bicarbonate appear to be the charge carrier. The involvement of anions is consistent with prestin's membership in the SLC26A family of anion transporters. Computational (informatic) analysis of the amino acid sequence of prestin and close family members indicate those regions of the proteins that are highly conserved in the prestins from different mammals. These include two sets of residues at the extracellular ends of transmembrane helices 1 and 2. Membrane electromotility is reduced (Rajagopalan et al., 2006) and prestin-associated charge movement is blocked by single point mutations in these residues (Zhang et al., 2007). Prestin-associated charge density is five times greater in high-frequency guinea pig OHCs than it is in low-frequency OHCs (Corbitt et al., 2012), which may reflect the greater force generation required at higher frequencies.

Prestin is an important component of the OHC membrane-based motor, but its precise role remains elusive. In contrast to the three- to fivefold increase in electromechanical force resulting from the presence of prestin, a greater than three order of magnitude increase in charge movement is observed. Its ability to modulate anion movement in and out of the membrane could be its most important role in the motor mechanism. There is no direct experimental or theoretical evidence that prestin is a motor molecule. Unlike well studied motor proteins such as myosin that can act independently in solution to generate force, prestin is unable to generate force in the absence of a membrane.

2.2.2.4 Membrane Material Properties Matter

The dependence of electromotility and prestin-associated charge movement on the material properties of the membrane has long been known. Changes in membrane tension shift the voltage dependence of both electromotility and the charge movement. The concentration of cholesterol in biological membranes modulates membrane mechanics and membrane–protein interactions (Khatibzadeh et al., 2012). The voltage/prestin-charge movement function can be shifted over a 100/mV range by adding or depleting cholesterol in the membrane (Rajagopalan et al., 2007; Sfondouris et al., 2008). Reduction of membrane cholesterol in the living cochlea eliminates distortion-product otoacoustic emissions (DPOAEs) while cholesterol enrichment results in a small (~3/dB) increase followed by the elimination of DPOAEs. Cholesterol depletion is the only manipulation known to increase cochlear electromechanics (Brownell et al., 2011). Immunofluorescent and immunogold labeling of prestin indicates that it is uniformly distributed in the lateral wall plasma membrane but prestin function is not (Corbitt et al., 2012) and is functionally most active in the middle (Takahashi & Santos-Sacchi, 2001) where the cholesterol concentration is lowest (Nguyen & Brownell, 1998). Membrane cholesterol levels decrease during OHC development and reach their minimal adult levels near the onset of hearing. The decrease spans the time subsurface cisternae appear and mature with a concomitant increase in OHC electromotility. Cholesterol reduction during development is not unique to the OHC, as depletion also occurs during chicken hair cell development. The nature of the lipid–protein interactions in the lateral wall membrane requires further exploration and may be related to gender differences in hearing sensitivity and cochlear amplification.

2.2.2.5 Turgor Pressure and Membrane Poration

A hydrostat is a mechanical structure formed from an elastic outer shell enclosing a pressurized core. The OHC is a cellular hydrostat with a modest cytoplasmic turgor (~1–2 kPa) (Brownell et al., 1985; Chertoff & Brownell, 1994; Ratnanather et al., 1996). This facilitates the rapid hydraulic communication of the mechanical force generated by the lateral wall plasma membrane to the ends of the cell. OHC electromotility diminishes and vanishes if the cell becomes flaccid. The requirement for intracellular turgor pressure places constraints on the OHC plasma membrane that are not shared by the other cells of the body. Most eukaryotic cells burst when their internal pressure is increased by even a small amount. The reinforcement provided by the trilaminate lateral wall and more specifically the cortical lattice (Oghalai et al., 1998) prevents this happening in the OHC. The maintenance of high turgor by plant and bacterial cell walls is associated with low water permeability through their membranes. It would be energetically costly if water could easily pass through their membranes; imagine trying to keep a leaky balloon full of air. The OHC plasma membrane shares the low water permeability of plant and bacterial cells (Chertoff & Brownell, 1994; Ratnanather, et al., 1996). The water permeability is increased when the membrane is mechanically deformed either with fluid jets or through contact with micropipettes. When test solutions are applied to OHCs by pressure injection, the membrane is deformed by the jet and the associated turbulent flow. The strain-dependent water permeability is more than an order of magnitude greater than that measured with equally rapid fluid exchange under laminar flow conditions (Morimoto et al., 2002). The strain-dependent increase in water permeability may be attributed to mechanosensitive ion channels in the plasma membrane or to an increase of spontaneously formed meta-stable (10–100 ns) water defects in the plasma membrane lipid bilayer. It may also be related to the mechanically sensitive chloride permeability of the lateral wall plasma membrane that is postulated to maintain elevated chloride concentrations in the extracisternal space for improved prestin function (Rybalchenko & Santos-Sacchi, 2003).

2.2.2.6 OHC Stereocilia Bundle and Cochlear Amplification

Throughout the past 20 years there has been a running debate as to whether electrical filtering by the OHC membrane would allow the lateral wall motor to operate at high frequencies (Santos-Sacchi, 1989), and some have suggested the stereocilia bundle motor is the mammalian cochlear amplifier. The debate about whether active hearing is based on the bundle or somatic motor has fueled research and focused interpretations. Theoretical treatments suggest the bundle could generate high-frequency electromechanical force but there is no experimental evidence that it does. Many of the participants in this discussion have recently presented their opinions in a jointly authored paper (Ashmore et al., 2010). Current modeling efforts suggest that active force production by the OHC can involve both the bundle and the lateral wall but most of these models require the lateral wall motor assume the greater burden [see proceedings of the most recent Mechanics of Hearing Workshop (Shera & Olson, 2011)]. The debate will no doubt continue despite substantial evidence that electrical filtering is not a problem for the OHC.

The earliest indication that OHC time constants may not be an insurmountable problem comes from the fact that cochlear microphonics extend to the limits of hearing. OHCs are notoriously conductive, and their low input impedance would contribute to a short time constant. For most cells the input impedance of small cells is larger than that of bigger cells. Short, high-frequency OHCs, in contrast, have smaller input impedances than long, low-frequency OHCs (Housley & Ashmore, 1992). The paradoxical length/input impedance relation would benefit hearing as it would result in shorter time constants for high-frequency OHCs.

The molecular basis of the large conductance of the OHC basolateral membrane remains to be identified. The conductance increase with frequency was attributed to the increase of a potassium channel $I_{K,n}$ (so named because it was activated at strongly negative potentials). KCNQ4 is a high- conductance potassium channel found in hair cells and neurons that is activated at negative potentials. Some have considered it to be the same as or a component of $I_{K,n}$. However, KCNQ4 expression is greatest in the longer, low-frequency OHCs and negligible in the shortest (Beisel et al., 2000, 2005), which suggests that the conductance difference arises from either different channel types or from mechanisms that modulate channel availability. One possibility is that the channel(s) are modulated by the composition and nanomechanics of the membrane. Membrane cholesterol concentration modulates ion channel conductance in hair cells (Levic & Yamoah, 2011; Purcell et al., 2011). Coexpression of prestin and KCNQ4 in cultured cells shifts the current–voltage relationship in the negative direction (Chambard & Ashmore, 2005). Cholesterol concentration and prestin function are nonuniformly distributed along the length of the OHC lateral wall (Brownell et al., 2011; Corbitt et al., 2012) and prestin-associated charge density is five times greater in short hair cells (Corbitt et al., 2012), making direct or indirect interaction with ion channels more likely for short hair cells. Although the precise mechanism for preserving temporal information of the electrical signal in OHCs remains to be identified, it is of interest that KCNQ4 expression in IHCs and spiral ganglion cells is tonotopically appropriate because it increases with increasing frequency.

Even though the highly conductive OHC basolateral membranes might have allowed the passage of the 500 pA transduction and silent currents, whole-cell currents of this magnitude were not measured until recently. Our estimates on the transduction current and the resting state of the mechanoelectrical transduction channel were confirmed by perfusing the bundle with a low-calcium-concentration solution resembling that found in the endolymph in vivo (Johnson et al., 2011). Under these conditions the upper frequency roll off occurred beyond the best frequency of where the cells were obtained up to the 8 kHz place in the cochlea. In addition to replicating the ionic environment around the bundle, we have found that optimal mechanoelectrical transduction requires the electrical environment be replicated as well. Current injections that mimic the silent current enhance the mechanical response of the bundle over its performance in the absence of the current (Hakizimana et al., 2012). We anticipate that even more features of the OHC will be found to depend on the large standing current that washes through it.

The silent current is maintained by active ion transport mechanisms located in the stria vascularis and the marginal neurons (type IV fibrocytes, etc.) that make up a circulatory system that creates a potassium-rich endolymph, and the process has been the topic of models and reviews (Quraishi & Raphael, 2008; Mistrik & Ashmore, 2010; Patuzzi, 2011). As has been long appreciated, there is a physical separation of the ion transport mechanisms and their energetic requirements from the actual sensory processing that takes place in the hair cells. Precise temporal processing imposes the need for a large silent current. High-conductance transduction and ion channels are required for short time constant, low-impedance electrical properties. We will discover similar requirements in the central auditory pathways.

2.3 Central Processing Mechanisms

2.3.1 *Pre-1992—Establishing the Temporal Limits of Central Processing*

Although the temporal resolution of acoustic signaling in the auditory nerve is impressive, the central nervous system is challenged to maintain a precise temporal representation of sounds at least up to the level of the superior olive, and possibly higher. Consequently, the central auditory system has processing demands in the time domain that parallel those of the auditory periphery.

Before 1992, much of our understanding of how this processing takes place was based on single-unit studies of neurons, a limited number of in vivo intracellular recordings, and an emerging literature utilizing the relatively new methods of brain slice preparations. Each method provided insight into different aspects of the central temporal processing. Single-unit studies demonstrated that the phase locking present in the auditory nerve was also present in neurons of the anterior ventral cochlear nucleus (AVCN) that exhibited prepotentials (Kiang et al., 1962; Goldberg & Brownell, 1973; Bourk, 1976), in their axons in the trapezoid body (Brownell, 1975), in neurons of the medial superior olive (Goldberg & Brown, 1969), and in the medial nucleus of the trapezoid body (Guinan & Li, 1990). A significant limitation of extracellular recordings is the inability to provide unequivocal identification of cell types. Although various methods had been used (detailed histological reconstructions of electrode tracks to determine which regions were recorded, characterization of prepotentials that are associated with particular cell types, and antidromic identification of projection pathways), these often left unresolved ambiguities, particularly in the cochlear nucleus where cell populations are well mixed. A major push began in the mid-1970s to obtain intracellular recordings of the acoustic responses of morphologically identified neurons, as a way of providing definitive correlations between neuron function and structure. Early efforts without simultaneous cell identification yielded interesting insights (Romand, 1978), including the presence of small action potentials and a lack of sustained discharge when recording from cells in the anterior portion of the AVCN. Further studies (Rhode et al., 1983; Rouiller & Ryugo, 1984) provided the first reasonably sized data sets and put structure–function correlations on solid ground. The majority of the conclusions from these studies were consistent with the currently accepted correlations between cell types and their physiology. Subsequent studies employed these methods studying the axons of cochlear nucleus neurons in the trapezoid body (Spirou et al., 1990) and cells in the medial nucleus of the trapezoid body (Friauf & Ostwald, 1988). Intracellular studies of the medial superior olive, which consists of a planar array of cells and thus is more difficult to target for intracellular recordings in vivo, were not to appear until later.

The relatively new brain slice technique offered opportunities to study the cellular basis of central auditory processing from the perspectives of synaptic transmission and intrinsic excitability. These first studies revealed that cochlear nucleus neurons had unusual properties, including phasic firing, low input resistances, fast membrane time constants, small action potentials, and brief EPSPs (Oertel, 1983). The identification of phasically firing cells as bushy cells, and repetitively firing cells as stellate cells were made the following year utilizing intracellular injections of horseradish peroxidase (Wu & Oertel, 1984). The slice preparation of the cochlear nucleus was considered difficult at that time, especially in comparison to hippocampal slices that were in common use. Oertel's success may have been in part the result of performing the slicing at an elevated temperature rather than cooling the brain with iced artificial cerebrospinal fluid as is still the standard approach for most other brain regions. The small action potentials in the bushy cells echoed the earlier in vivo recordings (Romand, 1978). The short time constant, coupled with the brief EPSPs in the cells, seemed optimal for conveying fine timing information. However, the mechanism behind the short time constant and the pronounced rectification in the steady-state current–voltage relationship was not at all clear. Because these experiments were performed in current clamp with high resistance, sharp electrodes, it was not an easy task to identify and characterize the conductances that produce particular effects on the membrane voltage.

We performed the first voltage-clamp experiments on VCN neurons (Manis & Marx, 1991) using cells isolated from the guinea pig cochlear nucleus. The preparation largely eliminated dendritic and axonal compartments that can escape voltage clamp, and we were able to obtain sufficient spatial control of the membrane voltage to make accurate measurements of the conductances. The experiments revealed two cell classes. One class fired phasically, matching the pattern seen in bushy cells. These cells exhibited strong outward currents that were diminished by one potassium channel blocker, 4-amino pyridine (4-AP), but were insensitive to another blocker, tetraethyl ammonium (TEA). These potassium currents activated at a voltage below the resting potential. A major clue about the resting activation came from currents that followed hyperpolarizing pulses that were also sensitive to 4-AP, and whose voltage dependence matched that of currents activated by depolarization. The relatively negative voltage dependence, as compared to most "delayed rectifiers," and a high conductance explains key aspects of the intrinsic excitability of bushy cells. In particular, the low-voltage-activated potassium conductance is a major contributor to both a short membrane time constant and a low input resistance, properties that are essential for cells to process synaptic input in ways that preserve, or even improve, the timing of information conveyed in spiral ganglion cell (SGC) action potentials. We initially thought that these currents might be mediated by "M-currents" (Adams et al., 1982), owing to several similarities in their kinetics and voltage dependence. In particular, the time course of activation of the conductance from rest was best fit with a single exponential function, which is consistent with M-currents, but not with the typical sigmoidal time-dependent activation of other delayed rectifier channels. The importance of this observation did not become clear until later, however. Furthermore, the conductance shows only a modest amount of very slow inactivation, and thus is distinct from rapidly inactivating potassium currents (so-called "A currents"). The first model of the low-voltage-activated conductance (g_{KLT}) was developed from the kinetic analysis of the voltage-clamped currents, and the model confirmed the key features of the conductance. We noted that the low-voltage-activated conductance in the bushy cells was similar to one of the potassium currents in IHCs (Kros & Crawford, 1990), which, as mentioned previously, is also partially activated at the resting potential and speculated that similar conductances might be present in other auditory neurons that required precise timing, including the medial nucleus of the trapezoid body (MNTB) and medial superior olive (MSO). Our speculations regarding the expression in other auditory neurons were later confirmed by other laboratories.

The other class of isolated cells encountered in this study fired regularly to depolarizing currents and had a higher input resistance and a longer time constant. In voltage clamp, these cells lacked the low-voltage-activated current, although they expressed high-voltage-activated TEA-sensitive potassium currents with virtually identical kinetics to the 4-AP resistant portion of the outward current in bushy cells. We concluded these were stellate cells.

EPSPs in bushy neurons of the avian nucleus magnocellularis (Hackett et al., 1982) and mouse (Oertel, 1983) were found to be very brief and did not summate. These properties seemed appropriate for neurons perform as timing devices and coincidence detectors. It was necessary to use fast-flow techniques to rapidly change the concentration of glutamate on outside-out somatic patches to understand the kinetics of the receptors that produce such brief synaptic events (Raman & Trussell, 1992). The two key observations were that the receptor desensitization was extremely fast (a concentration-dependent time constant of ~1 ms at 22 °C), and that the receptors could desensitize in the presence of tens of micromolar glutamate. This rapid desensitization was faster than that seen in prior studies in the nervous system, and seemed to represent a specialization of synaptic transmission in certain auditory pathways to be both temporally precise and optimized for coincidence detection.

These early observations provided an explanation for how bushy neurons could provide precisely timed responses to auditory nerve input. The large multisite endbulb of Held (and the modified endbulbs on globular bushy cells), in concert with a very rapidly activating and desensitizing glutamate receptor and a low-voltage-activated potassium conductance in the postsynaptic neurons, all combine to provide the needed synaptic security, brevity and short postsynaptic integration window that could support precisely timed generation of action potentials. These studies also emphasized the important roles of cell-specific intrinsic conductances and the expression of specific receptor populations in the central auditory system in the processing of the acoustic environment.

2.3.2 A 20-Year Perspective as Viewed from the Cochlear Nucleus

There are four themes that permeate the work that followed these initial studies. The first is that the specific roles of different ion channels, their kinetics, and more recently their plasticity, were explored in much greater depth, using a variety of techniques and across different parts of the auditory system. The second is that the subunit composition of the receptors that mediate synaptic transmission and the channels that contribute to the intrinsic excitability were identified, in large part aided by the rapid advances arising form the cloning and expression of specific channels and receptors, as well as by the identification of toxins that can be used to discriminate among channels and receptors with different subunits. The third theme was an emerging stream of studies on synaptic transmission, driven in large part by a technical advance that allowed the presynaptic compartment of the calyx of Held to be voltage clamped. The fourth is that computational modeling based on detailed kinetic measurements, and in some cases incorporated into dynamic clamp systems, allowed an in-depth exploration of the contributions of the different aspects of the conductances. However, before discussing these specific elements of these advances, it is important first to consider the timing information that reaches the cochlear nucleus via the auditory nerve fibers.

2.3.2.1 Precision in the Auditory Nerve

After transduction in the hair cells and transmission across the peripheral synapse to the SGC dendrite, the acoustic environment is represented as spike trains. These spike trains carry all of the information about the sound energy that falls within the receptive field of the hair cell both in the precise timing of individual action potentials and in the rate of action potentials. The temporal precision of firing in the type I afferent fibers is the result of multivesicular release from the IHC ribbon synapses (Glowatzki & Fuchs, 2002). This multivesicular release creates a rapidly rising EPSP in the SGC dendrites that allows spikes to be initiated with a precision (standard deviation) of less than 50 μs.

The precision of firing is commonly estimated by considering the classical phase locking measurement of vector strength (Goldberg & Brown, 1969). Although quite useful, this measure can be misleading because it combines information from both the stimulus cycle time and the spike timing variance. A better estimate of the temporal precision of the nerve's representation can be calculated from the standard deviation of spike times in individual auditory nerve fibers (Avissar et al., 2007), or by examining the ensemble responses of individual fibers to repeated presentations of stationary broadband noise (Louage et al., 2004). Although vector strength is highest for low frequencies (less than about 500 Hz), the spike time precision needed to achieve this high vector strength is notably less at those frequencies than at higher frequencies. In the avian ear, the standard deviation of spikes is about 450 μs at 150 Hz versus 110 μs at 2100 Hz (Avissar et al., 2007), and in owls can be as small as 22 μs at 8–10 kHz (Koppl, 1997). Thus the precision of spiking is frequency dependent, with the most precise firing occurring at the upper limits of phase locking. A similar result is evident from shuffled autocorrelation analysis (SAC) of cat auditory nerve pike trains to stationary noise stimuli (Louage et al., 2004), where the half-width of the difference between the SAC and the cross-stimulus autocorrelograms (the "DIFCOR") follows the same trend. Spikes can also be entrained to the stimulus envelope in auditory nerve fibers (ANFs) with high characteristic frequencies above the classical phase-locking limit (Louage et al., 2004), although the precision of firing appears to be about five times less than near the phase-locking limit.

The upper limit for phase locking in the auditory nerve is typically considered to be about 5 kHz, based on measurements in cat (Johnson, 1980). More recent measurements suggest that there may be detectable phase locking up to 12 kHz in chinchilla (Recio-Spinoso et al., 2005), consistent with suggestions from modeling (Heinz et al., 2001). Phase locking up to at least 10 kHz has been reported in barn owl auditory nerve (Koppl, 1997). It has also been suggested that temporal fine structure information might be used to make discriminations for sounds limited to high frequencies (>5 kHz) (Moore & Sek, 2009), although it is unclear whether phase locking to the waveform per se is required or whether timing information in the sub-millisecond range is the cue for this discrimination. Nonetheless, the existence of such weak but very fast timing cues is consistent with multiple adaptations by the auditory system, including extremely short time constants in the cochlea so that the stimulus-dependent receptor potential exceeds the intrinsic membrane noise sufficiently to modulate transmitter release, temporally precise hair cell transmitter release, systematic and precisely timed transmission at the endbulb synapses in the cochlear nucleus, and neurons in the cochlear nucleus that can detect the coincidence of convergent inputs from a limited region of the basilar membrane (Carney, 1990).

At present, there is limited evidence that such high-frequency phase-locking information is represented at the output of the cochlear nucleus. Koppl (1997) noted that phase locking in the barn owl nucleus magnocellularis (NM) was on average less robust than in the nerve, but even so, some NM neurons have vector strengths approaching those of auditory nerve fibers at frequencies greater than 5 kHz (Ashida & Carr, 2010). Comparable observations from mammals are currently lacking.

There is also evidence that the central auditory system can improve on the temporal information present in the auditory nerve. For tones below about 500 Hz most bushy cells in the ventral cochlear nucleus exhibit phase locking that exceeds that in the auditory nerve (Joris et al., 1994a, b). Computational models (Rothman et al., 1993; Rothman & Young, 1996; Rothman & Manis, 2003c; Zhang & Carney, 2005) suggest that to achieve this kind of enhancement likely requires convergence of subthreshold inputs, as occurs in globular bushy cells (Spirou et al., 2005).

The precision of timing for auditory information at high frequencies is often assumed to be less important than at low frequencies. However, we would argue that this is not the case. The central auditory system has a daunting task of reconstructing the auditory environment into "auditory objects" on the basis of both spectral and temporal features (Shinn-Cunningham & Wang, 2008; Shamma et al., 2011), and for many stimuli across disparate regions of the basilar membrane. The temporal cues, in particular, are key in binding together the different acoustic elements, and so precision in their representation, even for high-frequency sound, is of paramount importance. Although the requirement for timing precision may be somewhat relaxed, for example, in the sub-millisecond time domain, rather than the tens of microsecond range needed for azimuthal sound localization, it is still stringent enough to require specialized mechanisms be employed across the entire tonotopic axis.

Furthermore, although the cues present in the discharge patterns of auditory nerve fibers may be "weak," central processing mechanism that integrate across multiple ANFs arising from hair cells spanning a small region of the cochlea will have access to additional cues (Carney, 1990) such as local phase structure. However, analysis of these cues is complicated by level-dependent changes in phase information, at least as evaluated at lower frequencies (Carlyon et al., 2012).

2.3.2.2 Presynaptic Mechanisms in the Cochlear Nucleus

To communicate the precise firing times to the central auditory neurons, it is necessary for synapses to have a reliable and temporally precise release that can match the nerve timing. Because synaptic release is stochastic (Ribrault et al., 2011), one solution to improve reliability is to have multiple release sites onto a target cell so that there is a high probability of release across multiple synapses immediately after the arrival of an action potential. Indeed, the central terminations of the auditory nerve, at least onto those cells that contribute to pathways involved in analyzing timing information, are often composed of multiple synaptic sites. These are the endings termed the "endbulbs of Held" and the "modified endbulbs of Held" (reviewed in Manis et al., 2011). An average of 155 synaptic sites was estimated from an ultrastructural analysis of a limited data set in rat (Nicol & Walmsley, 2002). A physiological estimate in rats yielded a mean of 142 sites (Oleskevich et al., 2000). In CBA and young adult DBA/2 J mice, a mean-variance analysis suggested that approximately 90 sites participate in functional transmission from a presumptive single ANF (Oleskevich et al., 2004; Wang & Manis, 2005). In P9–P11 C57BL/6 mice, capacitance measurements suggest that approximately 40 vesicles are released per action potential in 2.0 mM Ca^{2+} (Lin et al., 2011). The large number of release sites, coupled with large quantal size (~100 pA per event at −60 mV, corresponding to 1.6 nS) could produce an excitatory synaptic conductance change (EPSG) of approximately 50–75 nS at rest [the current–voltage relationship of auditory nerve α-amino-3-hydroxy-5-methyl-4-isoxazolepropionic acid (AMPA) EPSGs is not linear because of partial voltage-dependent block by polyamines (Gardner et al., 1999), significantly reducing the peak conductance under current clamp conditions]. In brain slice studies that use 2.0–2.5 mM Ca^{2+}, endbulb release probability is 0.5–0.6 (Oleskevich, et al., 2004; Wang & Manis, 2005), which suggests a quantal content per action potential of 50–70. However, in vivo estimates from the calyx of Held in the MNTB suggest a much lower "physiological" release probability (Lorteije et al., 2009), and a quantal content of about 20. Recordings in 1 mM Ca^{2+} (Wang & Manis, 2005) show EPSGs of about 8.3 nS at −60 mV, and would suggest a quantal content of only 5. There are three primary consequences of this relatively low quantal content. First, the in vivo EPSGs will be substantially smaller than what is typically reported in the literature. Second, as is the case at the MNTB (Borst & Soria van Hoeve, 2012), the synapses release only a small fraction of their available pool for each action potential, as appropriate for the relatively high spontaneous and driven rates of auditory nerve fibers. Third, the low release probability at "physiological" calcium levels also results in less rate-dependent synaptic depression. Consequently, each incoming action potential produces an EPSG that is less dependent on the prior activity, which linearizes the transmission from the nerve into the cochlear nucleus.

The rate of vesicle release at the onset of the EPSP is also a significant limiting factor in setting the precision of postsynaptic timing. While hair cells seem to utilize multivesicular release to activate the single dendritic synapse of each auditory nerve fiber, the central synapses take a "parallel" release approach. Here, it is likely (though unproven) that release is primarily univesicular at any given synaptic contact, but is multivesicular with respect to the innervating auditory nerve fiber. The precision of synchronous release across individual sites then determines the rate of rise of the synaptic conductance change, as well as its amplitude and duration. As we discuss in Section 3.2.3, the spike threshold of bushy cells in the VCN depends on the rate of rise of the membrane potential, and thus is influenced by the rate of rise of EPSPs. Variability of release times for single quanta, measured under conditions of low release probability so that single events can be identified, show a tight temporal clustering that can be represented by either a gamma or a log-normal distribution, with a deviation (at 31 °C) of about 50 µs (Isaacson & Walmsley, 1996). Interestingly, presynaptic recordings indicate that the action potentials at endbulb synapses are slightly narrower than at the calyx of Held at the same age (P9–P11), while the calcium channels involved in release at endbulb terminals appear to have slightly faster kinetics and less inactivation (Lin et al., 2011). The narrow action potential with a rapid falling phase will create a rapidly rising, brief calcium influx, and the rapid kinetics could also help close the channels quickly and reduce delayed exocytosis.

The data to date do not address whether the release rate is dependent on the cochlear origin of the ANF or the tonotopic position in central nuclei. As discussed in Section 3.2.1, the precision of timing in the auditory nerve depends on the characteristic frequency. In addition, the tonotopic gradients of ion channels and excitability expressed by SGCs (Adamson et al., 2002) may also participate in shaping the terminal action potential and regulate the presynaptic release of transmitter. A role for a variety of presynaptic ion channels in release has been convincingly demonstrated at the calyx of Held (Kim et al., 2010; Huang & Trussell, 2011; Kopp-Scheinpflug et al., 2011). Although the small size of the endbulbs presents a technical challenge, measurements of presynaptic channel function in endbulbs are feasible (Lin et al., 2011).

It is unknown whether the individual release sites (synapses) of each endbulb synapse exhibit any coordinated release timing, other than that provided by the invading action potential. Because each synapse, at least on a short time scale, must act as an independent compartment with respect to calcium influx and the cellular mechanisms involved in release, it is unlikely that there is any overarching regulation of the release process between terminals. However, it is possible that individual synapses associated with a terminal, or synapses with common targets share aspects of transmitter release features that are "tuned" according to their history of activity or by neuromodulatory systems. For example, the paired-pulse ratios of ANF synapses converging onto individual bushy cells tend to be more highly correlated than expected from the overall distribution, suggesting that some aspect of transmission onto individual bushy cells is coordinated (Yang & Xu-Friedman, 2012). Such shared synaptic dynamics could arise from shared auditory nerve fiber action potential shapes (based on a tonotopic gradient) that affect calcium influx, by modulation via retrograde communication from the target cell, or possibly simply by a shared history of activity and regulation by long-term homeostatic mechanisms. One possibility is that such release properties correlate with the spontaneous rates of the ANFs, similar to the morphological complexity of the endbulbs (Sento & Ryugo, 1989).

2.3.2.3 Postsynaptic Mechanisms in Cochlear Nucleus Bushy Cells

Consistent with the glutamate receptor subunit compositions that produce the briefest and most rapidly desensitizing and deactivating currents (Geiger et al., 1995), the AMPA receptors that are postsynaptic to the auditory nerve fibers are composed primarily of GluR3 and 4 subunits, with little expression of GluR2 subunits (Wang et al., 1998). These receptors open in the tens of microseconds, and close (from both desensitization and deactivation) with a single fast time constant of 160–200 µs at 31–34 °C. For comparison, the decay time constant of excitatory postsynaptic currents in neocortex is 1–3 ms (Hestrin, 1992).

Although the EPSPs can clearly depolarize the cell to action potential threshold, they are not so large as to provide an insurmountable safety factor. Inhibition plays an additional, and potentially very important, role in temporal coding. Before 1992, inhibition was demonstrated in the discharge patterns of bushy cell axons (Brownell, 1975) and the cells were known to receive both glycinergic inhibition and γ-aminobutyric acid-ergic (GABAergic) inhibition (Wenthold et al., 1986, 1987; Wu & Oertel, 1986). Subsequently, inhibition onto bushy cells has been shown to be able to suppress spike generation (Caspary et al., 1994; Kopp-Scheinpflug et al., 2002), and contributes to regulation of the firing rate, particularly for louder sounds. Inhibition, even when it is slow, can also help improve firing precision, by restricting spike generation to only the largest and fastest-rising EPSPs (Xie & Manis, 2013), and thus reduce the incidence of less precisely timed spikes. However, Gai and Carney (2008) did not observe a significant change in phase locking for low-frequency sounds when blocking inhibition. Although the strong synaptic weights of ANF synapses strongly bias spike initiation toward the incoming auditory nerve fiber activity, inhibition is among additional mechanisms are available to modify how the nerve input is integrated, both to improve spike timing and presumably to filter activity under control of systems that regulate attention and arousal.

Detailed voltage-clamp studies provided better kinetic information about the different potassium conductances in VCN neurons (Rothman & Manis, 2003a, b, c). One of the key observations from these studies was that the time course of g_{KLT} activation was sigmoidal, with an n^4 activation shape, which was evident only when activation of the conductance followed deep hyperpolarization. A side-by-side comparison of voltage-clamped bushy cells (Rothman & Manis, 2003a) and OHCs (e.g., Johnson et al., 2011) show remarkable similarities in the overall potassium currents. We incorporated the kinetic measurements into Hodgkin–Huxley style models to help clarify how the different conductances in the bushy cells contribute to their firing patterns. In particular, the amount and voltage-dependence of g_{KLT} regulates the transition from phasic to tonic firing. A high density of g_{KLT} channels (or a more negative half-activation voltage) allows neurons to fire only phasically in response to rectangular current pulses. A lower density of g_{KLT} (or a more positive half-activation voltage) permit cells to fire with one to three spikes in a short burst at the onset of a current pulse. A very low density of g_{KLT} permits tonic firing.

An interesting observation is that the magnitude of g_{KLT} that could be measured in isolated VCN neurons was highly variable but seems to form a continuous distribution [Figure 7 of Rothman & Manis (2003b); unpublished data from another ~140 cells studied by Manis and Marx (1991) shows a similar distribution]. A continuous distribution may seem initially at odds with the discrete classification of firing patterns or projection sites. However, repetitive firing patterns are regulated in part by the nonlinear dynamics of spike generation, and the presence of a critical availability of g_{KLT} can shift the amount of current needed to enter into a stable limit cycle. Thus, cells with roughly similar sets of conductances can have qualitatively different firing patterns with the same driving depolarization. Nonetheless, the subthreshold temporal integrative properties of those cells would be different. In addition, the firing patterns of cochlear nucleus neurons show variability (Typlt et al., 2012) that might be consistent with a variable expression of channel densities and afferent convergence patterns. From a theoretical standpoint, a distribution of response characteristics is less redundant and often is an efficient way to encode information.

Bushy cells of guinea pig (Manis & Marx, 1991; Rothman & Manis, 2003b), gerbil (Schwarz & Puil, 1997), mouse (Cao et al., 2007; Cao & Oertel, 2011), rat (Pal et al., 2004), and dog (Bal et al., 2009) all express g_{KLT}. Because the bushy cell pathway is phylogentically old, it is reasonable to ask whether the same mechanisms are present in non-mammalians. Neurons of the avian nucleus magnocellularis also express g_{KLT} (Reyes et al., 1994; Koyano et al., 1996; Rathouz & Trussell, 1998), as do neurons in the dorsal medullary nucleus of anurans (Yang & Feng, 2007). Some evidence suggests that there is a tonotopic gradient of g_{KLT} magnitude that parallels a gradient of synaptic convergence in avians (Fukui & Ohmori, 2004).

However, other conductances, the cell's dendritic structure, and the location of channels (such as the location of the sodium channels that initiate action potentials) all play a role in setting the discharge patterns. In addition to g_{KLT}, the hyperpolarization-activated cation channels (Cao et al., 2007; Cao & Oertel, 2011), the two-pore channels (Holt et al., 2006) contribute to subthreshold integration and possibly spike afterhyperpolarizations, and other channels, such as the sodium-activated potassium channels, may be present as well.

2.3.2.4 Postsynaptic Mechanisms in the MSO

Although limiting temporal summation is important for cells such as globular bushy cells, which receive convergent input (Liberman, 1991; Spirou et al., 2005), it may be even more important for binaural computations in the medial superior olive.

These neurons are thought to operate as coincidence detectors (Jeffress, 1948) and perform the key computations needed for the azimuthal localization of low-frequency sounds. The neurons of the MSO appear to express the same sets of channels as bushy cells (Svirskis et al., 2002; Chirila et al., 2007; Mathews et al., 2010), although the total conductance levels are different. Nonetheless, the channels are open at rest and contribute significantly to the resting conductance, just as in bushy neurons. One important observation in the MSO is that action potentials have a very limited ability to invade the cell body (Scott et al., 2007). Although the cell body sodium channel density is similar to that in other neurons (Scott et al., 2010) the inactivation curve is shifted quite negative, so that few channels are available to be activated from rest. Nonetheless, these channels appear to contribute a voltage dependence to subthreshold EPSPs that facilities coincidence detection from the two ears.

2.3.2.5 General Roles for Low-Voltage-Activated K^+ Channels in the Auditory Brain Stem

How does each channel type contribute to transmitting timing information? In both bushy and MSO cells, g_{KLT} is already partially conducting at rest (Manis & Marx, 1991). When viewed from the conceptual framework of a standard Hodgkin–Huxley model, this means that across a population of channels, some of the channels are distributed in the open state (all four gates open), some are in the penultimate closed state (three gates open, one closed), while fewer are in the remaining three closed states (two gates open, two closed; one gate open, three closed, and no gates open). Thus, when the cell depolarizes with incoming excitatory synaptic input, a significant population of channels in the penultimate closed state are available to open immediately (e.g., without a delay imposed by the channel traversing multiple closed states before it opens). In this way, the cells can rapidly increase their membrane conductance and decrease their time constant. Furthermore, following even a brief EPSP, g_{KLT} contributes to an after hyperpolarization while it deactivates (Rothman & Manis, 2003b). Thus, the conductance also provides a kind of feed-forward inhibition that is faster than what could be provided by a local recurrent inhibitory network. In addition, the increased conductance provided by g_{KLT} in concert with the EPSG shortens the membrane time constant. At rest, the measured membrane time constant of bushy cells is very short at 0.5–1.5 ms (Francis & Manis, 2000; Rothman & Manis, 2003b; Cao et al., 2007), but the combined shunt from g_{KLT} and the EPSG would dynamically shorten the time constant even more and hasten the return of the membrane potential toward rest at the closing rate of the receptors. This has been directly demonstrated in the MSO (Mathews et al., 2010). The role of this feed-forward channel-mediated inhibition also reduces the excitability of the cell and curtails supernumerary spikes (Gittelman & Tempel, 2006). The net result is an improvement of the precision of timing of synaptic events at the next stage, or in the case of the MSO, a narrowing of the window for coincidence detection.

The hyperpolarization-activated currents are also partly open at rest, and so contribute to the integrative functions of bushy cells and MSO neurons. These channels tend to operate more slowly than g_{KLT}, and also are potential targets of neuromodulatory systems. However, prolonged depolarization (such as with sustained sounds) will tend to close these channels. Because their equilibrium potential is determined by a combined sodium and potassium permeability (typically around −45 mV), closing the channels will tend to hyperpolarize the cells, given that the remaining g_{KLT} and two-pore channels have equilibrium potentials at E_k, or about −90 mV. As such, I_h channels will tend to counter the time-averaged excitatory synaptic input to the cells, and act as a kind of "gain control." The activation state of I_h has been shown to contribute to coincidence detection in the MSO, in part by setting the resting potential and in part by engaging or disengaging g_{KLT} (Khurana et al., 2012). A similar role may exist in bushy cells, where modeling suggests that modulation of I_h could enhance monaural coincidence detection (Rothman & Manis, 2003c).

2.3.2.6 Other Exceptional Timing Functions in Auditory Pathways

There are other signs that the central processing is particularly sensitive to temporal information. The spike-timing plasticity seen at parallel fiber synapses onto the dendrites of DCN pyramidal cells and cartwheel cells has a very narrow interaction window that may be the narrowest in the brain (Tzounopoulos et al., 2004). DCN pyramidal cells can also report relatively fine temporal information that is transmitted in membrane voltage fluctuations in their spike trains (Street & Manis, 2007) with a sub-millisecond precision. Even at the auditory cortex, there appear to be mechanisms that constrain temporal interactions. The spike timing long-term potentiation window in auditory cortex is bounded by long-term depression windows on both the canonical negative side (spike preceding EPSP) and surprisingly on the positive side, where EPSPs precede spikes by ~50 ms (Rao, 2011). These observations suggest that an unusual set of mechanisms are present in auditory cortex that structure synaptic plasticity rules based on temporal interactions within a relatively narrow domain. Although such plasticity rules have important theoretical implications for cellular and network learning rules, these mechanisms operate on a slower time scale than the faster mechanisms required for temporal processing in the 10's of microseconds.

2.3.2.7 The Metabolic Costs of Mechanisms Enabling High Temporal Precision

Bushy cells (and likely by extension, octopus cells, MNTB neurons, and MSO or nucleus laminaris neurons) have a "standing conductance" that biases the "transduction" of synaptic conductance changes into a different dynamic regime. This conductance is created by g_{KLT} channels, in parallel with I_h and leak channels. The conductance is associated with ionic currents across the membrane that may enter and leave in different cellular compartments. The net effect is quite similar to that in the hair cells, in that the ion channels open a rest create a system that is biased to respond rapidly to incoming signals.

Let us consider for a moment the magnitude of the transmembrane current, and the amount of potassium, that a bushy cells "leaks" over time through g_{KLT}, two-pore, and I_h channels, when the cells is at rest. g_{KLT} has a half-activation of ~ −50 mV with a slope of 5.8 [guinea pig; –47 to −53 mV (Manis & Marx, 1991; Rothman & Manis, 2003a)]. Assuming an average resting potential of −62 mV (−64.4 mV, ICR mouse (Cao et al., 2007); –59 mV, Sprague–Dawley rat (Francis & Manis, 2000), and taking into consideration the slow inactivation of the channels, this leads to a fractional activation at rest of 4.8 %. The total g_{KLT} conductance in the guinea pig is approximately 171 nS (Rothman & Manis, 2003b), so the resting conductance is ~8.2 nS. Assuming that E_k is −77 mV (taken for the Rothman model and associated experimental measurements), we compute a "standing current" through these channels of approximately 64 pA, based on guinea pig data. The current is likely smaller in mouse neurons, where g_{KLT} averages 80.8 ns, with a shallower slope factor of 10.2 and a $V_{0.5}$ of −37.6 mV (Cao et al., 2007). In either case, the resting efflux of potassium ions is large, corresponding to 5.8×10^{11} ions per second for g_{KLT} channels for the guinea pig. Other channels (leak potassium channels and I_h channels) that are also conducting at rest will increase this value. The loss of K^+ in neurons is primarily balanced by the Na^+,K^+-ATPase (which exchanges three internal Na^+ ions for two K^+ ions, while consuming one molecule of ATP). If the maintenance of internal K^+ is solely dependent on this pump, then 1.82×10^8 molecules of ATP are consumed per second to keep the cell from slowly depolarizing in the absence of any spiking or synaptic input. This is somewhat larger than the estimate of approximately 10^7 ATP needed to produce just one action potential (Attwell & Laughlin, 2001; Hallermann et al., 2012), and might contribute to the relatively high metabolism noted in auditory brain stem structures. In a bushy cell model, using a wild-type $Na_v1.1$ sodium channel (Barela et al., 2006), we calculate that 6×10^6 ATP are needed for a single action potential. Thus, it would appear that the need for speed, as implemented by utilizing low-voltage-activated potassium channels, comes at a significant metabolic cost.

At the resting potential, however, the net current is zero, indicating that there must be a conductance that balances g_{KLT}. A candidate for the balancing conductance is I_h channels. The magnitude of g_{KLT} and I_h at rest appears to be "balanced" across different cell types in the cochlear nucleus (Cao & Oertel, 2011), and a pharmacological separation of the currents in octopus cells suggested that 0.6–1.2 nA flow through g_{KLT} channels at rest. The total g_{KLT} conductance in mouse octopus cells is at least ~510 nS (Bal & Oertel, 2001), several times larger than in mouse bushy cells (81 nS). Assuming that the ratio of the g_{KLT} and g_H conductances is about the same in bushy cells, the resting current through g_{KLT} in bushy cells could be 95–180 pA. This is smaller than that estimated in hair cells (500 pA), but is still substantial. The conductance contributed by two-pore potassium channels has not been measured in the VCN. However, the residual conductance remaining after blocking g_{KLT} in isolated cells that do not exhibit I_h suggest that it is likely to be a small fraction of the total conductance normally present at rest (Rothman & Manis, 2003b).

These observations then raise a question about how central auditory neurons handle the metabolic demands of maintaining a low input resistance and short time constant. Bushy cells are unusual, in that they receive most of their synaptic input on their cell bodies, and have a short, relatively unbranched dendritic tree, with the exception of an elaborate tuft of stringy dendrites that adorn the short primary dendrites. Ultrastructural analysis showed that the distal dendrites of cat bushy neurons have a high density of mitochondria, as compared to the proximal dendrites (Cant & Morest, 1979). In contrast to the cell body, the dendrites of bushy cells are sparsely innervated (Gomez-Nieto & Rubio, 2011). Although the distal dendrites may be involved in synaptic integration that contributes to temporal processing, a low density of synaptic inputs does not explain the need for many mitochondria in the dendrites. Thus, these dendrites may also participate in a distinct process that requires a substantial local source ATP. The preceding calculations suggest that with significant standing outward potassium current, Na^+,K^+ ATPases should play an important role in maintaining the intracellular ionic balance in bushy cells. The low-voltage-activated K^+ channels are located at the soma, although their density elsewhere on the cell surface is unknown. The soma is also covered with the endbulb as well as inhibitory synapses (Spirou et al., 2005), leaving little surface area to place pumps. In addition, many of the astrocytic processes around the soma envelop the synapses themselves rather than being directly opposed to the somatic membrane (Gulley et al., 1978; Cant & Morest, 1979), and so are likely primarily involved in regulating the synaptic microenvironment, including retrieving glutamate released both by the high spontaneous rates of some auditory nerve fibers and by the high rates of release that occur during acoustic stimulation.

We hypothesize that the membrane pumps and other exchange systems that replenish potassium in bushy cells are located in the distal dendrites. The large surface area of the dendritic tuft (brush) would provide an optimal arrangement for transferring potassium ions into the bushy cells from extracellular spaces and astrocytic sources. In this context, it is interesting that the dendrites of groups of bushy cells in cat tend to cluster together. Such clusters of dendrites are illustrated in the drawings in Figures 3–12, 3–27, and 3–28 in Lorente de Nó (1981), although an important caveat is that the drawings were made from relatively thick sections and so the apparent "clustering" does not necessarily reflect interdigitation of the dendrites. An interesting speculation is that these dendritic domains serve as a K^+ refilling station, working in conjunction with local astrocytes. An alternative arrangement, seen in rodents, could be that the dendritic processes that are located adjacent to other bushy cell soma (Gomez-Nieto & Rubio, 2011) serve as a way of siphoning K^+ away from one cell and into another. Along these lines, it is interesting that other auditory neurons that utilize low-voltage-activated potassium conductances, such as octopus cells, MNTB neurons (Kuwabara & Zook, 1991; Smith et al., 1998), and medial superior olivary neurons (Schwartz, 1977; Henkel & Brunso-Bechtold, 1990; Smith, 1995; Chirila et al., 2007; Rautenberg et al., 2009) also have distal dendritic tufts. Schwartz (1977) specifically mentions the presence of mitochondria in beaded areas of secondary and tertiary dendrites, recapitulating the pattern seen in bushy cells. Recognizing that such tufts are not unique to cells with g_{KLT}, these observations can be taken only as suggestive evidence in support of our hypothesis. It is also possible that these dendritic mitochondria are present for another purpose than maintaining K^+ homeostasis. In the neurons of the MNTB, the distal dendrites appear to have sodium channels that help boost the action potentials that are initiated in the axon initial segment (Leao et al., 2008). However, the sodium channels found in the dendrites of these cells appear to have unusually slow inactivation, which would lead to a larger sodium influx and could trigger the Na^+,K^+ ATPase, thereby assisting in the recirculation of K^+. Little of the α3 Na^+,K^+ ATPase was detected by immunohistochemistry in any location other than the cell body and axons in this study. However, a different isoform or even exchange mechanism could be utilized.

The hypothesis that bushy cells might handle their metabolic demands by a separation of ion channels and the mechanisms that maintain the electrochemical gradients has parallels in the photoreceptors in the retina (Hagins et al., 1970; Baylor et al., 1984; Yan & Matthews, 1992). Here, a non-uniform distribution of channels generates a circular extracellular current that flows from one part of the cell to another in the absence of stimulation, and biases the synaptic release machinery. Similarly, as discussed earlier, the separation of metabolic demands from the transduction and synaptic mechanisms in hair cells involves an extracellular flow, primarily of K^+ ions.

2.4 Synthesis and Summary

The remarkable sensitivity and speed of the auditory system depends on adaptations at the membrane and cellular level. Although before 1992, a number of important discoveries were made that provided insight into the mechanisms needed to meet the timing demands of auditory processing, the exploration of these mechanisms and the discovery and elucidation of others in the last 20 years has deepened our appreciation of how these molecular tools are employed. Over the past two decades there has been substantial progress in understanding the structural diversity and mechanisms involved at the organ and cellular level. We have found that detection relies on OHC electromechanics to provide sensitivity, frequency selectivity, and intensity invariant temporal precision. The mechanism requires the OHC have low input impedance for a short electrical time constant. The high conductivity necessitates the large standing current generated by the stria vascularis. Similar short time constant requirements for the IHC, SGC, and brain stem auditory neurons involved in transmitting and processing the temporal information impose similar requirements for a high conductance and short time constants. Surprisingly, some of the neurons have the structural specializations that would support the generation of a standing current for use by the cell. In addition, there is diversity in channel and protein function along the cochlear partition, as well as centrally. In the cochlea, this diversity is closely tied to the tuning of the basilar membrane, as was first elucidated in avian and turtle hair cells. However, centrally (including in the spiral ganglion) the transformed representation of sound into spike trains places a different set of demands on the processing mechanisms. Although these demands are related to the tonotopic organization and the limitations of the representation of sound features in action potentials with finite refractory periods, they are focused instead on optimizing the use of the available timing information across at least three synapses. As noted earlier, the larger functional role of reconstructing the auditory environment and "objects" from information across the entire cochlear partition extends the timing problem, however relaxed, even to very high frequencies.

We have discussed some of the specific membrane proteins (motor elements, channels, pumps) that participate in allowing the auditory system to perform feats such as localizing sounds based on time cues in the microsecond range. Although the cochlear hair cells and the brain stem neurons that participate in these challenging tasks are very different, the mechanisms across cell type share a number of parallel features. We are confident that even these have not been fully explored. We have also touched on the elegant synaptic mechanisms that allow information to be transmitted across chemical synapses in ways that preserve precise timing information. We are only beginning to unravel how the synapses function, as new technical approaches allowing better visualization, precise manipulation, and high-resolution measurements of synaptic transmission are developed and applied.

2.5 Future Directions

Although substantial progress has been made in the past two decades in understanding the cellular diversity and specializations that enable some auditory neurons to process timing information on a microsecond time scale, there remain a number of unanswered questions and several controversies.

2.5.1 Stereocilia

The molecular identity of the mechanoelectrical transduction channel located at the stereociliar tips has not been identified. It could be a protein, a peptide, or lipid pore or a combination of all of these. Once identified the coupling between it and the tip link will need to be clarified. Although progress has been made in identifying a protein that contributes to the coupling between the OHC stereocilia and the tectorial membrane, further work will be required to identify how the nanomechanics of the tectorial membrane contribute to stereocilia deflection. Stereociliar nanoelectromechanics must be explored to understand the impact of the silent current on bundle deflection. The signaling mechanisms that lead to differences in stereocilia length during development and maintaining them throughout life have not yet been addressed.

2.5.2 OHC Soma

The OHC is unusual both structurally and functionally and there is no shortage of unanswered questions. The subsurface cisternae is as novel a structure as electromotility is a function, although its role (as well as that of the related caniculi reticulum) has yet to be revealed. So called "pillars" span the extracisternal space from the circumferentially oriented f-actin in the cortical lattice to the plasma membrane, but their molecular identity and mechanical properties are unknown. The osmolyte that maintains OHC turgor has not been identified. Given the importance of membrane cholesterol in prestin and voltage gated ion channel function it would useful to have a precise knowledge of its concentration in the lateral wall, synaptic, and apical membranes. Other phospholipids can also affect protein function and little is known about their presence in the different membrane domains. OHC length is tonotopically and radially (rows 1–3) specific but the signaling mechanism that determines it during development and maintains it during aging is unknown.

2.5.3 IHC

One of the longest standing questions in cochlear function is the mechanism by which the IHC is stimulated. The free standing nature of its stereocilia bundle makes contact deflection by the tectorial membrane difficult and most models favor a hydraulic coupling with the fluids in the subtectorial space but experimental evidence for this possibility has not yet been obtained. Even more mysterious are the response differences in neurotransmitter release depending on where the synapse is located in the cell. There is little or no spontaneous release at synapses located on the modiolar quadrant and high spontaneous release at synapses located away from the modiolus. The mechanism responsible is difficult to imagine if the IHC is essentially isopotential.

2.5.4 Auditory Nerve

The representation of timing information in the auditory nerve has been investigated for more than 50 years. However, as our knowledge of cochlear mechanisms grows, new questions have arisen and some fundamental issues have been readdressed. The upper frequency of the fine structure of the acoustic waveform that is presented to neurons in the cochlear nucleus is certainly limited and depends on species.

However, observations in the literature suggest that there might be some information available for frequencies >5 kHz. It is not clear whether this information can be effectively used by the central auditory system. Similarly, the representations of temporal cues related to envelope in high-frequency SGCs are important.

2.5.5 Bushy Cells

Although the major ion channels have been identified, there are other classes of channels that are expressed in bushy cells, including sodium activated potassium channels, additional delayed rectifier channels, and two-pore potassium channels, whose roles are less well understood. Because these conductances are generally smaller than the ones identified so far, and models have not yet required them in order to explain basic bushy cell responses to auditory nerve input, their roles may be tied to sensory adaptation or descending control. Another fundamental problem is that of auditory nerve fiber convergence from different spontaneous rate groups onto cochlear nucleus cells, including bushy cells. This question is related to the integration of information across the dynamic range of hearing, as well as to the processing of various cues, such as stimulus phase, that change with stimulus intensity. It is also has important implications for the capabilities of hearing after both temporary and permanent hearing loss scenarios that may differentially affect the availability of information from different spontaneous rate groups of ANFs. Although there is evidence in the literature to support some segregation in central projections of spontaneous rate groups, as well as convergence onto certain cell classes, concrete demonstrations remain elusive.

Recently, it has been demonstrated that potassium channel expression and function in the MNTB depends on the recent history of sensory input (Strumbos et al., 2010), and is also dependent on the peripheral hearing status, suggesting that the auditory system adapts to both its short-term and long-term environmental drive. With respect to hearing loss, more subtle effects are present in bushy cells of the VCN. Other evidence suggests that auditory nerve fibers adapt their dynamic range and firing rates to the structure of the sensory input (Wen et al., 2009, 2012; Carlyon et al., 2012). How does the central system adapt to these changes and "interpret" inputs whose representation of the environment no longer holds a 1:1 relationship to the physical stimulus? How does the perceptual performance change as channel expression adapts to the sensory environment, or is there a compensatory mechanism that maintains constancy?

The use of low-voltage-activated potassium channels to set the input resistance and time constant of the hair cells and neurons is one clear parallel between the hair cells and some central neurons. That the voltage-gated channels are biased to be partly activated at rest presents metabolic challenges that are met in different ways in the cochlea and the central system, yet which may share some general features, and certainly molecular constituents. Although intriguing observations have been made, the roles of relatively sparse distal dendritic synaptic inputs on auditory neurons that have strong somatic synapses, and the purpose of dendritic mitochondria located distally from key ion channels and receptors remain speculative and challenging to investigate.

The role(s) of inhibition in the central pathways that process timing information remains controversial. Although some evidence suggests that precisely timed and brief inhibition is critical to sharpen coincidence detection in the MSO (Brand et al., 2002), it is clear that only this is not the only way that inhibition works at other sites in brain stem circuits that process timing information (Xie & Manis, 2013).

Descending inputs from upper auditory nuclei and nonauditory sources to the cochlear nucleus are poorly understood, and difficult to study. Descending input can shape sensory representations based on prior experience or expectations in ways that provide perceptual advantages in a challenging sensory environment. Studies so far have identified some basic cellular mechanisms, including the intrinsic excitability of neurons that are frequently the targets of neuromodulatory systems. For example, norepinephrine can improve temporal precision of VCN neurons in bats (Kossl & Vater, 1989), and a simple model implementation of one likely mechanism suggests that the same kind of improvement can occur in bushy cells (Rothman & Manis, 2003c). The specific roles of descending systems are now amenable to in vitro and in vivo analysis using optogenetic approaches, and such studies will undoubtedly reveal new aspects of sensory processing that differ from the static views that are currently available.

Acknowledgments This work was supported by the Baylor College of Medicine Jake and Nina Kamin Chair of Otorhinolaryngology and Communicative Sciences and NIH/NIDCD grants R01 DC000354 and DC002775 (WEB) and the Univeristy of North Carolina at Chapel Hill Thomas J. Dark Distinguished Research Professorship and NIH/NIDCD grants R01 DC004551 and DC009809 (PBM).

References

Adams, P. R., Brown, D. A., & Constanti, A. (1982). M-currents and other potassium currents in bullfrog sympathetic neurones. *The Journal of Physiology*, 330, 537–572.

Adamson, C. L., Reid, M. A., Mo, Z. L., Bowne-English, J., & Davis, R. L. (2002). Firing features and potassium channel content of murine spiral ganglion neurons vary with cochlear location. *The Journal of Comparative Neurology*, 447(4), 331–350.

Ashida, G., & Carr, C. E. (2010). Effect of sampling frequency on the measurement of phase-locked action potentials. *Frontiers in Neuroscience*, 4(100), 1–9.

Ashmore, J. (2008). Cochlear outer hair cell motility. *Physiological Review*, 88(1), 173–210.

Ashmore, J., Avan, P., Brownell, W. E., Dallos, P., Dierkes, K., Fettiplace, R., Grosh, K., Hackney, C. M., Hudspeth, A. J., Julicher, F., Lindner, B., Martin, P., Meaud, J., Petit, C., Santos-Sacchi, J., & Canlon, B. (2010). The remarkable cochlear amplifier. *Hearing Research*, 266(1–2), 1–17.

Ashmore, J. F. (1987). A fast motile response in guinea-pig outer hair cells: The cellular basis of the cochlear amplifier. *The Journal of Physiology*, 388, 323–347.

Ashmore, J. F., & Brownell, W. E. (1986). Kilohertz movements induced by electrical stimulation in outer hair cells isolated from the guinea pig cochlea. *The Journal of Physiology*, 376, 49P.

Attwell, D., & Laughlin, S. B. (2001). An energy budget for signaling in the grey matter of the brain. *Journal of Cerebral Blood Flow and Metabolism*, 21(10), 1133–1145.

Avissar, M., Furman, A. C., Saunders, J. C., & Parsons, T. D. (2007). Adaptation reduces spike-count reliability, but not spike-timing precision, of auditory nerve responses. *The Journal of Neuroscience*, 27(24), 6461–6472.

Bal, R., & Oertel, D. (2001). Potassium currents in octopus cells of the mammalian cochlear nucleus. *Journal of Neurophysiology*, 86(5), 2299–2311.

Bal, R., Baydas, G., & Naziroglu, M. (2009). Electrophysiological properties of ventral cochlear nucleus neurons of the dog. *Hearing Research*, 256(1–2), 93–103.

Barela, A. J., Waddy, S. P., Lickfett, J. G., Hunter, J., Anido, A., Helmers, S. L., Goldin, A. L., & Escayg, A. (2006). An epilepsy mutation in the sodium channel SCN1A that decreases channel excitability. *The Journal of Neuroscience*, 26(10), 2714–2723.

Baylor, D. A., Matthews, G., & Nunn, B. J. (1984). Location and function of voltage-sensitive conductances in retinal rods of the salamander, *Ambystoma tigrinum. The Journal of Physiology*, 354, 203–223.

Beisel, K. W., Nelson, N. C., Delimont, D. C., & Fritzsch, B. (2000). Longitudinal gradients of KCNQ4 expression in spiral ganglion and cochlear hair cells correlate with progressive hearing loss in DFNA2. *Brain Research. Molecular Brain Research*, 82(1–2), 137–149.

Beisel, K. W., Rocha-Sanchez, S. M., Morris, K. A., Nie, L., Feng, F., Kachar, B., Yamoah, E. N., & Fritzsch, B. (2005). Differential expression of KCNQ4 in inner hair cells and sensory neurons is the basis of progressive high-frequency hearing loss. *The Journal of Neuroscience*, 25(40), 9285–9293.

Békésy, G. von (1952). DC resting potentials inside the cochlear partition. *Journal Acoustical Society of America*, 24(1), 72–76.

Borst, J. G., & Soria van Hoeve, J. (2012). The calyx of held synapse: From model synapse to auditory relay. *Annual Review of Physiology*, 74, 199–224.

Bourk, T. R. (1976). *Electrical responses of neural units in the anteroventral cochlear nucleus of the cat.* Ph.D. dissertation, Massachusetts Institute of Technology.

Brand, A., Behrend, O., Marquardt, T., McAlpine, D., & Grothe, B. (2002). Precise inhibition is essential for microsecond interaural time difference coding. *Nature*, 417(6888), 543–547.

Breneman, K. D., Brownell, W. E., & Rabbitt, R. D. (2009). Hair cell bundles: Flexoelectric motors of the inner ear. *PLoS One*, 4(4), e5201.

Brownell, W. E. (1975). Organization of the cat trapezoid body and the discharge characteristics of its fibers. *Brain Research*, 94(3), 413–433.

Brownell, W. E. (1982). Cochlear transduction: An integrative model and review. *Hearing Research*, 6(3), 335–360.

Brownell, W. E. (1983). Observations on a motile response in isolated outer hair cells. In W. R. Webster & L. M. Aitken (Eds.), *Mechanisms of hearing* (pp. 5–10). Monash: Monash University Press.

Brownell, W. E. (1984). Microscopic observation of cochlear hair cell motility. *Scanning Electron Microscopy*(Pt 3), 1401–1406.

Brownell, W. E. (1990). Outer hair cell electromotility and otoacoustic emissions. *Ear and Hearing*, 11(2), 82–92.

Brownell, W. E. (2006). The piezoelectric outer hair cell. In R. A. Eatock, R. R. Fay, & A. N. Popper (Eds.), *Vertebrate hair cells* (pp. 313–347). New York: Springer Science+Business Media.

Brownell, W. E., & Kachar, B. (1986). Outer hair cell motility: A possible electro-kinetic mechanism. In J. B. Allen, J. L. Hall, A. E. Hubbard, S. T. Neely & A. Tubis (Eds.), *Peripheral auditory mechanisms* (pp. 369–376).New York: Springer-Verlag.

Brownell, W. E., Zidanic, M., & Spirou, G. A. (1986). Standing currents and their modulation in the Cochlea. In R. A. Altschuler, D. W. Hoffman, & R. P. Bobbin (Eds.), *Neurobiology of hearing: The cochlea* (pp. 91–107). New York: Raven Press.

Brownell, W. E., Manis, P. B., Zidanic, M., & Spirou, G. A. (1983). Acoustically evoked radial current densities in scala tympani. *Journal Acoustical Society of America*, 74(3), 792–800.

Brownell, W. E., Bader, C. R., Bertrand, D., & de Ribaupierre, Y. (1985). Evoked mechanical responses of isolated cochlear outer hair cells. *Science*, 227(4683), 194–196.

Brownell, W. E., Qian, F., & Anvari, B. (2010). Cell membrane tethers generate mechanical force in response to electrical stimulation. *Biophysical Journal*, 99(3), 845–852.

Brownell, W. E., Jacob, S., Hakizimana, P., Ulfendahl, M., & Fridberger, A. (2011). Membrane cholesterol modulates cochlear electromechanics. *Pflugers Archive: European Journal of Physiology*, 461(6), 677–686.

Cant, N. B., & Morest, D. K. (1979). The bushy cells in the anteroventral cochlear nucleus of the cat. A study with the electron microscope. *Neuroscience*, 4(12), 1925–1945.

Cao, X. J., & Oertel, D. (2011). The magnitudes of hyperpolarization-activated and low-voltage-activated potassium currents co-vary in neurons of the ventral cochlear nucleus. *Journal of Neurophysiology*, 106(2), 630–640.

Cao, X. J., Shatadal, S., & Oertel, D. (2007). Voltage-sensitive conductances of bushy cells of the mammalian ventral cochlear nucleus. *Journal of Neurophysiology*, 97(6), 3961–3975.

Carlyon, R. P., Long, C. J., & Micheyl, C. (2012). Across-channel timing differences as a potential code for the frequency of pure tones. *Journal of the Association for Research in Otolaryngology: JARO*, 13(2), 159–171.

Carney, L. H. (1990). Sensitivities of cells in anteroventral cochlear nucleus of cat to spatiotemporal discharge patterns across primary afferents. *Journal of Neurophysiology*, 64(2), 437–456.

Caspary, D. M., Backoff, P. M., Finlayson, P. G., & Palombi, P. S. (1994). Inhibitory inputs modulate discharge rate within frequency receptive fields of anteroventral cochlear nucleus neurons. *Journal of Neurophysiology*, 72(5), 2124–2133.

Chambard, J. M., & Ashmore, J. F. (2005). Regulation of the voltage-gated potassium channel KCNQ4 in the auditory pathway. *Pflugers Archive: European Journal of Physiology*, 450(1), 34–44.

Chertoff, M. E., & Brownell, W. E. (1994). Characterization of cochlear outer hair cell turgor. *American Journal of Physiology*, 266(2 Pt 1), C467–479.

Chirila, F. V., Rowland, K. C., Thompson, J. M., & Spirou, G. A. (2007). Development of gerbil medial superior olive: Integration of temporally delayed excitation and inhibition at physiological temperature. *The Journal of Physiology*, 584(Pt 1), 167–190.

Corbitt, C., Farinelli, F., Brownell, W. E., & Farrell, B. (2012). Tonotopic relationships reveal the charge density varies along the lateral wall of outer hair cells. *Biophysical Journal*, 102(12), 2715–2724.

Dallos, P., Popper, A. N., & Fay, R. R. (Eds.). (1996). *The cochlea* (Vol. 8). New York: Springer-Verlag.

Davis, H. (1983). An active process in cochlear mechanics. *Hearing Research*, 9(1), 79–90.

Davis, H., Tasaki, I., & Goldstein, R. (1952). The peripheral origin of activity, with reference to the ear. *Cold Spring Harbor Symposium Quantitative Biology*, 17, 143–154.

Dieler, R., Shehata-Dieler, W. E., & Brownell, W. E. (1991). Concomitant salicylate-induced alterations of outer hair cell subsurface cisternae and electromotility. *Journal of Neurocytology*, 20(8), 637–653.

Evans, B. N. (1990). Fatal contractions: ultrastructural and electromechanical changes in outer hair cells following transmembraneous electrical stimulation. *Hearing Research*, 45(3), 265–282.

Francis, H. W., & Manis, P. B. (2000). Effects of deafferentation on the electrophysiology of ventral cochlear nucleus neurons. *Hearing Research*, 149(1–2), 91–105.

Frank, G., Hemmert, W., & Gummer, A. W. (1999). Limiting dynamics of high-frequency electromechanical transduction of outer hair cells. *Proceedings of the National Academy of Sciences of the USA*, 96(8), 4420–4425.

Friauf, E., & Ostwald, J. (1988). Divergent projections of physiologically characterized rat ventral cochlear nucleus neurons as shown by intra-axonal injection of horseradish peroxidase. *Experimental Brain Research*, 73(2), 263–284.

Fukui, I., & Ohmori, H. (2004). Tonotopic gradients of membrane and synaptic properties for neurons of the chicken nucleus magnocellularis. *The Journal of Neuroscience*, 24(34), 7514–7523.

Gai, Y., & Carney, L. H. (2008). Influence of inhibitory inputs on rate and timing of responses in the anteroventral cochlear nucleus. *Journal of Neurophysiology*, 99(3), 1077–1095.

Gardner, S. M., Trussell, L. O., & Oertel, D. (1999). Time course and permeation of synaptic AMPA receptors in cochlear nuclear neurons correlate with input. *The Journal of Neuroscience*, 19(20), 8721–8729.

Geiger, J. R., Melcher, T., Koh, D. S., Sakmann, B., Seeburg, P. H., Jonas, P., & Monyer, H. (1995). Relative abundance of subunit mRNAs determines gating and Ca^{2+} permeability of AMPA receptors in principal neurons and interneurons in rat CNS. *Neuron*, 15(1), 193–204.

Ghaffari, R., Aranyosi, A. J., & Freeman, D. M. (2007). Longitudinally propagating traveling waves of the mammalian tectorial membrane. *Proceedings of the National Academy of Sciences of the USA*, 104(42), 16510–16515.

Gittelman, J. X., & Tempel, B. L. (2006). Kv1.1–containing channels are critical for temporal precision during spike initiation. *Journal of Neurophysiology*, 96(3), 1203–1214.

Glowatzki, E., & Fuchs, P. A. (2002). Transmitter release at the hair cell ribbon synapse. *Nature Neuroscience*, 5(2), 147–154.

Gold, T. (1948). Hearing. II. The physical basis of the action of the cochlea. *Proceedings of the Royal Society London B: Biological Science*, 135, 492–498.

Goldberg, J. M., & Brown, P. B. (1969). Response of binaural neurons of dog superior olivary complex to dichotic tonal stimuli: some physiological mechanisms of sound localization. *Journal of Neurophysiology*, 32(4), 613–636.

Goldberg, J. M., & Brownell, W. E. (1973). Discharge characteristics of neurons in anteroventral and dorsal cochlear nuclei of cat. *Brain Research*, 64, 35–54.

Gomez-Nieto, R., & Rubio, M. E. (2011). Ultrastructure, synaptic organization, and molecular components of bushy cell networks in the anteroventral cochlear nucleus of the rhesus monkey. *Neuroscience*, 179, 188–207.

Guinan, J. J., Jr., & Li, R. Y. (1990). Signal processing in brainstem auditory neurons which receive giant endings (calyces of Held) in the medial nucleus of the trapezoid body of the cat. *Hearing Research*, 49(1–3), 321–334.

Gulley, R. L., Landis, D. M., & Reese, T. S. (1978). Internal organization of membranes at end bulbs of Held in the anteroventral cochlear nucleus. *The Journal of Comparative Neurology*, 180(4), 707–741.

Hackett, J. T., Jackson, H., & Rubel, E. W. (1982). Synaptic excitation of the second and third order auditory neurons in the avian brain stem. *Neuroscience*, 7(6), 1455–1469.

Hagins, W. A., Penn, R. D., & Yoshikami, S. (1970). Dark current and photocurrent in retinal rods. *Biophysical Journal*, 10(5), 380–412.

Hakizimana, P., Brownell, W. E., Jacob, S., & Fridberger, A. (2012). Sound-induced length changes in outer hair cell stereocilia. *Nature Communications*, 3(1094), 10 pages.

Hallermann, S., de Kock, C. P., Stuart, G. J., & Kole, M. H. (2012). State and location dependence of action potential metabolic cost in cortical pyramidal neurons. *Nature Neuroscience*, 15(7), 1007–1014.

Hallworth, R., & Jensen-Smith, H. (2008). The morphological specializations and electromotility of the mammalian outer hair cell. In G. A. Manley, R. R. Fay, & A. N. Popper (Eds.), *Active processes and otoacoustic emissions* (Vol. 30, pp. 145–189). New York: Springer Science+Business Media.

Heinz, M. G., Colburn, H. S., & Carney, L. H. (2001). Rate and timing cues associated with the cochlear amplifier: level discrimination based on monaural cross-frequency coincidence detection. *The Journal of the Acoustical Society of America*, 110(4), 2065–2084.

Henkel, C. K., & Brunso-Bechtold, J. K. (1990). Dendritic morphology and development in the ferret medial superior olivary nucleus. *The Journal of Comparative Neurology*, 294(3), 377–388.

Hestrin, S. (1992). Developmental regulation of NMDA receptor-mediated synaptic currents at a central synapse. *Nature*, 357(6380), 686–689.

Holt, A. G., Asako, M., Duncan, R. K., Lomax, C. A., Juiz, J. M., & Altschuler, R. A. (2006). Deafness associated changes in expression of two-pore domain potassium channels in the rat cochlear nucleus. [Research Support, N.I.H., Extramural]. *Hearing Research*, 216–217, 146–153.

Housley, G. D., & Ashmore, J. F. (1992). Ionic currents of outer hair cells isolated from the guinea-pig cochlea. [Research Support, Non-U.S. Gov't]. *The Journal of Physiology*, 448, 73–98.

Huang, H., & Trussell, L. O. (2011). KCNQ5 channels control resting properties and release probability of a synapse. *Nature Neuroscience*, 14(7), 840–847.

Isaacson, J. S., & Walmsley, B. (1996). Amplitude and time course of spontaneous and evoked excitatory postsynaptic currents in bushy cells of the anteroventral cochlear nucleus. *Journal of Neurophysiology*, 76(3), 1566–1571.

Jeffress, L. A. (1948). A place theory of sound localization. *Journal of Comparative and Physiological Psychology*, 41(1), 35–39.

Johnson, D. H. (1980). The relationship between spike rate and synchrony in responses of auditory-nerve fibers to single tones. *Journal Acoustical Society of America*, 68(4), 1115–1122.

Johnson, S. L., Beurg, M., Marcotti, W., & Fettiplace, R. (2011). Prestin-driven cochlear amplification is not limited by the outer hair cell membrane time constant. *Neuron*, 70(6), 1143–1154.

Joris, P. X., Smith, P. H., & Yin, T. C. (1994a). Enhancement of neural synchronization in the anteroventral cochlear nucleus. II. Responses in the tuning curve tail. *Journal of Neurophysiology*, 71(3), 1037–1051.

Joris, P. X., Carney, L. H., Smith, P. H., & Yin, T. C. (1994b). Enhancement of neural synchronization in the anteroventral cochlear nucleus. I. Responses to tones at the characteristic frequency. *Journal of Neurophysiology*, 71(3), 1022–1036.

Kachar, B., Brownell, W. E., Altschuler, R., & Fex, J. (1986). Electrokinetic shape changes of cochlear outer hair cells. *Nature*, 322(6077), 365–368.

Kemp, D. T. (1978). Stimulated acoustic emissions from within the human auditory system. *Journal Acoustical Society of America*, 64(5), 1386–1391.

Khatibzadeh, N., Gupta, S., Farrell, B., Brownell, W. E., & Anvari, B. (2012). Effects of cholesterol on nano-mechanical properties of the living cell plasma membrane. *Soft Matter*, 8(32), 8350–8360.

Khurana, S., Liu, Z., Lewis, A. S., Rosa, K., Chetkovich, D., & Golding, N. L. (2012). An essential role for modulation of hyperpolarization-activated current in the development of binaural temporal precision. *The Journal of Neuroscience*, 32(8), 2814–2823.

Kiang, N. Y., Watanabe, T., Thomas, E. C., & Clark, L. F. (1962). Stimulus coding in the cat's auditory nerve. Preliminary report. *Transactions of the American Otological Society*, 50, 264–283.

Kim, J. H., Kushmerick, C., & von Gersdorff, H. (2010). Presynaptic resurgent Na^+ currents sculpt the action potential waveform and increase firing reliability at a CNS nerve terminal. *The Journal of Neuroscience*, 30(46), 15479–15490.

Kopp-Scheinpflug, C., Dehmel, S., Dorrscheidt, G. J., & Rubsamen, R. (2002). Interaction of excitation and inhibition in anteroventral cochlear nucleus neurons that receive large endbulb synaptic endings. *The Journal of Neuroscience*, 22(24), 11004–11018.

Kopp-Scheinpflug, C., Steinert, J. R., & Forsythe, I. D. (2011). Modulation and control of synaptic transmission across the MNTB. *Hearing Research*, 279(1–2), 22–31.

Koppl, C. (1997). Phase locking to high frequencies in the auditory nerve and cochlear nucleus magnocellularis of the barn owl, Tyto alba. *The Journal of Neuroscience*, 17(9), 3312–3321.

Kossl, M., & Vater, M. (1989). Noradrenaline enhances temporal auditory contrast and neuronal timing precision in the cochlear nucleus of the mustached bat. *The Journal of Neuroscience*, 9(12), 4169–4178.

Koyano, K., Funabiki, K., & Ohmori, H. (1996). Voltage-gated ionic currents and their roles in timing coding in auditory neurons of the nucleus magnocellularis of the chick. *Neuroscience Research*, 26(1), 29–45.

Kros, C. J., & Crawford, A. C. (1990). Potassium currents in inner hair cells isolated from the guinea-pig cochlea. *The Journal of Physiology*, 421, 263–291.

Kuwabara, N., & Zook, J. M. (1991). Classification of the principal cells of the medial nucleus of the trapezoid body. *The Journal of Comparative Neurology*, 314(4), 707–720.

Leao, R. N., Leao, R. M., da Costa, L. F., Rock Levinson, S., & Walmsley, B. (2008). A novel role for MNTB neuron dendrites in regulating action potential amplitude and cell excitability during repetitive firing. *The European Journal of Neuroscience*, 27(12), 3095–3108.

Levic, S., & Yamoah, E. N. (2011). Plasticity in membrane cholesterol contributes toward electrical maturation of hearing. *The Journal of Biological Chemistry*, 286(7), 5768–5773.

Liberman, M. C. (1991). Central projections of auditory-nerve fibers of differing spontaneous rate. I. Anteroventral cochlear nucleus. *The Journal of Comparative Neurology*, 313(2), 240–258.

Lin, K. H., Oleskevich, S., & Taschenberger, H. (2011). Presynaptic Ca^{2+} influx and vesicle exocytosis at the mouse endbulb of Held: a comparison of two auditory nerve terminals. *The Journal of Physiology*, 589(Pt 17), 4301–4320.

Lorente de Nó, R. (1981). *The primary acoustic nuclei*. New York: Raven Press.

Lorteije, J. A., Rusu, S. I., Kushmerick, C., & Borst, J. G. (2009). Reliability and precision of the mouse calyx of Held synapse. *The Journal of Neuroscience*, 29(44), 13770–13784.

Louage, D. H., van der Heijden, M., & Joris, P. X. (2004). Temporal properties of responses to broadband noise in the auditory nerve. *Journal of Neurophysiology*, 91(5), 2051–2065.

Ludwig, J., Oliver, D., Frank, G., Klocker, N., Gummer, A. W., & Fakler, B. (2001). Reciprocal electromechanical properties of rat prestin: The motor molecule from rat outer hair cells. *Proceedings of the National Academy of Sciences of the USA*, 98(7), 4178–4183.

Manis, P. B., & Brownell, W. E. (1983). Synaptic organization of eighth nerve afferents to cat dorsal cochlear nucleus. *Journal of Neurophysiology*, 50(5), 1156–1181.

Manis, P. B., & Marx, S. O. (1991). Outward currents in isolated ventral cochlear nucleus neurons. *The Journal of Neuroscience*, 11(9), 2865–2880.

Manis, P. B., Xie, R., Wang, Y., Marrs, G. S., & Spirou, G. A. (2011). The Endbulbs of Held. In L. O. Trussell, A. N. Popper & R. R. Fay (Eds.), *Synaptic mechanisms in the auditory system* (Vol. 41, pp. 61–94). New York: Springer Science+Business Media.

Manley, G. A., Fay, R. R., & Popper, A. N. (2008). *Active processes and otoacoustic emissions in hearing* (Vol. 30). New York: Springer Science+Business Media.

Mathews, P. J., Jercog, P. E., Rinzel, J., Scott, L. L., & Golding, N. L. (2010). Control of submillisecond synaptic timing in binaural coincidence detectors by K(v)1 channels. *Nature Neuroscience*, 13(5), 601–609.

Mistrik, P., & Ashmore, J. F. (2010). Reduced electromotility of outer hair cells associated with connexin-related forms of deafness: An in silico study of a cochlear network mechanism. *Journal of the Association for Research in Otolaryngology : JARO*, 11(4), 559–571.

Moore, B. C., & Sek, A. (2009). Sensitivity of the human auditory system to temporal fine structure at high frequencies. *The Journal of the Acoustical Society of America*, 125(5), 3186–3193.

Morimoto, N., Raphael, R. M., Nygren, A., & Brownell, W. E. (2002). Excess plasma membrane and effects of ionic amphipaths on mechanics of outer hair cell lateral wall. *American Journal of Physiology*, 282(5), C1076–1086.

Mountain, D. C. (1980). Changes in endolymphatic potential and crossed olivocochlear bundle stimulation alter cochlear mechanics. *Science*, 210(4465), 71–72.

Nguyen, T. V., & Brownell, W. E. (1998). Contribution of membrane cholesterol to outer hair cell lateral wall stiffness. *Otolaryngology-Head and Neck Surgery*, 119(1), 14–20.

Nicol, M. J., & Walmsley, B. (2002). Ultrastructural basis of synaptic transmission between endbulbs of Held and bushy cells in the rat cochlear nucleus. *The Journal of Physiology*, 539(Pt 3), 713–723.

Oertel, D. (1983). Synaptic responses and electrical properties of cells in brain slices of the mouse anteroventral cochlear nucleus. *The Journal of Neuroscience*, 3(10), 2043–2053.

Oghalai, J. S., Patel, A. A., Nakagawa, T., & Brownell, W. E. (1998). Fluorescence-imaged microdeformation of the outer hair cell lateral wall. *The Journal of Neuroscience*, 18(1), 48–58.

Oleskevich, S., Clements, J., & Walmsley, B. (2000). Release probability modulates short-term plasticity at a rat giant terminal. *The Journal of Physiology*, 524 Pt 2, 513–523.

Oleskevich, S., Youssoufian, M., & Walmsley, B. (2004). Presynaptic plasticity at two giant auditory synapses in normal and deaf mice. *The Journal of Physiology*, 560(Pt 3), 709–719.

Pal, B., Rusznak, Z., Harasztosi, C., & Szucs, G. (2004). Depolarization-activated K^+ currents of the bushy neurones of the rat cochlear nucleus in a thin brain slice preparation. *Acta Physiologica Hungarica*, 91(2), 83–98.

Patuzzi, R. (2011). Ion flow in stria vascularis and the production and regulation of cochlear endolymph and the endolymphatic potential. *Hearing Research*, 277(1–2), 4–19.

Purcell, E. K., Liu, L., Thomas, P. V., & Duncan, R. K. (2011). Cholesterol influences voltage-gated calcium channels and BK-type potassium channels in auditory hair cells. *PLoS One*, 6(10), e26289.

Quraishi, I. H., & Raphael, R. M. (2008). Generation of the endocochlear potential: A biophysical model. *Biophysical Journal*, 94(8), L64–66.

Rabbitt, R. D., Ayliffe, H. E., Christensen, D., Pamarthy, K., Durney, C., Clifford, S., & Brownell, W. E. (2005). Evidence of piezoelectric resonance in isolated outer hair cells. *Biophysical Journal*, 88(3), 2257–2265.

Rabbitt, R. D., Clifford, S., Breneman, K. D., Farrell, B., & Brownell, W. E. (2009). Power efficiency of outer hair cell somatic electromotility. *PLoS Computational Biology*, 5(7), e1000444.

Rajagopalan, L., Patel, N., Madabushi, S., Goddard, J. A., Anjan, V., Lin, F., Shope, C., Farrell, B., Lichtarge, O., Davidson, A. L., Brownell, W. E., & Pereira, F. A. (2006). Essential helix interactions in the anion transporter domain of prestin revealed by evolutionary trace analysis. *The Journal of Neuroscience*, 26(49), 12727–12734.

Rajagopalan, L., Greeson, J. N., Xia, A., Liu, H., Sturm, A., Raphael, R. M., Davidson, A. L., Oghalai, J. S., Pereira, F. A., & Brownell, W. E. (2007). Tuning of the outer hair cell motor by membrane cholesterol. *The Journal of Biological Chemistry*, 282(50), 36659–36670.

Raman, I. M., & Trussell, L. O. (1992). The kinetics of the response to glutamate and kainate in neurons of the avian cochlear nucleus. *Neuron*, 9(1), 173–186.

Rao, D. (2011). *Neuromodulation of intrinsic and synaptic plasticity in auditory cortex.* Ph.D. thesis, The University of North Carolina, Chapel Hill.

Raphael, R. M., Popel, A. S., & Brownell, W. E. (2000). A membrane bending model of outer hair cell electromotility. *Biophysical Journal*, 78(6), 2844–2862.

Rathouz, M., & Trussell, L. (1998). Characterization of outward currents in neurons of the avian nucleus magnocellularis. *Journal of Neurophysiology*, 80(6), 2824.

Ratnanather, J. T., Zhi, M., Brownell, W. E., & Popel, A. S. (1996). Measurements and a model of the outer hair cell hydraulic conductivity. *Hearing Research*, *96*(1–2), 33–40.

Rautenberg, P. L., Grothe, B., & Felmy, F. (2009). Quantification of the three-dimensional morphology of coincidence detector neurons in the medial superior olive of gerbils during late postnatal development. *The Journal of Comparative Neurology*, *517*(3), 385–396.

Recio, A., & Rhode, W. S. (2000). Basilar membrane responses to broadband stimuli. *Journal Acoustical Society of America*, *108*(5 Pt 1), 2281–2298.

Recio-Spinoso, A., Temchin, A. N., van Dijk, P., Fan, Y. H., & Ruggero, M. A. (2005). Wiener-kernel analysis of responses to noise of chinchilla auditory-nerve fibers. *Journal of Neurophysiology*, *93*(6), 3615–3634.

Reyes, A. D., Rubel, E. W., & Spain, W. J. (1994). Membrane properties underlying the firing of neurons in the avian cochlear nucleus. *The Journal of Neuroscience*, 14(9), 5352.

Rhode, W. S. (1971). Observations of the vibration of the basilar membrane in squirrel monkeys using the Mossbauer technique. *Journal Acoustical Society of America*, 49(4), Suppl 2:1218–1231.

Rhode, W. S., Oertel, D., & Smith, P. H. (1983). Physiological response properties of cells labeled intracellularly with horseradish peroxidase in cat ventral cochlear nucleus. *The Journal of comparative neurology*, 213(4), 448–463.

Ribrault, C., Sekimoto, K., & Triller, A. (2011). From the stochasticity of molecular processes to the variability of synaptic transmission. *Nature reviews. Neuroscience*, 12(7), 375–387.

Romand, R. (1978). Survey of intracellular recording in the cochlear nucleus of the cat. *Brain Research*, 148(1), 43–65.

Rothman, J. S., & Young, E. D. (1996). Enhancement of neural synchronization in computational models of ventral cochlear nucleus bushy cells. *Auditory Neuroscience*, 2, 47–62.

Rothman, J. S., & Manis, P. B. (2003a). Kinetic analyses of three distinct potassium conductances in ventral cochlear nucleus neurons. *Journal of Neurophysiology*, 89(6), 3083–3096.

Rothman, J. S., & Manis, P. B. (2003b). Differential expression of three distinct potassium currents in the ventral cochlear nucleus. *Journal of Neurophysiology*, 89(6), 3070–3082.

Rothman, J. S., & Manis, P. B. (2003c). The roles potassium currents play in regulating the electrical activity of ventral cochlear nucleus neurons. *Journal of Neurophysiology*, 89(6), 3097–3113.

Rothman, J. S., Young, E. D., & Manis, P. B. (1993). Convergence of auditory nerve fibers onto bushy cells in the ventral cochlear nucleus: Implications of a computational model. *Journal of Neurophysiology*, 70(6), 2562–2583.

Rouiller, E. M., & Ryugo, D. K. (1984). Intracellular marking of physiologically characterized cells in the ventral cochlear nucleus of the cat. *The Journal of Comparative Neurology*, 225(2), 167–186.

Rybalchenko, V., & Santos-Sacchi, J. (2003). Cl- flux through a non-selective, stretch-sensitive conductance influences the outer hair cell motor of the guinea-pig. *The Journal of Physiology*, 547(Pt 3), 873–891.

Santos-Sacchi, J. (1989). Asymmetry in voltage-dependent movements of isolated outer hair cells from the organ of Corti. *The Journal of Neuroscience*, 9(8), 2954–2962.

Santos-Sacchi, J., & Dilger, J. P. (1988). Whole cell currents and mechanical responses of isolated outer hair cells. *Hearing Research*, 35(2–3), 143–150.

Schwartz, I. R. (1977). Dendritic arrangements in the cat medial superior olive. *Neuroscience*, 2(1), 81–101.

Schwarz, D. W., & Puil, E. (1997). Firing properties of spherical bushy cells in the anteroventral cochlear nucleus of the gerbil. *Hearing Research*, 114(1–2), 127–138.

Scott, L. L., Hage, T. A., & Golding, N. L. (2007). Weak action potential backpropagation is associated with high-frequency axonal firing capability in principal neurons of the gerbil medial superior olive. *The Journal of Physiology*, 583(Pt 2), 647–661.

Scott, L. L., Mathews, P. J., & Golding, N. L. (2010). Perisomatic voltage-gated sodium channels actively maintain linear synaptic integration in principal neurons of the medial superior olive. *The Journal of Neuroscience*, 30(6), 2039–2050.

Sento, S., & Ryugo, D. K. (1989). Endbulbs of held and spherical bushy cells in cats: Morphological correlates with physiological properties. *The Journal of Comparative Neurology*, 280(4), 553–562.

Sewell, W. F. (1984). The effects of furosemide on the endocochlear potential and auditory-nerve fiber tuning curves in cats. *Hearing Research*, 14(3), 305–314.

Sfondouris, J., Rajagopalan, L., Pereira, F. A., & Brownell, W. E. (2008). Membrane composition modulates prestin-associated charge movement. *The Journal of Biological Chemistry*, 283(33), 22473–22481.

Shamma, S. A., Elhilali, M., & Micheyl, C. (2011). Temporal coherence and attention in auditory scene analysis. *Trends in Neurosciences*, 34(3), 114–123.

Shehata, W. E., Brownell, W. E., & Dieler, R. (1991). Effects of salicylate on shape, electromotility and membrane characteristics of isolated outer hair cells from guinea pig cochlea. *Acta Oto-Laryngologica*, 111(4), 707–718.

Shera, C. A. (2001). Intensity-invariance of fine time structure in basilar-membrane click responses: implications for cochlear mechanics. *The Journal of the Acoustical Society of America*, 110(1), 332–348.

Shera, C. A., & Olson, E. S. (2011). *What fire is in mine ears: Progress in auditory biomechanics* (Vol. 1403). Williamstown, MA: AIP Conference Proceedings

Shinn-Cunningham, B. G., & Wang, D. (2008). Influences of auditory object formation on phonemic restoration. *The Journal of the Acoustical Society of America*, 123(1), 295–301.

Smith, P. H. (1995). Structural and functional differences distinguish principal from nonprincipal cells in the guinea pig MSO slice. *Journal of Neurophysiology*, 73(4), 1653–1667.

Smith, P. H., Joris, P. X., & Yin, T. C. (1998). Anatomy and physiology of principal cells of the medial nucleus of the trapezoid body (MNTB) of the cat. *Journal of Neurophysiology*, 79(6), 3127–3142.

Spector, A. A., Brownell, W. E., & Popel, A. S. (2003). Effect of outer hair cell piezoelectricity on high-frequency receptor potentials. *The Journal of the Acoustical Society of America*, 113(1), 453–461.

Spector, A. A., Popel, A. S., Eatock, R. A., & Brownell, W. E. (2005). Mechanosensitive channels in the lateral wall can enhance the cochlear outer hair cell frequency response. *Annals of Biomedical Engineering*, 33(8), 991–1002.

Spirou, G. A., Brownell, W. E., & Zidanic, M. (1990). Recordings from cat trapezoid body and HRP labeling of globular bushy cell axons. *Journal of Neurophysiology*, 63(5), 1169–1190.

Spirou, G. A., Rager, J., & Manis, P. B. (2005). Convergence of auditory-nerve fiber projections onto globular bushy cells. *Neuroscience*, 136(3), 843–863.

Street, S. E., & Manis, P. B. (2007). Action potential timing precision in dorsal cochlear nucleus pyramidal cells. *Journal of Neurophysiology*, 97(6), 4162–4172.

Strumbos, J. G., Polley, D. B., & Kaczmarek, L. K. (2010). Specific and rapid effects of acoustic stimulation on the tonotopic distribution of Kv3.1b potassium channels in the adult rat. *Neuroscience*, 167(3), 567–572.

Svirskis, G., Kotak, V., Sanes, D. H., & Rinzel, J. (2002). Enhancement of signal-to-noise ratio and phase locking for small inputs by a low-threshold outward current in auditory neurons. *The Journal of Neuroscience*, 22(24), 11019–11025.

Takahashi, S., & Santos-Sacchi, J. (2001). Non-uniform mapping of stress-induced, motility-related charge movement in the outer hair cell plasma membrane. *Pflugers Archive: European Journal of Physiology*, 441(4), 506–513.

Typlt, M., Englitz, B., Sonntag, M., Dehmel, S., Kopp-Scheinpflug, C., & Ruebsamen, R. (2012). Multidimensional characterization and differentiation of neurons in the anteroventral cochlear nucleus. *PLoS One*, 7(1), e29965.

Tzounopoulos, T., Kim, Y., Oertel, D., & Trussell, L. O. (2004). Cell-specific, spike timing-dependent plasticities in the dorsal cochlear nucleus. *Nature Neuroscience*, 7(7), 719–725.

Wang, Y., & Manis, P. B. (2005). Synaptic transmission at the cochlear nucleus endbulb synapse during age-related hearing loss in mice. *Journal of Neurophysiology*, 94(3), 1814–1824.

Wang, Y. X., Wenthold, R. J., Ottersen, O. P., & Petralia, R. S. (1998). Endbulb synapses in the anteroventral cochlear nucleus express a specific subset of AMPA-type glutamate receptor subunits. *The Journal of Neuroscience*, 18(3), 1148–1160.

Weitzel, E. K., Tasker, R., & Brownell, W. E. (2003). Outer hair cell piezoelectricity: Frequency response enhancement and resonance behavior. *The Journal of the Acoustical Society of America*, 114(3), 1462–1466.

Wen, B., Wang, G. I., Dean, I., & Delgutte, B. (2009). Dynamic range adaptation to sound level statistics in the auditory nerve. *The Journal of Neuroscience*, 29(44), 13797–13808.

Wen, B., Wang, G. I., Dean, I., & Delgutte, B. (2012). Time course of dynamic range adaptation in the auditory nerve. *Journal of Neurophysiology*, 108(1), 69–82.

Wenthold, R. J., Zempel, J. M., Parakkal, M. H., Reeks, K. A., & Altschuler, R. A. (1986). Immunocytochemical localization of GABA in the cochlear nucleus of the guinea pig. *Brain Research*, 380(1), 7–18.

Wenthold, R. J., Huie, D., Altschuler, R. A., & Reeks, K. A. (1987). Glycine immunoreactivity localized in the cochlear nucleus and superior olivary complex. *Neuroscience*, 22(3), 897–912.

Wu, S. H., & Oertel, D. (1984). Intracellular injection with horseradish peroxidase of physiologically characterized stellate and bushy cells in slices of mouse anteroventral cochlear nucleus. *The Journal of Neuroscience*, 4(6), 1577–1588.

Wu, S. H., & Oertel, D. (1986). Inhibitory circuitry in the ventral cochlear nucleus is probably mediated by glycine. *The Journal of Neuroscience*, 6(9), 2691–2706.

Xie, R. & Manis, P.B. (2013). Target-specific IPSC kinetics promote temporal processing in auditory parallel pathways. *The Journal of Neuroscience*, 33(4), 1598–1614.

Yan, K., & Matthews, G. (1992). Blockers of potassium channels reduce the outward dark current in rod photoreceptor inner segments. *Visual Neuroscience*, 8(5), 479–481.

Yang, H., & Xu-Friedman, M. A. (2012). Emergence of coordinated plasticity in the cochlear nucleus and cerebellum. *The Journal of Neuroscience*, 32(23), 7862–7868.

Yang, S., & Feng, A. S. (2007). Heterogeneous biophysical properties of frog dorsal medullary nucleus (cochlear nucleus) neurons. *Journal of Neurophysiology*, 98(4), 1953–1964.

Zhang, R., Qian, F., Rajagopalan, L., Pereira, F. A., Brownell, W. E., & Anvari, B. (2007). Prestin modulates mechanics and electromechanical force of the plasma membrane. *Biophysical Journal*, 93(1), L07–09.

Zhang, X., & Carney, L. H. (2005). Response properties of an integrate-and-fire model that receives subthreshold inputs. *Neural Computation*, 17(12), 2571–2601.

Zheng, J., Shen, W., He, D. Z., Long, K. B., Madison, L. D., & Dallos, P. (2000). Prestin is the motor protein of cochlear outer hair cells. *Nature*, 405(6783), 149–155.

Zidanic, M., & Brownell, W. E. (1989). Low-frequency modulation of the intracochlear potential field: An energy source for the cochlear amplifier. *Il Valsalva*, 65(Suppl. 1), 20–27.

Zidanic, M., & Brownell, W. E. (1990). Fine structure of the intracochlear potential field. I. The silent current. *Biophysical Journal*, 57(6), 1253–1268.

Zidanic, M., & Brownell, W. E. (1992). Fine structure of the intracochlear potential field. II. Tone-evoked waveforms and cochlear microphonics. *Journal of Neurophysiology*, 67(1), 108–124.

Chapter 3
Human Auditory Cortex: In Search of the Flying Dutchman

John F. Brugge

John F. Brugge

J.F. Brugge (✉)
Department of Neuroscience, University of Wisconsin–Madison, Madison, WI 53706, USA

Department of Neurosurgery, University of Iowa, Iowa City, IA 52242, USA
e-mail: brugge.johnf@gmail.com

A.N. Popper and R.R. Fay (eds.), *Perspectives on Auditory Research*, Springer Handbook of Auditory Research 50, DOI 10.1007/978-1-4614-9102-6_3,

3.1 Introduction: Beginnings

> I believe that excessive admiration for the work of great minds is one of the most unfortunate preoccupations of intellectual youth.
>
> —Ramon y Cajal, *Advice for a Young Investigator*, 1897/1999

There are some things you hear that just stick in your mind. Sometime in my first year in grad school an eminent visiting neuroanatomist (who will remain unnamed) proclaimed to a group of us grad students that research was essentially a series of shipwrecks. Whether this was meant to warn or scare us or maybe to steer us into banking or law careers I don't know, but it certainly grabbed our attention. Whatever the intention, it was not good news, and we soon learned the hard truth of this. We also discovered that every so often we steer clear of the rocks only to veer off in directions we never would have imagined.

I came to the University of Wisconsin as a postdoc in 1963, drawn there by a group of neurophysiologists (the term "neuroscience" had not yet worked its way into our lexicon) led by Clinton Woolsey. Postdocs were given enormous freedom then to pursue whatever questions they thought interesting and important and to work with whoever would take them in. I was lucky enough to be taken in by Jerzy Rose and Joe Hind, two of the great pioneers of auditory neuroscience. Jerzy, known for his sage remarks, once commented that the thalamus could be likened to the Flying Dutchman, which one may recall was the ghost ship that many may had heard of but few had ever seen. The same might have been said in 1963 for auditory cortex.

One of my first experiments at Wisconsin was a single neuron study of auditory cortex of the cat. I had never before operated on a cat nor had I ever seen an action potential; it was a thrill and, to continue a metaphor, the sirens called. Over these past 50 years I've had the great fortune to work with extraordinarily talented colleagues on a winding road that took us up and down the auditory pathways. Rick Reale is one of those, and for more than 30 years I've been enormously privileged to have him as a colleague, teacher, and friend. Nearing the end of that road, and ignoring the fact that according to legend the sighting of Rose's phantom ship is considered to be a nasty omen portending doom, I found myself having come full circle from that first cat experiment to be offered the opportunity to help take on one of the great challenges in neuroscience, probing the workings of auditory cortex of humans.

The story of auditory cortex as we know it today can be traced to the late 19th and early 20th centuries with such intellectual giants as Paul Flechsig, Santiago Ramon y Cajal, Cécile and Oskar Vogt, Alfred Campbell, Korbinian Brodmann, and Constantine von Economo, among others, laying out in detail the cyto- and myeloarchitecture of what was then believed to be auditory cortex of the mammalian (including human) brain. But from there further progress slowed to a crawl, for even though these early scientists had hunches about the functionality of their anatomically identified cortical fields, they had few experimental tools to test how right or wrong those hunches might be. Indeed, until the 1940s there was no more than a scattering of research reports on auditory cortex, and many of these were clinical case studies of deficits associated with cortical lesions.

Advances in electrophysiology in the 1920s ushered in a revolution in neurophysiology. This revolution was somewhat late in reaching auditory cortex, but the pace would quicken with the advent of electroencephalography (EEG) in the 1930s. And it would propel research on human and nonhuman auditory cortex along two separate paths.

Just as World War II was breaking out, Pauline and Hallowell Davis, working between the private Tuxedo Park (New York) laboratory of polymath Alfred Loomis and Harvard Medical School, carried out the first systematic studies of auditory evoked potentials recorded from the human scalp. From then onward, functional studies of human auditory cortex came to rely heavily on scalp recording of neuroelectric activity. The field was to receive a jolt in 1964 when W. Grey Walter and his colleagues working in Bristol, England, reported auditory event-related potentials that included early deflections securely time locked to the onset of sensory events but also later and more labile stimulus related activity—the contingent negative variation—associated with perceptual and cognitive processes. Electroencephalography, along with later developed magnetoencephalography (MEG) and functional imaging, continue to be mainstay approaches to functional studies of human auditory cortex.

Shortly after auditory evoked potential recordings were reported, Clint Woolsey and Edward Walzl, working at Johns Hopkins, recorded voltage deflections directly from the cortical surface of both cat and monkey evoked by focal stimulation of the osseous spiral lamina and in doing so mapped for the first time the cochleotopic organization of auditory cortex. To this day, experiments on auditory cortex of nonhuman mammals have relied almost entirely on invasive approaches, including recording of single neuron action potentials and local field potentials, controlled lesions, and anatomical tracing of neural connectivity.

Over the next 50 years these two paths of auditory cortical research—invasive studies in lab animals and noninvasive studies in human subjects—rarely intersected, and when they did it was only for brief periods of time. But that has changed dramatically in the past 10 years or so, with several research groups now engaged in direct recording from auditory cortex of human neurosurgical patients. This work is bridging the human–animal gap, adding interpretive power to mainline noninvasive approaches and providing new insight into cortical mechanisms that underlie human auditory sensory processing, perception, and cognition.

3.2 Tools of the Trade

> But lo! Men have become the tools of their tools.
>
> —Henry David Thoreau, *Walden*, 1854

We can only imagine the surprise on the face of that great British electrophysiologist Edgar (Lord) Adrian when, having placed the tip of an electrolyte-filled glass micropipette electrode on the optic nerve of a toad, he first realized the noise that a sprang from loudspeakers attached to his recording amplifier was in fact the neuronal discharge related directly to the toad's eye tracking his movements. As important as it was (and still is) to studies of the synaptic and membrane properties of auditory cortical neurons, the micropipette electrode turned out not to be the tool of choice for mapping the functional organization of auditory cortex or for studying the coding properties of its neurons.

Before the mid-1950s functional mapping of auditory cortex was typically carried out using a metal macroelectrode having a terminal ball of 0.5–1 mm diameter, which was brought into direct contact with the cortical pial surface. The electrode typically was held in a micromanipulator to allow systematic placement of the ball at 1-mm intervals, in a grid-like fashion. To speed things up Woolsey devised an array of nine contacts, which was plugged into a bank of nine preamplifiers that fed nine oscilloscope screens. We called it the "old nine-channel." The auditory evoked potentials displayed on an oscilloscope face were photographed on 35 mm film. Images were later transferred to 35 mm photographic paper to be cut out and pasted in their respective locations on a large drawing of that animal's major cortical gyri and sulci. This approach worked well then for animals with relatively large cortical auditory areas (e.g., cat and dog), and a technically updated version of it works just as well today for humans.

Around the time Woolsey was generating tonotopic maps of cat auditory cortex, Rose was discovering the close relationships that exist between tonotopy and underlying cytoarchitecture and medial geniculate projections. Although that work was based on Nissl cytoarchitecture and retrograde degeneration of thalamic neurons following cortical lesions, it set the bar high for all future studies of structure–function relationships in auditory cortex. Moreover it demanded higher spatial resolution for the functional maps and more sensitive ways of delineating anatomical boundaries and connectivity.

After learning from Woldring and Dirken in 1950 that spontaneous spike activity could be recorded from the cortex through the cut end of an insulated wire, there was a flurry of activity to fabricate microelectrodes from materials other than glass that would be robust enough to penetrate the cortex repeatedly, record extracellular action potentials from single neurons or neuronal clusters for hours at a time, and leave a visible scar that could be seen later in stained tissue sections under the microscope. Fitting the bill were indium (Dowben & Rose, 1953), tungsten (Hubel, 1957), stainless steel (Green, 1958), and platinum-iridium (Wolbarsht et al., 1960). Whereas a macroelectrode recorded the summed activity of an enormous population of neurons beneath the 0.5–1 mm^2 patch of tissue covered by the electrode contact, a sharpened metal microelectrode was capable of recording from clusters of a few neurons at dozens of different loci within each mm^2, and within each of these mm^2 patches could be found a highly differentiated projection pattern.

Nearly 25 years after Woolsey and Walzl mapped auditory cortex of the monkey and 20 years after Rose and Woolsey reported on structure–function relationships in the auditory forebrain of the cat, Mike Merzenich and I took up fine grain microelectrode mapping of tonotopic organization of auditory cortex of the macaque monkey, relating the maps obtained to underlying cytoarchitecture. Those results stood for another 20 years until Ann Morel, working in Jon Kaas' lab at Vanderbilt

University, confirmed and extended our findings, and using neuronal tracers related the tonotopic maps obtained to corticocortical and thalamocortical connectivity. Shortly thereafter, Troy Hackett and Jon Kaas, in a tour de force of modern neuroanatomy, added much to the fundamental knowledge of cellular architecture and neuronal connectivity that frames our current model of auditory cortical organization in human and nonhuman primates.

Theodor Meynert, the German-Austrian anatomist of the late 19th century considered by many the father of cortical cytoarchitectonics, paid particular attention to that dense network of fibers within the cortical mantle and attempted to associate the connections these axons make with mental processes. He also had a great influence on Karl Wernicke, who made clinical observations following brain lesions and drew circuit diagrams representing what he believed to be the underlying anatomical connections interrupted by the lesion that could account for the clinical manifestations, including and especially the complex language disorder that bears his name. Henry Head later derisively dismissed him as a "diagram maker," a moniker that many of us would proudly wear today.

Until the 1950s the only methods available for tracing neural pathways experimentally were the Marchi method, which relied on anterograde degeneration of myelinated axons following surgical destruction of their cell bodies, and retrograde degeneration, the method used by Rose in his structure–function study of cat auditory cortex. All of this was to change dramatically when, in 1951 Walle Nauta, then a researcher at the Walter Reed Army Research Institute, introduced the modified reduced silver method of staining degenerating axons that came to bear his name. This method, with subsequent refinements, remained the way experimental tract tracing was carried out until the advent in the 1970s of techniques based on axonal transport of molecules made visible under the light and electron microscopes. The creative application of these new and powerful molecular tracer methods, combined with detailed microelectrode mapping, gave us the current structure–function model of primate auditory cortex. These experimental tract-tracing approaches used so effectively in laboratory animals for obvious reasons cannot be used in humans. Indeed, until the advent of noninvasive MR diffusion tensor imaging (DTI) of white matter in the living brain and the use of lipophilic carbocyanine dyes in fixed brain tissue, anatomical tracing of connections in human cortex was carried out by blunt dissection in autopsy specimens or by microscopic examination of serial tissue sections stained for myelinated axons. In the past decade or so a new chapter has opened up to study functional connectivity in the human cortex using an old methodology.

Initially, animal-based auditory cortical mapping by evoked potentials or multiunit activity and the tracing of anatomical connectivity patterns were to a large extent qualitative exercises that were both time and labor intensive. Questions of how single neuron or neuronal assemblies encode a sound in trains of action potentials were essentially out of reach, and the same held true for complex processes unveiled by neuroelectric signals recorded noninvasively. The development of a digital computer that could be placed in the hands of the individual investigator was to change all of that.

When I arrived at Wisconsin a large mainframe computer, located in the physics department and serving the entire university, was the only tool available (if you don't count a Burroughs mechanical calculator) for analyzing rapidly accumulating electrophysiological data. I (along with most other electrophysiologists of my generation) learned to program in Fortran and laboriously to create a stack of punched cards, which then was the primary medium for entering lines of program code and experimental data. I trudged back and forth regularly to submit my card stacks, being careful not to fold, bend, spindle, or mutilate them and later (usually the next day) to retrieve the printed fan-folded output, the pages of which were about the size of pages from *The New York Times*. But fortunes soon changed when my colleagues Joe Hind and Dan Geisler were chosen to participate in a program set up by the National Institutes of Health (NIH) to evaluate a relatively small digital computer designed with the laboratory scientist in mind.

Conceived around 1960 by Wesley Clark at MIT's Lincoln Laboratory, the Laboratory Instrument Computer, the LINC, would soon come to transform the bioscience research laboratory. It was easy to program, to communicate with, and to maintain. It was relatively cheap, no larger than a refrigerator, and able to control experiments directly and process biological signals online. Consider the prevailing vision of the computer of the future that appeared in a *Popular Mechanics* magazine article by Andrew Hamilton just a decade earlier: "Where a calculator like the ENIAC is equipped with 18,000 vacuum tubes and weights 30 tons, computers in the future may have only 1,000 vacuum tubes and perhaps only weigh 1 1/2 tons." What the LINC required was not technical innovation but a radically different way of thinking about how a computer could meet the needs of the laboratory scientist. As a footnote to LINC (and auditory neuroscience) history, Wes Clark and grad student Charlie Molnar first demonstrated the use of the LINC in the lab of Robert Livingston, who was then scientific director of the National Institute of Neurological Disorders and Blindness (NINDB) and the National Institute of Mental Health (NIMH), and a major supporter of the LINC program. Working in his lab at the time, and on the auditory system of the cat, was postdoc Arnold Starr. Within hours after being set up the LINC would collect data from one of Arne's cats and another revolution in neurophysiology would begin. At a celebration of the 20th anniversary of the LINC Samuel Rosenfeld recounted Livingston's reaction at that time: "it was such a triumph that we danced a jig right there around the equipment."

Joe and Dan spent the summer of 1963 at MIT building, testing, and programming a LINC and shipping it back to Wisconsin, where it, along with several others of its kind that were later obtained, became the workhorses of our auditory labs for more than 25 years. Not long after the first LINC was installed, telephone cables strung between the surgical operating rooms of the Wisconsin General Hospital with our labs several floors below where the LINC was located were put to work. Neurologist Gastone Celesia and neurosurgeon (and first cousin) Flavio Puletti began recording systematically for the first time auditory evoked activity directly from the exposed temporal lobe of the human brain, during neurosurgical procedures, controlling the LINC remotely via a 12-button Touch Tone telephone. The LINC used all of its 1 K of memory to compute the averaged auditory evoked potential, a feat first performed on scalp recordings by Dan Geisler as a grad student at MIT just a few years earlier. Celesia and Puletti went on to describe the fundamental temporal features of the evoked waveforms recorded from auditory cortex deep within the lateral fissure and on the free lateral surface of the superior temporal gyrus (STG). All this was all going on right in front of me, but more than 35 years had to go by before I would come to understand and appreciate fully the importance of this pioneering work. This was also one of those rare times when studies of human auditory cortex intersected those of nonhumans. As ships passing in the night, we wouldn't get to see this happen again until the 1980s, when Otto Creutzfeldt and George Ojemann teamed up at the University of Washington to record, again for the first time, the responses of single neurons in human auditory cortex to speech using those same tungsten microelectrodes so commonly and effectively used in invasive studies in lab animals.

3.3 Enter the Modern Era

Cats and monkeys, monkeys and cats; all human life is there.

—Henry James, *The Madonna of the Future*, 1873

To see where our field stood 20 or so years ago we need only browse through the nine volumes of *Cerebral Cortex* edited by Alan Peters and E. G. Jones. Just two chapters (118 pages) were devoted to auditory cortex, one reviewing literature on structure-function relationships in nonhumans the other on the cellular makeup of human auditory cortex as seen in Golgi stained material. The chapters are included in Volume 4 published in 1985 and entitled *Association and Auditory Cortices*. Visual cortex got its own volume while auditory cortex was found relegated to the "miscellaneous" category. Five years later, Lindsay Aitkin came to the rescue with a small but influential book, *Auditory Cortex: Structural and Functional Bases of Auditory Perception* (1990), which even today is well worth keeping on the same shelf as *The Auditory Cortex*, edited by Jeff Winer and Christoph Schreiner, and the SHAR volume, Human Auditory Cortex, edited by David Poeppel, Tobias Overath, Arthur Popper, and Richard Fay.

It was also just about 25 years ago that the National Institute on Deafness and Other Communication Disorders (NIDCD) received its authorization as an independent agency of the NIH and by the early 1990s was reviewing grant applications and funding hearing research. The agency soon found its footings and the impact on auditory research became enormous. By the end of the 1980s the midwinter meeting of the Association for Research in Otolaryngology (ARO) had gained momentum to become *the* place for auditory system researchers to present, discuss, and argue their latest findings, all over a cold beer. In 2000 the ARO launched its own journal to stand beside 22-year-old *Hearing Research*. We're now seeing more national and international symposia devoted to human auditory cortex or to a comparative view from animal and human based studies. The field of auditory neuroscience has taken its place at the table, and with it auditory cortex.

Faster computers, proliferating anatomical tracer techniques, more powerful signal processing algorithms—these are just a few of the technical advances of the past 20 years or so that have accelerated the pace of research on auditory cortex. We can now outline with considerable confidence some of the fundamental features of structural and functional organization of auditory cortex gained both from invasive studies in laboratory animals and noninvasive studies in humans.

The current structure–function model of "classic" auditory cortex, which has evolved from microelectrode mapping and correlative neuronal tracer studies in a variety of laboratory mammals but especially monkeys, envisions multiple auditory fields connected with one another and with the auditory thalamus so as to facilitate serial and parallel hierarchical processing of acoustic information. While exercising some interpretive restraint we may identify auditory fields functionally by a family of response characteristics exhibited by their constituent neurons, a "physiological fingerprint," if you will. These response characteristics, displayed mainly by those neurons in the thalamo-recipient layers III and IV, include the shape and sharpness of frequency tuning, relationships of discharge rate to changes in stimulus intensity, latency to stimulus onset, sensitivity to frequency modulated sounds, capacity to phase lock to the envelope of amplitude modulated sounds including those of running speech, and sensitivity to increasing sound complexity, to interaural localization cues of time and intensity, and to the location of sound in space. Synaptic and neuronal membrane properties of auditory cortical neurons that underlie thalamocortical transformations and auditory coding at the cellular level are becoming known. Single neuron and ensemble coding strategies are now better understood and can be related to human sensory processing and perception. Although it has long been thought that much of classic auditory cortex, and the core area in particular, is unisensory in function, we may have to change that view as a substantial portion of temporal auditory cortex may be multisensory, in one way or another. The original primate model of Hackett and Kaas has also been extended to include "auditory-related" areas of the parietal and frontal lobes, so named because they receive much if not all of their auditory input from temporal lobe auditory cortical fields. Drawing from a model of visual cortical organization, dual pathways from temporal auditory fields to auditory related cortex are posited to carry information about the identity (the "what") of a sound source along one leg and the location (the "where") of that source along another. This expanded model captures the major features of a highly distributed and interactive forebrain complex.

Within just a few years of employing the micromapping approach, details emerged on the functional organization of auditory cortex that had not been so evident in the macroelectrode maps. Particularly striking were the shapes and orientations of isofrequency contours and extent of cortex devoted to any given frequency band, which could differ greatly from one animal to the next of the same species (or even litter). Moreover, the boundaries and internal structure of tonotopic maps showed no straightforward relationship to gross gyral patterns, which themselves exhibited considerable intersubject variability. There was confusion over whether cortical tonotopy even existed, a situation that arose in part from attempts to derive a common tonotopic map based on microelectrode recording data pooled from experiments on individual animals using common gyral landmarks as spatial references. Once it was recognized that pooling of data only blurred details of functional maps, the practice was begun of deriving maps from hundreds of cortical recording sites in a single experiment. Christoph Schreiner and his colleagues no doubt hold the record in this regard. It didn't take long to realize that individual differences in auditory map structure were related environmental factors, and so with the groundbreaking work of Mike Merzenich, Dexter Irvine, Nina Kraus, and Norm Weinberger and their coworkers, a new field of study was ushered in—auditory cortical plasticity in adults. What were we thinking just 30 years or so ago when we accepted the scientific wisdom of the day that whereas the young brain may undergo environmentally driven change, the adult brain had finished its work and we were essentially stuck with what we got?

While knowledge of auditory cortical functional organization and information processing based on invasive studies was advancing on one front, new knowledge was also emerging from noninvasive EEG and MEG studies of human auditory cortex pioneered by Terry Picton and Christo Pantev, among others. Both EEG and MEG capture the temporal structure of cortical activity evoked by an acoustic stimulus, but they are saddled with the well-known "inverse problem," meaning that the sources of cortical active foci contributing to the surface recorded neuroelectric signals are ambiguous. Here, then, are experimental tools that give important information on the underlying temporal processing of an acoustic event, but that are not suited to telling us exactly where over the cortical surface those events are taking place.

The landscape of human brain research was to undergo a dramatic change with the introduction of magnetic resonance imaging (MRI) as a safe, noninvasive method of providing unsurpassed anatomical detail of the structure of the living human brain. Soon thereafter functional MRI and positron emission tomography (PET), which visualize changes in regional blood flow and cellular metabolism associated with changes in local cortical activity, were unleashed. Functional images provide a global view of cortical activity and the spatial resolution to localize small areas of cortex activated by a sound. Now and for the first time investigators had in their hands tools that could address some of the same questions addressed in invasive lab animal studies, and perhaps even more importantly, answer questions related to perception and cognition not easily approachable by EEG or MEG. For auditory physiologists and cognitive neuroscientists there was optimism in the air. But the auditory system tends to operate over a wide range of time scales, some measured as short as milliseconds, and because changes in blood flow and cellular metabolism on which functional imaging methods are based are relatively slow processes measured in seconds, they are incapable of capturing the rapid events that occur in the natural environment, including those critical to speech perception. In other words, functional imaging could help tell us where on cortex neural activity was taking place, but it couldn't tell us anything about the short-time-dependent neural processes associated with that activity.

One solution was to combine functional imaging and neuroelectric recording, thereby taking advantage of the spatial resolution of the former and temporal resolution of the latter. With functional imaging in the experimental tool box, knowledge of where and how complex sound, including speech, is represented in auditory cortex has been greatly advanced, and in the hands of Jeff Binder, Tim Griffiths, Greg Hickok, Ingrid Johnsrude, David Poeppel, Sophie Scott, and Robert Zatorre, among others, a new functional model has emerged. Sounds of every degree of complexity, from tones and noise to running speech, have been employed. As in the monkey, there is a spectrotemporal representation within an auditory core, although there is still some dispute over how many tonotopic maps are found there and how these maps relate to the gross and microscopic anatomy of Heschl's gyrus and surrounding cortex.

Phase locking to temporal modulations, including the envelope of running speech, is attributed to the core field, which is consistent with what we know of the core field from intracranial studies in humans and microelectrode recording in monkeys. Beyond the core things get a bit fuzzy. As many as six belt fields have been identified histochemically, but attempts to determine their respective functional roles have been met with mixed results. A transformation from the extraction of acoustic parameters by auditory core to a phonetic representation of speech points to auditory belt fields, and this may extend to the lateral free surface of the STG. Staying with a general hierarchical framework, an interface between phonetic and lexical processing is moved to cortex around the superior temporal sulcus and middle temporal gyrus, with further lexical processing engaging perhaps inferior temporal and basal temporal cortex. Dual pathways to prefrontal cortex, now shown also arising from temporal cortex beyond the STG, are postulated to play additional roles in mapping "sound meaning" and "sound-to-articulation" to prefrontal cortex. And then we need to contend with issues of cerebral lateralization that have been with us since Paul Broca's descriptions 140 years ago of language deficits following frontal lobe lesions. Here also we may depart from monkeys.

3.4 Been There, Done That

> If I have seen further it is by standing on the shoulders of Giants.
>
> —Isaac Newton in a letter to Robert Hooke, 1675

Well before the modern era multiple auditory cortical fields in humans and a variety of laboratory animals had been described and functions ascribed to them. Alfred Walter Campbell, in his monumental 1905 monograph, *Histological Studies on the Localization of Cerebral Function*, identified multiple areas of the human temporal lobe based on cellular architecture, and taking into account the extant experimental, clinicopathological, and anatomical evidence posited a structure/function scheme that has a familiar ring to us today. He summarized his findings thusly: "The first and perhaps the most firmly-grounded conclusions derived from this research is, that the area of cortex laid down for the primary reception of simple auditory stimuli is that which I have mapped out and described as mainly covering the transverse temporal gyri of Heschl" (p. 149). A century later his "audito-sensory area" has been renamed (primary and primary-like auditory cortex, auditory core cortex) but we haven't found a better way of describing it. Campell went on to describe a "skirt" or "belt" of cortex around this core field, suggesting that it is "specialized for the elaboration of primary stimuli" and may have a "psychic function."

We may often think of adult brain plasticity as a modern concept, but the early thinkers were well aware of the phenomenon. Korbinian Brodmann, in his 1909 monograph, *Localization in the Cerebral Cortex*, refers to recovery of function following brain damage as the principle of "functional replacement" and adds that "clinical pathology also recognizes such cortical plasticity." Santiago Ramon y Cajal went even further, noting in his *Advice for a Young Investigator*: "When one reflects on the ability that humans display for modifying and refining mental activity related to a problem under serious examination, it is difficult to avoid concluding that the brain is plastic and goes through a process of anatomical and functional differentiation, adapting itself progressively to the problem" (p. 61). So even as the early auditory cortical anatomists published their static maps, as we do today, they knew full well that at some structural level they were chasing a moving target. This is not meant to distract from all the progress made since then, but simply to remind ourselves that we weren't the first ones there. In the words of Yogi Berra, "It's déjà vu all over again."

3.5 Convergence

Mind the Gap

—Heard and seen in the London Underground

Invasive anatomical and electrophysiological studies in lab animals and architectural and noninvasive neuroelectric recording and functional imaging studies in humans have run parallel courses. To reconcile findings from these two bodies of research and attempt to arrive at a pan-primate model we tend to work under the assumption that the basic organization of the mammalian brain is largely invariant across species. Because certain anatomical and physiological features of auditory cortex are shared between humans and nonhuman primates the current primate model based largely on monkey data has become an attractive starting point for understanding the functional organization of auditory cortex in the human. Homologies between auditory cortical fields of humans and our near primate relatives are still unclear though, with the possible exception of the primary and primary-like core. Celesia and Puletti and, later, Creutzfeldt and Ojemann, started to bridge the human–nonhuman gap, and although they were far ahead of their time they opened the door enough to give us a glimpse of what could be done. Catherine Liegeois-Chauvel and her colleagues picked up on this in the 1990s, recording auditory evoked activity directly from temporal lobe cortex in neurosurgical patients. They described in some detail the evoked waveforms recorded from different auditory cortical fields, relating them to the waveforms typically recorded from the scalp. With the structure–function framework for a primate auditory forebrain model unfolding around that time as well, my research life took a sharp and unexpected turn.

3.6 Transitioning

It ain't over till it's over.

—Yogi Berra, *The Yogi Book*, 2010

For some years, my colleagues, Rick Reale and Rick Jenison, and I had been focusing attention on the mechanisms by which neurons in auditory cortex (of the cat) encode the direction from which sounds arise from different locations in space. The work promised to spill over into the current millennium to a time when society would expect me to "retire." Anthony Trollope laid it on the line rather bluntly in his novel entitled *The Fixed Period*: "It is self-evident that at sixty-five a man has done all he is fit to do." Our lexicon continues to evolve as new words and terms are added and old ones get the axe. This year we saw the Concise Oxford English Dictionary put its seal of approval on "sexting," "retweet," and "mankini." Destined for the ash heap was "cassette tape," which apparently created uproar from a very upset music historian in Dallas. If I had my way we would add to the ash heap the word "retirement," and I suspect that few scientists would disagree. To retire, according to The American Oxford Dictionary (AOD) is to "leave one's job and cease to work, typically upon reaching the normal age for leaving employment." Fair enough, but the AOD adds that athletes—and here we could include scientists as well—may "be compelled to leave their job when they no longer play competitively." Ouch! Equally offensive, of course, is a British stand-in for "retirement," the otherwise perfectly good word "redundant," which we learn is "the state of being no longer needed or useful." We still need a word to describe this state of our professional lives, but I don't know what it should be.

In any event, a surprise was waiting for me around the bend. While wandering about at the 1996 Society For Neuroscience meeting I dropped by a poster in front of which stood a young neurosurgeon from the University of Iowa, Matthew Howard. Howard claimed to have a passion for understanding the workings of human auditory cortex and quite brazenly claimed he had devised a tool to do just that: a multicontact "hybrid" depth electrode capable of recording action potentials from neurons deep in the temporal lobe, in awake neurosurgical patients. I was aware of some of the technical challenges he faced, having myself carried out single-unit experiments on auditory cortex of awake monkeys. Needless to say I was skeptical both of the electrode and of the surgeon who invented it and claimed to have used it successfully. To my surprise and delight I was proven wrong on both counts. Having put studies of the primate auditory cortex on the back burner some years earlier with every expectation of returning to them I now saw before me that chance, but with what I considered at first blush just a bigger monkey. I now understand what Ursala LeGuin meant in her 1972 fantasy novel, *The Farthest Shore*, when the protagonist implores: "Try to choose carefully … when the great choices must be made. When I was young, I had to choose between the life of being and the life of doing. And I leapt at the latter like a trout to a fly. But each deed you do, each act, binds you to itself and to its consequences, and makes you act again and yet again. Then very seldom do you come upon a space, a time like this, between act and act, when you may stop and simply be. Or wonder who, after all, you are." It wasn't long before I along with my Wisconsin colleagues, Rick Reale, Rick Jenison, and Joe Hind, signed on to the nascent Iowa team of surgeon Howard, physiologist Igor Volkov, and engineer Daniel Noh. Mitch Steinschneider, who was studying the functional architecture of primary auditory cortex of the monkey, was engaged long distance. Having made that leap I now figured we had this big primate to work with, which would be far easier to handle than the nonhuman kind we dealt with some years back, and the neurosurgeons would do all the heavy lifting. This last supposition turned out to be correct, but of course my image of human-as-monkey was way off base. Had I given it a little more thought I would have realized immediately the great opportunity that was being presented to study the functional organization of a part of our brain that sets us apart from all other mammalian species.

The descriptions of the cyto-, myelo-, and chemo-architectonic fields of human auditory cortex and their locations on drawings of the brain simply cannot convey the full anatomical complexity of this structure. Anyone having examined the gross anatomy of the superior surface of the STG—the superior temporal plane—buried deep in the lateral fissure and hidden from view, would have to agree that this is without doubt the most complex anatomical landscape in the human brain. The superior temporal plane consists of one or more transverse gyri, obliquely oriented with respect to the lateral free surface of the STG. The most rostral of these is referred to as Heschl's gyrus (HG). In front of HG is the planum polare and behind the broad planum temporale. Adding to this complexity is the fact that, like many other mammalian brains, there is great intersubject anatomical variation, and it is impossible to predict with any accuracy the alignment of architecturally identified fields with gyri and sulci.

Though breaking new ground on a number of fronts, the recording methods used in modern human intracranial studies are very much like those employed in animal-based experiments since the 1950s. Auditory evoked activity is still recorded with arrays of metal macroelectrodes brought into contact with the pial surface or of cutoff microwires inserted into cortical tissue. Instead of the "old nine channel" of Woolsey's days we have an "old 256 channel" data acquisition system. The sea change since those early days has been the development of powerful computers (no more LINCs) that now enable us to collect enormous amounts of data from multiple cortical sites over a period of a week or more from chronically implanted arrays or during short epochs in the operating room. We continue to compute the averaged evoked potential, the old workhorse of auditory neurophysiology that captures the neural events precisely timed to the onset of a stimulus. Emphasis now is on oscillations of the kind captured by EEG recording and found embedded in the locally recorded activity. These oscillations range in frequency, usually between about 40 and 200 Hz, and may be time related but not necessarily precisely time locked to the stimulus. Powerful signal processing algorithms now allow us to pull from these recordings a wealth of information about the cortical processing of simple and complex sounds, including speech. Acoustic stimuli may be those commonly used in auditory cortical studies in lab animals, which permits direct comparison with results obtained in both invasive animal-based and noninvasive human-based studies carried out under similar experimental conditions. Speech utterances, from synthesized consonant–vowel segments to spoken running speech, are now routinely employed. Recordings are often carried out while the subject performs a behavioral task.

A hybrid depth electrode makes multiple contacts with cortex of HG, including the posteromedial portion postulated to be the auditory cortical core. This cortex responds with a short latency to the onset of a wide range of acoustic stimuli, and it is tonotopically organized. Responses are tightly time locked to trains of acoustic transients and to the envelope of amplitude-modulated signals. This cortex captures the onset time of consonant release and of voicing, thereby accurately representing voice onset time, and tracks the temporal modulation envelope of spoken sentences, which is critical for speech comprehension. Taken together these findings compare remarkably well with those obtained from auditory core in the awake macaque monkey by Mitch Steinschneider, Yon Fishman, and their colleagues and in the awake marmoset monkey by the Xiaoquin Wang lab. The results also align with those obtained from EEG and MEG studies. Although there are some unanswered questions regarding the number and anatomical orientation of primary and primary-like areas, this intersection of human and monkey experimentation leaves little doubt regarding homology between human and monkey core cortex. A possible departure from the monkey could be the evolution of a population of HG neurons, reported by an Israeli and UCLA research team, having frequency tuning curves far sharper than those typically recorded in the core field of monkey, or other mammals for that matter.

So far, so good: the fit to the monkey model holds reasonably well for the auditory core. But when we look beyond the core, on anterolateral HG and on the free surface of the STG, all bets seem to be off. Josef Rauschecker and his colleagues tell us that neurons in lateral belt cortex adjacent to the core in the rhesus monkey are particularly sensitive to complex sounds, especially monkey calls and human speech, which would be consistent with the primate hierarchical and combinatorial model. I have no doubt that this could be the case for the macaque monkey, but in our human subjects we see quite a different picture. Cortex on anterolateral HG immediately adjacent to the acknowledged auditory core is poorly responsive to any acoustic stimulus we throw at it, from pure tones to running speech. At best we find relatively nonspecific low-amplitude long-latency waveforms and late gamma activity. Activity recorded simultaneously on the posterolateral surface of the STG (area PLST) is just as surprising. One might predict from the current monkey model that this field would be particularly selective for speech, but this seems not to be the case. Here we find an auditory field that, like the auditory core, exhibits robust responsiveness to a wide range of acoustic stimuli, from tones to speech, and in this regard is consistent with functional imaging data. Stimuli are represented in highly dynamic ways both in the spatial distribution of active sites and the unfolding of activity over time. Yet, spectrotemporal representations remain robust enough to encode phonemes, which suggests we may be witnessing here acoustic-to-phoneme transformations. Here also the accurate representation of speech is highly attention dependent, and this area exhibits audiovisual speech interactions.

Critical to any model are the connections made between anatomically identified fields. Whether the network that interconnects fields of human auditory cortex supports a core-to-belt-to-parabelt flow diagram ascribed to the monkey is difficult to test in humans, but some data are available on this point. By stimulating one cortical site while recording the resulting evoked activity from all others, we find a short-latency (~2 ms) functional connection between the auditory core cortex and area PLST, which could agree with results from modern tract tracing in monkeys. Electrically stimulating PLST results in a focus of evoked activity in an area of ventral prefrontal cortex that corresponds closely with an auditory-related field in monkey receiving a direct projection from belt/parabelt temporal cortex. Whether this represents a pathway that maps the identity of a sound source or the meaning of speech onto ventral prefrontal cortex has yet to be tested directly. Perhaps this temporofrontal pathway is part of a feedback circuit that kicks in as speakers modulate their own voices. Jeremy Greenlee and his coworkers at the University of Iowa think that this may be so. Activity evoked on PLST by a subject's own voice under passive listening conditions is modulated when the subject actually speaks the same words. Here, again, there is correspondence with the results of Xiaoquin Wang and his colleagues showing modulation of auditory cortical unit activity related to a monkey's spontaneous vocalizations.

So, where does this all leave us? Coming full circle from those early days, a great deal has been learned about auditory cortex, but at each step there have been surprises that have forced us to rethink how this piece of the brain works. Perhaps Jerzy Rose's metaphor of the ghost ship does apply here. Just as we think we have a grasp of it, the fog rolls in. It seems quite clear that human auditory cortex exhibits a functional architecture that at its core is very similar to that of monkeys (and even other mammals for that matter). But beyond this core area there almost surely has been an evolutionary change that allocated cortical space for speech communication. Many people are in the hunt, and there's good reason to believe that with new ideas and with the emerging technology the fog will clear. It's going to take a while though, and for me it's now about gradually wrapping up and letting go.

3.7 End Game

> The degree of decadence that old age can impose upon a brain is very variable and cannot be calculated.
>
> —Paul Broca, 1861

After all these years how does one let go? William Osler was perhaps the most renowned and influential physician of his era. When he decided to retire from the medical faculty of Johns Hopkins University in 1905 he was but 55 years of age. He titled his farewell speech to his Hopkins colleagues *The Fixed Period*, a reference to Anthony Trollope's novel of the same title mentioned earlier. Osler argued that "effective, moving, vitalizing work of the world is done between the ages of 25 and 40—these 15 golden years of plenty" while men above 40 years of age are essentially useless. He further argued that as for those older than the age of 60 there would be "incalculable benefit…in commercial, political and professional life if, as a matter of course, men stopped working at this age." Jokingly he suggested that these "incalculable benefits" might follow from Trollope's "admirable scheme" in which men retire at the age of 67 for a year of quiet contemplation before a peaceful death by chloroform. This joke got Osler into hot water, and although he later explained himself, "to interpose a little ease," he refused to retreat from his belief "that the real work of life is done before the fortieth years and that after the sixtieth years it would be best for the world and best for the themselves if men rested from their labours."

At the time that Trollope wrote *The Fixed Period* researchers had to make their mark early, as most were lucky even to live to the age of 40 (by Osler's time that average upper age limit had budged by no more than about 8 years). Today there's a good chance that most of us can enjoy another 20–30 years. Child prodigies aside, most researchers don't attain their doctorate until the age of 30 and in the biosciences it's not unusual for a Ph.D. grad to go on to one, two, or even three postdocs stretching out over a number of years. Once landing a job and getting a lab set up they're 40 before that first NIH grant comes through (if it ever does). By the age of 60 they may be at the top of their game. I'm in the middle of my eighth decade and am encouraged to know that the upper bound for the span of life has been set at 122, apparently by a French woman who reportedly continued to ride her bicycle until her 100th birthday.

3.8 Looking Ahead

> Ah, but a man's reach should exceed his grasp, or what's a heaven for?
>
> —Robert Browning, "Andrea del Sarto," 1855

It is inevitable that senior scientists be asked what's down the road, as though we have some insider information not available to our younger colleagues. Here are a few thoughts.

A new functional model of auditory cortex is emerging that takes into account the uniqueness of human communication. Perhaps we should now refer to this cortex as "communication cortex." Trying to shoehorn functional data obtained from humans into a model based on anatomical and physiological studies in the monkey takes us only so far. It's not simply that we're big monkeys having big brains with lots of cerebral cortex (we do but so do whales and elephants). The new model has a framework with multiple interconnected fields, not unlike the current monkey model, but includes a wider swath of cortex a homolog of which may be underdeveloped or perhaps not even present in other primates. Cerebral lateralization, once it is better understood, becomes fully integrated into this model. A new model takes into account what Israel Nelken argues is a more important role for core cortex in analyzing the natural scene, considering the fairly high place it occupies in the overall auditory system hierarchy. This new model puts emphasis on neuronal connectivity, which we remember was the clarion call of Meynert more than a century ago when he argued that functional differences between cell complexes was based not on their cellular structure but on their connections. In the end, it may be that it is the neural connectivity in our brains that make each of us unique, and uniquely human. Whether the Human Connectome Project, which aims to map the connectivity of the human brain with imaging approaches, can pull this off is still an open question, but the thinking behind it is aimed in the right direction. The evolution of any new model of human auditory cortex will require the coordinated application of all the experimental tools in our arsenal.

We've paid far too little attention to the dynamics of auditory cortical organization and reorganization. There is no one "human auditory cortex." Auditory cortex of each of us is unique, and it changes on a moment-to-moment basis over a lifetime. Just where, how, and under what conditions these changes take place are actively being pursued and this will spill over to the next generations of auditory neuroscientists. Answering them is critical. We're finally coming to understand that restoring hearing in individuals with severe sensory hearing loss, by introducing a cochlear implant, and eventually perhaps by growing new hair cells, will require ways of retraining the an old auditory brain now being called on not only to hear but also to comprehend the output of a newly engineered receptor organ.

We need to explore new and improved ways of gaining access to the structure and function of human auditory cortex. Intracranial recording will remain in the hands of a few centers. At the cortical interface are relics from the 1950s: metal discs, cut-off wires, and sharpened needles. Currently, a cable to bulky electronic instruments tethers subjects. Electrodes can be displaced, and external noise is easily coupled to wires. Wires are paths for infection, and they can and do break. Electronic miniaturization, wireless transmission, and electrodes fabricated from conductive polymers are not far down the road. Increasing channel capacity will improve spatial resolution of not only intracranial but also EEG and MEG recordings. Higher magnetic field strengths well beyond those that meet clinical MRI needs will find their way into the research lab. As the recently announced BRAIN initiative (also known as the Brain Activity Map Project) and the Human Brain Project gain traction we may see a radical change in the experimental landscape over the next decade.

Finally, we'll need to continually renew in our ranks with bright new minds. These will be PhDs with years of postdoc experience and MDs typically with less research experience but with unique perspectives and orientations. Both are needed, but it's the physician-scientist who may be the endangered species here. I've had the privilege of being mentor to neurosurgical residents who are given two years of protected research time. These are some of the brightest and hardest working individuals I know, yet even for them becoming a physician-scientist at an academic medical center is a serious challenge. To be sure, there are certain intangibles that draw an individual to a life of science and others that steer a person to clinical medicine. For those medically trained who try to make a go of research, the first year or two out is make-or-break time even for the very best of them. Time and money are critical, but both are often in short supply. The competition for limited funds becomes fierce, with added pressure to generate clinical revenue (and support a salary).

The head of a clinical department once told me that if one of his faculty members wanted to act like a PhD he'd be paid like one. Enormous debt loads carried by many medical graduates only exacerbate the problem. It's not surprising that a budding physician-scientist may simply throw in the towel. The Howard Hughes Medical Institute has stepped up to create a Physician-Scientist Early Career Award program and the NIH continues to provide special funding paths, but these may not be enough.

3.9 The Last Word

> One last drink, please.
>
> —Jack Daniels' final words, 1911

Now after years in the research lab what do we tell these new people why we chose research in the first place and why they should consider doing the same? The late Lewis Thomas may have said it best: "Very few see science as the high adventure it really is, the wildest of all explorations ever undertaken by human beings, the chance to catch close views of things never seen before, the shrewdest maneuver for discovering how the world works."

References

Adrian, E. D. (1928). *Basis of sensation: The action of sense organs*. London: Christophers.

Aitkin, L. M. (1990). *Auditory cortex: Structural and functional bases of auditory perception*. London: Chapman and Hall.

Berra, Y. (2010). *The Yogi book*. New York: Workman.

Broca, P. (1861). Sur le volume et la forme du cerveau suivant les individus et suivant les races. *Bulletin Société d'Anthropologie Paris*, 2, 139–207. Taken from Gould, S. J. (1996). *The Mismeasure of Man*. New York: W. W. Norton.

Brodmann, K. (2006). *Localization of the brain*. This work appeared originally in 1909 in German under the title *Vergleichende Localisationslehre der Grosshirnrinde*. Leipzig: Barth. English translation by L. J. Garey. New York: Springer.

Cajal, S. R. (1995). *Histology of the nervous system of man and vertebrates*, Vol. II, Chapter 26. This work appeared originally in 1904 in Spanish under the title *Textura del sistema nervioso del hombre y de los vertebrados*. Madrid: Moya. English translation from the 1995 French edition by N. Swanson & L. W. Swanson. New York: Oxford.

Cajal, S. R. (1999). *Advice for a young Investigator*. This work was first published in 1897 and later appeared in 1923 under the title *Reglas y consejos sobre investigación científica: Los tónicos de la voluntad*. Madrid: CSIC. English translation by N. Swanson and L. W. Swanson. Cambridge, MA: MIT Press.

Campbell, A. W. (1905). *Histological studies of the localization of cerebral function*. Cambridge, UK: Cambridge University Press.

Celesia, G. G., & Puletti, F. (1969). Auditory cortical areas of man. *Neurology*, 19, 211–220.

Creutzfeldt, O., Ojemann, G., & Lettich, E. (1989). Neuronal activity in the human lateral temporal lobe. I. Responses to speech. *Experimental Brain Research*, 77, 451–475.

Davis, P. A. (1939). Effects of acoustic stimuli on the waking human brain. *Journal of Neurophysiology*, 2, 494–499.

Dowben, R. M., & Rose, J. E. (1953). A metal-filled microelectrode. *Science*, 118, 22–24.

Eggert, G. H. (1973). *Wernicke's work on aphasia: A sourcebook and review*. The Hague: Mouton. (This work includes a detailed review of Wernicke's contributions to aphasia and translation of his major works on the subject including *Der aphasische Symptomencomplex: Eine psychologiische Studie auf anatomisher Basis*).

Flechsig, P. (1898). Neue Untersuchengen über die Markbildung in den menschlichen Grosshirnlappen. *Neurologisches Centralblatt*, 17, 977–996.

Geisler, C. D., Frishkopf, L. S., & Rosenblith, W. A. (1958). Extracranial responses to acoustic clicks in man. *Science*, 128, 1210–1211.

Green, J. D. (1958). A simple microelectrode for recording from the centeral nervous system. *Nature*, 182, 962–962.

Hackett, T. A. (2003). Comparative anatomy of the primate auditory cortex. In A. A. Ghazanfar (Ed.), *Primate audition: Ethology and neurobiology* (pp. 199–219). Boca Raton, FL: CRC Press.

Hamilton, A. (1949). Brains that click. *Popular Mechanics*, 91, 162.

Hubel, D. H. (1957). Microelectrode for recording from single units. *Science*, 125, 549–550.

Kaas, J. H., & Hackett, T. A. (2005). Subdivisions and connections of auditory cortex in primates: A working model. In R. Konig, P. Heil, E. Budinger, & H. Scheich (Eds.), *Auditory cortex: A synthesis of human and animal research* (pp. 7–25). Mahwah, NJ: Lawrence Erlbaum.

LeGuin, U. K. (2001). *The farthest shore*. New York: Simon and Schuster.

Merzenich, M. M., & Brugge, J. F. (1973). Representation of the cochlear partition of the superior temporal plane of the macaque monkey. *Brain Res*earch, 50, 275–296.

Meynert, T. (1872). Vom Gehirne der Säugethiere. In S. Stricker (Ed.), *Handbuck der Lehre von den Geweben des Menschen und der Thiere*, Vol. II. Leipzig: Engelmann.

Morel, A., Garraghty, P. E., & Kaas, J. H. (1993). Tonotopic organization, architectonic fields, and connections of auditory cortex in macaque monkeys. *Journal of Comparative Neurology*, 335, 437–459.

Nauta, W. J., & Gygax, P. A. (1951). Silver impregnation of degenerating axon terminals in the central nervous system: (1) Technic, (2) Chemical notes. *Stain Technol*ogy, 26, 5–11.

Peters, A., & Jones, E. G., Eds. (1985). *Cerebral cortex: Association and auditory cortices*, Vol. 4. New York: Plenum.

Poeppel, D., Overath, T., Popper, A., & Fay, R., Eds. (2012). *The human auditory cortex*. New York: Springer Science+Business Media.

Rose, J. E., & Woolsey, C. N. (1949). The relations of thalamic connections, cellular structure and evocable electrical activity in the auditory region of the cat. *Journal of Comparative Neurol*ogy, 91, 441–466.

Rosenfeld, S. A. (1983). LINC: The genesis of a technological revolution. Seminar in Celebration of the Twentieth Anniversary of the LINC Computer. ACRE Amphitheater, National Institutes of Health, Bethesda, MD (for more on the history of the LINC go to http://www.digibarn.com/history/07-11-04-VCF10-LINC/index.html)

Thomas, L. (1995). *Late night thoughts on listening to Mahler's Ninth Symphony*. Middlesex: Penguin Books.

Thoreau, H. D. (1854). *Walden; or, life in the woods*. Boston: Ticknor and Fields.

Trollope, A. (1882). *The fixed period: A novel*. Edinburgh: William Blackwood.

Vogt, C., & Vogt, O. (1903). Zur anatomischen Gliederung des Cortex cerebri. *Journal für Psychologie und Neurologie*, 2, 160–180.

von Economo, C. (2009). *Cellular structure of the human cerebral cortex*. This work appeared originally in 1927 in French under the title *L'architecture cellulaire normale de l'ecorce cerebral*. Paris: Masson et Cie, and in German under the title *Zellaufbau der Grosshernrinde des Menschen. Zehn Vorlesungen mit 61 Abbildungen*. Berlin: Verlag Julius Springer. English translation by L. C. Triarhou. Basel: Karger.

Walter, W. G. (1964). The convergence and interaction of visual, auditory, and tactile responses in human nonspecific cortex. *Annals of the New York Academy of Sciences*, 112, 320–361.

Winer, J. A., & Schreiner, C. E., Eds. (2011). *The auditory cortex*. New York: Springer.

Wolbarsht, M. L., MacNichol, E. F., & Wagner, H. G. (1960). Glass insulated platinum microelectrode. *Science*, 132, 1309–1310.

Woldring, S., & Dirken, M. N. J. (1950). Spontaneous unit activity in the superficial cortical layers. *Acta Physiologica et Pharmacologica Neerlandica*, 1, 369–379.

Woolsey, C. N., & Walzl, E. M. (1942). Topical projection of nerve fibers from local regions of the cochlea to the cerebral cortex of the cat. *Bulletin of the Johns Hopkins Hospital*, 71, 315–344.

Woolsey, C. N., & Walzl, E. M. (1982). Cortical auditory area of *Macaca mulatta* and its relation to the second somatic sensory area (Sm II). Determination by electrical excitation of auditory nerve fibers in the spiral ossious lamina and by click stimulation. In C. N. Woolsey (Ed.), *Cortical organization: Auditory areas*, Vol. 3 (pp. 231–256). Clifton, NJ: Humana Press.

Chapter 4
From Cajal to the Connectome: Building a Neuroanatomical Framework for Understanding the Auditory System

Nell Beatty Cant

Nell Beatty Cant

N.B. Cant (✉)
Department of Neurobiology, Duke University Medical School,
Durham, NC 27710, USA
e-mail: nellcant@neuro.duke.edu

A.N. Popper and R.R. Fay (eds.), *Perspectives on Auditory Research*, Springer Handbook of Auditory Research 50, DOI 10.1007/978-1-4614-9102-6_4,

4.1 Neuroanatomy by Any Other Name…

New approaches for neuroanatomical study of the central nervous system have been invented and perfected for more than a century. Typically, the introduction of each new method is accompanied by a "considerable resurgence of interest in fundamental neuroanatomy" (Jones & Hartman, 1978, p. 215; Nauta, 1993). We find ourselves in just such a resurgence today, not only because of the development of extraordinary new techniques for studying neuronal circuitry, of which there are many, but also because of massive increases in affordable computer memory and storage capacity, wide availability of sophisticated image analysis technology, high-resolution imaging capabilities (at both the light and electron microscopic levels), and the expanding field of neuroinformatics, all of which have combined to open the door for analysis and dissemination of neuroanatomical data in ways only dimly imagined when the first SHAR volumes appeared in 1992. Although it has long been clear that understanding the functioning of the brain requires a detailed knowledge of the patterns of synaptic interactions among specific neuronal types (e.g., Morest, 1975), this fact has assumed a new cachet as more and more neuroscientists recognize the importance and utility of a detailed anatomical framework for organizing data and understanding function. The prevailing validation of the importance of neuroanatomy (although often disguised by catchy nomenclature—connectome!, projectome!, neurome!, synaptome!— cf. Lichtman & Sanes, 2008; DeFilipe, 2010) presents exciting opportunities for studies of the auditory system over the next few decades.

In the spirit of the invitation to contribute to this special volume of SHAR, I wish to consider some of the challenges and opportunities that confront auditory neuroanatomists as we look forward to the next few decades. A comprehensive review of progress in auditory neuroanatomy over the last 20 years and a complete account of the exciting new developments for analysis and dissemination of our data would require considerably more space than is available here. Rather, this chapter is devoted mainly to a discussion of some of the ways that members of the auditory research community might jump on the "connectome" bandwagon and begin to take optimal advantage of new technologies for organization of information. To support the points that I wish to make, I have endeavored to refer to interesting papers that themselves offer further entries into a large (and rapidly expanding) literature; many other relevant studies could have been cited.

4.2 Neuroanatomy of the Auditory System

4.2.1 Some Beginnings

In 1973, when I joined Kent Morest's group at Harvard Medical School and began my career as a neuroanatomist studying the auditory pathways, the methods available were relatively limited. A succinct portrayal of the state of neuroanatomy at that time was drawn by Sandy Palay—whose laboratory was just down the hall from Kent's and who was among the first to fully exploit electron microscopy to study the synaptic organization of the central nervous system (Palay, 1975). In Kent's laboratory the main methods in use were analysis of rapid Golgi preparations, giving us a sense of direct descent from Ramón y Cajal, and electron microscopy, presaging the current interest in "connectomics" at the level of fine structure (e.g., Kleinfeld et al., 2011). The few degeneration studies (cf. Nauta, 1993) in which I participated made me highly appreciative of the work of those intrepid artists who learned so much about the auditory pathways using this method (e.g., Diamond et al., 1969; Osen, 1972; Warr, 1982) and also made me deeply grateful to the pioneers who developed the new, more easily accomplished, tracing methods based on axonal transport that were just beginning to appear (LaVail, 1975; Jones & Hartman, 1978). Also in the 1970s and early 1980s, advances in immunocytochemical identification of neurotransmitter-related proteins were beginning to provide new ways to identify and classify synapses at both the light and electron microscopic levels (Wenthold et al., 1990). By the time the first two volumes in the SHAR series appeared (Popper & Fay, 1992; Webster et al., 1992), these new methods had had considerable impact both on our understanding of the anatomical organization of the auditory pathways and also on our ability to integrate studies of structure and function, a goal of basic auditory neuroscience from the beginning (e.g., Rose & Woolsey, 1949). The chapters in those initial volumes still afford a good introduction to many of the important issues in the anatomy and physiology of the auditory system.

4.2.2 The Knowledge Base Grows Apace

Since 1992, the now standard neuroanatomical and immunocytochemical techniques—greatly expanded and developed—have continued to lead to more detailed understanding of known auditory pathways and even to the discovery of previously unknown connections. A few examples of the latter include discoveries like the following: direct projections from the auditory cortex to nuclei in the brainstem (Feliciano et al., 1995); monosynaptic γ-aminobutyric acid-containing (GABAergic) projections from the inferior colliculus to the medial geniculate nucleus (Peruzzi et al., 1997); topographically organized projections from the nucleus of the brachium of the inferior colliculus to the superior colliculus (King et al., 1998); projections from the amygdala to the inferior colliculus (Marsh et al., 2002); projections from the dorsal cochlear nucleus to the medial geniculate nucleus (Malmierca et al., 2002); a "new" nucleus in the midbrain—the tectal longitudinal column—that interconnects with auditory nuclei in the brain stem (Saldaña et al., 2007; Viñuela et al., 2011); and two distinct GABAergic neuronal populations in the inferior colliculus, defined based on both the organization of their synaptic inputs and their projection targets (Ito et al., 2009). Many other examples of new discoveries could be cited (see reviews by Smith & Spirou, 2002, and Casseday et al., 2002, in SHAR Vol. 15,

Integrative Functions in the Mammalian Auditory Pathway, edited by Oertel, Fay, & Popper), but this list should suffice to support the argument that there is still plenty to be learned using the now "old" methods.

One of the most dramatic changes in auditory neuroscience in the past 20 years has been the explosion of research on the auditory forebrain of both experimental animals (Winer & Schreiner, 2011) and humans (Poeppel et al., 2012). Two additional notable developments include an increased emphasis on rodent and nonhuman primate models for auditory research (Malmierca, 2003; Romanski & Averbeck, 2009) and what amounts to something of a renaissance in studies using the electron microscope as investigators figure out new ways of incorporating techniques for identifying specific types of neurons and synaptic profiles (Gómez-Nieto & Rubio, 2009) and develop improved methods for accumulating and analyzing ultrastructural data (Hoffpauir et al., 2007). All of these trends are contributing to a rapidly expanding body of knowledge concerning the basic structure and function of the auditory system.

4.2.3 The Future Is Here

Complementing established methods, new methods for exploring the mysteries of the nervous system are arriving at a brisk pace. Questions about the functional organization of the auditory system, like those regarding all parts of the central nervous system, will be approached in the coming decades with powerful new methods of diverse kinds, many of which yield inherently neuroanatomical data. Numerous technical advances in imaging methods, both in light and electron microscopy (Denk & Horstmann, 2004; Wilt et al., 2009; Kleinfeld et al., 2011); genetic and molecular strategies for labeling specific neuronal types (Feng et al., 2000; Arenkiel & Ehlers, 2009; Zador et al., 2012—a barcode on every synapse!); optogenetic approaches, including the ability to control the activity of specific neuronal populations with light (Szobota & Isacoff, 2010; Fenno et al., 2011); high-resolution magnetic resonance imaging of the brains of small animals (Johnson et al., 2010)—this is a short list of only a few of the more highly touted developments since the turn of the 21st century.

Looking forward to the future of auditory neuroscience, one (and perhaps only one) prediction seems safe: floods of new information will continue to pour in, collected in a great variety of ways by investigators with expertise in widely diverse fields and overwhelming the ability of most of us to remain abreast of it. Now is the time as a community to think carefully about the implementation of new methods and standards for data collection and dissemination that will help us channel the flow of information. In particular, we will benefit from improved methods for integrating and representing data collected across diverse disciplines and laboratories, allowing deeper insights into the implications and meaning of the collective results.

4.3 Very Large Databases

The challenge of organizing the vast amounts of data collected in studies of the central nervous system has been widely recognized and discussed (Bjaalie, 2002; Van Essen, 2002; Koslow & Subramaniam, 2005 [a multi-authored book on diverse aspects of the subject]; Bota & Swanson, 2007; Mikula et al., 2007; Boline et al., 2008; Bohland et al., 2009; and many others). We have entered an age in which the information that was once contained in hundreds of thousands of books in musty library stacks can be stored electronically in the form of "very large databases" equipped with sophisticated tools for searching them for specific information. Because the demand for management of large amounts of data permeates every aspect of modern life, the science of very large databases is thriving (exemplified by the diversity of publications in a journal devoted to the subject, *The International Journal on Very Large Databases*, published by Springer). Many applications of the new tools are being applied to the study of the brain. However, the greatest success in using these tools will depend critically on the involvement of a wide community of scientists, including continued interaction between those in the field of neuroinformatics who build the bases and the anatomists, physiologists, and others who supply the data (cf. Leergaard et al., 2012). An important responsibility of the bench scientist is to ensure the quality and reliability of the data entered into whatever database systems are employed.

4.3.1 Finding Better Ways to Share Neuroanatomical Findings

4.3.1.1 The Frustration

From the point of view of sharing neuroanatomical findings, a major advance has been the development of tools that make it possible to share high-resolution images on the Internet. A source of frustration for neuroanatomists has always been the difficulty in conveying to others the detailed information about the organization of the brain that becomes evident only through study of hundreds of microscopic images. Until very recently, a relatively few figures had to suffice to represent neuroanatomical findings, and, even for those few figures, the loss of clarity and resolution inherent in the translation from the microscope to the printed page was lamentable. Today, it has become possible to share large image files online, including very high resolution images that appear almost as though the viewer were looking through a microscope. Importantly, however, the neuroanatomist is still challenged with developing the best ways to share the images in the most appropriate manner (e.g., involving expert analysis and subject to community standards of peer review).

4.3.1.2 A Solution: "Microscopy" Online

Two of the more elaborate websites that exemplify the tremendous promise of a Web-based approach for providing access to neuroanatomical data are the Allen Mouse Connectivity Project (connectivity.brain-map.org) and the Brain Architecture Project (Bohland et al., 2009; brainarchitecture.org/mouse). The images of serial sections through experimental brains (i.e., brains injected with neuroanatomical tracers) archived on these two websites are high resolution (the "magnification" can be varied from low to quite high), and they are often spectacularly beautiful. Cells and terminal fields labeled by the injections are easily visualized. The sites are easy to navigate and offer tools for manipulating the material in various ways, sometimes in three dimensions. In addition, annotation and links to the relevant literature for each case are provided. It will represent a remarkable step forward when all neuroanatomical cases are prepared for a similar sort of presentation; publication of small numbers of "typical" sections, often these days in postage-stamp size figures, could become a thing of the past. In light of the beauty and promise of such presentations, however, it is important to emphasize that crucial elements are currently missing from these two websites.

4.3.1.3 Not So Fast! (A Brief Digression)

First and foremost, the neuroanatomists themselves appear to be missing. Without expert interpretation of the archived cases, like that demanded in peer-reviewed publications, the images may be nice to look at, but they are not very useful for advancing understanding. It is fundamentally important that neuroanatomical expertise be brought to bear on the interpretation of the injection sites in each case. This is such an important problem (and is handled so casually on the two highly sophisticated websites cited) that it I am going to yield to the temptation to introduce a short sermon on the subject here. The need for interpretation of tracer injection sites lies in the fact that the *apparent* tracer injection sites that show up on histological sections as blobs of something—blobs of fluorescence, blobs of heavy accumulations of the reaction products used to visualize a tracer, blobs of labeled cell bodies—do not necessarily correspond to the area of *effective* tracer uptake. Indeed, the region of uptake for any neuroanatomical tracer can only be estimated because its appearance on histological sections affords no more than a snapshot in time (i.e., the time at which the animal was perfused). This was demonstrated convincingly when tracer techniques first became widely used (Hedreen & McGrath, 1977; Warr et al., 1981), but is too often ignored when putative uptake zones are delineated based on the apparent boundaries of the blobs. An added complication with some anterograde tracers (perhaps most or even all of them) is that they can be transported very effectively to all terminal fields arising from the branches of a particular neuron, both those that actually do arise from neurons located at the site of the injection but also through labeling of neurons (located elsewhere) that send a separate branch into the site (so-called "false anterograde" or "collateral" label;

Shneiderman & Henkel, 1985; Chen & Aston-Jones, 1998; see further discussion in Saldaña et al., 2009). Because of the many variables that can make the *effective* uptake site different (larger or smaller) than the *apparent* site, I agree with the conclusion of Warr et al. (1981, p. 232) that "establishing a rule for the visual identification of the actual transport site within an apparent injection area is, for all practical purposes, impossible." A further difficulty is that there is no a priori reason to assume that, with those tracers that move in both the anterograde and the retrograde directions, the effective uptake areas are the same for both directions.

Because of the uncertainty inherent in the localization of any given injection site, the key to interpretation of results using modern neuroanatomical tracers lies in a systematic sample of injection locations and comparisons of the results of complementary anterograde and retrograde experiments. Comparisons of the results of cases involving all of the nuclei involved in a particular pathway, combined with (1) comparisons of injection sites in similar but not identical locations and (2) a consideration of what is already known about the connections of those nuclei, can usually (although perhaps not always) lead to an unambiguous interpretation of the results of a tracing study. Therefore, in Web-based presentations of neuroanatomical cases, as in traditional publications, interpretation of the injection sites must be presented in the larger context of multiple cases and *not* on a case-by-case basis. As far as I can tell, none of these considerations has been taken into account in "defining" the injection sites on the two websites mentioned (as of January, 2013).

4.4 The Central Nervous System: Now Appearing in 3-D!

4.4.1 An Exemplary Model

The systematic development of a properly analyzed neuroanatomical database for a specific part of the nervous system is exemplified by the elegant work of Bjaalie and colleagues (Bjaalie, 2002 [highly recommended as a pithy introduction]; Bjaalie & Leergaard, 2005), which has the goal of providing "opportunities for harmonized data presentation in neuroinformatics databases" (Brevik et al., 2001, p. 319). A key feature of their approach is the development of a *brain-based* three-dimensional coordinate system for the particular region of interest (in their case, the pontine nuclei and associated structures). In many brain atlases, position is defined based on landmarks visible on the skull, most importantly the junction of the sagittal and coronal sutures known as the bregma (sometimes referred to as the bregma "point," although it is, at best, a very fuzzy point, Blasiak et al., 2010), and on a particular plane of section. Creation of a three-dimensional coordinate system for specific parts of the brain stem based on *local* brain-based landmarks represents a more flexible approach. Because of the emphasis on local landmarks, experimental cases (with the inevitable distortions associated with histological processing; Simmons & Swanson, 2009) can be related to a coordinate system much more directly than they could be related to a map with skull-based coordinates (the skull, of course, being long gone). In addition, assumptions about the plane of section are less critical when brain-based landmarks are used.

There are two major advantages of presenting neuroanatomical data in a three-dimensional coordinate framework such as that described by Bjaalie and Leergaard (2005). First, it allows comparisons of neuroanatomical data across cases and across laboratories and makes it relatively straightforward to present data in different planes of view (the importance of which is emphasized in a different context by Elias, 1971). As Bjaalie (2002) also points out, systematic representation of data in this way greatly facilitates discussions of nomenclature and definition of subdivisions. The second, and perhaps even more important, advantage of this approach is that the three-dimensional coordinate grid forms a framework or backbone for development of spatial database systems that can incorporate any type of information that can be localized (e.g., recordings from single units; see Section 4.5.1). In studies of the auditory system, an early forerunner of this approach was the development of a "block" model of the cochlear nucleus of the cat (Kiang et al., 1975). This three-dimensional model based on histological sections was computerized and used as a common reference frame for mapping anatomical and physiological data (e.g., Bourk et al., 1981). In terms of ease of use and accessibility, the model was before its time, but the data provided in the papers cited could be incorporated into any three-dimensional coordinate system of the cat's cochlear nucleus; the same would not be true of much of the published data that is not related to spatial location.

4.4.2 Auditory Nuclei in the Gerbil

In my own efforts to understand better the organization of the inferior colliculus (IC) and medial geniculate nucleus (MGN) and their connections in the gerbil, I have developed three-dimensional coordinate atlases of these two areas (IC: Cant & Benson, 2005; MGN: early version in Cant & Benson, 2007). Like Bjaalie and Leergaard (2005), I defined coordinate grids for these structures based on local landmarks. Even given the many potential sources of error in the construction of the atlases (cf. Cant & Benson, 2005), I find the comparisons across cases made possible by relating each one to the atlas remarkably informative (e.g., Cant & Benson, 2008; Cant, 2013), especially because it is relatively easy (although time-consuming) to reconstruct and visualize the images from each case in different planes of section. The mathematical simplicity of a Cartesian framework for representation of neuroanatomical results means that reconciliation of atlases developed in individual laboratories should be relatively straightforward as long as sufficient attention is paid to precise definitions of the coordinate axes. The ultimate goal would be a community consensus on standard reference frames for each auditory structure in each species commonly used in auditory research. The creation of such spatial reference frames combined with websites for representation of data as illustrated by the Allen Brain maps and the Brain Architecture projects described previously, are worthy goals for the immediate future.

4.5 Online Databases Will Repay the Efforts Involved to Build Them

The advances in analysis and representation of neuroanatomical data made possible by new technology are noteworthy enough for their own sake, but what is most compelling is the potential for improved correlations of structure and function. The great Norwegian neuroanatomist Alf Brodal, reflecting on his life's work (Brodal, 1975, pp. 123, 124), concluded, "After more than thirty years of occupation with this subject [neuroanatomy], I am more convinced than ever that a knowledge of the structure of the brain in its minutest details is a prerequisite for meaningful interpretations of observations in all other fields of the neurosciences . . . . the study of structure [is] meaningful only insofar as it contributes in some way to an understanding of function." Brodal summarizes what for me has always been the essential point of neuroanatomical studies. Although their pursuit makes for a most satisfying way to spend one's career, the ultimate reason for documenting the "minutest details" is to provide a supporting framework for the interpretation of functional studies.

4.5.1 Ways Are Needed to Facilitate Localization of Physiological Recording Sites

Development of multidimensional data repositories in which physiological results are tied to specific anatomical locations are feasible, but the considerable challenges associated with sufficiently precise localization of recording sites must be overcome. New methods for controlled placement of recording electrodes and postmortem recovery of recording sites offer hope that this difficult problem can be addressed (Sperka & Ditterich, 2011; Markovitz et al., 2012) so that physiological results can be more closely tied to the details of anatomical circuitry (cf. Loftus et al., 2010). As we continue the development of anatomical reference frames, we should also encourage new standards for localization of neurophysiological recording sites.

4.5.2 Community Organization

The tools for building an infrastructure for organizing and sharing data derived from experimental studies of the auditory system are available and are becoming increasingly accessible. Optimal use of these tools will require the cooperation and participation of the international auditory research community. Indeed, it seems to me that this community, with a good history of productive interactions and shared goals, is an excellent body to explore and implement the development of comprehensive, community-supported sets of databases devoted to a specific neuronal system. Of course, a properly constituted framework could ultimately be incorporated into database systems devoted to the entire brain (as exemplified by the large-scale initiatives described in Section 4.3.1.2). It seems more rational and manageable, however, to start with more limited goals. Even there, the challenges are formidable, and practical considerations abound.

The most important requirement is community buy-in and investment in the effort. This has several implications, perhaps the most important of which is that the development of large, sophisticated, online database systems that are continuously updated and maintained over the long term requires a serious and stable funding commitment. Consensus conferences such as those described in Bohland et al. (2009) and Kleinfeld et al. (2011) represent an effective way to bring together experts with diverse points of view to establish methods and standards for the most efficient use of funding sources. Similar conferences, devoted specifically to the auditory system and perhaps held on an ongoing basis, would provide a way to promote widespread input and to monitor progress. At the all-important level of individual investigators and laboratories, the goal of representing data from both neuroanatomical and neurophysiological studies (in the broadest sense of those terms) within a three-dimensional framework should be recognized and supported as a desirable priority. Of course, this is easier said than done and comes with additional financial requirements, as the work needed to achieve this goal is often tedious and time-consuming, requiring skilled technical assistance. Different groups will be invested in the compilation of data from different species and, for developmental studies, different age ranges within a species. This means that acquisition of some experimental cases might be desirable even when they seem to do little more than repeat what has already been done in another species. Such (non-innovative) repetition may be desirable when neuroanatomy is seen as valuable not only for its own sake but also because it provides a skeleton for building a data-organizing framework. (As usual, there are substantial funding implications.)

4.6 In Conclusion

What would Ramón y Cajal think were he to drop in to check on the state of neuroanatomy today? It was his opinion that, "No matter how exact and minute the verbal description may be, it will always be less clear than a good illustration" (Ramón y Cajal, 1897; translated into English, 2000, p. 132). Ramón y Cajal made full use of the most advanced techniques available to him and developed many improvements of his own. Imagine what he could have done with the wide range of approaches for study and representation of neuronal circuitry available to us today!

Even given the scope of the task and the difficulties of implementation, the invention of new ways of visualizing, sharing and comparing results of auditory research is a goal that seems guaranteed to repay itself many times over. It is impossible to predict the state of auditory research 20 years from now, but if the changes are as great as they have been over the past 20 years, it seems clear that our ways of collecting and presenting experimental results will be quite different. Indeed, we are already experiencing these changes, and now seems a most opportune time to co-opt excitement over the "connectome" and to begin to invest seriously in new ways to archive, share, and use auditory information.

References

Arenkiel, B. R., & Ehlers, M. D. (2009). Molecular genetics and imaging technologies for circuit-based neuroanatomy. *Nature*, 461, 900–907.

Bjaalie, J.G. (2002). Localization in the brain: New solutions emerging. *Nature Reviews Neuroscience*, 3, 322–325.

Bjaalie, J. G., & Leergaard, T. B. (2005). Three-dimensional visualization and analysis of wiring patterns in the brain: Experiments, tools, models, and databases. In S. H. Koslow & S. Subramaniam (Eds.), *Databasing the brain* (pp. 349–368). New York: Wiley-Liss.

Blasiak, T., Czubak, W., Ignaciak, A., & Lewandowski, M. H. (2010). A new approach to detection of the bregma point on the rat skull. *Journal of Neuroscience Methods*, 185, 199–203.

Bohland, J. W., Wu, C., Barbas, H., Bokil, H., Bota, M., Breiter, H. C., Cline, H. T., Doyle, J. C., Freed, P.J., Greenspan, R. J., Haber, S. N., Hawrylycz, M., Herrera, D. G., Hilgetag, C. C., Huang, Z. J., Jones, A., Jones, E. G., Karten, H. J., Kleinfeld, D., Kötter, R., Lester, H. A., Lin, J. M., Mensh, B. D., Mikula, S., Panksepp, J., Price, J. L., Safdieh, J., Saper, C. B., Schiff, N. D., Schmahmann, J. D., Stillman, B. W., Svoboda, K., Swanson, L. W., Toga, A. W., Van Essen, D. C., Watson, J. D., & Mitra, P. P. (2009). A proposal for a coordinated effort for the determination of brainwide neuroanatomical connectivity in model organisms at a mesoscopic scale. *PLoS Computational Biology*, 5, 1–9.

Boline, J., Lee, E-F., & Toga, A. W. (2008). Digital atlases as a framework for data sharing. *Frontiers in Neuroscience*, 2, 100–106.

Bota, M., & Swanson, L. W. (2007). Online workbenches for neural network connections. *Journal of Comparative Neurology*, 500, 807–814.

Bourk, T. R., Mielcarz, J. P., & Norris, B. E. (1981). Tonotopic organization of the anteroventral cochlear nucleus of the cat. *Hearing Research*, 4, 215–241.

Brevik, A., Leergaard, T. B., Svanevik, M., & Bjaalie, J. G. (2001). Three-dimensional computerised atlas of the rat brain stem precerebellar system: Approaches for mapping, visualization, and comparison of spatial distribution data. *Anatomy and Embryology*, 204, 319–332.

Brodal, A. (1975). The "wiring patterns" of the brain: Neuroanatomical experiences and their implications for general views of the organization of the brain. In F. G. Worden, J. P. Swazey, & G. Adelman (Eds.), *The Neurosciences: Paths of Discovery* (pp. 123–139). Cambridge, MA: The MIT Press.

Cant, N. B. (2013). Patterns of convergence in the central nucleus of the inferior colliculus of the Mongolian gerbil: Organization of inputs from the superior olivary complex in the low frequency representation. *Frontiers in Neural Circuits*, 7, 1–22 (Article 29).

Cant, N. B., & Benson, C. G. (2005). An atlas of the inferior colliculus of the gerbil in three dimensions. *Hearing Research*, 206, 12–27.

Cant, N. B., & Benson, C. G. (2007). Multiple topographically organized projections connect the central nucleus of the inferior colliculus to the ventral division of the medial geniculate nucleus in the gerbil, *Meriones unguiculatus*. *Journal of Comparative Neurology*, 503, 432–453.

Cant, N. B., & Benson, C. G. (2008). Organization of the inferior colliculus of the gerbil (*Meriones unguiculatus*): Projections from the cochlear nucleus. *Neuroscience*, 154, 206–217.

Casseday, J. H., Fremouw, T., & Covey, E. (2002). The inferior colliculus: A hub for the central auditory system. In D. Oertel, R. R. Fay, & A. N. Popper (Eds.), *Integrative functions in the mammalian auditory pathway* (pp. 238–318). New York: Springer Science+Business Media.

Chen, S., & Aston-Jones, G. (1998). Axonal collateral-collateral transport of tract tracers in brain neurons: False anterograde labeling and useful tool. *Neuroscience*, 82, 1151–1163.

DeFilipe, J. (2010). From the connectome to the synaptome: An epic love story. *Science*, 330, 1198–1201.

Denk, W., & Horstmann, H. (2004). Serial block-face scanning electron microscopy to reconstruct three-dimensional tissue nanostructure. *PLoS Biology*, 2, 1900–1909.

Diamond, I. T., Jones, E. G., & Powell, T. P. S. (1969). The projection of the auditory cortex upon the diencephalon and brain stem in the cat. *Brain Research*, 15, 305–340.

Elias, H. (1971). Three-dimensional structure identified from single sections. *Science*, 174, 993–1000.

Feliciano, M. E., Saldaña, E., & Mugnaini, E. (1995). Direct projections from the rat primary auditory neocortex to nucleus sagulum, paralemniscal regions, superior olivary complex and cochlear nuclei. *Auditory Neuroscience*, 1, 287–308.

Feng, G., Mellor, R.H., Bernstein, M., Keller-Peck, C., Nguyen, Q. T., Wallace, M., Nerbonne, J. M., Lichtman, J. M., & Sanes, J. R. (2000). Imaging neuronal subsets in transgenic mice expressing multiple spectral variants of GFP. *Neuron*, 28, 41–51.

Fenno, L., Yizhar, O., & Deisseroth, K. (2011). The development and application of optogenetics. *Annual Review of Neuroscience*, 34, 389–412.

Gómez-Nieto, R., & Rubio, M. E. (2009). A bushy cell network in the rat ventral cochlear nucleus. *Journal of Comparative Neurology*, 516, 241–263.

Hedreen, J. C., & McGrath, S. (1977). Observations on labeling of neuronal cell bodies, axons and terminals after injection of horseradish peroxidase into rat brain. *Journal of Comparative Neurology*, 176, 225–246.

Hoffpauir, B. K., Pope, B. A., & Spirou, G. A. (2007). Serial sectioning and electron microscopy of large tissue volumes for 3D analysis and reconstruction: A case study of the calyx of Held. *Nature Protocols*, 2, 9–22.

Ito, T., Bishop, D. C., & Oliver, D. L. (2009). Two classes of GABAergic neurons in the inferior colliculus. *Journal of Neuroscience*, 29, 13860–13869.

Johnson, G. A., Badea, A., Brandenburg, J., Cofer, G., Fubara, B., Liu, S., & Nissanov, J. (2010). Waxholm space: An image-based reference for coordinating mouse brain research. *NeuroImage*, 53, 365–372.

Jones, E. G., & Hartman, B. K. (1978). Recent advances in neuroanatomical methodology. *Annual Review of Neuroscience*, 1, 215–296.

Kiang, N. Y. S., Godfrey, D. A., Norris, B. E., & Moxon, S. E. (1975). A block model of the cat cochlear nucleus. *Journal of Comparative Neurology*, 162, 221–246.

Kleinfeld, D., Bharioke, A., Blinder, P., Bock, D. D., Briggman, K. L., Chklovskii, D. B., Denk, W., Helmstaedter, M., Kaufhold, J. P., Lee, W-C. A., Meyer, H. S., Micheva, K. D., Oberlaender, M., Prohaska, S., Reid, R. C., Smith, S. J., Takemura, S., Tsai, P. S., & Sakmann, B. (2011). Large-scale automated histology in the pursuit of connectomes. *Journal of Neuroscience*, 31, 16125–16138.

King, A. J., Jiang, Z. D., & Moore, D. R. (1998). Auditory brainstem projections to the ferret superior colliculus: Anatomical contribution to the neural coding of sound azimuth. *Journal of Comparative Neurology*, 390, 342–365.

Koslow, S. H., & Subramaniam, S., Eds. (2005). *Databasing the brain: From data to knowledge (Neuroinformatics)*. New York: Wiley-Liss.

LaVail, J. H. (1975). The retrograde transport method. *Federation Proceedings*, 34, 1618–1624.

Leergaard, T. B., Hilgetag, C. C., & Sporns, O. (2012). Mapping the connectome: Multi-level analysis of brain connectivity. *Frontiers in Neuroinformatics*, 6, 1–6.

Lichtman, J. W., & Sanes, J. R. (2008). Ome sweet ome: What can the genome tell us about the connectome? *Current Opinion in Neurobiology*, 18, 346–353.

Loftus, W. C., Bishop, D. C., & Oliver, D. L. (2010). Differential patterns of inputs create functional zones in central nucleus of inferior colliculus. *Journal of Neuroscience*, 30, 13396–13408.

Malmierca, M. S. (2003). The structure and physiology of the rat auditory system: An overview. *International Reviews in Neurobiology*, 56, 147–211.

Malmierca, M. S., Merchán, M. A., Henkel, C. K., & Oliver, D. L. (2002). Direct projections from cochlear nuclear complex to auditory thalamus in the rat. *Journal of Neuroscience*, 22, 10891–10897.

Markovitz, C. D., Tang, T. T., Edge, D. P., & Lim, H. H. (2012). Three-dimensional brain reconstruction of *in vivo* electrode tracks for neuroscience and neural prosthetic applications. *Frontiers in Neural Circuits*, 6, 1–14.

Marsh, R. A., Fuzessery, Z. M., Grose, C. D., & Wenstrup, J. J. (2002). Projection to the inferior colliculus from the basal nucleus of the amygdala. *Journal of Neuroscience*, 22, 10449–10460.

Mikula, S., Trotts, I., Stone, J. M., & Jones, E. G. (2007). Internet-enabled high-resolution brain mapping and virtual microscopy. *NeuroImage*, 35, 9–15.

Morest, D. K. (1975). Structural organization of the auditory pathways. In D. B. Tower & E. L. Eagles (Eds.), *The nervous system*, Vol. 3. *Human communication and its disorders* (pp. 19–29). New York: Raven Press.

Nauta, W. J. H. (1993). Some early travails of tracing axonal pathways in the brain. *Journal of Neuroscience*, 13, 1337–1345.

Osen, K. K. (1972). Projection of the cochlear nuclei on the inferior colliculus in the cat. *Journal of Comparative Neurology*, 144, 355–372.

Palay, S. L. (1975). An essay on neurocytology. In D. B. Tower (Ed.), *The nervous system*, Vol. 1. *The basic neurosciences* (pp. 13–18). New York: Raven Press.

Peruzzi, D., Bartlett, E., Smith, P. H., & Oliver, D. L. (1997). A monosynaptic GABAergic input from the inferior colliculus to the medial geniculate body in rat. *Journal of Neuroscience*, 17, 3766–3777.

Poeppel, D., Overath, T., Popper, A. N., & Fay, R. R., Eds. (2012). *The human auditory cortex*. New York: Springer Science+Business Media.

Popper A. N., & Fay, R. R., Eds. (1992). *The mammalian auditory pathway: Neurophysiology*. New York: Springer Science+Business Media.

Ramón y Cajal, S. (1897 [2000]). *Advice for a young investigator* (Swanson, N. & Swanson, L. W., Translators). Cambridge, MA: The MIT Press.

Romanski, L. M., & Averbeck, B. B. (2009). The primate cortical auditory system and neural representation of conspecific vocalizations. *Annual Review of Neuroscience*, 32, 315–346.

Rose, J. E., & Woolsey, C. N. (1949). The relations of thalamic connections, cellular structure and evocable electrical activity in the auditory region of the cat. *Journal of Comparative Neurology*, 91, 441–466.

Saldaña, E., Viñuela, A., Marshall, A. F., Fitzpatrick, D. C., & Aparicio, M-A. (2007). The TLC: A novel auditory nucleus of the mammalian brain. *Journal of Neuroscience*, 27, 13108–13116.

Saldaña, E., Aparicio, M-A., Fuentes-Santamaría, V., & Berrebi, A. S. (2009). Connections of the superior paraolivary nucleus of the rat: Projections to the inferior colliculus. *Neuroscience*, 163, 372–387.

Shneiderman, A., & Henkel, C. K. (1985). Evidence of collateral axonal projections to the superior olivary complex. *Hearing Research*, 19, 199–205.

Simmons, D. M., & Swanson, L. W. (2009). Comparing histological data from different brains: sources of error and strategies for minimizing them. *Brain Research Reviews*, 60, 349–367.

Smith, P. H., & Spirou, G. A. (2002). From the cochlea to the cortex and back. In D. Oertel, R. R. Fay, & A. N. Popper (Eds.), *Integrative functions in the mammalian auditory pathway* (pp. 6–71). New York: Springer Science+Business Media.

Sperka, D. J., & Ditterich, J. (2011). SPLASh: A software tool for stereotactic planning of recording chamber placement and electrode trajectories. *Frontiers in Neuroinformatics*, 5, 1–10.

Szobota, S., & Isacoff, E. Y. (2010). Optical control of neuronal activity. *Annual Review of Biophysics*, 39, 329–348.

Van Essen, D. C. (2002). Windows on the brain: The emerging role of atlases and databases in neuroscience. *Current Opinion in Neurobiology*, 12, 574–579.

Viñuela, A., Aparicio, M-A., Berrebi, A. S., & Saldaña, E. (2011). Connections of the superior paraolivary nucleus of the rat: II. Reciprocal connections with the tectal longitudinal column. *Frontiers in Neuroanatomy*, 5, 1–15.

Warr, W. B. (1982). Parallel ascending pathways from the cochlear nucleus: Neuroanatomical evidence of functional specialization. *Contributions to Sensory Physiology*, 7, 1–38.

Warr, W. B., Olmos, J. S., & Heimer, L. (1981). Horseradish peroxidase: The basic procedure. In L. Heimer, & M. J. RoBards (Eds.), *Neuroanatomical tract-tracing methods* (pp. 207–262). New York: Plenum Press.

Webster, D. B., Popper, A. N., & Fay, R. R. (Eds.) (1992). *The mammalian auditory pathway: Neuroanatomy*. New York: Springer-Verlag.

Wenthold, R. J., Altschuler, R. A., & Hampson, D. R. (1990). Immunocytochemistry of neurotransmitter receptors. *Journal of Electron Microscopy Technique*, 15, 81–96.

Wilt, B. A., Burns, L. D., Ho, E. T. W., Ghosh, K. K., Mukamel, E. A., & Schnitzer, M. J. (2009). Advances in light microscopy for neuroscience. *Annual Review of Neuroscience*, 32, 435–506.

Winer, J. A., & Schreiner, C. E., Eds. (2011). *The auditory cortex*. New York: Springer Science+Business Media.

Zador, A. M., Dubnau, J., Oyibo, H. K., Zhan, H., Cao, G., & Peikon, I. D. (2012). Sequencing the connectome. *PLoS Biology*, 10, 1–7.

Chapter 5
Recording from Hair Cells

Ruth Anne Eatock

Ruth Anne Eatock

R.A. Eatock (✉)
Department of Neurobiology, University of Chicago,
Chicago, IL 60637, USA
e-mail: eatock@uchicago.edu

A.N. Popper and R.R. Fay (eds.), *Perspectives on Auditory Research*, Springer Handbook of Auditory Research 50, DOI 10.1007/978-1-4614-9102-6_5,

5.1 Introduction

On the plane to Los Angeles for my graduate school interview, I poured over the Caltech Biology Annual Report and was drawn to a small abstract from Jim Hudspeth and David Corey, describing intracellular recordings from single hair cells in excised frog saccules. I was already intrigued by the inner ear—Geoff Manley had been teaching me how to infer inner ear functions from afferent nerve recordings—and the prospect of direct access to hair cells was thrilling. During my visit, Jim recorded from a hair cell while deflecting its hair bundle at 10 Hz; seeing the hair bundle move and hearing the rhythmic receptor potential on the audio monitor, I was enthralled. I still find joy in the mechanically gated response of a hair cell or its afferent fiber.

In the intervening decades, hair cells have been poked, prodded, sucked, electrified, calcium-imaged, and puffed-upon. The favored early models, frog saccular and turtle cochlear hair cells, proved impressively adaptable to new methods (or was it the skill of the investigators?), such that they became models for multiple key questions in neurobiology: sensory transduction, ion-channel signaling, and synaptic transmission. In each case, the hair cell performs better, or at least faster, than counterparts in other model systems. As the list of hair cells recorded from grew to include avian, mammalian, vestibular, and lateral line hair cells, more exotic properties emerged, such as the electromechanical response of the outer hair cell and nonquantal transmission at the type I hair cell–calyceal synapse.

5.1.1 The Seventies and Eighties

The first intracellular recordings from hair cells were made *in vivo* during natural stimulation (Harris et al., 1970; Mulroy et al., 1974; Russell and Sellick, 1977). Then Jim Hudspeth and David Corey (Hudspeth & Corey, 1977) and Robert Fettiplace and Andrew Crawford (Fettiplace & Crawford, 1978), inspired by experiments in the phototransduction field, began recording in tissue maintained in vitro in simple saline solutions. The greater access yielded much larger receptor potentials than had been recorded in vivo. Fettiplace and Crawford began with a preparation of the turtle auditory organ that included an intact middle ear, allowing for sound stimulation. Hudspeth and Corey, in contrast, excised the sensory epithelium and removed accessory structures to provide full visual access to the hair bundles. They could then directly manipulate individual mechanosensitive hair bundles (Fig. 5.1). Such recordings showed that the hair bundle has a functional polarity, the transduction channels open with positive bundle deflections and then adapt, the cascade is 100- to 1000-fold faster than photo- and chemotransduction, and the transduction current has a sigmoidal dependence on bundle deflection. Some of these properties could be inferred from in vivo extracellular potentials and afferent nerve recordings evoked by natural stimuli, but in vitro preparations provided high signal-to-noise ratios and access to intracellular mechanisms.

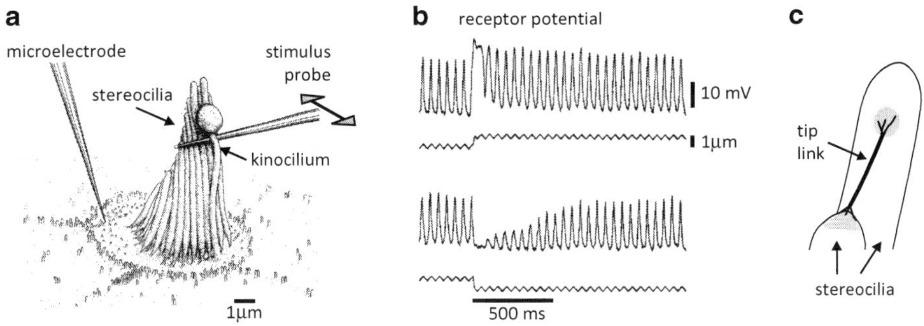

Fig. 5.1 Recording from hair cells in the frog saccule revealed transducer adaptation. (Modified from Eatock et al., 1987.) (**a**) Schematic, drawn by David Corey, showing a method for coupling a rigid probe to the hair bundle while recording transmembrane potential with a sharp microelectrode. (**b**) Transducer adaptation. Receptor potentials evoked in a frog saccular hair cell by a small triangle wave superimposed on positive (top) and negative (bottom) static deflections (steps). At the step onset, transduction was saturated, then adapted, such that responses to the triangle wave recovered as the operating range shifted toward the static position of the bundle. (**c**) Schematic of a tip link connecting two stereocilia in adjacent rows. See Kachar et al. (2000) for high-resolution micrographs

In the early 1980s, hair cells were among the first small cells to be recorded with the new tight-seal, whole-cell method (Lewis & Hudspeth, 1983; Ohmori, 1984). To do so, investigators developed ways to dissociate hair cell epithelia, breaking the tight junctions that hold cells in place and cleaning their membranes with enzymes. Patch clamping quickly became the method of choice by virtue of its impressive fidelity and low noise. As the quality of in vitro preparations, imaging, and recording improved, hair cells were seen to be more active manipulators of the transduced signal than previously imagined, exhibiting micromechanical changes associated with gating and adaptation (reviewed in Vollrath et al., 2007), electrical tuning by voltage-gated conductances (reviewed in Fettiplace & Fuchs, 1999; Art & Fettiplace, 2006), and, most unexpectedly, outer hair cell electromotility (reviewed in Brownell, 2006). Because in vitro preparations do not perfectly reproduce in vivo conditions, each new observation was challenged as possibly artifactual. How could a transduction mechanism sensitive to angstrom displacements of the basilar membrane survive being yanked out and pinned down in a dish? Surely those onset transients or oscillations reflect poor stimulus control? It turned out that hair cells tolerate the trauma of excision and dissociation remarkably well as, decades on, most of the basic observations of the first decade of in vitro hair-cell recordings hold.

5.1.2 The Nineties and Oughts

In the 1990s, investigators began applying the whole-cell patch clamp method to excised mammalian inner ear epithelia, beginning with organ of Corti (Kros et al., 1992) and utricular macula (Rüsch & Eatock, 1996; Géléoc et al., 1997) from mice.

Whereas hair cells from the established models—frog saccule, turtle cochlea, and chick cochlea—all had large Ca^{2+} and Ca^{2+}-gated K^+ conductances that electrically tune the receptor potential (Art & Fettiplace, 1987; Hudspeth & Lewis, 1988a, b; Fuchs et al., 1988), mammalian hair cells emphasized temporal fidelity by expressing large numbers of low-voltage-activated potassium (K_{LV}) channels (Correia & Lang, 1990; Housley & Ashmore, 1992). Adding these channels decreases membrane resistance and therefore membrane charging time and voltage gain. The marked attenuation of gain seemed counterintuitive at first and early reports of K_{LV} channels in type I hair cells were skeptically received. When Alfons Rüsch first joined my lab, his first recordings seemed to confirm his doubts, as neonatal mouse utricular hair cells showed no sign of K_{LV} channels; but as he pursued them to older ages, K_{LV} currents became more frequent and larger (Rüsch et al., 1998a). K_{LV} channels from different channel families have been reported in many neurons concerned about speed of signaling, such as cells in the auditory brain stem (Golding & Oertel, 2012).

Because molecular-genetic approaches to hair cell questions lagged biophysical approaches, sophisticated phenotype testing was available as soon as candidate genes and proteins began emerging (e.g., Siemens et al., 2004; Sollner et al., 2004). Working with mouse tissue allowed biophysicists to test the functions of candidate proteins in hair cells from genetically manipulated animals (Rüsch et al., 1998b; Kros et al., 2002). The zebrafish (*Danio rerio*) emerged as an alternative model for inner ear development and function (Whitfield et al., 1996; Nicolson et al., 1998), but the tiny size of larval hair cells made single-cell patch clamping daunting until Ricci et al. (2013) applied techniques refined on excised epithelia to a whole-larva preparation. Single-cell resolution has also been achieved in zebrafish hair cells by genetically engineering Ca^{2+} signals (Kindt et al., 2012).

Before the 1990s, what we knew about hair cell afferent transmission was derived largely from the output stage, afferent activity (Sewell, 1996). Then the hair cell was adopted as a model presynaptic terminal (reviewed in Fuchs & Parsons, 2006). The hair cell shares accessibility, size, glutamatergic transmission, and presynaptic ribbons with retinal photoreceptor and bipolar cells. But hair cell synapses are very fast and in special cases exhibit extraordinary speed-enhancing mechanisms such as the coincident release of many synaptic vesicles on a single afferent terminal (Glowatzki & Fuchs, 2002) or nonquantal transmission (Songer & Eatock, 2013).

The following sections highlight examples of how thinking on hair cell transduction, voltage-dependent signaling, and synaptic transmission has evolved since we began to record from single hair cells.

5.2 Mechanoelectrical Transduction

There are many critically important mechanotransducers in our bodies, but hair cells occupy pride of place because they serve hearing, a special sense of intense popular interest, and because they have been more accessible to recording electrodes than are, say, the distant terminals of dorsal root ganglion touch sensors.

They also appeal to the imagination: hair bundles are weirdly beautiful on the outside, and have been conceptualized as tiny motorized gear-works on the inside (e.g., Gillespie & Hudspeth, 1994).

In 1983, David Corey and Jim Hudspeth (Corey & Hudspeth, 1983) introduced a model of transduction gating that formalized in a simple way the mechanosensitivity of transduction in the hair bundle: Elastic elements (gating springs) apply force to the channels; positive bundle deflections stretch the springs, increasing the force and favoring channel opening, which reduces the stretch. Negative deflections reduce the slight resting stretch of the springs, favoring channel closing. Next, Jim Pickles and colleagues (Pickles et al., 1984) observed, in electron micrographs, interstereociliary linkages (tip links; Fig. 5.1c) that are oriented along the functional axis of the hair bundle (arrows in Fig. 5.1a). They proposed that the tip links are the gating springs, attaching to transduction channels at one or both ends of each tip link.

The gating spring/tip link model stimulated important experiments. As the model predicted, bundles experience a softening as the channels gate—the "gating compliance" (Howard & Hudspeth, 1988). This was detected by measuring bundle stiffness with flexible glass probes of calibrated stiffness. The same experiments showed that a slower decrease in bundle stiffness accompanies transducer adaptation (Howard & Hudspeth, 1987). When a hair bundle is deflected and then held in place by a probe (Fig. 5.1a), the transduction current rises rapidly, then decays (adapts) more slowly, while sensitivity to superimposed bundle deflections increases (Fig. 5.1b) and the bundle softens. To account for the transducer adaptation and bundle softening, Howard and Hudspeth envisioned an intra-stereociliary "motor" moving along the actin core of the stereocilium, dragging the attached transduction channel with it and reducing the stretch of gating springs. By removing the original stretch from the gating element, the adaptation process restores sensitivity to novel forces superimposed on the tonic force—particularly useful in vestibular organs, which are subject to slow and even static forces (Eatock et al., 1987). The motor idea became formalized as an elegant physical model centered on the tip link anchor point on the side of the tallest stereocilium in each linked pair. At the anchor point, the tip link (gating spring), transduction channel, a plaque of myosin-1c molecules, and the actin core of the stereocilium were serially connected (Gillespie & Hudspeth, 1994). This beautiful structure compellingly pulled together diverse observations on structure and function of the bundle, including the presence of actin, myosins, and calmodulin, and the Ca^{2+} dependence of adaptation. Over the next 20 years, it stimulated many experiments, and has evolved to accommodate new results.

5.2.1 *Adaptation Is Amplification*

Transducer adaptation in frog saccular and turtle cochlear hair cells resembled each other in some ways and not others; eventually it dawned on everyone that there were two kinds of adaptation. The early frog data focused on slow adaptation, and the early turtle data on fast adaptation. Both cell types have both, however, as do mouse utricular cells (Howard & Hudspeth, 1987; Wu et al., 1999; Vollrath & Eatock, 2003). Fast adaptation, with time constants of 0.1–10 ms, correlates in time with a brief stiffening of the hair bundle, preceding the 10-fold slower relaxation of slow adaptation (Howard & Hudspeth, 1987). The fast mechanism might be as simple as Ca^{2+} binding to the internal face of the channel protein and promoting a conformational change (Howard & Hudspeth, 1988; Crawford et al., 1989).

When investigators began to monitor hair bundle motion with photodiodes, they saw oscillatory bundle motions (Crawford & Fettiplace, 1985), indicating tuning of the transduction apparatus centered on the oscillation frequency. Adaptation is a damped form of the oscillations—transduction can be more adaptive or more tuned (resonant) depending on Ca^{2+} level (Ricci et al., 1998; Martin et al., 2003). Thus, vertebrate hair cells have versatile Ca^{2+}-dependent mechanisms that reduce sensitivity at low frequencies (adaptation) and enhance sensitivity at best frequencies (amplification) (Hudspeth et al., 2000).

Could such mechanisms drive sharp tuning at 50–100 kHz in bat and whale cochleas? It's possible, given that the transduction channels gate quickly enough to pass currents at such frequencies and that fast adaptation in rodent cochlear hair cells can have time constants <100 μs (Ricci et al., 2005; Jia et al., 2007). In the outer hair cell, stereociliary amplification could work in series with power amplification by prestin, a different kind of motor which drives electromotility (see comments by Fettiplace and Hackney and Martin and Hudspeth in Ashmore et al., 2010).

The cochlear amplification field has struggled frequently and creatively with the problem of how a cellular, voltage-dependent mechanism—electromotility—can operate at the remarkable high frequencies of the mammalian audiogram (Ashmore et al., 2010). Relief may be in sight; as in vitro methods have improved, the measured performance of the outer hair cell has improved. By using appropriate Ca^{2+} levels, mammalian temperature, high-speed stimulation, and high-quality whole-cell patch clamping, Johnson et al. (2011) showed that the membrane charging time is much faster than previously measured, raising the low-pass corner frequency for the receptor potential to the tens of kilohertz. Key factors are the low Ca^{2+} of cochlear endolymph, which unblocks the transduction channels, increasing their open probability at rest; body temperature, which increases transduction current; and K_{LV} channels, which reduce membrane charging time.

5.2.2 Transduction Channels: How, Where, What?

Tip links are now considered too stiff to be the gating springs. Kachar et al. (2000) noted that tip links imaged with high-resolution electron microscopy can hold deformations. Genetic analyses showed that the braids of the tip links comprise two kinds of calcium-dependent adhesion protein (cadherin) (Siemens et al., 2004; Sollner et al., 2004; Kazmierczak et al., 2007), extending from the upper and lower attachment points, respectively, and linking at their other ends. Mutations in either cadherin cause human deafness (El-Amraoui & Petit, 2010). Long cadherin domains bind multiple Ca^{2+} ions, and in the absence of Ca^{2+} they unravel, nicely explaining why tip links require a pinch of Ca^{2+} in the endolymph (>~10 μM). The parallel loss of both tip link integrity and transduction when Ca^{2+} chelators are added to the bath solution (Assad et al., 1991) had seemed strong support for the idea that tip links are gating springs. But the high stiffness of cadherins does not accord with the measured elasticity of gating springs (Sotomayor et al., 2005). Instead, tip links may be in series with the elastic gating springs, which might reside in the lipid bilayer itself or in something linked to or within the bilayer, such as an elastic component of the transduction channels.

Denk et al. (1995) used the new method of two-photon microscopy to address whether channels are located near opposite ends of the tip links on adjacent stereocilia (see Fig. 5.1c) by imaging Ca^{2+} entry through transduction channels at the level of individual stereocilia. The shortest stereocilia at the very back of the hair bundle staircase have attachments only near their tips, not along their sides. Thus, a rise of Ca^{2+} in the very back row would imply that channels are located at the stereociliary tips. A rise of Ca^{2+} in the tallest stereocilia, next to the kinocilium, would suggest that channels are located at side attachment points (although there are connections to kinocilia; see later in this section). A rise was seen in both stereociliary rows, suggesting that channels are located at both ends of each tip link.

Beurg et al. (2009) revisited the question in rat cochlear hair cells and saw no labeling in the tallest stereocilia. They concluded that Ca^{2+} enters only at the tips of the stereocilia and not at the side attachment points. The difference may depend on true differences between the hair bundles examined or on technical differences affecting spatial and temporal resolution. The hair bundles of rat inner hair cells have relatively thick stereocilia in a steep staircase, allowing better resolution of the tips of the stereocilia in adjacent rows. Also, advances in Ca^{2+} indicators and our understanding of intrastereociliary buffering permitted a better match between the working range of the Ca^{2+} signal and the physiological range of Ca^{2+} changes.

Ruling out transduction channels at the side tip link attachment site requires some re-thinking about how slow adaptation works. As illustrated in Gillespie and Hudspeth (1994), Ca^{2+} ions had been visualized as entering the stereocilium through transduction channels *at the side tip link attachments* and modulating interactions between myosin–calmodulin complexes connecting the transduction channels to actin filaments running the length of the stereocilium. This mechanism might be salvaged were Ca^{2+} to enter transduction channels at the stereociliary tip and drift down to the myosin–calmodulin complex near the side attachment point of the same stereocilium. In this way, tension at a channel in one stereocilium could modulate tension at channels in the lower stereocilium.

The mammalian inner hair cell's bundle is a special case in which the tallest stereocilia are free-standing and unconnected to a kinocilium—though present initially, kinocilia degenerate as cochlear hair cells mature. Thus, it remains possible that in the *frog* hair bundle, Ca^{2+} does enter the tallest stereocilia, which are connected firmly to the kinocilium (Fig. 5.1a). Stretching links between stereocilia and the kinocilium could activate channels at the tips of the tallest stereocilia and/or in the kinocilium itself (Kindt et al., 2012).

For a while, it was thought that the transduction channels were likely to be members of the TRP ion channel family, which has been implicated in sensory transduction in diverse cell types (reviewed in Gillespie et al., 2005). But TRPN1 does not exist in mammals, and despite promising localization, timing, and functional results, mice in which TRPA1 was knocked out were not deaf (Bautista et al., 2006; Kwan et al., 2006). A novel transduction channel candidate, *TMC-1* (transmembrane cochlear gene 1), emerged from a human deafness analysis (Kurima et al., 2002). Early on, the direct involvement of *TMC-1* in transduction appeared to be ruled out by normal transduction currents recorded from mice with mutations in *TMC-1* (Vreugde et al., 2002; Marcotti et al., 2006). More recent results with double knockouts of *TMC-1* and its close relative TMC-2 (Kawashima et al., 2011), however, indicate roles for both TMCs either as transduction channels or as another essential part of the mechanosensory apparatus. Both *TMC-1* and *TMC-2* are present in the developing cochlea, such that mouse mutants lacking functional *TMC-1* can transduce in early postnatal life via *TMC-2*—explaining the lack of biophysical phenotype in *immature* hair cells with *TMC-1* mutations (Vreugde et al., 2002; Marcotti et al., 2006). With maturation, *TMC-1* takes over completely in the cochlea but both proteins are still expressed in the vestibular system.

Since 1992, then, molecular genetics and biophysical imaging experiments have redrawn the cartoons of the mechanosensory transduction apparatus: transduction channel gating may be effected not by tip links but rather by springy elements that are part of the channel, the membrane that it is embedded in, or the cytoskeleton, or some combination; transducer adaptation, viewed in another light, causes amplification; and after some competitive back and forth on the relative importance of stereociliary amplification and outer hair cell electromotility, a civilized compromise has taken hold, with both mechanisms thought critical to the amazing ability of the mammalian cochlea to communicate high-frequency signals to the brain.

5.3 Receptor Potentials Are Unexpectedly Diverse

In the 1980s, microelectrode recordings gave way to whole-cell patch recordings for the characterization of voltage-gated conductances in the basolateral membranes of hair cells. The first hair cells studied—frog saccular, turtle cochlear, and chick cochlear—all have complements of voltage-gated calcium (Ca_V) channels and large numbers of Ca^{2+}-gated K (K(Ca)) channels that lead to sharp electrical tuning of the receptor potential. The tuning manifests as oscillations at the electrical best frequency in response to step displacements of the hair bundle: transduction current depolarizes the membrane, activating Ca_V current that further depolarizes the membrane but also activates K(Ca) current, which repolarizes the membrane (Hudspeth & Lewis, 1988b). As possibly the most elegant demonstration of how voltage-gated ion channels can shape signals on record, these descriptions entered the textbooks and were thought to apply to all hair cells.

As new inner ear preparations were developed in the 1990s, however, it became clear that hair cells were more eclectic than previously supposed. In fact, hair cells offer striking examples of how ion channel expression can diverge through ontogeny and phylogeny to enhance distinct signal features, notably, frequency *vs.* timing. Some hair cells are specialized for fast and linear voltage responses, rather than electrical tuning, by the expression of large K_{LV} conductances that are active at resting potential. This is especially true of outer hair cells in the mammalian cochlea, where the conductance is called $g_{K,n}$ (Housley & Ashmore, 1992) and type I hair cells of vestibular organs from birds, mammals, and reptiles, where the conductance is called $g_{K,I}$ or $g_{K,L}$ (Correia & Lang, 1990; Eatock & Hutzler, 1992; Brichta et al., 2002). The low-voltage-activated conductances are acquired somewhat late in hair cell maturation, along with other changes that dramatically affect excitability and bandwidth.

Immature hair cells go through at least two stages of ion channel configuration until the onset of hearing and eye opening and the maturation of vestibular reflexes (reviewed in Eatock & Hurley, 2003; Goodyear et al., 2006). The combination of high input resistance, fast inward conductances, and delayed outward conductances can produce spikes in immature cochlear hair cells (Marcotti et al., 2003) and electrical resonance in immature vestibular hair cells (Songer & Eatock, 2013). The spiking and resonances should boost transmitter release in the pre-hearing, pre-seeing, pre-mobile animal, when mechanical inputs are weak, and so promote downstream development. Then, to become fully mature, the cells drop some inward channels and add K channels, decreasing spiking and resonance but improving the speed and linearity of the receptor potential's representation of hair bundle deflections.

Recent work has highlighted the value of the low-voltage-activated hair cell conductances, $g_{K,n}$ and $g_{K,L}$, in mammalian hearing and vestibular function. Both conductances broaden bandwidth and speed up the voltage response. In outer hair cells, $g_{K,n}$ is needed to drive electromotility at high characteristic frequencies (see Johnson et al., 2011). Similarly, $g_{K,L}$ in type I hair cells improves response timing, reducing phase lags over a broad frequency range, as may be needed for reflexes to compensate for fast head motions (Songer & Eatock, 2013). These conductances may also be significant links in K^+ circulation pathways in the inner ear (Zdebik et al., 2009).

5.4 Hair Cell-to-Afferent Transmission Has Surprising Properties

By the 1990s, many aspects of hair cell-to-afferent transmission had been shown to fit the canonical view of excitatory transmission in the vertebrate central nervous system: Depolarization stimulates release of quanta (packets) of glutamate from synaptic vesicles (reviewed in Sewell, 1996). But already it was clear that hair cell transmission has special features, including presynaptic ribbons and some very large postsynaptic terminals, notably the chalices (calyces) around vestibular type I hair cells (reviewed in Eatock & Lysakowski, 2006). First described by Wersäll (1956), these large synapses are just beginning to reveal their secrets. Unlike the better-known *pre*synaptic calyces such as the calyx of Held in the auditory midbrain, vestibular afferent calyces are postsynaptic and smoothly continuous. The hand-in-glove proximity of type I cells and calyces has excited speculation that transmission is non-quantal (Yamashita & Ohmori, 1990) or retrograde (Sans & Scarfone, 1996).

Parsons et al. (1994) introduced whole-cell recordings of the increase in membrane capacitance produced when synaptic vesicles fuse with the hair cell membrane (exocytosis). Vesicle exocytosis occurs in stages with distinct time courses, corresponding neatly to different physical pools of transmitter (Lenzi et al., 1999; Moser & Beutner, 2000). Different time scales of synaptic adaptation may reflect serial depletion of the synaptic vesicle pools (Furukawa & Matsuura, 1978; Schnee et al., 2005). The dependence of vesicle fusion on Ca^{2+} goes through developmental stages that have been interpreted as a restriction of the volume of Ca^{2+} that controls exocytosis, allowing Ca^{2+} influx through one to several channels to drive fusion of a particular vesicle (Moser et al., 2006).

In the first patch-clamp recordings from afferent boutons on cochlear inner hair cells, Glowatzki and Fuchs (2002) discovered that inner hair cells can simultaneously release large numbers of vesicles simultaneously, more than other ribbon synapses. The simultaneous release of many quanta makes the summated excitatory postsynaptic currents both large and fast, which in turn drives spiking with high fidelity and low temporal jitter (Wittig & Parsons, 2008). These properties may be essential to the remarkable ability of the mammalian auditory system to follow sound frequencies (phase-lock) up to about 5 kHz and to detect minute interaural time differences.

Investigators have also patch-clamped the large but delicate type I-calyx synapse of amniote vestibular epithelia (Bonsacquet et al., 2006; Hurley et al., 2006; Rennie & Streeter, 2006; Songer & Eatock, 2013) (Fig. 5.2). The calyx may support both quantal and non-quantal transmission (Holt et al., 2007) (Fig. 5.2); the non-quantal transmission has remarkably short synaptic delay (Songer & Eatock, 2013). Like the type I hair cell it envelops, the postsynaptic calyx membrane expresses large numbers of K_{LV} channels (Lysakowski et al., 2011; Songer & Eatock, 2013) and HCN channels (Meredith et al., 2012). Together these reduce membrane charging time and hence the delay between excitatory postsynaptic current and spike generation. These channels are more abundantly present in the calyces of irregularly spiking afferents than of regularly spiking afferents, consistent with other measures that indicate that irregular afferents are specialized for speed of signaling (reviewed in Eatock & Songer, 2011). Blocking the afferents' K_{LV} channels makes firing more regular (Iwasaki et al., 2008; Kalluri et al., 2010), suggesting that highly irregular firing is a side effect of the high K_{LV} conductance. Note that K_{LV} channels are also important in mammalian cochlear hair cells (K_V7 channels, Kharkovets et al., 2000) and auditory afferents (K_V1 channels, Mo et al., 2002)—and that auditory afferents are also highly irregular.

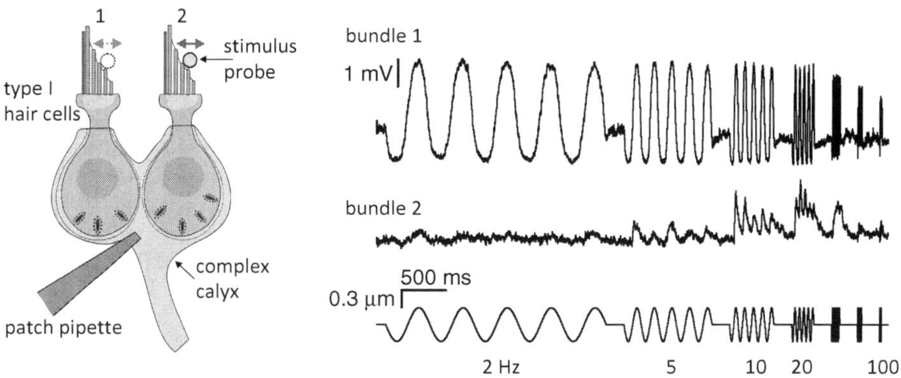

Fig. 5.2 Recording from calyceal afferent terminals on vestibular type I hair cells revealed complex synaptic transmission. (Modified from Songer & Eatock, 2013.) (**Left**) Schematic showing a micropipette recording from a complex calyx enveloping two hair cells; each hair bundle was stimulated one at a time. The stimulus probe (viewed end on) was pushed against the back of each bundle, one at a time; the same probe was moved between the two bundles. (**Right**) Postsynaptic potentials recorded as one hair bundle was stimulated, and then another, with a series of sinusoidal bursts at frequencies incrementing from 2 to 100 Hz. Stimulating one hair bundle evoked non-quantal responses; stimulating the other hair bundle evoked quantal responses from the same calyx ending. The non-quantal response to stimulation of bundle 1 had a wider bandwidth (broader tuning) than the quantal response to stimulation of bundle 2, in part because the response to bundle 1 had shorter latency

5.5 Concluding Remarks

Biophysical research on hair cells in the 1970s and 1980s was dominated by two model preparations: the frog sacculus and the turtle cochlea. These preparations bore remarkable similarities, promoting a sense that hair cells generally have similar adaptation and amplification mechanisms, ion channel expression, and synaptic machinery. In hindsight, the common properties of frog saccular and turtle cochlear hair cells may reflect their similar niches in mature hair-cell organs that respond selectively to frequencies from tens to hundreds of Hertz.

When the Springer Handbook of Auditory Research series began in 1992, we were exiting the heroic era of pure hair cell electrophysiology, which established many of the tools that we still use, and entering a more complex scene in which hair cell recordings share the stage with molecular biological, high-resolution anatomical, optical, and genetic approaches. Recordings diversified to birds, rodents, and zebrafish, vestibular and lateral line organs, and synaptic terminals, driven by curiosity as well as pressure for data from molecular-genetic model organisms. Naturally, all this activity expanded the range of known inner ear electrophysiology.

The artisanal skills for hair-cell recording require apprenticeship and persistence. The experiments can be arduous and low-throughput; although the basic equipment is not expensive when amortized over a research program's life time, the experiments are costly in man-hours. We keep doing them because they have magnificent signal-to-noise ratios, combine high spatial and temporal resolution, directly control the voltage-dependent mechanisms that make signals, and yield functional read-out that is intuitive because it is electrical and comprehensive because it is analogue.

Acknowledgments Work in my laboratory has been supported principally by the U.S. National Institutes of Deafness and Communication Disorders, but our first work on vestibular hair cells was stimulated by the Office of Naval Research; a scientific officer who'd heard Jim Hudspeth speak persuaded the admirals that they ought to fund hair cell research. The ONR also introduced me to Art Popper and Dick Fay, who originated and shepherded, at enormous effort, the Springer Handbook of Auditory Research. Their commitment to the encouragement and dissemination of research on sensory systems of the inner ear has benefitted me and my colleagues in countless ways.

References

Art, J. J., & Fettiplace, R. (1987). Variation of membrane properties in hair cells isolated from the turtle cochlea. *Journal of Physiology*, 385, 207–242.

Art, J. J., & Fettiplace, R. (2006). Contribution of ionic currents to tuning in auditory hair cells. In R. A. Eatock, R. R. Fay & A. N. Popper (Eds.), *Vertebrate hair cells* (pp. 204). New York: Springer Science + Business Media.

Ashmore, J., Avan, P., Brownell, W. E., Dallos, P., Dierkes, K., Fettiplace, R., Grosh, K., Hackney, C. M., Hudspeth, A. J., Julicher, F., Lindner, B., Martin, P., Meaud, J., Petit, C., Sacchi, J. R., & Canlon, B. (2010). The remarkable cochlear amplifier. *Hearing Research*, 266(1–2), 1–17.

Assad, J. A., Shepherd, G. M. G., & Corey, D. P. (1991). Tip-link integrity and mechanical transduction in vertebrate hair cells. *Neuron*, 7, 985–994.

Bautista, D. M., Jordt, S. E., Nikai, T., Tsuruda, P. R., Read, A. J., Poblete, J., Yamoah, E. N., Basbaum, A. I., & Julius, D. (2006). TRPA1 mediates the inflammatory actions of environmental irritants and proalgesic agents. *Cell*, 124(6), 1269–1282.

Beurg, M., Fettiplace, R., Nam J. H., & Ricci, A. J. (2009). Localization of inner hair cell mechanotransducer channels using high-speed calcium imaging. *Nature Neuroscience*, 12(5), 553–558.

Bonsacquet, J., Brugeaud, A., Compan, V., Desmadryl, G., & Chabbert, C. (2006). AMPA type glutamate receptor mediates neurotransmission at turtle vestibular calyx synapse. *Journal of Physiology*, 576(Part 1), 63–71.

Brichta, A. M., Aubert, A., Eatock, R. A., & Goldberg, J. M. (2002). Regional analysis of whole cell currents from hair cells of the turtle posterior crista. *Journal of Neurophysiology*, 88, 3259–3278.

Brownell, W. E. (2006). The piezoelectric outer hair cell. In R. A. Eatock, R. R. Fay, & A. N. Popper (Eds.), *Vertebrate hair cells* (pp. 313–347). New York: Springer Science + Business Media.

Corey, D. P., & Hudspeth, A. J. (1983). Kinetics of the receptor current in bullfrog saccular hair cells. *Journal of Neuroscience*, 3, 962–976.

Correia, M. J., & Lang, D. G. (1990). An electrophysiological comparison of solitary type I and type II vestibular hair cells. *Neuroscience Letters*, 116, 106–111.

Crawford, A. C., & Fettiplace, R. (1985). The mechanical properties of ciliary bundles of turtle cochlear hair cells. *Journal of Physiology*, 364, 359–379.

Crawford, A. C., Evans, M. G., & Fettiplace, R. (1989). Activation and adaptation of transducer currents in turtle hair cells. *Journal of Physiology*, 419, 405–434.

Denk, W., Holt, J. R., Shepherd, G. M. G., & Corey, D. P. (1995). Calcium imaging of single stereocilia in hair cells: Localization of transduction channels at both ends of tip links. *Neuron*, 15, 1311–1321.

Eatock, R. A., & Hutzler, M. J. (1992). Ionic currents of mammalian vestibular hair cells. *Annals of the New York Academy of Sciences*, 656, 58–74.

Eatock, R. A., & Hurley, K. M. (2003). Functional development of hair cells. In R. Romand & I. Varela-Nieto (Eds.), *Development of the auditory and vestibular systems 3: Molecular development of the inner ear* (pp. 389–448). San Diego: Academic Press.

Eatock, R. A., & Lysakowski, A. (2006). Mammalian vestibular hair cells. In R. A. Eatock, R. R. Fay, & A. N. Popper (Eds.), *Vertebrate hair cells* (pp. 348–442). New York: Springer Science + Business Media.

Eatock, R. A., & Songer, J. E. (2011). Vestibular hair cells and afferents: Two channels for head motion signals. *Annual Review of Neuroscience*, 34, 501–534.

Eatock, R. A., Corey, D. P., & Hudspeth, A. J. (1987). Adaptation of mechanoelectrical transduction in hair cells of the bullfrog's sacculus. *Journal of Neuroscience*, 7, 2821–2836.

El-Amraoui, A., & Petit, C. (2010). Cadherins as targets for genetic diseases. *Cold Spring Harbor Perspectives in Biology*, 2(1), a003095.

Fettiplace, R., & Crawford, A. C. (1978). The coding of sound pressure and frequency in cochlear hair cells of the terrapin. *Proceedings of the Royal Society of London B: Biological Sciences*, 203, 209–218.

Fettiplace, R., & Fuchs, P. A. (1999). Mechanisms of hair cell tuning. *Annual Review of Physiology*, 61, 809–834.

Fuchs, P. A., Nagai, T., & Evans, M. G. (1988). Electrical tuning in hair cells isolated from the chick cochlea. *Journal of Neuroscience*, 8(7), 2460–2467.

Fuchs, P. A., & Parsons, T. D. (2006). The synaptic physiology of hair cells. In R. A. Eatock, R. R. Fay & A. N. Popper (Eds.), *Vertebrate hair cells* (pp. 249–312). New York: Springer Science + Business Media.

Furukawa, T., & Matsuura, S. (1978). Adaptive rundown of excitatory post-synaptic potentials at synapses between hair cells and eighth nerve fibres in the goldfish. *Journal of Physiology*, 276, 193–209.

Géléoc, G. S., Lennan, G. W., Richardson, G. P., & Kros, C. J. (1997). A quantitative comparison of mechanoelectrical transduction in vestibular and auditory hair cells of neonatal mice. *Proceedings of the Royal Society of London B: Biological Sciences*, 264(1381), 611–621.

Gillespie, P. G., & Hudspeth, A. J. (1994). Pulling springs to tune transduction: Adaptation by hair cells. *Neuron*, 12(1), 1–9.

Gillespie, P. G., Dumont, R. A., & Kachar, B. (2005). Have we found the tip link, transduction channel, and gating spring of the hair cell? *Current Opinion in Neurobiology*, 15(4), 389–396.

Glowatzki, E., & Fuchs, P. A. (2002). Transmitter release at the hair cell ribbon synapse. *Nature Neuroscience*, 5(2), 147–154.

Golding, N. L., & Oertel, D. (2012). Synaptic integration in dendrites: Exceptional need for speed. *Journal of Physiology*, 590, 5563–5569.

Goodyear, R. J., Kros, C. J., & Richardson, G. P. (2006). The development of hair cells in the inner ear. In R. A. Eatock, R. R. Fay & A. N. Popper (Eds.), *Vertebrate hair cells* (pp. 20–94). New York: Springer Science + Business Media.

Harris, G. G., Frishkopf, L. S., & Flock, A. (1970). Receptor potentials from hair cells of the lateral line. *Science*, 167(3914), 76–79.

Holt, J. C., Chatlani, S., Lysakowski, A., & Goldberg, J. M. (2007). Quantal and nonquantal transmission in calyx-bearing fibers of the turtle posterior crista. *Journal of Neurophysiology*, 98(3), 1083–1101.

Housley, G. D., & Ashmore, J. F. (1992). Ionic currents of outer hair cells isolated from the guinea-pig cochlea. *Journal of Physiology*, 448, 73–98.

Howard, J., & Hudspeth, A. J. (1987). Mechanical relaxation of the hair bundle mediates adaptation in mechanoelectrical transduction by the bullfrog's saccular hair cell. *Proceedings of the National Academy of Science of the U.S.A.*, 84, 3064–3068.

Howard, J., & Hudspeth, A. J. (1988). Compliance of the hair bundle associated with gating of mechanoelectrical transduction channels in the bullfrog's saccular hair cell. *Neuron*, 1, 189–199.

Hudspeth, A. J., & Corey, D. P. (1977). Sensitivity, polarity, and conductance change in the response of vertebrate hair cells to controlled mechanical stimuli. *Proceedings of the National Academy of Sciences of the U.S.A.*, 74, 2407–2411.

Hudspeth, A. J., & Lewis, R. S. (1988a). Kinetic analysis of voltage- and ion-dependent conductances in saccular hair cells of the bull-frog, *Rana catesbeiana Journal of Physiology*, 400, 237–274.

Hudspeth, A. J., & Lewis, R. S. (1988b). A model for electrical resonance and frequency tuning in saccular hair cells of the bull-frog, *Rana catesbeiana*. *Journal of Physiology*, 400, 275–297.

Hudspeth, A. J., Choe, Y., Mehta, A. D., & Martin, P. (2000). Putting ion channels to work: Mechanoelectrical transduction, adaptation and amplification by hair cells. *Proceedings of the National Academy of Sciences of the U.S.A.*, 97, 11765–11772.

Hurley, K. M., Gaboyard, S., Zhong, M., Price, S. D., Wooltorton, J. R., Lysakowski, A., & Eatock, R. A. (2006). M-like K^+ currents in type I hair cells and calyx afferent endings of the developing rat utricle. *Journal of Neuroscience*, 26(40), 10253–10269.

Iwasaki, S., Chihara, Y., Komuta, Y., Ito, K., & Sahara, Y. (2008). Low-voltage-activated potassium channels underlie the regulation of intrinsic firing properties of rat vestibular ganglion cells. *Journal of Neurophysiology*, 100(4), 2192–2204.

Jia, S., Dallos, P., & He, D. Z. (2007). Mechanoelectric transduction of adult inner hair cells. *Journal of Neuroscience*, 27(5), 1006–1014.

Johnson, S. L., Beurg, M., Marcotti, W., & Fettiplace, R. (2011). Prestin-driven cochlear amplification is not limited by the outer hair cell membrane time constant. *Neuron*, 70(6), 1143–1154.

Kachar, B., Parakkal, M., Kurc, M., Zhao, Y., & Gillespie, P. G. (2000). High-resolution structure of hair-cell tip links. *Proceedings of the National .Academy of Science of the U.S.A.*, 97(24), 13336–13341.

Kalluri, R., Xue, J., & Eatock, R. A. (2010). Ion channels set spike timing regularity of mammalian vestibular afferent neurons. *Journal of Neurophysiology*, 105(4), 2034–2051.

Kawashima, Y., Geleoc, G. S., Kurima, K., Labay, V., Lelli, A., Asai, Y., Makishima, T., Wu, D. K., Della Santina, C. C., Holt, J. R., & Griffith, A. J. (2011). Mechanotransduction in mouse inner ear hair cells requires transmembrane channel-like genes. *Journal of Clinical Investigation*, 121(12), 4796–4809.

Kazmierczak, P., Sakaguchi, H., Tokita, J., Wilson-Kubalek, E. M., Milligan, R. A., Muller, U., & Kachar, B. (2007). Cadherin 23 and protocadherin 15 interact to form tip-link filaments in sensory hair cells. *Nature*, 449(7158), 87–91.

Kharkovets, T., Hardelin, J. P., Safieddine, S., Schweizer, M., El-Amraoui, A., Petit, C., & Jentsch, T. J. (2000). KCNQ4, a K^+ channel mutated in a form of dominant deafness, is expressed in the inner ear and the central auditory pathway. *Proceedings of the National Academy of Sciences of the USA*, 97(8), 4333–4338.

Kindt, K. S., Finch, G., & Nicolson, T. (2012). Kinocilia mediate mechanosensitivity in developing zebrafish hair cells. *Developmental Cell*, 23(2), 329–341.

Kros, C. J., Rüsch, A., & Richardson, G. P. (1992). Mechano-electrical transducer currents in hair cells of the cultured neonatal mouse cochlea. *Proceedings of the Royal Society of London B: Biological Sciences*, 249, 185–193.

Kros, C. J., Marcotti, W., van Netten, S. M., Self, T. J., Libby, R. T., Brown, S. D., Richardson, G. P., & Steel, K. P. (2002). Reduced climbing and increased slipping adaptation in cochlear hair cells of mice with Myo7a mutations. *Nature Neuroscience*, 5(1), 41–47.

Kurima, K., Peters, L. M., Yang, Y., Riazuddin, S., Ahmed, Z. M., Naz, S., Arnaud, D., Drury, S., Mo, J., Makishima, T., Ghosh, M., Menon, P. S., Deshmukh, D., Oddoux, C., Ostrer, H., Khan, S., Deininger, P. L., Hampton, L. L., Sullivan, S. L., Battey, J. F., Jr., Keats, B. J., Wilcox, E. R., Friedman, T. B., & Griffith, A. J. (2002). Dominant and recessive deafness caused by mutations of a novel gene, *TMC1*, required for cochlear hair-cell function. *Nature Genetics*, 30(3), 277–284.

Kwan, K. Y., Allchorne, A. J., Vollrath, M. A., Christensen, A. P., Zhang, D. S., Woolf, C. J., & Corey, D. P. (2006). TRPA1 contributes to cold, mechanical, and chemical nociception but is not essential for hair-cell transduction. *Neuron*, 50(2), 277–289.

Lenzi, D., Runyeon, J. W., Crum, J., Ellisman, M. H., & Roberts, W. M. (1999). Synaptic vesicle populations in saccular hair cells reconstructed by electron tomography. *Journal of Neuroscience*, 19(1), 119–132.

Lewis, R. S., & Hudspeth, A. J. (1983). Voltage- and ion-dependent conductances in solitary vertebrate hair cells. *Nature*, 304, 538–541.

Lysakowski, A., Gaboyard-Niay, S., Calin-Jageman, I., Chatlani, S., Price, S. D., & Eatock, R. A. (2011). Molecular microdomains in a sensory terminal, the vestibular calyx ending. *Journal of Neuroscience*, 31(27), 10101–10114.

Marcotti, W., Johnson, S. L., Rusch, A., & Kros, C. J. (2003). Sodium and calcium currents shape action potentials in immature mouse inner hair cells. *Journal of Physiology*, 552(Part 3), 743–761.

Marcotti, W., Erven, A., Johnson, S. L., Steel, K. P., & Kros, C. J. (2006). Tmc1 is necessary for normal functional maturation and survival of inner and outer hair cells in the mouse cochlea. *Journal of Physiology*, 574(Part 3), 677–698.

Martin, P., Bozovic, D., Choe, Y., & Hudspeth, A. J. (2003). Spontaneous oscillation by hair bundles of the bullfrog's sacculus. *Journal of Neuroscience*, 23(11), 4533–4548.

Meredith, F. L., Benke, T. A., & Rennie, K. J. (2012). Hyperpolarization-activated current (I (h)) in vestibular calyx terminals: Characterization and role in shaping postsynaptic events. *Journal of the Association for Research in Otolaryngology*, 13, 745–758.

Mo, Z. L., Adamson, C. L., & Davis, R. L. (2002). Dendrotoxin-sensitive K^+ currents contribute to accommodation in murine spiral ganglion neurons. *Journal of Physiology*, 542,763–768.

Moser, T., & Beutner, D. (2000). Kinetics of exocytosis and endocytosis at the cochlear inner hair cell afferent synapse of the mouse. *Proceedings of the National Academy of Sciences of the USA*, 97(2), 883–888.

Moser, T., Brandt, A., & Lysakowski, A. (2006). Hair cell ribbon synapses. *Cell and Tissue Research*, 326(2), 347–359.

Mulroy, M. J., Altmann, D. W., Weiss, T. F., & Peake, W. T. (1974). Intracellular electric responses to sound in a vertebrate cochlea. *Nature*, 249, 482–485.

Nicolson, T., Rüsch, A., Friedrich, R. W., Granato, M., Ruppersberg, J. P., & Nusslein-Volhard, C. (1998). Genetic analysis of vertebrate sensory hair cell mechanosensation: The zebrafish circler mutants. *Neuron*, 20(2), 271–283.

Ohmori, H. (1984). Studies of ionic currents in the isolated vestibular hair cell of the chick. *Journal of Physiology*, 350, 561–581.

Parsons, T. D., Lenzi, D., Almers, W., & Roberts, W. M. (1994). Calcium-triggered exocytosis and endocytosis in an isolated presynaptic cell: Capacitance measurements in saccular hair cells. *Neuron*, 13, 875–883.

Pickles, J. O., Comis, S. D., & Osborne, M. P. (1984). Cross-links between stereocilia in the guinea pig organ of Corti, and their possible relation to sensory transduction. *Hearing Research*, 15, 103–112.

Rennie, K. J., & Streeter, M. A. (2006). Voltage-dependent currents in isolated vestibular afferent calyx terminals. *Journal of Neurophysiology*, 95(1), 26–32.

Ricci, A. J., Wu, Y. C., & Fettiplace, R. (1998). The endogenous calcium buffer and the time course of transducer adaptation in auditory hair cells. *Journal of Neuroscience*, 18, 8261–8277.

Ricci, A. J., Kennedy, H. J., Crawford, A. C., & Fettiplace, R. (2005). The transduction channel filter in auditory hair cells. *Journal of Neuroscience*, 25(34), 7831–7839.

Ricci, A. J., Bai, J. P., Song, L., Lv, C., Zenisek, D., & Santos-Sacchi, J. (2013). Patch-clamp recordings from lateral line neuromast hair cells of the living zebrafish. *Journal of Neuroscience*, 33(7), 3131–3134.

Rüsch, A., & Eatock, R. A. (1996). A delayed rectifier conductance in type I hair cells of the mouse utricle. *Journal of Neurophysiology*, 76(2), 995–1004.

Rüsch, A., Lysakowski, A., & Eatock, R. A. (1998a). Postnatal development of type I and type II hair cells in the mouse utricle: Acquisition of voltage-gated conductances and differentiated morphology. *Journal of Neuroscience*, 18(18), 7487–7501.

Rüsch, A., Erway, L. C., Oliver, D., Vennstrom, B., & Forrest, D. (1998b). Thyroid hormone receptor beta-dependent expression of a potassium conductance in inner hair cells at the onset of hearing. *Proceedings of the National Academy of Sciences of the U.S.A.*, 95(26), 15758–15762.

Russell, I. J., & Sellick, P. M. (1977). Tuning properties of hair cells in the mammalian cochlea. *Nature*, 267, 858–860.

Sans, A., & Scarfone, E. (1996). Afferent calyces and type I hair cells during development: A new morphofunctional hypothesis. *Annals of the New York Academy of Sciences*, 781, 1–12.

Schnee, M. E., Lawton, D. M., Furness, D. N., Benke, T. A., & Ricci, A. J. (2005). Auditory hair cell-afferent fiber synapses are specialized to operate at their best frequencies. *Neuron*, 47(2), 243–254.

Sewell, W. F. (1996). Neurotransmitters and synaptic transmission. In P. Dallos, A. N. Popper, & R. R. Fay (Eds.), *The cochlea* (pp. 503–533). New York: Springer-Verlag.

Siemens, J., Lillo, C., Dumont, R. A., Reynolds, A., Williams, D. S., Gillespie, P. G., & Muller, U. (2004). Cadherin 23 is a component of the tip link in hair-cell stereocilia. *Nature*, 428(6986), 950–955.

Sollner, C., Rauch, G. J., Siemens, J., Geisler, R., Schuster, S. C., Muller, U., & Nicolson, T. (2004). Mutations in cadherin 23 affect tip links in zebrafish sensory hair cells. *Nature*, 428(6986), 955–959.

Songer, J. E., & Eatock, R. A. (2013). Tuning and timing in mammalian type I hair cells and calyceal synapses. *Journal of Neuroscience*, 33(8), 3706–3724.

Sotomayor, M., Corey, D. P., & Schulten, K. (2005). In search of the hair-cell gating spring elastic properties of ankyrin and cadherin repeats. *Structure.*, 13(4), 669–682.

Vollrath, M. A., & Eatock, R. A. (2003). Time course and extent of mechanotransducer adaptation in mouse utricular hair cells: comparison with frog saccular hair cells. *Journal of Neurophysiology*, 90(4), 2676–2689.

Vollrath, M. A., Kwan, K. Y., & Corey, D. P. (2007). The micromachinery of mechanotransduction in hair cells. *Annual Review of Neuroscience*, 30, 339–365.

Vreugde, S., Erven, A., Kros, C. J., Marcotti, W., Fuchs, H., Kurima, K., Wilcox, E. R., Friedman, T. B., Griffith, A. J., Balling, R., Hrabe, D. A., Avraham, K. B., & Steel, K. P. (2002). Beethoven, a mouse model for dominant, progressive hearing loss DFNA36. *Nature Genetics*, 30(3), 257–258.

Wersäll, J. (1956). Studies on the structure and innervation of the sensory epithelium of the cristae ampullares in the guinea pig; a light and electron microscopic investigation. *Acta Oto-Laryngologica, Supplementum*, 126, 1–85.

Whitfield, T. T., Granato, M., van Eeden, F. J., Schach, U., Brand, M., Furutani-Seiki, M., Haffter, P., Hammerschmidt, M., Heisenberg, C. P., Jiang, Y. J., Kane, D. A., Kelsh, R. N., Mullins, M. C., Odenthal, J., & Nusslein-Volhard, C. (1996). Mutations affecting development of the zebrafish inner ear and lateral line. *Development*, 123, 241–254.

Wittig, J. H., Jr., & Parsons, T. D. (2008). Synaptic ribbon enables temporal precision of hair cell afferent synapse by increasing the number of readily releasable vesicles: A modeling study. *Journal of Neurophysiology*, 100(4), 1724–1739.

Wu, Y. C., Ricci, A. J., & Fettiplace, R. (1999). Two components of transducer adaptation in auditory hair cells. *Journal of Neurophysiology*, 82(5), 2171–2181.

Yamashita, M., & Ohmori, H. (1990). Synaptic responses to mechanical stimulation in calyceal and bouton type vestibular afferents studied in an isolated preparation of semicircular canal ampullae of chicken. *Experimental Brain Research*, 80, 475–488.

Zdebik, A. A., Wangemann, P., & Jentsch, T. J. (2009). Potassium ion movement in the inner ear: Insights from genetic disease and mouse models. *Physiology*, 24, 307–316.

Chapter 6
Three Decades of Tinnitus-Related Research

Jos J. Eggermont

Jos J. Eggermont

J.J. Eggermont (✉)
Department of Physiology and Pharmacology, University of Calgary,
2500 University Drive N.W., Calgary, AB, Canada T2N 1N4

Department of Psychology, University of Calgary,
2500 University Drive N.W., Calgary, AB, Canada T2N 1N4
e-mail: eggermon@ucalgary.ca

A.N. Popper and R.R. Fay (eds.), *Perspectives on Auditory Research*, Springer Handbook of Auditory Research 50, DOI 10.1007/978-1-4614-9102-6_6,

6.1 Introduction

I will illustrate my tinnitus research and its relationship to use-dependent plasticity with a narrative centered on International Tinnitus Seminars (ITS) and Tinnitus Research Initiative (TRI) conferences that I attended over a time span of nearly three decades. I follow this by examining how the field of use-dependent plasticity impacted on my understanding of tinnitus. Finally, I speculate about the much-needed integration of mechanism and phenotypes of tinnitus that has to occur in future research and treatment.

6.2 Before SHAR

In 1983, the second ITS was held in New York City. I was at that time still living in the Netherlands and was invited to give a talk about potential physiological mechanisms for tinnitus. I had never studied tinnitus but had experienced it weekly in transient form following shooting practice during my military service. Thus I had at least an idea of how it sounded and knew one etiology, and decided to talk about "Tinnitus: Some thoughts about its origin" (Eggermont, 1984, p. 34); in hindsight I made some comments that were relevant:

> Sharply defined hearing losses, such as those caused by noise trauma or ototoxic drugs, often are accompanied by tinnitus. In such situations apparently normal parts of the cochlea are at the high frequency side bordered by abnormal parts. This leads to a reduced suppression effect at or near the cut-off frequency of the audiogram and may cause an artificial cooperative effect in the spontaneous activity at the edge of the audiogram. This propagates through the auditory system leading to a tinnitus sensation with a pitch corresponding to the cut-off frequency of the audiogram. … Inhibitory interaction between neurons in the auditory system starts from the level of the dorsal cochlear nucleus (…) and are found in all higher centers. In the dorsal cochlear nucleus type II/III interneurons act in an inhibitory way upon spontaneously active type IV neurons. One can conceive that, if type II/III neurons are becoming devoid of input from their afferents, then the spontaneous active type IV neurons are no longer inhibited. Los of inhibition might give rise to abnormally large spontaneous activity and become audible.

I also promoted (Eggermont, 1984, p. 33) interneuronal synchronization as a general phenomenon underlying any sound sensation, stimulus-induced or pathological:

> This [reorganization of firings] actually means that there are instantaneous rate changes in the individual nerve fiber firing patterns, but above all it means that activity patterns of small groups of nerve fibers become synchronized. This is a cooperative effect and it is now postulated that such cooperative effects in the central nervous system or auditory periphery may give rise to spontaneous sensations of sound, to tinnitus.

After moving to Calgary in 1986, the first tinnitus conference that I attended was the fourth ITS held in Bordeaux, France in 1991. At that time I was still leaning more toward a peripheral than central origin of tinnitus, as reflected in the title of the paper "On the pathophysiology of tinnitus: A review and a peripheral model" (Eggermont, 1990, p. 111), but was convinced that correlated neural activity was at the source of tinnitus.

> … I investigate the consequences of the assumption that tinnitus is the result of correlated neural activity in auditory nerve fibers under 'no sound' conditions. Two possible pathological conditions capable of causing this correlation are ephaptic excitation of one nerve fiber by neighboring nerve fibers and synchronization of the various synapses in individual hair cells. The first condition is likely to be found in cases suffering from acoustic neuroma where the myelin sheath of the auditory neurons is damaged. The second condition is attributed to a spontaneous excess influx of K^+ or Ca^{2+} -ions into the hair cell resulting in transient hair cell depolarization causing synchronous transmitter release at all hair cell synapses. This condition is postulated in noise trauma and ototoxic drug damage of the inner hair cell membrane. The model produces the excess of short interspike intervals found in auditory nerve fiber recordings in animal models of tinnitus as well as the theoretically required correlation in the activity of neighboring neurons.

6.3 After SHAR, Volume 1

Around 1992, I became more and more convinced that the locus of tinnitus was in the brain. This may in part have been due to my changing research interest toward adult auditory cortical plasticity which led me to envision tinnitus as "maladaptive" plasticity. We started simple by studying the effects of systemically applied salicylate on neural activity in cat auditory cortex, mainly because previous studies by Jastreboff et al. (1988) had shown that salicylate caused increased spontaneous firing rates (SFRs) in the inferior colliculus and that this correlated with the presence of tinnitus as deduced from behavioral tests in the same animals. Kentaro Ochi and Mutsumi Kenmochi (Ochi & Eggermont, 1996; Eggermont & Kenmochi, 1998), who at that time were post docs in my lab, found a clear dose-related effect of the tinnitus-inducing drug quinine on neural correlation strength in cat primary auditory cortex (AI) but very little for salicylate. This was largely the result of the low tolerance of cats for salicylate, which prohibited the use of a large dose. We presented this at the fifth ITS held in 1995 in Portland, Oregon. After more experiments that, in addition to AI, included recordings from secondary auditory cortex (AII) and anterior auditory field (AAF), our data (Eggermont & Kenmochi, 1998, p. 149) suggested:

> …(1) that both salicylate and quinine significantly increase spontaneous firing rates in AII. In AI and AAF, both quinine and salicylate reduced the spontaneous rate; (2) the effect of both drugs was to increase spontaneous rates for high CF sites and a tendency to decrease them for low CF sites; (3) the mean driven firing rates were not affected by either drug except for a decrease produced by quinine in AI; (4) changes in driven firing rate were positively correlated with changes in spontaneous firing rates.

Salicylate and quinine present a straightforward way of inducing tinnitus in animals and humans, but the cure is to stop taking the drug, so it can hardly be a model for continuous tinnitus in humans. We thus decided to switch to noise trauma as a more relevant model that would be useful in a search for neural correlates in humans. Hisashi Komiya and I presented some early work at the sixth ITS (1999) held in Cambridge, UK. We found that permanent but moderate hearing loss resulting from noise trauma in the cat caused, besides increased SFR and increased neural synchrony, a reorganization of the cortical tonotopic map in AI (Eggermont & Komiya, 2000). This was the first indication of the triad of electrophysiological substrates that commonly result from noise trauma. At that time it was well established that sound conditioning, such as exposing the animals to a non-traumatizing sound before inducing a noise trauma, reduces the amount of hearing loss it causes (Canlon et al., 1988). It was also noted (Fukushima et al., 1990) that unilateral removal of middle ear ossicles, resulting in an attenuation of environmental sound of approximately 50 dB over a wide frequency band (250 Hz to 8 kHz), after an acoustic trauma was associated with greater remaining hearing loss in that ear compared with that with the intact middle ear. This suggested that environmental sound stimulation immediately after a noise trauma might reduce or prevent permanent cochlear lesions. My post doc at that time, Arnaud Noreña, developed the notion that tinnitus is caused by an imbalance between excitatory and inhibitory activity in the central auditory nervous system resulting from a frequency-specific decrease in spontaneous and driven activity in the auditory nerve fibers following acoustic trauma. Based on all this, we speculated that using an enhanced acoustic environment (EAE), such as presenting sounds in the frequency range of the hearing loss, would balance this uneven output of the auditory nerve fibers. We found that post-trauma exposure to the EAE for 3 weeks or more prevented the triad of tonotopic map change, increased SFR, and increased neural synchrony from occurring (Noreña & Eggermont, 2005, p. 704):

> At least two different types of inhibitory mechanisms, phasic and tonic, could be involved in the tuning properties of the neurons … The tonic inhibition (driven by spontaneous activity of afferent inputs) is supposed to spread widely across frequency and to be proportional to the amount of spontaneous firing rate of excitatory inputs. When spontaneous firing rates at the auditory nerve level are reduced (caused by hearing loss; see above), this tonic inhibition is also reduced. As a consequence, inputs that were previously inhibited are "un-masked." This release from inhibition can unmask inputs at frequencies relatively remote from the hearing loss region. In summary, a decrease in spontaneous firing rate at a peripheral level induces a release from inhibition at central level, causing the emergence of new responses (i.e., reorganization of the tonotopic map)…. By stimulating the frequency region corresponding to the hearing loss, we compensated for the decrease in firing rates in the auditory nerve fibers caused by hearing loss and thereby prevented the cascade of central changes (release from inhibition) that would normally lead to cortical tonotopic map reorganization.

In the follow-up paper (Noreña & Eggermont, 2006, p. 559) we subsequently concluded:

> Noise-induced hearing loss causes a reorganization of the cortical tonotopic map and increased firing rate and neural synchrony. If acoustic stimulation with a spectrum corresponding to the frequency band of the hearing loss is provided after the trauma, however, the hearing loss is reduced, the tonotopic map is normal, and the spontaneous firing rate and synchrony are unchanged. Thus, if trauma-induced tinnitus is related to an increase in spontaneous firing rate or synchrony in AI, post-trauma stimulation might prevent the occurrence of tinnitus.

We also found that stimulation with a low-frequency EAE, such as with a frequency range covering the normal part of the cochlea, had little or no effect. We also found, but did not publish because we had only three animals in that experiment and never followed up on it, that applying an EAE 1 week after the end of the trauma had no effect. This suggests a critical period for recovery of the neurotoxic aspects of the hearing loss (in the high-frequency region basally from the region of hair cell loss). Potential consequences for posttrauma treatment of tinnitus in humans by tailored auditory environments are that its effectiveness will greatly decrease with time after the onset of tinnitus.

Around this time (2005), the eighth ITS was held in Pau, France. I talked about cortical tonotopic map reorganization and its implications for treatment of tinnitus (Eggermont, 2006, p. 12):

> If we presume that changed tonotopic maps are co-occurring with tinnitus, that aberrant thalamo-cortical loops can give rise to and maintain tinnitus, and that all these may originate from an imbalance of excitation and inhibition at the thalamo-cortical level … , the obvious treatment would be to restore that balance. This has been demonstrated to work immediately after the trauma by stimulating in the hearing loss region … by placing the animals in an enriched acoustic environment. This mimics in a sense equipping the animals with a hearing aid that only amplifies the sound in the hearing loss range. This suggests that in early stages of hearing loss, hearing aids that provide local increases in excitation may prevent local thalamic hyperpolarization, prevent cortical tonotopic map reorganization, and prevent tinnitus from occurring. It has been reported earlier that hearing loss following noise trauma is considerably larger after sound deprivation by unilateral removal of middle ear ossicles compared with the other side showing an intact middle ear … So hearing aids as well as enriched acoustic environments … may also help partial recovery from noise trauma.

Only recently did it became clear that this ubiquitous finding of tonotopic map reorganization in animals after noise trauma was not present in human tinnitus sufferers with modest hearing loss (Langers et al., 2012). It is important to note that Seki and Eggermont (2002) did not find tonotopic map changes in animals when noise-induced hearing losses were ≤25 db SPL despite the fact that there was an increased SFR for nearly all units from which they recorded. Thus, tonotopic map reorganization is not a necessary condition for tinnitus in general.

A series of conferences promoted by the TRI started to focus on the neuroscience aspects of tinnitus that we had emphasized in a well-cited review in *TINS* (Eggermont & Roberts, 2004). This paper reviewed the field of tinnitus research based on our experiences with a multidisciplinary and multicenter study on tinnitus in Canada. Our group combined investigations based on psychoacoustics (Ward and Baumann, 2009), human EEG and evoked potential recordings (Roberts et al., 2008), neural modeling studies (Dominguez et al., 2006), and animal electrophysiology (Eggermont, 2006); comprised universities from east (McMaster, Hamilton) to west (University of Calgary, University of British Columbia); and met at least once a year to discuss the field and sample wines from the two corresponding wine regions (Niagara and Okanagan).

I attended the third of the TRI conferences held in 2009 in Stresa, Italy and that featured a series of fresh approaches to tinnitus research. Studies presenting the amplification of peripheral SFR in the inferior colliculus, on the effect of noise trauma on the balance of excitation and inhibition in auditory cortex, and on the role of plasticity in the DCN, signaled a welcome boost to the research in tinnitus. Changes in EEG resting-state activity in patients with tinnitus were reported by de Ridder and colleagues and correlated gamma band activity in temporal cortex with tinnitus loudness and alpha band activity in prefrontal cortex with tinnitus distress. This represents a potentially important research direction for understanding the centralization of sustained tinnitus. At the fifth TRI conference, held in 2011 in Buffalo, NY, I addressed discordances between the proposed electrophysiological substrates of tinnitus, specifically increased cortical SFR, and conditioned responses and gap-startle reflex tests for salicylate-induced tinnitus. Either the behavioral tests do not reflect what we presume, that is the presence of tinnitus, or cortical SFR increase is not a substrate of tinnitus. This, and the impressive gains made in especially the last decade of tinnitus research, are extensively reviewed in my book *The Neuroscience of Tinnitus* (Eggermont, 2012) and in a SHAR volume on Tinnitus that I co-edited (Eggermont et al., 2012).

6.4 The Past Decade: Mechanisms for Tinnitus Without Hearing Loss

Around 2005, my interest started to diverge from cortical plasticity induced by traumatic hearing loss toward the effects of non-damaging (in the sense of not producing hearing loss) EAEs on auditory cortex in normal hearing adult cats. To our surprise, a 5-month continuous stimulation with a dynamic multi-frequency 4- to 20-kHz EAE presented at a level of 80 dB peak equivalent SPL in adult cats (Noreña et al., 2006) produced a nearly complete loss of response to this frequency range in AI. This resulted in a reorganization of this original 4- to 20-kHz part of AI so that neurons now had characteristic frequencies (CFs) with normal thresholds either to frequencies >20 kHz, <4 kHz, or were double tuned to frequencies from both ranges. Occasionally there were triple-tuned neurons to the original CF, and those from the high- and low-frequency regions, all with normal thresholds at the various CFs. Corroborating this were the normal auditory brain stem response (ABR) thresholds as compared to those in controls. We concluded (Noreña et al., 2006, p. 937):

> The present study shows that continuous and long-term stimulation from the end of the maturation period into adulthood can induce central changes similar to those caused by hearing loss. Namely, the representation of the frequencies of the chronically presented stimulus is markedly decreased and is replaced by an enlarged representation of the frequencies adjacent to the EAE spectrum. Notably, thalamus (LFPs) and cortex (multiunit responses) appear to become unspecialized for the EAE frequency band. It is not clear if these changes are permanent or reversible. To establish this, we would need to place the cats in a quiet environment for a sufficiently long time after the exposure to the EAE and then assess the tonotopic maps. It is also conceivable that associating the EAE with a reward could have changed the direction and extent of the map changes. Finally, presenting the

> EAE sounds interspersed with silent recovery periods could potentially prevent the occurrence of complete synaptic depression and change the central reorganization. Further studies will be needed to unravel this new form of representational plasticity caused by continuous stimulation with a spectrally enhanced acoustic environment.

We subsequently reported on changes to the sound frequency representation in auditory cortex following persistent exposure of mature cats to more moderate-level (<70 dB SPL), random, band-limited multi-frequency sounds. The effect was similar but weaker following an intermittent exposure (12 h-on/12 h-off) to the same EAE. No hearing loss was detected in ABR responses. The changes in cortex partially recovered over a 12-week time period (Pienkowski & Eggermont, 2009, p. 38):

> Following up to 12 weeks of (post-exposure) recovery in a quiet room shared with littermates, a more or less partial reversal of some of the exposure-induced changes occurred. Most notably, the proportion of units tuned to frequencies in the EAE band was restored to normal, as were their spectral bandwidths. Nevertheless, the tonotopic organization of AI persisted in a partially disrupted state at the end of our 12 week observation period.

Profound suppression was also observed after long-term, uninterrupted exposure to 4- to 20-kHz band-limited noise. With narrower tonal exposure bandwidths (an octave-band spanning 2- to 4-kHz suppression could extend an octave or more. For an EAE consisting of a pair of third-octave bands centered at 4 and 16 kHz) the region between 4 and 16 kHz was "filled in," so that the suppression looked very much like the one for a 4–20 kHz EAE. This suggested a large role for lateral inhibition effects. In the very long-term, passive EAE exposure in adult cats led gradually to a reorganization of the AI tonotopic map. The reorganization is reminiscent of that following restricted hearing loss: Neurons tuned to the EAE band, initially suppressed, eventually become tuned to higher and/or lower frequencies, with no evident decrease in sensitivity. There are potentially many real life consequences (Pienkowski & Eggermont, 2012, p. 312):

> Many people with normal or near-normal audiograms, especially among the elderly, have problems with speech intelligibility in noisy environments … We have suggested that at least some of these cases may be linked to noise exposure. The noise may be traumatic, leading to damage to cochlear structures and [spiral ganglion cells] SGCs without necessarily producing permanent absolute threshold shifts, at least not until later in life. The noise may also be nontraumatic yet lead to persistent changes in auditory cortical function even when the cochlea and lower brainstem remain structurally and functionally sound. Both types of exposure fall under the radar of present occupational noise standards, which aim only to prevent permanent increases in pure-tone thresholds. Another area of potential concern is sound exposure during early infancy. Perhaps most vulnerable are premature infants spending time in neonatal intensive care units. As might be expected given prevailing neonatal intensive care unit sound levels …, large-sample studies using DPOAEs and ABR found little evidence of increased risk of peripheral hearing loss. Nevertheless, the developing brain is in general considerably more plastic than the adult brain. Thus, plasticity of the developing auditory brain (or disruption of the normal developmental trajectory) can be triggered with a relatively shorter sound exposure period, and with more lasting effects, as demonstrated in animal studies. It is vital to note that plasticity induced by moderate-level noise in infants could delay or impair language development, although more work is needed to substantiate this risk.

Recordings of SFR in control and EAE cats in the Noreña et al. (2006) study were split into three groups according to the location of the electrode array along the postero-anterior axis: posterior (<10 % of the postero-anterior ectosylvian sulci distance, and including the low-frequency region in AI caudal from the posterior ectosylvian sulcus), middle, that is, the region with CFs in the EAE frequency range (10–70 % of distance) and anterior (>70 % of the postero-anterior sulci distance). The SFR was not significantly different between control and EAE cats for recordings with characteristic frequencies normally corresponding to the EAE spectrum (middle area). On the other hand, the SFR was significantly increased for recordings normally corresponding to characteristic frequencies below 4 kHz (posterior) and above 20 kHz (anterior). The synchrony between spontaneous firings was significantly elevated in EAE cats compared to controls in posterior, middle, and anterior areas. For control cats we found no significant difference in neural synchrony within the three areas of AI. However, in EAE cats the neural synchrony was significantly lower in the middle area compared with the posterior and anterior areas of AI.

We also observed a correlate of hyperacusis, steeper rate-intensity functions and lower neural thresholds, in the regions bordering the EAE frequency range, and reflecting the frequency-dependent gain changes. The gain increase could underlie the increased SFRs and increased synchrony. It is tempting to interpret these changes, especially in neural synchrony, as substrates for tinnitus without hearing loss.

6.5 Perspectives: Typology of Tinnitus and Use-Dependent Plasticity

Currently, clinical trials on tinnitus treatment generally have a negative or inconclusive outcome. This may be the result of using inhomogeneous populations by combining tinnitus patients across etiology or symptom. Tyler et al. (2008) used cluster analysis to identify four subgroups among tinnitus patients based on their symptoms: (1) constant distressing tinnitus, (2) varying tinnitus that is worse in noise, (3) tinnitus patients who can cope and whose tinnitus is not influenced by touch (somatic modulation), and (4) tinnitus patients who can cope but whose tinnitus is worse in quiet environments. For people with tinnitus, their etiologies and underlying biological substrates may be very different. At present we do not know whether there is a connection of these clusters to the etiology, nor do we know what differentiates the brains of these four classes of tinnitus. The degree of involvement of the limbic system is a likely factor but a definitive answer is lacking. In addition to the current use of questionnaires, it is critical to develop objective diagnostics such as "resting state" brain imaging to classify tinnitus and to evaluate its treatment outcomes, without which it would be difficult to conduct meaningful clinical trials. A typology of tinnitus, based on etiology as well as psychological aspects, co-occurrence with hyperacusis, depression, etc. is needed for targeted treatment. However, about half of tinnitus patients cannot identify a cause for their tinnitus. Etiology-based types of tinnitus are fairly obvious (for details see Eggermont, 2012).

6.5.1 Noise-Induced, Hearing-Loss-Related Tinnitus

Here the driving force to tinnitus is likely the imbalance in spontaneous and driven output of the cochlea along the frequency axis. Auditory nerve fibers in the region of the hearing loss have lower SFR and lower driven firing rate than those in the unaffected frequency regions. This results likely in a frequency-dependent gain change in the central nervous system, already present in the ventral cochlear nucleus, which drives increased SFR in the inferior colliculus. The increased SFR in the inferior colliculus becomes internalized after several weeks to months. Lower auditory input to the dorsal cochlear nucleus slowly enhances its dorsal column and trigeminal nerve inputs, and results in enhanced SFR in that can be passed on to downstream nuclei. Recalibration should initially be possible by equalizing auditory nerve output across frequency by presenting an EAE or by reducing the SFR in auditory nerve fibers. The big question is why fewer than 30 % of people with hearing loss (older than the age of 65) have tinnitus. This likely implies a central regulating mechanism, potentially in the limbic system.

6.5.2 Somatic-Induced Tinnitus Without Hearing Loss

This tinnitus can result from trauma in the C2–C4 region of the spinal cord (dorsal column), from dental problems, or from gaze changes (trigeminal). All of these changing inputs to the dorsal cochlear nucleus upset the balance of excitation and inhibition. They are likely the result of potential growth of new or strengthening of existing synaptic connections of the somatic inputs to the dorsal cochlear nucleus. It is questionable if recalibration using an EAE could work.

6.5.3 Tinnitus Without Hearing Loss

There are several possibilities for tinnitus without demonstrable hearing loss besides that of somatic origin (see Section 5.2). The hearing loss may remain undetected owing to limitation of clinical audiometry, either the highest frequency used is too low (8 kHz) or the octave separation is too wide to detect local dead regions (induced by inner hair cell ribbon synapse damage). Another possibility is a frequency-dependent disturbance in the central gain as described in Section 4, which leads to frequency-dependent increases in SFR and in neural synchrony. Presenting a targeted acoustic environment following exposure in both occupational and recreational settings can likely prevent this. It is also possible that there is a purely central neuromodulator or hormonal imbalance that affects inhibitory activity. This could potentially be targeted by pharmaceutical means.

A totally different group consists of patients that present co-occurrence of tinnitus with stress and/or depression. And these are typically the people who really suffer from tinnitus.

6.5.4 Stress-Related Tinnitus

Here the distinction is not based on a particular etiology but on the often co-occurrence of stress and tinnitus. It is often not clear if stress is causal to tinnitus or that tinnitus causes stress. Stress is a psychological problem resulting from stress-related pathways that may be affected by tinnitus. Stress should be treated first and likely reduces the complaints about tinnitus.

6.5.5 Tinnitus and Depression

There is a direct and long-term association between tinnitus severity and clinical depression. Hébert et al. (2012) examined the relationship between depression and tinnitus prevalence and severity in a representative sample of the general Swedish working population. They found that a decrease in depression was associated with a decrease in tinnitus prevalence, and even more markedly with tinnitus severity. Parallels in the pathophysiology of tinnitus and depression argue against depression as a result of tinnitus, but suggest a complex interplay between tinnitus and depression.

6.6 Tinnitus as a Neural Network Problem

An important common mechanism of bothersome tinnitus is likely the activity in positive feedback loops formed by stress and depression related pathways that turn the focus of attention to the tinnitus and engrain it in memory. Gating studies suggest modulation of auditory signals in the limbic areas of the brain that incorporate feedback loops either to the thalamus (thalamus–amygdala–nucleus accumbens–thalamic reticular nucleus–thalamus) or to the cortex (thalamus–amygdala–basal forebrain–cortex). This indicates that the thalamocortical system is crucial for the perception of tinnitus, but may be tuned out (in normal hearing subjects) by the nucleus acumbens, which is deficient at least in some tinnitus patients or by stimulating the caudate, which may function insufficiently in tinnitus.

Corticofugal feedback may also be an important factor in the perception of tinnitus. Magnetic dipole-source imaging suggests that tinnitus may be accompanied by a reorganization of the auditory cortical tonotopy. As we now know, this depends likely on the amount of hearing loss. The pattern of tonotopic map reorganization correlated with the subjective tinnitus strength and with the shift in the representation of tinnitus frequencies in the auditory cortex. Corticofugal feedback, induced by the tinnitus to which a person directs her or his attention, could enhance the processing of tinnitus-related frequencies and suppress the processing of surround frequencies in the brain stem and auditory midbrain. Therefore, this frequency-specific amplification by corticofugal feedback in subcortical areas might contribute to stabilizing the tinnitus percept, leading to the chronic form of tinnitus. Feedback loops tend to stabilize systems and not always in a positive way. In the long run, peripheral and central activity may enhance each other, and the result is that there is no particular site in the central auditory system that can be held solely responsible for tinnitus. Opening the loop by blocking connections, for example, by using drugs such as lidocaine, or by desynchronizing the activity of the nested loops, for example, by stimulation through a cochlear implant, by neuromoulation either electrically or by transcranial magnetic stimulation of the cortex, or even by deep brain stimulation are potential ways to alleviate tinnitus. However, given the current state of these potential treatments, it would be more effective using cognitive behavioral therapy or "tinnitus retraining therapy."

Acknowledgments My research in tinnitus and plasticity-related topics has been funded by the Canadian Institutes of Health Research, the Natural Sciences and Engineering Research Council of Canada, the Alberta Heritage Foundation for Medical Research, and the Campbell McLaurin Chair for Hearing Deficiencies.

References

Canlon, B., Borg, E., & Flock, A. (1988). Protection against noise trauma by pre-exposure to a low level acoustic stimulus. *Hearing Research*, 34, 197–200.

Dominguez, M., Becker, S., Bruce, I., & Read, H. (2006). A spiking neuron model of cortical correlates of sensorineural hearing loss: Spontaneous firing, synchrony, and tinnitus. *Neural Computation*, 18, 2942–2958.

Eggermont, J. J. (1984). Tinnitus: Some thoughts about its origin. *Journal of Laryngology and Otology*, Supplement 9, 31–37.

Eggermont, J. J. (1990). On the pathophysiology of tinnitus: A review and a peripheral model. *Hearing Research*, 48, 111–124.

Eggermont, J. J. (2006). Cortical tonotopic map reorganization and its implications for treatment of tinnitus. *Acta Oto-laryngologica Supplementum*, 556, 9–12.

Eggermont, J. J. (2012). *The neuroscience of tinnitus*. Oxford: Oxford University Press.

Eggermont, J. J., & Kenmochi, M. (1998). Salicylate and quinine selectively increase spontaneous firing rates in secondary auditory cortex. *Hearing Research*, 117, 149–160.

Eggermont, J. J., & Komiya, H. (2000). Moderate noise trauma in juvenile cats results in profound cortical topographic map changes in adulthood. *Hearing Research*, 142, 89–101.

Eggermont, J. J., & Roberts, L. E. (2004). The neuroscience of tinnitus. *Trends in Neuroscience*, 27, 676–682.

Eggermont, J. J., Zeng, F.-G., Popper, A. N., & Fay, R. R., Eds. (2012). *Tinnitus*. New York: Springer Science + Business Media.

Fukushima, N., White, P., & Harrison, R.V. (1990). Influence of acoustic deprivation on recovery of hair cells after acoustic trauma. *Hearing Research*, 50, 107–118.

Hébert, S., Canlon, B., Hasson, D., Magnusson Hasson, L.L., Westerlund, H., & Theorell T. (2012). Tinnitus severity is reduced with reduction of depressive mood—a prospective population study in Sweden. *PLoS One* 7(5), e37733.

Jastreboff, P. J., Brennan, J. F., Coleman, J. K., & Sasaki, C. T. (1988). Phantom auditory sensation in rats: An animal model for tinnitus. *Behavioral Neuroscience*, 102, 811–822.

Langers, D. M., de Kleine, E., & van Dijk, P. (2012). Tinnitus does not require macroscopic tonotopic map reorganization. *Frontiers in Systems Neuroscience*, 6, 2.

Noreña, A. J., & Eggermont, J. J. (2005). Enriched acoustic environment after noise trauma reduces hearing loss and prevents cortical map reorganization. *Journal of Neuroscience*, 25, 699–705.

Noreña, A. J., & Eggermont, J. J. (2006). Enriched acoustic environment after noise trauma abolishes neural signs of tinnitus. *NeuroReport*, 17, 559–563.

Noreña, A. J., Gourévitch, B., Aizawa, N., & Eggermont, J. J. (2006). Spectrally enhanced acoustic environment disrupts frequency representation in cat auditory cortex. *Nature Neuroscience*, 9, 932–939.

Ochi, K., & Eggermont, J. J. (1996). Effects of salicylate on neural activity in cat primary auditory cortex. *Hearing Research*, 95, 63–76.

Pienkowski, M., & Eggermont, J. J. (2009). Long-term, partially-reversible reorganization of frequency tuning in mature cat primary auditory cortex can be induced by passive exposure to moderate-level sounds. *Hearing Research*, 257, 24–40.

Pienkowski, M., & Eggermont, J. J. (2012). Reversible long-term changes in auditory processing in mature auditory cortex in the absence of hearing loss induced by passive, moderate-level sound exposure. *Ear & Hearing*, 33, 305–314.

Roberts, L. E., Moffat, G., Baumann, M., Ward, L. M., & Bosnyak, D. J. (2008). Residual inhibition functions overlap tinnitus spectra and the region of auditory threshold shift. *Journal of the Association for Research in Otolaryngology*, 9, 417–435.

Seki, S., & Eggermont, J. J. (2002). Changes in cat primary auditory cortex after minor-to-moderate pure-tone induced hearing loss. *Hearing Research*, 173, 172–186.

Tyler, R., Coelho, C., Tao, P., Ji, H., Noble, W., Gehringer, A., & Gogel, S. (2008). Identifying tinnitus subgroups with cluster analysis. *American Journal of Audiology*, 17, S176–184.

Ward, L. M., & Baumann, M. (2009). Measuring tinnitus loudness using constrained psychophysical scaling. *American Journal of Audiology*, 18, 119–128.

Chapter 7
The Sense of Hearing in Fishes

Richard R. Fay

Richard R. Fay

R.R. Fay (✉)
Marine Biological Laboratory, Woods Hole, MA 02543, USA
e-mail: rfay@luc.edu

A.N. Popper and R.R. Fay (eds.), *Perspectives on Auditory Research*, Springer Handbook of Auditory Research 50, DOI 10.1007/978-1-4614-9102-6_7,

7.1 The Early Years

When I was a graduate student in fish bioacoustics (1966–1970), the entire field could be summarized in three scholarly books: *Marine Bioacoustics*, Vols. 1 and 2, edited by William N. Tavolga (Tavolga, 1964, 1967) and *Lateral Line Detectors*, edited by Phyllis Cahn (Cahn, 1967); a chapter titled "The evolution of vertebrate hearing" by Willem van Bergeijk (van Bergeijk, 1967); and a handful of earlier papers by von Frisch and Stetter (1932), Dijkgraaf (1950), Pumphrey (1950), Poggendorf (1952), Kleerekoper and Chagnon (1954), Harris and van Bergeijk (1962), Tavolga and Wodinski (1963), Harris (1964), Tavolga (1967), and Jacobs and Tavolga (1967, 1968) (but also see Chapter 14 by Hawkins for a discussion of more European-oriented research). Needless to say, I devoured this literature with great excitement. There were a few behavioral audiograms and a few studies of frequency and intensity discrimination. That was it.

At that time, an important issue was the "near field" as presented by Harris and van Bergeijk (1962) in a paper that made everyone in the field stop and think. We now appreciate, and some of us did then as well, that all experiments carried out in tanks in the lab were probably flawed if the near field was not taken into account (measured). This was because all of them were conducted in the near field where particle motion dominated the sound field (see Chapter 14 by Hawkins) and may have been responsible for stimulating the lateral line system as well as the auditory system via two routes to the ears.

As I look back on the field and my role in it, I think van Bergeijk was the most important influence on my early thinking. Van Bergeijk erroneously believed that "hearing" was dependent solely on detecting sound pressure, so that the only fish species that could be said to hear by means of their ears were the "ostariophysines," which had Weberian ossicles (Weber, 1820) connecting the swim bladder and inner ear or something comparable in herrings, mormyrids, and possibly gouramis. The erroneous assumption was that sound pressure alone stimulated the ear and that all other "hearing" in fish was mediated by the lateral line system ("acoustico-lateralis hypothesis") detecting near-field disturbances. Of course, van Bergeijk was well aware that particle motion existed throughout the far field as well, but he believed that it was too small to be detected using the ears. He believed that any response of the otolith organs directly to particle motion, even within the near field, would be more like "vertigo" than hearing.

I was skeptical of this acoustico-lateralis hypothesis from the beginning but had no way of designing experiments that could test it. So I chose to study the goldfish (*Carassius auratus*, an "ostariophysine," now called an otophysan), with Weberian ossicles that could hear by van Bergeijk's definition, but I first focused on the response to the near field specifically by using a dipole sound source in the laboratory. If all laboratory studies on hearing in fish were subject to the near-field "contamination" of the sound field, then why not investigate the near-field contributions to hearing directly?

Although this approach didn't address the question of the lateral line system and hearing (partially answered by others much later), the choice of the goldfish had the ultimate advantage in that this species has very sensitive hearing but doesn't communicate using vocalizations; they are mute as far as anyone knows. So the question of what goldfish listen to immediately came up. My biologist colleagues studying hearing in fish focused some attention on sound production and hearing because the one signal we were sure fish listen to are conspecific vocalizations. My rather compulsive and detailed look at goldfish over the years was always motivated, in part, by the question of what they listen to. I didn't have the ready answer I would have had if I had instead studied a vocal fish. This was, in a way, liberating for me.

The first experiment I did in 1969 (Fay, 1969a) while I was a research assistant to Paul Smith at the U.S. Navy Submarine Base, Groton CT, used a homemade dipole sound source to study sound detection, masking, and sound source localization and in addition introduced the method of stimulus generalization to investigate the sense of hearing in goldfish. Oddly, this experiment helped define my entire research program over the years even though the experiment and the data obtained from it were almost completely uninterpretable, primarily because of the inadequate measurements of the sound field and my failure to disentangle the contributions of the ears and lateral line system.

At about this time, I became a Ph.D. student in E. G. Wever's Auditory Research Laboratory at Princeton University. Wever was famous at the time for two books, *Theories of Hearing* (Wever, 1949) and *Physiological Acoustics* (Wever & Lawrence, 1954). In the first book, he reviewed and presented auditory theory in a historical context and put forward the "volley theory" of frequency processing. In the second book, Wever and Lawrence summarized their and others' research on the cochlear potential (discovered by Wever & Bray, 1930) and its use in clinical and basic research on hearing in mammals and humans. Wever was a very kind gentleman, entirely focused on whatever interested him at the time and tireless in the lab. He was a "hands-off" mentor and advisor. He had become interested in hearing in reptiles (a decades-long project culminating in a thick book titled *The Reptile Ear* [Wever, 1978]) and spent each day measuring cochlear potentials and doing whole-head histological analysis on lizards and other "subjects," including sea turtles (Ridgway et al., 1979) and dolphins (e.g., Wever et al., 1971). One day in 1968, some reporters from *Newsweek* came to visit and were somewhat taken aback to learn that Wever had been studying reptiles almost exclusively for more than 10 years, having already established his reputation in otological research in the 1930s to 1950s. Wever's comment to them was, "You go where the scientific questions take you." He seemed perfectly satisfied to let me do whatever I wanted, with the resources of the entire lab at my disposal, including a "private" lab with a built-in anechoic chamber and a fully equipped and manned machine shop and carpentry shop. When the lab was first built, Wever had a 1-foot diameter, 10-meter "tube" built underground for future experiments with fishes. It's unfortunate that it was never used. This was one of the "paradise" labs that I have had the good fortune to be associated with. One of the resources in the very well equipped electronics shop was Jerry Palin, the electrical engineer, who was extremely helpful and taught me to make instruments out of op-amps and various digital chips. This was Wever's tradition—to make whatever instruments you needed.

7.2 Princeton and Hawaii

While at Wever's lab (part of the Psychology Department), we had many visitors assisting Wever in the reptile project. Two from Princeton's Biology Department were Mark Konishi and Geoff Manley. Sam Ridgeway and Carl Gans, also from Princeton, were frequent visitors. Yehudah Werner from the Hebrew University of Jerusalem in Israel came to the lab often to pursue lizard hearing with Wever. Of course, Jim Simmons (bats) was a graduate student with me, and Jim McCormick (dolphins) and Jim Saunders (cats) had just graduated from the Psychology Department's Ph.D. program. While a graduate student, I tinkered a bit with lizard cochlear potentials but got discouraged when I saw how labile and complex the cochlear potentials were because of the bidirectional hair cell orientations in geckos and other lizards (see comment by Manley in this volume on the same issue).

My primary project was to determine an audiogram for goldfish (Fay, 1969b), the starting point for any further study. I also determined frequency discrimination thresholds for the goldfish (Fay, 1970a), continued to develop the stimulus generalization method for studying amplitude-modulation perception in goldfish, and obtained evidence that goldfish perceived periodicity pitch much as humans do (Fay, 1970b, 1972).

Georg von Békésy visited Wever's lab in 1970 (Wever had translated and edited von Békésy's famous book *Experiments in Hearing* [1960]), and from that meeting, von Békésy offered me a postdoc in Hawaii, where he had founded the Laboratory of Sensory Sciences with help from Hawaiian Telephone. So, within two months of being offered the job, we left for Honolulu. Von Békésy told me to work on the statocyst organs of "crabs" (he meant any crustacean) instead of fish, inspired by the early work by Kreidl (1893). Von Békésy envisioned raising crayfish in sand with iron grains and then stimulating the statocysts using computer-controlled electromagnets. At that time, von Békésy was interested primarily in Mach bands in vision and lateral inhibition in general as an explanation for the mystery of how the sharp tuning could be explained given the rather broad cochlear excitation patterns. So I applied myself to the crayfish (Fay, 1973, 1975) but didn't get to the ultimate experiment before von Békésy died in 1972.

At this point, I drifted back to my first love—hearing in fishes. I investigated the masking effects in goldfish and observed a function that was a linear function of log frequency, very similar to that in humans, and just like the near-power functions of frequency I saw when measuring frequency discrimination thresholds. I marveled at these two humanlike functions and started to wonder what else made goldfish hearing similar to human hearing. Of course, the human behavior was explained pretty well in terms of cochlear patterns (Greenwood, 1961), but goldfish had no analog of the basilar membrane. Maybe the psychophysics reflected a very general requirement for signal processing and was not a "constraint" of basilar membrane mechanics. Or, more specifically, the basilar membrane and the otolith organs of hearing were both adaptations for optimal signal processing independent of vocal communication. This reinforced my interest in the question of what goldfish listen to. If the answer was something as specific as fish vocalizations, it was hard to explain these very humanlike (very general) functions. I began to seriously doubt the "matched filter" hypothesis touted by some neuroethologists.

M. E. ("Jeff") Bitterman from the Psychology Department was a major researcher at the time in von Békésy's lab. He was known as a great comparative psychologist (American style) and a behaviorist in the classic tradition. Bitterman had a great influence on me and was a good friend. I remember Jeff lecturing Art Popper and me about the misuse of the term "natural." He was quite skeptical of the ethologists, some of whom had been highly critical of Jeff and behaviorism in general.

Art Popper was doing experiments on fish hearing at the time, and was one of the first people I met in Honolulu. He was an assistant professor in the Zoology Department, but soon after my arrival, he joined von Békésy's lab and did his research there. Our first collaboration was investigating shark hearing at Eniwetok Atoll in the Marshall Islands with Al Tester, a famous shark expert at the University of Hawaii, and his colleague Jim Kendall, a retired professor from City College of New York. We originally fantasized recording single units from the saccular nerve or from the macula neglecta of the sharks we would catch in Eniwetok ("the most shark-infested waters on the face of the earth" according to a frog-man film from the 1950s that we saw at Eniwetok with some WWII frogmen in the audience). We caught a blacktip reef shark (*Carcharinus melanopterus*, about 5 feet long) from a beach strewn with 50-caliber shells from the Battle of Eniwetok in 1944. ("Don't put them in your pocket!" was the advice from the Atomic Energy Commission because atomic bombs were tested there from 1948 to 1958 and the shells were radioactive.) Once reality forced itself upon us (we had no previous experience recording single units of the auditory nerve), we switched to recording potentials from the ear ("microphonics") in response to a vertical vibrator on the head and got results consistent with stimulation of the macula neglecta via the parietal fossa on the top of the head (Fay et al., 1974). While at Eniwetok for three weeks, Art and I hatched our first joint writing project, a review of fish hearing to be published in the *Journal of the Acoustical Society of America* (Popper and Fay, 1973).

Art and I also collaborated on several saccular potential studies while in Hawaii. One was a study of the effects of swim bladder deflation on the auditory response in goldfish (Fay & Popper, 1974). We used an air-filled standing wave tube with the fish in a plastic bag full of water to demonstrate that the swim bladder gave sound pressure sensitivity to the goldfish. After deflation, the maximal response shifted in the standing wave with a displacement maximum at the fish, indicating that goldfish do not become deaf without a swim bladder but respond to particle motion only once the swim bladder is deflated.

7.3 To Loyola University Chicago

After von Békésy died in 1972, the lab tried to continue his NIH grant without him but failed. I took a job in 1974 at the Bowman Gray School of Medicine in Winston-Salem, North Carolina, with Jim McCormick (formerly at Wever's lab) in the

Otolaryngology Department. I spent my time writing grant proposals but decided that a medical school environment was not for me. After about 11 months, I was offered an associate professor position in the Psychology Department at Loyola University Chicago. So we went to Chicago in 1975 and I remained at Loyola until retirement in 2011. I was very fortunate to be able to join the Parmly Hearing Institute once Bill Yost became director in 1976. Here, I spent about 36 years in my lab, essentially doing research pretty much alone but supported intellectually by Bill and the others at Parmly. Parmly, including Bill Yost and several postdocs and graduate students, made up another paradise laboratory in which I was left free to do whatever I wanted. Yost was not a boss in the traditional sense but a highly supportive colleague. Because the Institute was privately endowed and did not report to the Arts and Sciences dean, we were essentially ignored by the university and nobody interfered with our research program. All we had to do was cover several undergraduate and graduate courses. Coupled with the fact that we occupied a rather unpretentious but large space in the basement of our building, this contributed greatly to our contentedness and made for a stable and happy lab where we were perfectly free to pursue our particular interests as long as we had grant funding. We were fortunate to have had continuous grant funding from 1975 through 2011.

Sheryl Coombs came to my lab in 1981 as a postdoc from Art Popper's lab at Georgetown University to work on an NSF grant on directional hearing in fishes. She remained an important colleague and collaborator and later a Biology Department faculty member and a world expert and authority on the lateral line system. She moved to Bowling Green State University in Ohio in 2005. Sheryl and I did many experiments together while she was at Loyola, the most recent being a study showing that the goldfish ear (saccule) responds to a nearby dipole sound source at 50 Hz and sound pressure (Coombs et al., 2010).

7.3.1 Research Program

7.3.1.1 Time and Frequency Domain Processing

Research began with psychophysical experiments on goldfish and soon thereafter continued with single-unit studies on the goldfish auditory nerve. Principally, I determined to investigate the sense of hearing in goldfish using psychophysical methods to define and answer the interesting questions that arose and electrophysiological methods to help determine the responsible neural mechanisms for the behavior. The physiological experiments, in turn, often demanded further psychophysical studies. In the back of my mind, always, was the question of what goldfish could be listening to.

I first investigated the perception of amplitude-modulated signals and obtained temporal modulation transfer functions (TMTFs) for noise and tones (Fay, 1980). The noise TMTF is different from the human TMTF primarily, I believe, because the bandwidth of goldfish hearing is narrow (up to 1 or 2 kHz), so the function is generally too low to show the expected low-pass characteristic.

One important thread in the research program was on various aspects of intensity discrimination, particularly the effects of signal duration, studied using psychophysics and electrophysiology (single units of the saccular nerve), in both cases analyzed according to the methods of the theory of signal detection (TSD; Fay, 1985; Fay and Coombs, 1992). I think these studies were important in applying TSD to the psychophysical results using classical respiratory conditioning and to the single-unit neurophysiological results.

In general, the results showed that the psychophysics of intensity processing in the goldfish is remarkably similar to that in humans despite differences between the structures and functions of the cochlea and the otolithic saccule of goldfish. We observed that Weber's law approximately holds for both humans and goldfish for pulsed noise and tones but that Weber's law misses widely for continuous tones in fish. Both species show the same degree of temporal summation in intensity discrimination, with identical slopes as a function of signal duration (e.g., Florentine, 1986). These experiments on duration effects in intensity discrimination followed a paper Sheryl Coombs and I wrote on temporal summation in sound detection (Fay & Coombs, 1983) in which, among other things, we got masked thresholds for tones of various frequency for single units of the saccular nerve and in psychophysics. We found that the thresholds for masked tones fell precisely at the S/N of the critical masking ratio and precisely at the point in the rate-level functions for saccular fibers where the addition of the tone to the noise first caused spike rate increments. All saccular fibers phase-lock to tones, and the vector strength at which spike rate increments first occur is 0.5. In other words, there is significant phase-locking well below the behavioral S/N threshold that apparently is not available for use in signal detection.

It had become generally accepted that phase-locking information is all the fish had for auditory processing, so this finding was surprising to me. I began to take seriously an alternative hypothesis that the peripheral frequency tuning of saccular fibers (i.e., their spike-rate response) was more important in audition than previously thought. This motivated a series of experimental studies focusing on peripheral frequency selectivity (e.g., Fay & Ream, 1986; Fay, 1997). We found that the goldfish has at least two or three differently tuned frequency-selective channels in the saccular nerve. In the end, I decided that the issue of "rate" versus time coding in the goldfish is as complex and controversial as it is for other vertebrates. Earlier, we got evidence that frequency-discrimination thresholds could be accounted for by the temporal error (jitter) with which saccular fibers phase-locked to tones (Fay, 1978). So the question persists today, as it does for humans and other mammals, regarding the relative importance of processing in the frequency ("rate") or time ("phase-locking") domains. The goldfish is like other vertebrates in this respect.

7.3.1.2 Spike Rate Suppression

We observed two types of suppression among saccular nerve fibers: single-tone suppression (suppression of spontaneous activity) and two-tone rate suppression. Single-tone suppression remains controversial because it is said not to occur in mammals (cf. Henry & Lewis, 1992) and because it is impossible to be certain that the spike activity that is nominally "spontaneous" is not caused by noise or otherwise evoked. But this type of background activity can be suppressed by tones in a subset of low-frequency tuned, spontaneously active saccular fibers (Fay, 1990). In goldfish, two-tone rate suppression (TTRS) is the suppression of tone-evoked activity in low-frequency tuned fibers caused by a second tone of different frequency (Fay, 1990, 1991). In mammals, suppression arises from basilar membrane mechanics, presumably determined by the active cochlear amplifier, but in goldfish, the explanation is not yet clear. In goldfish, the suppressive effects seem to arise either at or central to the hair cell synapse, possibly at the spike-initiating zone of saccular fibers (Hill et al., 1989). In any case, although the responsible mechanisms are apparently completely different, the phenomena of peripheral TTRS in fish and mammals are very similar and may have a similar function in determining frequency selectivity in the peripheral nerve channels.

In the goldfish midbrain, suppression is a major phenomenon that appears to shape the frequency selectivity of auditory nerve cells. We investigated this with my graduate student Zhongmin Lu (Lu & Fay, 1996). In response to a two-tone stimulus, the frequency-selective response areas of the majority of midbrain cells are determined by two-tone interaction (TTI) that is probably a combination of peripheral suppression and central inhibition. The observed TTI sharpens the frequency-selective responses and sometimes creates "islands" of inhibition between sharp excitatory peaks in the brain. This inhibition may also be responsible for peristimulus time histogram (PSTH) shapes that have been described as "onset," "chopper," "build-up," and "pauser" units seen in the mammalian cochlear nucleus (Lu & Fay, 1993). It is likely that peripheral TTRS has served as a sort of model for the evolution of additional inhibitory interactions in the midbrain and thalamus (Lu and Fay, 1995). The TTRS, central inhibition, and PSTH shapes observed in the goldfish brain are very much like comparable response patterns in mammals and other vertebrates, although, as noted, some of the responsible mechanisms may be quite different. It appears that the goldfish auditory system is adapted for very general auditory signal-processing functions, as are the auditory systems of most vertebrates investigated (see Chapter 24 by Pollak). In general, although the peripheral physiology of goldfish (and other species as well) seems somehow "special" in fishes, central neurophysiology (and behavior) seems more generally "vertebrate" or general. This underscores the question of not only what goldfish are listening to but also the broader question of what all vertebrates are listening to. Perhaps the answer as it relates to goldfish is the same as that for all vertebrates.

7.3.1.3 Ripple Noise Processing

Ripple noise (RN) is a signal made up of the sum of noise and a delayed replica of itself that has been used to investigate pitch processing (e.g., Yost et al., 1978). The pitch of RN is equal to $1/T$ Hz, where T is the time delay used to create the stimulus. Bill Yost and I were naturally interested in asking how goldfish respond to

RN because the dominant theory was that, for human listeners, the perception of pitch was computed in the time domain (with a process perhaps like autocorrelation) whereas fish are thought to be "naturals." The shallow underwater environment that goldfish inhabit is reverberant with sound reflections from the bottom and surface, so perhaps goldfish made use of RN processing to measure and generally assess their environment.

I studied RN processing in goldfish, with experiments on masking by RN and discrimination of RN delay differences (Fay et al., 1983). I also looked carefully at the distributions of interspike intervals (ISIs) in saccular units stimulated with RN. I found that goldfish can discriminate between RN with about a 6% change in delay between 1/T values of 100 and 800 Hz. The thresholds for human listeners are only slightly below the goldfish thresholds and the best ones are in the same range of $1/T$ values. The ISI distributions for these same stimuli had robust peaks at T seconds in most saccular fibers. I noted that the frequency-discrimination high thresholds for pure tones are nearly identical to the $1/T$ discrimination thresholds, yet the spike rate differences occurring for these two different types of stimuli are much greater for pure tones than for RN. Therefore, I hypothesized that RN processing in goldfish was based on the processing of peaks in the ISI distributions in the time domain and not on spike rate differences.

The RN experiment described earlier did not allow me to determine whether or not RN produces a pitch percept in fish as it does in human listeners. Much later, I used a stimulus-generalization method to help answer questions about the pitch of RN in goldfish (Fay, 2005). Although several attempts were made to demonstrate it, goldfish did not behave in generalization experiments as if RN evokes a pitch percept as is does for human listeners. Fish were first conditioned to a harmonic series with a fundamental frequency of 100 Hz and were then tested for their generalization to RN and iterated RN of various delays (T values). No generalization was observed at any T value, indicating that the two stimuli produced perceptions that had little or nothing in common. On the one hand, this makes some sense because noise and a mixture of tones are very different stimuli (e.g., the tone complex is a deterministic signal and the noise is a stochastic one), so the goldfish would not be expected to perceive them as similar. On the other hand, I hypothesized that when the two stimuli were constructed to have a common pitch, generalization would occur. But it didn't. Perhaps the generalization method is too heavily influenced by the overall differences in the perceptions of the stimuli to indicate whether or not they had a pitch in common. But perhaps the goldfish does not have a perception of pitch at all. This last alternative is consistent with some of the other results (see Section 3.1.4) from the pitch perception experiments (Fay, 2005).

7.3.1.4 Stimulus Generalization

I used a stimulus-generalization method in the very first experiment I did in 1969. I used it again in 1970 and 1972 but after that didn't use it until 1992. About this time, I became even more interested in what the method could reveal about auditory perception in goldfish and the limits of the discrimination methods in psychophysics. After 1992, I did generalization studies almost exclusively.

The stimulus-generalization paradigm is an old one in psychology (e.g., Pavlov, 1927) and has been investigated in various contexts in every decade since that time. Stimulus generalization has been defined as "the behavioral fact that a conditioned response formed to one stimulus may also be elicited by other stimuli which have not been used in the course of conditioning" (Hilgard & Marquis, 1940). Applied to hearing in the goldfish, the basic experiment is to first establish a classically conditioned response (respiratory suppression) to one sound (about 40 conditioning trials in one session). Later that day or on the next day, various sound stimuli are presented, without reinforcement, and the strength of the response produced is measured. Typically, eight test stimuli are presented, four times each, in random order. This experiment was first done by conditioning the fish to a 40-Hz tone and then testing them with eight novel frequencies above and below 40 Hz (Fay, 1969a). Remarkably, the percent generalization (with the average response to 40 Hz defined as 100%) reached 0% (no change in respiration) at 20 Hz and 70 Hz. In other words, animals conditioned to 40 Hz did not respond at all to about one octave above and below 40 Hz. I interpreted this in terms of the perceived similarity between these tones of different frequency. At one octave above and below the conditioning frequency, the test tones produced perceptions that had nothing in common with those evoked by the conditioning frequency. Note that there is no explicit discrimination training involved; the goldfish behaves in this discriminating way without training, or "naturally." In other words, generalization methods can reveal what simple and complex sounds "sound like" to goldfish.

Stimulus generalization was used to study pure-tone frequency differentiation (Fay, 1970b) and the perception of amplitude-modulated tones (Fay, 1972). The generalization functions around tones from 40 Hz to 1600 Hz can all be described as sharply tuned, with responses from 100% at the conditioning frequency to about 0% one octave above and below the conditioning frequency. I interpreted this as an indication that goldfish "know" all about the sound spectrum (or frequency), as if they possessed pitch perception, and that frequency is a most salient aspect of tone perception. When goldfish are conditioned to sinusoidally amplitude-modulated tones, they generalized very little to unmodulated tones and vice versa, except when the modulated test tone was presented at the same rate as the pure-tone conditioning stimulus, suggesting the existence of periodicity pitch in goldfish.

In 1992, the generalization method was used again to study what was then called "analytic listening." Fish were conditioned to a mixture of two tones (166 Hz and 724 Hz) and tested for generalization to single tones, including the frequencies used in conditioning. They responded robustly only when the generalization test frequency was at 100 Hz and 724 Hz, demonstrating that goldfish listened analytically; they were aware of the two mixed frequencies and didn't hear the mixture as a unique entity (i.e., an unresolved chord).

I investigated the segregation of two simultaneous tones as a function of the frequency separation between the tones (Fay, 2009). In general, a comparison of saccular unit tuning curves, tone-generalization frequency selectivity, and two-tone segregation reinforced the view that for goldfish and possibly other fishes, it appears to be central computations based on a rather crude peripheral frequency selectivity that account for the fundamental aspects of source segregation. For humans and some other vertebrates, the analysis of the spectrum at the periphery seems to be adequate to account for many aspects of source segregation. The absence of a fine peripheral frequency selectivity and tonotopicity in fishes has not been an important limitation on the sense of hearing or its performance in source segregation.

Generalization was used again in a study of the perception of ramped and damped tones in goldfish, motivated by a finding by Roy Patterson (Fay et al., 1996). For humans, repeated tone bursts with a gradual rise and rapid fall (ramps) are perceived quite differently from bursts reversed in time (damps; e.g., Patterson, 1994). Ramps, but not damps, are judged to have a tonal component in perception by human listeners despite having the identical long-term spectrum as damps. I was interested in determining whether goldfish perceived ramps and damps in the same way. Goldfish conditioned to pure tones generalized significantly more to ramps than to damps, indicating that the goldfish behaved as if ramps were perceived as more tonelike than damps. The behavioral and neurophysiological data from goldfish with respect to ramps and damps support the idea that ramp/damp discrimination is based in the processing time interval information. The auditory image model of Patterson (1994) also identifies time interval processing as likely underlying ramp/damp perception. In any case, the perceptual phenomena and their underlying processes in goldfish and human listeners seem to have much in common.

7.3.2 Auditory Scene Analysis and What Fish Listen to

Auditory scene analysis (ASA) is an important yet conceptually simple hypothesis that is used in the analysis of sound perception in humans and other vertebrates (Bregman, 1990). The first chapter of Bregman's (1990) book is a brilliant, general essay on human sound perception that has been very influential for students of animal hearing as well. Essentially, the rationale for ASA for humans applies equally well for all vertebrate animals (and possibly invertebrates as well). In the end, ASA refers to the requirement that individual sources of sound be segregated in perception according to the likely sources as physical objects or events and arrayed as a "scene" of sources without necessarily recognizing the identity of the sources or source location. A good analogy is the determination of the various musical instruments playing in an orchestra (without being able to identify the specific instrument) while listening to a monophonic recording. Reading this essay, which deals exclusively with human perception, I became convinced that ASA must be a capacity of all sensory systems for all organisms and for all time. In other words, it is a fundamental capacity for any hearing at all and is the reason that auditory brains have "all those neurons." In fact, I believe that vertebrate brains evolved the way they did because the problem of ASA had to be solved first, and the solution was complex and required interactions among millions of cells. We all take ASA for granted (and therefore some don't see that it is a necessary capacity for hearing), but it is the most difficult and complex fundamental task that auditory systems must accomplish (Popper and Fay, 1993). Automatic speech recognition is easy compared to ASA, as artificial intelligence researchers know well.

So I assumed that fishes must have the capacity for ASA. The following simplistic insight from Stephen Hawking (1988), combined with the recognition that ASA applied to all organisms, was the answer to my original question of what fish listen to: everything that they could hear, not just vocalizations. Hawking stated, "… in any population of self-reproducing organisms, there will be … differences [that] will mean that some individuals are better able than others to *draw the right conclusions about the world around them and to act accordingly*. These individuals will be more likely to survive and reproduce and so their pattern of behavior and thought will come to dominate" (Hawking, 1988, p. 12). In other words, goldfish hear to be aware of the various sources so they could know as much as possible about the world, allowing them to behave appropriately. This is what their brains are for. Without source segregation and ASA, hearing would be of little value. So I believe that goldfish listen to the total soundscape to detect predators and prey, orient to environmental features, and tell them where they are so that they can "draw the right conclusions about the world." This is a very simple, but somewhat vague, conclusion about hearing.

I demonstrated ASA in two studies (Fay, 1998, 2000) using stimulus generalization as the method. Goldfish were conditioned to a mixture of two simultaneous pulse trains differing in both repetition rate (19 and 85 pulses per second [pps]) having different spectral profiles (238-Hz and 625-Hz center frequencies, respectively). Then, two groups of fish were tested for generalization to single-pulse trains (either the low-frequency or high-frequency pulse) at a variety of different pulse rates from 19 through 85 pps. The two groups showed very different generalization functions despite receiving identical stimuli during conditioning. When tested with the low-frequency pulse, the function was clearly tilted toward the 10-pps pulse rate (where more generalization and thus greater stimulus similarity occurs), whereas in the group tested with the high-frequency pulse, the generalization function was tilted oppositely toward the 85-pps pulse rate. These results demonstrated that the frequency region of the pulse and the repetition rate were associated; goldfish were aware of both the pulse rate and pulse frequency region, independently, during conditioning. This is evidence that the two pulse trains were segregated in perception, a principal requirement for auditory scene analysis. This ability of goldfish is analogous to our ability to hear individual instruments playing in an orchestra. Note that, in this experiment, the locations of the sources were indeterminate (the loudspeaker was located below the animal). Although sound source localization is not required for ASA, it is likely that spatially separated sources would have added to the effectiveness of ASA.

7.3.3 Conclusions

My focus on the goldfish has revealed an auditory system and sense of hearing that is very mammal-like. Ed Burns used to kid me at Acoustical Society meetings before there was an Animal Bioacoustics Technical Committee. Whenever the psychoacoustics people made a new observation about human hearing, Ed would say, "Of course, goldfish can do this—right?" I used to claim that goldfish hearing was very much like human hearing. But I have since reconsidered this statement and decided that human hearing is very goldfishlike. After all, we have inherited an auditory system that originated with fish (Popper and Fay, 1993). At the same time, Günter Ehret (a neuroethologist) once "thanked me" for "turning a fish into a mammal." He obviously disapproved. I think both of these statements reflect my attitude and design of my experiments. I was one of the very few people trained as a psychologist in a field made up primarily of biologists and ethologists. For 100 years, psychologists have been known for studying rat behavior. But they are not interested in rat behavior per se; they are interested in animal models of human behavior and are looking at what rats and humans have in common. I have also been told (Arthur Popper, personal communication) that the goldfish is the "wet rat" of fish hearing research. So it is natural that I would be looking at what goldfish and humans have in common and not at what adaptations made them different (see the essay by Bullock, 1992 on this topic).

My answer to the question of what fish listen to is therefore a psychologist's answer. In any case, I now think I know what the answer likely is. They are listening to the soundscape (Fay, 2009), like most animals on earth. This is a very vague answer, particularly considering that we are profoundly ignorant about natural soundscapes and the possible sounds and sources that probably have biological significance to fish and all vertebrate animals. But I have come to this conclusion on the basis of three conceptions: that the fish brain is similar in morphotype to all vertebrate brains (e.g., McCormick and Hernandez, 1996); that the goldfish, at least, has the processing power to analyze complex soundscapes (e.g., this chapter); and that the processing of fish vocalizations couldn't be a determining factor in the evolution of their hearing (e.g., Ladich, 1999).

The auditory brain of goldfish, and all fishes, is organized in the same way as in other vertebrates. They all have various branches of the auditory/vestibular nerve that project to multiple, first-order nuclei in the medulla (primarily the descending octaval nucleus [DON]). Another medullar nucleus, the secondary octaval nucleus (SON), has been identified in many fishes and may be analogous to the inferior olivary nucleus. Both of these medullar nuclei project to the midbrain (torus semicircularis), which, in turn, projects to several nuclei in the thalamus (primarily the central posterior nucleus [CPN]). The thalamic nuclei project to multiple forebrain nuclei that are less well studied. Whether or not each of these analogous nuclei is homologous to mammalian or avian nuclei is at least controversial and, in the end, impossible to determine now. But the physiology of brain stem nuclei in fish tends to show cellular response properties very similar to those seen in mammals, birds, and reptiles despite the fact that there have been very few studies in fishes (e.g., Lu & Fay, 1993, 1995, 1996). What, other than brain size, could be the rationale for believing that the sense of hearing is limited, reduced, or impoverished in fishes? A possible answer could be that fishes don't have a cochlea or a basilar papilla. This chapter has attempted to demonstrate that a cochlea is not necessary for complex hearing functions, at least at low frequencies.

Goldfish, at least, possess the sense of hearing required to process the various sources likely making up a fish's soundscape (this chapter). In particular, they seem to be able to process the acoustic spectrum (e.g., Fay, 1992); they can process reverberations and reflections and various temporal patterns (e.g., Fay, 1994); and some species, at least, can locate the sources of sound (Zeddies et al., 2011) using binaural hearing. The general biological significance of the soundscape has not been studied or defined, however. There are at least two reasons for this. The first is that the underwater soundscape potentially contains sounds consisting of particle motion stimuli that are not measured conventionally in acoustics but that stimulate the ears of all fishes (e.g., Popper and Fay, 2011). Second, particle motion sensitivity extends to extremely low (infrasonic) frequencies (e.g., Sand and Karlsen, 1986) that are probably important in the fish's soundscape (e.g., Sand et al., 2008). Until the technology improves, we will not be able to understand fully the underwater soundscape's role in the acoustic ecology of fishes.

Some of the work described in this chapter shows that goldfish are probably very good at processing the soundscape, as are, presumably, humans and most other animals. This is an alternative to the view that animal vocalizations have determined or contributed to the adaptations of the characteristics of auditory systems throughout the evolution of vertebrates. However, it remains nearly true that vocalization sounds are the only sounds that have known biological significance. This arises from our ignorance of the other possible sounds and sources that probably have biological significance to fish and all vertebrate animals: the soundscapes that bathe all organisms as "acoustic daylight." The notion of environmental soundscapes as most probably important sources of information to the organism is suggested here. Environmental information exists to be exploited for appropriate behavior with respect to audible sound sources and events, and fish have the capacity to exploit it for general orientation. It seems logical to assume that this is what fish and other species listen to.

References

Bregman, A. (1990). *Auditory scene analysis: The perceptual organization of sound.* Cambridge, MA: MIT Press.

Bullock, T. (1992). Comparisons of major and minor taxa reveal two kinds of differences: "lateral" adaptations and "vertical" changes in grade. In D. Webster, R. R. Fay, & A. N. Popper (Eds.), *The evolutionary biology of hearing* (pp. 15–19). New York: Springer Verlag.

Cahn, P. H., Ed. (1967). *Lateral line detectors.* Bloomington: Indiana University Press.

Coombs, S. L., Fay, R. R., & Elepfandt, A. (2010). Dipole source detection and tracking by the goldfish auditory system. *Journal of Experimental Biology*, 213, 3536–3547.

Dijkgraaf, S. (1950). Untersuchungen uber die Functionen das Ohrlabyrinths bei Meeresfischen. *Physiologia Comparata et Oecologia*, 2, 81–106.

Fay, R. R. (1969a). Auditory sensitivity of the goldfish within the near acoustic field. *U.S. Naval Submarine Medical Center, Submarine Base, Groton, Connecticut, Report No. 605*, 1–11.

Fay, R. R. (1969b). Behavioral audiogram for the goldfish. *Journal of Auditory Research*, 9, 112–121.

Fay, R. R. (1970a). Auditory frequency discrimination in the goldfish (*Carassius auratus*). *Journal of Comparative and Physiological Psychology*, 73, 175–180.
Fay, R. R. (1970b). Auditory frequency generalization in the goldfish (*Carassius auratus*). *Journal of the Experimental Analysis of Behavior*, 14, 353–360.
Fay, R. R. (1972). Perception of amplitude-modulated auditory signals by the goldfish. *Journal of the Acoustical Society of America*, 52, 660–666.
Fay, R. R. (1973). Multisensory interaction in control of eye-stalk rotation response in the crayfish (*Procambarus clarkii*). *Journal of Comparative and Physiological Psychology*, 84, 527–533.
Fay, R. R. (1975). Dynamic properties of the compensatory eye-stalk rotation response of the crayfish (*Procambarus clarkii*). *Comparative Biochemistry and Physiology*, 51, 101–103.
Fay, R. R. (1978). Phase-locking in goldfish saccular nerve fibres accounts for frequency discrimination capacities. *Nature*, 275, 320–322.
Fay, R. R. (1980). Psychophysics and neurophysiology of temporal factors in hearing by the goldfish: Amplitude modulation detection. *Journal of Neurophysiology*, 44, 312–332.
Fay, R. R. (1985). Sound intensity processing by the goldfish. *Journal of the Acoustical Society of America*, 78, 1296–1309.
Fay, R. R. (1990). Suppression and excitation in auditory nerve fibers of the goldfish, *Carassius auratus. Hearing Research*, 48, 93–110.
Fay, R. R. (1991). Masking and suppression in auditory nerve fibers of the goldfish, (*Carassius auratus*). *Hearing Research*, 55, 177–187.
Fay, R. R. (1992). Analytic listening by the goldfish. *Hearing Research*, 59, 101–107.
Fay, R. R. (1994). Perception of temporal acoustic patterns by the goldfish (*Carassius auratus*). *Hearing Research*, 76, 158–172.
Fay, R. R. (1997). Frequency selectivity of saccular afferents of the goldfish revealed by revcor analysis. In E. R. Lewis, G. R. Long, R. F. Lyon, P. M. Narins, C. R. Steele, & E. Hecht-Poinar (Eds.) *Diversity in auditory mechanics* (pp. 69–75).Singapore: World Scientific Publishers.
Fay, R. R. (1998). Auditory stream segregation in goldfish (*Carassius auratus*). *Hearing Research*, 120, 69–76.
Fay, R. R. (2000). Frequency contrasts underlying auditory stream segregation in goldfish. *Journal of the Association for Research in Otolaryngology*, 1, 120–128.
Fay, R. R. (2005). Pitch perception in goldfish. *Hearing Research*, 205, 7–20.
Fay, R. R. (2009). Soundscape and the sense of hearing in fish. *Integrative Zoology*, 4, 26–32.
Fay, R. R., & Popper, A. N. (1974). Acoustic stimulation of the ear of the goldfish (*Carassius auratus*). *Journal of Experimental Biology*, 61–260.
Fay, R. R., & Coombs, S. (1983). Neural mechanisms in sound detection and temporal summation. *Hearing Research*, 10, 69–92.
Fay, R. R., & Ream, T. J. (1986). Acoustic response and tuning in saccular nerve fibers of the goldfish (*Carassius auratus*). *Journal of the Acoustical Society of America*, 79, 1883–1895.
Fay, R. R., & Coombs, S. L. (1992). Psychometric functions for level discrimination and the effects of signal duration in the goldfish (*Carassius auratus*): Psychophysics and neurophysiology. *Journal of the Acoustical Society of America*, 92, 189–201.
Fay, R. R., Kendall, J. I., Popper, A. N., & Tester, A. L. (1974). Vibration detection by the macula neglecta of sharks. *Comparative Biochemistry and Physiology*, 47, 1235–1240.
Fay, R. R., Yost, W. A., & Coombs, S. (1983). Psychophysics and neurophysiology of repetition noise processing in a vertebrate auditory system. *Hearing Research*, 12, 31–55.
Fay, R. R., Chronopoulos, M., & Patterson, R. D. (1996). The sound of a sinusoid: Perception and neural representations in the goldfish (*Carassius auratus*). *Auditory Neuroscience*, 2, 377–392.
Florentine, M. (1986). Level discrimination of tones as a function of duration, *Journal of the Acoustical Society of America*, 79, 792–798.
Frisch, K. von, & Stetter, H. (1932). Untersuchungen uber den Sitz des Gehörsinnes bei der Elritze. *Zeitschrift fur vergleichende Physiologie*, 17, 686–801.

Greenwood, D. (1961). Critical bandwidth and the frequency coordinates of the basilar membrane. *Journal of the Acoustical Society of America*, 33, 1344–1356.
Harris, G. G. (1964). Considerations on the physics of sound production by fishes. In W. N. Tavolga (Ed.), *Marine bio-acoustics* (pp. 233–248). Oxford: Pergamon Press.
Harris, G. G., & van Bergeijk, W. A. (1962). Evidence that the lateral line organ responds to near field displacements of a sound source in water. *Journal of the Acoustical Society of America*, 34, 1831–1841.
Hawking, S. (1988). *A brief history of time*. Toronto: Bantam Books, p. 12.
Henry, K., & Lewis, E. (1992). One-tone suppression in the cochlear nerve of the gerbil. *Hearing Research*, 63, 1–6.
Hilgard, E. R., & Marquis, D. G. (1940). *Conditioning and learning*. New York: Appleton-Century.
Hill, K., Stange, G., Gummer, A., & Mo, J. (1989). A model proposing synaptic and extra-synaptic influences on the response of cochlear nerve fibers. *Hearing Research*, 39, 75–90.
Jacobs, D. W., & Tavolga, W. N. (1967). Acoustic intensity limens in the goldfish. *Animal Behaviour*, 15, 324–335.
Jacobs, D. W., & Tavolga, W. N. (1968). Acoustic frequency discrimination in the goldfish. *Animal Behaviour*, 16, 67–71.
Kleerekoper, H., & Chagnon, E. (1954). Hearing in fish, with special reference to *Semotilus atromaculatus atromaculatus* (Mitchell). *Journal of the Fisheries Research Board of Canada*, 11,130–152.
Kreidl, A. (1893). Weitere Beitrage zur Physiologie des Ohrlabyrinthes: Versuche an Krebsen. *Sitzungsberichteder Kaiserlichen Akadamie der Wissenschajt, Wien, Mathematische-Naturwissenschaftliche Klasse*, 102, 149–174.
Ladich, F. (1999). Did auditory sensitivity and vocalization evolve independently in otophysan fishes? *Brain Behavior and Evolution*, 53, 288–304.
Lu, Z., & Fay, R. R. (1993). Acoustic response properties of single units in the torus semicircularis of the goldfish, *Carassius auratus*. *Journal of Comparative Physiology*, 173, 33–48.
Lu, Z., & Fay, R. R. (1995). Acoustic response properties of single units in the central posterior nucleus of the thalamus of the goldfish (*Carassius auratus*). *Journal of Comparative Physiology*, 176, 747–760.
Lu, Z., & Fay, R. R. (1996). Two-tone interaction in auditory nerve fibers and midbrain neurons of the goldfish, *Carassius auratus*. *Auditory Neuroscience*, 2, 257–273.
McCormick, C. A., & Hernandez, D. V. (1996). Connections of octaval and lateral line nuclei of the medulla in the goldfish, including the cytoarchitecture of the secondary octaval population in goldfish and catfish (Part 1 of 2). *Brain Behavior and Evolution*, 47,113–125.
Patterson, R. (1994). The sound of a sinusoid: Time-interval models. *Journal of the Acoustical Society of America*, 96, 1419–1428.
Pavlov, I. P., (1927). *Conditioned reflexes*. Translated by G. V. Anrep. London: Oxford University Press.
Poggendorf, D. (1952). Die absoluten Hörschwellen des Zergwelses und Beiträge zur Physik des Weverschen Apparates der Ostariophysen. *Zeitschrift für vergleichende Physiologie*, 34, 222–257.
Popper, A. N., & Fay, R. R. (1973). Sound detection and processing by teleost fishes: A critical review. *Journal of the Acoustical Society of America*, 53, 1515–1529.
Popper, A. N., & Fay, R. R. (1993). Sound detection and processing by fish: Critical review and major research questions. *Brain Behaviour and Evolution*, 41, 14–38.
Popper, A. N., & Fay, R. R. (2011). Rethinking sound detection by fishes. *Hearing Research*, 273, 25–36.
Pumphrey, R. J. (1950). Hearing. In *Physiological mechanisms in animal behaviour. Symposia of the Society for Experimental Biology*, 4, 3–18.
Ridgway, S., Wever, E. G., Mcormick, J. G., Palin, J., & Anderson, J. (1979). Hearing in the giant sea turtle, *Chelonia mydas*. *Proceedings of the National Academy of Sciences of the USA*, 64, 884–890.

Sand, O., & Karlsen, H. E. (1986). Detection of infrasound by the Atlantic cod. *Journal of Experimental Biology*, 125, 197–204.

Sand, O., Karlsen, H., & Knudsen, F. (2008). Comment on "Silent research vessels are not quiet." *Journal of the Acoustical Society of America*, 121, EL145–EL150. *Journal of the Acoustical Society of America*, 123, 1831–1833.

Tavolga, W. N., Ed. (1964). *Marine bio-acoustics,* Vol I. Oxford: Pergamon Press.

Tavolga, W. N., Ed. (1967). *Marine bio-acoustics*, Vol II. Oxford: Pergamon Press.

Tavolga, W. N., & Wodinski, J. (1963). Auditory capacities in fishes: Pure tone thresholds in nine species of marine teleosts. *Bulletin of the American Museum of Natural History*, 126, 177–240.

van Bergeijk, W. A. (1967). The evolution of vertebrate hearing. In W. D. Neff (Ed.), *Contributions to sensory physiology* (pp. 1–49). New York: Academic Press.

von Békésy, G. (1960). *Experiments in hearing*. New York: McGraw-Hill.

Weber, E. H. (1820). *De Aure et Auditu Hominis et Animalium.* Pars I. *De Aure Animalium Aquatilium* (p. 134). Leipzig: Gerhard Fleischer.

Wever, E. G. (1949). *Theory of hearing*. New York: John Wiley & Sons.

Wever, E. G. (1978). *The reptile ear*. Princeton, NJ: Princeton University Press.

Wever, E.G., & Bray, C.W. (1930). *Auditory Nerve Impulses*. Science 71(1834), 215

Wever, E. G., & Lawrence, M. (1954). *Physiological acoustics*. Princeton, NJ: Princeton University Press.

Wever, E. G., McCormick, J. G., Palin, J., & Ridgway, S. H. (1971). Cochlea of the dolphin, *Tursiops truncatus*: The basilar membrane. *Proceedings of the National Academy of Sciences of the USA*, 68, 2708–2711.

Yost, W., Hill, R., & Perez-Falcon, T. (1978). Pitch and pitch discrimination of broadband signals with rippled power spectra. *Journal of the Acoustical Society of America*, 63, 1166–1173.

Zeddies, D., Fay, R. R., Gray, M. D., Alderks, P. W., Acob, A., & Sisneros, J. A. (2011). Local acoustic particle motion guides sound source localization behavior in the plainfin midshipman fish (*Porichthys notatus*). *Journal of Experimental Biology*, 215, 152–160.

Chapter 8
A Quarter-Century's Perspective on a Psychoacoustical Approach to Loudness

Mary Florentine

Mary Florentine

M. Florentine (✉)
Bouvé College of Health Science and College of Engineering,
Northeastern University, SPLA Department,
360 Huntington Avenue, Boston, MA 02115, USA

Department of Electrical and Computer Engineering, Institute for Hearing,
Speech and Language, Northeastern University, 360 Huntington Ave.,
Boston, MA 02115, USA
e-mail: m.florentine@neu.edu

A.N. Popper and R.R. Fay (eds.), *Perspectives on Auditory Research*, Springer Handbook of Auditory Research 50, DOI 10.1007/978-1-4614-9102-6_8,

8.1 Introduction

Although most people think they understand the concept of loudness and have a clear opinion about it, few people have a basic understanding of the various issues related to loudness (Florentine, 2011). The study of loudness is strewn with conceptual errors and methodological pitfalls. Anyone who truly wants to understand loudness—and perception, in general—must have a broad knowledge across scientific disciplines, an open mind, and a high level of emotional intelligence. The latter attribute may seem humorous, and it is, but it is also meant in a deeply serious way. One must be interested in discovery, have an ability to accept harsh criticism, and not be emotionally attached to any theory.

We are limited by our senses with which we study perception. We assume that our perceptions are accurate, that others perceive the physical world in nearly—if not exactly—the same way, and that the words used by others to describe a perception always have the same meaning as we do when we use them. A wide body of research indicates that this is not true.

The study of perception is difficult because we have preconceived ideas about our own experience. While mulling over a question regarding auditory perception, I find it helpful to ask myself the same question as it applies to another modality. For example, binaural loudness constancy is a phenomenon by which listening binaurally in a room is about as loud as listening monaurally. One of my colleagues asked me, "Isn't your discovery of binaural loudness constancy an illusion?" It made me think of an analogy with binocular brightness. As pointed out by Sivonen and Ellermeier (2011), the world is not brighter when you look at something with two eyes than when you look with only one eye. In the same way, the world is not any louder when you listen to a talker with two ears in a room than when you listen with only one ear (Epstein & Florentine, 2009, 2012; Florentine & Epstein, 2010). For a classroom demonstration, see Florentine and Epstein (2012a). I next thought about perceptual constancy of object size—a house at a distance has a small size on our retinas, but we do not perceive a dollhouse, we perceive a house of typical size at a distance. Finally, I switched to the perception of time as unfolding continuously as we move forward with a clock. Enlightening experiments provide compelling evidence that the physical world does not have a simple correspondence with our perception of time. Does that mean that our perception of time is an illusion and, if it is, are *all* of our perceptions illusions? Sometimes I wonder what is beyond the limits of our perceptions, but that is not within the realm of this book.

My current purpose is to pass on to you (the reader) some thoughts that I have acquired over several decades using a psychoacoustical approach to understand what is known as "loudness" both in and out of the laboratory. My approach to writing this chapter is unlike almost all of my scientific published works; it provides a more personal view as I might relate it to a student in my laboratory, and gives only a brief overview of the many factors that influence loudness. A scholarly approach to loudness with many references can be found in the various chapters written by international authors in the book *Loudness* (Florentine et al., 2011; SHAR 37).

By drawing together authors from Australia, Europe, Japan, and the United States to write edited chapters on various aspects related to the study of loudness, we tried to provide a global network of information about loudness in one volume of the SHAR series.

We like to think of science as continually progressing forward in time, but that is not how science works. Discovery has many faulty starts; information is forgotten and rediscovered when the *Zeitgeist* is ready for it. Loudness and psychoacoustics are no exceptions to this rule. I am concerned that much of what we have learned about psychoacoustics may be forgotten when my generation passes on. Most young scholars are pursuing areas in neuroscience in which few people have a deep understanding of perception, which takes considerable time to acquire. Although it is not possible to know everything, it is important to have an awareness of bodies of knowledge that may be useful. Loudness not only changes with stimulus factors (level, time, frequency, spectrum, and the presence of other sounds); it also changes with various cognitive factors, such as context, memory, meaning, multisensory interactions, procedures (how sounds are presented and how they are measured), cross-cultural differences, and the general psychological and physical state of the listener.

There is clearly much to learn about loudness, and this chapter can provide only a brief glimpse into some important issues. The first section defines a psychoacoustical approach to loudness with a bit of history, gives examples of two approaches that I find useful, and concludes with the importance of using accurate terminology. The second section describes some aspects of the complex nature of loudness, and perception in general. The third section summarizes some elements of loudness measurement (i.e., some theoretical, empirical, and practical constraints). The fourth section gives a brief history of loudness at Northeastern University in Boston, MA. The final section concludes with some suggestions for future work.

8.2 Definitions, Approaches, and the Importance of Terminology

Various people have defined loudness in different ways. After editing a book on loudness with Art Popper and Dick Fay, writing two encyclopedic entries, and working in the area for several decades, I prefer this one: Loudness is the *primary* psychological correlate of physical level. The word "primary" is important because loudness also changes with a host of parameters, as do other perceptual experiences. Loudness is a subfield of psychoacoustics (a.k.a. auditory perception or auditory sensation and perception).

The term "psychoacoustics" is widely used today, even in commercial sales. The first use of the term, as far as Lawrence E. Marks and I know, was in 1901 by M. F. Larroque in a paper entitled, "Études de psycho-acoustique," which translates to "Studies on psycho-acoustics." The aim of Larroque's paper was to explore a connection between perception of roughness (tremolo) and brightness with the acoustics of bowed and brass instruments, respectively. The English word "psychoacoustics" came into wider use in 1940 when the U.S. Air Force funded S. S. Stevens to experimentally address problems arising from high noise levels in aircraft (Miller, 1975). According to Bertram Scharf—a student of Stevens and one of my teachers and co-investigators—this financial support led to the founding of the Psycho-Acoustics Laboratory (PAL) at Harvard University, employing about 50 people. The founding of the PAL issued in a period of applied research in humans, primarily on communication in noisy environments with distorted acoustic signals and on the effects of intense noise on humans. From its very inception, psychoacoustics was broadly defined to address how auditory sensation and perception ("psych") depends on physical parameters of sound ("acoustics").

In 1901, Larroque wrote that the wavering vibrational sound ("tremolo digiti") contributed to a strong shaking of the nervous system of the listener. This was the first example of the close connection between psychoacoustics and the physiology of hearing, which even today are commonly viewed as separate fields. Thirty-four years later, Stevens, Davis, and Lurie (1935) were able to provide insight into Helmholtz's Place Theory when they showed that a lesion on a guinea pig's basilar membrane varied with the frequency at which hearing loss was greatest. The article by Stevens and colleagues appeared around the same time that von Békésy was investigating the traveling wave and the tonotopic representation of the basilar membrane, work for which he later received a Nobel Prize (Nobel Foundation, 1961). It seems obvious that an understanding of physiology is important to understand perception fully, because alterations in the physiology affect perception.

As an academic discipline today, psychoacoustics remains broadly defined and is a study of the relation between the physical, acoustic stimulus and the psychological response to sound. In the broadest definition of psychoacoustics, it includes all sounds: environmental sounds (soundscapes of nature, cities, etc.), spoken words, music, laughs, cries, and so forth. Accordingly, loudness—a subarea of psychoacoustics—includes the loudness of all sounds studied with "experimental" and "applied" approaches. These terms are in quotes because I believe that it is unwise to make an artificial division between experimental and applied research. These terms come with a lot of baggage and an air of intellectual elitism that impedes understanding. Both approaches span a continuum and fluidity of the approach facilitates understanding perception under conditions in which most perceptions occur (i.e., in daily environments).

Just as loudness is a subfield of psychoacoustics, psychoacoustics is a subfield of psychophysics, which is a study of the relationship between stimuli in the physical world that impinge on our senses and our perception of them. The study of psychoacoustics/psychophysics includes what I have referred to as microscopic and macroscopic approaches (Florentine, 2013). A microscopic approach entails strict stimulus control, usually varies only one physical parameter at a time, and is performed with headphones or loudspeakers in a sound-treated test chamber. A macroscopic approach entails a complex stimulus environment and has greater potential for confounding variables affecting the dependent variable. Both approaches are valid and used in psychoacoustics, but the favorite of funding reviewers in the

United States, United Kingdom, and much of Europe is the microscopic approach. Data obtained with a microscopic approach are more "controlled" and appear more "clean" than those obtained with the macroscopic approach, whereas data obtained with a macroscopic approach provide more potential for understanding how sounds are perceived in ecologically valid environments. Assumptions about perception made from data obtained with a microscopic approach may not be valid when tested with a macroscopic approach. In other words, just because a tone presented to two ears of a listener is louder than the same tone presented to only one ear in a laboratory setting with headphones *does not* mean that the listener will have the same binaural loudness summation in an ecologically valid environment. This underscores the importance of ecologically valid research.

Thus, a combination of both the microscopic and the macroscopic approaches seem best to me—just as we need detailed and global views of our data. I like the Native American analogy of mouse vision and eagle vision; I have adapted this concept to how I approach research questions. Both views are necessary, but it is unwise to be trapped in one view for too long. I try to alternately soar up to eagle vision to view the wider implications and connections among other aspects of a perception that I am examining (after all, it's all connected at some level), and then glide down to assess the details of the subject matter.

Details are important. Most mistakes in auditory research—and perhaps in other fields as well—are made at a basic level of initial assumptions. One cannot build a foundation of understanding on pillars of faulty assumptions. For example, it is often incorrectly assumed that (1) a sound can be discriminated and understood if it is heard and (2) a listener has normal audition if pure-tone absolute thresholds are within the normal range. Compelling evidence indicates that these assumptions are true for some, but not all, listeners. The same holds true for loudness. For example, it is often incorrectly assumed that (1) loudness grows at the same rate for listeners with the same pure-tone absolute thresholds, and (2) loudness grows at a normal rate if pure-tone absolute thresholds are within normal limits. Data clearly indicate that loudness can grow at different rates in different listeners, and loudness may not be normal even if pure-tone absolute thresholds are within the normal range.

Because the language with which we study perception limits us, the thorny problem of terminology is briefly addressed. It is essential to use well-defined terminology that clearly separates terms used to connote the physical attributes of sound and the psychological attributes of sound. For example, the term "loudness" is often incorrectly used to connote the physical level of a sound. When referring to the physical level of a sound, the term "sound level" should be used and "loudness" should be used when referring to the primary perceived attribute of sound level. Otherwise, it makes it difficult to understand what is meant. Another important reason for using clear terminology is that there is no one-to-one correspondence between physical acoustics of a sound and our perception and cognition of it. For example, in an elegant little experiment by S. S. Stevens (1934) the principle of independent invariance was used to show that changes in only two physical parameters (sound level and frequency) create four different percepts (loudness, pitch, volume, and density). This is just one example of why it is essential to keep a clear distinction between terms connoting the physical and the psychological attributes of sound.

8.3 The Complex Nature of Sound Perception

Loudness is a pervasive and complex issue and one that should be examined from a range of perspectives. Loudness is often viewed as a one-dimensional concept in theory and research, but it is multidimensional as experienced in daily life. This should not surprise us; at some level everything is probably connected—at least from an eagle's view of nature and matter.

In addition to acoustic stimulus variables, a host of variables affect loudness, such as physiological and psychological states of the listeners. Loudness is clearly in the ear of the beholder. For example, imagine this scenario: An ambulance with siren at maximum output is trying to get through rush-hour traffic. Assume that the sound level produced by the siren is the same at each person's ears in the surrounding cars (which it is surely not, but assume this anyway for this example). A person in one car with an average level of alertness may perceive a very loud siren. A person in another car, who had been listening to the radio at full blast, may perceive the siren to be not as loud. A passenger in a third car, who is listening on a cellphone, may perceive the siren as more annoying than loud. A person with a hearing loss may hear the siren as soft, whereas a deaf person may not even hear the siren. Loudness, therefore, can be influenced by the physical and psychological states of listeners. Hearing loss, attention, and previous sound exposure (auditory fatigue, adaptation, context, etc.) affect loudness. Ample data in the literature support this claim (for specific studies, see Florentine et al., 2011).

Whereas loudness, annoyance, and perceived noise are correlated percepts, they are clearly separate percepts. Music provides an excellent example of this. In a cross-cultural study initiated and coordinated by Seiichiro Namba and Sonoko Kuwano in the 1980s, an international group of scientists came together to address issues related to the perceptions of environmental sounds, as well as social factors related to community noise problems. Some differences were found among the participating countries, but there were also some universal similarities. In one study, six types of sounds (aircraft noise, construction noise, music, road traffic noise, speech, and train noise) were prerecorded and presented at different levels to listeners in China, Japan, and the United States. Results showed a clear correlation between loudness and annoyance; as the sounds became louder they also became more annoying. Music was the only sound that could be very loud and not annoying. (For details of this experiment and data figure, see Fastl & Florentine, 2011, Fig. 8.2, p. 203). Of course, this occurs only if the music is experienced as pleasing. If a person does not like a particular type of music, it can become very annoying. As a general rule, well-liked music is enjoyed at greater loudness than other types of music. There is also a general rule about the loudness of unpleasant music, which is perceived as noise and best tolerated at low levels.

The topic of what constitutes the psychological attribute "noise" has been debated for more than a century. An analogy to personal possessions may be useful. Imagine two college roommates sharing a room in a residence hall. Each student has the same types of possessions, such as clothing, bedding, computers and other electronic devices, books, food, and dishes. From the view of an objective observer, both sides of the room look similar. The student on one side of the room, however, may look across the room and perceive an unsightly mess, whereas the student's own possessions are viewed as cherished belongings. Accordingly, if you are making sounds or enjoying them, it is not noise to you. The old joke is "noise is something your neighbor makes" (see Berendt et al., 1976 for a somewhat dated, but wise and practical guide to noise control from members of the National Bureau of Standards, USA).

Not all loud sounds are bad; some are good and even desirable. Some warning signals may be very loud, but most people understand the importance of these signals and respond to these loud sounds differentially, depending on the assumed meaning of the signal. The meaning associated with the same sound can be different in different countries and impact perception in different ways. For example, garbage trucks in Tokyo, Japan and ice cream trucks in Boston, MA, USA look very similar. They look clean in blue and white colors and they have writing on them. Both types of trucks play a cheerful carillon-bell sound. Having a similar visual and auditory stimulus with different meanings in different countries can lead to mistakes. A person raised in Boston can fly to Tokyo and hear the sounds of a garbage truck and wonder why the Japanese people eat ice cream so early in the morning! Therefore, a loud sound is better tolerated if it is associated with a useful purpose. Keeping Tokyo clean is a useful purpose; ice cream in the morning is not.

The multisensory nature of perception related to loudness has been revealed by a host of experiments. Low-level noise bursts are rated louder when they are heard in the presence of lights than in their absence (Odgaard et al., 2004). In another experiment, loudness of a white noise depended on the type of picture that was paired with the noise; when paired with a picture of a waterfall, the noise was judged as softer and more pleasant than when the same noise was judged without the picture. Studies of audio-visual interactions with loudness are more numerous than studies of audio-tactile interactions, but there are clear audio-tactile interactions. For example, subjects were asked to engage in hand-rubbing behavior while they listened to recorded sounds of hand rubbing at various levels. When the level was reduced from moderate to soft loudness, the subjects reported that their hands felt less rough. (For more examples of multisensory interactions and references, see Fastl & Florentine, 2011).

8.4 Approaches to Measuring Loudness

In addition to the traditional experiment, there are two methods used to acquire information about perceptions that I find especially helpful: observation and introspection. Each of these methods has its uses and limitations, but together they are a formidable set of learning tools. For example, if we want to know whether our voice is perceived as soft or loud to a person listening to us, we have only to observe them. If our voice sounds too soft, the listener may lean toward us in order to increase the signal's reception level, focus intently on our face for speech-reading cues, and/or make head adjustments to move the better ear into an advantageous position for receiving our output. When sounds are too loud, listeners may grimace, move away from the sound source, and/or try to limit the sound from reaching their ears. Most people provide a trained observer with abundant cues to assess whether a sound is too loud or too soft, although some people provide clearer cues than others. A young boy may simply cover his ears with his hands when he perceives uncomfortably loud sounds, whereas a middle-aged woman—trying to hide her hearing loss—may nod her head in agreement in an inappropriate manner. Some researchers dismiss the method of observation as unscientific, but I have found it to be one of the essential tools in discovery.

Another rarely used and almost-forgotten method is introspection, which I learned from Eberhard Zwicker while he was my doctoral-dissertation advisor. Every Friday afternoon at the Institüt für Electroakustik (Institute for Electroacoustics) at the Technische Universität München (Technical University of Munich), Professor Dr.-Ing. Zwicker made his rounds to talk with each of his doctoral students. He encouraged us to carefully listen to sounds and to develop an ability to describe them. This is not as easy as it sounds; it required much reflective time listening that I argued could be better spent accomplishing other tasks. Now, I am grateful to him for that guidance because this skill has served me well when used as a tool to acquire information. For example, years after completing my doctoral dissertation, I was listening to pulsed tones near my absolute threshold and comparing the sounds of tones at two different frequencies. At one frequency my threshold was normal; as I slowly increased this tone's level, I heard a sound become audible as very, very soft. At the other frequency my threshold was elevated due to a mild cochlear hearing loss; as I slowly increased the tone's level I heard a sound become audible as somewhat soft, but not very, very soft. In other words, loudness at threshold was different for the two tones and it was greater for the frequency with an elevated threshold. After checking that this was not a microstructure issue, the concept of softness imperceptions was born (Florentine & Buus 2001, 2002; Florentine et al., 2004). Later, Andrzej Miśkiewicz (2004) also used introspection to describe his softness imperceptions in a brief publication entitled, "From pianissimo to mezzopiano."

The method of introspection also opened my mind to other aspects of perception and the inadequacies of language to transmit perceptual experiences. For example, the same word "soft" can refer to different perceptions when used by different people. When people with hearing losses describe a sound as "very soft" it may mean that it is the softest sound that they are able to hear (i.e., closest to the absolute thresholds) or it may indicate an ordinal difference—that one sound is soft relative to another.

Although observation and introspection provide useful information, the gold standard is the traditional experiment. Using the taxonomy of Buus (2002), methods of experimentation include the mode of stimulus presentation, the listener's task, measurement strategy, and datum definition. In each area of the method with which we measure loudness, there are potential pitfalls. Two broad types of measurement methods are equal loudness matching and scaling methods. It is essential that the chosen method of measurement meet the basic requirement of yielding internally consistent measurements. Because there is no objective measure of loudness, measurements of loudness should employ methods of converging evidence. For example, a matrix design can be used in which sounds are matched in loudness to themselves and to each of the other sounds; the resulting data can then be examined (for an example, see Florentine et al., 1978). Examination should reveal that the data conform to two principles: an ordinal indicant of relative loudness and transitivity of the loudness. An ordinal indicant means that if A has a measured loudness greater than that of B, then A is louder than B, and B is softer than A. Further, whenever A and B are equally loud, the system must assign them the same loudness value. Transitivity means that if the acoustic signal A is as loud as signal B, and B is as loud as C, then A must be as loud as C. It takes considerably more time to include measures of internal consistency, but without them measurement bias can be hiding in the data. Careful measurements show that most people are consistent when judging loudness.

Although there is no perfect method with which to measure loudness, there are some methods that are much more biased than others and some that are clearly founded on faulty assumptions. Methodological pitfalls can be avoided by taking these limitations into account. See Marks and Florentine (2011) for the theoretical, empirical, and practical constraints on loudness measurement and see Arieh and Marks (2011) for ways in which context affects loudness and loudness judgments.

Even with the best methods to study loudness, we must approach the topic with an open mind to the complex nature of perception. Because scientists like to reduce information and look for one or two causes, there is a risk of our views becoming too simplistic. When our theories are too narrow, they inhibit our ability to see more broadly and obtain understanding. I fell prey to this erroneous line of reasoning when I first discovered softness imperception. At the time, it seemed so reasonable to me that almost all people with hearing losses of primarily cochlear origin heard a greater loudness at absolute threshold in a frequency range in which they had elevated thresholds than in a frequency range in which they had normal absolute thresholds (i.e., softness imperceptions)—just as I experienced. Søren Buus and I (2001; Florentine & Buus, 2002) modeled data from five listeners with hearing losses of primarily cochlear origin and our loudness model fit the individual data, supporting the concept of softness imperception. In hindsight, there was one listener whose data did not fit as well as the others, but this observation was lost in the excitement of the discovery. At the time, I believed that rapid growth of loudness probably did not occur in hearing losses of cochlear origin, especially since our data showed little difference in loudness growth near threshold in listeners with normal hearing and listeners with hearing losses (Buus & Florentine, 2002).

At the 2003 meeting of the American Auditory Society, Brian C. J. Moore gave a lecture and stated from the podium that he did not believe in softness imperceptions, because he had a hearing loss and did not experience it. When I heard his statement, I was surprised. We both used our personal experience to make assumptions about the perceptions of others. Brian Moore's comment forced me to sharpen my thinking and obtain more data. Today, I believe that we were both right and wrong.

The most current view of softness imperception can be found in Marozeau and Florentine (2007). It shows that loudness-growth functions encompass a wide range of shapes from rapid growth of loudness to softness imperceptions, and neither model can account for all the data.

8.5 Loudness Work at Northeastern University

Research environments can support or extinguish innovative ideas. Northeastern University has sustained an outstanding interdisciplinary and collegial environment to promote a psychoacoustical approach to loudness and loudness-related topics over several decades with tenured faculty who directed nationally funded research laboratories across three colleges: Liberal Arts (Bertram Scharf & Adam Reeves), Engineering (Søren Buus & Michael Epstein), and Health Sciences (Michael Epstein & Mary Florentine). The work in the research laboratories has been aided by close contact with the on-campus audiology clinic, directed by Sandra Cleveland. Other investigators, who have not worked at Northeastern University, are often confused by the organization and function of the laboratories. The next four paragraphs give a brief history.

About 50 years ago, Bertram Scharf left S. S. Steven's laboratory at Harvard University to start his Auditory Perception Laboratory in the Psychology Department in the College of Liberal Arts at Northeastern University. He brought the first NIH grant on loudness to Northeastern University. Since that time, at least one of the laboratories has been funded through NIH to work on loudness or loudness-related questions. The second research laboratory—which became known as the Communication Research Laboratory, now in the College of Health Sciences—was established in 1980 when I left Nat Durlach and Lou Braida's laboratory in the Research Laboratory of Electronics at MIT for a faculty position at Northeastern University. When a faculty position opened in the Department of Electrical and Computer Engineering at Northeastern University in 1986, Søren Buus left David M. Green's laboratory at Harvard University to establish a third research laboratory (i.e., the Hearing Research Laboratory) at Northeastern University's College of Engineering. Although the Communication Research Laboratory and the Hearing Research Laboratory were in different colleges, the laboratories were next door to one another and the doors were open for interdisciplinary research long before "interdisciplinary" became a buzz word. Students in all three colleges took courses in psychoacoustics and worked together on the same projects. We all participated in colloquia every week, which hosted many local, national, and international speakers.

A number of postdoctoral research associates worked in the Communication Research Laboratory for one-to-two-year appointments, including Robert Carlyon, Michael Epstein, Michelle Hicks, Renier Kortekaas, Jeremy Marozeau, Peter Marvit, Andrzej Miśkiewicz, Hannes Müsch, Bärbel Nieder, Andrew Oxenham, Eva Wagner, Linda Welsh, and Tilmann Zwicker. A number of colleagues visited to work on joint projects for various amounts of time, including Eberhard Zwicker, Hugo Fastl, Georg Klump, Torben Poulsen, Sonoko Kuwano, Seiichiro Namba, Georges Canévet and others.

At the height of operation, it had more than 20 members. I have been asked a number of times how we were able to work together with so many different types of people and strong personalities. It is true that we did not always agree on theory or interpretation of data, and we sometimes argued in our perfectly imperfect ways, but we all sought to understand auditory perception, had a sense of humor, and wanted to get along with one another. We often used positive humor and several enjoyable activities to reduce stress, such as the annual Halloween party when we would wear costumes of our own design or the one afternoon each year when we would take a few hours to engage in some sightseeing activity in Boston (e.g., sailing on the Charles River or viewing Boston from the top of the Custom House).

The disagreement made us better scientists and moved the research forward. My favorite example occurred one night in Wellesley, Massachusetts after dinner, when Eberhard Zwicker, Søren Buus, and I were debating different theories about the workings of the auditory system. (It involved signal-detection vs. excitation-pattern modeling.) We had three people and three theories! Eberhard Zwicker placed the palm of his hand firmly on the table and announced, "Now, we design the experiment!" After some hard work, we designed an experiment that we all agreed the outcome of which would resolve the issue. The result was a synthesis of the notion of excitation patterns with signal-detection theory to model intensity discrimination and was based on Florentine and Buus (1981). The experiments showed that discrimination must be performed by integration of information across all stimulated frequency-selective channels. The model permitted the development of the first quantitative model of auditory perception in a nonhuman species (the starling, *Sturnus vulgaris*), tying together a wide variety of physiological and behavioral data for that species (Buus et al., 1995). The integration of information across independent frequency bands embodied in both models has been used by other authors in the development of physiologically based models of perception.

When Bertram Scharf closed his Auditory Perception Laboratory after more than 40 years of continuous operation, he continued working as an Emeritus researcher in the other laboratories on campus—including Adam Reeves' Perception and Attention Laboratory, where he and Adam Reeves started doing loudness work. Rhona Hellman, who had worked in Bertram Scharf's laboratory as an Adjunct Research Professor, moved to the Communication Research Laboratory. Another major change came in 2003 when Søren Buus became unable to work due to illness; his Hearing Research Laboratory and the Communication Research Laboratory were administratively combined without loss of personnel or space. Members of the hearing-research community at Northeastern University came together in an energy-charged collaborative effort to keep the research on track. With an education in both engineering and audiology, Michael Epstein was uniquely qualified to take a post-doctoral Research Associate position in the Communication Research Laboratory and move the work forward. In fact, he did such an excellent job that he was hired as a faculty member in 2006 and established his NIH-funded Auditory Modeling and Processing Laboratory (AMP Lab); he recently received tenure and his laboratory is thriving. One of the primary aims of his lab is to bridge the gap between physiological representations of loudness and perceptual loudness in human listeners and he is uniquely qualified to do this. He also seeks to improve loudness models and amplification strategies used in hearing aids and cochlear implants.

Now that the AMP Lab has taken off, it seems like a good time to bring my laboratory (the Communication Research Laboratory) to a close. Generous colleagues have offered me use of their laboratory facilities should I need them. Because of my long career and the pace of my work with two ongoing NIH research grants, I have amassed data that only I can sort out. Not wanting to risk the possibility of sudden illness or death with my research in disarray and of use to no one, I have decided to pause and put my research in order. The plan is to use the time freed up from running my lab to use writing up the mounds of data acquired over the years and completing unfinished projects. This should also provide time to make a smooth transition into other types of loudness research, and to allow for new growth of leadership in loudness and other areas. After coediting the SHAR loudness book and bringing almost 25 years of continuous NIH research to an orderly close as PI, my interest in loudness is as strong as ever, but I think it best for me to take a more supportive role. Transitioning to a more supportive role may be difficult, but I believe that it is an important right of passage in the life cycle of a scientist—or, at least, it is for me. I think it odd that much attention is paid to early career build-up, but discussions of the late stages of careers are often treated as taboo. I think this is unwise; information needs to be transferred or it may be lost.

Over the past several decades, work performed at the Communication Research Laboratory at Northeastern University using a psychoacoustical approach to loudness has brought us to a better understanding of loudness among normal listeners and listeners with hearing losses of primarily cochlear origin (Florentine, 2009). Four areas of inquiry give examples of our approach: (1) investigations of individual differences in loudness functions among normal listeners and listeners with different types of hearing losses, (2) investigations and models of the relationship between temporal and spectral integration of loudness and the loudness function, (3) investigations of how context affects loudness, and (4) loudness in the laboratory and in ecologically valid environments. These areas of inquiry have important theoretical implications for understanding loudness among normal listeners and listeners with different types of hearing losses, as well as how loudness is perceived in daily environments and how to improve the design of hearing aids, cochlear implants, and various assistive-listening devices. As is the case in most of science, knowledge has progressed in incremental steps with moments of sudden enlightenment. Noteworthy areas are summarized in the remainder of this section.

8.5.1 Investigations of Individual Differences in Loudness Functions Among Normal Listeners and Listeners with Different Types of Hearing Losses

Attitudes toward individual differences and variability in loudness data have changed very much over the years. When I published data from my doctoral dissertation (Florentine & Zwicker 1979; Florentine et al., 1980), it was almost impossible to publish individual data. Reviewers viewed individual differences as variability in the measurements, something that needed to be averaged out. Almost all studies published around that time grouped listeners with sensorineural hearing losses together. My data clearly showed that individual listeners were consistent in their judgments (i.e., small intrasubject variability). There were, however, considerable differences among the data for normal listeners, and even more variation among listeners with hearing losses of primarily cochlear origin (i.e., large intersubject variability). The reviewers insisted that the data be grouped and, as a consequence, we grouped the data into six groups: normal, conductive, noise-induced, otosclerotic without elevated bone conduction, otosclerotic with elevated bone conduction, and degenerative. Loudness modeling was successfully applied only to the noise-induced hearing loss group, because it was clear that listeners with other types of hearing losses behaved differently and not enough was known about the physiology to make reasonable assumptions.

At the time of this writing, more knowledge is available and the general consensus supports the importance of maintaining the integrity of data in broad diagnostic groups, but much work is still needed in loudness modeling (Marozeau, 2011). Most hearing researchers agree that a young person with a noise-induced hearing loss does not process information in the same manner as an older person with presbycusis. I still believe that it is important to *maintain the integrity of individual data* because two people with similar audiograms can have different loudness-growth functions (Hellman, 1994; Florentine et al., 1997). Further, important insights regarding context effects have been gained from examining the individual loudness data from listeners with normal hearing (Florentine & Epstein, 2012b). We should treasure these individual differences because they are highly likely to reveal important mechanisms that contribute to loudness. (Carefully recording individual differences worked for Gregor Mendel. Remember his pea experiments?)

To compare the data from normal listeners and listeners with different types of hearing losses, an understanding of loudness at and near threshold is essential. Although many authors have proposed loudness functions according to which loudness at threshold is zero (for references and discussion, see Florentine & Buus, 2002), actual measurements in individual listeners indicate that loudness at threshold is greater than zero for normal listeners (Buus et al., 1998). In other words, if a sound is heard, it has a percept of loudness associated with it. Buus and colleagues used a clever paradigm to examine the form of the loudness function in normal listeners at threshold and low levels. They derived loudness functions from loudness-matching experiments between equally loud tones and tone complexes (i.e., spectral integration of loudness). Listeners matched the loudness of tone complexes composed of subthreshold components with pure tones above threshold. Results show that a tone complex composed of components at threshold can easily be heard and can be relatively loud. The slopes of the loudness functions at and near threshold are consistent with loudness functions obtained with a wide variety of methods and yield a slope close to unity. A slope close to unity indicates a positive loudness at threshold in accord with the most recent loudness standard (ANSI S3.4, 2007). Some of the most surprising data and modeling indicate that loudness grows at a normal (or close to normal) rate near the elevated thresholds of listeners with cochlear hearing losses and that loudness at threshold may be greater than normal in *some* listeners (Buus & Florentine, 2002; Marozeau & Florentine, 2007).

When it became clear that loudness at threshold has a positive value, it seemed logical that loudness at threshold may have a value that differs among individuals. My own introspection experiments (described earlier) provided insight that loudness at threshold could be greater in a frequency range of hearing loss than in a frequency range of normal hearing. Other evidence came from absolute-threshold measurements of many listeners with hearing losses using a clinical finger-raising method (i.e., "raise your finger when you hear a beep"). Normal listeners usually raised their fingers very slowly as the beeps approached their thresholds, and sometimes they did not complete a full finger extension that I interpreted as expressing a percept so soft that it was almost inaudible. Some other listeners raised their fingers in a quick, full extension even at levels approaching their absolute thresholds, which I interpreted as indicating that they clearly heard a beep. Using the method of behavioral observation, there appeared to be clear differences, and it also seemed reasonable that various disturbances in the cochlea that cause hearing losses could be related to different changes in loudness at threshold.

Because loudness at threshold is very difficult to measure, a reaction time (RT) paradigm that correlates with loudness was used. [Much can be learned by using correlates of loudness (Epstein, 2011).] Simple RTs for tones have been shown to correspond closely to equal loudness: the louder the sound, the faster the RT. Another advantage of this method is that measurements of RTs are possible even when tones are set at threshold (for review, see Wagner et al., 2004). In this paradigm, 200-ms tones of various levels were presented to listeners with high-frequency cochlear hearing losses and their task was to press a key as soon as they hear a sound. Two frequencies were used for each listener. One frequency was chosen to have normal absolute thresholds or a mild hearing loss; the other was chosen to have a moderate-to-severe hearing loss. Results for six listeners with cochlear hearing losses consistently showed faster RTs to tones at and near threshold for the frequency with elevated threshold than for the frequency with normal or near-normal threshold (Florentine et al., 2005). Normal controls showed an effect of frequency in some listeners, but the effect was not large enough to account for the difference attributable to hearing loss (Epstein & Florentine, 2006a). This finding provides strong support for the concept that some listeners with hearing losses of primarily cochlear origin have a greater loudness than normal listeners at threshold. In addition to the RT data from six impaired listeners described above, RT data from 22 impaired listeners and equal-loudness balances from 13 of the 22 listeners were obtained to check the relationship between RT and loudness (Florentine et al., 2004). Agreement is good in trained listeners.

The preponderance of the data at and near threshold provided clear evidence against the pervasive and long-held notion that all listeners with hearing losses of primarily cochlear origin show abnormally rapid loudness growth near their elevated thresholds (i.e., the original definition of recruitment). It also indicates that most hearing scientists have been using a faulty theoretical framework for over 60 years by assuming that all impaired listeners with losses of primarily cochlear origin perceive loudness growth in a similar manner. If loudness at threshold is greater than normal in some listeners with cochlear hearing loss, they have a reduced dynamic range, not only in terms of SPL, but also in terms of loudness. This phenomenon is known as "softness imperception" (i.e., the inability to hear a range of low loudnesses that can be heard by normal listeners).

The new concept of softness imperception has very important theoretical and practical implications. It agrees with recent knowledge about basilar-membrane mechanics (Epstein et al., 2006), and it provides a scientific basis for the design of hearing aids that apply expansion to low-level sounds to ensure that only sounds for which normal loudness is above the impaired listener's elevated threshold for loudness are amplified to audibility (e.g., Blamey, 2005). It also provides a basis for understanding the common observation that wide-dynamic-range-compression hearing aids have optimal compression rates that are considerably smaller than those derived from mapping the physical dynamic range of the impaired ear into that of the normal ear. This is because results from some listeners indicate that the reduction of physical dynamic range is accompanied by a reduction in the subjective dynamic range of loudness. Implications of softness imperception may also apply to cochlear implants. When I told Margo Skinner of our discovery of softness imperceptions over box-lunch sandwiches at a meeting of the Association of Auditory Research in Phoenix, AZ, she became very excited and said, "I think that's what could be happening with some of our cochlear-implant patients!" Unfortunately, the exceptionally kind, delightfully bright, and dedicated scholar Margo Skinner passed away before we had time to collaborate on the experiments. Although I have not worked with cochlear-implant patients yet, I have seen some data that indicate that this may be true for some individuals with cochlear implants.

There are important individual differences in loudness growth functions of normal listeners and those with hearing losses of primarily cochlear origin. Some of these differences have been obscured by methodological pitfalls in measuring loudness (Marks & Florentine, 2011; SHAR 37), context effects (Epstein, 2007; Arieh & Marks, 2011; SHAR 37), differences in psychophysical procedures (Epstein & Florentine, 2006b; Silva & Florentine, 2006), and averaging data from individuals who exhibited clearly different responses (Marozeau & Florentine, 2007). Marozeau and Florentine (2007) reanalyzed data in the literature from individual listeners to test two loudness-growth models: rapid growth (a.k.a. recruitment) and softness imperception. Five different studies using different methods to obtain individual loudness functions were used: absolute magnitude estimation, cross-modality matching with string length, categorical loudness scaling, loudness functions derived from binaural loudness summation, and loudness functions derived from spectral summation of loudness. Results from each of the methods show large inter-listener differences. Individual loudness-growth functions encompass a wide range of shapes from rapid growth to softness imperception. These results indicate that neither theory (classical recruitment or softness imperception) accounts for all the data. It is clear that some of the impaired listeners deviate markedly from the average, further supporting the importance of individual data and indicating that group data do not accurately represent the behavior of all impaired listeners.

8.5.2 Investigations and Models of the Relationship Between Temporal and Spectral Integration of Loudness and the Loudness Function

Apparent paradoxes in loudness data sets provided enjoyable and productive hours in which Søren Buus and I mulled over ideas on the upper edge of our abilities at the Acoustics Laboratory of the Technical University of Denmark. Several six-month visits with Torben Poulsen gave us the reflective time to read the literature, reanalyze published and unpublished data, and apply several models to large sets of data. One puzzling question was how to reconcile two sets of loudness data from normal listeners. One set of data clearly indicated that the amount of temporal integration for loudness varies nonmonotonically with level and is greatest at moderate levels (Florentine et al., 1996, 1998; Buus et al., 1997). [This holds true for both monaural and binaural loudness summation (Marozeau et al., 2006; Whilby et al., 2006; Marozeau & Florentine, 2009).] The second set of data indicated that the loudness functions for short and long noises (Stevens & Hall, 1966) and tones are parallel when plotted on a log scale as a function of level (Epstein et al., 2001; Epstein & Florentine, 2005). In other words, there is a constant vertical distance between loudness functions for short- and long-duration tones, which is known as the equal-loudness-ratio hypothesis (for review see Epstein & Florentine, 2005). A logical deduction came into focus: If the vertical distance between loudness functions for short and long sounds is independent of level, then the loudness-growth functions for these stimuli must be shallower at moderate levels than at low and high levels. When we went back and examined the literature, we could see shallower functions at moderate levels in some individual data. Straight lines through the data points in graphs, however, obscured the visual representation of the data!

Computer modeling of large amounts of data in the literature ensued to answer this question. For example, polynomial fitting procedures reveal orderly deviations from a simple power function that are important (e.g., Marozeau et al., 2006). Careful examination of the loudness growth function reveals that it is shallower at moderate levels than at low and high levels (for review, see Florentine & Epstein, 2006). This nonlinearity is consistent with masking and peripheral nonlinearity (Buus & Florentine, 2001; Oxenham & Bacon, 2004; SHAR 17; Epstein et al., 2006).

This new discovery that the loudness function is less steep at moderate levels than at low and high levels was combined with data that have been around for decades. The older data indicate that the loudness function becomes steep as it approaches threshold (Hellman & Zwislocki, 1961; also more recent data Marks, 1979; Canévet et al., 1986; Buus et al., 1998). Near threshold, the average slope is about unity or slightly larger. As level increases, the slope decreases as the function approaches moderate levels (see Buus et al., 1997; Buus & Florentine, 2002). These two deviations to the power law led to a non-stationary point of inflection law [or an inflected exponential (INEX) law] that appears to be the best description of currently available data for normal listeners. For review of the INEX law, see Florentine and Epstein (2006).

8.5.3 Investigations of How Context Affects Loudness

A new understanding has been reached regarding ways that context affects loudness and loudness judgments. Previously, context effects were often regarded as unknown sources of variability. One of the most interesting and pervasive context effects is known as induced loudness reduction (ILR); it is also known as the slippery-context effect (Marks, 1992) and loudness recalibration (Marks, 1994; Arieh & Marks, 2001). It is a phenomenon by which a preceding higher-level tone (an inducer tone) reduces the loudness of a lower-level tone (a test tone). The strength of this effect depends on the following factors: tone levels, frequency separation between the inducer and test tone, duration of inducer and test tone, time separation between inducer and test tone, number of exposures to the inducer, and individual differences (for review see Epstein, 2007).

The mechanism of ILR is still a matter of debate, but any hypotheses about the basis for ILR will have to take into account discoveries at the Communication Research Laboratory regarding binaural hearing and large spectral effects. For example, Nieder et al. (2007) measured induced loudness reduction induced by a contralateral tone. The ILR of a weaker tone caused by a preceding stronger tone was measured with both tones in the same ear (ipsilateral ILR) and also in opposite ears (contralateral ILR). When the tone duration was 200 ms, the loudness reduction averaged 11 dB under ipsilateral ILR and 6 dB under contralateral ILR. When the duration was 5 ms, ILR was 8 dB for both the ipsilateral and contralateral conditions. For each duration, ipsilateral and contralateral ILR were strongly correlated (r around 0.80). Regarding spectral effects, ILR can occur even when the frequency separation between the inducer and test tone is wider than four equivalent rectangular bandwidths (Marozeau & Epstein, 2008).

In two elegant little experiments, Epstein and Gifford (2006) showed that the majority of ILR studies have used an experimental paradigm that results in an underestimation of the amount of ILR, because the level of the comparison tone in the baseline condition tends to be substantially higher than in the experimental condition. Because of this difference, exposure to the baseline condition immediately prior to the experimental condition causes an unintended ILR for the comparison tone. Therefore, it is highly likely that loudness data in the literature are confounded by ILR.

The discovery of ILR has implications for psychophysical procedures. For example, the marked reduction in loudness under ILR, which persists across frequency and over many minutes, means that care must be taken in the sequence of sound presentation. Some of the measurement differences in the literature could result from differences in psychophysical procedures. To gain insight into this possibility, Silva and Florentine (2006) used four adaptive two-interval, two-alternatives-forced-choice procedures to obtain equal-loudness matches between 5- and 200-ms 1-kHz tones as a function of level for each of six normal listeners. The procedures differed primarily in the sequence in which the stimuli were presented. The variations tested included: the ordering of stimuli by amplitude across blocks of trials (both increasing and decreasing amplitudes), randomizing the order across those blocks, and randomizing the order within blocks. The random-within-block procedure yielded a significantly greater amount of temporal integration than the other three procedures. The results show significant differences in temporal integration measurements at moderate levels for the same listeners across different procedures. Therefore, although there are individual differences among listeners in the amount of temporal integration measured across paradigms, the choice of paradigm also affects the amount of temporal integration measured at moderate levels. It is likely that ILR is responsible for the differences among the psychophysical procedures. Given the large intersubject differences in the amount of ILR (Epstein, 2007), loudness functions and loudness growth may vary substantially depending on stimulus order and context. This is likely to influence measurements in the laboratory as well as the clinic, which could influence differences in hearing-aid programming depending on the evaluation procedure.

8.5.4 Loudness in the Laboratory and in Ecologically Valid Environments

Generalizations from controlled laboratory experiments often fall apart when extended to daily environments in which there are more complex stimulus networks. I experienced this myself while teaching a group of engineers in a classroom at the Acoustics Laboratory at the Technical University of Denmark in the 1970s. I wanted to give the students an example of binaural loudness summation (i.e., a sound heard with two ears is louder than a sound heard with only one ear).

The students were asked to sit in their usual seats and directly face me. I recited memorized passages while attempting to keep my voice at a constant level that was typical for my lectures. The students' task was to estimate the loudness of my voice while listening with both ears compared to the loudness of my voice while they were blocking one ear by pressing on a tragus with an index finger. They were encouraged to make several observations for each of the two conditions before making a judgment. The students' subjective reports indicated that the loudness of speech changed only a negligible amount, if at all!

I was very excited about this outcome and told several researchers. Their response to me was to cite the literature. Of course, I knew the literature, which was why I was so excited. Despite my efforts, they dismissed the idea. In fact, laboratory experiments using headphones did indicate that a tone presented binaurally is louder than the same tone presented monaurally (Fletcher & Munson, 1933). It had been generally assumed that the binaural-to-monaural loudness ratio is equal to two for dichotic tones at the same loudness [i.e., a tone presented to two ears is twice as loud as a tone presented to only one ear. (For a review, see Sivonen & Ellermeier, 2011)]. Although Scharf and Fishken (1970) found a lower ratio than double loudness with two ears (they found an increase of only 1.7), they still assumed that the amount of binaural loudness summation with pure-tone stimuli applied to complex stimuli in daily environments.

It became clear to me that it would take too much of my energy to push this idea forward and I proceeded on to research other aspects of hearing. There were so many new ideas to pursue. I did, however, continue to use the classroom demonstration of binaural loudness summation under real-world conditions and collect data from various lectures (see Florentine & Epstein, 2012a). Since 1975, these demonstrations have been repeated more than 38 times to groups of students in many different classrooms in several countries. The results from these demonstrations are about the same if the students have not been told prior to the demonstration to expect a doubling or large increase in loudness. This classroom demonstration has been published (see, Florentine and Epstein, 2012a).

Recently, Michael Epstein and I have formally studied this phenomenon, which is known as Binaural Loudness Constancy, BLC—an almost complete absence of binaural loudness summation. [Note that BLC is not the same as loudness constancy with distance in which loudness remains relatively constant while sound source distance is varied (for review, see Sivonen & Ellermeier, 2011)]. Our first traditional experiment compared the loudness of three stimuli [tones, recorded spondees, and monitored live voice (MLV) spondees]. The stimuli were presented monaurally and binaurally to normal listeners, who judged their loudness. Statistical analysis indicated that (1) the binaural-to-monaural loudness ratio is significantly smaller for speech from a visually present talker than for recorded speech and tones, (2) the binaural-to-monaural loudness ratio is significantly smaller for loudspeaker presentation than for earphone presentation, and (3) the binaural-to-monaural loudness ratio is smallest for speech from a visually present talker presented via loudspeakers than any of their other test conditions. This experiment was very difficult to publish because the results were strongly questioned, but it was finally published in *Ear and Hearing* (Epstein & Florentine, 2009).

Our second paper (Epstein & Florentine, 2012) built on these earlier findings and tested the following hypothesis: Speech from the same talkers presented under more ecologically valid conditions results in a smaller binaural-to-monaural loudness ratio than speech presented without visual cues and/or presented via headphones. To provide a condition that had more ecological validity—while being experimentally defined—loudspeaker presentation of speech with visual cues was chosen. This condition was compared to the same stimuli presented in three other conditions: without visual cues presented via a loudspeaker, without visual cues presented via headphones, and with visual cues presented via headphones. Results show that the binaural-to-monaural loudness ratio was significantly less for speech with visual cues presented via a loudspeaker than for stimuli with any other combination of test parameters (i.e., speech without visual cues presented via both headphones and loudspeakers, and speech presented with visual cues via headphones). These results indicate that the loudness of a visually present talker in daily environments is little affected by switching between binaural and monaural listening.

In summary, three experiments (one introspection and two traditional) support the importance of ecological validity in loudness research, which could change how perception of loudness is understood. The lesson learned from the story of the acceptance of this new idea is an old one: *Zeitgeist* is important.

8.6 Looking Toward the Future

The past quarter-century has been especially fruitful in the area of loudness research. Although significant progress has been made, our understanding of loudness is still unfolding. There is no comprehensive theory that explains all phenomena related to the perception of loudness and our knowledge of the physiology of loudness is just starting to unfold (e.g., Epstein et al. 2006; Florentine & Heinz, 2009; Epstein, 2011). There are, however, areas that are primed for new discoveries, as predicted in the SHAR Loudness book (Florentine et al., 2011). These predictions are already being borne out through combining a psychoacoustical approach to loudness with other academic areas of study. It is highly likely that over the next quarter-century: (1) there will be an understanding of the physiological basis of loudness; (2) individual differences in loudness of listeners with normal hearing and hearing losses will be understood, resulting in better rehabilitation of people with hearing losses; (3) loudness context effects will be widely acknowledged—the gap between loudness in the laboratory and in daily environments will be better understood; and (4) new models will be developed that can predict individual differences in loudness among normal listeners and listeners with hearing losses, as well as predict the average perception of loudness for large groups of listeners in various daily environments. Prospects for understanding loudness are bright and the *Zeitgeist* is with us!

Acknowledgments I am very grateful for my students, colleagues, continuous funding from the National Institutes of Health for about a quarter century, and time to reflect while visiting with colleagues. Michael Epstein and Julia B. Florentine made helpful comments on this manuscript.

References

ANSI-S3.4 (2007). American National Standard Procedure for the Computation of Loudness of Steady Sounds. New York: American National Standards Institute.

Arieh, Y., & Marks, L. E. (2001). Recalibration of loudness: Sensory vs. decisional processes. In E. Sommerfield, R. Kompass, & T. Lachmann (Eds.), *Fechner Day*. Berlin: Pabst Science Publishers.

Arieh, Y., & Marks, L. E. (2011). Measurement of loudness, part II: Context effects. In M. Florentine, R. Fay, & A. Popper (Eds.), *Loudness* (pp. 17–56). New York: Springer.

Berendt, R. D., Corliss, E., & Ojalvo, M. (1976). *Quieting: A practical guide to noise control. National Bureau of Standards Handbook # 119*. Washington, DC: US Environmental Protection Agency, Office of Noise Abatement and Control.

Blamey, P. J. (2005). Adaptive dynamic range optimization (ADRO). *Trends in Amplification*, 9, 77–98.

Buus, S. (2002). Psychophysical methods and other factors that affect the outcome of psychoacoustic measurements. In L. Tranebjærg, J. Christensen-Dalsgaard, T. Andersen, & T. Poulsen (Eds.), *Genetics and the function of the auditory system: Proceedings of the 19th Danavox Symposium* (pp. 183–225). Copenhagen, Denmark: Holmens Trykkeri, The Danavox Jubilee Foundation.

Buus, S., & Florentine, M. (2001). Modifications to the power function for loudness. In E. Sommerfeld, R. Kompass, & T. Lachmann, (Eds.), *Fechner Day 2001* (pp. 236–241). Berlin: Pabst Science Publishers.

Buus, S., & Florentine, M. (2002). Growth of loudness in listeners with cochlear hearing losses: Recruitment reconsidered. *Journal of the Association for Research Otolaryngology*, 3, 120–139.

Buus, S., Klump, G. M., Gliech, O., & Langemann, U. (1995). An excitation-pattern model for the starling (Sturnus vulgaris). *Journal of the Acoustical Society of America*, 98, 112–124.

Buus, S., Florentine, M., & Poulsen, T. (1997). Temporal integration of loudness, loudness discrimination, and the form of the loudness function. *Journal of the Acoustical Society of America*, 101, 669–680.

Buus, S., Müsch, H., & Florentine, M. (1998). On loudness at threshold. *Journal of the Acoustical Society of America*, 104, 399–410.

Canévet, G., Hellman, R. P., & Scharf, B. (1986). Group estimations of loudness in sound fields. *Acustica*, 60, 277–282.

Epstein, M. (2007). An introduction to induced loudness reduction. *Journal of the Acoustical Society of America*, 122, EL74–80.

Epstein, M. J. (2011). Correlates of loudness. In M. Florentine, R. Fay, & A. Popper (Eds.), *Loudness* (pp. 89–108). New York: Springer Science+Business Media.

Epstein, M., & Florentine, M. (2005). A test of the equal-loudness-ratio hypothesis using cross-modality matching functions. *Journal of the Acoustical Society of America*, 118, 907–913.

Epstein, M., & Florentine, M. (2006a). Reaction time to 1- and 4-kHz tones as a function of sensation level in listeners with normal hearing. *Ear and Hearing*, 27, 424–429.

Epstein, M., & Florentine, M. (2006b). Loudness of brief tones measured by magnitude estimation and loudness matching. *Journal of the Acoustical Society of America*, 119, 1943–1945.

Epstein, M., & Gifford, E. (2006). A potential carry-over effect in the measurement of induced loudness reduction. *Journal of the Acoustical Society of* America, 120, 305–309.

Epstein, M., & Florentine, M. (2009). Binaural loudness summation for speech and tones presented via earphones and loudspeakers. *Ear and Hearing*, 30, 234–237.

Epstein, M., & Florentine, M. (2012). Binaural loudness summation for speech with and without visual cues presented via earphones and loudspeaker. *Journal of the Acoustical Society of America*, 131, 3981–3988.

Epstein, M., Florentine, M., & Buus, S. (2001). Measuring loudness of long and short tones using magnitude estimation. In E. Sommerfield, R. Kompass, & T. Lachmann (Eds.), *Fechner Day 2001* (pp. 231–235). Berlin: Pabst Science Publishers.

Epstein, M., Marozeau, J., & Florentine, M. (2006). Basilar-membrane activity and loudness. In D. E. Kornbrot, R. M. Msetfi, & A. W. MacRae (Eds.), *Fechner Day 2006: Proceedings of the 22nd Annual Meeting of the International Society for Psychophysics* (pp. 181–186). St. Albans, U.K.: The International Society for Psychophysics.

Fastl, H., & Florentine, M. (2011). Loudness in daily environments. In M. Florentine, R. Fay, & A. Popper (Eds.), *Loudness* (pp. 199–222). New York: Springer Science+Business Media.

Fletcher, H., & Munson, W. A. (1933). Loudness, its definition, measurement and calculation. *Journal of the Acoustical Society of America*, 5, 82–108.

Florentine, M. (2009). Advancements in psychophysics lead a new understanding of loudness in normal hearing and hearing loss. In M. A. Elliot, S. Antonijević, S. Berthaud, P. Mulcahy, C. Martyn, B. Bargery, & H. Schmidt (Eds.), *Fechner Day 2009: Proceedings of the 25th Annual Meeting of the International Society for Psychophysics* (pp. 83–88). Galway: Snap Printing.

Florentine, M. (2011). Loudness. In M. Florentine, R. Fay, & A. Popper (Eds.), *Loudness* (pp. 1–15). New York: Springer Science+Business Media.

Florentine, M. (2013). Importance of microscopic and macroscopic psychoacoustical approaches: An example from binaural loudness constancy. In *Proceedings of Inter-Noise 2013, Innsbruck, Austria*, Institute of Noise Control Engineering (SS68, 1228, invited).

Florentine, M., & Zwicker, E. (1979). A model of loudness summation applied to noise-induced hearing loss. *Hearing Research*, 1, 121–132.

Florentine, M., & Buus, S. (1981). An excitation-pattern model for intensity discrimination. *Journal of the Acoustical Society of America*, 70, 1646–1654.

Florentine, M., & Buus, S. (2001). Cochlear hearing loss reduces the dynamic range for loudness: Implications for hearing aids. *Proceedings of the 17th International Congress of Acoustics*, Volume IV (5B16.02).

Florentine, M., & Buus, S. (2002). Evidence for normal loudness growth near threshold in cochlear hearing loss. In L. Tranebjaerg, J. Christensen-Dalsgaard, T. Andersen, & T. Poulsen (Eds.), *Genetics and the function of the auditory system* (pp. 411–426). Denmark: GN ReSound.

Florentine, M., & Epstein, M. (2006). To honor Stevens and repeal his law for the auditory system. In *Proceedings of the 22nd Annual Meeting of the International Society for Psychophysics* (pp. 37–42). England: St. Albans.

Florentine, M., & Heinz, M. (2009). Audition: Loudness. In B. E. Goldstein (Ed.), *Encyclopedia of perception* (pp. 145–151). Thousand Oaks, CA: SAGE.

Florentine, M., & Epstein, M. (2010). Ecological loudness: Binaural loudness constancy. In: *Proceedings of the International Congress of Acoustics*, Sydney, Australia.

Florentine, M., & Epstein, M. (2012a). Measures of the ecological loudness of speech. In T. Dau, M. L. Jepsen, T. Poulsen, & J. C. Dalsgaard (Eds.), *Speech perception and auditory disorders: 3rd International Symposium on Auditory and Audiological Research* (EAN 9788799001330, pp. 87–94). Denmark: The Danavox Jubilee Foundation.

Florentine, M., & Epstein, M. (2012b). Insights gained from Rhona Hellman's loudness functions for individual listeners. In C. Burroughs, & S. Conlon (Eds.), *Proceedings of Inter-Noise 2012*, New York City, NY, USA, Institute of Noise Control Engineering.

Florentine, M., Buus, S., & Bonding, P. (1978). Loudness of complex sounds as a function of the standard stimulus and the number of components. *Journal of the Acoustical Society of America*, 64, 1036–1040.

Florentine, M., Buus, S., Scharf, B., & Zwicker, E. (1980). Frequency selectivity in normally-hearing and hearing-impaired observers. *Journal of Speech and Hearing Research*, 23, 113–132.

Florentine, M., Buus, S., & Poulsen, T. (1996). Temporal integration of loudness as a function of level. *Journal of the Acoustical Society of America*, 99, 1633–1644.

Florentine, M., Buus, S., & Hellman, R. (1997). A model of loudness summation applied to high-frequency hearing loss. In W. Jesteadt (Ed.), *Modeling sensorineural hearing loss* (pp. 187–197). Mahwah, NJ: Erlbaum.

Florentine, M., Buus, S., & Robinson, M. (1998). Temporal integration of loudness under partial masking. *Journal of the Acoustical Society of America*, 104, 999–1007.

Florentine, M., Mumby, S., & Cleveland, S. (2004). FYI: Softness imperceptions. *Hearing Health*, 20, 31–34.

Florentine, M., Buus, S., & Rosenberg, M. (2005). Reaction-time data support the existence of softness imperception in cochlear hearing loss. In S. McAdams, A. Cheveigne, L. Collet, & D. Pressnitzer. (Eds.), *Auditory signal processing: Physiology, psychoacoustics and models*. New York: Springer Science+Business Media.

Florentine, M., Popper, A., & Fay, R. (2011). *Loudness*. New York: Springer Science+Business Media.

Hellman, R. P. (1994). Relation between the growth of loudness and high-frequency excitation. *Journal of the Acoustical Society of America*, 96, 2655–2663.

Hellman, R. P., & Zwislocki, J. J. (1961). Some factors affecting the estimation of loudness. *Journal of the Acoustical Society of America*, 33, 687–694.

Larroque, M. F. (1901). Études de psycho-acoustique. *Comptes Rendus Hebdomadaires des Séances de l'Académie des Sciences*, 132, 821.

Marks, L. E. (1979). A theory of loudness and loudness judgments. *Psychological Review*, 86, 256–285.

Marks, L. E. (1992). The contingency of perceptual processing: Context modifies equal-loudness relations. *Psychological Science*, 3, 285–291.

Marks, L. E. (1994). Recalibrating the auditory system: The perception of loudness. *Journal of Experimental Psychology*, 20, 382–396.

Marks, L. E., & Florentine, M. (2011). Measurement of loudness, part I: Methods, problems, and pitfalls. In M. Florentine, R. Fay, & A. Popper (Eds.), *Loudness* (pp. 17–56). New York: Springer Science+Business Media.

Marozeau, J. (2011). Models of loudness. In M. Florentine, R. Fay, & A. Popper (Eds.), *Loudness* (pp. 261–284). New York: Springer Science+Business Media.

Marozeau, J., & Florentine, M. (2007). Loudness growth in individual listeners with hearing losses: A review. *Journal of the Acoustical Society of America*, 122, EL81–87.

Marozeau, J., & Epstein, M. (2008). Induced loudness reduction as function of frequency difference between test tone and inducer. *Perception and Psychophysics*, 70, 725–731.

Marozeau, J., & Florentine, M. (2009). Testing the binaural equal-loudness-ratio hypothesis with hearing-impaired listeners. *Journal of the Acoustical Society of America*, 126, 310–317.

Marozeau, J., Epstein, M., Florentine, M., & Daley, B. (2006). A test of the binaural equal-loudness-ratio hypothesis for tones. *Journal of the Acoustical Society of America*, 120, 3870–3877.

Miller, G. A. (1975). *Stanley Smith Stevens 1906-1973 – A biographical memoir*. Washington DC: National Academy of Sciences.

Miśkiewicz, A. (2004). From pianissimo to mezzopiano. *Hearing Health*, 20, 31.

Nieder, B., Buus, S., Cazals, Y., & Scharf, B. (2007). Loudness reduction induced by a contralateral tone. *Journal of the Acoustical Society of America*, 122, 35–37.

Nobel Foundation (1961). *The Nobel Prize in Physiology or Medicine 1961*. Retrieved from http://www.nobelprize.org/nobel_prizes/medicine/laureates/1961/index.html.

Odgaard, E. C., Arieh, Y., & Marks, L. E. (2004). Brighter noise: Sensory enhancement of perceived loudness by concurrent visual stimulation. *Cognitive, Affective, & Behavioral Neuroscience*, 4(2), 127–132.

Oxenham, A. J., & Bacon, S. P. (2004). Psychophysical manifestations of compression: Normal-hearing listeners. In S. P. Bacon, R. R. Fay, & A. N. Popper (Eds.), *Compression: From cochlea to cochlear implants* (pp. 62–106). New York: Springer Science+Business Media.

Scharf, B., & Fishken, D. (1970). Binaural summation of loudness reconsidered. *Journal of Experimental Psychology*, 86, 374–379.

Silva, I., & Florentine, M. (2006). Effect of adaptive psychophysical procedure on loudness matches. *Journal of the Acoustical Society of America*, 120, 2124–2131.

Sivonen, V. P., & Ellermeier, M. (2011). Binaural loudness. In M. Florentine, R. Fay, & A. Popper (Eds.), *Loudness* (pp. 169–198). New York: Springer Science+Business Media.

Stevens, J. C., & Hall, J. W. (1966). Brightness and loudness as a function of stimulus duration. *Perception of Psychophysiology*, 1, 319–327.

Stevens, S. S. (1934). The attributes of tones. *Proceedings of the National Academy of Science USA*, 20, 457–459.

Stevens, S. S., Davis, H., & Lurie, M. H. (1935). Localization of pitch perception on the basilar membrane. *Journal of General Psychology*, 13, 297–315.

Wagner, E., Florentine, M., Buus, S., & McCormack, J. (2004). Spectral loudness summation and simple reaction time. *Journal of the Acoustical Society of America*, 116, 1681–1686.

Whilby, S., Florentine, M., Wagner, E., & Marozeau, J. (2006). Monaural and binaural loudness of 5- and 200-ms tones in normal and impaired hearing. *Journal of the Acoustical Society of America*, 119, 3931–3939.

Chapter 9
Nonsyndromic Deafness: It Ain't Necessarily So

Thomas B. Friedman and Sheikh Riazuddin

Thomas B. Friedman and Sheikh Riazuddin

"It ain't necessarily so" is a song about doubt in George and Ira Gershwin's opera *Porgy and Bess*.

T.B. Friedman (✉)
Porter Neuroscience Research Center, 35A Convent Drive, Bethesda, MD 20892-3729, USA
e-mail: friedman@nidcd.nih.gov

S. Riazuddin
Centre of Excellence in Molecular Biology, University of the Punjab, Lahore 54550, Pakistan

Allama Iqbal Medical Research Centre, Jinnah Hospital, Lahore 54550, Pakistan
e-mail: riazuddin@aimrc.org

A.N. Popper and R.R. Fay (eds.), *Perspectives on Auditory Research*, Springer Handbook of Auditory Research 50, DOI 10.1007/978-1-4614-9102-6_9,

9.1 Introduction

Improvement in medical technology and a deeper understanding of human disease has lowered the clinical threshold for detecting many disorders, gradually enhancing our ability to identify medically relevant anomalies. Pathology is detected earlier, and with greater precision and finer definition. For instance, diagnoses of diabetes, hypertension, impaired vision, and inner ear abnormalities are occurring at an earlier age because of improved clinical awareness and detection. Concurrently, clinical phenotyping coupled to molecular genetic breakthroughs such as exome and genomic sequencing are accelerating the identification of potentially causative genetic variants that exert a large effect in human disorders. Studies of hereditary hearing loss from different perspectives have benefitted from these technical and conceptual developments.

The inner ear has many highly specialized, delicate, and intricate structures that are functionally unique and necessary for normal hearing. Assuming that unique structures require highly specialized proteins, perhaps limited in expression to the inner ear, it would not have been surprising that inherited abnormalities of such molecular machinery made of proteins expressed exclusively in the inner ear might be expected to cause hearing loss and no other accompanying disorder. When this happens it is referred to as nonsyndromic deafness. Many different recessive and dominant mutant alleles of a variety of genes have been associated with nonsyndromic deafness. However, the assumption that the expression of the "nonsyndromic deafness genes" is largely limited to the inner ear (or at the least not widely expressed in other organ systems) is not strictly correct. It is true that the inner ear is physiologically an exquisitely sensitive structure, composed of startlingly beautiful architectural arrangements of cells that transduce sound to signals then sent to the brain. However, the expression of the majority of macromolecules exploited for this remarkably complex development of the inner ear appears to have been genetically hijacked during the evolution of the auditory system from alternative functions that have remained in many other tissue types. This broad expression of many deafness genes was worrisome for us and offers a cautionary note when assuming a particular hearing loss is nonsyndromic.

Congenital and early onset hearing loss is a common neurosensory deficit that may be associated with other disorders. There are many syndromes that include hearing loss as one feature of a pleiotropic phenotype. The online database Mendelian Inheritance of Man (http://www.ncbi.nlm.nih.gov/sites/entrez?db=omim) lists hundreds of entries that include deafness alone or as one manifestation of a variety of syndromes. For example, Usher syndrome and Jervell and Lange–Nielsen syndrome are disorders involving the eye and heart, respectively, in addition to impaired hearing. These two syndromes are mentioned because progressive loss of vision due to retinitis pigmentosa (RP) may not be obvious in young children, and the life-threatening cardiac conduction defect of Jervell and Lange–Nielsen syndrome is sometimes discovered only posthumously. For a deaf child checking the electrical activity of the heart by an electrocardiogram (EKG) is not yet the standard of care. An EKG is a quick, easy and inexpensive test and in our opinion should become routine during a pediatric examination of a deaf person.

In contrast to such syndromic forms of hearing loss, nonsyndromic deafness is not accompanied by additional medically significant features, as is often believed at the time of ascertainment and reported in the literature. How might medically relevant issues accompanying hearing loss have been overlooked or ignored? We think that questions addressed to the subjects or relatives may be too narrowly focused on hearing ability. RP or prolongation of the QT cardiac conduction interval (delayed repolarization of the heart detected by an EKG test) are hidden from view, as are many other possible additional clinically relevant features. Hearing loss may be the tip of a clinical iceberg. Lacking a comprehensive physical examination, a strongly held belief that the diagnosis is nonsyndromic deafness may become a self-fulfilling prophecy. Fooling oneself is easy. In science, always doubt you got it right.

Study subjects are often visited where they dwell, which can be a considerable distance from medical staff and not allow for a wide-ranging physical exam to completely characterize the actual phenotype of the study subjects. But even more important, the underlying function of the causative disease gene responsible for hearing loss is likely to be unknown early in a study, further frustrating characterization of the full phenotype. Without knowledge of gene function and expression pattern in the various organ systems, a clinician investigator would need to be thinking of an inordinately large number of relevant questions about organs and tissues throughout the body.

Nearly 70 different nonsyndromic deafness genes have been reported to date (http://dnalab-www.uia.ac.be/dnalab/hhh/). How many of these different nonsyndromic deafness disorders, in reality, are syndromic remains to be determined. Early in a study of hereditary deafness, this can be an inherently difficult issue to resolve. For all these reasons, classification of human hereditary hearing loss as nonsyndromic should be considered provisional until there is adequate understanding of the normal function and expression pattern of a deafness gene that can then guide a focused clinical evaluation.

Functional studies can be technically difficult, especially when a gene expresses multiple mRNA splice isoforms, which is common for eukaryotic genes. Meticulous functional studies are also expensive and resource intensive. As a consequence, crucial mechanistic information about gene function that might inform a clinician about where to focus a physical examination often follows many years after disease gene identification. And that has meant a delay between disease gene discovery and expanded clinical insight concerning hitherto unrecognized syndromic manifestations.

A few published examples of mistaken assignment of hereditary deafness as nonsyndromic are described briefly, including an instance from our own recent work. In this chapter we also emphasize that most of the studies of hereditary hearing loss were initiated with the ascertainment of affected subjects from families where marriages often occur within the ethnic group (endogamy), from community isolates, or from large consanguineous families living almost exclusively outside of Western countries. In many countries study subjects often live far from sophisticated medical resources. Family ascertainment is tough work that is the bedrock for identification of the many novel mutated genes responsible for this monogenic disorder, and is underappreciated. In addition, ascertaining families segregating hereditary disorders is not hypothesis driven research and attracts support with difficulty. At times our own personal funds have subsidized this critically important first step of locating large families for molecular genetic studies.

9.2 Nonsyndromic Deafness

In the 1980s restriction fragment length polymorphism (RFLP) markers allowed genetic mapping of Mendelian disorders, although the process was sluggish by today's standards. Studies were initiated to map the chromosomal locations (loci) for human syndromic and nonsyndromic deafness genes. To achieve statistical significance, genetic linkage studies required large families with several affected and unaffected participants, and achieved stronger results when parents and grandparents were included. In Western countries there have been many opportunities to ascertain large families segregating nonsyndromic deafness as an autosomal dominant trait (DFNA). However, large families segregating autosomal recessive deafness (DFNB) are rare in the United States and Europe.

Large families or community isolates segregating recessive deafness were first studied in Bali, India, Saudi Arabia, Iran, Turkey, Tunisia, and Pakistan. Given these locations, it was uncertain if the mutated deafness genes, once discovered, would have relevance in Western countries. During this time, there was a concern that families ascertained in remote locations would reveal mutant genes that would turn out to be private to particular ethnic groups, and it was possible that the knowledge gained would contribute little clinically relevant information toward an understanding of deafness in the United States or Europe. In the absence of molecular genetic data it was not possible to refute such concerns. However, it seemed likely that basic knowledge about how hearing happens would expand through functional studies of the genes causing deafness, even if the mutations discovered were private and their clinical relevance limited to non-Western populations.

This matter was settled when it was shown that hereditary deafness in Western countries was caused by mutant alleles of nearly all the same deafness genes identified through studies of families in non-Western countries. This was also true for deafness genes segregating in community isolates and even in families living in a centuries-old Balinese village. It was the best of all possible outcomes. The vast majority of mutated genes associated with recessive deafness segregating in these families ascertained from remote areas were also found among individuals with hearing loss worldwide. For example, mutations of *MYO15A* encoding unconventional myosin XVA (Probst et al., 1998; Wang et al., 1998) were first discovered through studies of deaf and hearing individuals living in Bengkala, Indonesia (Friedman et al., 1995), and are now reported to be associated with recessive deafness in North and South America, the Middle East, Asia, and Europe (Nal et al., 2007) (Table 9.1).

Table 9.1 Selected genes underlying human syndromic and nonsyndromic deafness mentioned in this chapter[a]

Syndrome	Locus	Chromosomal location	Gene	Comments
Chudley–McCullough (CMCS)	*DFNB32* *DFNB82*	1p13.3	*GPSM2*	G-protein signaling modulator, unknown function in the ear
Deafness–dystonia–optic neuropathy	*MTS*	Xq22.1	*TIMM8A*	Essential for import of proteins into the mitochondrion inner membrane
Jervell and Lange–Nielsen	*JLNS1*	11p15.5-p15.4	*KCNQ1*	Potassium voltage-gated channel, KQT-like subfamily, member 1, mediates secretion of K^+ from stria vascularis marginal cells into scala media
Perrault	*DFNB81*	19p13.3	*CLPP*	Chambered protease, highly conserved mitochondrial protein
	PRLTS1	5q23.1	*HSD17B4*	17-β-hydroxysteroid dehydrogenase IV, enzyme involved in peroxisomal fatty acid β-oxidation
	PRLTS2	5q31.3	*HARS2*	Histidyl tRNA synthetase, highly conserved mitochondrial protein
	LARS2	3p21.31	*LARS2*	Leucyl-tRNA synthetase, mitochondrial protein
Usher, type 1	*USH1B*	11q13.5	*MYO7A*	Myosin VIIA, actin-based motor required for stereocilia function
	USH1C	11p15.1	*USH1C*	Harmonin, scaffolding protein required for stereocilia function
	USH1D	10q22.1	*CDH23*	Cadherin 23, component of the stereocilia tip link
	USH1F	10q21.1	*PCDH15*	Protocadherin 15, component of the stereocilia tip link
	USH1G	17q25.1	*USH1G*	Sans, scaffolding protein required for stereocilia function
	USH1J	15q25.1	*CIB2*	Calcium- and integrin-binding protein 2
Nonsyndromic deafness				
	DFNB1	13q12	*GJB2*	Connexin 26, gap junctions for intercellular communication
	DFNB3	17p11.2	*MYO15A*	Myosin motor protein 15 essential for stereocilia elongation and maintenance
	DFNB12	10q21-q22	*CDH23*	Cadherin 23, component of the tip link
	DFNB72	19p13.3	*GIPC3*	GAIP C-terminus-interacting protein 3, essential for survival of hair cells and spiral ganglion
	DFNB81	19p13.3	*CLPP*	Chambered protease, highly conserved mitochondrial protein

[a]Modified from Tables 1 and 3 of Griffith and Friedman (2014)

As of 2012, approximately 100 nonsyndromic recessive deafness loci have been genetically mapped using consanguineous families ascertained in non-Western countries. There are several reasons why India and Pakistan in particular have been troves of immense scientific value for geneticists studying monogenic disorders. Schools for the hearing impaired are good starting points to identify affected members of large families. In many countries in Asia and the Middle East, individuals marry relatives, often their first cousins. Generations of consanguineous marriages or endogamy within a community bring together mothers and fathers who are carriers (heterozygotes) of the same recessive mutations. Their progeny have a one in four chance of being homozygous for a particular recessive mutant allele. In a sufficiently large family, the location of the causative allele can be mapped to a chromosome using a genome-wide homozygosity mapping strategy (Friedman et al., 1995), and the gene eventually identified. Among many consanguineous families there is a tradition of large sibships and members frequently live close to one another. A shared environment helps to rule out major contributions of extrinsic factors to a phenotype. In addition, family members often are genuinely interested in understanding the reasons for their hearing loss and generously participate in basic research studies when direct benefits are not promised and evidence-informed therapies are not close at hand.

9.3 Rhetoric and Reality

Human nonsyndromic deafness can be caused by any one of a great many different large-effect mutant genes. A skeptic might ask about other clinically relevant issues considered and ruled out before accepting an assertion that a hearing impaired individual has no other co-segregating features indicative of a known or novel syndrome. Already there are a few reports where the initial supposition of nonsyndromic deafness was just the beginning of an evolving diagnosis.

9.3.1 X-Linked Nonsyndromic Deafness DFN1 Is Deafness–Dystonia–Optic Neuropathy Syndrome

X-linked genes account for only a few percent of all cases of inherited hearing loss. In 1960, a four-generation Norwegian family was described that segregated early-onset, progressive, nonsyndromic neurosensory deafness consistent with an X-linked pattern of inheritance (Mohr & Mageroy, 1960) and designated DFN1 (X-linked nonsyndromic deafness). There was an initial clinical evaluation of the seven affected males in this family, all of whom were described as just having hearing loss. However, when this family was restudied, extensive intrafamilial phenotypic variation was observed among affected males including progressive loss of vision, mental deterioration, dystonia, ataxia, and hip fractures in addition to the postlingual progressive hearing loss (Tranebjaerg et al., 1995). Some female carriers also seemed to manifest focal dystonias (Swerdlow & Wooten, 2001). The hearing loss component of this syndrome (DDON; deafness–dystonia–optic neuropathy syndrome; previously named Mohr–Tranebjaerg syndrome, MTS; MIM 304700) was due to degeneration of cochlear neurons observed in temporal bone specimens (Bahmad et al., 2007) (Table 1).

Linkage analysis was conducted and the *DDON* locus was mapped to Xq22 (Tranebjaerg et al., 1995). Subsequently, a variety of pathogenic mutations of *TIMM8A* (translocase of the *i*nner *m*itochondrial *m*embrane *8A*) were shown to cause DDON (Jin et al., 1996; Engl et al., 2012). *TIMM8A* mediates selective import of proteins from the cytosol to the inner mitochondria membrane (Koehler et al., 1999; Roesch et al., 2002). DDON syndrome is therefore a disorder of defective mitochondrial protein import. But the pathophysiology that accounts for the pleiotropy of DDON remains somewhat of a mystery when there are apparently phenotypically normal tissues and organs that also have substantial respiratory demands. An inner ear conditional *Timm8a1* mutant mouse or other animal models of DDON syndrome should provide insight. Nearly 20 years ago, the thesis that Tranebjaerg and co-authors stressed was the necessity of implementing detailed clinical evaluations of study subjects when a novel deafness gene is identified.

9.3.2 Nonsyndromic Deafness DFNB82 Is Chudley–McCullough Syndrome

A large Palestinian family segregating profound deafness as a recessive trait was ascertained in the West Bank and the phenotype was genetically mapped to a novel locus on chromosome 1p (Shahin et al., 2010) (Table 1). Subsequently, a homozygous nonsense mutation (p.Arg127X) of *GPSM* was reported as the cause of hearing loss DFNB82 initially reported to be nonsyndromic (Walsh et al., 2010). A Turkish family also segregating deafness was reported to be homozygous for a p.Gln562X allele of *GPSM* (Yariz et al., 2012). Subsequently, recessive mutations of *GPSM* were reported to be associated with Chudley–McCullough syndrome (CMS; MIM 604213). CMS is characterized by hearing loss and a prominent although partial failure in the development of the corpus callosum, an anatomic defect of the brain that appears not to translate into any obvious or consistently contemporaneous abnormalities. However, in two of the twelve affected subjects pharmacologically controlled seizures were present, a disorder known to be associated with corpus callosum defects (Doherty et al., 2012).

Because CMS subjects were found to have two mutant alleles of *GPSM in trans*, Doherty and coauthors used brain imaging to re-examine some of the deaf DFNB82 subjects. The Palestinian affected individuals and three Turkish deaf subjects all had brain abnormalities consistent with CMS. In these subjects there were no obvious developmental or behavioral deficits that might have indicated the possibility of additional clinically relevant features beyond hearing loss, the phenotype used for ascertainment. How many other study subjects reported to have nonsyndromic deafness actually have a new syndromic form of hearing loss that was overlooked or ignored?

9.3.3 Perrault Syndrome Not Nonsyndromic Deafness (DFNB81)

It was easy to assume hearing loss is nonsyndromic when we didn't ask spot-on questions. The study subject is certainly not at fault. Who would have thought there was a relationship between hearing loss and infertility when we genetically mapped deafness segregating in Pakistani family PKDF291? Family PKDF291 has four deaf female siblings, two normal hearing sibs, and normal hearing parents (Rehman et al., 2011). After discovering the pathogenic mutation causing deafness in this family, but before publishing this observation, William Newman in Manchester, UK identified a different mutation of the *CLPP* gene in a family segregating Perrault syndrome, a genetically heterogeneous disorder (Jenkinson et al., 2013). Perrault syndrome is a sex-influenced, autosomal recessive disorder characterized by sensorineural hearing loss in males and females, and gonadal dysgenesis only in females (Pierce et al., 2011). Armed with this information, the next step was to revisit members of the DFNB81 family and ask questions about fertility and request relevant tests for the levels of serum follicle-stimulating hormone (FSH), luteinizing hormone (LH), progesterone, and estradiol. Affected females of family PKDF291 were found to have the hormonal aberrations expected in Perrault syndrome. Interestingly, if the affected members of family PKD291 were instead males, nonsyndromic deafness would have correctly described the phenotype. On the other hand, females homozygous for the mutation had no menses and in effect are postmenopausal at a very young age (Jenkinson et al., 2013). The functional nexus between hearing loss and ovarian dysgenesis in Perrault syndrome remains an open question.

9.4 Allelic Mutations Can Cause Nonsyndromic or Syndromic Deafness

There are now several published examples of different mutations of the same gene causing syndromic deafness or nonsyndromic deafness, a heretical supposition a few decades ago. Such clinical heterogeneity results from the variable downstream impact of different mutations in the same gene. The severity of the phenotype is presumably directly related to the degree of gene disablement, while genetic modifier variants in the background (Riazuddin et al., 2000; Schultz et al., 2005) and environmental factors may also have a significant impact on the fully evolved phenotype, for better or for worse. A few examples illustrate this point.

Usher syndrome accounts for the majority of inherited human deaf-blindness. The defining features of the most severe presentation of Usher syndrome are hearing loss, peripheral vestibular areflexia, and progressive loss of vision due to retinitis pigmentosa (RP) that becomes noticeable in the second decade of life. Usher syndrome is inherited as a recessive disorder. Even knowing the molecular details of a mutated Usher syndrome gene may make it challenging to predict precisely the ultimate phenotype in young hearing impaired human subjects.

Christine Petit's group genetically mapped the nonsyndromic deafness locus DFNB12 to chromosome 10q21-22 (Chaib et al., 1996) while that same year, Richard Smith and colleagues (Wayne et al., 1996) reported that an Usher syndrome locus (*USH1D*) genetically mapped to roughly the same chromosomal interval. Our group then reported that some alleles of *CDH23* encoding cadherin 23 can cause only nonsyndromic deafness DFNB12 while other presumably more disabling mutations of *CDH23* cause Usher syndrome 1D (Bork et al., 2001). This observation is now well supported by reports from other investigators. A similar genotype–phenotype correlation was found for mutations of some of the other Usher syndrome genes including *USH1C*, *PCDH15*, and *CIB2*. Our working hypothesis is that loss of gene function (null mutation) results in Usher syndrome while hypomorphic alleles spare vision but cause hearing loss (Schultz et al., 2011). Why vision is maintained and hearing lost with some hypomorphic alleles of Usher syndrome genes is an unanswered, experimentally challenging question.

Without an eye examination by an ophthalmologist, the early signs of RP can be easily overlooked in a juvenile. A young deaf person with two predicted pathogenic missense mutations of an Usher gene may wish to know if vision will also be lost. The following examples illustrate this situation. Nonsyndromic deafness in the Ashkenazi Jewish population is predominantly caused by one particular recessive founder mutation (c.167delT) of *GJB2* (Morell et al., 1998). This frameshift mutation ablates the function of connexin 26 and is associated with well-documented nonsyndromic deafness DFNB1.

In the Ashkenazi Jewish community there is also a recessive founder mutation (p.R245X) of *PCDH15* that appears invariably to cause the most severe form of Usher syndrome (Ben-Yosef et al., 2003). In a young person, homozygosity for p.R245X may be incompletely diagnosed as nonsyndromic deafness (Brownstein et al., 2004). As an example, the underlying reasons for hearing loss were sought for 20 young Ashkenazi Jewish adolescents who were previously diagnosed with nonsyndromic deafness that was known to not be due to the c.167delT mutation of *GJB2* or any other mutant allele of *GJB2*. Subsequently, when the *PCDH15* gene was sequenced, 2 of the 20 children were found to be homozygous for p.R245X, foreshadowing loss of vision. The diagnosis of deafness was certainly correct, but incomplete, while early indications of impending vision loss due to RP was overlooked. The standard of care for young deaf *GJB2*-negative Ashkenazi Jewish children should now include an ophthalmological evaluation even if there is no family history of Usher syndrome.

Worldwide, pediatricians, otolaryngologists, and audiologists are more alert for the possibility of Usher syndrome in young individuals of any ethnicity with hearing loss, especially among children who exhibit delayed independent ambulation (walking), a condition possibly indicative of a vestibular dysfunction. Early indications of RP can be detected as a small deviation from a normal electroretinogram (ERG) in a young person. An ERG can detect electrical responses from photoreceptors (rods and cones) in the eye and other cell types of the retina. Timely habilitation of hearing through cochlear implantation of otherwise profoundly deaf Usher syndrome children often provides sufficient hearing when later loss of vision precludes sign language and lip reading.

An obvious question arose as to the phenotypic consequence of a person who has one DFNB12 mutation of *CDH23* on one chromosome and an Usher syndrome allele of *CDH23* on the other chromosome (referred to as the *trans*-configuration). Will this person develop RP? Is a nonsyndromic deafness allele of *CDH23* phenotypically recessive or phenotypically dominant to an Usher syndrome allele of *CDH23*? Although the number of subjects in the study was small, Schultz and co-authors addressed this question by characterizing the phenotype of individuals who were compound heterozygotes for an Usher allele and a DFNB12 allele of *CDH23*. The conclusion from this study is that such persons did not have a vestibular dysfunction nor did they develop RP. However, they were deaf. Therefore, a DFNB12 allele is phenotypically dominant to an USH1 allele in the eye and vestibular system. Although at present there is no assay for cadherin 23 function in the retina, it was proposed that one DFNB12 mutation of *CDH23* provides sufficient residual cadherin 23 function for normal retinal and vestibular labyrinth function but, because the subjects were deaf, this is inadequate for inner ear function. The genotype–phenotype relationship for mutations of *CDH23* is reason for optimism that a therapy can be developed to prevent or slow the progression of RP as partial function of defective cadherin 23 seems to be sufficient to preserve vision (Schultz et al., 2011).

9.5 Summary

We predict that there will be more examples of incompletely diagnosed nonsyndromic deafness in which the hearing loss, upon closer clinical examination, will be just one feature of a more complex phenotype. Intellectual flexibility and continuous conversation between subjects, basic scientists and clinicians will help to clarify these situations over time. In part, clues will come from in vitro functional analyses of deafness genes as well as detailed in vivo studies of mouse models. Admittedly, mouse models do not always recapitulate the full phenotypic spectrum of human syndromic forms of deafness. Although not necessarily so, one tipoff that the mutant phenotype may be more complex than just hearing loss alone is a wide pattern of expression of the wild type gene beyond the auditory system. Surprisingly, nearly ubiquitous expression of several presumably nonsyndromic deafness genes appears to be the rule rather than the exception (Schultz et al., 2009).

Our own recent experience, and the published examples described in the preceding text, suggests to us that quick pronouncements of the nonsyndromic nature of an inherited hearing loss might better be tempered with a degree of uncertainty commensurate with the depth of the initial clinical investigation, which is often sparse. In our case, the focus has been on hearing and vision, and thus the phenotype was nonsyndromic deafness or Usher syndrome. In the fullness of time, how many of the nearly hundred genes associated with nonsyndromic deafness will turn out to actually cause a syndromic form of hearing loss? Before revisiting affected individuals, a goal will be to empower collaborating clinicians with focused biological insight about pathologic and normal "deafness gene" functions from comprehensive studies of cognate mouse models. Germane clinical data beyond the auditory system can then be gathered. Correct and complete science is partial repayment for the generosity of human subjects in our studies and a prerequisite for contributions to the body of published knowledge that withstands the test of time.

Acknowledgments We thank Dennis Drayna, Andrew J. Griffith, and Julie Schultz for their comments and suggestions. SR is supported by Higher Education Commission, Islamabad, Pakistan. Intramural research fund DC000039-16 to TBF is from the National Institute on Deafness and Other Communication Disorders at the National Institutes of Health, Bethesda, MD.

References

Bahmad, F., Merchant, S. N., Nadol, J. B., & Tranebjaerg, L. (2007). Otopathology in Mohr-Tranebjaerg syndrome. *The Laryngoscope*, 117, 1202–1208.

Ben-Yosef, T., Ness, S. L., Madeo, A. C., Bar-Lev, A., Wolfman, J. H., Ahmed, Z. M., Desnick, R. J., Willner, J. P., Avraham, K. B., Ostrer, H, Oddoux, C., Griffith, A. J., & Friedman, T. B. (2003). A mutation of PCDH15 among Ashkenazi Jews with the type 1 Usher syndrome. *New England Journal of Medicine*, 348, 1664–1670.

Bork J. M., Peters, L. M., Riazuddin, S., Bernstein, S., Ahmed, Z. M., Ness, S. L., Polomeno, R., Ramesh, A., Schloss, M., Srisailpathy, C. R. S., Wayne, S., Bellman, S., Desmukh, D., Der Kaloustian, V. M., Lalwani, A., Riazuddin, S., Bitner-Glindzicz, M., Nance, W., Liu X-Z., Wistow, G., Smith R. J. H., Griffith A. J., Wilcox, E., Friedman, T. B., & Morell, R. J. (2001). Usher syndrome 1D and nonsyndromic autosomal recessive deafness DFNB12 are caused by allelic mutations of the novel cadherin-like gene CDH23. *The American Journal of Human Genetics*, 68, 26–37.

Brownstein, Z., Ben-Yosef, T., Dagan, O., Frydman, M., Abeliovich, D., Sagi, M., Abraham, F. A., Taitelbaum-Swead, R., Shohat, M., Hildesheimer, M., Friedman, T. B., & Avraham K. B. (2004). The R245X mutation of PCDH15 in Ashkenazi Jewish children diagnosed with nonsyndromic hearing loss foreshadows retinitis pigmentosa. *Pediatric Research*, 55, 995–1000.

Chaib, H., Place, C., Salem, N., Dode, C., Chardenoux, S., Weissenbach, J., el Zir, E., Loiselet, J., & Petit, C. (1996). Mapping of DFNB12, a gene for a non-syndromal autosomal recessive deafness, to chromosome 10q21-22. *Human Molecular Genetics*, 5, 1061–1064.

Doherty, D., Chudley, A. E., Coghlan, G., Ishak, G. E., Innes, A. M., Lemire, E. G., Rogers, R. C., Mhanni, A. A., Phelps, I. G., Jones, S. J., Zhan, S. H., Fejes, A. P., Shahin, H., Kanaan, M., Akay, H., Tekin, M., Triggs-Raine, B., & Zelinski, T. (2012). GPSM2 mutations cause the brain malformations and hearing loss in Chudley-McCullough syndrome. *American Journal of Human Genetics*, 90, 1088–1093.

Engl, G., Florian, S., Tranebjaerg, L., & Rapaport, D. (2012). Alterations in expression levels of deafness dystonia protein 1 affect mitochondrial morphology. *Human Molecular Genetics*, 21, 287–299.

Friedman, T. B., Liang, Y., Weber, J. L., Hinnant, J. T., Barber, T. D., Winata, S., Arhya, I. N., & Asher, J. H., Jr. (1995). A gene for congenital, recessive deafness DFNB3 maps to the pericentromeric region of chromosome 17. *Nature Genetics*, 9, 86–91.

Griffith, A. J., & Friedman, T. B. (2014). Hereditary hearing loss. In J. B. Snow, Jr. & P. A. Wackym (Eds.), Ballenger's otorhinolaryngology head and neck surgery, 18th ed. Beijing: People's Medical Publishing House-USA.

Jenkinson, E. M., Rehman, A. U., Tom Walsh, T., Clayton-Smith, J., Lee K., Robert J., Morell R. J., Drummond M. C., Khan S. N., Naeem M. A., Rauf B., Billington N., Schultz J. M., Urquhart, J. E., Lee, M. K., Berry, A., Hanley, N. A., Mehta, S., Cilliers, D., Clayton, P. E., Kingston, H., Miriam, J.. Smith, M. J., Warner, T. T., Black, G. C., Trump, D., Davis, J. R. E., Ahmad, W., Leal, S. M., Riazuddin S., King, M-C., Friedman, T. B., & Newman, W. G. (2013). Perrault syndrome is caused by recessive mutations in *CLPP*, encoding a mitochondrial ATP-dependent chambered protease. *The American Journal of Human Genetics*, 92, 605–613.

Jin, H., May, M., Tranebjaerg, L., Kendall, E., Fontan, G., Jackson, J., Subramony, S. H., Arena, F., Lubs, H., Smith, S., Stevenson, R., Schwartz, C., & Vetrie, D. (1996). A novel X-linked gene, DDP, shows mutations in families with deafness (DFN-1), dystonia, mental deficiency and blindness. *Nature Genetics*, 14, 177–180.

Koehler, C. M., Leuenberger, D., Merchant, S., Renold, A., Junne, T.., & Schatz, G. (1999). Human deafness dystonia syndrome is a mitochondrial disease. *Proceedings of the National Academy of Sciences of the USA*, 96, 2141–2146.

Mohr, J., & Mageroy, K. (1960). Sex-linked deafness of a possibly new type. *Acta Genetica et Statistica Medica*, 10, 54–62.

Morell, R. J., Kim, H. J., Hood, L. J., Goforth, L., Friderici, K., Fisher, R., Van Camp, G., Berlin, C. I., Oddoux, C., Ostrer, H., Keats, B., & Friedman, T. B. (1998). Mutations in the connexin 26 gene (*GJB2*) among Ashkenazi Jews with nonsyndromic recessive deafness. *New England Journal of Medicine*, 339, 1500–1505.

Nal, N., Ahmed, Z. M., Erkal, E., Alper, O. M., Luleci, G., Dinc, O., Waryah, A. M., Ain, Q., Tasneem, S., Husnain, T., Chattaraj, P., Riazuddin, S., Boger, E., Ghosh, M., Kabra, M., Morell, R. J., & Friedman, T. B. (2007). Mutational spectrum of *MYO15A*: The large N-terminal extension of myosin XVA is required for hearing. *Human Mutation*, 28, 1014–1019.

Pierce, S. B., Chisholm, K. M., Lynch, E. D., Lee, M. K., Walsh, T., Opitz, J. M., Li, W., Klevit, R. E., & King, M. C. (2011). Mutations in mitochondrial histidyl tRNA synthetase HARS2 cause ovarian dysgenesis and sensorineural hearing loss of Perrault syndrome. *Proceedings of the National Academy of Sciences of the USA*, 108, 6543–6548.

Probst, F. J., Fridell, R. A., Raphael, Y., Saunders, T. L., Wang, A., Liang, Y., Morell, R. J., Touchman, J. W., Lyons, R. H., Noben-Trauth, K., Friedman, T. B., & Camper, S. A. (1998). Correction of deafness in shaker-2 mice by an unconventional myosin in a BAC transgene. *Science*, 280, 1444–1447.

Rehman, A. U., Gul, K., Morell, R. J., Lee, K., Ahmed, Z. M., Riazuddin, S., Ali, R. A., Shahzad, M., Jaleel, A. U., Andrade, P. B., Khan, S. N., Khan, S., Brewer, C. C., Ahmad, W., Leal, S. M., & Friedman, T. B. (2011). Mutations of *GIPC3* cause nonsyndromic hearing loss DFNB72 but not DFNB81 that also maps to chromosome 19p. *Human Genetics*, 130, 759–765.

Riazuddin, S., Castelein, C. M., Ahmed, Z. M., Lalwani, A. K., Mastroianni, M. A., Naz, S., Smith, T. N., Liburd, N. A., Friedman, T. B., Griffith, A. J., & Wilcox, E. R. (2000). Dominant modifier DFNM1 suppresses recessive deafness DFNB26. *Nature* Genetics, 26, 431–434.

Roesch, K.., Curran, S. P., Tranebjaerg, L., & Koehler, C. M. (2002). Human deafness dystonia syndrome is caused by a defect in assembly of the DDP1/TIMM8a-TIMM13 complex. *Human Molecular Genetics* 11, 477–486.

Schultz, J. M., Yang, Y., Caride, A. J., Filoteo, A. G., Penheiter, A. R., Lagziel, A., Morell, R. J., Mohiddin, S. A., Fananapazir, L., Madeo, A. C., Penniston, J. T., & Griffith, A. J. (2005). Modification of human hearing loss by plasma-membrane calcium pump PMCA2. *New England Journal of Medicine*, 352, 1557–1564.

Schultz, J. M., Khan, S. N., Ahmed, Z. M., Riazuddin, S., Waryah, A. M., Chhatre, D., Starost, M. F., Ploplis, B., Buckley, S., Velasquez, D., Kabra, M., Lee, K., Hassan, M. J., Ali, G., Ansar,

M., Ghosh, M., Wilcox, E. R., Ahmad, W., Merlino, G., Leal, S. M., Friedman, T. B., & Morell, R. J. (2009). Noncoding mutations of HGF are associated with nonsyndromic hearing loss, DFNB39. *American Journal of Human Genetics*, 85, 25–39.

Schultz, J. M., Bhatti, R., Madeo, A. C., Turriff, A., Muskett, J. A., Zalewski, C. K., King, K. A., Ahmed, Z. M., Riazuddin, S., Ahmad, N., Hussain, Z., Qasim, M., Kahn, S. N., Meltzer, M. R., Liu, X. Z., Munisamy, M., Ghosh, M., Rehm, H. L., Tsilou, E. T., Griffith, A. J., Zein, W. M., Brewer, C. C., & Friedman, T. B. (2011). Allelic hierarchy of CDH23 mutations causing non-syndromic deafness DFNB12 or Usher syndrome USH1D in compound heterozygotes. *Journal of Medical Genetics*, 48, 767–775.

Shahin, H., Walsh, T., Rayyan, A. A., Lee, M. K., Higgins, J., Dickel, D., Lewis, K., Thompson, J., Baker, C., Nord, A. S., Stray, S., Gurwitz, D., Avraham, K. B., King, M. C., & Kanaan, M. (2010). Five novel loci for inherited hearing loss mapped by SNP-based homozygosity profiles in Palestinian families. *European Journal of Human Genetics*, 18, 407–413.

Swerdlow, R. H., & Wooten, G. F. (2001). A novel deafness/dystonia peptide gene mutation that causes dystonia in female carriers of Mohr-Tranebjaerg syndrome. *Annals Neurology*, 50, 537–540.

Tranebjaerg, L., Schwartz, C., Eriksen, H., Andreasson, S., Ponjavic, V., Dahl, A., Stevenson, R. E., May, M., Arena, F., & Barker, D. (1995). A new X linked recessive deafness syndrome with blindness, dystonia, fractures, and mental deficiency is linked to Xq22. *Journal of Medical Genetics*, 32, 257–263.

Walsh, T., Shahin, H., Elkan-Miller, T., Lee, M. K., Thornton, A. M., Roeb, W., Abu Rayyan, A., Loulus, S., Avraham, K. B., King, M-C., & Kanaan, M. (2010). Whole exome sequencing and homozygosity mapping identify mutation in the cell polarity protein GPSM2 as the cause of nonsyndromic hearing loss DFNB82. American *Journal of Human Genetics*, 87, 90–94.

Wang, A., Liang, Y., Fridell, R. A., Probst, F. J., Wilcox, E. R., Touchman, J. W., Morton, C. C., Morell, R. J., Noben-Trauth, K., Camper, S. A., & Friedman, T. B. (1998). Association of unconventional myosin MYO15 mutations with human nonsyndromic deafness DFNB3. *Science*, 280, 1447–1451.

Wayne, S.., Der Kaloustian, V. M., Schloss, M., Polomeno, R., Scott, D. A., Hejtmancik, J. F., Sheffield, V. C., & Smith, R. J. (1996). Localization of the Usher syndrome type ID gene (Ush1D) to chromosome 10. *Human Molecular Genetics*, 5, 1689–1692.

Yariz, K. O., Walsh, T., Akay, H., Duman, D., Akkaynak, A. C., King, M. C., & Tekin, M. (2012). A truncating mutation in GPSM2 is associated with recessive non-syndromic hearing loss. *Clinical Genetics*, 81, 289–293.

Chapter 10
Evolving Mechanosensory Hair Cells to Hearing Organs by Altering Genes and Their Expression: The Molecular and Cellular Basis of Inner Ear and Auditory Organ Evolution and Development

Bernd Fritzsch and M.-Dominique Crapon de Caprona

Bernd Fritzsch

M.-Dominique Crapon de Caprona

B. Fritzsch (✉) • M.-D. Crapon de Caprona
Department of Biology, College of Liberal Arts, University of Iowa,
Iowa City, IA 52242, USA
e-mail: bernd-fritzsch@uiowa.edu; Sloughi@gowebway.com

A.N. Popper and R.R. Fay (eds.), *Perspectives on Auditory Research*, Springer Handbook of Auditory Research 50, DOI 10.1007/978-1-4614-9102-6_10,

10.1 Introduction

After having discovered in the early 1980s that some amphibians are electroreceptive and that this novel sense had been dismissed over 100 years by many of the great pioneers in developmental biology (Fritzsch, 1981; Munz et al., 1982; Fritzsch & Wahnschaffe, 1983), including Nobel Prize winners, we were looking for another challenge. Major questions that fascinated us some 30 years ago were the apparently unresolved origin and adaptive diversification of animal sensory systems such as the eye or the ear, including the evolution of the human auditory and speech system that enables us to communicate. That problem was eloquently pointed out some 150 years ago for the eye: "To suppose that the eye with all its inimitable contrivancies for adjusting the focus to different distances, for admitting different amounts of light, and for correction of spherical and chromatic aberration, could have been formed by natural selection, seems, I freely confess, absurd in the highest possible degrees" (Darwin, 1859).

Unfortunately, Darwin chose for his example a sensory system whose evolution is difficult to trace in vertebrates, as all vertebrates (except for certain cave-dwelling fish) have eyes that show limited evolutionary changes in the retina structure, eye muscles, and motor innervation (Fritzsch, 1991; Fritzsch & Glover, 2007). In contrast to eyes, the auditory sensory system has long been known for its transformative changes in the middle ear, which turn bones associated with chewing into middle ear ossicles used for sound conduction. This evolutionary transformation is possible through ontogenetic alteration of bone precursors that have lost their original function to acquire a novel function (Reichert, 1837). Ideas about the evolution of middle ear ossicles were combined with beliefs that the ear serves only an auditory function. Only in the late 19th century were the vestibular functions of the ear discovered and eventually the "vestibular first" hypothesis was formulated, arguing that during evolution a purely vestibular ear acquires an additional auditory function (de Burlet, 1934). Despite these early interests in auditory evolution, the ear has not become as much a part of mainstream evolutionary theory during the last 20 years as has the eye. We aim here to show that vertebrate ear evolution provides an outstanding example to understand how humble cellular beginnings can be transformed into the most highly ordered cellular assembly of the mammalian body: an organ capable of extracting specific frequencies over a wide range to enjoy, in extreme cases, perfect pitch perception.

Research in the 1960s and 1970s concentrated on the physiology of the auditory system, adding little to the evolutionary origin of the vertebrate mechanosensory cell, the ear, and the auditory system, beyond repeating older speculations. To boot, most of the data presented in the 1970s were not put into an evolutionary context following cladistic principles to order the data against increasingly better known phylogenetic relationship between vertebrate taxa. Still, during the past 40 years many novel aspects pertaining to the evolution of cells, ears, and auditory organs have been uncovered, providing new insights. It is an interesting coincidence for this chapter that some initial summaries of earlier work (Fritzsch, 1992) were published in one of the first books in the SHAR series of Springer books edited by some of the editors of the current volume. Other chapters in SHAR series books summarize data on efferent development (Simmons et al., 2011), central auditory system development and evolution (Grothe et al., 2004), and theoretical considerations of auditory system evolution (Fritzsch, 1999). More recent conceptual work summarized earlier hypotheses in the context of our current understanding of the cellular basis of ear development and evolution (Pan et al., 2012). Readers are referred to these more detailed summaries so we can maintain here the flow of the narrative without overdue disruptions by citations.

We focus here on the conceptual aspects underlying all of this work (Wagner, 2011): The quest is to understand how sensory novelties arise in vertebrates (or any animal) through molecular evolution followed by natural selection, precisely as anticipated by Darwin 150 years ago. Following the central dogma of biology, namely that information flows from genes through messenger RNA to proteins, decoding the evolution of an organ must ultimately establish the genetic basis for the developmental alterations that result in observed adult changes. We narrate in this chapter how our understanding evolved and what the open questions are. We propose that ultimately the inherited developmental changes need to be explained at a mechanistic, molecular level to understand how newly evolved adult morphologies can be selected for, stabilizing genetic changes in subsequent generations.

Progress in each of the three themes of this overview (hair cell, ear, and auditory system evolution) can be divided into three, partially overlapping phases: (1) an initial phase characterized by the search for the evolutionary pattern of adult changes underlying each novelty; (2) a second phase characterized by the arrangement of adult evolutionary changes as a series of ontogenetic changes; and (3) a third, still ongoing, phase analyzing the molecular basis toward an understanding of evolutionary alterations of gene sequence and expression (and thus signaling) to modify the outcome of development and thus generate the selected adult morphology.

In parallel with our efforts to understand evolution and development of these systems, others have clarified the functional aspects of these systems. Their efforts have added considerably to the depth of our current understanding of the physical and physiological basis of selection pressure acting on development, producing different adult morphologies that define the auditory and vestibular system (Lewis & Fay, 2004).

10.2 From Single Cells to Organs: The Evolution of Hair Cells

Since the earliest descriptions of the vertebrate "hair cell" with its asymmetric staircase of stereocilia next to a kinocilium, the hair cell of the lateral line system and inner ear has served as the prototype of highly specialized cellular receptors unique to vertebrates. How it evolved was not even discussed, the research in the 1960s and 1970s being more concerned with the physiological understanding of this cell's unique features, not shared by any other cell type in either vertebrates or invertebrates.

The first insight into the apparent similarity between the vertebrate hair cells and its possible ancestral cells we are aware of was provided in a review by Jørgensen (1989). This review specified that the apparently unique vertebrate mechanosensory hair cell might have evolved from other cell types that did not share the stereotyped staircase arrangement of stereocilia, the hallmark of vertebrate hair cells (Jørgensen, 1989). Three findings helped to trigger this novel concept.

First, the newly discovered "hair cells" of electroreceptive organs showed a highly variable pattern of microvilli and kinocilia and intense discussions about the significance of this variability was rampant in the 1980s. Because these organs were believed by us and others to be derived from mechanosensory hair cells containing neuromast-like organs, these cells forced others and us to abandon the strict rule of mechanosensory hair cell organization as the only acceptable vertebrate "hair cell." Once the morphological variation of a theme within vertebrates was accepted as belonging to a single cell type, it was a logical step for all of us to expand that relaxed morphological concept to nonvertebrate sensory cells of mostly unknown function, and Jørgensen (1989) was the first to formalize these insights.

Second, developmental studies showed that the staircase arrangement of stereocilia, so characteristic of the adult vertebrate mechanosensory hair cell, did not develop as such. Rather, all hair cells start out in development with a central kinocilium surrounded by microvilli (Fig. 10.1). Only later in development, under the influence of a planar cell patterning signal, is the kinocilium moved into an eccentric position. During this developmental reorganization other, yet to be identified signals turn microvilli into actin-filled stereocilia of variable length and diameter. This plastic cellular morphology during development implied to us that evolutionary changes in hair cell organization are also possible. Like developmental changes, evolutionary changes could sequentially modify the developmental program to evolve the seemingly invariable adult structure of the vertebrate mechanosensory hair cell out of a morphologically different precursor cell.

Third, continuing work on nonvertebrates identified hair cell–like cells with more or less eccentric kinocilia and multiple microvilli of variable diameter. For example, jellyfish cells sensitive to mechanical stimulation may have a central kinocilium surrounded by somewhat variable stereocilia. These cells either trigger the release of nettle capsules, or may function as lateral line like organs, or form a complex statocyst for gravistatic sensing. Obviously, mechanosensation with somewhat asymmetric microvilli surrounding a kinocilium evolved long before bilaterian animals such as vertebrates evolved (Fig. 10.1). More importantly, these cells already were aggregated occasionally into complex organs serving the same functions known for vertebrate mechanosensors: perception of water movement across the surface (e.g., lateral line organs), or perceiving gravity mediated displacement (e.g., the gravistatic receptors of the vestibular part of the vertebrate ear).

Combined, these data painted for us an evolutionary picture suggesting a progressive morphological transformation of an ancestral mechanosensory cell into the vertebrate hair cell with its highly directional mechanosensation. We also concluded that development seemingly repeated evolution during mammalian hair cell development: hair cells start out with a central kinocilium surrounded by microvilli like

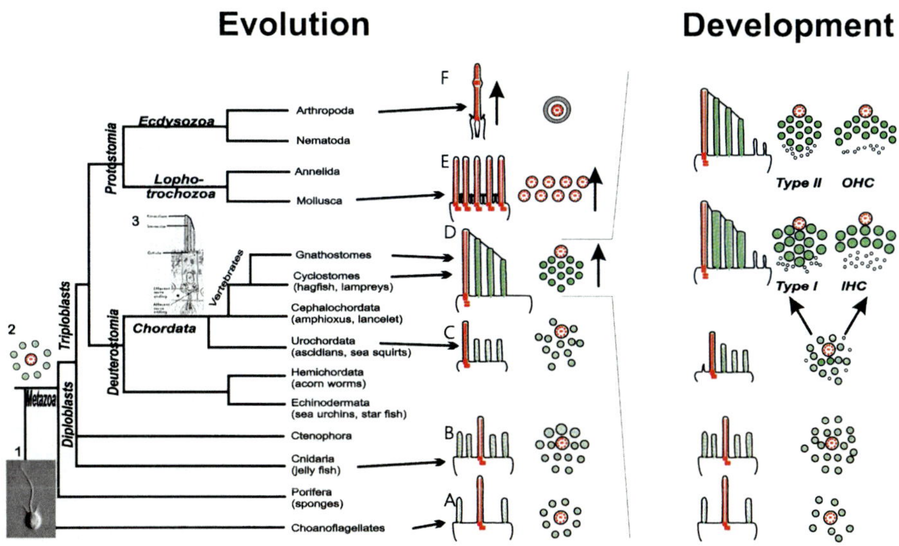

Fig. 10.1 Evolution of mechanosensory cells. Kinocilia (red) and microvilli (light blue) of known or suspected mechanosensory cells in various eukaryotic unicellular (1) and multicellular (3) organisms are shown. Orthologs of structural genes relevant for mechanosensation or for development of polarity such as actin, tubulin, rare myosins, cadherins, espin, β- catenin, and *Wnt* genes and several transcription factors are known in diploblasts (2) and various triploblasts and are thus likely ancestral to vertebrates. Note that the single-celled ancestor of all multicellular animals, the choanoflagellates (1), has a single kinocilium surrounded by microvilli with actin filaments (**A**). In some diploblasts, the central kinocilium is surrounded by an asymmetric assembly of microvilli, potentially providing directional sensitivity (**B**). Among deuterostomes, urochordates have sensory cells with a kinocilium and asymmetrically arranged microvilli. Vertebrates are unique with a highly polarized, organ-pipe assembly of actin-rich stereocilia, connected via tip links with each other. Mammalian hair cells develop their stereocilia in a process that starts with a central kinocilium surrounded by few microvilli. As the number of microvilli increases, the kinocilium moves into an off-center position and eventually toward one end of the developing hair cell. Microvilli in front of the moving kinocilium become reduced and eventually all disappear. In contrast, microvilli trailing the kinocilium grow in length, thickness, and actin content to turn into stereocilia. As the kinocilium reaches its acentric position, the polarity matures, with the longest stereocilia being next to the kinocilium. Development diverges through unknown molecular means to generate the four different hair cells found in the mammalian sensory epithelia. Type I and inner hair cells develop thick stereocilia and the characteristic bundles which are C-shaped for inner hair cells. Other vestibular hair cells develop as Type II and outer hair cells with thinner stereocilia. In addition, the organization of the stereocilia in the outer hair cells forms a characteristic M shape, with the kinocilium in the inflection of the M. Later in development small microvilli all but disappear and the kinocilium is resorbed in inner and outer hair cells. It appears that ampullary electroreceptive cells could be viewed as developmentally truncated mechanosensory hair cells that adopt a different phenotype without stereocilia development (Duncan & Fritzsch, 2012)

the single cell ancestor of animals, the choanoflagellates. Only later in development does the kinocilium move into an eccentric position, followed by the actin accumulation in microvilli, turning them into stereocilia. We and others also provided data suggesting that electroreceptive hair cells are derived from the same placodal material as mechanosensory hair cells. However, electrosensory "hair cells" never differentiate the microvilli into stereocilia. Some electrosensory hair cells look to us like vertebrate mechanosensory hair cell arrested in early development.

Although this perspective emerged for us in the late 1980s, it remained "non-mainstream" because the evidence left room for alternative explanations. Such alternative explanations revolved around invoking multiple convergent evolution of somewhat similar looking cell types. In part the arguments focused on issues that are unrelated to homology, namely the potential function of the cells in question. Conceptually, if we are to accept the organization of the stereocilia and kinocilium of a typical vertebrate mechanosensory hair cell as the divide between vertebrates and nonvertebrates, we generate an unresolvable conundrum that leads to the conclusion that vertebrate hair cells evolved de novo as they are now, without any trace in nonvertebrate ancestors. In contrast, if we accept our relaxed interpretation outlined in the preceding paragraphs, we can identify cells with microvilli and a central kinocilium as obvious developmental and likely evolutionary precursors of vertebrate hair cells (such as single-celled choanoflagellates). We propose that transformative developmental changes can evolve the nondirectional (mechanosensory) cell of ancestral pre-vertebrates into the directionally sensitive mechanosensory cell of vertebrates.

Our assumption is in line with the evolution at large, which seemingly transforms cells from humble beginnings into cells with increasingly complex morphologies such as the apical tuft of mechanotransducing stereocilia. Interestingly, beyond the comparative and developmental argument across phyla and, within vertebrates, across organs, there are also the very real data on the "lateral line" of hagfish. This "lateral line" has sensory cells with a kinocilium surrounded by microvilli, much like many sensory cells in invertebrates. Unfortunately, even these data can be interpreted either as an atavism of ancestral conditions or as a novel feature of unknown functional significance (Braun & Northcutt, 1997). It could even indicate that these cells are not mechanosensors at all but rather electroreceptive "hair cells" owing to the similarity with amphibian ampullary electroreceptors (Fritzsch & Wahnschaffe, 1983). Mammalian hair cells can be transformed into such cells after loss of certain genes (Jahan et al., 2012). We interpret this as an atavism, arresting mechanosensory hair cell development in hagfish at the level equivalent to adult cells in other organs or animals.

We realized already 15 years ago that breaking this stalemate of opinions that developed over the past 20 years required a novel approach. This approach needed to go beyond the plausibility of morphological transformations. This approach needed to establish molecular similarity to consolidate or refute the apparent morphocline that we proposed leading from single-cell ancestors to the mammalian hair cell. Such a novel approach was provided during the last 20 years by the molecular biology revolution. These novel bioinformatics and molecular technologies could identify genes as homologous based on their sequence similarity. These emerging techniques allowed researchers also to establish the essential functions for these genes, mostly through loss of function approaches (the gene knockout technique) or gain of function approaches (knockin of one gene into the locus of another gene or overexpressing a gene using promoter fragments). The accelerating technical breakthroughs in which we participated has identified multiple genes that are expressed in morphologically different sensory cells across phyla. In the past, certain sensory cells, identifiable as being related by the expression of closely related genes, would not have been homologized based on their different morphologies. For example, the fly chordotonal sensory cells and the mammalian hair cell are morphologically very dissimilar but express multiple conserved genes (Fritzsch et al., 2010; Senthilan et al., 2012).

Foremost among the genes that we propose to be useful to establish molecular homology are all genes needed for hair cell differentiation. We propose that such critical genes connect the development of mechanosensory cells with their evolution. In our opinion, this conservation of developmentally essential and homologous genes indicate what has been phrased as "deep molecular homology" in other systems such as the limbs (Shubin et al., 2009) and the eyes (Gehring, 2011). In essence, we propose that all sensory cells that share the molecular developmental module found in vertebrates, flies (Senthilan et al., 2012), or other animals evolved from a common ancestral cell, possibly identical with the single-celled ancestor of all multicellular animals (Fig. 10.1).

The primary gene initiating mechanosensory hair cell development in vertebrates is Atonal homolog 1 (*Atoh1*). *Atoh1*-related genes are expressed in mechanosensory cells across phyla and are essential for their differentiation. Further, these homologous genes can be experimentally exchanged bidirectionally between flies and mammals to rescue normal hair cell/mechanosensory cell development in the inner ear or the Johnston's organ (the fly hearing organ) but also in other cells that express this gene such as cerebellar granule cells. Exchanging an essential gene needed for either mammalian or fly mechanosensory cell development shows that either can function to initiate development in a foreign cellular context but differentiates hair cells only in the context of expression in hair cell precursors. In our opinion, this indicates that despite morphological dissimilarities of the adult cell, the genetic context of the cellular development is somewhat conserved between these cells. Therefore, each transcription factor can drive the development in either cell type, albeit toward an outcome that is defined by the different molecular context within which each gene is signaling. In other words, *Atoh1/atonal* functions as general differentiation signals and specificity in terms of hair cell/chordotonal organ cell differentiation is provided by the molecular context in a given cell.

We therefore tested this alternative explanation, which implies that any gene of the *Atoh1* family of transcription factors can initiate hair cell differentiation and that it is the context of other factors that specifies the outcome. Generating a novel mouse line, we could recently dismiss this possible alternative explanation (Jahan et al., 2012). Specifically, we replaced *Atoh1* with the closely related mammalian transcription factor *Neurog1* instead of the fly *atonal* gene. *Neurog1* is needed for neuronal formation in ear development and affects hair cells only indirectly through a presumed clonal relationship of neurons with hair cells. However, in contrast to the rescue of hair cell development by replacing the mammalian *Atoh1* gene with the fly ortholog *atonal*, replacing it with this closely related mammalian gene does not lead to either neuronal or hair cell differentiation of the mammalian hair cell precursors. Instead, cells remain for a short while partially differentiated (central kinocilia surrounded by multiple microvilli) before they die. In our opinion, this demonstrates that the fly *atonal* gene has indeed conserved signal equivalence to the mammalian *Atoh1* gene and is its true ortholog. In contrast, at least one mammalian member of the Atoh family of genes, *Neurog1*, does not have the capacity to initiate hair cell differentiation.

We therefore feel justified to suggest that these experiments have established that sensory cells expressing these homologous genes can be considered to be derived from the same ancestral cell type. This may also imply that these ancestral cells may have been mechanosensitive, as transient expression of *Atoh1* leads only to cellular differentiation without development of mechanosensory capacity (Pan et al., 2012). Indeed, further support for such an idea comes from data showing transformation of neurons into hair cells when *Atoh1* is continuously misexpressed in ganglion neurons due to removal of the *Atoh1* expression inhibition via another transcription factor, *Neurod1*. Combined, these data suggest to us that *Atoh1* differentiates hair cells only in the context of ear-derived cells (either hair cells or neurons). In contrast, in other tissue lacking unknown additional factors specific to the neurosensory cells of the ear and the fly hearing system, *Atoh1* helps to differentiate neurons. Thus *Atoh1* is necessary but not sufficient to differentiate hair cells as many cells that express *Atoh1* do not differentiate as hair cells, not even in the ear.

Although this experimental fact is already a strong argument for both molecular and cellular homology across phyla despite morphological dissimilarities, more recent data showed that specific micro RNAs (miRs) are also uniquely associated with sensory cells across phyla. These small, noncoding RNAs have emerged as a major driver for cellular development and their sequence conservation across phyla can be extremely high. For example, miR-124 is a neuronal specific miR that has identical nucleotide sequences in all triploblasts, suggesting that neurons (and miR-124) evolved in bilaterian ancestors only once. Moreover, all neurons require this miR for their development, and misexpression of this miR can convert mesoderm-derived fibroblasts into neurons. It is important to note that absence of all micro RNAs disrupts neurosensory cell development in mice and fish, whereas other cell types are less affected by the absence of miRs. Much like the "neuro-miR" 124, sensory cell specific miRs exist and are needed for hair cell development. Moreover, even a single nucleotide change in the 21 nucleotide long sequence of one of these hair cell specific miRs can cause deafness, further underlining the critical role these highly conserved regulators of cellular development have. The highly conserved sequence also highlights that these miRs have deep molecular homology across phyla and indicates that sensory cells expressing these hair cell specific miRs may be as homologous across phyla as are neurons of bilaterian animals based on the uniform expression of miR-124. Importantly, both hair cell and neuronal miRs are expressed in the neurons of the inner ear. However, hair cells express only hair cell miRs, but not neuronal miRs.

We suggest that the ability of *Atoh1* to convert neurons into hair cells is due to hair cell related miRs, also found in inner ear neurons. In contrast, converting hair cell precursors into neurons with a neuronal specific transcription factor will not work in our opinion. This should be so because no neuronal specific miRs are expressed in hair cells, thus blocking the differentiation of hair cells as neurons. Interestingly, once a certain degree of development is achieved, some misexpressed genes can apparently maintain hair cells. Despite the fact that these genes cannot initiate hair cell differentiation, they can stabilize viability, indicating that there are differences in the gene networks that initiate differentiation and maintain hair cells.

During the past 10 years, many other genes have been identified in flies and mice in which mutations result in severe disruption of normal development if not loss of sensory cells. Furthermore, many animals have family members of these genes and at least some are co-expressed in their sensory cells. Unfortunately, no studies have yet introduced jellyfish genes of the *Atoh1/atonal* family into a mammalian locus to check the ability of those genes with a very different sequence to functionally replace the mammalian *Atoh1* gene. Given that all transcription factors will cooperate in complex networks, such an experiment could identify the stability of networks against mutational perturbation, an essential feature for innovations to occur (Wagner, 2011). Indeed, transformational identity of a developmental gene network across phyla is the underlying molecular cause of morphological homology. Mechanosensory hair cells are in our opinion an excellent example for this basic biological concept.

Although the details of the cellular and molecular evolution of vertebrate hair cells still leave some room for alternative explanations, we are convinced that the emerging consensus is already clear: (Mechano)sensory cells form not only a morphological continuum across phyla but do so also at the molecular level with multiple interacting genes belonging to evolving developmental modules. Gene replacement studies demonstrated that these orthologous genes are functionally conserved in a yet to be fully defined developmental module providing deep molecular homology across phyla. What remains to be shown is how the interactions of these genes with other partners, and the evolution of downstream genes, changed to result in the variations observed in the adult cellular morphologies across phyla. Future research will have to establish the evolutionary changes not only in single genes. It will also have to establish how these changes alter the network of interacting transcription factors through qualitative and quantitative changes. Given that some of the genes identified thus far may regulate hundreds of downstream genes, the complexity of these gene networks and their robustness, and inherent "evolvability" (Wagner, 2011) require extensive future work.

One of the most puzzling aspects of the sensory cell evolution will be to reconcile the apparent molecular differences among animals (different numbers of bHLH genes, radical differences in miRs) with the apparent high level of morphological conservation between mechanosensory cells in coelenterates, tunicates, and to a certain extent, vertebrates. Based on these discrepancies it appears that we still have ways to go before we can reconcile the emerging concepts of cellular and molecular deep homologies across phyla with a molecular, mechanistic explanation.

As an aside, the early promise of one of those genes as a potential master gene for hair cell development (*Atoh1*) led to the idea that expression of this gene might suffice to regenerate hair cells in mammal. However, initial success using viral transfection or molecular manipulations have thus far not resulted in stable differentiation of hair cells. Obviously, as already shown by overexpression of *Atoh1* in frog development some 15 years ago, without proper molecular context *Atoh1* may generate neurons but not hair cells. Even if hair cells are generated they are of vestibular instead of cochlear variety or hair cells die soon after their development. Understanding the evolution of the minimal developmental gene network needed to differentiate the vertebrate mechanosensory hair cell will not only have ultimate, heuristic value to understand mammalian hair cell evolution through a molecularly altered development. Understanding hair cell evolution at a molecular level will also define a minimally essential gene network needed to differentiate hair cells as the proximate gain of this endeavor. Future work will have to establish how molecularly similar developmental processes lead to different hair cell types with different physiological properties and different level of susceptibility to ototoxic drugs or sound. It is only a complete understanding of the developmental gene network that will guarantee the development of, for example, inner hair cells precisely at the right position in the organ of Corti to functionally regenerate a lost organ of Corti, including long term viable mechanotransducing hair cells. Current attempts to reconstitute a lost organ of Corti are nowhere near a solution to this topological problem.

10.3 From Hair Cells to Ears, Lateral Line Neuromasts, and Vitalli's Organ

Obviously, evolving a vestibular ear requires the prior evolution of mechanosensory cells since it seems in our view logically inconceivable that an ear as an empty vesicle without mechanosensory hair cells evolved first, followed by the evolution of mechanosensory hair cells. To sidestep this problem, others argued in the past that the simpler neuromasts might be the precursor of the inner ear. However, this hypothesis does not explain how the mechanosensory hair cells of lateral line organs might have evolved in the first place. As indicated above, hair cell development seems to go through stages closely resembling adult cells found in other phyla, indicating that hair cell development may recapitulate critical steps in its evolution during development (Fig. 10.1).

Developmental studies have revealed that neuromasts develop after the ear. If neuromasts evolved before the ear, this would suggest a significant heterochronic shift in the developmental time table, developing the evolutionarily older lateral line organs after the evolutionarily more recent ear. In contrast to this 'lateral line first' hypothesis, we propose that it is equally possible that mechanosensory hair cells evolved as single cells first. Organ formation, including lateral line neuromast and inner ear formation, followed later in evolution. In the preceding discussion we outlined the argument that all mechanosensory cells share what has been dubbed a deep molecular homology (Shubin et al., 2009). In our opinion, this argument about the evolution of the hair cells before the vertebrate ear aligns ear evolution closely with what is now known about the evolution of another major sense, the eye.

Much like hair cells, similar issues of cellular and molecular homology have been raised for photoreceptors. Photoreceptors are classically divided into morphologically unique cell types which appeared to be also molecularly distinct. However, recent evidence suggests that molecular intermediates to these various cell types exist, thus forcing the consideration that very dissimilar looking photoreceptors are indeed morphologically divergent variations of a single ancestral photoreceptor (Arendt, 2003). In fact, it has been proposed that all eyes derive from a single pair of cells, a hypothetical receptor cell and the accompanying pigment cell (Gehring, 2011). We propose that all mechanosensory organs derive from a pair of cells, a mechanosensory cell and a possibly accompanying supporting cell.

Like the possibly homologous mechanosensory cells, all morphologically and molecularly rather divergent photoreceptors are now considered by some to be homologous. In addition, molecular evidence suggested that all "eyes" across phyla could indeed be homologous and share a developmental molecular program around the *Pax6* transcription factor. This is in stark contrast to the considerable morphological dissimilarity that has traditionally been interpreted as evidence for multiple parallelisms of eye evolution in animals. *Pax6* is not only expressed in various developing eyes but replacement of *Pax6* across phyla suggests some degree of functional preservation. For example, mammalian *Pax6* can organize fly specific eyes with structurally different eyes and photoreceptors (Gehring, 2011). We agree with the emerging unifying concept that suggests a deep homology of photoreceptors and eyes. It remains to be seen how exactly molecular similarities in at least a set of essential eye developmental genes are mechanistically tied into the developmental programs of eyes that have been modified across vertebrates to generate eyes with very different ocular muscle systems, lens accommodation, and retina organization (Fritzsch, 1991; Lamb, 2013).

We recently proposed a deep homology of a set of transcription factors for mechanosensory cells and the various gravistatic organs in which they are embedded (Bouchard et al., 2010), aligning hair cell and ear evolution with photoreceptor and eye evolution. As with the morphologically dissimilar photoreceptors in morphologically dissimilar eyes, morphologically dissimilar mechanosensory cells are also assembled into morphologically dissimilar receptor organs.

These morphological differences have been used for a long time to suggest multiple parallel and independent evolution of statocysts and mechanosensory cells (for review see Markl, 1974). As with the eyes, the arguments raised for statocysts were extremely compelling and built on the obvious morphological differences of various organs. However, looking at it from a larger perspective, these arguments for parallel evolution of statocysts may ultimately be as flawed as proposing that multicellular organisms could not have arisen from the same single-celled precursor because they are morphologically so dissimilar. Yet, the molecular evidence suggests that there was once a single-celled ancestor that shared a set of molecular features with modern animals such as flies, mice, and worms. We propose that morphologically dissimilar mechanosensory organs evolved independently out of a common precursor using morphological diversification from a simple ancestor cell while grouping evolving mechanosensory cells into morphologically dissimilar organs. As with morphologically dissimilar eyes and photoreceptors, the search for a molecular unifying gene equivalent to *Pax6* was needed to break through the very convincing argument centered on morphology.

In contrast to *Pax6* which remained a single gene in all animals, the ear specific *Pax* gene multiplied into three genes, *Pax2*, *5*, and *8*. Indeed, this multiplication/diversification happened after the *Pax6/2* genes split. It has been shown that a single *Pax* gene is expressed in both eyes and statocysts of jellyfish, indicating that even such dissimilar organs may have a deep homology based on certain development organizing transcription factors that allow sensory organs to form.

Eye evolution was seemingly invariably tied into a single transcription factor, *Pax6*. However, ear evolution proved to be more complex. First, knocking out either *Pax2*, *5*, or *8* alone did not abolish ear development. Indeed, *Pax5* and *Pax8* loss had no obvious effects on embryonic ear development, and the late effects of *Pax8* could be connected to the thyroid gland, absent in *Pax8* null mice. In contrast, *Pax2* mutations generated ear defects that were, however, associated with the evolutionary novel cochlea, not with the ancestral vestibular system. This seemingly indicated that *Pax2* is associated with the late evolution of the mammalian auditory system, not what one would expect from an evolutionary early association of this gene with a statocyst evolution. In addition, *Pax8* and *Pax2* were also expressed in kidneys, thus complicating identification of *Pax2/8*-positive, presumed homologous precursor organs across phyla. In essence, it remained an open question throughout the past 20 years whether *Pax2/5/8* played in the ear an equally important role during evolution as *Pax6* in the eye.

To solve that problem required complex mutants that have neither *Pax2* nor *Pax8*. Such double null mutants were eventually generated by us and showed that *Pax8* and *Pax2* together are needed to develop an ear past the otic vesicle (Bouchard et al., 2010). The cochlea, previously described as missing in *Pax2* null mutants, was identified as a sack expanded into the brain cavity, apparently causing earlier work on this mutant to claim lack of cochlea formation. We suggest that early expressed *Pax8* drives the early vestibular development, but its absence can be compensated for by the later expressed *Pax2*. However, *Pax2* has evolved a unique and new function in mammalian auditory organ development. Unclear is still how ear and kidney evolution relate to each other as both depend on *Pax2/8* and several other coexpressed factors (*Gata3*, *Eya4*). The present data suggest that *Pax2/8* may indeed be a part of the unifying molecular principles, providing deep molecular homology of diverse statocysts (including the vertebrate vestibular ear) across phyla. However, the molecular data will likely never be as compelling as the *Pax6* case for eye evolution because of the early gene multiplication of the ancestral gene into three paralogous genes, *Pax2/5/8*.

Irrespective of the uncertainty regarding *Pax* genes in ear evolution, one of the basic principles of ear development has been well characterized molecularly. The development of the dorsolateral placodes give rise to the ear and, when present, lateral line organs, ampullary electroreceptors and Vitalli's organs. Molecular data suggest that all share certain gene expression, but also differ in others. In each of these cases of independent placodes it is now clear that they give rise to both neurons and hair cells, possibly from multipotent progenitor cells shifting cell specification over time (O'Neill et al., 2012; Pan et al., 2012). Be that as it may, it appears at the moment equally likely to us that neuromasts, ears, and Vitalli's organs represent independent molecular transformations of the general epidermal capacity to form hair cell–bearing organs and associated neurons, with neither being ancestral to the other but each evolving independently. Combined with multiple points of critique on the original octavolateralis hypothesis (Duncan & Fritzsch, 2012), it appears best to abandon this concept and instead begin to dissect molecularly how each of these various organ types could have arisen independently, separated in space and time and yet always building on the ancestral molecular machinery that forms the mechanosensory/electrosensory hair cell.

Experimentally, it would be appropriate to expand on the existing data by inducing the developing frog ectoderm through forced gene expression to transform into different mechnosensory organs using transcription factors uniquely associated with specific inner ear organs, lateral line organs, or Vitalli's organ. Demonstrating thereby molecular similarities in organ development would consolidate in our view the idea that molecules can be recruited in evolution to alter development. Essentially we propose that existing genes needed to differentiate mechanosensory hair cells can be bundled to be expressed in restricted areas (in vertebrates referred to as placodes) to differentiate into distinct sensory organs via distinct sets of placodes that share common molecular modules (Grocott et al., 2012).

10.4 From Vestibular Ears to Tetrapod Hearing Organs: Toward the Molecular Basis of Organ of Corti Evolution

The preceding overviews highlighted the molecular basis of mechanosensory hair cell development and evolution and the aggregation of neurosensory precursors into placodes. These cells derived from placods were transformed in the course of evolution into various mechanosensory organs of vertebrates, including the vestibular ear. In this section we discuss what we see as the stepwise morphological transformation of a vestibular ear into the mammalian hearing organ.

Over the past 100 years a uniform agreement has emerged that the mammalian hearing organ, the organ of Corti of the cochlea, is most likely a transformed vestibular sensory organ. However, controversies remain as to when in vertebrate evolution this transformation occurred. It is also unclear how the evolution of the terrestrial middle ear of tetrapods was associated with the inner ear evolution to minimize the impedance mismatch in the air–water transition, moving the stapes footplate into the newly evolved oval window.

The discussion has to be put into the context that since the late 1960s, it has become clear that impedance mismatch of sound conductance in air as compared to the fluid-filled ear, can best be overcome in water through any association of the ear with a gas-filled resonator such as the swim bladder or lung. The last 20 years have confirmed and extended previous observations indicating morphologically distinct sound pressure receptions in various bony fish (Fritzsch, 1999; Ladich & Popper, 2004). This suggested that associations between lungs/swim bladder and ear to enhance sound pressure reception happened multiple times but was always associated with modifications of a gravistatic receptor (the utricle). The most striking example of an obvious parallelism is the apparent transformation of the utricle into a hearing organ only in herrings (Fritzsch, 1999; Ladich & Popper, 2004).

Given this multitude of hearing organ development in bony fish, it was unavoidable that similar multitudes of hearing organ evolution would be speculated for amphibians—the group of animals that represented the water to land transition. In contrast to these speculations about multiple evolutionary events to evolve hearing organs among amphibians, which dominated the field until the late 1980s, our analysis has identified these alleged parallelisms to be morphological variations on a common theme. We provided evidence that the basilar papilla at the orifice of the lagenar recess becomes the organ of Corti and the amphibian specific translocation of parts of the neglected papilla to become the amphibian papilla (Fritzsch & Wake, 1988). Indeed, most recent summaries of this problem all agree that the evolution of the tetrapod hearing organ, the basilar papilla, happened early in tetrapods and was associated with the formation of a lagenar recess, allowing a physical segregation of a sensory patch between the lagenar macula and the saccular macula (Manley & Clack, 2004; Fritzsch et al., 2013). Opinions differ with the interpretation of the ears of two taxa forming the outgroup of tetrapods among sarcopterygians, the lungfish and the coelacanth.

This discussion around the ear of non-tetrapod sarcopterygians largely revolves around the still unresolved taxonomic position of lungfish and *Latimeria*, the only two representatives of the sarcopterygian lineage other than tetrapods. Whereas some data suggest that lungfish may be more closely related to tetrapods than *Latimeria* is, it seems to be most appropriate at the moment to treat this taxonomic problem as an unresolved trisomy. Looking at the ear data alone it is clear that the assertion already presented 150 years ago (Retzius, 1881), that lungfish ears show shared derived features with sharks and rays (a unique utricular recess found only in these taxa), is not easy to reconcile with regressive evolution of the lungfish ear. Such a regressive evolution is the preferred explanation of proponents of the lungfish/tetrapod affinity, if this unique shared feature of lungfish and shark ears is discussed at all. Likewise, there is a detailed similarity of lagena macula in the saccular recess with an identical polarity and arrangement of hair cells between lungfish and basic actinopterygians, presumably presenting a shared primitive character of unclear significance (Fritzsch et al., 2013). These data agree with the assumption that the lungfish ear may represent a primitive similarity to non-sarcopterygians based on retention of those shared features instead of being a highly derived regressive convergence in features not found in other cases of regressive evolution of the ear such as among limbless amphibians.

In contrast to lungfish, we previously showed that *Latimeria* possesses at the entrance of the lagenar recess a sensory epithelium that in several features resembles the tetrapod basilar papilla, including innervation and association with a perilymphatic space to a round window (Fritzsch, 1992). Although it is formally possible that this sensory epithelium evolved in parallel to that of tetrapods, such an assumption contrasts with the obvious parallel evolution of auditory sensory epithelia and their association with perilymphatic space among actinopterygian fish. The last 100 years showed that pressure difference reception such as hearing among actinopterygian fish evolved multiple times in parallel, coupling different endorgans in different ways to the gas-filled swim bladder to enable sound pressure reception (Fritzsch, 1999; Ladich & Popper, 2004). These differences among hearing organs of bony fish make it unlikely that a detailed similarity evolved twice and independently at the base of sarcopterygians, in *Latimeria*, and in tetrapods. In our opinion, rejecting this detailed similarity while assuming that the lungfish ear devolved to look convergently like basic actinopterygians, and even chondrichthyans, makes for a complicated hypothesis.

Thus, a sister taxa relationship of lungfish and tetrapods, with *Latimeria* being the sarcopterygian outgroup, requires complex assumptions to reconcile ear morphology with taxonomy. A more parsimonious explanation for ear morphology is that the lungfish ear is primitively similar to ancestral ears. In contrast, the similarity of the ears of *Latimeria* and tetrapods reflects a single evolutionary change, indicating a common ancestry: The basilar papilla at the orifice of the lagenar recess evolved only once in the common sarcopterygian ancestor of *Latimeria* and tetrapods (Fig. 10.2).

Pending full resolution of the unresolved trisomy of lungfish, *Latimeria* and tetrapods, we propose to use for the time being the more parsimonious assumption for ear evolution. Irrespective of this dispute, it is generally agreed now that the organ of Corti in the coiled therian cochlea is a transformed basilar papilla that evolved in tetrapod ancestors (Manley & Clack, 2004; Fritzsch et al., 2013). The transformation of the basilar papilla into the organ of Corti entailed several important changes: (1) The lagena was either lost or integrated into the expanding organ of Corti, allowing the elongation and coiling, including the coiling of the spiral ganglion. (2) The more or less uniform tetrapod hair cells evolved into two discrete types with a distinct distribution, the inner and outer hair cells. (3) A unique association of some sensory neurons with either inner or outer hair cells evolved, forming the type I spiral ganglion neurons to inner hair cells and the type II ganglion cells to outer hair cells. (4) As an additional association with efferent innervations of these two types of hair cells, the lateral and medial olivo-cochlear fibers evolved.

Research in the 1990s highlighted for the first time the detailed organization of the basilar papilla in monotremes, the only mammals that have no coiled cochlea and a lagena at the tip of the lagena recess. Importantly, monotreme mammals have two types of hair cells in the basilar papilla, but they are not organized into a single row of inner and three rows of outer hair cells, except for the most basal tip of basilar papilla. These data suggest that evolution of inner and outer hair cells (and presumably the associated spiral ganglion neuron types) evolved in early mammals. In contrast, the elongation of the organ of Corti, the formation of a single row of inner hair cells, and complete segregation of two types of afferents to two types of hair cells evolved together with the coiling of the cochlear duct in therian mammalian ancestors. This elongation and reorganization was possibly facilitated by the loss of

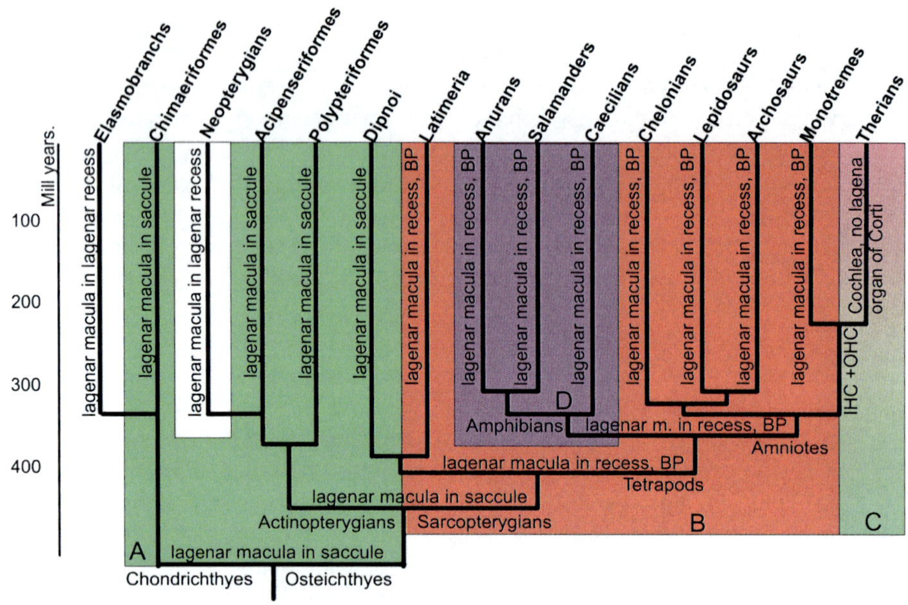

Fig. 10.2 A consensus diagram of multiple cladistic analyses pointing out the likely relationship between different sarcopterygian and actinopterygian taxa and in different shades the major morphological changes. Note that lungfish resemble basal actinopterygians and chondrichthyes, in particular ratfish, in great detail in that the lagenar macula is together with the saccular macula in the saccular recess (highlighted in light green, **A**). The derived conditions of tetrapods, apparently shared with the coelacanth *Latimeria*, is the possession of a lagenar recess with the lagenar macula at the tip and the formation of a basilar papilla and a perilymphatic sac to a round window in the posterior wall of the otocyst (highlighted in light red, **B**). Some derived chondrichthyes and actinopterygians also have a lagenar recess with a lagenar macula in it (shown in white). All amphibians have a shared derived character, the amphibian papilla in its own recess (highlighted in light lilac, **D**). Salamanders and frogs have lost the neglected papilla and some salamanders and cecilians have lost the basilar papilla, which may be in its own recess that comes off the lagenar recess. Some caecilians have lost the lagenar macula but retain the lagenar recess devoid of any hair cells or innervation. Ancestral mammals evolved a basilar papilla with two types of hair cells arranged in multiple rows, inner and outer hair cells (IHC, OHC). Therian mammals either lost or transformed the lagenar macula into the apex of the organ of Corti in a greatly elongated lagenar recess, now referred to as the cochlea or cochlear duct (shown in shades of light green to light red, **C**). (From Fritzsch et al., 2013)

the lagenar macula or its integration into the expanding organ of Corti. This lack of a lagena, combined with the coiled extension of this duct, requires renaming the ancestral lagenar recess of *Latimeria* and tetrapods, including monotreme prototherian mammals, into the cochlear duct of therian mammals.

The last 10 years have provided a wealth of molecular data with some potential meaning for this reorganization to form the lagenar recess, segregate the basilar papilla from the lagenar macula, change development to form a perilymphatic space associated with the basilar papilla/organ of Corti, and transform variable sets of hair cells in the tetrapod basilar papilla into the mammalian set of inner and outer hair cells. Only recently has there been some insight into the possible molecular regulation of inner and outer hair cell differentiation. It is possible that changes in types of hair cells reflect systematic variation in timing and level of expression of the general hair cell differentiation factor of mechanosensory hair cells, *Atoh1*. Experimental transformation of outer to inner hair cells indicates that level and duration of *Atoh1* expression might be tightly regulated to achieve specific hair cell type differentiation in the correct position (Pan et al., 2012).

The differentiation of inner and outer hair cells and the association of the electromotility protein Prestin with outer hair cells require novel transcription factors to be associated with the developing cochlea/organ of Corti. Among those factors is *Pax2*, an ancient transcription factor that plays a unique role in cochlea development. *Pax8* can compensate vestibular development in *Pax2* null mice but apparently not organ of Corti development (Bouchard et al., 2010). It is possible that this is simply related to the late development of the cochlea relative to vestibular sensory epithelia. The loss of *Pax8* expression prior to the need for a *Pax* gene in cochlear development makes *Pax2* unique in that respect. Alternatively, *Pax2* has changed its sequence so that it has a unique function. Although insertions of *Pax5* into *Pax8* have shown that these two orthologs are functionally equivalent (Bouchard et al., 2010), such an experiment needs to be made with a knockin of *Pax8* into the *Pax2* locus. Should *Pax8* be able to compensate for *Pax2* we have to consider that any *Pax2/5/8* protein is needed for cochlear development and evolution. This would also indicate that *Pax2/8* has been early on associated with ear development and evolution, but only *Pax2* was recruited to a new role in the most derived part of the vertebrate ear, the mammalian cochlea development through expression changes. It is important to realize that either temporal shift of expression through mutations in the promoter region or changes in the protein coding part of the *Pax2* gene may be associated with this therian novelty. It also needs to be stressed that although *Pax2* is necessary for neurosensory development of the cochlea, there is formation of the cochlear duct, indicating that duct formation and neurosensory development are coupled through the *Pax2* gene.

A second gene important for cochlear neurosensory development, but not for vestibular neurosensory development, is *Gata3*. This gene not only is expressed in the cochlear duct, delaminating spiral ganglion neurons and olivo-cochlear efferents, but it is also essential for cochlear neurosensory development. Most interesting is that the lack of *Gata3* can result in a short cochlear duct without neurosensory cell development. Thus, *Gata3* is indeed a transcription factor directly involved in yet to be understood aspects of cochlea neurosensory development. Obviously, further studies of the molecular interactions of genes regulated by *Pax2* and *Gata3* will be central for a detailed understanding of the molecular evolution of the organ of Corti and its innervation.

Finally, a gene widely expressed in the developing ear, *Lmx1a*, affects the segregation of the organ of Corti from the saccule, thus clarifying once and for all that the organ of Corti/basilar papilla segregates during development from the saccular epithelium. This validates the evolutionary segregation proposed in the preceding text for the basilar papilla. Whether or not modifications in *Lmx1a* are the only changes needed to induce the segregation of basilar papilla from the saccular and lagenar macula needs to be tested by manipulating the ear development of animals that have all three of these sensory epithelia.

The road ahead will see soon the complete sequence of all sarcopterygian genomes to resolve the current trisomy of lungfish, *Latimeria*, and tetrapods. These data will decide if the most parsimonious explanation of otic features reflects indeed the evolutionary history of the sarcopterygian ear or if we have to accept more complicated scenarios of both progressive and regressive parallelism that convergently evolved the *Latimeria* ear into a tetrapod look-alike ear while at the same time devolving the lungfish ear into an actinopterygian/chondrichthyan look-alike ear. No matter the outcome of this quest, it is now commonly accepted that the tetrapod basilar papilla evolved only once in the last common ancestor and that the mammalian organ of Corti is a transformed basilar papilla sitting in a modified lagenar recess referred to in therian mammals as the cochlear duct. We have begun to understand molecular aspects of different hair cell type development and molecular mechanisms needed to make the unique organ of Corti neurosensory cells and segregate them from saccular cells, but details still need to be worked out. At least, a conceptual start has been made that may prove useful in our quest to understand the absolute inability of hair cell proliferation in the adult mammalian organ of Corti compared to the tetrapod basilar papilla. We propose that the structural changes related to the high degree of supporting cell differentiation, needed for the function of the organ of Corti, disables proliferation and transdifferentiation that can restore hearing in nonmammalian vertebrates. Our inability to cure the hearing loss currently plaguing more elderly people worldwide than any other medical problem, and disabilities in communication due to deafness at a time many people need the ability the most, may be a consequence of unique cellular differentiation needed for a mammalian organ of Corti.

10.5 Summary

The last 20 years have seen significant steps forward in the quest to understand the evolution of hair cells, and their afferent and efferent innervation in the context of ear evolution overall as well as the segregation of auditory sensory system from the ancestral vestibular ear. Old paradigms presented by the more than 100-year-old "octavolateralis hypothesis" have been largely replaced by a more molecular driven concept that aims to explain the vertebrate ear evolution as a series of progressively transformed ontogenies. This scientific progress has accelerated in recent years as a result of the increasing availability of molecular tools. This chapter provides a snapshot of these insights but also a warning that future progress will likely face the same winding road as in the past. How many more surprises are waiting to be discovered remains to be seen. The past has demonstrated that in the ear we certainly should not tune out novel and seemingly outlandish findings but literally keep an open ear, even if at first glance findings might be counterintuitive and out of tune with the perceived reality believed by many. Like other scientific progress, understanding hair cell, ear, and cochlear evolution was driven by theoretical insights and technical progress, to be united to challenge previously held views.

Acknowledgments This work has been supported by grants from National Aeronautics and Space Administration, National Institutes of Health, the German Science Foundation, and funds from Office of the Vice Provost for Research of the University of Iowa.

References

Arendt, D. (2003). Evolution of eyes and photoreceptor cell types. *International Journal of Developmental Biology*, 47(7–8), 563–571.

Bouchard, M., Busslinger, M., Xu, P., De Caprona, D., & Fritzsch, B. (2010). PAX2 and PAX8 cooperate in mouse inner ear morphogenesis and innervation. *BMC Developmental Biology*, 10, 89.

Braun, C. B., & Northcutt, R. G. (1997). The lateral line system of hagfishes (Craniata: Myxinoidea). *Acta Zoologica*, 3, 247–268.

Darwin, C. (1859). *On the origin of species by means of natural selection or the preservation of favoured races in the struggle for life*. London: Murray.

de Burlet, H. M. (1934). Vergleichende Anatomie des statoakustischen Organs. a) Die innere Ohrsphäre. In L. Bolk, E. Göppert, E. Kallius & W. Lubosch (Eds.), *Handbuch der Vergleichenden Anatomie der Wirbeltiere* (Vol. 2, pp. 1293–1432). Berlin: Urban and Schwarzenberg.

Duncan, J. S., & Fritzsch, B. (2012). Evolution of sound & balance perception: Innovations that aggregate single hair cells into the ear and transform a gravistatic sensor into the organ of Corti. *Journal of Anatomy*, 295, 1760–1774.

Fritzsch, B. (1981). The pattern of lateral-line afferents in urodeles: A horseradish-peroxidase study. *Cell and Tissue Research*, 218(3), 581–594.

Fritzsch, B. (1991). Ontogenetic clues to the phylogeny of the visual system. In P. Bagnoli & W. Hodos (Eds.), *The changing visual system* (pp. 33–49). London: Plenum Press.

Fritzsch, B. (1992). The water-to-land transition: Evolution of the tetrapod basilar papilla, middle ear and auditory nuclei. In D. B. Webster, R. R. Fay, & A. N. Popper (Eds.), *The evolutionary biology of hearing* (pp. 351–375). New York: Springer-Verlag.

Fritzsch, B. (1999). Hearing in two worlds: Theoretical and realistic adaptive changes of the aquatic and terrestrial ear for sound reception. In R. R. Fay & A. N. Popper (Eds.), *Comparative hearing: Fish and amphibians*. (pp. 15–42). New York: Springer-Verlag.

Fritzsch, B., & Wahnschaffe, U. (1983). The electroreceptive ampullary organs of urodeles. *Cell and Tissue Research*, 229(3), 483–503.

Fritzsch, B., & Wake, M. H. (1988). The inner ear of gymnophione amphibians and its nerve supply: A comparative study of regressive events in a complex sensory system. *Zoomorphology*, 108, 210–217.

Fritzsch, B., & Glover, J. C. (2007). Evolution of the deuterostome central nervous system: An intercalation of developmental patterning processes with cellular specification processes. In J. H. Kaas (Ed.), *Evolution of nervous systems* (Vol. 2, pp. 1–24). Oxford: Academic Press.

Fritzsch, B., Eberl, D. F., & Beisel, K. W. (2010). The role of bHLH genes in ear development and evolution: Revisiting a 10-year-old hypothesis. *Cell and Molecular Life Sciences*, 67(18), 3089–3099.

Fritzsch, B., Pan, N., Jahan, I., Duncan, J. S., Kopecky, B. J., Elliott, K. L., Kersigo, J., & Yang, T. (2013). Evolution and development of the tetrapod auditory system: An organ of Corti-centric perspective. *Evolution & Development*, 15, 63–79.

Gehring, W. J. (2011). Chance and necessity in eye evolution. *Genome Biology and Evolution*, 3, 1053–1066.

Grocott, T., Tambalo, M., & Streit, A. (2012). The peripheral sensory nervous system in the vertebrate head: A gene regulatory perspective. *Developmental Biology*, 370(1), 3–23.

Grothe, B., Carr, E. C., Casseday, J. H., Fritzsch, B., & Köppl, C. (2004). The evolution of central pathways and their neural processing patterns. In G. A. Manley, A. N. Popper, & R. R. Fay (Eds.), *Evolution of the vertebrate auditory system*. (pp. 289–359). New York: Springer Science+Business Media.

Jahan, I., Pan, N., Kersigo, J., Calisto, L. E., Morris, K. A., Kopecky, B., Duncan, J. S., Beisel, K. W., & Fritzsch, B. (2012). Expression of *Neurog1* instead of *Atoh1* can partially rescue organ of Corti cell survival. *PLoS One*, 7(1), e30853.

Jørgensen, J. M. (1989). Evolution of octavolateralis sensory cells. In S. Coombs, P. Goerner, & H. Muenz (Eds.), *The mechanosensory lateral line: Neurobiology and evolution* (pp. 99–115.). New York: Springer Verlag.

Ladich, F., & Popper, A. N. (2004). Parallel evolution of fish hearing organs. In G. A. Manley, A. N. Popper, & R. R. Fay (Eds.), *Evolution of the vertebrate auditory system* (pp. 95–127). New York: Springer Science+Business Media.

Lamb, T. D. (2013). Evolution of phototransduction, vertebrate photoreceptors and retina. Progress in Retinal and Eye Research, 36, 52–119.

Lewis, E. R., & Fay, R. R. (2004). Environmental variables and the fundamental nature of hearing. In G. A. Manley, A. N. Popper, & R. R. Fay (Eds.), *Evolution of the vertebrate auditory system* (pp. 27–54). New York: Springer Science+Business Media.

Manley, G. A., & Clack, J. A. (2004). An outline of evolution of vertebrate hearing organs. In G. A. Manley, A. N. Popper, & R. R. Fay (Eds.), *Evolution of the vertebrate auditory system* (pp. 1–26). New York: Springer Science+Business Media.

Markl, H. (1974). The perception of gravity and of angular acceleration in invertebrates. In H. H. Kornhuber (Ed.), *Handbook of sensory physiology* (Vol. VI/1: *Vestibular system*, pp. 17–74). Berlin: Springer-Verlag.

Munz, H., Claas, B., & Fritzsch, B. (1982). Electrophysiological evidence of electroreception in the axolotl Siredon mexicanum. *Neuroscience Letters*, 28(1), 107–111.

O'Neill, P., Mak, S.-S., Fritzsch, B., Ladher, R. K., & Baker, C. V. H. (2012). The amniote paratympanic organ develops from a previously undiscovered sensory placode. *Nature Communications*, 3, 1041.

Pan, N., Kopecky, B., Jahan, I., & Fritzsch, B. (2012). Understanding the evolution and development of neurosensory transcription factors of the ear to enhance therapeutic translation. *Cell and Tissue Research*, 349, 415–432.

Reichert, C. (1837). Ueber die Visceralbogen der Wierbelthiere im allgemeinen und deren Metamorphose bei den Saeugethieren und Voegeln. *Acta Anatomica & Physiologica*, 120–222.

Retzius, G. (1881). *Das Gehörorgan der Wirbeltiere* (Vol. I.). Stockholm: Valin.

Senthilan, P. R., Piepenbrock, D., Ovezmyradov, G., Nadrowski, B., Bechstedt, S., Pauls, S., Winkler, M., Mobius, W., Howard, J., & Gopfert, M. C. (2012). *Drosophila* auditory organ genes and genetic hearing defects. *Cell*, 150(5), 1042–1054.

Shubin, N., Tabin, C., & Carroll, S. (2009). Deep homology and the origins of evolutionary novelty. *Nature*, 457(7231), 818–823.

Simmons, D., Duncan, J., Crapon de Caprona, D., & Fritzsch, B. (2011). Development of the inner ear efferent system. In D. K. Ryugo, R. R. Fay, & A. N. Popper (Eds.), *Auditory and vestibular efferents* (pp. 187–216.). New York: Springer Science+Business Media.

Wagner, A. (2011). *The origins of evolutionary innovations*. Oxford: Oxford University Press.

Chapter 11
The Implications of Discharge Regularity: My Forty-Year Peek into the Vestibular System

Jay M. Goldberg

Jay M. Goldberg

J.M. Goldberg (✉)
Department of Pharmacological & Physiological Sciences,
University of Chicago, Chicago, IL 60637, USA
e-mail: jgoldber@bsd.uchicago.edu

A.N. Popper and R.R. Fay (eds.), *Perspectives on Auditory Research*, Springer Handbook of Auditory Research 50, DOI 10.1007/978-1-4614-9102-6_11,

11.1 Background

I was trained in auditory neurophysiology, first in ablation studies by W. D. "Dewey" Neff in the Biopsychology Section at the University of Chicago, then in single-neuron microelectrode recording methods by Joseph E. Hind and Jerzy E. Rose in the Laboratory of Neurophysiology at the University of Wisconsin. On finishing my postdoctoral training, I moved back to the University of Chicago as an Assistant Professor of Physiology. As my first independent research project, I had been studying binaural mechanisms in the superior olive with a graduate student, Paul Brown, when César Fernández, a Professor of Otolaryngology, introduced me to the vestibular system and suggested that we collaborate. Having finished two papers on the superior olive (Goldberg & Brown, 1968, 1969), I had come to a convenient time for a pause in my auditory research and thought that a peek into another system might be instructive. The peek turned into more than 40 years of research. In this essay, I have taken advantage of Arthur Popper and Richard Fay's invitation to write an account that "might be very personal."

An autobiographical essay can easily become self-indulgent. An author has to decide whether such an effort would be of use to others, and so might be published rather than being placed in a drawer. My reasons for writing this chapter for publication are as follows. The scientific literature emphasizes the orderly progress of knowledge. But research advances by fits and starts, with crucial clues coming not only from well-thought-out experiments, but also from experiments undertaken for purposes that, in retrospect, might be ill-conceived. In this chapter I will emphasize those intellectual way stations, both accidental and well conceived, that drove my own research. The hope is that such a telling would amuse my contemporaries, who are largely unaware of the personal details, as well as profit younger scientists, who may have been taught, I think unwisely, that science is a strictly logical enterprise. My research has been a very personal adventure that at key points took advantage of seeming missteps. In this adventure, I have often been reminded of a paraphrase by my mentor, Jerzy Rose, of a quote by Benjamin Franklin: "Experience is a poor teacher, but is the only way a fool can learn."

11.2 My Introduction to the Vestibular System

During the first few years after my return to the University of Chicago in 1965, I would occasionally talk to César Fernández (Fig. 11.1), usually over a cup of coffee. During one such occasion, we discussed our mutual interest in lower auditory pathways and decided to collaborate on the following neuroanatomical project. I had observed that fibers from the medial nucleus of the trapezoid body (MNTB) pass through the medial superior olive (MSO) on their way to the lateral superior olive (LSO), where they were known to provide a powerful inhibitory input (Boudreau & Tsuchitani, 1968). The question was whether some of the fibers might

Fig. 11.1 A photograph of César Fernández taken in 1965. The setting in his laboratory before it was modified for the study of the vestibular nerve

synapse in the MSO, thus providing inputs for the inhibition Paul Brown and I had seen in the latter nucleus. To address the issue, fiber degeneration studies were to be done. This required that stereotaxic lesions be placed in the MNTB, which, after some discussion, we decided would be done in Fernández's rather than in my laboratory. The decision was based on Fernández feeling more comfortable in familiar surroundings. While we were doing the surgery, I noticed a rather large apparatus in the distance and was informed that it was an animal rotator. Its performance, which Fernández demonstrated, was impressive; for example, it could accurately track sinusoidal commands. On further inquiry, Fernández told me that there were only a handful of such devices in existence and that there were no contemporary studies of the discharge of the vestibular nerve. I was intrigued and asked for a collection of reprints that would summarize the current state of research on the vestibular system.

A reading of the reprints revealed that, except for the studies of Lowenstein and Sand in the isolated elasmobranch labyrinth (Lowenstein & Sand, 1936, 1940a, b), most of the studies, particularly those in mammals, were unreliable. Reminiscent of the ambiguity posed by the classic Galambos' papers in the auditory system (Galambos & Davis, 1946), it was unclear whether the published recordings were in the vestibular nerve or vestibular nuclei. Two papers illustrate the situation. Gernandt (1949) described a variety of response patterns that were unlike the Lowenstein and Sand results, but similar to recordings in the vestibular nuclei subsequently made by Duensing and Schaefer (1958). The other study, that by Adrian (1943), resembled the Lowenstein and Sand findings, but as was emphasized by Adrian himself, his recordings were in the brain and so could have been from central neurons and/or from the central processes of vestibular nerve fibers.

Fernández and I decided to begin a study of the vestibular nerve, which required that we build a laboratory from scratch. This led to my neglecting my auditory studies and the young people in my laboratory, William Brownell, Mario Ruggero, and Eric Young. Fortunately, they hardly needed my help. Each of them finished their projects, published their results (Ruggero, 1973; Brownell, 1975; Young & Brownell, 1976), and subsequently went on to distinguished careers in auditory neuroscience.

I sometimes wonder whether I would have become interested in the vestibular system had we done the stereotaxic lesions in my laboratory, in which case I might never have seen the rotator. As for the project we abandoned, later studies showed that MNTB fibers provide inputs to the MSO (Adams & Mugnaini, 1990) and that these might be crucial in the handling of binaural processing in that nucleus (Hassfurth et al., 2010).

11.3 Discharge Characteristics of Vestibular Nerve Fibers

Fernández and I began recording from the squirrel monkey vestibular nerve, concentrating first on afferents innervating the semicircular canals (SCCs) and then on otolith (OTO) fibers. Six papers were published, three on the SCCs (Fernández & Goldberg, 1971; Goldberg & Fernández, 1971a, b) and three on the OTOs (Fernández & Goldberg, 1976a, b, c). The results settled several questions including directional properties, resting discharge, response dynamics, and response diversity. Three more recent reviews can be recommended for placing these results in a contemporary context (Goldberg, 2000; Lysakowski & Goldberg, 2004; Eatock & Songer, 2011).

At the time of these studies, the cat was the animal of choice in neurophysiological research. Yet, we chose the squirrel monkey for our studies. The rationale for the choice was the shallow internal meatus in the monkey, which allowed easy access to the vestibular nerve. My mentors, Hind, Rose, and their colleagues, had exploited this feature to record auditory nerve discharge (Hind et al., 1967; Rose et al., 1967). There was a distinct possibility that Gernandt (1949) had recorded from central neurons. Because of this possibility, we wanted to be certain that we were recording from peripheral neurons and a shallow meatus ensured this.

Our reasoning was specious. As exemplified by the pioneering work of Nelson Kiang in the cat auditory nerve (Kiang, 1965), one can gently retract the brain stem and cerebellum away from the meatus to get a clear view of the eighth nerve. Even so, the choice of the squirrel monkey proved to be fortuitous because the preparation of choice for central studies was becoming the alert, behaving rhesus monkey (Miles 1974; Fuchs & Kimm, 1975). Recordings from the vestibular nerve in the alert rhesus monkey led Keller (1976) and Louie and Kimm (1976) to conclude that the discharge properties in that preparation were similar to those in the anesthetized squirrel monkey, implying that our results could be used as a benchmark for the alert studies. Eventually, recordings were made in the cat

(Tomko et al., 1981a, b) and, except for somewhat lower discharge rates, were again similar to those in the squirrel monkey.

We continued to use the squirrel monkey through the mid-1980s. When we started, unconditioned squirrel monkeys could be imported from animal dealers in Colombia for the incredibly low price of $25. By the mid-1980s the price for squirrel monkeys had risen to $750, a cost that might be justified in studies involving chronic recordings, but not acute experiments. So we turned to a rodent model. The choice was made by Fernández without consulting me. One day he had a chinchilla cadaver on his dissecting table and was devising an approach to the vestibular nerve. The price of a chinchilla was $75. These animals were readily available because breeders were happy to dispose of animals whose coats were less than optimal.

11.3.1 Directional Properties

Fibers innervating a given SCC had identical directional properties, conforming to those found by Lowenstein and Sand (1940a), but not by Gernandt (1949). The directional properties were precisely those predicted by Ewald (1892) in his First and Second Laws and by the morphological polarization of hair bundles in the three cristae (Lowenstein & Wersäll, 1954). By correlating ultrastructure with physiology, the latter workers concluded that bundle deflections toward the kinocilium always increased discharge, while oppositely directed deflections decreased it.

Whereas the SCCs monitor the angular forces acting on the head, the two OTOs, the utricular (UM) and saccular maculae (SM), monitor linear forces. At the time we began our studies, we knew the disposition of the two maculae and the morphological polarization of their hair bundles (Lindeman, 1969). The plane of the UM has a predominantly horizontal orientation, whereas the SM plane is oriented in a parasagittal plane. Unlike the situation for each crista, where bundle polarizations are uniform, those in a macula are oppositely directed across a reversal line lying within a specialized zone, the striola (Lindeman, 1969) or, as recently discovered for the UM of mice and rats (Li et al., 2008; Schweizer et al., 2009), just outside the striola, but parallel to it. The zones to either side of the striola are referred to as the extrastriola. In the UM, polarizations in either extrastriola are arranged in a semicircular fan. Those in the main part of the SM, while reversing direction across the striola, are aligned in a near vertical orientation.

Recordings showed that individual OTO units were responsive to linear forces lying within the plane of its macula, but not those oriented perpendicular to that plane. So, as was suggested by earlier workers (deVries, 1950; Trincker, 1962), shearing forces were effective, whereas compressional forces were not. The ineffectiveness of compressional forces extended to their not affecting the responses to simultaneously applied shearing forces.

Unit recordings were consistent with the disposition of the maculae and of its hair bundles. We concluded from these recordings that the UM monitored forces broadly disposed in the horizontal plane, whereas the SM added the third or vertical dimension.

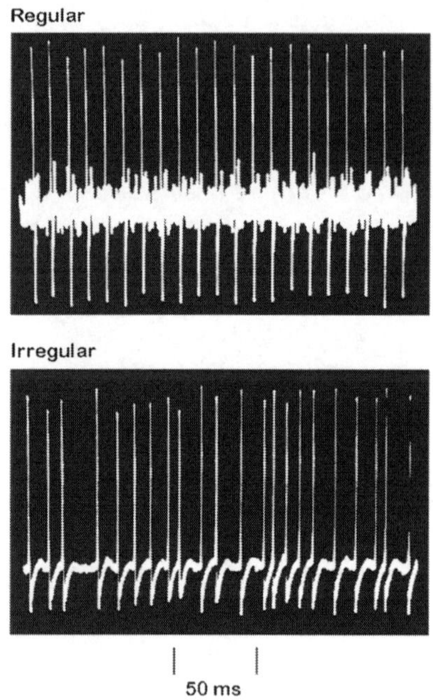

Fig. 11.2 Resting activity, two superior-canal units, squirrel monkey. Although both units have similar rates near 90 spikes/s, they differ in their discharge regularity. (From Goldberg & Fernández, 1971b)

11.3.2 Resting Discharge

As was anticipated by Lowenstein and Sand (1936), almost all SCC afferents had an appreciable resting discharge in the absence of angular accelerations. In the monkey, resting rates approached and could even exceed 100 spikes/s (Fig. 11.2). As concluded by Lowenstein (1956), the presence of a background discharge offers three advantages: (1) it allows afferents to respond bidirectionally; (2) it effectively eliminates a sensory threshold; and (3) it provides a massive input to the brain stem and cerebellum that contributes to postural tone.

A resting discharge is defined as the activity occurring in the absence of stimulation. Insistence on this definition might lead to the conclusion that the ever-presence of terrestrial gravity would preclude estimating a resting discharge for OTO units on earth. Because the directional properties of an OTO unit are summarized by a polarization vector, the response to tilts about pitch or roll axes is a sinusoidal function about a constant rate. The latter rate can be taken as the resting rate because its constancy implies a lack of response to linear forces.

11.3.3 Response Dynamics

Based on direct observations of the cupula, Steinhausen (1933) described the dynamics of the SCC in terms of a second-order, over-damped linear differential equation, the so-called torsion-pendulum model. Our vestibular nerve recordings

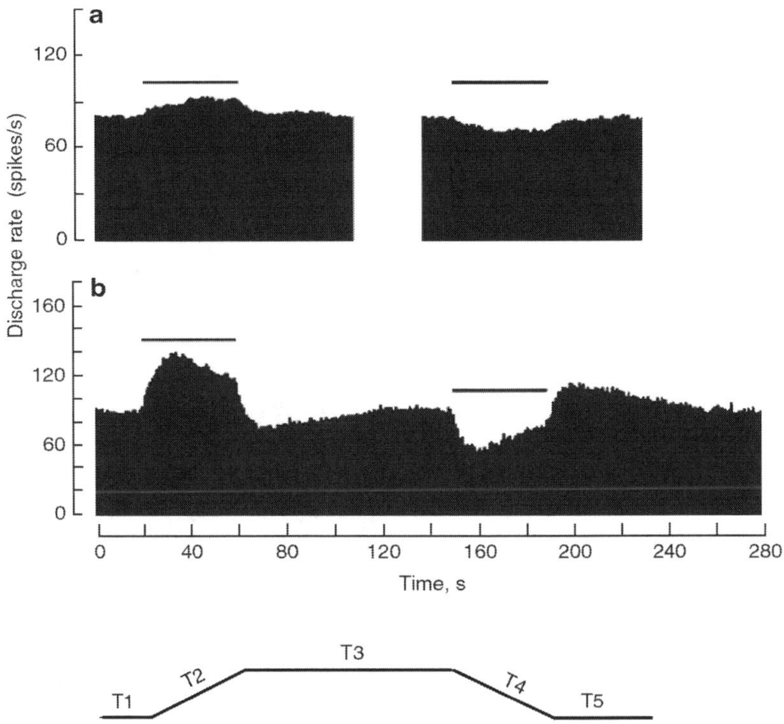

Fig. 11.3 Responses of two semicircular-canal units to velocity trapezoids (bottom trace). (**a**) A regular unit shows per-acceleratory response increases and post-acceleratory decreases, both exponential in form, consistent with the torsion-pendulum model. (**b**) An irregular unit shows deviations from the torsion-pendulum model, including per-acceleratory response declines and post-acceleratory secondary responses. (From Goldberg & Fernández, 1971a)

confirmed that the model described SCC response dynamics with two exceptions, a slow, low-frequency adaptation (Fig. 11.3b) and a high-frequency velocity sensitivity (Fig. 11.4a, c) (Fernández & Goldberg, 1971; Goldberg & Fernández, 1971b). Responses of OTO units consist of a combination of a tonic component, proportional to linear force, and a phasic component, proportional to the rate of force application (Fernández & Goldberg, 1976c).

As presented in older textbooks (see, e.g., Ruch & Patton, 1965), the response dynamics may be an impediment to an appreciation of the vestibular system. There, the emphasis was on responses to low-frequency maneuvers that lead to a persistent response not following head motion and, as such, may be considered nonveridical (Fig. 11.5b, c). This was the emphasis in the 1950s possibly because clinical tests emphasized such motion paradigms.

While in graduate school, I was an avid student of almost all neural systems, but I could not understand how the vestibular system could function when the SCCs provided such apparently false information. It was only when Fernández and I were well into our vestibular research that I understood the situation.

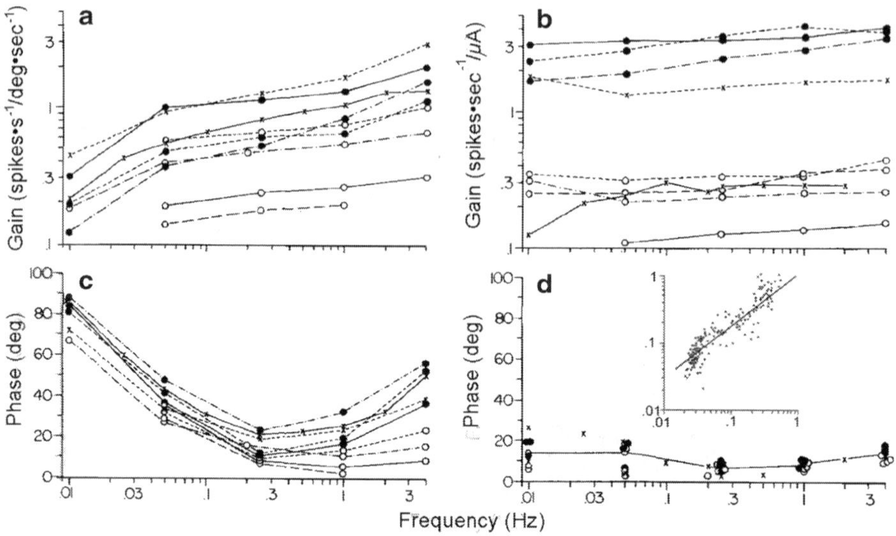

Fig. 11.4 Bode plots, semicircular canal units. All graphs include the same collection of units including regular (○, CV* < 0.10), intermediate (×, 0.10 ≤ CV* ≤ 0.30), or irregular (●, CV* > 0.30). (**a**, **c**) Rotational responses. Frequency-dependent gain increases (**a**) and high-frequency phase leads (**c**) increase the more irregular the discharge. (**b**, **d**) Galvanic responses also show gains that increase the more irregular the discharge; dynamic effects, including frequency-dependent gain increases (**b**) and phase leads, (**d**) are much smaller than those for rotational responses. Inset, power-law relation between galvanic sensitivity (ordinate) and CV* (abscissa). (**a–d** from Goldberg et al., 1982. Inset from Goldberg et al., 1984)

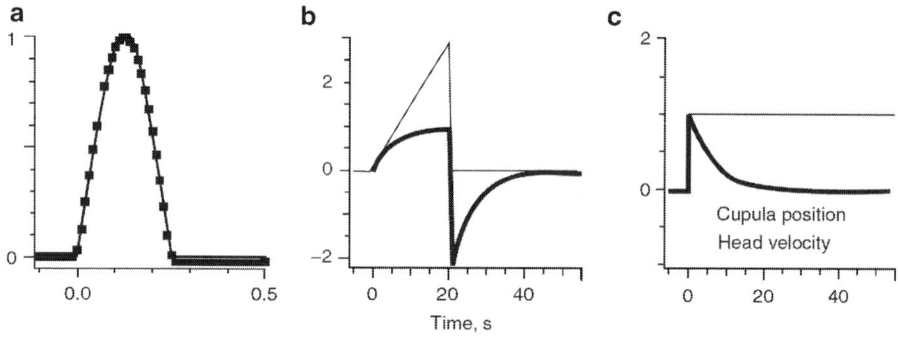

Fig. 11.5 Responses calculated from torsion-pendulum equation (bold lines and squares) to various angular-velocity profiles (thin lines). Responses to (**a**) a 2 Hz half-wave of a sinusoid; (**b**) a 20-s acceleration step followed by a brief deceleration; and (**c**) a velocity step. Only in (**a**) does the response follow head velocity. (From Goldberg et al., 2012)

During everyday life, virtually all head movements are brief, bidirectional, and in a mid-frequency range (Fig. 11.5a). Under such circumstances, responses precisely follow head motion. Realization of the importance of mid-band motions leads to the conclusion that the vestibular system monitors head velocity and, as such, is well designed. Low-frequency stimulation can occur during vehicular traffic or the playground stunt of sudden stopping after prolonged rotation. Such maneuvers can lead to vertigo and even motion sickness (Money, 1970; Reason & Brand, 1975). The importance of rapid head movements in vestibular function is reflected in the change in clinical testing of SCC function, which classically was done by the Bárány chair maneuver (Fig. 11.5b; Camis, 1930) but now is accomplished by the head thrust procedure in which the accuracy of vestibular-evoked eye movements following rapid head rotations is determined (Halmagyi et al., 2008). Caloric irrigation remains the one common low-frequency test in part because it allows the separate examination of each labyrinth.

11.3.4 Response Diversity and Discharge Regularity

Most sensory systems show diversity in the handling of information, with different afferents emphasizing distinct aspects of the external sensory world. The vestibular system is unexceptional in this regard. Neurons differ in the regularity of spacing of their action potentials (Fig. 11.2). As compared to their regular counterparts, irregular units have larger gains, calculated as the ratio of response amplitude to stimulus amplitude (Fig. 11.4a). Irregular SCC units also show larger deviations from the torsion-pendulum model (Figs. 11.3b, 11.4a, c) (Goldberg & Fernández, 1971b) and irregular OTO units show larger phasic, as compared to tonic, response components (Fernández & Goldberg, 1976c).

Perhaps our most original finding was that the diversity of afferent classes was related to discharge regularity. Although there had been no precedent in the physiological literature for the finding of diversity, it could have been anticipated in morphological studies of afferent branching patterns. As described most clearly by Lorente de Nó (1926), vestibular afferents could be distinguished into thick fibers innervating the central zone and thin fibers supplying the peripheral zone of each neuroepithelium. The thick fibers terminate as calyx endings around type I hair cells, while the thin fibers enter an intraepithelial plexus. Medium-sized fibers could provide a mixed innervation consisting of a thick branch ending as a calyx and thinner branches entering the plexus. The silver nitrate stains used by Lorente de Nó were nonselective, which made it impossible to trace the complete terminal arbors of individual fibers. Only later was it determined by ultrastructural studies that the calyx ending was distinct from its type I hair cell(s), and that the intraepithelial branches ended as bud-shaped or bouton endings on type II hair cells (Wersäll, 1956). Still later, modern neuroanatomical methods allowed the complete arbors of vestibular afferents to be reconstructed (Fernández et al., 1988, 1990, 1995).

11.4 Discharge Regularity and Galvanic Sensitivity

Until now, the narrative has followed chronological order. The next major topic that Fernández and I explored was the efferent vestibular system (EVS) (Goldberg & Fernández, 1980). As research on the EVS has recently been reviewed (Holt et al., 2011),

I will not cover it here, but rather use it to describe how we were led to study the responses to galvanic currents. We have noted that in their responses to natural stimulation, irregular afferents have larger responses than do regular afferents. The same was true of the responses of afferents to electrical stimulation of the EVS; efferent responses were much larger in irregular units regardless of the vestibular organ they innervated. For reasons that are now not entirely clear, I favored the notion that the differences in efferent responses with discharge regularity reflected the EVS input to the two afferent classes, rather than intrinsic properties of the afferents.

A test of the idea was to introduce artificial inputs that might stimulate all afferents equally. The most obvious such input was a galvanic current. Others had studied galvanic responses (Lowenstein, 1955; Lifschitz, 1973), but these authors were unconcerned with afferent diversity. We learned how to deliver galvanic currents through chlorided silver wires implanted in the perilymphatic space of the vestibule. My hypothesis was that regular and irregular afferents would respond similarly to the currents. How wrong could I be! As became obvious with the testing of a few units, the galvanic responses of irregular units were qualitatively larger (Fig. 11.4b and inset) and, in this respect, were similar to the differences in rotational responses (Fig. 11.4a) (Goldberg et al., 1982).

The result disappointed me greatly. Fernández and I were hoping to finish our EVS experiments, which had been going on for a long time. We supposed that a confirmation of our hypothesis would be a step in that direction. No such luck. I went home to lick my wounds. But almost immediately on settling down for the evening, I realized that the galvanic results were incredibly important for two reasons. First, they showed unequivocally that intrinsic factors were a major determinant of the gain differences between regular and irregular afferents. Second, an intrinsic difference in excitability with discharge regularity was to be expected theoretically. About 15 years earlier, Dan Geisler and I had published a leaky integrate-and-fire model with a noisy input. The model predicted that post-spike recovery determined both discharge regularity and sensitivity to depolarizing inputs (Geisler & Goldberg, 1966).

The next day Fernández and I dropped everything else that we were doing and began work on galvanic responses, which would occupy us for several years. In these efforts, a graduate student, Charley Smith, joined us. Three papers were published, two experimental (Goldberg et al., 1982, 1984) and one theoretical (Smith & Goldberg, 1986).

We needed to quantify the relation between galvanic sensitivity and discharge regularity. A suitable statistic for the latter is the coefficient of variation (CV) characterizing the distribution of interspike intervals. The fact that CV varies with mean interval prompted us to devise a method to transform the CV over a wide range of intervals to a CV*, a normalized CV appropriate to a standard mean interval, 15 ms in the case of mammals. This index, which was first developed in our efferent studies (Goldberg & Fernández, 1980), has been almost universally adopted in the vestibular literature (see, e.g., Ramachandran & Lisberger, 2006). In our case, the galvanic response versus CV* relation could be fit by a power law with an exponent of 0.80, equivalent to a 16-fold variation in galvanic sensitivity across the afferent population (Fig. 11.4b and inset).

As the currents were delivered via the perilymphatic space, they could have affected discharge by acting on each afferent or on the hair cells it innervates. Responses to short shocks clearly arise postsynaptically as latencies are too short to accommodate a synaptic delay between hair cells and the afferent (Goldberg et al., 1984). Evidence as to the origin of responses to longer galvanic steps are more ambiguous (for discussion, see Goldberg et al., 1984; Highstein et al., 1996; Aw et al., 2008); nevertheless, the flat phases of Fig. 11.4d suggest that these responses also arise postsynaptically. A likely site of action for both short shocks and longer steps is the spike encoder in the afferent terminal, by which is meant the set of conductances that convert synaptic depolarization into a spike train (Goldberg et al., 1984). The main conclusion of the galvanic studies, that irregular afferents have distinctively high sensitivities, was confirmed in recent studies of the turtle posterior crista in which intra-axonal recordings near the crista allowed a comparison between the modulation of the afferent depolarization evoked by mechanical stimulation of the SCC and spike discharge Goldberg and Holt (2013).

Short shocks, delivered at several distinct times after naturally occurring spikes, were used to chart the post-spike recovery of excitability. Regular units had much slower and deeper recoveries than did irregular units. Results were explained by an update of the Geisler model. In the newer model (Smith & Goldberg, 1986), discharge regularity was jointly determined by post-spike recovery and synaptic noise. Confirmation of this conclusion comes from the aforementioned turtle studies (Goldberg & Holt, 2013).

Differences in galvanic sensitivity were exploited to investigate the central projections of regular and irregular fibers. One method involved the intracellular recordings of excitatory postsynaptic potentials (EPSPs) from secondary neurons in the vestibular nuclei. This was done in collaboration with Stephen Highstein and Richard Boyle (Goldberg et al., 1987; Highstein et al., 1987; Boyle et al., 1992). The growth of secondary EPSPs with peripheral shock strength was used to estimate the profile of regular and irregular inputs received by impaled secondary neurons. We concluded that most secondary neurons receive a mixed input from both afferent classes. A similar conclusion was reached by Sato and Sasaki (1993), who used intra-axonal dye fills to trace the central projections of regular and irregular fibers.

Many of the differences between regular and irregular afferents could be explained by the postsynaptic spike encoders of irregular afferents being especially sensitive to depolarizing inputs. The differences so explained included the gains of SCC and OTO afferents to natural stimulation (Goldberg & Fernández, 1971b; Fernández & Goldberg, 1976c), the increase in discharge resulting from electrical stimulation of the EVS (Goldberg & Fernández, 1980), as well as galvanic responses (Goldberg et al., 1982, 1984). The two kinds of SCC afferents also differ in their response dynamics. This is exemplified in irregular SCC afferents having more conspicuous high-frequency gain enhancements (Fig. 11.4a) and phase leads (Fig. 11.4c) (Goldberg & Fernández, 1971b). In contrast to gain variations, encoder differences cannot explain differences in response dynamics (Fig. 11.4a), as these effects were absent in galvanic responses (Fig. 11.4b, d) (Goldberg et al., 1982). Dynamic differences might involve the presynaptic depletion of neurotransmitter, neurotransmitter receptor mechanisms involved in the generation of postsynaptic voltages, or postsynaptic adaptation triggered by spiking activity (reviewed in Highstein et al., 2005; Eatock & Songer 2011; Goldberg et al., 2012).

The galvanic experiments remain among my favorite projects. One reason is that we stumbled into these studies inadvertently, yet recognized their potential importance almost immediately. Though an elaborate set of experiments ensued, the essential idea was arrived at the night I was licking my wounds. A second reason is that experiments kept coming up that led to important, interpretable results. This was so even though the approach was crude, as currents from a distant electrode could have affected discharge in several ways. For this reason, there was no guarantee that the experiments should have worked; yet they did so with little difficulty.

That the Geisler model was important in devising the experiments was especially gratifying. The model had been developed to explain interspike-interval distributions in the superior olive (Goldberg et al., 1964). In the latter study, I learned how to analyze discharge regularity, which proved essential in our later vestibular studies. While the superior-olive study was technically adept, it gave me little personal satisfaction as it led to few functional insights. The moral I draw from this experience is that even a scientific cul de sac can be profitable if it enlarges one's technical skills.

Finally, galvanic stimulation has become increasingly important in human studies as indicated by PubMed listing almost 500 references on this subject over the last 20 years. The subject has been reviewed (Fitzpatrick & Day, 2004; St George & Fitzpatrick, 2011). Our findings have provided a foundation for this work.

11.5 Discharge Regularity and Innervation Patterns

It remained to determine the relation between discharge properties and the morphology of afferent terminations. In the first studies to address this issue, the discharge regularity of afferents was correlated with their fiber diameters as estimated from conduction times (CTs) (Goldberg & Fernández, 1977; Yagi et al., 1977). Thick and thin afferents were irregularly and regularly discharging, respectively. Fibers with intermediate CV*s had a wide range of CTs such that the larger the inferred fiber diameter, the more irregular the discharge. Given the difference in fiber diameters of afferents innervating central and peripheral zones (Lorente de Nó, 1926; Wersäll, 1956), results were consistent with the hypothesis that central calyx fibers were irregular and peripheral bouton fibers were regular.

Several questions remained. From Lorente de Nó's description, intermediate fibers might provide a mixed calyx and bouton (dimorphic) innervation, but such fibers made up a much larger fraction of our sample than implied by ultrastructural studies, which suggested that dimorphic fibers were rare (Wersäll, 1956). In addition, it was unclear what determined the difference between the most regular and irregular of the presumed dimorphic fibers. To study these problems, we used the then recently developed intra-axonal labeling methods (Baird et al., 1988; Goldberg et al., 1990). Here, an afferent is impaled, its physiology characterized,

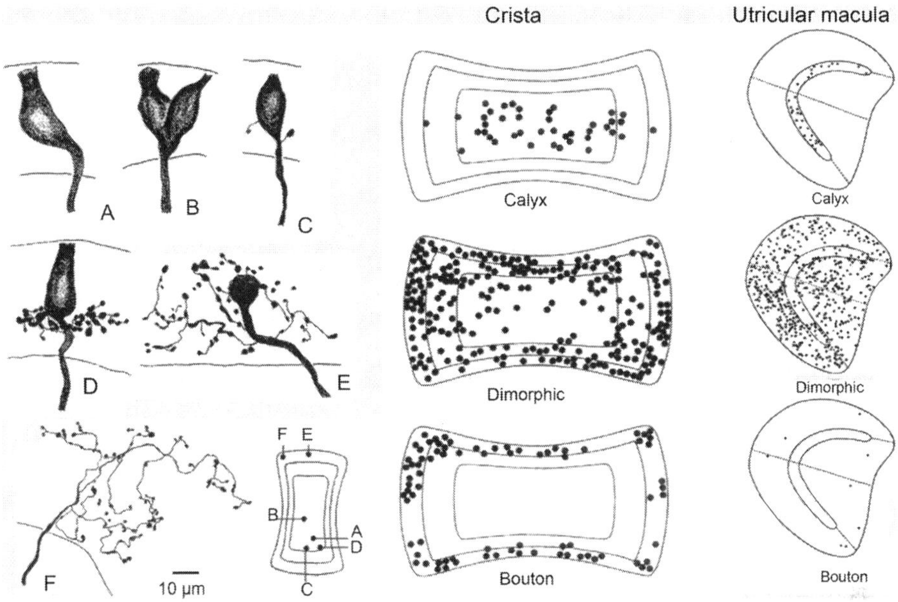

Fig. 11.6 Left: Camera-lucida drawings of six afferent terminals from the cristae, including two calyx units (**a**, **b**), three dimorphic units (**c**–**e**), and a bouton unit (**f**); locations of units indicated on standard map, right of **f**. Right: Standard maps of the cristae and the utricular macula showing the distribution of the three unit types. (From Fernández et al., 1988, 1990)

and it is then marked by the injection of horseradish peroxidase, biocytin or some other intracellular label, which travels from the injection site to the neuroepithelium. After histological processing, the labeled fiber is recovered and can be traced, if one is lucky, to its termination. Because the impaling of axons is far from routine, and is especially difficult in small-diameter fibers, there is a bias favoring thicker fibers. For this reason, intra-axonal labeling was supplemented by the extracellular labeling of large numbers of fibers (Fernández et al., 1988, 1990, 1995).

The extracellular labeling studies were revealing. Three kinds of afferents innervate the vestibular organs of mammals (Fig. 11.6a–f). Calyx fibers are typically unbranched and give rise to single calyx endings innervating from one to three type I hair cells (Fig. 11.6a, b). Bouton fibers branch profusely to innervate several type II hair cells in a territory typically extending 20–50 μm in all directions (Fig. 11.6f). Dimorphic fibers consist of a thicker branch, giving rise to a calyx ending, and thinner collaterals ending as boutons (Fig. 11.6c–e). Calyx fibers are restricted to central/striolar zones and bouton fibers to peripheral/extrastriolar zones (Fig. 11.6, maps on right). Dimorphic fibers, contrary to previous suggestions, are by far the most numerous fiber type and are found throughout each neuroepithelium, including both central (striolar) and peripheral (extrastriolar) zones. Calyx fibers have the thickest axons and bouton fibers the thinnest axons.

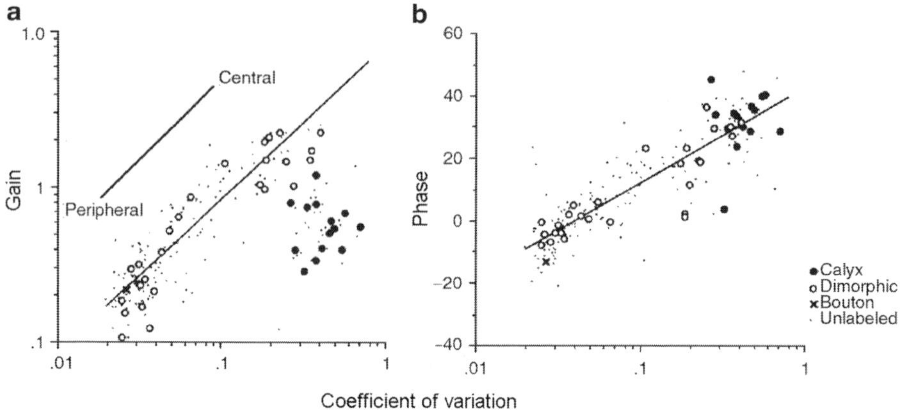

Fig. 11.7 Semicircular canal units. (**a**) Gain (spikes s^{-1}/deg s^{-1}) and (**b**) phase (deg) versus coefficient of variation (CV*) for dye-filled units (large symbols; see key) and unlabeled units (dots). (From Baird et al., 1988)

Based on dye fills of physiologically characterized afferents, calyx and central/striolar dimorphic fibers are irregularly discharging, phasic afferents. Among central/striolar axons, calyx fibers are slightly more irregular and slightly more phasic than dimorphic fibers. Peripherally located dimorphic and bouton fibers are regularly discharging and tonic. The overall results for the SCC are illustrated by plotting rotational gain and phase at 2 Hz as a function of CV* for dye-filled units (Fig. 11.7). Two populations are seen. The first consists of dimorphic and bouton fibers whose rotational gains increase with CV*. The second group consists of calyx afferents characterized by the most irregular discharge, the largest phase leads and rotational gains that are three to five times lower than those of irregular dimorphic fibers. Remarkably, despite their low rotational gains, the galvanic sensitivity of calyx afferents is unexceptional in being quite high, conforming to the power law for the entire population. This last result implies that the spike encoders of calyx afferents are very sensitive; rather, their low rotational gains must be the result of a low synaptic input provided by type I hair cells. Attempts to correlate gains with the numbers of endings or of release sites have been unsuccessful. One suspects that the low gain of calyx units is related to the distinctive electrophysiology of type I hair cells, which is dominated by a large, slow, low-voltage current (I_{KL}) (Correia & Lang, 1990; Rüsch & Eatock, 1996). Another distinctive set of currents with similar properties exists in the calyx ending and may contribute (Chatlani & Goldberg, 2010; Eatock & Songer, 2011).

In the UM, the results were similar to those in the SCC. There was an increase in linear-force gains and phase leads with CV*. The one difference concerned calyx units, which had low gains in the SCC, but high gains in the UM. It has been speculated that the large head rotations achieved during rapid gaze shifts (Armand & Minor, 2001; Liao et al., 2005) could be handled only by the low gains of SCC calyx afferents (Goldberg et al., 2012). In a similar vein vein (Lysakowski & Goldberg, 2008), the lack of low-gain irregular units in the UM may reflect the more modest demands placed on the OTOs by the linear forces occurring during everyday life (MacDougall & Moore 2005). Although this teleological reasoning is comforting, it is no substitute for the need to explain the differences between SCC and OTO calyx afferents in biophysical terms.

There has been only one study in the mammalian crista where the morphology and physiology of dye-filled units could be compared (Baird et al., 1988), Yet, extracellular studies exist in a variety of species—squirrel monkey (Lysakowski et al., 1995), chinchilla (Baird et al., 1988), rhesus monkey (Ramachandran & Lisberger, 2006) and mouse (Yang & Hullar, 2007)—where the pattern of gains and phases from extracellularly recorded units (Fig. 11.7, small symbols) conforms to that of intracellularly recorded, dye-filled units in the chinchilla (Fig. 11.7, large symbols).

Studies have also been done in several non-mammalian species in which both the morphology and physiology of individual SCC afferents have been characterized, including toadfish (Boyle et al., 1991), frogs (Honrubia et al., 1989), lizards (Schessel et al., 1991) and turtles (Brichta & Goldberg, 2000) (reviewed in Lysakowski & Goldberg, 2004). There are cross-species differences within the vertebrate scale. For example, amniotes (reptiles, birds and mammals) have type I hair cells and calyx endings, while non-amniotes (fish and amphibians) do not (Wersäll & Bagger-Sjöbäck, 1974; Lysakowski, 1996). Despite such differences, there are common themes in all species studied. Both regular and irregular units exist side by side and the two fiber types differ in many of the same features as in mammals, including innervation zones, fiber caliber, rotational gains and response dynamics (Lysakowski & Goldberg, 2004).

11.6 Discharge Regularity and Depolarization Sensitivity

Based on the study of galvanic responses (Goldberg et al., 1984), on intracellular recordings of afterhyperpolarizations (AHPs) and miniature (mEPSPs; Highstein & Politoff, 1978; Schessel et al., 1991; Goldberg & Holt, 2013), and on the Smith and Goldberg (1986; Goldberg & Holt, 2013) model, we conclude that discharge regularity jointly depends on post-spike recovery and synaptic noise. How these factors interact is illustrated by simulations of the Smith–Goldberg model (Fig. 11.8), which closely mirror actual recordings (Goldberg & Holt, 2013). For both a regular (Fig. 11.8a, c) and an irregular unit (Fig. 11.8b, d), simulations include a fixed background rate, which is then changed by the addition of a 1 mV depolarization. In the regular unit, where there are large, slow afterhyperpolarizations (AHPs) and small synaptic potentials (miniature EPSPs, mEPSPs), firing is largely determined when AHPs cross threshold and is regular because the AHPs are deterministic. The irregular unit has fast AHPs and large mEPSPs. Because the latter occur on a flat baseline that remains below threshold, discharge only occurs when mEPSPs cross threshold. As the timing of mEPSPs is random, discharge is irregular.

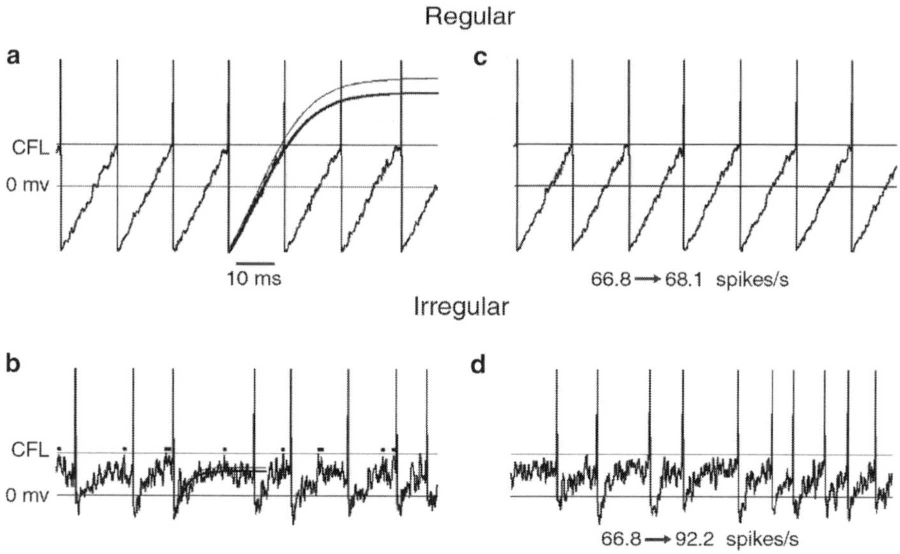

Fig. 11.8 Integrate-and-fire simulations, Smith and Goldberg (1986) model for a regular unit (above) and an irregular unit (below). In both cases, synaptic inputs were set to produce identical discharge rates of 66.8 spikes/s (**a**, **b**), to which was added a 1 mV depolarization (**c**, **d**). CFL, critical firing level. The depolarization slightly speeds up a prominent AHP in the regular unit and renders effective several previously ineffective mEPSPs (dots in b) the irregular unit; as a result, the depolarization results in a much smaller discharge rate increase in the regular, as compared to the irregular, unit. (From Goldberg et al., 2012)

Addition of the depolarization leads in the regular unit to a slight shortening of the AHP and a slight increase in firing rate of 1.3 spikes/s. For the irregular unit, the same depolarization leads to a small shift of the baseline with the result that several hitherto ineffective mEPSPs now cross threshold and cause spikes; the increase in rate is 15.4 spikes/s or more than 10 times that of the regular unit.

11.7 Discharge Regularity and Information Transmission

Two themes are seen in all vertebrates studied to date (reviewed in Lysakowski & Goldberg, 2004). (1) Regular and irregular afferents exist side by side in individual vestibular organs. (2) The two kinds of afferents differ in their response dynamics, which is tonic in regular units and more phasic in irregular units. Were the two features causally linked, this might explain the close association between them. But as we have seen in Fig. 11.4, this is not the case at least in the mammalian crista (Goldberg et al., 1982). To explore this issue further, we consider the encoding of information by the SCCs. There have been two studies of information transmission in vestibular organs, one in the rhesus monkey (Sadeghi et al., 2007a) and the second in the turtle (Rowe & Neiman, 2012). Neither study is ideal for our purposes;

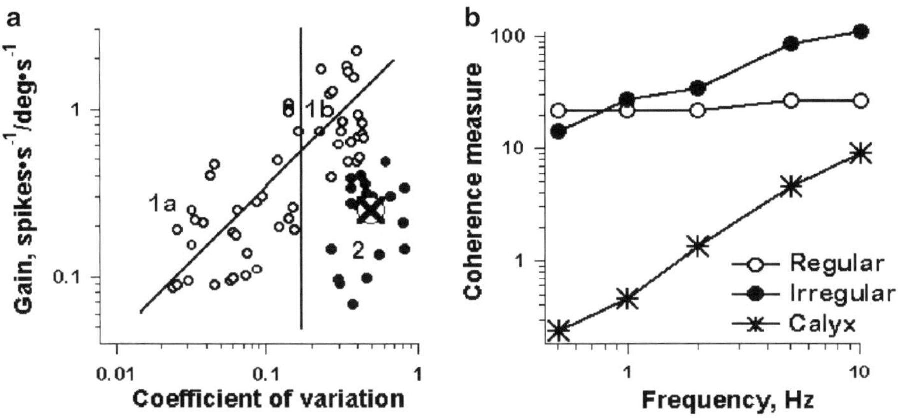

Fig. 11.9 (**a**) Gains versus CV* for horizontal semicircular-canal units, rhesus monkey. The units have been divided into regular (1a), high-gain irregular (1b), and low-gain irregular (2) units; ⊗ depicts the mean values of gains and CV* for group 2. (**b**) Plots the ratios (Gain/CV^2) for the three groups. The ratios are proportional to *Co*(*f*), the coherence between unit discharge and pseudo-random gaussian rotation. (Data in **a** from Ramachandran and Lisberger, 2006)

the first because of a possibly unfortunate sampling of irregular units, the other because low background rates in the turtle limit conclusions about the information encoding of regular afferents to a narrow band of stimulus frequencies.

An adequate sample of units is presented in Fig. 11.9a from the rhesus monkey (Ramachandran & Lisberger, 2006) and consists of the same two groups we have already seen in Fig. 11.7: (1) a group extending from regular units (1a) to high-gain irregular units (1b); and (2) low-gain irregular units (2). Subgroups 1a and 1b are, somewhat arbitrarily separated at CV*=0.18 and, based on labeled units (Fig. 11.7), are made up of bouton and dimorphic units in the peripheral zone (1a) and dimorphic units in the central zone (1b). Group 2 consists of calyx units. As a measure of information transmission, the coherence [*Co*(*f*)] between a band-limited Gaussian stimulus and spike response is calculated and leads to

$$MI(f) = -\log_2\left[1 - Co(f)\right] \tag{11.1}$$

where *MI*(*f*) is the lower-bound estimate of mutual information, expressed as a function of frequency (*f*) and measured in bits/(s Hz). *Co*(*f*)is approximated from so-called linear response theory (LRT) (Lindner et al., 2005; Rowe & Neiman, 2012)

$$Co(f) = \left[\frac{G(f)}{CV}\right]^2 \frac{\sigma^2}{2\bar{r}f_b} \tag{11.2}$$

where *G*(*f*) is the conventional gain, *CV* is the coefficient of variation and σ^2 is the total power in the bandwidth of the Gaussian stimulus. The quantity $\tilde{r}\,.CV^2$ is an approximation for the power spectral density (PSD) that holds provided that the stimulus frequency band (f_b) is much smaller than the resting discharge $(\bar{r})$, a condition that holds for mammalian afferents. In addition, when $Co(f) \ll 1$, *MI*(*f*)is proportional to *Co*(*f*).

Figure 11.9b plots $[G(f)/CV]^2$ versus f for the three groups. The quantity plotted is proportional both to $Co(f)$ and $MI(f)$. Regular units have a flat gain and a low CV, which results in a high $Co(f)$ throughout the frequency bandwidth. High-gain irregular units have a high CV and a high gain that increases with frequency; its $Co(f)$ starts below that for regular units, but crosses the latter near 1 Hz. Low-gain irregular units have the most irregular discharge, a very low gain at low frequencies and the most phasic response dynamics with the most prominent high-frequency gain enhancement; this combination leads to an especially steep $Co(f)$ function that, by extrapolation, crosses the regular $Co(f)$ function only near 20 Hz.

These results suggest that the two kinds of group 1 units may differ in their ability to process information throughout the frequency spectrum with regular and high-gain irregular units doing better at low and high frequencies, respectively. A similar conclusion was reached in the turtle study, although the inability of regular afferents to encode higher frequencies reflects the low resting rates in the turtle (Rowe & Neiman, 2012). As for group 2 (calyx) units, we have already suggested that they function to encode the very rapid head rotations accompanying rapid gaze shifts. A second possibility is that calyx afferents are specialized to encode very high frequency head rotations; along those lines, although there is little energy above 20 Hz, high-frequency components can still influence head trajectories (Armand & Minor, 2001).

Sadeghi et al. (2007a) concluded that mammalian regular units were more efficient in encoding for stimulus frequencies up to 20 Hz. Notice that we arrived at a similar conclusion when regular and low-gain irregular units were compared. In the Sadeghi et al. paper, no distinction was made between the two classes of irregular units. A simple explanation for their results is that there was an oversampling of low-gain units in their irregular group. The presumed lack of high-gain irregular units is consistent with a paucity of these units in a published graph of gain versus CV* (Sadeghi et al., 2007b).

The superior ability of high-gain irregular units to encode higher frequencies depends on their phasic response dynamics, in particular on their high-frequency gain enhancements. We have already emphasized that discharge regularity and response dynamics are not causally linked. The present reasoning suggests that the two attributes are functionally tethered by the necessity to encode information over a broad frequency range. But for the reasoning to be persuasive requires that an irregular discharge provide unique advantages in the encoding of high-frequency stimulation.

Here, we note that the superior performance of high-gain irregular units at high frequencies depends on the gains of these units showing a threefold gain enhancement as frequency increases from 0.5 to 10 Hz. The required increase in spike rate depends on the sensitivity of the spike encoder and on quantal size. Encoder sensitivity is 5–10 times higher in irregular than in regular units (Goldberg et al., 1984; Goldberg & Holt, 2013). In addition, the size of quantal mEPSPs is 5–10 times smaller in regular units. For a regular unit to achieve a threefold gain enhancement would require its hair cells to increase their quantal rate fiftyfold as compared to hair cells supplying irregular afferents. Given their low encoder sensitivities and small mEPSP sizes, the synaptic machinery of regular units might be incapable of meeting such a demand. Stated another way, high-frequency gain enhancements may require the high encoder sensitivities and large mEPSPs of irregular afferents.

11.8 A Case History: SCCs Can Respond to Linear Forces

This section was included in early drafts of the chapter, but was removed because it was not germane to our understanding of discharge regularity. It is included here as a striking example of the sometimes extralogical ways that "a fool can learn."

During everyday life, the SCCs do not respond to linear forces, yet they can be rendered sensitive, for example, during the caloric test. A caloric stimulus leads to a response because it creates an inhomogeneity in the density along the endolymphatic ring, which when acted on by gravity results in a convective movement of endolymph and a cupular displacement. Here, the head is positioned to maximize convection before warm or cold fluid is introduced into the external ear. An alternative is to set up a static thermal gradient and then change the magnitude of the convective current by altering the position of the head and of the endolymphatic ring relative to gravity. The latter mechanism explains how SCC afferents can give vigorous responses when the head is tilted to different positions. Here, the thermal gradient would be produced by the difference between room and body temperature.

As obvious as this mechanism is, I did not appreciate it until the possibility was suggested late one afternoon by a graduate student, William Abend. My first impulse was to dismiss the suggestion as, in our preparation at that time, the exposure of the vestibular nerve and the temporal bone was covered by a plastic cylinder filled with mineral oil. Surely, as I explained to Abend, the cylinder would insulate the labyrinth from room air. Eric Young happened by and I confidently explained the situation to him, emphasizing the insulating properties of the plastic chamber. Young simply walked me over to a window and invited me to place my finger on the glass pane, which was cold to the touch because of the typically frigid air of a Chicago December. I was crestfallen to realize that glass (or plastic for that matter) was a poor insulator and that a caloric response could be the basis for the gravity responses. The realization was particularly galling as Fernández and I had devoted a lot of effort to characterizing the response in the hopes of understanding its etiology.

Having regained my composure, I telephoned Fernández and explained the situation to him. We agreed to meet the next morning, a Saturday. At that meeting, we designed a set of experiments to test the new hypothesis. The basic idea was to measure thermal gradients with thermistors placed at the top and bottom of the chamber and to control the gradients by passing dc currents through a coil of wire fixed to the top of the fluid column. The thermistors were lent to us by an otolaryngology colleague, Leonard Proctor, whose research involved the caloric response. Within a few weeks we had assembled the equipment and had done enough experiments to be convinced that thermal gradients were a potential cause of the SCC gravity responses. We published a paper that included procedures to distinguish SCC and OTO units (Goldberg & Fernandez, 1975).

Another group had discovered the linear-force responses of SCC units without realizing the possibility that they could be artifacts due to the exposure of the labyrinth (Estes et al., 1975). Our paper dampened the enthusiasm in the field for the response, with the exception of Manning Correia (Perachio & Correia, 1983). Correia was dissatisfied with our conclusion that the response was an artifact. In several conversations, I tried to convince him that the only way to settle the issue was to prevent exposure and that this was easily achieved by recording vestibular nerve discharge in the alert, intact preparation. Eventually, such recordings were made (Correia et al., 1992; Somps et al., 1994) with the results that SCC units no longer showed linear-force responses when exposure was minimized.

I end this section with an anecdote. Our paper was published in 1975. Some 30 years later I was a guest of Kathleen Cullen, McGill University. Along with her graduate student, Soroush Sadeghi, we were recording from vestibular nerve fibers in the alert monkey and I was pleased to note the absence of a tilt response in SCC units. At one point, we decided to anesthetize the animal with ketamine. Within a few minutes of an intramuscular injection, the SCC unit we were studying developed a large response on being tilted to new positions. This is hardly surprising. As is now well known, circulating drugs can be differentially absorbed by the crista and the overlying cupula and render the SCC sensitive to tilts. A well-studied example of this is positional alcohol nystagmus (PAN) (Money et al., 1965; Money & Myles, 1974; Fetter et al., 1999). When my colleagues saw the tilt response, they became excited and wondered whether the phenomenon should be studied. My response was: "Been there, done that."

11.9 Current and Future Directions

I now list several problems that are the subjects of current and, possibly, of future research.

- The type I hair cell and its calyx ending are unique structures, each characterized by the presence of large, low-voltage activated conductances. In the case of the hair cell, its large conductance limits the ability of transducer currents to result in presynaptic depolarization and in neurotransmitter release. This has led to the suggestion that depolarization is enhanced by the accumulation of K^+ ions in the intercellular space (Goldberg, 1996). Evidence for such accumulation has been obtained (Holt et al., 2007; Lim et al., 2011; Contini et al., 2012). But the relative importance of accumulation and more conventional neurotransmission remains to be determined.
- There are a whole host of ion channels, neurotransmitter receptors, and accessory molecules in the calyx ending (Lysakowski et al., 2011). Only a few ion channels have been characterized in ganglion cells (Iwasaki et al., 2008; Kalluri et al., 2010) and calyx endings (Dhawan et al., 2010; Meredith et al., 2011, 2012). How do the several entities shape synaptic transmission and repetitive discharge?

- Differences in response dynamics between regular and irregular units may involve several presynaptic and postsynaptic mechanisms (Highstein et al., 2005; Eatock & Songer, 2011; Goldberg et al., 2012). The role of each of these mechanisms needs clarification. There are also differences in the spike gains of calyx units in the SCC (Baird et al., 1988; Goldberg et al., 2012) and OTO (Fernández & Goldberg, 1976c), which are low in the former and high in the latter. We offered a rather tortured explanation for the difference (Fernández and Goldberg, 1976c). The problem deserves further study, as a biophysical basis for the difference may shed light on calyx function.
- The efferent vestibular system (EVS) has been reviewed in a previous SHAR volume (Holt et al., 2011). Although we know much about the peripheral neuroanatomy and actions of the EVS, the function of the system remains a mystery. Studies in decerebrate (Plotnik et al., 2002) and alert animals (Sadeghi et al., 2009) have shown that intense head rotations can lead to efferent-mediated alterations in the discharge of vestibular afferents, but the significance of such effects is unclear.
- The central projections of regular and irregular afferents have been studied both anatomically and physiologically (review: Goldberg, 2000). There is a convergence onto individual secondary neurons of both kinds of afferents (Sato & Sasaki, 1993; Goldberg 2000), but it is unclear whether the separate projections have distinctive functions. One possibility that needs to be investigated is that irregular afferents, because their transient responses are phase led, mediate the fast transmission responsible for the incredibly short latency of the vestibulo-ocular reflex (Huterer & Cullen, 2002). Of the paradigms that have been used to examine the roles of different afferent contingents, the most promising has been a so-called functional ablation in which irregular afferents are silenced by the brief introduction of bilateral anodal galvanic currents (Minor & Goldberg, 1991; Angelaki & Perachio, 1993). Previous functions tested in this way lacked the high-frequency components that information theory suggests are targeted by irregular afferents.
- Traditionally, the vestibular system has been viewed as a set of reflexes that ensure gaze and postural stability. While these functions are important, they should not obscure the fact that the system involves complex transformations that allow it to distinguish, for example, tilt from translation and voluntary from imposed head movements (Angelaki & Cullen, 2008). Such transformation involve the cerebellum (Angelaki et al., 2010; Yakusheva et al., 2007; Brooks & Cullen, 2009), which also participates in the adaptive plasticity of the vestibulo-ocular reflex (du Lac et al., 1995; Jörntell & Hansel, 2007). In addition, the vestibular system projects throughout the neuraxis, including ascending connections to the neocortex, where it interacts with visual and other inputs in the computation of visual flow (Angelaki et al., 2011). A vestibular projection to the hippocampal gyrus is crucial in the firing of spatial-coding cells (Taube, 2007). As these examples illustrate, the vestibular system, rather than being confined to reflex stabilization, participates in a wide variety of functions and is a major component of the proprioceptive sixth sense (Goldberg et al., 2012). Study of these more general functions is in its infancy.

References

Adams, J. C., & Mugnaini, E. (1990). Immunocytochemical evidence for inhibitory and disinhibitory circuits in the superior olive. *Hearing Research*, 49, 281–296.

Adrian, E. D. (1943). Discharge from vestibular receptors in the cat. *Journal of Physiology*, 101, 389–407.

Angelaki, D. E., & Perachio, A. A. (1993). Contribution of irregular semicircular canal afferents to the horizontal vestibuloocular response during constant velocity rotation. *Journal of Neurophysiology*, 69(3), 996–999.

Angelaki, D. E., & Cullen, K. E. (2008). Vestibular system: The many facets of a multimodal sense. *Annual Review of Neuroscience*, 31, 125–150.

Angelaki, D. E., Yakusheva, T. A., Green, A. M., Dickman, J. D., & Blazquez, P. M. (2010). Computation of egomotion in the macaque cerebellar vermis. *Cerebellum*, 9(2), 174–182.

Angelaki, D. E., Gu, Y., & DeAngelis, G. C. (2011). Visual and vestibular cue integration for heading perception in extrastriate visual cortex. *Journal of Physiology*, 589(4), 825–833.

Armand, M., & Minor, L. B. (2001). Relationship between time- and frequency-domain analyses of angular head movements in the squirrel monkey. *Journal of Computational Neuroscience*, 11, 217–239.

Aw, S. T., Todd, M. J., Aw, G. E., Weber, K. P., & Halmagyi, G. M. (2008). Gentamicin vestibulotoxicity impairs human electrically evoked vestibulo-ocular reflex. *Neurology*, 71, 1776–1782.

Baird, R. A., Desmadryl, G., Fernández, C., & Goldberg, J. M. (1988). The vestibular nerve of the chinchilla. II. Relation between afferent response properties and peripheral innervation patterns in the semicircular canals. *Journal of Neurophysiology*, 60(1), 182–203.

Boudreau, J. C., & Tsuchitani, C. (1968). Binaural interaction in the cat superior olive S segment. *Journal of Neurophysiology*, 31, 442–454.

Boyle, R., Carey, J. P., & Highstein, S. M. (1991). Morphological correlates of response dynamics and efferent stimulation in horizontal semicircular canal afferents of the toadfish, *Opsanus tau*. *Journal of Neurophysiology*, 66(5), 1504–1521.

Boyle, R., Goldberg, J. M., & Highstein, S. M. (1992). Inputs from regularly and irregularly discharging vestibular nerve afferents to secondary neurons in squirrel monkey vestibular nuclei. III. Correlation with vestibulospinal and vestibuloocular output pathways. *Journal of Neurophysiology*, 68(2), 471–484.

Brichta, A. M., & Goldberg, J. M. (2000). Morphological identification of physiologically characterized afferents innervating the turtle posterior crista. *Journal of Neurophysiology*, 83(3), 1202–1223.

Brooks, J. X., & Cullen, K. E. (2009). Multimodal integration in rostral fastigial nucleus provides an estimate of body movement. *Journal of Neuroscience*, 29(34), 10499–10511.

Brownell, W. E. (1975). Organization of the cat trapezoid body and the discharge characteristics of its fibers. *Brain Research.*, 94(3), 413–433.

Camis, M. (1930). *The physiology of the vestibular apparatus*, translated by R.S. Creed. Oxford: Clarendon Press.

Chatlani, S., & Goldberg, J. M. (2010). Whole-cell recordings from calyx endings in the turtle posterior crista. *Association for Research in Otolaryngology Abstract*, 937, 702–703.

Contini, D., Zampini, V., Tavazzani, E., Maggistretti, J., Russo, G., Prigioni, I., & Masetto, S. (2012). Intercellular K^+ accumulation depolarizes type I vestibular hair cells and their associated afferent nerve calyx. *Neuroscience*, 227, 232–246.

Correia, M. J., & Lang, D. G. (1990). An electrophysiological comparison of solitary type I and type II vestibular hair cells. *Neuroscience Letters*, 116, 106–111.

Correia, M. J., Perachio, A. A., Dickman, J. D., Kozlovskaya, I. B., Sirota, M. G., Yakushin, S. B., & Beloozerova, I. N. (1992). Changes in monkey horizontal semicircular canal afferent responses after spaceflight. *Journal of Applied Physiology*, 73(2 Suppl), 112S–120S.

deVries, H. (1950). The mechanics of the labyrinthine otoliths. *Acta Oto-Laryngologica (Stockholm)*, 38, 262–273.

Dhawan, R., Mann, S. E., Meredith, F. L., & Rennie, K. J. (2010). K^+ currents in isolated vestibular afferent calyx terminals. *Journal of the Association for Research in Otolaryngology*, 11, 463–476.

Duensing, F., & Schaefer, K.-P. (1958). Die Aktivität einzelner Neurone im Bereich der Vestibulariskerne bei Horizontalbeschleunigungen unter besonderer Berücksichtigung des vestibulären Nystagmus. *Archiv für Psychiatrie und Nervenkrankheiten*, 198, 225–252.

du Lac, S., Raymond, J. L., Sejnowski, T. J., & Lisberger, S. G. (1995). Learning and memory in the vestibulo-ocular reflex. *Annual Review of Neuroscience*, 18, 409–441.

Eatock, R. A., & Songer, J. E. (2011). Vestibular hair cells and afferents: Two channels for head motion signals. *Annual Review of Neuroscience*, 34, 501–534.

Estes, M. S., Blanks, R. H., & Markham, C. H. (1975). Physiologic characteristics of vestibular first-order canal neurons in the cat. I. Response plane determination and resting discharge characteristics. *Journal of Neurophysiology*, 38, 1232–1249.

Ewald, J. R. (1892). *Physiologische Untersuchungen ueber das Endorgan des Nervus Octavus*. Wiesbaden: J.F. Bergmann.

Fernández, C., & Goldberg, J. M. (1971). Physiology of peripheral neurons innervating semicircular canals of the squirrel monkey. II. Response to sinusoidal stimulation and dynamics of peripheral vestibular system. *Journal of Neurophysiology*, 34(4), 661–675.

Fernández, C., & Goldberg, J. M. (1976a). Physiology of peripheral neurons innervating otolith organs of the squirrel monkey. I. Response to static tilts and to long-duration centrifugal force. *Journal of Neurophysiology*, 39(5), 970–984.

Fernández, C., & Goldberg, J. M. (1976b). Physiology of peripheral neurons innervating otolith organs of the squirrel monkey. II. Directional selectivity and force-response relations. *Journal of Neurophysiology*, 39(5), 985–995.

Fernández, C., & Goldberg, J. M. (1976c). Physiology of peripheral neurons innervating otolith organs of the squirrel monkey. III. Response dynamics. *Journal of Neurophysiology*, 39(5), 996–1008.

Fernández, C., Baird, R. A., & Goldberg, J. M. (1988). The vestibular nerve of the chinchilla. I. Peripheral innervation patterns in the horizontal and superior semicircular canals. *Journal of Neurophysiology*, 60(1), 167–181.

Fernández, C., Goldberg, J. M., & Baird, R. A. (1990). The vestibular nerve of the chinchilla. III. Peripheral innervation patterns in the utricular macula. *Journal of Neurophysiology*, 63(4), 767–780.

Fernández, C., Lysakowski, A., & Goldberg, J. M. (1995). Hair-cell counts and afferent innervation patterns in the cristae ampullares of the squirrel monkey with a comparison to the chinchilla. *Journal of Neurophysiology*, 73(3), 1253–1269.

Fetter, M., Haslwanter, T., Bork, M., & Dichgans, J. (1999). New insights into positional alcohol nystagmus using three-dimensional eye-movement analysis. *Annals of Neurology*, 45, 216–223.

Fitzpatrick, R. C., & Day, B. L. (2004). Probing the human vestibular system with galvanic stimulation. *Journal of Applied Physiology*, 96, 2301–2316.

Fuchs, A. F., & Kimm, J. (1975). Unit activity in vestibular nucleus of the alert monkey during horizontal angular acceleration and eye movement. *Journal of Neurophysiology*, 38, 1140–1161.

Galambos, R., & Davis, H. (1946). Action potentials from single auditory-nerve fibers? *Science*, 108, 513.

Geisler, C. D., & Goldberg, J. M. (1966). A stochastic model of the repetitive activity of neurons. *Biophysics Journal*, 6, 53–69.

Gernandt, B. (1949). Response of mammalian vestibular neurons to horizontal rotation and caloric stimulation. *Journal of Neurophysiology*, 12, 173–184.

Goldberg, J. M. (1996). Theoretical analysis of intercellular communication between the vestibular type I hair cell and its calyx ending. *Journal of Neurophysiology*, 76, 1942–1957.

Goldberg, J. M. (2000). Afferent diversity and the organization of central vestibular pathways. *Experimental Brain Research*, 130(3), 277–297.

Goldberg, J. M., & Holt, J. C. (2013). Discharge regularity in the turtle posterior crista: comparisons between experiment and theory. *J Neurophysiol*, 110(12), 2830–2848.

Goldberg, J. M., & Brown, P. B. (1968). Functional organization of the dog superior olivary complex: An anatomical and electrophysiological study. *Journal of Neurophysiology*, 31(4), 639–656.

Goldberg, J. M., & Brown, P. B. (1969). Response of binaural neurons of dog superior olivary complex to dichotic tonal stimuli: Some physiological mechanisms of sound localization. *Journal of Neurophysiology*, 32(4), 613–636.

Goldberg, J. M., & Fernández, C. (1971a). Physiology of peripheral neurons innervating semicircular canals of the squirrel monkey. I. Resting discharge and response to constant angular accelerations. *Journal of Neurophysiology*, 34(4), 635–660.

Goldberg, J. M., & Fernández, C. (1971b). Physiology of peripheral neurons innervating semicircular canals of the squirrel monkey. III. Variations among units in their discharge properties. *Journal of Neurophysiology*, 34(4), 676–684.

Goldberg, J. M., & Fernández, C. (1975). Responses of peripheral vestibular neurons to angular and linear accelerations in the squirrel monkey. *Acta Oto-Laryngologica*, 80, 101–110.

Goldberg, J. M., & Fernández, C. (1977). Conduction times and background discharge of vestibular afferents. *Brain Research*, 122(3), 545–550.

Goldberg, J. M., & Fernández, C. (1980). Efferent vestibular system in the squirrel monkey: Anatomical location and influence on afferent activity. *Journal of Neurophysiology*, 43(4), 986–1025.

Goldberg, J. M., Adrian, H. O., & Smith, F. D. (1964). Response of neurons of the superior olivary complex of the cat to acoustic stimuli of long duration. *Journal of Neurophysiology*, 27, 706–749.

Goldberg, J. M., Fernández, C., & Smith, C. E. (1982). Responses of vestibular-nerve afferents in the squirrel monkey to externally applied galvanic currents. *Brain Research*, 252(1), 156–160.

Goldberg, J. M., Smith, C. E., & Fernández, C. (1984). Relation between discharge regularity and responses to externally applied galvanic currents in vestibular nerve afferents of the squirrel monkey. *Journal of Neurophysiology*, 51, 1236–1256.

Goldberg, J. M., Highstein, S. M., Moschovakis, A. K., & Fernández, C. (1987). Inputs from regularly and irregularly discharging vestibular nerve afferents to secondary neurons in the vestibular nuclei of the squirrel monkey. I. An electrophysiological analysis. *Journal of Neurophysiology*, 58(4), 700–718.

Goldberg, J. M., Desmadryl, G., Baird, R. A., & Fernández, C. (1990). The vestibular nerve of the chinchilla. V. Relation between afferent discharge properties and peripheral innervation patterns in the utricular macula. *Journal of Neurophysiology*, 63(4), 791–804.

Goldberg, J. M., Wilson, V. J., Cullen, K. E., Angelaki, D. E., Broussard, D. M., Büttner-Ennever, J., Fukushima, K., & Minor, L. B. (2012). *The Vestibular system: A sixth sense*. New York: Oxford.

Halmagyi, G. M., Weber, K. P., Aw, S. T., Todd, M. J., & Curthoys, I. S. (2008). Impulsive testing of semicircular canal function. *Progress in Brain Research*, 171, 187–194.

Hassfurth, B., Grothe, B., & Koch, U. (2010). The mammalian interaural time difference detection circuit is differentially controlled by GABA B receptors during development. *Journal of Neuroscience*, 30, 15–27.

Highstein, S. M., & Politoff, A. L. (1978). Relation of interspike baseline activity to the spontaneous discharges of primary afferents from the labyrinth of the toadfish, Opsanus tau. *Brain Research*, 150(1), 182–187.

Highstein, S. M., Goldberg, J. M., Moschovakis, A. K., & Fernández, C. (1987). Inputs from regularly and irregularly discharging vestibular nerve afferents to secondary neurons in the vestibular nuclei of the squirrel monkey. II. Correlation with output pathways of secondary neurons. *Journal of Neurophysiology*, 58(4), 719–738.

Highstein, S. M., Rabbitt, R. D., & Boyle, R. (1996). Determinants of semicircular canal afferent response dynamics in the toadfish, Opsanus tau. *Journal of Neurophysiology*, 75(2), 575–596.

Highstein, S. M., Rabbitt, R. D., Holstein, G. R., & Boyle, R. D. (2005). Determinants of spatial and temporal coding by semicircular canal afferents. *Journal of Neurophysiology*, 93(5), 2359–2370.

Hind, J. E., Anderson, D. J., Brugge, J. F., & Rose, J. E. (1967). Coding of information pertaining to paired low-frequency tones in single auditory nerve fibers of the squirrel monkey. *Journal of Neurophysiology*, 30, 794–816.

Holt, J. C., Chatlani, S., Lysakowski, A., & Goldberg, J. M. (2007). Quantal and non-quantal transmission in calyx-bearing fibers of the turtle posterior crista. *Journal of Neurophysiology*, 98, 1083–1101.

Holt, J. C., Lysakowski, A., & Goldberg, J. M. (2011). The efferent vestibular system. In D. K. Ryugo, R. R. Fay, & A. N. Popper (Eds.), *Auditory and vestibular efferents*. New York: Springer, 135–186.

Honrubia, V., Hoffman, L. F., Sitko, S., & Schwartz, I. R. (1989). Anatomic and physiological correlates in bullfrog vestibular nerve. *Journal of Neurophysiology*, 61, 688–701.

Huterer, M., & Cullen, K. E. (2002). Vestibuloocular reflex dynamics during high-frequency and high-acceleration rotations of the head on body in rhesus monkey. *Journal of Neurophysiology*, 88, 13–28.

Iwasaki, S., Chihara, Y., Komuta, Y., Ito, K., & Sahara, Y. (2008). Low-voltage activated potassium channels underlie the regulation of intrinsic firing properties of rat vestibular ganglion cells. *Journal of Neurophysiology*, 100, 2192–2204.

Jörntell, H., Hansel C. (2006). Synaptic memories upside down: bidirectional plasticity at cerebellar parallel fiber-Purkinje cell synapses. *Neuron*, 52(2), 227–238.

Kalluri, R., Xue, J., & Eatock, R. A. (2010). Ion channels set spike timing regularity of mammalian vestibular afferent neurons. *Journal of Neurophysiology*, 104, 2034–2051.

Keller, E. L. (1976). Behavior of horizontal semicircular canal afferents in alert monkey during vestibular and optokinetic stimulation. *Experimental Brain Research*, 24(5), 459–471.

Kiang, N. Y. S. (1965) Discharge patterns of single fibers in the cat's auditory nerve. *M.I.T. Research Monograph No.* 35. Cambridge, MA: MIT Press.

Li, A., Xue, J., & Peterson, E. H. (2008). Architecture of the mouse utricle: Macular organization and hair bundle heights. *Journal of Neurophysiology*, 99, 18–33.

Liao, K., Kumar, A. N., Han, Y. H., Grammer, V. A., Gedeon, B. T., & Leigh, R. J. (2005). Comparison of velocity waveforms of eye and head saccades. *Annals of the New York Academy of Sciences*, 1039, 477–479.

Lifschitz, W. S. (1973). Responses from the first order neurons of the horizontal semicircular canal in the pigeon. *Brain Research*, 63, 43–57.

Lim, R., Kindig, A. E., Donne, S. W., Callister, R. J., & Brichta, A. M. (2011). Potassium accumulation between type I hair cells and calyx terminals in mouse crista. *Experimental Brain Research*, 210, 607–621.

Lindeman, H. H. (1969). Studies on the morphology of the sensory regions of the vestibular apparatus. *Ergebnisse Anatomie Entwicklungsgeschite*, 42, 1–113.

Lindner, B., Chacron, M. J., & Longtin, A. (2005). Integrate-and-fire neurons with threshold noise: A tractable model of how interspike interval correlations affect neuronal signal transmission. *Physical Review E*, 72, 021911.

Lorente de Nó, R. (1926). Etudes sur l'anatomie et la physiologie du labyrinthe de l'oreille et du VIIIe nerf. Deuxiéme partie. Quelques données au sujet d l'anatomie des organes sensoriels du labyrinthe. *Travaux du Laboratoire de Recherches Biologiques de l'Université de Madrid*, 24, 53–153.

Louie, A. W., & Kimm, J. (1976). The response of 8th nerve fibers to horizontal sinusoidal oscillation in the alert monkey. *Experimental Brain Research*, 24, 447–457.

Lowenstein, O. (1955). The effect of galvanic polarization on the impulse discharge from sense endings in the isolated labyrinth in the thornback ray (*Raja clavata*). *Journal of Physiology*, 127, 104–117.

Lowenstein, O. (1956). Peripheral mechanisms of equilibrium. *British Medical Bulletin*, 12(2), 114–118.

Lowenstein, O., & Sand, A. (1936). The activity of the horizontal semicircular canal of the dogfish, *Scyllium canalicula. Journal of Experimental Biology*, 13, 416–428.

Lowenstein, O., & Sand, A. (1940a). The mechanisms of the semicircular canal. A study of the response of single-fiber preparations to angular accelerations and to rotation at constant speed. *Proceedings of the Royal Society of London. Series B: Biological Sciences*, 129, 256–275.

Lowenstein, O., & Sand, A. (1940b). The individual and integrated activity of the semicircular canals of the elasmobranch labyrinth. *Journal of Physiology*, 99, 89–101.

Lowenstein, O., & Wersäll, J. (1954). A functional interpretation of the electron microscopic structure of the sensory hairs in the cristae of the elasmobranch, *Raja clavata. Nature*, 184, 1807–1810.

Lysakowski, A. (1996). Synaptic organization of the crista ampullaris in vertebrates. *Annals of the New York Academy of Sciences*, 781, 164–182.

Lysakowski, A., & Goldberg, J. M. (2004). Morphophysiology of the vestibular periphery. In S. M. Highstein, A. N. Popper, & R. R. Fay (Eds.), *The vestibular system* (pp. 57–152). New York: Springer-Verlag.

Lysakowski, A., & Goldberg, J. M. (2008). Ultrastructural analysis of the cristae ampullares in the squirrel monkey (Saimiri sciureus). *Journal of Comparative Neurology*, 511, 47–64.

Lysakowski, A., Minor, L. B., Fernández, C., & Goldberg, J. M. (1995). Physiological identification of morphologically distinct afferent classes innervating the cristae ampullares of the squirrel monkey. *Journal of Neurophysiology*, 73, 1270–1281.

Lysakowski, A., Gaboyard-Niay, S., Calin-Jageman, I., Chatlani, S., Price, S. D., & Eatock, R. A. (2011). Molecular microdomains in a sensory terminal, the vestibular calyx ending. *Journal of Neuroscience*, 31, 10101–10114.

MacDougall, H. G., & Moore, S. T. (2005). Marching to the beat of the same drummer: The spontaneous tempo of human locomotion. *Journal of Applied Physiology*, 99, 1164–1173.

Meredith, F., Li, G. Q., & Rennie, K. J. (2011). Postnatal expression of an apamin-sensitive K(Ca) current in vestibular calyx terminals. *Journal of Membrane Biology*, 244, 81–91.

Meredith, F. L., Benke, T. A., & Rennie, K. (2012). Hyperpolarization-activated current (I(h)) in vestibular calyx terminals: Characterization and role in shaping postsynaptic events. *Journal of the Association of Research in Otolaryngology*, 13, 745–758.

Miles, F. A. (1974). Single unit firing patterns in the vestibular nuclei related to voluntary eye movements and passive body rotation in conscious monkeys. *Brain Research*, 71, 215–224.

Minor, L. B., & Goldberg, J. M. (1991). Vestibular-nerve inputs to the vestibulo-ocular reflex: A functional-ablation study in the squirrel monkey. *Journal of Neuroscience*, 11(6), 1636–1648.

Money, K. E. (1970). Motion sickness. *Physiological Reviews*, 50, 1–39.

Money, K. E., & Myles, W. S. (1974). Heavy water nystagmus and effects of alcohol. *Nature*, 247, 404–405.

Money, K. E., Johnson, W. H., & Corlett, B. M. (1965). Role of the semicircular canals in positional alcohol nystagmus. *American Journal of Physiology*, 208, 1065–1070.

Perachio, A. A., & Correia, M. J. (1983). Responses of semicircular canal and otolith afferents to small angle static head tilts in the gerbil. *Brain Research*, 280, 287–298.

Plotnik, M., Marlinski, V., & Goldberg, J. M. (2002). Reflections of efferent activity in rotational responses of chinchilla vestibular afferents. *Journal of Neurophysiology*, 88(3), 1234–1244.

Ramachandran, R., & Lisberger, S. G. (2006). Transformation of vestibular signals into motor commands in the vestibulo-ocular reflex pathways of monkeys. *Journal of Neurophysiology*, 96(3), 1061–1074.

Reason, J. T., & Brand, J. J. (1975). *Motion sickness*. New York: Academic Press.

Rose, J. E., Brugge, J. F., Anderson, D. J., & Hind, J. E. (1967). Phase-locked response to low-frequency tones in single auditory nerve fibers of the squirrel monkey. *Journal of Neurophysiology*, 30, 769–793.

Rowe, M. H., & Neiman, A. B. (2012). Information analysis of posterior canal afferents in the turtle, *Trachemys scripta elegans. Brain Research*, 1434, 226–242.

Ruch, T. C., & Patton, H. D. (1965) *Physiology and biophysics*. Philadelphia: Saunders.

Ruggero, M. A. (1973). Response to noise of auditory nerve fibers in the squirrel monkey. *Journal of Neurophysiology*, 36(4), 569–587.

Rüsch, A., & Eatock, R. A. (1996). A delayed rectifier conductance in type I hair cells of the mouse utricle. *Journal of Neurophysiology*, 76, 995–1004.
Sadeghi, S. G., Chacron, M. J., Taylor, M. C., & Cullen, K. E. (2007a). Neural variability, detection thresholds, and information transmission in the vestibular system. *Journal of Neuroscience*, 27(4), 771–781.
Sadeghi, S. G., Minor, L. B., & Cullen, K. E. (2007b). Response of vestibular-nerve afferents to active and passive rotations under normal conditions and after unilateral labyrinthectomy. *Journal of Neurophysiology*, 97, 1503–1514.
Sadeghi, S. G., Goldberg, J. M., Minor, L. B., & Cullen, K. E. (2009). Efferent-mediated responses in vestibular nerve afferents of the alert macaque. *Journal of Neurophysiology*, 101, 988–1001.
Sato, F., & Sasaki, H. (1993). Morphological correlations between spontaneously discharging primary vestibular afferents and vestibular nucleus neurons in the cat. *Journal of Comparative Neurology*, 333(4), 554–566.
Schessel, D. A., Ginzberg, R., & Highstein, S. M. (1991). Morphophysiology of synaptic transmission between type I hair cells and vestibular primary afferents. An intracellular study employing horseradish peroxidase in the lizard, *Calotes versicolor. Brain Research*, 544, 1–16.
Schweizer, F. E., Savin, D., Luu, C., Sultemeier, D. R., & Hoffman, L. F. (2009). Distribution of high-conductance calcium-activated potassium channels in rat vestibular epithelia. *Journal of Comparative Neurology*, 517, 134–145.
Smith, C., & Goldberg, J. M. (1986). A stochastic afterhyperpolarization model of repetitive activity in vestibular afferents. *Biological Cybernetics*, 54, 41–51.
Somps, C. J., Schor, R. H., & Tomko, D. L. (1994). *Vestibular afferent responses to linear accelerations in the alert squirrel monkey*. Moffett Field, CA: Ames Research Center.
St George, R. J., & Fitzpatrick, R. C. (2011). The sense of self-motion, orientation and balance explored by vestibular stimulation. *Journal of Physiology*, 589, 807–813.
Steinhausen, W. (1933). Über die Beobachtung der Cupula in den Bogengangsampullen des Labyrinthes des lebende Hechts. *Pflugers Archiv fur Die Gesamte Physiologie des Menschen und der Tiere*, 232, 500–512.
Taube, J. S. (2007). The head direction signal: Origins and sensory-motor integration. *Annual Review of Neuroscience*, 30, 181–207.
Tomko, D. L., Peterka, R. J., & Schor, R. H. (1981a). Responses to head tilt in cat eight nerve afferents. *Experimental Brain Research*, 41, 216–221.
Tomko, D. L., Peterka, R. J., Schor, R. H., & O'Leary, D. P. (1981b). Response dynamics of horizontal canal afferents in barbiturate-anesthetized cats. *Journal of Neurophysiology*, 45, 376–396.
Trincker, D. (1962). The transformation of mechanical stimulus into nervous excitation by the labyrinthine receptors. *Society for Experimental Biology*, 16, 289–316.
Wersäll, J. (1956). Studies on the structure and innervation of the sensory epithelium of the cristae ampullaris in the guinea pig. A light and electron microscopic investigation. *Acta Oto-Laryngologica Supplementum*, 126, 1–85.
Wersäll, J., & Bagger-Sjöbäck, D. (1974). Morphology of the vestibular sense organ. In H. H. Kornhuber (Ed.), *Vestibular system: Basic mechanisms* (pp. 123–170). Handbook of Sensory Physiology, Vol. 6, part I. Berlin: Springer-Verlag.
Yagi, T., Simpson, N. E., & Markham, C. H. (1977). The relationship of conduction velocity to other physiological properties of the cat's horizontal canal neurons. *Experimental Brain Research*, 30, 587–600.
Yakusheva, T. A., Shaikh, A. G., Green, A. M., Blazquez, P.M., Dickman, J. D., Angelaki, D. E. (2007). Purkinje cells in posterior cerebellar vermis encode motion in an inertial reference frame. *Neuron*, 54(6), 973–985.
Yang, A., & Hullar, T. E. (2007). Relationship of semicircular canal size to vestibular-nerve afferent sensitivity in mammals. *Journal of Neurophysiology*, 98(6), 3197–3205.
Young, E. D., & Brownell, E. D. (1976). Responses to tones and noise of single cells in dorsal cochlear nucleus of unanesthetized cats. *Journal of Neurophysiology*, 39(2), 282–300.

Chapter 12
Aging, Hearing Loss, and Speech Recognition: Stop Shouting, I Can't Understand You

Sandra Gordon-Salant

Sandra Gordon-Salant

S. Gordon-Salant (✉)
Department of Hearing and Speech Sciences, University of Maryland,
College Park, MD 20742, USA
e-mail: sgsalant@umd.edu

A.N. Popper and R.R. Fay (eds.), *Perspectives on Auditory Research*, Springer Handbook of Auditory Research 50, DOI 10.1007/978-1-4614-9102-6_12,

12.1 Introduction

My interest in age-related hearing loss began in 1984, as my father (age 75) exhibited significant noise-induced hearing loss but refused to wear a hearing aid. How frustrating for an audiologist to be unable to convince a parent to use amplification that would benefit him significantly. Eventually, his additional age-related hearing loss progressed to the degree that he used his hearing aid routinely, but he continued to experience difficulty in understanding speech in all but quiet face-to-face situations. In later years, my mother complained of specific challenges listening to accented English on televised broadcasts. These and other complaints of my elders inspired me to investigate the nature of age-related hearing loss and specifically, the factors underlying older listeners' speech recognition problems.

Is hearing loss an inevitable consequence of the aging process? Are humans destined to say, "I can hear you but I can't understand what you are saying"? Communication with others is fundamental to human existence and is the essence of social interaction; untreated decline in auditory abilities with age often results in social isolation and the appearance of dementia. A great deal is now known about the range of anatomical and physiological changes in the auditory system that occurs with age, but the functional consequences for communication and the benefit of assistive technology and rehabilitation are currently an intense area of scientific inquiry. How well an individual older listener understands speech in degraded situations, including poor acoustic environments and the speech of less-than-ideal talkers, depends not only on auditory capacity but also on cognitive and multisensory integration abilities. Also of great interest are lifestyles that young and middle-aged adults can embrace to promote healthy auditory aging to prevent progressive hearing loss with advancing age and preserve the ability to process distorted speech signals or speech in noisy environments. This chapter briefly reviews that state of knowledge on age-related hearing loss from 20 years ago (ca. 1992) and highlights some significant findings in recent years that have crystallized the current view of mechanisms underlying presbycusis (defined as hearing loss associated with the aging process) and its functional consequences. The chapter culminates in suggestions for emerging areas of research aimed at unraveling the persistent and interrelated issues of understanding and improving older listeners' difficulty perceiving speech in everyday communication situations.

12.2 Historical Perspective

12.2.1 Epidemiology

In 1992, the overall prevalence rate of hearing loss among those 65 years and older was generally accepted as approximately 30% in the United States. It was well known that older men exhibited poorer hearing sensitivity than older women, especially in the higher frequencies, and the typical audiogram was mild-to-moderate in degree with a gradually sloping configuration. Data from the hearing aid industry suggested that the acquisition rate of hearing aids among the older hearing-impaired population was approximately 23%, and surveys suggested that about half of these hearing aids remained in the drawer. At that time, ear-level analog devices were common, although hybrid analog-digital technology was beginning to emerge. The quality of amplified sound, especially in noise, and the stigma associated with hearing loss were often cited as the most common reasons for rejecting hearing aids.

12.2.2 Models of Presbycusis

The classic work of Harold Schuknecht (1974) identified four forms of presbycusis that were based on human temporal bone studies and corresponding audiometric profiles: (1) sensory, associated with loss of cochlear hair cells particularly in the basal turn; (2) metabolic, attributed to atrophy of the stria vascularis; (3) cochlear conductive, theorized as a stiffening of the basilar membrane; and (4) neural, ascribed to deterioriation of neurons comprising the auditory branch of cranial nerve (CN) VIII. Each form of presbycusis was associated with specific auditory manifestations, although Schuknecht and Gacek (1993) recognized that multiple forms of presbycusis can coexist with an additive effect on auditory thresholds. Others suggested the existence of "conductive" presbycusis, thought to be associated with elastic changes to the tympanic membrane, fixation of the ossicles, arthritic changes to muscles and ligaments, and Eustachian tube dysfunction (Glorig & Davis, 1961). Despite these possible anatomical changes, a small proportion of older people exhibit conductive hearing loss (Cruickshanks et al., 1998). There was also evidence of deterioration of the central auditory pathways in humans (Kirikae et al., 1964). Willott (1991) theorized that damage to the central auditory pathways could result either from direct deterioration of central structures (called the central effects of biological aging [CEBA]) or from the subsequent, retrograde effects of peripheral pathology on central nervous system (CNS) structures (called the central effect of peripheral pathology [CEPP]). Anatomical studies of aging human brains indicate a decrease in the volume of nuclei in the central auditory nervous system, a decrease in the number of dendrites and dendritic spines in the central auditory pathways, and considerable variability in the surviving neuronal population (Willott, 1991) lending some support to the theory of central biological aging. Nevertheless, strong evidence from animal studies (C57 mouse) also supports secondary effects of peripheral pathology, especially as a result of high-frequency sensorineural hearing loss (Willott, 1991).

12.2.3 Speech Understanding Performance

Prior to 1992, the effects of auditory aging were quantified by a performance comparison between younger listeners with normal hearing and older listeners with hearing loss on speech recognition measures conducted in quiet, noise, and reverberant environments. Differences between groups were typically ascribed to aging, but differences in signal audibility could also have accounted for group effects. One noteworthy exception was a study by Dubno et al. (1984), which showed that older listeners performed more poorly than younger listeners with matched audiograms, and hearing-impaired listeners performed more poorly than normal-hearing listeners on a low-context sentence recognition test presented in a multitalker babble background. The authors interpreted this finding as reflecting an audibility factor that limits performance for hearing-impaired listeners and a distortion factor that limits performance for older listeners; these two factors may combine in certain signal and noise conditions to produce excessively poor performance among older hearing-impaired listeners. A notable finding in this study is that the availability of contextual cues had a dramatic effect on the performance of older listeners, reducing the age differences considerably.

Early investigations also examined the effects of aging on recognition of time-compressed speech, a simulation of natural, rapid speech (e.g., Wingfield et al., 1985). These investigations attempted not only to quantify anecdotal reports of older listeners' difficulty understanding rapid speech, but also to investigate a prominent theory of cognitive aging that there is a decline in speed of sensory and perceptual encoding with increasing age (Salthouse, 1996). Related studies examined perception of reverberant speech by younger and older listeners (e.g., Helfer & Wilber, 1990). Findings from these investigations generally showed that older listeners performed much more poorly than younger listeners, but again, differences in auditory sensitivity between younger and older groups could have contributed significantly to observed group effects. All of these early studies revealed large variability in the performance of older listeners; few of them examined differences in cognitive abilities that may have contributed to individual differences. Finally, investigations in this era were confined to assessing unimodal, auditory-only speech recognition performance.

12.3 Key Findings in Recent Years

12.3.1 Epidemiology

Hearing loss among the older population in the United States is more widespread than reported previously, with an overall prevalence rate converging at 46% among those 48 years and older (Cruickshanks et al., 1998). Using a pure-tone average in the speech frequencies (500, 1000, 2000, and 4000 Hz) exceeding 25 dB hearing level (HL) in the better ear to identify hearing loss, Agrawal et al. (2008) reported a prevalence rate of 49% among 60- to 69-year-olds and Lin et al. (2011) reported a prevalence rate of 63.1% among those 70 years of age and older. Applying these hearing loss prevalence rates to the population older than 65 years (U.S. Census Bureau, 2011) indicates that there are approximately 20 million senior citizens with significant hearing loss in the United States today, and by 2035, there will be approximately 38 million. These data underscore the imperative to find realistic solutions to the communication problems related to aging and hearing loss. Also reported in recent epidemiologic studies is the frequency of hearing aid use among the older population, stratified by degree of hearing loss. In the population older than 70 years, hearing aids were used by 40% of those with a moderate hearing loss but only by 3.4% of those with mild hearing loss (Lin et al., 2011). Individuals with a moderate or greater degree of hearing loss report significant hearing handicap (Weinstein & Ventry, 1983); thus, a take-up rate of only 40% by older people with moderate hearing losses continues to reflect poor acceptance of hearing aids.

12.3.2 Models of Presbycusis

Animal models of age-related hearing loss have revealed three consistent age-related changes in the auditory periphery, and these vary by the specific animal model. In the Mongolian gerbil (*Meriones unguiculatus*), a mammal that exhibits hearing sensitivity similar to that of humans, the principal change in cochlear anatomy is atrophy of the stria vascularis, which in turn reduces the endocochlear potential (Schmiedt, 2010). The resulting hearing loss is mild in degree in the low frequencies, sloping to a moderate-to-severe hearing loss in the high frequencies, which is remarkably similar to the audiometric configuration reported for older humans with prebycusis (Schmiedt, 2010). Deterioration of stria vascularis is also a prominent change in Fischer 344 rats (Syka, 2010). In the C57 mouse model of age-related hearing loss, however, the primary locus of change in the auditory periphery is loss of outer hair cells in the basal turn of the cochlea (Sprongr et al., 1997). All animal models consistently show widespread shrinkage and loss of spiral ganglion cells of cranial nerve (CN) VIII with increasing age (Schmiedt, 2010). Unlike humans, animals in these studies are raised in a quiet environment with a well-controlled diet, suggesting that these observed age-related changes are not associated with acquired insult to the auditory periphery. The work of Kujawa and Liberman (2006) suggests that in humans, exposure to intense noise at a young age initiates not only a rapid deterioration of outer hair cells but also a slow process of progressive deterioration of CN VIII trunks, which may manifest as accumulated damage to spiral ganglion cells among older adults.

There is evidence of neural deterioration with age at every nucleus and the connecting pathways throughout the central auditory nervous system (Willott, 1991). A reduction in neurochemical inhibitors with age has also been observed at many of the nuclei of the central auditory nervous system, including the dorsal cochlear nucleus, inferior colliculus, and auditory cortex (e.g., Caspary et al., 1990). Finally, altered physiologic responses with age in processing brief temporal gaps have been recorded in CBA mice in the inferior colliculus (Walton et al., 1998). Taken together, these findings from animal models confirm that auditory aging occurs independently of lifestyle (noise exposure, diet, ototoxicity), deterioration of the stria vascularis and CN VIII are prominent peripheral changes, a reduction in neural inhibition is widespread throughout the central auditory nervous system, and deficits in auditory temporal processing at central levels accompany the aging process.

12.3.3 Factors Contributing to Speech Understanding Problems

In the intervening 20 years, a number of studies have examined the sources of individual variability in speech recognition performance among large cohorts of older listeners. These investigations typically applied a multifactorial approach to identify a set of measures (auditory sensitivity, auditory processing, cognitive) that predicts speech recognition performance in quiet, noise, and other forms of degradation. Invariably, these studies identified hearing sensitivity as the most important factor that contributes to speech understanding performance in quiet and noise (e.g., Humes et al., 1994), and this finding has been replicated in cross-sectional designs as well. Nevertheless, the variance in performance accounted for by hearing sensitivity approximated 40% to 60% across studies, indicating that there are additional factors that contribute to older listeners' diminished speech recognition performance.

Auditory temporal processing refers to the ability to detect or discriminate brief acoustic signals or those presented at a rapid rate. Older listeners consistently show poorer auditory temporal processing than younger listeners with comparable hearing sensitivity for tonal signals and/or noise bursts on measures of gap detection (Schneider et al., 1994), duration discrimination (Fitzgibbons & Gordon-Salant, 1995), and sequence timing (Fitzgibbons & Gordon-Salant, 2001). Recent evidence (Grose et al., 2006) also indicates that the age-related decline in auditory temporal processing is evident by middle age. The ability to process brief acoustic cues is inherent in nearly every speech recognition task, but is stressed when speech materials are time compressed. Older listeners with and without hearing loss have inordinate difficulty accurately recognizing speech that is time compressed by 50% or more, particularly when few contextual cues are available in sentence-length speech materials (Wingfield et al., 1985; Gordon-Salant & Fitzgibbons, 1993). They also show poorer performance when only a phrase of a sentence is time-compressed, unlike younger listeners (Gordon-Salant & Fitzgibbons, 2004). This performance decline is ascribed, in part, to deficits in accurately perceiving the brief consonant phonemes in time-compressed speech (Gordon-Salant & Fitzgibbons, 2001) and in part to age-related slowing of information processing (Wingfield et al., 1985). Age-related declines in sequential working memory have been linked to deficits in recognition of time-compressed sentences with and without context (Vaughan et al., 2006). Moreover, older listeners show less tolerance than younger listeners for speech presented in noise when the speech is presented at increasing rates, reflecting the combined effects of reduced inhibition of distracting sounds and age-related slowing of information processing (Tun, 1998). Watching television, in which programming is often time compressed, can be particularly difficult for older listeners, although use of closed captioning significantly improves television viewing (Gordon-Salant & Callahan, 2009).

Other speech recognition paradigms have been utilized to reveal the importance of cognitive skills to speech understanding in less-than-ideal conditions. Older people retain their knowledge of the language to help them overcome some of the challenges to listening in poor acoustic environments. For example, recognition of speech in noise or time-compressed speech by older listeners approaches performance levels of younger listeners when contextual cues are available (Gordon-Salant & Fitzgibbons, 1997). In contrast, adding a memory task to the speech recognition task has a more adverse effect on performance by older than younger listeners (Pichora-Fuller et al., 1995), and requiring listeners to identify a target speech signal in the presence of dichotic or monotic competing speech signals with varying cues to target identity presents a more difficult task for older than younger listeners in divided attention but not selective attention conditions (Humes et al., 2006).

A more recent trend is to examine age effects in a dual-task paradigm in which recognition of speech is the primary task and a memory task for read materials (for example) is a secondary task (Gosselin & Gagne, 2010). Cognitive resources are limited at any moment in time and are often reduced in older people. In the dual-task paradigm, utilizing cognitive resources for one challenging task such as a memory task reduces the resources available for processing degraded or noisy speech. As a result, performance is consistently poorer for older than younger listeners. These types of challenges also require greater perceptual effort, which likely characterizes the impact of listening to speech in everyday challenging situations for older people with an impaired auditory system. In sum, older listeners have greater difficulty than younger listeners in accurately recognizing a message that is spoken at a fast rate or in noise, due in large part to reduced audibility of the speech signal, the consumption of cognitive resources during these challenging tasks, and the greater perceptual effort required to follow a distorted signal.

12.3.4 Training to Improve Speech Understanding

Individuals with high frequency sensorineural hearing loss, typical of presbycusis, experience difficulty detecting and discriminating acoustic cues that are important for distinguishing place of articulation (e.g., *pie* vs. *tie*). The talker's cues that are available on the face provide place of articulation information and thus are complementary to the cues that are provided acoustically. Investigations have confirmed the substantial benefit afforded by providing visual cues in addition to auditory cues for hearing-impaired listeners in noise (e.g., Bernstein & Grant, 2009), although the magnitude of the benefit may be reduced with age (Tye-Murray et al., 2010). Nevertheless, a key aspect of any training program is to increase awareness and encourage use of all supplemental cues that may be available, including visual and semantic contextual cues.

A rich literature is now emerging on the benefits of training for detection and discrimination of auditory signals on tasks of frequency discrimination, temporal-interval discrimination, AM rate discrimination, sound-source localization, and signal detection in noise (for a review, see Wright & Zhang, 2009). Many of the improvements in auditory abilities generalize to related signals and tasks. Tedious training to discriminate a single auditory cue may not be necessary; rather, exposure to the cue interspersed with a cue discrimination paradigm with feedback has been shown to produce significant improvements (Wright et al., 2010). The application of these types of training modules with older people has not yet been implemented systematically; however, the significant benefit and generalization of training in temporal-interval discrimination (Wright et al., 1997) may have specific implications for improving the auditory temporal processing abilities of older listeners. Some issues that may require refinement with the geriatric population are the time-course of learning, the generalization of learning, and fatigue.

12.4 Current and Future Directions

12.4.1 Demographics

Where should researchers focus their efforts to improve our understanding of the receptive communication problems of older people and translate this knowledge into effective rehabilitative techniques and technology to improve communication? How can we communicate better with our family and friends as we all grow older? Undoubtedly, changing demographics and lifestyles in our society dictate the imperative to examine the prevalence of different degrees of hearing loss, hearing aid use, benefit, and training for older people in the upper decades of life (80s, 90s, 100s!). Comparable data are lacking for middle-aged adults, although the onset of progressive age-related hearing loss and auditory processing deficits begins in middle age. Researchers must understand better the onset and natural progression of declines in auditory capabilities throughout the adult life span so that efforts to arrest further progression through training or use of sensory aids can be implemented at critical intervals. This approach is fundamental toward preventing the consequences if presbycusis as we know it today.

12.4.2 Speech Recognition Performance for Real-World Degraded Signals

Although a great deal is now known about speech recognition in noisy environments by older listeners with hearing loss (e.g., Humes & Dubno, 2010), very little is known about the difficulties these listeners encounter in perceiving speech altered by other forms of degradation encountered every day. These other forms of speech degradation (described below) are more subtle and may not be recognized as contributing to the listener's perceptual problems. Nevertheless, such signal alterations may contribute substantially to the limited benefit that older listeners report from hearing aids in real-world settings. In particular, because older listeners experience deficits in auditory temporal processing (Fitzgibbons & Gordon-Salant, 1996), it is likely that alterations in the temporal characteristics of speech signals have a substantial impact on speech intelligibility by this group.

One form of degraded speech that is prevalent in today's society is accented speech. Approximately 23 % of the population speaks a native language other than English; many of these speakers provide services to elderly individuals (Shin & Kominski, 2010). Alterations in the acoustic characteristics of English by accent vary with the native language of the talker, however, many of the alterations involve the duration of speech segmental cues especially in Spanish-accented English. Deficits in recognizing Spanish-accented English monosyllabic words by younger and older listeners are attributed to errors for consonant contrasts cued by timing information (Gordon-Salant et al., 2010a, b). Moreover, age-related differences in recognition of Spanish-accented English are particularly prominent in noise. The findings suggest that difficulties in understanding Spanish-accented English are primarily associated with poor perception of temporal information in consonants, especially by older listeners with hearing loss (Gordon-Salant et al., 2010a). In addition, differences in stress and timing are known to exist between native English and Spanish-accented English in part because Spanish is a syllable-timed language that is perceived with equal stress for each syllable, whereas English is a stress-timed language with equal time between stressed syllables. The impact of these deviations on older listeners' performance has not been investigated.

The typical cues used by listeners to separate a target talker from competing speech or noise and hence improve speech recognition may be compromised by talker accent, age, and hearing loss. In younger listeners with normal hearing, a background composed of broadband, modulated noise (i.e., energetic masking) produces less interference than a background of multiple talkers (i.e., energetic + informational masking) (Carhart et al., 1969). Similarly, listeners take advantage of differences in voice pitch and speech rate between the target talker vs. the competing speech masker (Brungart, 2001; Gordon Salant & Fitzgibbons, 2004). The foregoing studies examined masking release for native English talkers (for the target and background speech). We recently investigated the use of these cues by younger and older listeners when the target and background talkers varied in English accent (Gordon-Salant et al., 2013). The hypothesis was that speech produced by native and Spanish-accented speakers of English could create a difference in relative timing between the target and background talkers; these timing differences potentially would serve as a cue to separate the target from the background. When the target talker was a native speaker of English, younger and older listeners with normal hearing were able to use a difference in accent between the target and background talkers to separate the two. However, listeners were unable to take advantage of any cues to speech segregation when the talker had a pronounced Spanish accent.

A tentative interpretation is that Spanish-accented English requires considerable effort to understand, and distracts listeners from taking advantage of typical cues for speech segregation.

Another form of alteration in the temporal cues in speech occurs with variations in talker rate, both across talkers and within the same talker. Most simulations of fast speech with time compression employed uniform time compression throughout the spoken message. However, talkers may vary the speed of their speech naturally even within a single sentence, for example, when they get excited or emphasize a point. Older listeners exhibit significant decline in recognition performance when a single phrase is time compressed, unlike younger listeners (Gordon-Salant & Fitzgibbons, 2004). Thus, older people are at a disadvantage in perceiving speech even when a brief segment of a message is spoken at a fast rate. Variations in natural speech rate between different talkers may contribute to some of the difficulties that older listeners with hearing loss report when participating in a group conversation. Sommers (1997) showed that trial-to-trial variations in talker rate produced significantly poorer performance among older hearing-impaired listeners than uniform talker rate, underscoring that older listeners with hearing loss have more difficulty in making moment-to-moment perceptual adjustments in this dimension. The Perceptually Robust English Sentence Test Open-Set (PRESTO) is composed of recordings by numerous talkers of both genders who speak with varying rates and dialects. These speech stimuli appear to be sensitive to the effects of aging and correlate with age-related cognitive decline (Pisoni et al., 2013). They may be particularly valuable to assess the impact of natural variations in talker speed on older hearing-impaired listeners' ability to process dynamically changing spoken materials.

12.4.3 Auditory–Visual Speech Perception

Real-world speech communication entails exposure to visual information as well as auditory information. The visual information may be supportive or detrimental to speech recognition, and hence may be viewed as providing a continuum of benefits and limitations. These benefits and limitations of visual information may be dependent on listener hearing status and age.

When the talker's face is visible, speech recognition in noise by hearing-impaired listeners improves dramatically because the speech information conveyed on the face is complementary to the speech information that is available through an impaired auditory system (Bernstein & Grant, 2009). Although older listeners benefit from the auditory + visual (AV) presentation of speech relative to A-only, they may not derive as much benefit as do younger listeners (Tye-Murray et al., 2010). This may be related, in part, to age-related decline in speechreading ability (Tye-Murray et al., 2007) or multimodal integration (Spehar et al., 2008).

Processing of AV stimuli may be compromised when there is asynchrony between the stimuli presented in the two modes. AV asynchrony is apparent in some television programs in which the auditory signal may lag or precede the presentation of the visual signal. Because older listeners show slowed auditory processing (Fitzgibbons & Gordon-Salant, 1996), it is possible that AV asynchrony is created for AV information or that preexisting asynchronous AV information is even more temporally mismatched among older people. Younger listeners exhibit a temporal window over which they can adequately integrate asynchronous AV speech information of −30 ms to +170 ms (re: auditory lag; van Wassenhove et al., 2007). There is no effect of age on the ability to *detect* asynchrony in AV stimuli (Başkent & Bazo, 2011), but recent pilot work in our lab (unpublished) suggests that aging affects the ability to *recognize* asynchronous AV speech signals and reduces the temporal integration window. Perception of accented asynchronous AV stimuli or asynchronous AV stimuli in noise may overwhelm the sensory, perceptual, and cognitive capacity of older hearing-impaired listeners, and could be a key factor underlying older listeners' difficulty following television programs, using hearing aids, or understanding accented speech.

One critical feature of nearly every communication setting that is taken for granted is the presence of visual information that is unrelated to the target speech message. For example, individuals' facial expressions while talking to others, a television program, and foot traffic as part of the visual scene can shift a listener's attention away from the primary communication task. As a result, these visual distracters may negatively impact speech recognition, particularly if the listener is monitoring information conveyed by the visual distracter (such as the latest baseball score on television). The addition of single visual distracters to auditory masking can decrease speech recognition performance by younger and older listeners with normal hearing (Gordon-Salant et al., 2011). However, the impact of visual distraction on the performance of older listeners with hearing loss or during a dual-task paradigm has not been investigated. Literature from auditory, cognitive, and visual neuroscience generally indicates that the ability to perform a target task in the presence of competing stimuli deteriorates with aging. This age-related decrease in performance appears to be related to decline in sensory and perceptual processes and their interaction with cognitive decline, suggesting that older hearing-impaired listeners will experience excessive difficulty in the presence of visual distraction. An understanding of the human and environmental factors that act to reduce the older listener's ability to parse relevant from irrelevant information in everyday auditory and visual scenes is fundamental to improving communication for this growing segment of the population.

12.4.4 *Cognitive Load*

The normal aging process entails a gradual and progressive decline in hearing sensitivity and cognitive abilities for most of us. The cognitive aging literature consistently shows that aging is accompanied by a decline in working memory (Baddeley, 1996), attention (Craik & Byrd, 1982), and speed of processing (Salthouse, 1996). The impact of specific dimensions of cognitive decline on auditory performance of older people in everyday communication situations is the subject of intense investigation.

It is reasonable to assume that listening to a spoken message distorted either by environmental conditions (noise), the talker (accent or rate), or the listener's hearing loss requires more cognitive resources to process the signal accurately. Given that an individual possesses a finite store of cognitive resources at any one moment in time and that some of these cognitive resources may be more limited in older than younger people, older listeners with hearing loss are expected to experience a greater cognitive load and hence may require greater listening effort to recognize speech in difficult listening conditions (Pichora-Fuller, 2003). Unfortunately, older listeners do not report greater cognitive load during increasingly difficult speech recognition tasks in noise (Larsby et al., 2005), perhaps reflecting an age-related bias toward underreporting communication problems.

A dual-task paradigm is a more objective technique to determine the impact of increased cognitive load on auditory performance. In this paradigm, performance for each single task is assessed, followed by performance in the two simultaneous tasks. The shift in performance on the primary task in the dual-task paradigm compared to the single-task paradigm indicates the impact of the additional cognitive load associated with the dual-task paradigm. Older participants perform much more poorly than younger listeners on a primary speech recognition task while also engaged in a secondary memory or pattern recognition task (Gosselin & Gagne, 2011). These results are interpreted as reflecting the reduced cognitive resources of older people that must be divided between the two tasks, under the assumption that there is a limited capacity of resources at any point in time.

12.4.5 New Directions for Hearing Aid Signal Processing

Directional microphones, noise-reduction algorithms, feedback management, digital programming, multichannel compression, and advanced connectivity to other communication devices are all remarkable advances that have been incorporated into contemporary hearing aids over the last decade. Laboratory studies indicate that older listeners with hearing loss benefit from amplification in quiet and steady-state noise conditions (Humes et al., 2002). Why, then, do older people with hearing loss largely reject them? The answer is undoubtedly multifactorial, including degree of hearing loss, personality factors, cost, appearance, etc. (Jenstad & Moon, 2011; Lin et al., 2011), but may also include perceived limited benefit of hearing aids in difficult communication situations. Although noise reduction algorithms attempt to attenuate the background noise, this is not accomplished easily if the background noise is composed of a speech signal that has comparable spectral and temporal properties as the target speech signal to be amplified. Another problem in successful use of hearing aids by older listeners is the combination of reduced spectral resolution, due to the sensorineural hearing loss, coupled with reduced auditory temporal processing associated with age. Souza and her colleagues (e.g., Souza, 2000) have confirmed that an age-related deficit in the use of temporal cues has an impact on older listeners' ability to process temporal-envelope cues in speech with compression amplification and/or across channels. These findings have critical implications for an older listener's success in perceiving hearing-aid processed speech in compression amplification conditions.

Current hearing aid signal processing algorithms do not alter the speech signal to accommodate the auditory temporal processing deficits of older listeners and the resulting difficulty in understanding rapid or variable-rate speech. One method that holds promise is to enhance selectively the duration of very brief consonants in ongoing (fast) speech (Gordon-Salant et al., 2007). However, significant time expansion of consonants may create asynchronous auditory and visual information. To rectify this potential asynchrony, it may be possible to compress the duration of vowels and pauses in natural speech while expanding the duration of consonants, as previous research has shown that selective time compression of vowels and pauses has little impact on performance (Gordon-Salant & Fitzgibbons, 2001). The implementation of this type of automatic signal processing method has yet to be developed.

12.4.6 New Models of Adaptation and Training

The established hierarchy of auditory training progresses from simple sound discrimination to more demanding complex speech identification paradigms, and forms the basis of traditional aural rehabilitation strategies. Principles of exposure, learning, and cognitive training have recently been applied to investigations of auditory learning and communication function (e.g., Wright & Zhang, 2009). For example, paradigms that present targeted acoustic cues to listeners with correct answer feedback interspersed with exposure to the same acoustic cues have been shown to promote auditory learning (Wright et al., 2010). Short-term exposure to accented English improves intelligibility accuracy and speed of processing for this type of distorted speech for younger listeners (Clarke & Garrett, 2004) and older listeners (Gordon-Salant et al., 2010c). Cognitive training paradigms focus on memory and speed of processing training; older listeners demonstrate significant gains in cognitive-specific measures but do not necessarily transfer improvements to functional activities (Ball et al., 2002). To date, the efficacy of these exclusively cognitive-based training paradigms for improving performance on auditory measures has not been reported.

The accessibility of home computers, videogames, and smartphones affords an unsurpassed opportunity to deliver auditory training paradigms easily to the end user. Indeed, several software programs and games have become widely available (e.g., LACE, Neurotone; Brain Fitness Program, Posit Science, etc.). A recent investigation by Anderson and her colleagues (Anderson et al., 2013) showed significant benefit of the Brain Fitness Program, which is a home-based computerized training program for older adults. The 40-hour training program, administered over the course of 8 weeks, emphasized auditory-based cognitive training, in which listeners discriminated slowed consonant–vowel (CV) syllables in isolation and various linguistic contexts; the CVs were compressed in duration as listeners exhibited performance improvement. Outcome measures demonstrated increments in sentence recognition in noise, improved memory and speed of processing, and faster neural timing, indicating that focused cognitive-auditory training has the potential to induce neural plasticity and reduce some of the speech understanding problems experienced by older people. Contemporary auditory training programs focus more on listening skills than cognitive skills, per se. One particularly promising program, the Speech Perception Assessment and Training System (SPATS; Miller et al., 2007), combines adaptive training with common speech syllable constituents (onsets, vowel nuclei, offsets) to sharpen spectrotemporal processing of specific syllabic segments, together with sentence-level training in which linguistic cues are stressed to improve bottom-up and top-down processing in quiet and noise. Initial evidence suggests that adults with hearing aids and cochlear implants gain significant benefit from SPATS training (Miller et al., 2008). An extensive multisite clinical trial to evaluate the efficacy of an intensive SPATS protocol for adult hearing aid users is currently underway.

Another relevant area of research derives from studies of the impact of long-term exposure to certain types of acoustic signals on signal processing and speech understanding. Long-term musical training has a significant benefit on speech recognition in noise as well as on latency and amplitude of auditory evoked potentials on matched groups of younger listeners with normal hearing (Parbery-Clark et al., 2009). Musical training also mitigates the expected decline in speech recognition performance in noise associated with aging (Parbery-Clark et al., 2012). Lifelong exposure to rapid speech appears to eliminate the expected age-related decline in recognizing time-compressed speech in quiet and in noise. For example, blind participants, who listen to recorded materials at fast playback rates for long periods of time, exhibit equivalent recognition scores for time-compressed speech as young listeners with comparable hearing sensitivity (Gordon-Salant & Friedman, 2011). Taken together, these findings suggest that age-related decline in speech recognition is not inevitable, and that significant experience with music and/or challenging speech materials can reduce or eliminate the problem. This type of long-term auditory training appears to require early and consistent implementation. Future investigations may be directed at the benefits of training at later stages of life and/or implementation during middle age as a form of prevention.

12.5 Summary

Research over the last 20 years has increased our understanding of the changes in the peripheral and central auditory nervous system that accompany the aging process and the import of these changes on functional communication in everyday settings. In addition to reduced audibility, older listeners exhibit deficits in auditory temporal processing that appear to limit the ability to understand rapid speech, accented speech (especially Spanish-accented English), and some forms of hearing-aid processed speech. Speech recognition in everyday face-to-face scenarios can be enhanced or diminished by the spectrum of visual information that is available in real-world settings. An analysis of the auditory and visual scene, as well as the sensitivity of older listeners with hearing loss to these varying AV cues, should be a priority in future investigations. Normal decline in working memory, speed of processing, selective attention, and other cognitive abilities also has a significant impact on speech recognition performance by older listeners in everyday challenging listening tasks, especially those involving dual simultaneous tasks.

Difficulty in hearing and understanding speech are not inevitable consequences of the aging process, however. Hearing aids do provide significant benefit to those with hearing loss, although signal processing adaptations for older listeners in certain conditions are still needed. Research has shown that lifestyle choices, implemented earlier in adulthood, can minimize some of the difficulties in speech understanding. Auditory, cognitive, and musical training, as well as exposure to rapid or accented speech stimuli all have the potential to improve communication function for seniors. The high prevalence rate of significant hearing loss among those older than 65 years coupled with the longer expected lifespan suggests that older people and their families will be seeking answers to the older listener's unique combination of auditory and cognitive deficits. It behooves researchers to identify and evaluate effective solutions aimed at our older hearing-impaired population, thereby enabling all of us to remain engaged and functioning members of society with the passage of time.

References

Agrawal, Y., Platz, E. A., & Niparko, J.K. (2008). Prevalence of hearing loss and differences by demographic characteristics among US adults. *Archives of Internal Medicine*, 168, 1522–1530.

Anderson, S., White-Schwoch, T., Parbery-Clark, A., & Kraus, N. (2013). Reversal of age-related neural timing delays with training. *Proceedings of the National Academy of Sciences of the USA*, 110, 4357–4362.

Baddeley, A.D. (1996). Exploring the central executive. *The Quarterly Journal of Experimental Psychology*, 49A, 5–28.

Ball, K., Berch, D. B., Helmers, K. F., Jobe, J. B., Leveck, M. D., Marsiske, M., Morris, J. N., Rebok, G.W., Smith, D. M., Tennstedt, S. L., Unverzagt, F. W., & Wiillis, S. L. (2002). Effects of cognitive training interventions with older adults. *JAMA*, 288, 2271–2281.

Başkent, D., & Bazo, D. (2011). Audiovisual asynchrony detection and speech intelligibility in noise with moderate and severe sensorineural hearing impairment. *Ear and Hearing*, 32, 582–592.

Bernstein, J., & Grant, K. (2009). Auditory and auditory-visual intelligibility of speech in fluctuating maskers for normal-hearing and hearing-impaired listeners. *Journal of the Acoustical Society of America*, 125, 3358–3372.

Brungart, D. S. (2001). Informational and energetic masking effects in the perception of two simultaneous talkers. *Journal of the Acoustical Society of America*, 109, 1101–1109.

Carhart, R., Tillman, T. W., & Greetis, E. S. (1969). Release from multiple maskers: Effects of interaural time disparities. *Journal of the Acoustical Society of America*, 45, 411–418.

Caspary, D. M., Raza, A., Lawhorn Armour, B. A., Pippin, J., & Arnerić, S. P. (1990). Immunocytochemical and neurochemical evidence for age-related loss of GABA in the inferior colliculus: Implications for neural presbycusis. *Journal of Neuroscience*, 10, 2363–2372.

Clarke, C. M., & Garrett, M. F. (2004). Rapid adaptation to foreign-accented English. *Journal of the Acoustical Society of America*, 116, 3647–3658.

Craik, F. I. M., & Byrd, M. (1982). Aging and cognitive deficits: The role of attentional resources. In F. I. M. Craik & S. Trehub, S. (Eds). *Aging and cognitive processes* (pp. 191–211). New York: Plenum Press.

Cruickshanks, K. J., Wiley, T. L., Tweed, T. S., Klein, B. E., Klein, R., Mares-Perlman, J. A., & Nondahl, D. M.. (1998). Prevalence of hearing loss in older adults in Beaver Dam, Wisconsin, *American Journal of Epidemiology*, 148, 879–886.

Dubno, J. R., Dirks, D. D., & Morgan, D. E. (1984). Effects of age and mild hearing loss on speech recognition in noise. *Journal of the Acoustical Society of America*, 76, 87–96

Fitzgibbons, P. J., & Gordon-Salant, S. (1995). Duration discrimination with simple and complex stimuli: Effects of age and hearing sensitivity. *Journal of the Acoustical Society of America*, 98, 3140–3145.

Fitzgibbons, P., & Gordon-Salant, S. (1996). Auditory temporal processing in elderly listeners. *Journal of the American Academy of Audiology*, 7, 183–189.

Fitzgibbons, P., & Gordon-Salant, S. (2001). Aging and temporal discrimination in auditory sequences, *Journal of the Acoustical Society of America*, 109, 2955–2963.

Glorig, A., & Davis, H. (1961). Age, noise and hearing loss. *Annals of Otology, Rhinology, and Laryngology*, 70, 556–574.

Gordon-Salant, S., & Callahan, J. (2009). The benefits of hearing aids and closed captioning for television viewing by older adults with hearing loss. *Ear and Hearing*, 30, 458–465.

Gordon-Salant, S., & Fitzgibbons, P. (1993). Temporal factors and speech recognition performance in young and elderly listeners. *Journal of Speech and Hearing Research*, 36, 1276–1285.

Gordon-Salant, S., & Fitzgibbons, P. (1997). Selected cognitive factors and speech recognition performance among young and elderly listeners. *Journal of Speech, Language, Hearing Research*, 40, 423–431.

Gordon-Salant, S., & Fitzgibbons, P. (2001). Sources of age-related difficulty for time- compressed speech, *Journal of Speech, Language, and Hearing Research*, 44, 709–719

Gordon-Salant, S., & Fitzgibbons, P. (2004). Effects of stimulus and noise rate variability on speech perception by younger and older adults. *Journal of the Acoustical Society of America* 115, 1808–1817.

Gordon-Salant, S., & Friedman, S. A. (2011). Recognition of rapid speech by blind and sighted older adults. *Journal of Speech, Language, and Hearing Research*, 54, 622–631.

Gordon-Salant, S., Fitzgibbons, P., & Friedman, S. (2007). Recognition of time-expanded speech by young and elderly listeners with normal hearing and mild-to-moderate hearing loss. *Journal of Speech, Language, and Hearing Research*, 50, 1181–1193.

Gordon-Salant, S., Yeni-Komshian, G. H., & Fitzgibbons, P. J. (2010a). Recognition of accented English in quiet by younger normal-hearing listeners and older listeners with normal hearing and hearing loss. *Journal of the Acoustical Society of America*, 128, 444–455.

Gordon-Salant, S., Yeni-Komshian, G. H., & Fitzgibbons, P. J. (2010b). Perception of accented English in quiet and noise by younger and older listeners. *Journal of the Acoustical Society of America*, 128, 3152–3160.

Gordon-Salant, S., Yeni-Komshian, G. H., Fitzgibbons, P. J., & Schurman, J. (2010c). Short-term adaptation to accented English by younger and older listeners. *Journal of the Acoustical Society of America*, 128, EL200–EL204.

Gordon-Salant, S., Ferruggiaro, A., Cohen, J. I., & Rainsford, J. (2011). Effect of visual distracters on auditory-visual speech perception in noise. American Auditory Society meeting, Scottsdale, March 2–6.

Gordon-Salant, S., Yeni-Komshian, G. H., Fitzgibbons, P. J., Cohen, J. I., & Waldroup, C. (2013). Recognition of accented and unaccented speech in different maskers by younger and older listeners. *Journal of the Acoustical Society of America*, 134, 618–627.

Gosselin, P. A., & Gagne, J.-P. (2010). Use of a dual-task paradigm to measure listening effort. *Canadian Journal of Speech-Language Pathology and Audiology*, 34(1), 43–51.

Gosselin, P. A., & Gagne, J.-P. (2011). Older adults expend more listening effort than young adults recognizing speech in noise. *Journal of Speech, Language, and Hearing Research*, 54, 944–958.

Grose, J. H, Hall, J. W. III, & Buss, E. (2006). Temporal processing deficits in the pre-senescent auditory system. *Journal of the Acoustical Society of America*, 119, 2305–2315.

Helfer, K. A., & Wilber, L. A. (1990). Hearing loss, aging, and speech perception in reverberation and noise, *Journal of Speech and Hearing Research*, 33, 149–155.

Humes, L. E., & Dubno, J. R. (2010). Factors affecting speech understanding in older adults. In S. Gordon-Salant, R. D. Frisina, A. N. Popper, & R. R. Fay (Eds.), *The aging auditory system* (pp. 211–258). New York: Springer Science+Business Media.

Humes, L. E., Watson, B. U., Christensen, L. A., Cokely, C. G., Halling, D. C., & Lee, L. (1994). Factors associated with individual differences in clinical measures of speech recognition among the elderly, *Journal of Speech and Hearing Research*, 37, 465–474.

Humes, L., Wilson, D., Barlow, N., & Garner, C. (2002). Changes in hearing aid benefit following 1- or 2 years of hearing aid use by older adults. *Journal of Speech, Language, and Hearing Research*, 45, 772–782.

Humes, L., Lee, J. H., & Coughlin, M. P. (2006). Auditory measures of selected and divided attention in younger and older adults using single-talker competition. *Journal of the Acoustical Society of America*, 120, 2926–2937.

Jenstad, L., & Moon, J. (2011). Systematic review of barriers and facilitators to hearing aid uptake in older adults. *Audiology Research*, 1, 91–96.

Kirikae, I., Sato, T., & Shitara, T. (1964). A study of hearing in advanced age. *Laryngoscope*, 74, 205–220.

Kujawa, S., & Liberman, C. (2006). Acceleration of age-related hearing loss by early noise exposure: Evidence of a misspent youth. *Journal of Neuroscience*, 26, 2115–2123.

Larsby, B., Hallgren, M., Lyxell, B., & Arlinger, S. (2005). Cognitive performance and perceived effort in speech processing tasks: Effects of different noise backgrounds in normal-hearing and hearing-impaired subjects. *International Journal of Audiology*, 44, 131–143.

Lin, F. R., Thorpe, R., Gordon-Salant, S., & Ferrucci, L. (2011). Hearing loss prevalence and risk factors among older adults in the United States. *The Journals of Gerontology A:, Biological Sciences and Medical Sciences*, 66, 582–590.

Miller, J. D., Watson, C. S., Kewley-Port, D., Sillings, R., Mills, W. B., & Burleson, D. F. (2007). SPATS: Speech Perception and Assessment Training System. *Journal of the Acoustical Society of America*, 122, 3063.

Miller, J. D., Watson, C. S., Kistler, D., Preminger, J., & Wark, D. (2008). Training listeners to identify the sounds of speech II: Using SPATS software. *Hearing Journal*, 61, 29–33.

Parbery-Clark, A., Skoe, E., Lam, C., & Kraus, N. (2009). Musician enhancement for speech-in-noise. *Ear and Hearing*, 30, 653–661.

Parbery-Clark, A., Anderson, S., Hittner, E., & Kraus, N. (2012). Musical experience offsets age-related delays in neural timing. *Neurobiology of Aging*, 33, 1483.e1–1483.e4.

Pichora-Fuller, M. K. (2003). Cognitive aging and auditory information processing. *International Journal of Audiology*, 42, S26–S32.

Pichora-Fuller, M. K., Schneider, B. A., & Daneman, M. (1995). How young and old adults listen to and remember speech in noise. *Journal of the Acoustical Society of America*, 97, 592–608.

Pisoni, D. B., Tamati, T. N., Faulkner, K. F., & Twiggs, T. (2013). Neurocognitive factors underlying individual differences in speech recognition on PRESTO: New findings and clinical applications. Presentation at the Aging and Speech Communication 5th International and Interdisciplinary Research Conference, October 6–9, Indiana University, Bloomington, IN.

Salthouse, T. A. (1996). The processing speed theory of adult age differences in cognition. *Psychological Review*, 103, 403–428.

Schmiedt, R. A. (2010). The physiology of cochlear presbycusis. In S. Gordon-Salant, R. D. Frisina, A. N. Popper, & R. R. Fay (Eds.), *The aging auditory system*. New York: Springer Science+Business Media.

Schneider, B. A., Pichora-Fuller, M. K., Kowalchuk, B., & Lamb, M. (1994). Gap detection and the precedence effect in young and old adults. *Journal of the Acoustical Society of America*, 95, 980–991.

Schuknecht, H. F. (1974). *Pathology of the ear*. Cambridge, MA: Harvard University Press.

Schuknecht, H. F., & Gacek, M. R. (1993). Cochlear pathology in presbycusis. *Annals of Otology, Rhinology, and Laryngology*, 102, 1–16.

Shin, H. B., & Kominski, R. A. (2010). *Language use in the United States:* 2007. American Community Survey Reports, ACS-127. Washington, DC: U.S. Census Bureau.

Sommers, M. S. (1997). Stimulus variability and spoken word recognition. II. The effects of age and hearing impairment. *Journal of the Acoustical Society of America*, 101, 2278–2288.

Souza, P. (2000). Older listeners' use of temporal cues altered by nonlinear amplification. *Journal of Speech, Language & Hearing Research*, 48, 661–674.

Spehar, B., Tye-Murray, N., & Sommers, M. (2008). Intra- versus inter-modal integration in young and older adults. *Journal of the Acoustical Society of America*, 123, 2855–2866.

Sprongr, V. P., Flood, D. G., Frisina, R. D., & Salvi, R. J. (1997). Quantitative measures of hair cell loss in CBA and C57BL/6 mice throughout their life spans. *Journal of the Acoustical Society of America*, 101, 3546–3553.

Syka, J. (2010). The Fischer 344 rat as a model of presbycusis. *Hearing Research*, 264, 70–78.

Tun, P. (1998). Fast noisy speech: Age differences in processing rapid speech with background noise. *Psychology and Aging*, 13, 424–434.

Tye-Murray, N., Sommers, M., & Spehar, B. (2007). The effects of age and gender on lipreading abilities. *Journal of the American Academy of Audiology*, 18, 883–892.

Tye-Murray, N., Sommers, M., Spehar, B., Myerson, J., & Hale, S. (2010). Aging, audiovisual integration, and the principle of inverse effectiveness. *Ear and Hearing*, 31, 636–644.

U.S. Census Bureau (2011). *The Older Population:* 2010. Available at http://www.census.gov/prod/cen2010/briefs/c2010br-09.pdf (Accessed September 1, 2012).

van Wassenhove, V., Grant, K. W., & Poeppel, D. (2007). Temporal window of integration in auditory-visual speech perception. *Neuropsychologia*, 45, 598–607.

Vaughan, N., Storzbach, D., & Furukawa, I. (2006). Sequencing versus nonsequencing working memory in understanding rapid speech by older listeners. *Journal of the American Academy of Audiology*, 17, 506–518.

Walton, J. P., Frisina, R. D., & O'Neill, W. E. (1998). Age-related alteration in neural processing of silent gaps in the central nucleus of the inferior colliculus in the CBA mouse model of presbycusis. *Journal of Neuroscience*, 18, 2764–2776.

Weinstein, B. E., & Ventry, I. M. (1983). Audiometric correlates of the Hearing Handicap for the Elderly. *Journal of Speech and Hearing Disorders*, 48, 379–384.

Willott, J. F. (1991). Central physiological correlates of ageing and presbycusis in mice. *Acta Oto-Laryngologica Supplementum*, 476, 153–156.

Wingfield, A., Poon, L.W., Lombardi, L., & Lowe, D. (1985). Speed of processing in normal aging: Effects of speech rate, linguistic structure, and processing time. *Journal of Gerontology*, 40, 579–585.

Wright, B. A., & Zhang, Y. (2009). A review of the generalization of auditory learning. *Philosophical Transactions of the Royal Society B: Biological Sciences*, 364, 301–311.

Wright, B. A., Buonomano, D. V., Mahncke, H. W., & Merzenich, M. M. (1997). Learning and generalization of auditory temporal-interval discrimination in humans. *Journal of Neuroscience*, 17, 3956–3963.

Wright, B. A., Sabin, A. T., Zhang, Y., Marrone, N., & Fitzgerald, M. B. (2010). Enhancing perceptual learning by combining practice with periods of additional sensory stimulation. *The Journal of Neuroscience*, 30, 12868–12877.

Chapter 13
Cochlear Mechanics, Otoacoustic Emissions, and Medial Olivocochlear Efferents: Twenty Years of Advances and Controversies Along with Areas Ripe for New Work

John J. Guinan Jr.

John J. Guinan Jr.

J.J. Guinan Jr. (✉)
Eaton Peabody Lab, Massachusetts Eye and Ear Infirmary, Boston, MA 02114, USA

Department of Otology and Laryngology, Harvard Medical School, Boston, MA 02115, USA
e-mail: jjg@epl.meei.harvard.edu

A.N. Popper and R.R. Fay (eds.), *Perspectives on Auditory Research*, Springer Handbook of Auditory Research 50, DOI 10.1007/978-1-4614-9102-6_13,

13.1 Introduction

There has been great progress in many areas of auditory neuroscience over the last 20 years. Here I cover areas that I have contributed to, principally with work involving medial olivocochlear (MOC) efferents. Progress in cochlear mechanics and micromechanics, areas with exciting new developments, are considered in Section 2. Progress in otoacoustic emissions (OAEs), and their use in studying MOC effects, are considered Sections 3 and 4. Finally, MOC function and its role in hearing are considered in Section 5. In each of these areas I start from where the field was 20 years ago, review developments over the past 20 years, outline present problems that need work, and in some cases, speculate about what will be found.

13.2 Cochlear Mechanics

13.2.1 Active Mechanisms and "Cochlear Amplification"

In the early 1990s, it was known that there was an active mechanism in the cochlea that increased basilar membrane (BM) motion at low sound levels near the local best frequency. This was called "cochlear amplification." Some investigators held that there was no true "amplification"; instead, the active process changed BM impedance so as to sharpen the resonance without the traveling wave receiving any increase in energy. Over the past 20 years, almost all researchers have come to accept that cochlear amplification of BM motion involves cycle-by-cycle injection of energy into the traveling wave. One support for this view is the existence of spontaneous OAEs (SOAEs) and the demonstration that the amplitude versus time structure of SOAEs cannot be explained by noise filtered by a sharp passive resonance (Shera, 2003). Another support was the measurement of BM motion coupled with model calculations that showed that the real part of the BM impedance is negative basal to the traveling wave peak, which means that energy is injected into the traveling wave by BM motion (de Boer & Nuttall, 2000).

13.2.2 What Is the Motor for Mammalian Cochlear Amplification?

In the early 1990s OHC somatic motility was well known and the idea of calcium-activated stereocilia motility was just coming into focus. Later in the decade, a controversy developed over which was the motor element for mammalian cochlear amplification. The main arguments for stereocilia motility were: (1) Amplification in nonmammalian hearing organs is due to stereocilia motility, and by Occam's razor, the same mechanism should explain amplification in mammals; and (2) OHC capacitance shunts the OHC receptor current and makes the OHC time constant too slow for somatic-motility to amplify at high frequencies. The first argument is not compelling, but the second one seems so until it is examined in detail. OHC time constants were measured with OHCs in a dish, which results in the loss of the piezoelectric coupling of OHC electric potentials to organ of Corti mechanics. With OHCs in their normal position, energy flowing into the OHC capacitance produces motion of the organ of Corti through OHC somatic motility. OHC piezoelectric coupling of the OHC electric and mechanical domains can increase the OHC time constant by changing the OHC electric circuit. Recently, it has been found that OHCs, in situ, have faster time constants than in a dish because of increased basolateral conductance from voltage-activated membrane channels. Although this increased conductance shortens the OHC membrane time constant, it does not increase OHC motion because, for any given OHC receptor current, increasing the OHC basolateral conductance *decreases* the resulting receptor voltage at frequencies below the time constant and produces no change at frequencies above the time constant where the OHC capacity dominates. This shows that the OHC time constant, by itself (i.e., without considering the gain of the overall system), says very little about how high a frequency can be amplified by OHC somatic motility. It is interesting that the voltage-activated membrane channels decrease OHC receptor potentials at frequencies below the OHC's characteristic frequency (CF), but with little receptor potential attenuation at the OHC's CF (Johnson et al., 2011). This again indicates that the increased OHC conductance has little or no effect on cochlear amplification at CF. What then, is the purpose of this decrease in the OHC receptor potential? One possibility is to reduce motion from OHC motility at frequencies below CF, thereby increasing frequency selectivity and/or decreasing acoustic trauma.

A key thing missing in these analyses is that OHC somatic motility is in a negative feedback relationship with the bending of the OHC stereocilia. Upward (toward scala vestibuli) BM movement moves the reticular lamina (RL) tectorial-membrane (TM) complex upward and bends OHC stereocilia in the excitatory (depolarizing) direction, which causes OHC contractions. These OHC contractions pull down the RL–TM complex and *decrease* the OHC stereocilia bending that caused the OHC contraction, and so there is negative feedback. The negative feedback loop makes it possible for the system to amplify at much higher frequencies than the OHC time-constant corner frequency (Lu et al., 2006).

It is curious that for all of the focus on the supposed high-frequency limitations of OHC somatic motility, there has been almost no attention to the high-frequency limitations of calcium-based stereocilia motility. This issue is typically dismissed by saying that cochlear microphonic shows that OHC mechanoelectric transduction channels are able to open and close at rates up to the highest frequencies heard by mammals. Although this is certainly true, much more than transduction channel opening and current flow is required for there to be stereocilia motility at high frequencies. To produce stereocilia fast motility at, say 50 kHz, calcium must bind and unbind at a site that changes the stereocilia channel-open probability at a 50-kHz rate. Calcium ion flow into a region with calcium ion binding sites is subject to buffered diffusion and the time constant involved is limited by both the calcium binding and unbinding rate constants at this site, as well as by the rates of calcium ion flow into and out of the stereociliar regions adjacent to the calcium binding site. Calcium is at a very low concentration in endolymph so a high-affinity calcium binding site is required, yet the binding site must have forward and reverse binding rates that are *both* extremely high. This is an area that could benefit from direct investigation and by comparison to known affinities and binding rates at other calcium binding sites. My guess is that stereocilia motility has been found in mammalian OHCs only up to a few kilohertz, not simply because of technical limitations, but because the slowness of calcium binding limits stereocilia motility at higher frequencies.

From experiments in which Prestin has been knocked out or modified, it is now well accepted that Prestin provides the main motor for mammalian cochlear amplification. A question remains of whether calcium-activated stereocilia motility still has a role. At high frequencies (e.g., >3 kHz), it seems likely that calcium-activated stereocilia motility has no role because of limitations on the calcium binding/unbinding speed, as explained earlier. However, at low frequencies, calcium-activated stereocilia motility may still play an important role both in OHCs and in inner hair cells (IHCs).

13.2.3 Cochlear Macromechanics: The Apex Is Different from the Base

In the early 1990s mechanical data from the apex showed little evidence of nonlinearity and tuning that was much broader in the apex than in the base. These data, however, were from excised preparations and were open to question. Data from the apex of live preparations became available later in the 1990s and confirmed the wide tuning and also showed a small degree of nonlinearity. This, together with evidence that AN two-tone suppression is much weaker in the apex than in the base, have suggested that the BM motion amplifier has much less gain in the apex than in the base.

It is difficult to compare mechanical data from the apex versus the base because the cochlear opening in the base is in scala tympani and allows viewing of the BM, whereas the cochlear opening in the apex is in scala vestibuli and allows viewing of the structures at the top of the organ of Corti and of the BM only with very large (and potentially damaging) openings. Other problems in apical measurements are the need to seal the hole in the cochlea to prevent a fast-wave artifact, and the lack of a metric for local cochlear health that is comparable to the tone-pip-evoked compound action potentials (CAPs) used in the base. A newly developed neural measure, the auditory nerve overlapped waveform (ANOW; Lichtenhan et al., 2013) provides information on neural sensitivities in this low-frequency range and should help to determine when an apical preparation has good thresholds. Much more work is needed to provide an adequate understanding of cochlear macromechanics in the apex.

13.2.4 Cochlear Micromechanics

The dominant view in the early 1990s was that BM motion caused the organ of Corti from the BM to the RL to rotate about the foot of the inner pillar cells, whereas the TM rotated around its insertion in the modiolus. These motions produced shear between the RL and the TM that deflected OHC and IHC stereocilia. The prevalent view was that the structures of the organ of Corti moved in phase so that excitation of IHCs followed BM motion with at most a small difference in frequency dependence. In particular, because IHC stereocilia are not attached to the TM, they are deflected by fluid motion and, at low frequencies, follow fluid velocity. One hypothesis was that the TM vibrated along its radial axis with a resonant frequency that was below the local BM best frequency and thereby produced the drive to OHC stereocilia that led to BM motion amplification.

However, even in the 1990s, there was evidence that could not be reconciled with the simple view that the drive to IHCs closely follows BM motion. Several labs showed that AN fiber responses had phase reversals as sound level increased, but reversals were not seen in BM motion (Liberman & Kiang, 1984; Ruggero et al., 1996). Further, although many investigators thought that OHC somatic motility produced BM motion amplification, the implication that tissue motion at the top and bottom of the OHC must be different was generally ignored.

The two decades from 1992 to 2012 saw a blossoming of cochlear micromechanical measurements that showed that the structures of the organ of Corti did not all vibrate in phase (see Guinan, 2012, for references). When electrical stimuli elicited OHC contractions, the RL was pulled toward the BM, which locally squeezed the fluid within the organ of Corti, causing the Hensen cells and the arcuate zone of the BM to bulge out, as well as causing fluid flow along the tunnel of Corti. OHC contractions tilted the reticular lamina about the head of the pillars so that when the RL over the OHCs moved down, the RL over the IHCs moved up. In contrast, for the same OHC contractions, TM movement was in phase over both OHCs and IHCs. Thus, over IHCs, RL and TM movement were antiphasic and OHC contractions caused the RL–TM gap to be squeezed. Measurements in excised TMs showed that the TM could carry a longitudinal traveling wave of radial vibrations, but did not show evidence of a TM radial resonance. In mice in which a TM protein was genetically altered, there was a small loss of threshold and cochlear tuning became *sharper*. Excised TM's from mice with the same genetic alteration showed a sharper decay of the TM traveling wave. Together, these results indicate a strong role for TM traveling waves in setting the sharpness of cochlear tuning.

The above measurements were made in excised cochleae without cochlear amplification, but recent measurements of BM and RL motion in the same cochlea were made in a live preparation with good thresholds. These showed the usual pattern of BM motion nonlinearity at low levels near the local best frequency and a somewhat similar pattern at the RL. However, at low levels the RL motion was more than twice as large as BM motion, leaving no doubt that BM and RL movement are not even close to identical. These measurements represent a tremendous technical advance. Now needed are measurements of TM radial motion relative to RL radial motion in a live preparation with good thresholds, so that the drive to the OHC stereocilia can be determined.

Stereocilia movement has been measured by sound-synchronized confocal microscopy and addition of a dye that made the stereocilia visible. The results showed a phase relationship between IHC and OHC stereocilia motion consistent with the long accepted view that OHC stereocilia are driven by the RL–TM shear and are displacement sensitive, whereas IHCs are driven by fluid motion and, at low frequencies, are velocity sensitive. Surprisingly, these measurements also indicated that OHC stereocilia length changed during a sound cycle. At high frequencies, friction and the mass of the fluid within the RL–TM gap must hold the RL–TM distance fixed. However, at low frequencies, it appears that the RL–TM distance over OHCs can vary due to elasticity in the OHC stereocilia and/or their attachment to the TM and RL (Guinan, 2012).

13.2.5 The Mechanical Drive to IHC Stereocilia

Auditory-nerve responses show a variety of properties that cannot be explained with the classic view of cochlear mechanics. In 1992 it was already known that AN responses to low-frequency tones can reverse phase at high sound levels. In the decades since 1992, it was found that the AN initial peak (ANIP) response to clicks reversed at high levels, but the BM response to clicks, as measured in the cochlear base, did not reverse. Further, MOC stimulation inhibited the ANIP response at sound levels below the phase reversal, but not at levels above the reversal. In contrast, MOC stimulation did not inhibit the first peak of BM motion at any sound level. Other mysterious AN behavior was found in AN responses to tones an octave or more below CF (tail frequencies). Ablating OHCs caused AN tail thresholds to be more sensitive and to reverse in phase. Further, MOC stimulation inhibited cat, tail-frequency AN responses by up to 15 dB. In contrast, BM responses an octave or more below the local BF grow linearly and are not inhibited by MOC stimulation. Although these anomalous AN results could not be explained by the prevailing view that BM motion excited AN fibers simply by RL–TM shear, there was no coherent view of how these AN responses were produced.

A key breakthrough in understanding the excitation of IHC stereocilia was that changes in the RL–TM gap would produce fluid flow in the gap that could deflect IHC stereocilia (Steele & Puria, 2005; Nowotny & Gummer, 2006; Guinan, 2012). From a detailed analysis of the fluid flows from various RL–TM gap changes, I have identified three drives, in addition to the classic "SHEAR" drive, that are expected to deflect IHC stereocilia (Guinan, 2012). They are: (1) OHC-MOTILITY: Upward BM motion causes OHC somatic contraction that tilts the RL, compresses the RL–TM gap over IHCs, and expands the RL-TM gap over OHCs, thereby producing an outward (away from the modiolus) radial fluid flow. (2) TM-PUSH: For upward BM motion, the force that moves the TM compresses the RL–TM gap over OHCs, causing inward radial fluid flow past IHCs. (3) CILIA-SLANT: Motions that produce large tilting of OHC stereocilia squeeze the supra-OHC RL–TM gap and cause inward radial flow past IHCs. For upward BM motion, IHC stereocilia are deflected in the *excitatory* direction by SHEAR and OHC-MOTILITY drives, but in the *inhibitory* direction by TM-PUSH and CILIA-SLANT drives. Combinations of these drives can explain (1) the reversal at high sound levels of AN initial peak (ANIP) responses to clicks, and medial olivocochlear (MOC) inhibition of ANIP responses below, but not above, the ANIP reversal, (2) dips and phase reversals in AN responses to tones in cats and chinchillas, (3) hypersensitivity and phase reversals in tuning-curve (TC) tails after OHC ablation, and (4) MOC inhibition of tail-frequency AN responses (Guinan, 2012). The ability of these IHC drives to explain previously anomalous AN data provides strong, although indirect, evidence that these drives (or something like them) exist and are significant. Overall, the success of these hypotheses argues that changes in the RL–TM gap produce fluid flows that drive IHC stereocilia, and marks the beginning of a paradigm shift in understanding of how the cochlea works at low frequencies.

An important implication of the preceding analysis is that the OHC-MOTILITY drive provides another mechanism, along with BM motion amplification, that uses active processes to enhance cochlear output. Traditional "cochlear amplification," which I now call "BM motion amplification," although large (>40 dB) in the base, appears to be small (≤10 dB) in the apex. Consistent with this, apical turn mechanical measurements show only a small nonlinearity. Nonetheless, psychophysical experiments at low, apical-turn frequencies show evidence for a strong nonlinearity, which implies that there is significant mechanical amplification in the apex (Lopez-Poveda et al., 2003). I hypothesize that the OHC-MOTILITY drive to IHCs produces most of the apical amplification (Guinan, 2012).

13.2.6 The Mechanisms by Which MOC Efferents Change Cochlear Mechanics

Activation of the MOC–OHC synapse increases the OHC basolateral conductance and hyperpolarizes OHCs, both of which reduce BM cochlear amplification. The synaptic increase in OHC basolateral conductance shunts OHC receptor current, thereby reducing the OHC voltage change. The OHC hyperpolarization may move the operating point of the OHC-length versus OHC-voltage function so that less motion is produced by a given voltage change. It seems likely that both of these factors reduce OHC somatic motility. In contrast, there is no known mechanism by which the MOC synapse would reduce cochlear amplification from OHC stereocilia. An OHC hyperpolarization would slightly increase the driving voltage for the OHC receptor current, and slightly increase calcium entry into the stereocilia. However, both of these are in the wrong direction for reducing BM motion amplification. The lack of a known mechanism by which MOC stimulation reduces amplification from stereocilia motility is another argument against stereocilia motility as the source of BM motion amplification.

Classic MOC efferent effects build up and decay with time constants in the range of 100 ms. In the mid-1990s, a slower MOC effect was discovered that builds up and decays over tens of seconds. By measuring MOC-induced changes in BM motion, we (Cooper & Guinan, 2003) found that fast and slow MOC effects both reduced BM motion but produced BM phase changes in opposite directions. This showed that MOC fast and slow effects originate from different mechanical changes in

OHCs. A hypothesis that fits the data is that the MOC fast effect is due to the OHC hyperpolarization and change in basolateral conductance (both change on the time scale of the fast effect) and that the MOC slow effect may be due to a slow decrease in OHC stiffness.

13.3 Otoacoustic Emissions

13.3.1 Understanding the Generation of OAEs

In the early 1990s OAEs were classified by the stimulus used to evoke them and their origin was poorly understood. Stimulus frequency OAEs (SFOAEs) were the least understood OAEs until the work of Zweig and Shera (1995) showed that SFOAEs are due to coherent reflection of energy at the traveling wave peak. In the 1990s it was shown that distortion product OAEs (DPOAEs) originate from two places along the basilar membrane, one near the F2 cochlear place and the other from the distortion product place, for example, at the 2F1–F2 place (F1 < F2 are the primary tones that evoke the DPOAEs). Later it was shown that the DPOAE component place is less important than the mechanism by which DPOAE energy is sent backward in the cochlea. Near the F2 place, energy is sent backward by a local distortion of cochlear properties caused by the traveling wave itself. This "distortion" source produces a DPOAE component whose phase changes little with frequency. In contrast, near the distortion-product place, 2F1–F2 energy is turned backward by coherent reflection and has the long group delay of coherent reflection. The different phase versus frequency profiles of these components produces beating between the components that is called "DPOAE fine structure."

Classifying OAEs using a taxonomy based on the two mechanisms by which energy flow is turned backward in the cochlea has helped to show the relationships among the OAEs (Shera & Guinan, 1999). For instance, transient-evoked OAEs (TEOAEs), for example from clicks (CEOAEs), are produced by the same coherent reflection mechanism as SFOAEs, and both share many properties. By 1992, spontaneous OAEs (SOAEs) had been identified as being due to multiple internal reflections, but it was not until the early 2000s that a compelling demonstration was provided that mammalian SOAEs are amplitude-stabilized standing waves (Shera, 2003). That is, SOAEs are energy that keeps reflecting back and forth within the cochlea (reflected at the apical end by coherent reflection and at the basal end by the impedance discontinuity from the stapes), each time being amplified, with the energy that leaks out backward through the middle ear producing the SOAE. Thus, SOAEs are an emergent property of BM motion amplification and cochlear reflections, and do not require a small number of OHCs that spontaneously oscillate.

Several aspects of the theory embodied in the OAE taxonomy have been disputed. Ren (2004) claimed that the energy that produces a DPOAE is carried backward in the cochlea by fast acoustic pressure waves and not by a slow reverse BM traveling wave. The reverse-fast-wave hypothesis was proposed because, in an experiment in which 2F1–F2 distortion products were measured in BM motion at two positions in the cochlea, only a short-latency forward 2F1–F2 BM traveling wave was seen and not a 2F1–F2 reverse BM traveling wave (Ren, 2004). No direct evidence for a reverse fast wave was presented; the reverse-fast-wave was postulated as a way to account for the data. Many subsequent papers have refuted this claim. In particular, models show that at the measurement points of the Ren experiment, the forward 2F1–F2 traveling wave is expected to be larger than the reverse 2F1–F2 traveling wave so there is no necessity to assume there is a reverse fast wave (e.g., Sisto et al., 2011). The Ren paper, and many others, incorrectly assume that the 2F1–F2 backward traveling wave originates at the F2 place. However, the 2F1–F2 backward wave originates over a region basal to the F2 place because only from this region do the wavelets from each place arrive in phase at the stapes. Near the F2 place the BM phase changes rapidly so that 2F1–F2 wavelets from this region are out of phase at the stapes and cancel (see Shera & Guinan, 2007 for a good exposition of this).

Another dispute is whether coherent reflection theory adequately explains SFOAEs at low sound levels. First, coherent reflection theory is expected to explain SFOAEs only at *low* levels, because at *high* levels local distortion can produce same-frequency distortion that swamps coherent reflection. Siegel et al. (2005) claimed that low-level SFOAEs could not be explained by coherent reflection because SFOAE group delays are too short. At frequencies above approximately 3 kHz, SFOAE group delays in chinchillas are <2 times longer than BM delays estimated from AN-response group delays, and coherent reflection theory was said to predict a factor of 2. However, an exact application of coherent reflection theory yields a SFOAE/BM delay ratio slightly less than 2, and the match of theory and data for the chinchilla are excellent (Shera et al., 2010). On the other hand, SFOAE group delays at low frequencies are sometimes actually less than a single estimated BM delay. In this frequency region, SFOAE data show a fine structure that appears to come from interference between SFOAE sources with long and short group delays. The long-delay source fits with coherent reflection, but the short-delay source may not, and its origin is unknown. In the low-frequency region, AN TCs have tips with long group delays and side lobes with short group delays. This implies that there are two cochlea motions that have different group delays. Perhaps these motions produce the long- and short-delay SFOAE components. One speculation is that the long-delay SFOAE component, the TC main lobe, and the post-ANIP click response are due to the traveling-wave *displacement*, whereas the short-latency SFOAE component, the TC side lobes, and the ANIP response all come from a micro-mechanical response to the traveling-wave *pressure*, which has a shorter group delay than the traveling wave displacement. Whether or not this speculation is true, it appears that in the low-frequency region, coherent reflection theory is not the whole answer (Shera & Guinan, 2003).

Another point of dispute is about the cochlear region over which SFOAEs originate. In coherent reflection theory, SFOAEs originate predominantly from the peak region where the traveling wave is large. In contrast, Siegel et al. (2003) argued that SFOAEs have major sources in the basal cochlea where best frequencies are an octave, or more, above the SFOAE frequency. This conclusion was based on finding large residuals from "suppressor" tones at high frequencies which, supposedly, show SFOAE components that originate from the regions "suppressed" by the suppressor tones. However, an alternate explanation is that the suppressor creates a disturbance in cochlear impedance at the suppressor-frequency cochlear place, and this disturbance generates a backward-directed wave at the SFOAE frequency that is present only when the suppressor tone is present. This is exactly what is predicted in a standard cochlear model when suppressor frequencies are much greater than the SFOAE frequency (Shera et al., 2004). I was the first to present suppressor data showing large SFOAE residuals at high suppressor frequencies (Guinan, 1990), and I originally gave the same interpretation as Siegel et al. (2003). However, I never published this as a journal paper because I became aware of the alternate interpretation and this explanation fit better with other data. Choi et al. (2008) argued that a substantial part of SFOAEs originates far basal to the traveling wave peak; however, this result was due to a model that assumed a pattern of irregularities weighted against producing SFOAEs from the traveling wave peak. Recently, Lichtenhan (2011) found that low-frequency "bias" tones produce effects on SFOAEs and on CAPs from tone bursts at the SFOAE frequency that together indicate that the SFOAEs originate from the same cochlear region as the CAPs, which argues strongly (at least for frequencies >2 kHz) that SFOAEs primarily originate from the region where the traveling wave is big, as predicted by coherent reflection. In summary, the data indicate that most of the SFOAE originates near the peak of the traveling wave, but do not rule out that a fraction originates basal to the peak, particularly for frequencies <2 kHz.

13.3.2 Using OAEs to Reveal Cochlear Properties

SFOAEs have group delays that are several times longer in humans than in small experimental animals such as cats, guinea pigs and chinchillas. From the long human SFOAE group delays, and the connection between long delays and narrow tuning in filters, we concluded that cochlear tuning is much sharper in humans than in cats, guinea pigs, and chinchillas, and this was confirmed by new psychophysical results (Shera et al., 2002). However, Ruggero and Temchin (2005, 2007) argued that cochlear tuning and traveling-wave delays are the same in humans and experimental animals, with the primary evidence being (1) that similar delays in BM motion were measured in human and animal cadavers, (2) the assertion that coherent reflection does not account for SFOAE delays in chinchillas, and (3) that forward-masking experiments done in animals do not match animal tuning curves. However, over most of the frequency range considered, the chinchilla data fit very well with coherent reflection theory (see Section 3.1). Further, observations in dead animals ignore the role that active processes play in live animals. In addition, the cited animal forward-masking experiments used older, less accurate methods than we used in humans. Finally, Ruggero and Temchin (2005, 2007) provided no explanation for the uncontested fact that SFOAE delays are longer in humans than in small experimental animals. Recently, Joris et al. (2011) measured both AN TCs and SFOAE delays in macaque monkeys and found values that are intermediate between those of humans and small animals, which confirms the ability of SFOAE delays to predict cochlear tuning. Overall, the conclusion that humans have sharper cochlear tuning than cats, guinea pigs, and chinchillas appears to be correct. This may be because tuning sharpness, measured in distance along the cochlea, is similar across species, and the cochlea is longer in humans than in small experimental animals (Shera et al., 2010).

13.4 Measuring MOC Effects Using Changes in OAEs

The early 1990s saw the first measurements of MOC effects in humans using OAEs (see Guinan, 1996, 2011 for detailed reviews that include references). MOC effects on DPOAEs had been measured in animals a decade earlier and provided the earliest strong evidence that MOC efferents produce mechanical changes in the cochlea. In these experiments, efferent effects were assayed by measuring the changes in DPOAE amplitudes. However, DPOAE amplitude provides an inaccurate measure of MOC-induced changes because, as noted in Section 3.1, DPOAEs are due to the mixing and interference of distortion and reflection components. If, for instance, a measurement is made at a frequency at which these sources cancel, then an MOC inhibition of one component will cause a release from cancellation and an *increase* in DPOAE amplitude, even though the underlying change is a *decrease* in a DPOAE component. The cure for this problem is to separate the two DPOAE source components and to measure the MOC effects on each. This has recently been done for DPOAEs in humans and revealed that MOC inhibition is larger on the reflection component than on the distortion component. Separation of DPOAEs into their components is complex and requires a great deal of data, but is required if accurate MOC measurements are to be made using DPOAEs in humans.

Several other techniques have been used, with mixed success, to quantify MOC effects using changes in DPOAEs (see Guinan, 2011 for more detail). Measuring MOC-induced changes in DPOAEs near a cancellation point magnifies the DPOAE change and provides a useful measure of MOC effects in small experimental animals but not in humans. Similarly, measurements of DPOAE fast adaptation works well in some, but not all, experimental animals. Its validity in humans cannot be checked and is unknown.

Many of the problems in measuring MOC effects using DPOAEs are not present with TEOAEs. TEOAEs are produced by the same coherent reflection process that produces SFOAEs, and, as long as the sound levels are kept low, distortion-source TEOAE components do not add to, and interfere with, the reflection-source TEOAE. TEOAE measurements work better in humans than in experimental animals for two reasons: (1) Reflection source OAEs are generally smaller in animals than in humans, and (2) coherent reflection group delays are much shorter in animals than in humans. The short TEOAE delays in animals make it difficult to separate the TEOAE from the ringing of the sound that evokes the TEOAE.

In my earliest work using OAEs to measure MOC effects, I used SFOAEs because (1) they are the most frequency specific OAE, (2) SFOAEs are large in humans, and (3) the sound that evokes SFOAEs is a less potent elicitor of MOC activity than any other sound used to elicit OAEs. However, SFOAEs have disadvantages. We found that MOC-induced SFOAE changes depend strongly on the exact frequency tested with patterns that were idiosyncratic to the individual subject (Backus & Guinan, 2007). Averaging MOC effects across a narrow range of frequencies removed these idiosyncrasies, but this requires more averaging time. We do not understand how MOC effects can change substantially for a frequency change as small as 20 Hz, but they did. One possibility is that activation of MOC synapses on OHCs changes the pattern of cochlear irregularities that provide the reflection points for SFOAE generation. Understanding this phenomenon is an area that needs work. Another disadvantage of SFOAEs is that measuring them is far more sensitive than other OAEs to changes in the position of acoustic assembly in the ear canal. Because of these disadvantages, in our recent experiments we have measured MOC-induced changes in CEOAEs instead of SFOAEs.

No matter which OAE is used, the most common shortcoming in reported measurements of MOC effects on OAEs is inadequate averaging. Many OAE studies use an OAE measurement if its amplitude is 6 dB, or more, above the noise floor. This may be adequate for a simple OAE measurement, but the measurement of an MOC effect involves taking the difference between two measurements. In this case, the signal is the OAE difference, not the OAE, so the correct application of a 6 dB criterion is to have the OAE difference be 6 dB above the noise level. This requires that each OAE must have a signal-to-noise ratio (SNR) that is far higher than 6 dB (Guinan, 2006).

13.5 Medial Olivocochlear Efferent Function

13.5.1 MOC Effects in Humans

Early OAE studies of MOC efferent effects in humans, for technical reasons, used only contralateral sounds and measured MOC effects by the change produced in DPOAEs or CEOAEs. The change in these emissions was called "contralateral suppression," which is unfortunate because it is an ambiguous term. Suppression (such as in two-tone-suppression) can also be produced by acoustic crosstalk. When a contralateral effect is clearly attributable to MOC synapses on OHCs, a better, correct, unambiguous term would be "contralateral inhibition."

There have been two important technical issues in using OAEs to measure MOC effects: (1) distinguishing MOC effects from middle-ear-muscle (MEM) effects and (2) MOC and MEM activity elicited by the stimulus used to evoke the OAE (Guinan, 2006). The MEM problem is most acute using TEOAEs, because clicks at the typical 50/s rate are a potent elicitor of both MOC and MEM activity. In particular, early work using 70 dB SPL clicks to elicit TEOAEs almost certainly had severe contamination by MEM effects because sound near the MEM threshold in one ear lowers the MEM threshold in the other ear. To use TEOAEs and avoid these problems, the click level should be kept to 60 dB SPL or lower, and the click rate should be less than 50/s.

Despite there being methodological issues in some papers, the early work using OAEs to study MOC effects in humans showed, for the most part, that MOC effects in humans are similar to MOC effects in experimental animals. One difference was that measured MOC effects were much smaller in humans than in animals (a few dB vs. 20 dB). A major source of this difference is that in animals high-rate shocks were used to activate MOC fibers, whereas in humans MOC activation was by sound that had to be presented at 60 dB SPL, or lower, to avoid eliciting MEM activity. Another difference is that in animals, MOC effects were often measured using AN responses, whereas in humans, MOC effects were measured by changes in OAEs. To compare neural and OAE effects, we did a study in cats comparing MOC effects on DPOAEs and on tone-pip-evoked CAPs, and found that MOC effects on CAPs were the same or larger (sometimes four times larger) than on DPOAEs. Because the DPOAE distortion component originates basal to the F2 place (see Section 3.1), it receives less amplification than responses at the F2 place and would be expected to show less MOC inhibition. Also, we did not separate the DPOAEs into reflection and distortion components. More work is needed with all OAE types to show how well MOC-induced changes in OAEs correspond to the MOC-induced inhibition of neural responses. Note that the change in neural response is the physiologically important MOC effect, not the change in OAEs.

The main reason why most measurements of MOC effects have been done using contralateral elicitors, but not ipsilateral elicitors, is that an elicitor sound in the OAE- measurement ear can suppress (by two-tone suppression, not synaptic inhibition) the OAE when the frequency content and timing of the ipsilateral elicitor and OAE-evoking sound overlap. Ipsilateral MOC effects can be measured by using an elicitor that has no frequency overlap with the OAE probe sound (e.g., a notched noise and an SFOAE-evoking tone centered in the notch), or by measuring the MOC-induced change in the OAE just after the termination of the elicitor sound (two-tone suppression decays almost instantaneously whereas MOC inhibition decays over hundreds of milliseconds).

These techniques have revealed interesting properties of the contralateral, ipsilateral, and bilateral MOC acoustic reflexes (see Guinan, 2011). Bilateral noise elicited MOC effects that equaled, on average, the addition of the effects of the separate ipsilateral and contralateral noises. The ratio of the MOC effects elicited by ipsilateral versus contralateral noise depended on the bandwidth of the noise. For narrowband noise, the ipsilateral reflex was about twice as strong as the contralateral reflex. This factor of two was expected from animal data because (1) the contralateral reflex is mediated by uncrossed MOC fibers and the ipsilateral reflex is mediated by crossed MOC fibers (it is a double-crossed reflex) and (2) experimental animals show about twice as many crossed MOC fibers as uncrossed MOC fibers. Surprisingly, wide-band noise, which elicits the strongest MOC reflexes, produces ipsilateral and contralateral inhibitions that are equal! This result indicates that, in determining the amplitudes of the reflexes, the peripheral innervation ratio is not the only thing that is important; the neural properties of the MOC brain stem circuitry are also important. One speculation is that the reflexes are kept equal when they are large to preserve binaural sound localization.

Another way in which human results have differed from the expectation from animal work is in how well focused MOC effects are along the frequency axis. In cats and guinea pigs, tuning curves from MOC fibers are only slightly wider than tuning curves from AN fibers, and the projection of MOC fibers onto the cochlear sensory epithelium is tonotopic. These MOC-reflex properties led to the expectation that the MOC acoustic reflex is frequency specific, for example, a 5-kHz tone would elicit MOC activity that mostly affected the 5-kHz cochlear place. In contrast, measurements in humans show a complex pattern, with MOC inhibition extending over a wider frequency range than expected and with peak effects sometimes offset a half octave above or below the elicitor frequency (Lilaonitkul & Guinan, 2012). Surprisingly, the pattern of efferent effects along the cochlea depended on which OAE metric was used and showed a different pattern for amplitude versus phase of the change in SFOAEs. Some confidence in these unexpected results is provided by data from cats that showed a complex pattern of MOC effects across AN fibers with different CFs (Warren & Liberman, 1989) that was similar to the change in SFOAE (ΔSFOAE) metric. However, the disparity of the MOC effects on SFOAE magnitude versus phase is unexplained and indicates a lack of understanding of exactly how MOC efferents change OAEs. This is another area that needs work.

13.5.2 *The Role of MOC Efferents in Hearing*

In the early 1990s the list of putative roles of MOC efferents in hearing was to (1) extend the dynamic range of hearing, (2) make it easier to hear signals in noise, (3) help provide selective attention, and (4) reduce acoustic trauma. The list is the same today, although it is now clearer that the first three roles are interconnected. The evidence that MOC efferents extend the dynamic range of hearing is not much different now than in 1992 and will not be considered further.

13.5.2.1 MOC Activity Makes It Easier to Hear Signals in Noise

Over the past 20 years, the ways in which MOC activity produces antimasking have come into better focus (see Guinan, 2011 for more detail). One way MOC activity produces unmasking is by increasing the SNR of the BM response to sound. MOC inhibition reduces BM motion amplification more for low-level sounds (e.g., for a background noise) than for high-level sounds (e.g., for a signal). The MOC increase in the SNR of the BM response makes the signal easier to hear. Another way MOC activity produces unmasking is by reducing adaptation at the IHC–AN synapse. This increases the ability to detect changes in transient sounds in a continuous background noise. Background noise that excites AN responses depletes the pool of neurotransmitter vesicles in IHCs. MOC activity reduces the response to the noise and the depletion of IHC vesicles which, in turn, increases the dynamic range of the IHC output and makes changes in transient sounds easier to hear.

A variety of experiments have been done with the goal of determining whether MOC activity actually produces unmasking. Typically these studies gather data across a range of subjects and seek to determine if an individual's ability to detect a sound signal in noise correlates with the individual's efferent activation. In most experiments, the ability to hear a signal in noise is measured by a psychophysical test, and MOC activation is measured by the inhibition of an OAE produced by contralateral sound, with the two tests done separately. These studies have produced a variety of results with some finding a correlation and others finding none. In some studies, the OAE averaging was inadequate to measure the MOC effect in all subjects accurately, which would reduce any correlation. However, a more basic problem is that the psychophysical and OAE tests have almost always been done separately, and the MOC activation during the psychophysical test may be quite different from the MOC activation in the OAE test. An overall assessment of this literature suggests that MOC efferents do produce unmasking. However, the diversity of results and lack of patterns across studies makes it difficult to understand exactly when, and by how much, MOC efferents make it easier to hear signals in noise. Furthermore, attention may change MOC activation and may be central to understanding how much MOC efferents help in hearing signals in noise

13.5.2.2 MOC Activity and Selective Attention

Modulation of MOC effects by selective attention is another area that has produced a variety of results. Overall, the results make it clear that attention to auditory and/or visual targets can modulate MOC activity, presumably via descending pathways from the cortex to the MOC centers in the brain stem. However, MOC activity was increased in some experiments and decreased in others, and it is difficult to discern a pattern in the conditions that accounts for the different results.

MOC activation during a task has been expected to be focused by attention on the frequency of the tone, but no focusing has been convincingly found. Perhaps this is because broad-band masking noise has typically been used. Would the MOC activation be focused if the masking noise were narrow band (i.e., more focused than the somewhat focused MOC activation from the narrow-band noise itself)? If so, how would this depend on the relationship of the noise band to the frequency of the tone being discriminated?

To make progress in understanding auditory attention and its role in unmasking, the efferent effects should be measured while the subject is doing the psychophysical task. Also, because the efferent effect may be modulated by the difficulty of the task, task difficulty should be held constant when making comparisons. The lack of equating difficulty across tasks may be the main reason why the literature shows such diverse results.

13.5.2.3 MOC Activity Reduces Acoustic Trauma

By the early 1990s there was evidence that efferent activity reduces acoustic trauma, and in the ensuing 20 ears that evidence has greatly increased. Recent evidence indicates that sounds at moderate levels that do not produce permanent threshold shifts, can, nonetheless, produce long-persisting changes in the cochlea that lead to degeneration of high-threshold AN fibers (Kujawa & Liberman, 2009). This loss of AN fibers might be the cause of degraded speech discrimination in noise often found in older subjects. These findings make it all the more important to understand how efferents reduce acoustic trauma and to develop fast, accurate ways to measure MOC effects in humans so as to identify subjects with weak MOC reflexes who may be particularly susceptible to trauma.

In guinea pigs, efferent-mediated protection from acoustic overexposure has been found to be related to the production of an MOC slow effect. The slow effect, in turn, may be due to a reduction in OHC stiffness, but how this might protect from acoustic trauma is not clear. Because the MOC slow effect appears to adapt away for long duration sounds, but MOC protection exists for these long duration sounds, it appears that the MOC fast effect is also involved in MOC trauma protection.

The MOC fast effect has been thought to change cochlear responses only at low sound levels, so how it might reduce acoustic trauma from high-level sounds has been a mystery. Measurements of MOC fast effects show that there is, at most, a small-amplitude reduction ($\ll$1 dB) in the BM motion response to tones of 90 dB SPL and higher. However, a new analysis reveals that there are large MOC-induced changes in BM motion at high sound levels, but these are largely phase changes (Cooper & Guinan, 2011). It is not clear how these changes might relate to the reduction of acoustic trauma, but they do show that MOC fast effects can produce substantial changes in the mechanical response of the cochlea at high sound levels.

13.6 Final Thoughts

The past 20 years has seen many changes in our understanding of the cochlea, and the next 20 years promises to be just as exciting. Hopefully, we will gain an understanding of the cochlear amplification of BM motion in the cochlear base. There should also be substantial progress in understanding the cochlear apex which is clearly different than the base and seemingly more complex. I also hope for progress in understanding how IHCs are excited. Current evidence shows a much more complicated picture than the classical view that IHC stereocilia are deflected only by RL–TM shear (Guinan, 2012). I expect that the MOC efferents, which produce reversible, physiologic changes in cochlear mechanics, will play an important role in uncovering the secrets of the cochlea, as they have in the past. Finally, experiments in which efferent mechanical changes are monitored during the performance of psychophysical tasks should provide definitive, quantitative knowledge of how and when MOC efferents benefit hearing.

Acknowledgments This work was supported by NIH NIDCD RO1 000235 and RO1 005977.

References

Backus, B. C., & Guinan, J. J., Jr. (2007). Measurement of the distribution of medial olivocochlear acoustic reflex strengths across normal-hearing individuals via otoacoustic emissions. *Journal of the Association for Research in Otolaryngology*, 8(4), 484–496.

Choi, Y. S., Lee, S. Y., Parham, K., Neely, S. T., & Kim, D. O. (2008). Stimulus-frequency otoacoustic emission: Measurements in humans and simulations with an active cochlear model. *Journal of the Acoustical Society of America*, 123(5), 2651–2669.

Cooper, N. P., & Guinan, J. J., Jr. (2003). Separate mechanical processes underlie fast and slow effects of medial olivocochlear efferent activity. *Journal of Physiology*, 548(1), 307–312.

Cooper, N. P., & Guinan, J. J., Jr. (2011). Efferent insights into cochlear mechanics. In C. A. Shera & E. S. Olson (Eds.), *What fire is in mine ears: Progress in auditory biomechanics* (Vol. 1403, pp. 396–402). Melville, NY: American Institute of Physics.

de Boer, E., & Nuttall, A. L. (2000). The mechanical waveform of the basilar membrane. II. From data to models—and back. *Journal of the Acoustical Society of America*, 107(3), 1487–1496.

Guinan, J. J., Jr. (1990). Changes in stimulus frequency otoacoustic emissions produced by two-tone suppression and efferent stimulation in cats. In P. Dallos, C. D. Geisler, J. W. Matthews & C. R. Steele (Eds.), *Mechanics and biophysics of hearing* (pp. 170–177). New York: Springer-Verlag.

Guinan, J. J., Jr. (1996). The physiology of olivocochlear efferents. In P. J. Dallos, A. N. Popper & R. R. Fay (Eds.), *The cochlea* (pp. 435–502). New York: Springer-Verlag.

Guinan, J. J., Jr. (2006). Olivocochlear efferents: Anatomy, physiology, function, and the measurement of efferent effects in humans. *Ear and Hearing*, 27(6), 589–607.

Guinan, J. J., Jr. (2011). Physiology of the medial and lateral olivocochlear systems. In D. K. Ryugo, R. R. Fay, & A. N. Popper (Eds.), *Auditory and vestibular efferents* (pp. 39–81). New York: Springer Science+Business Media, LLC.

Guinan, J. J., Jr. (2012). How are inner hair cells stimulated? Evidence for multiple mechanical drives. *Hearing Research*, 292(1–2), 35–50.

Johnson, S. L., Beurg, M., Marcotti, W., & Fettiplace, R. (2011). Prestin-driven cochlear amplification is not limited by the outer hair cell membrane time constant. *Neuron*, 70(6), 1143–1154.

Joris, P. X., Bergevin, C., Kalluri, R., Mc Laughlin, M., Michelet, P., van der Heijden, M., & Shera, C. A. (2011). Frequency selectivity in Old-World monkeys corroborates sharp cochlear tuning in humans. *Proceedings of the National Academy of Sciences of the USA*, 108(42), 17516–17520.

Kujawa, S. G., & Liberman, M. C. (2009). Adding insult to injury: Cochlear nerve degeneration after "temporary" noise-induced hearing loss. *Journal of Neuroscience*, 29(45), 14077–14085.

Liberman, M. C., & Kiang, N. Y. S. (1984). Single-neuron labeling and chronic cochlear pathology. IV. Stereocilia damage and alterations in rate- and phase-level functions. *Hearing Research*, 16, 75–90.

Lichtenhan, J. T. (2011). Effects of low-frequency biasing on otoacoustic and neural measures suggest that stimulus-frequency otoacoustic emissions originate near the peak region of the traveling wave. *Journal of the Association for Research in Otolaryngology*, 13, 17–28.

Lichtenhan, J. T., Cooper, N. P., & Guinan, J. J., Jr. (2013). A new auditory threshold estimation technique for low frequencies: Proof of concept. *Ear and Hearing*, 34(1), 42–51.

Lilaonitkul, W., & Guinan, J. J., Jr. (2012). Frequency tuning of medial-olivocochlear-efferent acoustic reflexes in humans as functions of probe frequency. *Journal of Neurophysiology*, 107(6), 1598–1611.

Lopez-Poveda, E. A., Plack, C. J., & Meddis, R. (2003). Cochlear nonlinearity between 500 and 8000 Hz in listeners with normal hearing. *Journal of the Acoustical Society of America*, 113(2), 951–960.

Lu, T. K., Zhak, S., Dallos, P., & Sarpeshkar, R. (2006). Fast cochlear amplification with slow outer hair cells. *Hearing Research*, 214(1–2), 45–67.

Nowotny, M., & Gummer, A. W. (2006). Nanomechanics of the subtectorial space caused by electromechanics of cochlear outer hair cells. *Proceedings of the National Academy of Sciences of the USA*, 103(7), 2120–2125.7–

Ren, T. (2004). Reverse propagation of sound in the gerbil cochlea. *Nature Neuroscience*, 7(4), 333–334.

Ruggero, M. A., & Temchin, A. N. (2005). Unexceptional sharpness of frequency tuning in the human cochlea. *Proceedings of the National Academy of Sciences of the USA*, 102(51), 18614–18619.

Ruggero, M. A., & Temchin, A. N. (2007). Similarity of traveling-wave delays in the hearing organs of humans and other tetrapods. *Journal of the Association for Research in Otolaryngology*, 8(2), 153–166.

Ruggero, M. A., Rich, N. C., Shivapuja, B. G., & Temchin, A. N. (1996). Auditory-nerve responses to low-frequency tones: Intensity dependence. *Auditory Neuroscience*, 2, 159–185.

Shera, C. A. (2003). Mammalian spontaneous otoacoustic emissions are amplitude-stabilized cochlear standing waves. *Journal of the Acoustical Society of America*, 114(1), 244–262.

Shera, C. A., & Guinan, J. J., Jr. (1999). Evoked otoacoustic emissions arise by two fundamentally different mechanisms: A taxonomy for mammalian OAEs. *Journal of the Acoustical Society of America*, 105(2), 782–798.

Shera, C. A., & Guinan, J. J., Jr. (2003). Stimulus-frequency-emission group delay: A test of coherent reflection filtering and a window on cochlear tuning. *Journal of the Acoustical Society of America*, 113(5), 2762–2772.

Shera, C. A., & Guinan, J. J., Jr. (2007). Mechanisms of mammalian otoacoustic emission. In G. A. Manley, B. L. Lonsbury-Martin, A. N. Popper, & R. Fay (Eds.), *Active processes and otoacoustic emissions* (pp. 305–342). New York: Springer Science+Business Media.

Shera, C. A., Guinan, J. J., Jr., & Oxenham, A. J. (2002). Revised estimates of human cochlear tuning from otoacoustic and behavioral measurements. *Proceedings of the National Academy of Sciences of the USA*, 99(5), 3318–3323.

Shera, C. A., Tubis, A., Talmadge, C. L., & Guinan, J. J., Jr. (2004). The dual effect of "suppressor" tones on stimulus-frequency otoacoustic emissions. *Association for Research in Otolaryngology Abstracts*, 27, Abs. 776.

Shera, C. A., Guinan, J. J., Jr., & Oxenham, A. J. (2010). Otoacoustic estimation of cochlear tuning: Validation in the chinchilla. *Journal of the Association for Research in Otolaryngology*, 11(3), 343–365.

Siegel, J. H., Temchin, A. N., & Ruggero, M. (2003). Empirical estimates of the spatial origin of stimulus-frequency otoacoustic emissions. *Association for Research in Otolaryngology Abstracts*, 26, Abstract 679.

Siegel, J. H., Cerka, A. J., Recio-Spinoso, A., Temchin, A. N., van Dijk, P., & Ruggero, M. (2005). Delays of stimulus-frequency otoacoustic emissions and cochlear vibrations contradict the theory of coherent reflection filtering. *Journal of the Acoustical Society of America*, 118(4), 2434–2443.

Sisto, R., Moleti, A., Botti, T., Bertaccini, D., & Shera, C. A. (2011). Distortion products and backward-traveling waves in nonlinear active models of the cochlea. *Journal of the Acoustical Society of America*, 129(5), 3141–3152.

Steele, C. R., & Puria, S. (2005). Force on inner hair cell cilia. *International Journal of Solids and Structures*, 42, 5887–5904.

Warren, E. H., III, & Liberman, M. C. (1989). Effects of contralateral sound on auditory-nerve responses. II. Dependence on stimulus variables. *Hearing Research*, 37, 105–122.

Zweig, G., & Shera, C. A. (1995). The origin of periodicity in the spectrum of evoked otoacoustic emissions. *Journal of the Acoustical Society of America*, 98(4), 2018–2047.

Chapter 14
Examining Fish in the Sea: A European Perspective on Fish Hearing Experiments

Anthony D. Hawkins

Anthony D. Hawkins

A.D. Hawkins (✉)
Loughine, Kincraig, Blairs, Aberdeen AB12 5YT, UK
e-mail: a.hawkins@btconnect.com

A.N. Popper and R.R. Fay (eds.), *Perspectives on Auditory Research*, Springer Handbook of Auditory Research 50, DOI 10.1007/978-1-4614-9102-6_14,

14.1 Introduction

In the late 1960s, at the beginning of my research career, when the Beatles and Janis Joplin reigned supreme, almost nothing was known of the hearing abilities of those important fishes that lived in the sea and supplied the major fisheries. Yet the role of sound in fish capture and fishing tactics had become of intense interest (Chapman & Hawkins, 1969). Funding to explore fish hearing had become available from fisheries departments, and several European groups were embarking on new research projects.

I was fortunate enough to work with some of the talented European scientists who undertook these new experiments into fish hearing. Although their work may now be largely forgotten, it is instructive to remind ourselves of their findings.

In parallel with this pioneering work on fish hearing in Europe, scientists elsewhere, and especially in the United States, were pursuing similar interests with other species. Because their work has already been well described in recent reviews (e.g., see Chapter 7 by Fay and Chapter 25 by Popper), I will describe only those studies carried out in northern Europe.

14.2 Earlier State of Knowledge

What did we know about fish hearing back in the 1960s? Griffin (1950) and Lowenstein (1957) had reviewed the very earliest accounts of the hearing abilities of fish. They had concluded that fish could hear, and that sounds played an important part in their behavior. Two sensory systems were suggested as the acoustic receptors: the paired labyrinth organs of the head (the inner ears) and the lateral line system of the head and trunk. Some of the earlier experiments presenting sounds to fish had yielded negative results but Karl von Frisch had pointed out (von Frisch, 1936) that most of these studies relied on the fish exhibiting fright reactions. The sounds presented had little biological significance; they were generated using whistles, bells, and tuning forks and a lack of response was not surprising. Karl von Frisch subsequently received the Nobel Prize in Physiology and Medicine in 1973, along with Nikolaas Tinbergen and Konrad Lorenz, for his work on the behavior of animals. He and his colleagues at the University of Munich developed and applied completely new methods for examining animal behavior. Among them were behavioral conditioning methods whereby fish were trained to respond in an easily defined way to sounds. By a series of elegant experiments, many of them involving the elimination of parts of sensory organs, von Frisch and his colleagues established that the fish ear did serve as a hearing organ (von Frisch & Stetter, 1932; Dijkgraaf, 1949).

It had been proposed that within the fish ear it was the movement of the dense calcareous otoliths, relative to patches of sensory hair cells, that mediated the detection of sounds (Pumphrey, 1950; de Vries, 1950). The tissues of the head were acoustically transparent to sound in water. The dense otoliths would lag behind oscillations of the head in a sound field and the resultant relative movements would stimulate the hair cells. Willem van Bergeijk (1964, 1967) stressed, however, that the magnitude of the motion would be very small and that there was little evidence that the otolith organs alone would detect a sound wave of even moderate amplitude. Others had already suggested that the effectiveness of a motion sensitive end organ would be greatly increased by the presence of a coupling to a gas-filled sac like the swim bladder (Poggendorf, 1952; de Vries, 1956; Alexander, 1966). The gas, being more compressible than the surrounding medium, would change volume in response to a sound, causing the swim bladder wall to generate greater amplitudes of motion at the ear. In many fishes there were indeed special anatomical linkages between the swim bladder and the inner ear, catalogued by Jones and Marshall (1953).

Van Bergeijk (1964, 1967) emphasized that hearing in fish depended on the detection of sound pressure. The back and forth particle motion accompanying the sound wave was too small to be detected by the unaided otolith organs, unless the fish were very close to a sound source. He concluded, however, that fish with a single receptor, the swim bladder, would be unable to detect the direction of a sound source. In terrestrial vertebrates two spaced ears are necessary to determine direction, analyzing differences in the amplitude, phase, and time of arrival. The acoustic properties and dimensions of the fish head would preclude localization through binaural differences. Moreover, a single sound pressure detector would respond equally to sounds from all directions. Van Bergeijk opined that directional hearing would be possible only very close to a sound source, with the lateral line as the only organ capable of providing directional information. This view stood in contrast to that of Dijkgraaf (1963), who had proposed that the lateral line served mainly to detect and locate moving objects at short range on the basis of current-like water disturbances. The lateral line system was not engaged in the detection of propagated sonic or infrasonic sound waves. Confusion over a possible auditory role of the lateral line has lasted until the present day. Evidence now strongly supports Dijgraaf's view of the lateral line as an independent sensory system, detecting local movements of the surrounding water.

Early studies of the hearing abilities of fish had shown great variability in sensitivity to sounds. Experiments on the same species often gave very different results. Comparison of hearing thresholds obtained from the goldfish by different scientists showed differences of up to 60 dB (a factor of 10^3) at some frequencies (Hawkins, 1973). The chief reasons for these differences lay in the different acoustic conditions under which the experiments were conducted. Griffin (1950) had already emphasized that if the functioning of the fish ear were to be understood fully then more rigorous quantitative acoustic measurements would be required. He, and others (Parvulescu, 1964), pointed to the pitfalls in carrying out experiments in small tanks and specifying the sounds solely in terms of sound pressure. The propagated back-and-forth motion of the component particles of the medium accompanying a sound, and designated as the particle velocity, displacement or acceleration, was also important.

Measurement of sound pressure alone is valid only in the rather special case of a propagated sound wave in a free sound field. Real sounds may deviate from this ideal in two important respects. First, the point of measurement may be close to the source and in the region of sound spreading, where large particle motions may accompany quite small sound pressures (Pumphrey, 1950; Harris & van Bergeijk, 1962; Harris, 1964). Second, the medium may not be homogeneous and may be bounded by interfaces with media of different acoustic properties. Parvulescu (1964) emphasized that the majority of small aquarium tanks were effectively completely surrounded by air, resulting in any loudspeaker immersed in the tank close to the fish producing very high levels of particle motion.

At that time, and even to the present day, it was commonplace to work on the hearing of fish in small water tanks in the laboratory. Parvulescu had suggested that mounting a loudspeaker in air over a small tank could generate sound pressures accompanied by relatively small particle motions. He was careful to point out that this principle was applicable only if the sound waves in air were in phase at different points around a thin-walled tank. For most tanks of practical dimensions this would only be the case at rather low frequencies, but for many workers this method provided an attractive and deceptively simple way of presenting measured sound pressures to fish.

Another approach suggested by Parvulescu (1964) was to generate the acoustic field within a tubular tank, shorter than the wavelength of the sounds to be presented, and fitted with a loudspeaker at each end. By driving the loudspeakers with signals of similar amplitude, but differing phase, it would be possible to control the ratio of sound pressure to particle motion at the center of the tube. Poggendorf (1952) had been the first to apply these principles in examining the hearing of the catfish *Ictalurus nebulosus*, although in his case the tube was open at one end. Later, Hawkins and MacLennan (1976) were to describe in detail the characteristics of such a standing wave tube, and to apply it in hearing experiments.

A more difficult approach, but one with many attractions to those of us working in marine fisheries, was to carry out experiments on fish in a large body of water, such as the sea itself, where the fish actually lived. There, the effects of reflecting boundaries would largely be eliminated and, by changing the distance of the animal from the sound source, the ratio of sound pressure to particle motion could be varied. This was the method adopted independently and concurrently by workers in three European countries: Norway, the Netherlands, and Scotland.

14.3 Moving into the Sea: The First Steps

Per Enger, at the University of Oslo, Norway, was already at the center of research on fish hearing. Per had studied single unit activity in the fish auditory system and had demonstrated responses to sound from the auditory nerve of the bullhead *Cottus scorpius* (Enger, 1963). He had also shown differences in the responses of fish exposed to sounds from loudspeakers in water and in air, following the guidance from Parvulescu. He had concluded that particle motion might be a better parameter than sound pressure in determining auditory thresholds in some species (Enger, 1966).

In 1967, seeking improved acoustic conditions; Per Enger took the first step towards performing experiments in the sea together with Rolf Andersen from the Institute of Marine Research in Bergen. Atlantic cod (*Gadus morhua*) and bullhead (sculpin) were held at different distances from an underwater loudspeaker in the sea, and microphonic potentials recorded from the ear in response to sounds (Enger & Andersen, 1967). In the cod, the amplitudes of these potentials, originating in the hair cells of the ear, were related to the measured sound pressures and were independent of distance. In the bullhead, potentials could be recorded only within 1 m of the loudspeaker.

Enger and Andersen concluded that the Atlantic cod, a fish with a swim bladder, was able to detect sound pressure. The bullhead, without a swim bladder, could detect only the large particle motions found close to the source (within the acoustic near field). Enger and Andersen concluded, as van Bergeijk had done, that the swim bladder was essential for hearing in the far field.

If van Bergeijk were right, then fish with swim bladders, such as the cod and herring, would not be able to discriminate and respond to sounds from different directions. This suggestion ran contrary to the experience of fishermen and was an important issue in considering whether fishing vessels and research vessels were being detected and avoided by fish. Many of us working in fisheries research thought he must be wrong.

Kjell Olsen, a fisheries biologist working at the Institute of Marine Research, decided to look at the behavior of herring *Clupea harengus* in cages in the sea. He demonstrated clear directional responses by herring schools to the playback of noise (Olsen, 1969). At about the same time, Nelson and Gruber (1963) and Myrberg et al. (1969), also working in the sea, had concluded that sharks, which lacked a swim bladder, could detect and orient to sounds in the far field.

Kjell later encouraged a Dutch group of scientists to visit Norway to perform experiments on hearing in fish in the sea. Arie Schuif, the leader of the Dutch group, was a student of Professor Sven Dijkgraaf at the University of Utrecht. With help from Kjell and others at the Institute of Marine Research an experiment was set up beneath a raft in a Norwegian fjord at the island of Sotra. Arie's aim was to investigate whether fish could discriminate between sounds from different directions. Arie and his students confirmed that Ballan wrasse (*Labrus bergylta*) were able to detect a change in direction (Schuijf et al., 1972). This was to be the first of several key papers from the Dutch group.

At the same time, in Scotland, I was working with Colin Chapman at the Marine Laboratory in Aberdeen, the Scottish fisheries research institute (Fig. 14.1). I had studied sound production in fish as an undergraduate, and had moved to Scotland to investigate fish sounds as part of my PhD. We had successfully recorded sounds from spawning haddock (*Melanogrammus aeglefinus*), and had published this work in the journal *Nature*. We now turned to examining hearing in fish.

We were convinced that experiments in tanks were not the way to proceed. Our hearing experiments were going to be carried out in the sea on commercially important species of fish. Colin had found a site by the side of Loch Torridon, a sheltered fjord on the west coast, where relatively deep water could be reached within a short distance of the shore. He then built a tower on the seabed, with loudspeakers moored at different distances and in different angular positions (Fig. 14.2).

The remoteness and rugged nature of the site created many logistical difficulties. Torridon is one of the most beautiful areas of Scotland, with high mountains

Fig. 14.1 Tony Hawkins (center) and Colin Chapman (right) explaining their work at the Loch Torridon site to their boss, Bill Hemmings

towering above an enclosed arm of the sea (Fig. 14.3). A rough track to the site had first to be improved to allow access to the shore. A garden shed was constructed to house electronic equipment, and a diesel generator installed. Boat moorings were laid. We soon discovered that one of the great advantages of the site was that fish and shellfish were present in abundance. Fish could be caught on hand-lines and held in tanks and in cages in the sea ready for use in experiments. The shellfish could be gathered and eaten! Colin and his wife Margaret held legendary scallop parties at their cottage by the loch.

Although very temporary in nature, the Torridon laboratory was to remain in operation for more than 10 years. We were known locally as "the fishery boys" and eventually became part of that isolated rural community. We traveled west every summer in heavily laden vehicles, taking with us our families, pets, and a plentiful supply of the whisky that formed the local currency. Most of our local friends and helpers were from the McKenzie clan, and most of the men had been named Donald. They had to be distinguished by their nicknames: Donnie London (born in London); Donnie Merchant (owner of the tiny hardware store), and Donnie the Bus (driver of the school bus). Without their help and friendship our experimental work would not have been possible.

Our first experiments were focused on determining auditory thresholds for fish of several species. The fish were caught at shallow depth (<10 m) to prevent damage from expansion of the swim bladder (as a fish rises the gas expands and may damage the bladder and other organs). The fish were allowed to recover, placed in the sea in a cage on top of the tower, and left for 24 hours to equilibrate. A calibrated hydrophone obtained from the UK Admiralty Research Laboratory was placed

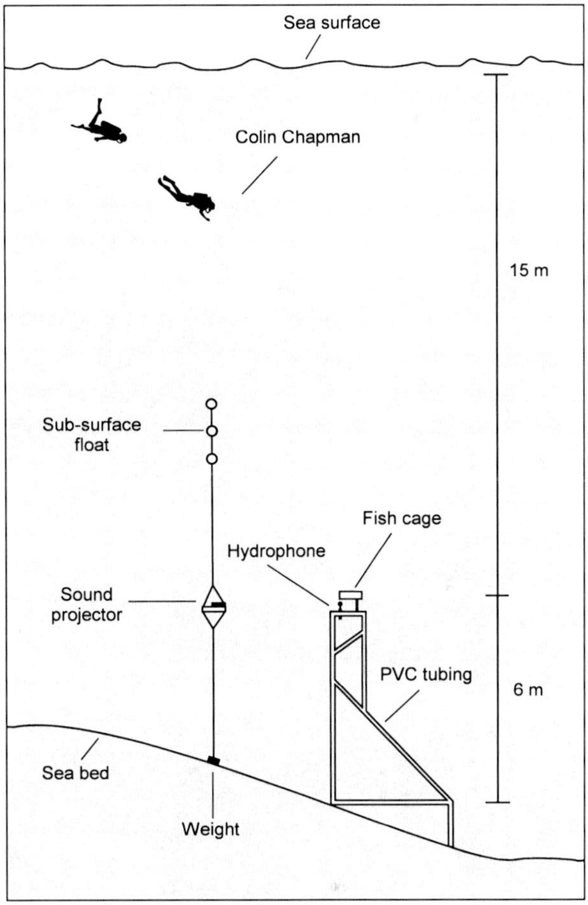

Fig. 14.2 The Loch Torridon acoustic range, with the fish placed on a tower in mid water for sound playback experiments

beneath the head of the fish (Fig. 14.4), and a number of US Navy J9 sound projectors placed at different distances, each supported on an aluminum platform anchored to the seabed and buoyed up by sub-surface floats.

We used a cardiac conditioning technique (Otis et al., 1957) that allowed auditory thresholds to be determined with the fish submerged below the sea surface and out of sight. The electrocardiogram of the fish was monitored with a small metal electrode. The fish was confined in a narrow open mesh cage on top of the underwater tower and the electrode plugged into a cable running 200 m to the shore. A pure tone stimulus was presented for the duration of four normal heartbeats, followed by a mild electric shock. The conditioned response consisted of a delay in one or more heartbeats following the onset of the sound. Once a clear positive response had been established the sound level was progressively lowered with each positive response and raised with each negative response. Karl von Frisch himself had developed this "staircase" method for threshold determination.

Fig. 14.3 View over the Loch Torridon acoustic range with two students planning an experiment. The mountains of Liathach and Beinn Alligin are in the distance

Fig. 14.4 A fish in its cage being placed on top of the Loch Torridon tower, with two hydrophones in position. Fouling organisms, including tubeworms, had to be removed from the hydrophones regularly

Hearing thresholds were determined at a number of frequencies for four species from the cod family: the haddock, Atlantic cod, European pollack (*Pollachius pollachius*), and ling (*Molva molva*) (Chapman, 1973; Chapman & Hawkins, 1973). Interestingly, the thresholds obtained from cod in the sea were much lower than those obtained previously by Udo Buerkle (1967) and Kjell Olsen (1969) working with this same species in laboratory tanks. However, conditions in Loch Torridon were much quieter. Only a few small fishing vessels passed by each day and the ambient noise levels proved to be quite low. We noted, however, that small variations in the ambient noise level, as wind and weather conditions changed, resulted in changes to the auditory thresholds. Detection of the stimulus was being masked by sea noise. These results underlined the importance of obtaining quiet conditions for hearing experiments, and demonstrated the wisdom of working under conditions that fish normally experienced.

We set out to examine whether fish were responding to the measured sound pressure, or to particle motion. Sounds were presented to the fish from sources at different distances, following the method introduced by Enger and Andersen (1967). For cod, thresholds at frequencies between 60 and 160 Hz were largely independent of sound source distance. At lower frequencies, the thresholds were lower when the source was very close to the fish, where particle motion amplitudes were higher. We concluded that the auditory system of the cod was effectively sensitive to sound pressure at higher frequencies in the far field, but responded to particle motion at low frequencies close to the source.

Our experiments at Loch Torridon were extended to include several other species. Olav Sand, a graduate student who had previously been working with Per Enger in Oslo, joined us at Torridon. Olav was both young and strong, and he proved a great asset, especially when 40-gallon drums of diesel fuel for the generator had to be manhandled up the hill to the site. Attention turned to measuring the hearing abilities of two species of flatfish, the plaice *Pleuronectes platessa* and the dab *Limanda limanda*, both of them lacking a gas-filled swim bladder. The sound stimulus in these experiments was varied in two ways (Chapman & Sand, 1974). First, sound projectors were placed at different distances from the fish to vary the ratio between sound pressure and particle motion. Second, the effect of sound radiation from a gas-filled balloon was examined by placing it close to the head of the fish. Actually, it was not a balloon, as the subsequent paper said, but an air-inflated condom, chosen for its closer resemblance to a swim bladder.

Sound pressure thresholds were much higher for the plaice and dab than they were for the gadoid species and their frequency range did not extend as high. Their thresholds were not affected by changes in the ambient sea noise. There were, however, very clear differences between sound pressure thresholds obtained at different distances. If threshold values were recalculated in terms of particle motion, then they were independent of distance. It was clear that that unaided otolith organs in the absence of a swim bladder were sensitive to particle motion rather than sound pressure. Although thresholds were high when expressed as sound pressures, the particle displacement thresholds were exceedingly low, that is, less than 0.1 nm.

Such high sensitivity to particle motion enabled detection of sounds in the far field, as well as the near field. Van Bergeijk had been wrong!

An especially exciting result was that the presence of a gas-filled condom close to the head of the dab resulted in much lower thresholds and a more extended frequency range. This experiment provided strong evidence of the potential role of gas-filled bodies such as the swim bladder in augmenting hearing. The result also confirmed the need for great care in carrying out experiments in a complex acoustic environment. It showed that the presence of any reflecting gas body or surface could produce large particle motions that would stimulate the ear of a particle motion sensitive fish.

Hawkins and MacLennan (1976), in laboratory experiments, later provided further evidence that the unaided ear in flatfish was sensitive to particle motion. Microphonic potentials were detected from the ear of the plaice in response to tone stimuli generated in a standing wave tank where the ratio of sound pressure to particle motion could be carefully controlled and measured. The potentials were evoked only to stimulation by particle motion.

14.4 The Acoustic Properties of the Swim Bladder

The results from these field experiments aroused our interest in the role of the swim bladder as an accessory hearing organ in fish. Many discussions took place while we tramped together across the heather at Torridon or sat together in the local pub over a dram or two. We designed a series of new experiments to elucidate more fully the role of the swim bladder in hearing.

We set out to measure the sound field re-radiated by the swim bladders of intact living cod (Sand & Hawkins, 1973). A technique for doing this had previously been applied at Loch Torridon by two associates, Brian McCartney and Ron Stubbs from the National Institute of Oceanography. They had an interest in developing low frequency sonar systems to estimate the size of fish and had set out to measure the resonance frequencies of fish swim bladders. Their procedure involved measuring the sound pressures generated by a wide-band source (a J11 loudspeaker), first in the free field and then close to a fish. To monitor the re-radiated sound the live fish was placed inside a large, ring-shaped, piezo-electric transducer.

Our experiments, using the same technique, were carried out from a raft moored in a deep, slate quarry, flooded with seawater, at the Island of Seil in Argyll, where calm and stable conditions prevailed. Observations were made on cod maintained at different depths to examine the effect of hydrostatic pressure on the acoustic properties of the organ.

We were not sure what to expect from these experiments. Poggendorf (1952) and de Vries (1956) had assumed a close correspondence between the swim bladder and a free gas bubble in water. They modeled the gas bladder as a simple mass/spring system, where the spring factor was provided by the low elastic modulus of the contained gas, and the mass resulted from the high inertia of the surrounding water. If a gas bubble is exposed to sound pressures of varying frequency but constant amplitude it pulsates, the response reaching a maximum at the resonance frequency. The resonance frequency depends upon the volume of gas inside the bubble, the hydrostatic pressure, the shape, and other factors.

Previous acoustical studies had indicated that the swim bladder of a fish did behave like a free bubble (Hersey & Backus, 1954; Andreeva, 1964). As the fish moved up and down in the water the volume of the swim bladder and the resonance frequency changed in a predictable manner. Alexander (1959) had demonstrated in the laboratory that the swim bladder of many physoclist fishes (species with closed swim bladders) followed the gas laws closely. Within certain limits the volume of the swim bladder was inversely proportional to the pressure. We anticipated, however, that if fish was moved only slowly from one depth to another, allowing time for adaptation to each depth, then comparison with the behavior of a constant volume of gas might be more relevant.

We discovered that the swim bladder in the cod did resemble a free bubble very closely in its acoustic behavior when the fish was subjected to large and rapid depth changes. However, it did not behave acoustically like a free bubble for moderate changes in depth. Once placed at a particular depth, and allowed to adapt, the swim bladder slowly changed in its properties to provide a much higher resonance frequency and much heavier damping than a free bubble of the same volume. This maintenance of a higher resonance frequency did not appear to be associated with any excess pressure developing within the organ. In separate experiments we were able to show that the internal hydrostatic pressure in depth-adapted cod was similar to that in the surrounding water (Sand & Hawkins, 1974).

We concluded that the changes on adaptation were associated with an increase in the shear modulus of the tissues surrounding the swim bladder; that is, they resulted from changes in the tone of the abdominal muscles. A lightly damped swim bladder resonant at a frequency falling within the hearing range of the fish would provide maximum advantage in terms of converting incident sound pressures into particle motion, but it would also cause the relative sensitivity to different frequencies to change with depth. With a resonance frequency maintained well above the hearing range of the fish, and with increased damping, sensitivity to different frequencies would be independent of depth.

From modeling the behavior of the swim bladder (Sand & Hawkins, 1973) we suggested that the particle motion re-radiated from the swim bladder would exceed the particle motion in the absence of the organ only at frequencies above about 30–50 Hz. There would be little auditory gain at the very lowest frequencies. It was evident, however, that the gas in the swim bladder of the cod had a positive effect on hearing sensitivity for all frequencies from a lower transition frequency to the upper frequency limit of hearing. The presence of the swim bladder also extended the audible frequency range. Yet in the cod there was no specialized physical link between the swim bladder and the ear.

Subsequently, Olav Sand re-joined Per Enger in Norway and provided direct evidence of an auditory function of the swim bladder in this species (Sand & Enger, 1973). Microphonic potentials were recorded from the ears of cod during exposure to sounds in a Norwegian fjord, while the swim bladder was inflated and deflated.

Swim bladder volume had no effect on the microphonic potentials at 100 Hz, but had a marked effect at higher frequencies. At 300 Hz, the sound pressure necessary to evoke microphonic potentials just above the electric background noise was about 20 dB (10 times) higher for the empty bladder compared to the full bladder.

Was this the case for all fishes with swim bladders? Our next step was to examine another group of fishes: the salmonids.

14.5 The Salmon and Its Kind

Salmonid fish also have a swim bladder, but they are physostomes. That is, unlike the cod family, where the swim bladder is closed, there is a connection (the pneumatic duct) between the bladder and the digestive tract. The organ is also placed further back in the abdomen than in the cod. Was it possible that in the salmonids the swim bladder was less important in hearing?

Olav Sand, working with Gunnar Sundnes in Salangen Fjord in northern Norway, investigated the acoustics of the swim bladder in intact living Arctic charr *Salvelinus alpinus* at different depths (Sundnes & Sand, 1975). The acoustical behavior was similar to that of a free bubble of gas, and varied with depth in a similar manner. However, because of shape and tissue effects, the resonance frequency of the charr swim bladder was about 40 % higher than for a comparable free bubble. The most interesting finding, however, was that the mechanism found in the cod, where the resonance frequency was maintained at a high value, was not present in the charr.

Back in Scotland I joined with Alastair Johnstone, the youngest of the fishery boys and a great angling enthusiast, to study the hearing abilities of Atlantic salmon *Salmo salar* (Hawkins & Johnstone, 1978). The hearing of salmon and its close relative the trout had been the subject of much debate, especially among anglers.

In the sea, salmon responded only to low-frequency tones (below 380 Hz). The fish were relatively insensitive compared to the cod, their audiograms more closely resembling those obtained from the plaice and dab. Masking of the thresholds did not take place under natural conditions of sea noise, but could be imposed by transmitting high levels of noise. Use was again made of the near field effect to expose the fish to different ratios of sound pressure to particle motion. As with the dab and plaice lower sound pressure thresholds were obtained closer to the source, but the estimated particle motion thresholds remained constant with distance, confirming that the salmon was sensitive to particle motion rather than sound pressure. The swim bladder was not involved in hearing in the salmon.

Parallel experiments in a laboratory tank were especially revealing. We made use of Parvulescu's suggestion that a sound field characterized by small particle motions could be obtained by imposing changes in the air pressure surrounding the tank. Conversely, very large particle motions could be obtained with a submerged loudspeaker. Sound pressure thresholds determined for salmon in the tank differed greatly under these two different conditions. The thresholds were much higher (by >30 dB) with a loudspeaker in air. Again, it was evident that the fish were responding only to particle motion. The experiment also underlined the extreme care that had to be taken in interpreting the results of experiments in a small tank in the laboratory. The use of a loudspeaker in air to determine sound pressure thresholds would greatly underestimate the hearing abilities of any fish sensitive to particle motion.

14.6 Masking

In our initial experiments on cod it had become clear that auditory thresholds to pure tones increased as the level of ambient noise increased. Sounds were readily masked by sea noise.

Masking indicates an inability to separate the signal and noise. However, not all the frequency components of the noise may cause masking. For human subjects it has long been known that a pure tone signal is masked most effectively by noise components at the same and similar frequencies. Fletcher (1940) applied the term "critical band" to the frequency span of noise that is effective. He introduced the analogy of an auditory filter, which can be tuned to the frequency of the stimulus and effectively eliminates noise at remote frequencies.

We set out to investigate whether an analogue of the critical band existed in the cod (Hawkins & Chapman, 1975). We used two different techniques to examine the effects of noise. First, bands of noise of different width were transmitted, and the threshold for detection of a pure tone stimulus determined as the band was progressively narrowed. The signal-to-noise ratio remained constant until a point was reached where the signal-to-noise ratio declined. Only frequencies close to that of the pure tone were responsible for the masking.

In the second experiment, pure tone thresholds were determined in the presence of a succession of narrow bands of noise (10 Hz wide), centered at different frequencies. Masking was most pronounced when the center frequency of the noise band coincided with the frequency of the signal. A small shift in the center frequency of the noise band above or below the frequency of the pure tone resulted in a sharp decline in masking. Again, the experiment confirmed the existence of a critical band for masking.

Similar experiments were later carried out with salmon. Critical bands in the salmon were wider than those for the cod at any given frequency.

In mammals and birds the critical band has been interpreted in terms of the mechanical response of the cochlea of the inner ear (von Békésy, 1960). There is no obvious form of mechanical frequency analyzer within the fish ear comparable to the cochlea. It seems likely that frequency selectivity in the fish auditory system is mediated within the central nervous system (CNS).

The full implications of masking have yet to be investigated fully for fishes. It is evident that it is the level of ambient sea noise that largely determines detection distances for sounds by the cod. Recent concern over the impact of man-made noise upon the environment has raised a number of queries over the effects of additional noise in terms of masking signals of interest to fish, including any effects upon their ability to detect the calls of other fish or the sounds of prey or predators.

14.7 Directional Hearing

One of the main priorities for all three European groups in carrying out hearing experiments in the sea was to investigate directional hearing in fish.

At Loch Torridon, Colin Chapman (1973) had noted that the masking effect of noise on the detection of a pure tone by cod was reduced when the masking noise was transmitted from a separate loudspeaker, spatially separated from the signal projector. In humans a similar improvement in signal detectability is attributed to binaural differences in the timing and amplitude of sounds reaching the two ears. Van Bergeijk had rejected this mechanism for fish.

Chapman and Johnstone (1974) repeated the experiment with cod using four sound projectors, allowing a wider range of separation between signal and masker. For angles greater than 10° there was a significant decrease in the mean threshold: noise ratio of about 7 dB.

Experiments were then carried out where cod and haddock were conditioned to a short period of switching of a pulsed tone from one loudspeaker to another at a different angle of azimuth. The fish readily responded to the switching when the loudspeakers were separated by 20° or more. The sound amplitude required for discrimination of the larger angles (40°–90°) in terms of measured sound pressure was proportional to frequency.

Colin and Alistair proposed that the otolith organs were involved in directional discrimination, through the detection of particle motion. Although a swim bladder was present, and used to detect sound pressure in these fishes, it did not appear to interfere with the ability of the fish to discriminate sounds from different directions.

Wersall et al. (1965) had earlier found orderly hair cell orientation patterns in the otolith organs of the burbot *Lota lota* (a freshwater cod). On a short visit to the Marine Laboratory in Aberdeen by Per Enger, debate between the Norwegian and Scottish teams over the possible mechanisms of directional hearing in fish resulted in the suggestion that we should investigate the directional properties of the ear by vibrating a fish in different directions.

An impromptu experiment was mounted over a weekend. Microphonic potentials were recorded from the haddock ear. The fish was mounted within a plastic tube filled with water on a vibration table, consisting of a rotatable metal slab resting upon a foam-rubber bed. The slab was driven back and forth by an electromagnetic vibrator. The amplitude of the potentials proved to be a function both of the stimulus strength and of the direction of vibration (Enger et al., 1973).

In many of the fish examined, the maximum microphonic amplitude was recorded at an azimuth angle of 0° (i.e., when the fish was subjected to vibration along its long axis). A minimum was recorded at an azimuth angle of 90–100° (i.e., when the fish was subjected to vibration from the side). In some cases, however, the optimal sensitivity was obtained for other vibration directions. We concluded that different groups of hair cells within the otolith organs showed different patterns of directional sensitivity when stimulated by vibration.

Olav Sand (1974) later followed up this approach by recording microphonic potentials from different parts of the ear in perch (*Perca fluviatilis*) as a function of vibration in the horizontal and vertical plane. Again, the amplitude of the microphonic potentials was a function of the vibration direction. Maximal responses in each sacculus were obtained when this direction deviated about 20° from the long axis of the fish, which is approximately parallel to the long axis of the sacculi. In recordings from the lagena, greatest sensitivity was to vertical vibrations. Olav concluded that fish might be able to detect the sound direction with even higher accuracy in the vertical than in the horizontal plane.

Others subsequently took up the technique of whole body vibration. Polar diagrams of the directional sensitivity of primary auditory afferents in fish were first presented by Hawkins and Horner (1981), who recorded from the saccular and utricular branches of the auditory nerve in Atlantic cod during whole-body vibrations in the horizontal plane. These and experiments by others showed that the primary auditory afferents gave directional response patterns similar to the cosine response functions of single hair cells, indicating that each afferent contacts a population of hair cells with the same directional orientation. The otolith organs were sufficiently sensitive to respond to the levels of particle motion associated with sounds of normal amplitude in the far field of a source. Strong phase locking to the stimulus shown by the primary auditory afferents indicated that information on stimulus phase was being conveyed to the CNS.

While these initial laboratory investigations of the directional properties of the sensory maculae were being pursued, Arie Schuijf (1975) was carrying out further experiments on directional hearing in cod in the sea. He trained cod to indicate the active one of two alternative sound projectors by swimming to either of two opposing corners of a cage in return for a food reward. He observed that a fish with a completely intact lateral line system, but with unilaterally severed saccular and lagenar nerves, was no longer capable of directional detection in the far field of a source. Directional hearing in fish required both the ears.

Arie followed Enger et al. (1973) in suggesting that discrimination of direction was based on the directional sensitivity of the sensory hair cells. It was likely that the fish brain could determine the direction of particle motion of the incident sound by a process of vector weighing, comparing the inputs from different regions of the sensory maculae. Current models of directional hearing in fish are still based on this idea. Horner et al. (1980), a student at Aberdeen, subsequently demonstrated binaural interaction in the CNS of the cod by recording single unit activity in the torus semicircularis of the brain during sound stimulation.

An explanation for directional hearing in fish was emerging. The directional information conveyed by the particle motion within a sound wave could be extracted from the incident sound by comparison of the outputs of differently orientated groups of hair cells.

There were two problems with this explanation. Detection of the axis of particle motion is not sufficient for determining the direction of a sound source. Because the particle motion takes place alternately toward the source and away from it there is an inherent bidirectionality or 180° ambiguity in the response of a simple vector weighing system. The fish would not be able to discriminate between two sound sources 180° apart.

Fig. 14.5 Arie Schuijf, standing on the raft, and a student, working at the Loch Torridon acoustic range

The second problem was that despite the otoliths being sensitive to particle motion, the auditory thresholds in a fish like the cod were determined by sound pressure at all but the lowest frequencies. The direct input to the otolith organs from the incident sound was supplemented by the indirect input from the pulsations of the gas-filled swim bladder. Such a single indirect input, carrying no directional information per se, might be expected to interfere with directional detection in at least part of the ear.

Later, working with his student Rob Buwalda (Fig. 14.5), Arie Schuijf showed that the fish could discriminate sound waves traveling toward the head from those traveling toward the tail (Schuijf & Buwalda, 1975). Essential to this discrimination, however, was preservation of the appropriate phase relationship between particle motion and sound pressure. Phase reversal of the acoustic pressure in the travelling wave caused a 180° reversal of the directional response. Far from interfering with the directionality of the ear, the indirect input of sound pressure appeared to be essential for unambiguous directional detection.

In a seminal paper (Schuijf, 1976), Arie outlined his phase model of directional hearing. He proposed that directional detection might be thought of as two distinct, but not necessarily unrelated processes: determination of the axis of particle motion by vector weighing and removal of the remaining 180° ambiguity by analysis of the phase relationship between sound pressure and particle motion.

Further experiments on directional hearing followed at Loch Torridon. We showed that cod were able to discriminate between sound sources in the median vertical plane (Hawkins & Sand, 1977). Cod could also discriminate between pure tones emitted alternately from two aligned sound projectors at different distances (Schuif & Hawkins, 1983). In both cases the cod was superior to man. This difference in auditory ability may be associated with differing habitats; the fish lives in a three-dimensional medium whereas humans are restricted to a surface.

Fig. 14.6 Arie Schuijf (left) and Rob Buwalda (right) on a fishing expedition. Olav Sand, a much keener fisherman, managed to be in the background in both pictures

Finally, we carried out experiments to test the validity of the phase model in three-dimensional space (Buwalda et al., 1983). The demanding experiments were carried out beneath a raft at Loch Torridon (Fig. 14.6) and required a complex configuration of fish and loudspeakers. The results showed that cod could discriminate between two sources of low-frequency sound positioned opposite one another in the median vertical plane. The cod was capable of ambiguity-free detection in three-dimensional space. A number of alternative versions of a phase detection model were then presented to account for the results.

Shortly after these experiments Arie Schuijf and Rob Buwalda took early retirement and ceased their scientific work. The departure of these very able and innovative scientists brought an end to experiments on directional hearing in fish in a free sound field. In a parting remark, their 1983 paper emphasized that future research in this field would rely heavily on independent and accurate control of relevant acoustic variables. Experiments on hearing in the laboratory were not ruled out, but it was pointed out that it would be necessary to control the amplitude, phase, and direction of particle motion and sound pressure at the position of the fish, a warning that has seldom been heeded.

14.8 Current State of Knowledge

By the mid-1980s experimental work on the hearing abilities of fish by the three different European teams had ceased. The Loch Torridon acoustic range was closed and most of the workers who had been actively involved in the experiments moved on to other fields.

Much progress had been made over quite a short period of time. The hearing sensitivities of a number of key species had been determined under conditions in which the properties of the sound stimuli could be properly measured or estimated. The audiograms established for these species still stand today, alongside those obtained under much poorer acoustic conditions, and are regularly used in environmental statements to estimate sound detection distances.

The experiments had demonstrated the role of the swim bladder in hearing, and had enabled the acoustic properties of swim bladders to be determined. The masking of sounds by background noise in the sea had been examined and the existence of frequency filtering mechanisms for improving signal detection in the presence of noise had been confirmed. It had become clear that a fish like the cod, with a swim bladder, could discriminate between sounds from different directions and different distances, and models had been developed to explain these abilities. Together with parallel experimental work carried out in North America (reviewed in this volume in Chapter 7 by Fay and Chapter 25 by Popper) these studies had set out the fundamental features of fish hearing.

Major gaps still remain. Most of these early studies on directional hearing in fish focused on detecting changes in sound direction or sound distance, rather than detection of the actual location of a sound source. Mechanisms of hearing within the ear and within the CNS are still not well understood. It is not clear how directional information in the incident particle motion is protected against masking by the amplified particle motions radiating from the swim bladder. Mechanisms for processing, separating, and comparing the representation of sound pressure and particle motion within the CNS remain to be elucidated. There have been a number of healthy and valid criticisms of the phase model, but satisfactory alternatives have not yet come forward.

Much of the early experimental work has now been forgotten and is seldom referred to. There is a tendency in science for earlier papers, even those that underpin current thought, to be forgotten. The ideas they promulgated are summarized, not always accurately, in reviews, which are then re-reviewed. There is a resemblance to the game of telephone whispers, in which one person gives a message to another, which is passed through a line of people until the last player announces the message. Errors accumulate in the retelling. It is now very difficult to obtain copies of the original papers. It costs almost 40 US dollars for those without access to library facilities to read them.

Where are we now? The recent expansion of offshore industrial activities has led to renewed concern about the impact of man-made noise upon marine animals (e.g., Popper & Hawkins, 2012). Sound travels well in the sea and the influence of underwater noise can be pervasive. The effects can range from mild and insignificant to severe and lasting. We need more and better information about the hearing abilities of fish and other marine animals.

Current lack of understanding is affecting our ability to properly evaluate and mitigate effects of man-made sound on marine ecosystems, making it difficult to take informed management decisions. It is important, however, that we learn from the past. Griffin had warned in the 1950s that if the functioning of the fish ear was to be understood fully then more rigorous quantitative acoustic measurements were required. Our experiments in the sea had then confirmed the importance of working under appropriate acoustic conditions. That lesson must not be forgotten.

References

Alexander, R. McN. (1959). The physical properties of the swimbladders of fish other than Cypriniformes. *Journal of Experimental Biology*, 36, 347–355.

Alexander, R. McN. (1966). Physical aspects of swimbladder function. *Biological Reviews*, 41, 141–176.

Andreeva, I. B. (1964). Scattering of sound by air bladders of fish in deep sound-scattering ocean layers. *Soviet Physics: Acoustics*. 10, 17–20.

Buerkle, U. (1967). An audiogram of the Atlantic cod, *Gadus morhua* L. *Journal of the Fisheries Research Board of Canada*, 24, 2309–2319.

Buwalda, R. J. A., Schuijf, A., & Hawkins, A. D. (1983). Discrimination by the cod of sound from opposing directions. *Journal of Comparative Physiology A*, 150, 175–184.

Chapman, C. J. (1973). Field studies of hearing in teleost fish. *Helgoländer wiss Meeresunters*, 24, 371–390.

Chapman, C. J., & Hawkins, A.D. (1969). The importance of sound in fish behaviour in relation to capture by trawls. *FA0 Fisheries Reports*, 621,717–729.

Chapman, C. J., & Hawkins, A. D. (1973). A field study of hearing in the cod, *Gadus morhua* L. *Journal of Comparative Physiology*, 85, 147–167.

Chapman, C. J., & Johnstone, A. D. F. (1974). Some auditory discrimination experiments on marine fish. *Journal of Experimental Biology*, 61, 521–528.

Chapman, C. J., & Sand, O. (1974). Field studies of hearing in two species of flatfish *Pleuronectes platessa* (L.) and *Limanda limanda* (L.) (family Pleuronectidae). *Comparative Biochemistry and Physiology A*, 47, 371–385.

Dijkgraaf, S. (1949). Untersuchungen uber die Functionen des Ohrlabyrinths bei Meeresfischen. *Physiologia Comparata Oecologia*, 2, 81–106.

Dijkgraaf, S. (1963). The functioning and significance of the lateral-line organs. *Biological Reviews*, 38, 51–105.

Enger, P. S. (1963). Single unit activity in the peripheral fish auditory system of a teleost fish, *Acta Physiologica Scandinavica*, 59 (Supplement 210), 948.

Enger, P. S. (1966). Acoustic threshold in goldfish and its relation to sound source distance. *Comparative Biochemistry and Physiology*, 18, 859–868.

Enger, P. S., & Andersen, R. (1967). An electrophysiological field study of hearing in fish. *Comparative Biochemistry and Physiology*, 33, 517–525.

Enger, P. S., Hawkins, A. D., Sand, O., & Chapman, C. J. (1973). Directional sensitivity of saccular microphonic potentials in the haddock. *Journal of Experimental Biology*, 59, 425–434.

Fletcher, H. (1940). Auditory patterns. *Reviews in Modern Physics* 12(1), 47–65.

Frisch, K. von (1936). Uber den Gehorsinn der Fische. *Biological Reviews*, 11, 210–246.

Frisch, K. von, & Stetter, R. H. (1932). Untersuchungen über den Sitz des Gehörsinnes bei der Elritze. *Zeitschrift für vergleichende Physiologie*, 17, 686–801.

Griffin, D. R. (1950). Underwater sounds and the orientation of marine animals—A preliminary survey. *Technical Report Number 3, Project NR 162–429*. Office of Naval Research and Cornell University.

Harris, G. G. (1964). Considerations on the physics of sound produced by fishes. In W. N. Tavolga (Ed.), *Marine bio-acoustics* (pp. 233–247). Oxford: Pergamon Press.

Harris, G. G., & van Bergeijk, W. A. (1962). Evidence that the lateral line organ responds to near-field displacements of sound sources in water. *Journal of the Acoustical Society of America*, 34(12), 1831–1841.

Hawkins, A. D. (1973). The sensitivity of fish to sounds. *Oceanography Marine Biology Annual Review*, 11, 291–340.

Hawkins, A. D., & Chapman, C. J. (1975). Masked auditory thresholds in the cod (*Gadus morhua* L.). *Journal of Comparative Physiology*, 103, 209–226.

Hawkins, A. D., & MacLennan, D. N. (1976). An acoustic tank for hearing studies on fish. In A. Schuijf & A. D. Hawkins (Eds.), *Sound reception in fish* (pp. 149–170). Amsterdam: Elsevier.

Hawkins, A. D., & Sand, O. (1977). Directional hearing in the median vertical plane by the cod. *Journal of Comparative Biology and Physiology A*, 122, 1–8.
Hawkins, A. D., & Johnstone, A. D. F. (1978). The hearing of the Atlantic salmon (*Salmo salar*). *Journal of Fish Biology*, 13, 655–673.
Hawkins, A. D., & Horner, K. (1981). Directional characteristics of primary auditory neurons from the cod ear. In W. N. Tavolga, A. N. Popper, & R. R. Fay (Eds.), *Hearing and sound communication in fishes* (pp. 311–328). New York: Springer-Verlag.
Hersey, J. B., & Backus, R. H. (1954). New evidence that migrating gas bubbles, probably the swim-bladders of fish, are largely responsible for scattering layers on the continental rise south of New England. *Deep Sea Research*, 1, 190–191.
Horner, K., Sand, O., & Enger, P. S. (1980). Binaural interaction in the cod. *Journal of Experimental Biology*, 85, 323–333.
Jones, F. R. H., & Marshall, N. B. (1953). The structure and functions of the teleostean swimbladder. *Biological Reviews*, 28, 16–83.
Lowenstein, O. (1957). The sense organs. 2. The acoustic-lateralis system. In M. E. Brown (Ed.), *The physiology of fishes* (pp. 155–186). New York: Academic Press.
Myrberg, A. A., Jr., Banner, A., & Richard, J. D. (1969). Shark attraction using a video-acoustic system. *Marine Biology*, 2, 264–276.
Nelson, D. R., & Gruber, S. H. (1963). Sharks: Attraction by low-frequency sounds. *Science*, 142, 975–977.
Olsen, K. (1969). Directional responses in herring to sound and noise stimuli. *International Council for the Exploration of the Sea*, B20, 8 pp.
Otis, S., Cerf, L. S., & Thomas, G. J. (1957). Conditioned inhibition of respiration and heart rate in the goldfish. *Science*, 126, 263–264.
Parvulescu, A. (1964). Problems of propagation and processing. In W. N. Tavolga (Ed.), *Marine bio-acoustics* (pp. 87–100). Oxford: Pergamon Press.
Poggendorf, D. (1952). Die absoluten Hörschwellen des Zwergwelses (*Amiurus nebulosus*) und Beiträge zur Physik des Weberschen Apparates der Ostariophysen. *Zeitschrift für vergleichende Physiologie*, 34, 222–257.
Popper, A. N., & Hawkins, A. D., Eds. (2012). *The effects of noise on aquatic life*. New York: Springer Science+Business Media.
Pumphrey, R. J. (1950). Hearing. In *Physiological Mechanisms in Animal Behaviour.* Symposium Society of Experimental Biology, 4, 31–38.
Sand, O. (1974). Directional sensitivity of microphonic potentials from the perch car. *Journal of Experimental Biology*, 60, 881–899.
Sand, O., & Enger, P. S. (1973). Evidence for an auditory function of the swimbladder in the cod. *Journal of Experimental Biology*, 59, 405–414.
Sand, O., & Hawkins, A. D. (1973). Acoustic properties of the cod swim bladder. *Journal of Experimental Biology*, 58, 797–820.
Sand, O., & Hawkins, A. D. (1974). Measurements of swimbladder volume and pressure in the cod. *Norwegian Journal of Zoology*, 22, 31–34.
Schuijf, A. (1975). Directional hearing of cod (*Gadus morhua*) under approximate free field conditions. *Journal of Comparative Physiology*, 98, 307–332.
Schuijf, A. (1976). The phase model of directional hearing in fish. In A. Schuijf & A. D. Hawkins (Eds.), *Sound reception in fish* (pp. 63–86). Amsterdam: Elsevier.
Schuijf, A., & Buwalda, R. J. A. (1975). On the mechanism of directional hearing in cod (*Gadus morhua* L.) *Journal of Comparative Physiology*, 98, 333–343.
Schuif, A., & Hawkins, A. D. (1983). Acoustic distance discrimination by the cod. *Nature*, 302, 143–144.
Schuijf, A., Baretta, J. W., & Wildschut, J. T. (1972). A field investigation on the discrimination of sound direction in *Labrus bergylta* (Pisces: Perciformes). *Netherlands Journal of Zoology*, 22, 81–104.

Sundnes, G., & Sand, O. (1975). Studies of a physostome swimbladder by resonance frequency analyses. *ICES Journal of Marine Science*, 36, 176–182.

van Bergeijk, W. A. (1964). Directional and nondirectional hearing in fish. In: W. N. Tavolga (Ed.), *Marine Bio-acoustics* (pp. 281–299). Oxford: Pergamon Press,

van Bergeijk, W. A. (1967). The evolution of vertebrate hearing. In W. D. Neff (Ed.), *Contributions to sensory physiology*, Vol. 2 (pp. 1–49). New York: Academic Press.

von Békésy, G. (1960). *Experiments in hearing*. New York: McGraw-Hill.

Vries, H. de (1950). The mechanics of the labyrinth otoliths. *Acta Oto-Laryngologica*, 38, 262–273.

Vries, H. de (1956). Physical aspects of the sense organs. *Progress in Biophysics and Biophysical Chemistry*, 6, 207–264.

Wersall, J., Flock, A., & Lundquist, P. G. (1965). Structural basis for directional sensitivity in cochlear and vestibular sensory receptors. *Cold Spring Harbor Symposium in Quantitative Biology*, 30, 115–132.

Chapter 15
The Behavioral Study of Mammalian Hearing

Henry E. Heffner and Rickye S. Heffner

Rickye S. Heffner and Henry E. Heffner

H.E. Heffner (✉) • R.S. Heffner
Department of Psychology, University of Toledo,
2801 West Bancroft, Toledo, OH 43606, USA
e-mail: hheff@adelphia.net; rheffner5@mac.com

A.N. Popper and R.R. Fay (eds.), *Perspectives on Auditory Research*, Springer Handbook of Auditory Research 50, DOI 10.1007/978-1-4614-9102-6_15,

15.1 Introduction

Modern auditory research had its beginnings in the late 1940s, at which time a general review of auditory research could be accomplished in four chapters (Stevens, 1951). The subsequent expansion of auditory research led to the Springer Handbook of Auditory Research series that began to appear in 1992. It is interesting to consider what advances have been made since then and whether the advances have been new discoveries or refinements of previous ones.

Our study of the auditory abilities of mammals started in the laboratory of Bruce Masterton, first at Vanderbilt University and then at Florida State University. It began as the study of auditory cortex using the ablation-behavior procedure (Masterton, 1997). Coming from Irving Diamond's laboratory at Duke University, Bruce was interested in the evolution of auditory cortex and wanted to observe the effect of auditory cortex lesions in animals approximating the human phyletic lineage: opossum (marsupial), hedgehog (insectivore), tree shrew (classified at that time as a primate), and bushbaby (prosimian). The first step was to establish the preoperative auditory abilities of these species beginning with their audiograms (Masterton et al., 1969). It soon became apparent, however, that the animals varied in their high-frequency hearing, and that this variation was systematically related to the availability of cues to localize sound. This was an unexpected finding that sent the lab on an exploration of the audiograms and later sound-localization thresholds in a larger sample of mammalian species.

The following is a description of two areas of research: The first is the comparative approach to understanding the selective pressures on mammalian hearing and sound localization; the second is the role of auditory cortex, as determined by the ablation-behavior procedure, in absolute sensitivity, sound localization, and more complex discriminations. Underlying both of these areas of research is a third—the development of behavioral tests for determining the auditory abilities of animals—and it is there that we begin.

15.2 The Evolution of Animal Psychophysics

The advances in animal psychophysical procedures have been largely conceptual. Although advances in electronics and computers have been helpful in generating auditory stimuli and recording behavioral responses, modern behavioral procedures for testing animals could have been conducted with the equipment available over half a century ago; computers make it all easier and certainly more compact, but the procedures could have been instrumented with the relay racks that were in use at that time. Instead it was the advances in behavioral conditioning techniques and refinements in the contingencies of reinforcement that led to faster and more accurate ways to determine an animal's sensory abilities.

15.2.1 The Early Years

The period prior to 1992 was one in which new animal psychophysical procedures were being developed and old ones improved. In the 1950s, there were two behavioral procedures employed by W. D. Neff and his students for testing hearing in animals (Neff et al., 1975). One was a sound-localization procedure in which a cat was placed in a start box facing two or more goal boxes that contained food; a buzzer was sounded over the one goal box that was unlocked and the cat was trained to go to the source of the sound to obtain the food. After it finished eating, it was picked up by the experimenter, who was in the test room, and returned to the start box for another trial. The other procedure used a double grill box in which an animal was trained to move from one compartment to the other when it detected a sound, or a change in an ongoing sound, to avoid a shock delivered through the floor bars. Although both procedures worked with cats, there were some limitations. For example, the sound-localization box could be used only with tame animals that could be picked up and returned to the start box, and of course, the person doing this could also present distractions and potential cues. With the double grill box, an animal's head position in the sound field varied introducing error in the measurement of absolute thresholds.

By the 1960s, new animal psychophysical procedures were appearing, many of which are described in the book, *Animal Psychophysics*, edited by W. C. Stebbins (1970). The book contains two chapters on the method of conditioned suppression, one by Barbara Ray, the other by James Smith. The conditioned suppression procedure differed from others in that instead of training an animal to make a response when it hears a particular sound, the animal was trained to make a steady response, such as licking a water spout, but to stop (suppress) responding when it heard a sound that signaled shock. This is a cognitively simple task, much like that of an animal at a water hole that stops drinking when it detects danger. Moreover, by having an animal respond by licking a water spout, its head was fixed within the sound field, making precise measurement of the sound at its ears possible.

We adopted conditioned suppression in the 1960s, having learned of it from Jim Smith at Florida State University. Since then we have made several modifications to simplify training and accelerate testing, the details of which can be found elsewhere (Heffner & Heffner, 1995; Heffner et al., 2006). Among the changes were (1) allowing an animal to make steady contact with a water or food spout instead of requiring it to make discrete licks on the spout which some species find difficult to do, (2) shortening the trial duration from 10 s to 2–3 s, (3) lowering the response cost by requiring an animal to break contact with the spout for only 150–200 ms, (4) changing how false positives were incorporated into the performance measure, and (5) making the shock avoidable. This last change, in which an animal avoids the shock by breaking contact with the spout when it hears a sound, greatly reduced the number of shocks the animal actually received thus increasing the number of trials that could be obtained in a session. However, it also changed the procedure from conditioned suppression to an avoidance task, one that behaviorists refer to as "discriminated punishment," and we have long struggled to find a good name for it.

Because the key feature is that an animal stop or suppress what it is doing when it detects a signal, we currently refer to it as "conditioned suppression/avoidance" (Heffner et al., 2012). Naming aside, it was this application of conditioned suppression that allowed us not only to test many different species of animals, but, because it is such a robust procedure, also enabled us to test hearing in animals with brain lesions and debilitating genetic defects.

During the early years we also explored procedures that did not use electric shock but relied instead on positive reinforcement. The first was Neff's two-choice procedure that we initially used for sound localization. We automated that procedure by having an animal lick a start spout to center its head between the loudspeakers (a "ready" or "observing" response), and using water reward that was automatically dispensed from a water spout under the active loudspeaker. The animal returned to the observing spout on its own thus eliminating the experimenter from the test room. In addition to sound localization, the two-choice procedure has also been used to determine detection and discrimination abilities. In each case, animals made an observing response to indicate their readiness to perform the discrimination. For sound localization, the animal responded left or right to left and right sounds, respectively; for detection, the animal responded to one side if it detected a sound and to the other side if it did not; for auditory discriminations, the animal would respond left to one type of sound and right to another, permitting the testing of fairly complex distinctions, such as between different categories of vocalizations. Correct responses were rewarded with either food or water and errors were punished with a short wait or time out. The two-choice procedure worked well with cats, monkeys, dogs, and an elephant. However, there were some species that did not perform consistently well on the two-choice sound-localization task; specifically, some horses and cattle did not perform at a high level when tested on easily discriminated angles and none of the rats we tested would consistently perform above 90% correct even at large angles of separation.

The second reward procedure was a go/no-go procedure that we used to determine the audiograms of horses and cattle; this procedure was patterned after those developed by John Dalland, who obtained the first behavioral audiograms of bats, and Bill Stebbins who tested hearing in monkeys (Dalland, 1965; Stebbins, 1970). In our tests, an animal was required to place its mouth on an observing plate to initiate testing and to contact a reporting plate with its mouth when it detected a sound. Correct detections were rewarded with water while false positives were punished with an error time out. Although the procedure worked fairly well, false positives easily intruded, especially if an animal was rewarded for making a chance response to a subthreshold tone; in the case of a pony, it was necessary for the experimenter to stand behind the animal with a switch, which quickly eliminated false positives.

15.2.2 The Past 20 Years

Since 1992, there have been both refinements and new developments in animal psychoacoustic procedures, some of which can be found in *Methods in Comparative*

Psychoacoustics (edited by Klump et al., 1995). There are three that we would like to note. The first is the use of the reflex inhibition procedure, which has been used to obtain auditory thresholds in a variety of animals including amphibians for which there are currently no operant procedures (e.g., Simmons & Moss, 1995). The reflex inhibition procedure is based on the observation that the magnitude of an animal's startle response to a stimulus, such as a loud sound or electric shock, can be reduced by preceding the startle stimulus with a low-level sound. Absolute thresholds are obtained by reducing the amplitude of the preceding sound until it no longer has a detectable effect on the magnitude of the startle response as compared to its magnitude when there is no preceding sound. The great advantage of this procedure is that the animals do not require any training beyond acclimation to the test box. However, because operant procedures often show that it is necessary for an animal to "learn to listen" to low-level sounds, it is likely that an audiogram obtained with the startle reflex will be less sensitive than those obtained with operant procedures.

A second advance has been the recognition of species differences in tolerating the negative consequences of errors when performing a sensory discrimination. For example, when testing monkeys, cats, and dogs in a two-choice auditory discrimination, rewarding correct responses with food or water and punishing errors with a short error time out of 3–5 s is often sufficient to maintain a high level of performance. Rats and hamsters, on the other hand, do not find an error time out sufficiently aversive and are content with the thinner reward schedule that accompanies errors and the consequent delays. However, we now know that if errors are punished by even mild shock, rats and hamsters will perform a two-choice auditory task at asymptotic levels as high as that of any other animal. As a result, it is now possible to use the two-choice procedure on these species to do equivalence testing (Heffner, 2011), a procedure that cannot easily be done with conditioned suppression. In short, there must be aversive consequences for making errors. For some animals, a short error time out is psychologically aversive enough to cause them to minimize their errors. For others, it is necessary to add a physically aversive stimulus to obtain good performance.

The third area in which there have been new advances has been the application of behavioral tests for detecting tinnitus in animals (for a review, see Heffner & Heffner, 2012). Such tests are inherently difficult because, unlike other auditory tests, the experimenter does not have direct control over the stimulus of interest; that is, there is no guarantee that administering a tinnitus-inducing agent to an animal will actually cause it to have tinnitus because significant individual variation in susceptibility to tinnitus has been observed in humans. Other difficulties arise from the need to tease out the effects of tinnitus from the effects of other auditory changes that often accompany it such as hearing loss and hyperacusis. The tests for tinnitus can be divided into three general types. The first has been to train animals to discriminate the presence of sound from its absence, such as training them to stop drinking when a background sound is turned off. The animals are then given a tinnitus-inducing agent, such as salicylate or exposure to loud sound, and tested in the absence of any physical sound to determine if they behave as though a sound is still present. The second type of test involves looking for interactions between tinnitus and physical sounds—interactions in which tinnitus might fill in a gap in a sound that is similar in pitch to the tinnitus or in which the tinnitus might change the perception of a physical sound. The third has been to train an animal on a left–right sound-localization task, expose one ear to a loud sound, and test to see if the animal responds as though it hears a sound in the exposed ear when no physical sounds are presented. The various tinnitus test procedures differ in the degree to which their validity has been assessed. One type of validation is to determine how the animals would perform when tinnitus is simulated by physical sounds. Another is to consider whether alternative explanations of the results are plausible such as the hearing loss and hyperacusis that are also caused by tinnitus-inducing agents. Finally, and most fundamentally, we should ask whether a particular test would detect tinnitus in humans.

15.3 Comparative Mammalian Hearing

The simplicity, accuracy and reliability of behavioral tests of hearing made it possible to study hearing in an astonishing variety of mammals. Species that differ in their size (bats to elephants), body configuration (horse to primate), motor abilities (blind mole rats, squirrels, mice with genetic movement disorders), and lifestyle (underground in deserts to aquatic) have been tested (Heffner & Heffner, 1998, 2003). The most interesting findings so far have centered on simple pure-tone sensitivity (the audiogram) and sound localization.

15.3.1 The Early Years

In the late 1960s we came to recognize that the variation in mammalian high-frequency hearing was linked to sound localization. Because both binaural cues for sound location, the difference in the time of arrival of a sound at the two ears and the difference in the frequency-intensity spectra reaching the two ears, are affected by head size, animals with small heads may have smaller binaural cues available to them. Time delays can be well below 100 μs, and, because small heads and pinnae do not block low frequencies as effectively as they block higher ones, small animals must hear higher frequencies than large animals to use the spectral-difference cue. Defining head size functionally as the maximum difference in the time of arrival of a sound at the two ears (i.e., the maximum time difference available to an animal), it was found that the high-frequency hearing limit correlated closely with functional head size. Indeed, this relationship accommodates species ranging in size from wild mice and bats to humans and elephants. However, with the testing of additional species, some exceptions began to appear. For example, there were some mammals that did not use the binaural spectral-difference cues and others that did not hear as high as predicted by their functional head size. These exceptions had to be explained.

The sound-localization tests given to the various species typically consisted of two types. The first was the determination of the minimal audible angle for noise bursts from azimuthal locations to the left and right of the animal's midline.

The second test examined the localization of pure tones—the ability to localize low-frequency tones indicating that an animal could use the binaural phase cue (a subset of the binaural time cue), and the ability to localize high-frequency tones indicating that it could use the binaural intensity-difference cue (a subset of the binaural spectral-difference cue). At the time, we expected that all animals would localize sound as accurately as the physical locus cues available to them permitted, so we were surprised when horses and cattle proved to be far less accurate than much smaller animals such as laboratory rats. We also found that these large animals could not localize high-frequency tones, indicating that they made little or no use of the binaural spectral-difference cue; yet they still heard high-frequency sounds as predicted by the relationship between functional head size and high-frequency hearing. Both of these discoveries had to be explained.

15.3.2 The Past 20 Years

Recent research has answered the questions concerning both high-frequency hearing and sound localization. It has also revealed large variation in mammalian low-frequency hearing.

15.3.2.1 High-Frequency Hearing

The relationship between functional head size and high-frequency hearing was initially based on only seven species; with an increase of the number of species in the sample by almost 10-fold, the relationship continues to account for about 80% of the variance in high-frequency hearing (Heffner & Heffner, 2008). The original explanation for this relationship was based on the view that high frequencies are necessary for sound localization because they provide the binaural spectral-difference cue. This is supported by the observation that filtering out high frequencies from a broadband noise reduces the performance of animals such as monkeys, humans, and chinchillas, on discriminating left from right sound sources. However, there are other mammals that are either partially or completely unable to localize high-frequency pure tones, indicating that their ability to hear high frequencies cannot be accounted for by the need to use the binaural spectral-difference cue. Specifically, the Indian elephant and domestic goat are unable to localize pure tones in the upper end of their hearing range and domestic pigs, horses, and cattle cannot localize tones that are too high to be localized using the binaural phase cue. However, it soon became apparent that these animals required high-frequency hearing to localize sound using pinnae cues (Heffner & Heffner, 2008).

Although the role of the pinna in sound localization has long been known, most human research has focused on the binaural locus cues, often using headphones that eliminated any contribution from the pinna. However, the work of Robert Butler and others had demonstrated that the directionality of the pinna not only provides the primary cues for vertical localization and for preventing front–back confusions, the pinnae also provide effective cues for localizing sound in the horizontal plane. Building on this work, we found that horses, which do not use the binaural spectral-difference cue, nevertheless require high frequencies to use pinnae cues for front–back localization.

A marked exception to the relationship between functional head size and high-frequency hearing was the discovery that subterranean mammals, specifically the pocket gopher, naked mole rat, and blind mole rat, did not hear nearly as high as their small functional head sizes predicted. Subsequent testing revealed that these animals also could not localize sound. Indeed, they are not only virtually unable to distinguish left sounds from right sounds, but they also lack pinnae and therefore are not under selective pressure to hear high frequencies to make front–back distinctions (Heffner & Heffner, 2008). Evidently, sound localization is of little use to animals living underground in a one-dimensional world. The observation that mammals that do not hear high frequencies as predicted by their functional head size also lack the ability to localize sound further supports the contention that high-frequency hearing is closely linked in mammals to sound localization.

Appreciating the importance of high-frequency hearing for generating pinna locus cues has implications for the evolution of the mammalian ear. One of the hallmarks of mammals is the three-boned middle ear, which appears to have evolved to enable them to hear high-frequency sounds. Indeed, fossils are often identified as mammalian based on the presence of the mammalian ear. However, it seems likely that the high-frequency mammalian ear evolved in conjunction with the pinnae, which improve left–right localization, allow for better localization within the lateral hemifield of sounds including those so faint that they are audible in only one ear, and reduce front–back confusions. This implies that birds, at least those that have not evolved a pinna-like facial ruff like the barn owl, are probably unable to determine whether a sound is coming from in front or behind them.

Another question that arose is whether selective pressure for echolocation has caused bats to increase their high-frequency hearing beyond that required for passive sound localization. Comparing bats with other mammals, it appears that echolocating bats do hear about 0.7 octaves higher than predicted for a similar-size non-echolocating mammal. Interestingly, non-echolocating bats do not appear to have extended their high-frequency hearing at all beyond that expected for passive sound localization.

15.3.2.2 Sound Localization

The discovery that horses and cattle, with their relatively large heads and pinnae, did not localize sound as accurately as much smaller animals came as a surprise. Until then, we had believed that animals would localize sound as accurately as permitted by the physical cues available to them. Only after several replications on additional individual animals and using different test procedures did we come to accept that these species were poor localizers.

In looking to explain this unexpected variation in mammalian sound-localization acuity, we examined seven potential factors (Heffner & Heffner, 1992). The first three were functional head size, trophic level (the degree to which a species is predatory or itself preyed upon), and activity cycle (whether a species is nocturnal, diurnal, or crepuscular). The other four were visual factors: visual acuity, width of the binocular visual field, width of the panoramic visual field, and width of the field of best vision. Examining visual factors proved interesting and required one of us (RH) to learn how to process retinal whole mounts to estimate visual acuity and width of the field of best vision—this latter we defined anatomically as the horizontal width (in degrees) of the retinal area encompassing ganglion cell densities at least 75% of maximum.

The results of our multiple correlation study revealed that sound-localization acuity is most closely related to the width of the field of best vision. Indeed, the current correlation coefficient is $r = .89$. We have interpreted this to mean that the primary function of sound localization is to direct the eyes to the source of a sound. Just how accurate sound localization must be to do this depends on the width of an animal's field of best vision. Animals with a narrow field of best vision, such as humans, require good sound-localization acuity to direct their eyes so that the image of the sound source falls within that narrow region, which in the case of humans is our fovea. Animals with broad fields of best vision do not require good localization acuity to direct their gaze as their field of best vision can encompass nearly the entire horizon as is the case with visual streaks.

One question that arises is how bats fit the relationship between sound-localization acuity and width of the field of best vision, given the exquisite ability of most bats to use echolocation to detect and even discriminate objects in their environment. Because bats are thought to essentially replace vision with echolocation, it was conceivable that vision and sound localization might have become decoupled. As it turns out, however, the ability of bats to *passively* localize sound is not unusual—they localize sound as expected based on the width of their field of best vision. Thus, their development of echolocation has not detectably affected their passive sound-localization ability, at least among those bats examined so far.

A final discovery has to do with the use of the two binaural locus cues (Heffner & Heffner, 2003). Over the years it has become apparent that the binaural locus cues are not both used universally. A small number of mammals do not use the binaural time cue and others do not use the binaural spectral-difference cue, and the subterranean rodents appear to use neither. Moreover, among the mammals that use the binaural time cue in the form of the phase cue, there is systematic variation in the highest frequency at which they can use it. For example, cattle appear able to use the binaural phase cue up to 500 Hz whereas the Jamaican fruit bat can use it up to 6.3 kHz, a span of more than 3 octaves. Moreover, this variation in the upper limit of binaural phase is closely related to the maximum time difference available to an animal such that the smaller the available time difference, the higher the upper limit of binaural phase ($r = -0.85$). This is because the smaller an animal's head size, the higher the frequencies for which the binaural phase cue remains unambiguous, although other factors may be involved. Because the use of the binaural phase cue presumably requires phase locking in the auditory system, one might expect there to be species differences in phase locking with smaller species phase locking to higher frequencies than larger animals.

15.3.2.3 Low-Frequency Hearing

The range of variation in mammalian low-frequency hearing is now known to be greater than that for high-frequency hearing. Using the lowest frequency audible at a level of 60 dB, low-frequency hearing limits extend from 17 Hz (Indian elephant) to 10.3 kHz (little brown bat), a range of 9.2 octaves, almost twice the 4.7-octave range in high-frequency hearing limits (Heffner et al., 2001). Moreover, the distribution of low-frequency hearing limits appears to be bimodal with some mammals falling into a group with good low-frequency hearing (i.e., those that hear 125 Hz and lower), and others forming a group with poor low-frequency hearing (i.e., those that do not hear below 250 Hz). There is a loose relationship between high- and low-frequency hearing such that animals with good high-frequency hearing often have poor low-frequency hearing. However, high-frequency hearing accounts for less than half of the variance in low-frequency hearing and there are many species from several different orders, including rodents and carnivores, that are quite sensitive to both high and low frequencies—with audiograms in some cases spanning over 13 octaves. There is currently no selective pressure that has been proposed to account for these unusual features of low-frequency hearing.

15.4 Auditory Cortex

The use of ablation-behavior experimentation to study auditory cortex dates back to the late 19th century after it was found that sensory and motor functions could be localized to different parts of the cortex. The modern study of this field began in the late 1940s with W. D. Neff and his students being major contributors (Masterton, 1997). Improvements in the behavioral test procedures discussed earlier have helped advance this area of auditory research.

15.4.1 The Early Years

The view that the cerebral cortex is, if not the seat of the soul, at least the seat of consciousness led 19th century researches to believe that ablation of auditory cortex would abolish the ability to respond to sound. Although initial studies indicated that this was so, other studies failed to find "cortical deafness" and in spite of sporadic reports of cortical deafness in human patients, animal researchers were unanimous in dismissing the possibility. Thus, when we found that ablation of auditory cortex in macaques caused a substantial hearing loss (Heffner, 2005), it was months before we dared tell colleagues of our discovery. That other researchers failed to find cortical deafness in animals is because they were working with cats and it has only been observed in primates so far (humans and macaques).

The role of auditory cortex in discriminating frequency has also been studied, motivated by the discovery that it is tonotopically organized. Results showed that ablation of auditory cortex does not abolish the ability to discriminate frequency, but it does increase discrimination thresholds.

Perhaps Neff's most famous finding was that ablation of auditory cortex abolishes the perception of locus. Further, this appears to be a perceptual deficit because an animal with bilateral auditory cortex lesions is able to discriminate left sounds from right sounds, but no longer associates a sound with a location in space. This finding has stood unmodified since Neff first reported his discovery in 1948 (Heffner & Heffner, 2003).

One other discovery, which was difficult to classify as sensory or perceptual, was the finding in the 1980s that bilateral ablation of auditory cortex abolishes the ability of Japanese macaques to discriminate two forms of their coo vocalizations. On one hand, it suggested an aphasia-like deficit in macaques following auditory cortex ablation. However, because the coos differed acoustically with one rising in frequency and the other falling, it was possible that the monkeys had a sensory deficit that affected their ability to determine if a sound was changing in frequency.

15.4.2 The Past 20 Years

Recent behavioral studies of auditory cortex have both refined previous discoveries and made new ones.

15.4.2.1 Cortical Hearing Loss

It had been known that the cortical hearing loss that occurs in macaques following bilateral ablation of auditory cortex—a loss that may begin as a complete inability to respond to sound—shows substantial recovery during the first 1–2 months postoperatively, though the animals still have a moderate hearing loss. However, longitudinal studies have shown that recovery continues 3–5 years after surgery with thresholds returning to normal levels at low frequencies and to near normal levels in the midrange of the animals' audiograms. There are at least two possible explanations for the recovery of hearing. One is that cortical areas outside auditory cortex are mediating the function of auditory cortex in detecting sound. Another possibility is that the hearing loss is due to the disruption of the lower auditory centers caused by the sudden loss of descending cortical input and that thresholds improve as the lower centers adapt to the loss. One way to investigate these possibilities would be to examine the remaining cortical areas and the lower auditory centers using electrophysiological and functional MRI techniques to determine how their functions change as a result of ablation of auditory cortex.

It was also discovered that unilateral ablation of auditory cortex in macaques results in a hearing loss in the ear opposite the lesion. The hearing loss is not permanent and thresholds quickly recover to normal or near-normal levels in a matter of weeks. This effect, which was found by testing each ear independently using earphones, can help explain some of the initial effects of damage to auditory cortex in humans.

15.4.2.2 Intensity Discrimination

Although early studies did not suggest a role for auditory cortex in discriminating changes in the intensity of a sound, recent studies have indicated that auditory cortex ablation does affect intensity discrimination. Specifically, although bilateral ablation in macaques may result in at most a slight increase in thresholds for detecting an *increase* in intensity, it results in a large threshold increase for detecting a *decrease* in the intensity of a sound. Currently, there is no theory to explain this result, although it brings to mind the Neff Neural Model that animals without auditory cortex can detect an increase in neural activity.

15.4.2.3 Frequency Discrimination

The ability to discriminate frequency has classically been tested by training animals to discriminate a train of tone pips of the same frequency from a train of tone pips that alternate in frequency. Thresholds obtained by reducing the difference between the two frequencies indicate that ablation of auditory cortex results in a small but consistent increase in thresholds in both monkeys and cats. However, presenting tone pips is only one way to test frequency discrimination; another way is to modulate the frequency to determine an animal's ability to detect when a steady tone is replaced by one that is changing in frequency. One common way is to train an animal to discriminate tones that are rising in frequency from those that are falling in frequency, that is, frequency ramps.

Frequency ramps are of special interest to auditory researchers for several reasons. First, many neurons in auditory cortex are sensitive to the direction of frequency change; a cell that responds to a rising frequency ramp may not respond to a falling ramp over the same frequency range, a discovery reported by Whitfield and Evans in 1965. Second, it is relatively easy for animals to learn to discriminate rising from falling ramps, suggesting that is a more natural discrimination than discriminating tone pips, which is more difficult for an animal to learn. Finally, discriminating rising from falling frequency ramps forms the sensory basis for discriminating the two forms of the Japanese macaque coo calls, a discrimination that is abolished in macaques by auditory cortex lesions. Indeed, frequency modulation is an important component of both language and echolocation, making this acoustic feature very common in nature.

The major discovery in this field was that auditory cortex lesions abolish the ability of monkeys to determine if a sound is changing in frequency (Heffner, 2005). Specifically, they can no longer discriminate a steady tone from one that is either rising or falling in frequency. However, to demonstrate this deficit, it is necessary to randomize the frequency of the steady tone from trial to trial to prevent animals from performing the discrimination on the basis of absolute frequency. This finding has two interesting consequences. First, it indicates that a deficit in the ability to discriminate the coo vocalizations, shown by Japanese macaques following auditory cortex ablation, is a sensory deficit. Second, it is an example of an electrophysiological observation that correctly identified a function of auditory cortex; whereas the view that auditory cortex might be necessary for frequency discrimination, based on the finding of tonotopic maps, turned out not to be true, the discovery by Whitfield and Evans that some auditory neurons were responsive to the direction of a change in frequency did foreshadow the discovery that auditory cortex is necessary for detecting if a sound is changing in frequency.

15.4.2.4 Functional Differences Between Areas of Auditory Cortex

Auditory cortex can be divided into different areas on the basis of the electrophysiological and anatomical properties leading to the question of whether different areas have different behavioral functions. Based on electrophysiological studies, it has been suggested that, in macaques, the identification of complex sounds is processed in the rostral portion of auditory cortex and that the localization of sound is processed in the caudal portion. Indeed, this view has been at least partially supported by ablation studies. Specifically, the ability of macaques to determine if a sound is changing in frequency is abolished by removal of either the rostral or core portions of auditory cortex, but not by removal of the caudal portion. On the other hand, the ability to localize sound is impaired (though not completely abolished) by removal of the caudal or core portions of auditory cortex, with the caudal lesion resulting in the largest impairment, but ablation of the rostral portion has no effect. Thus, it appears that we are making progress in determining the behavioral functions of the subareas of auditory cortex.

15.5 Future Perspectives

There are a number of directions in which future research can go, depending on the technical skills and interests of investigators in different disciplines.

15.5.1 The Comparative Study of Hearing

There are unanswered questions and unexplored areas in our knowledge of the hearing abilities of animals. With regard to mammals, one question concerns the wide variation in low-frequency hearing limits and their remarkable bimodal distribution, for which there is currently no explanation. Is low-frequency hearing the result of a single source of selective pressure, as seems to be the case for sound localization driving high-frequency hearing, or is it due to adaptations to specific conditions in which low-frequency hearing is used for different functions by different species? On the other hand, perhaps some mammals do not hear low frequencies because they would interfere with the reception of high-frequency sounds that are important to them. An area currently receiving little attention is the ability of mammals to resolve differences in intensity and frequency; one reason for this is that the standard procedure has been to train animals to indicate whether a train of tone pips is alternating in frequency or intensity, a task that most animals find difficult. There may yet be interesting findings of evolutionary relevance in these abilities; the use of tests in which a sound is modulated (rather than discrete tone pips) may reveal important species differences. Finally, there is the unexplored area of auditory perception, which includes the ability of animals to recognize objects, usually other animals, by the sounds they make.

The auditory abilities of other vertebrate classes also await exploration. In the case of birds, although high-frequency hearing shows little variation, there may be significant variation in low-frequency hearing, as suggested by the fact that pigeons, and perhaps other birds, are able to hear lower-frequency sounds than humans (infrasound). Even less is known about the behavioral hearing abilities of amphibians and reptiles, which, with the anatomical variety of their ears, make relevant subjects for both physiological and evolutionary theory.

Finally, the results of anatomical and neurophysiological studies are sometimes used to infer the sensory abilities of species whose hearing has not been studied behaviorally. Behavioral assessment in these species would help understand the significance of the results of those studies.

15.5.2 Behavioral Study of Auditory Cortex

There are also many directions in which ablation-behavior studies can reveal more about auditory cortex, of which two are mentioned here. The first has to do with the species variation in the effect of cortical ablation on hearing. It is well established that removal of auditory cortex has little effect in the rat, a greater effect in cats where sound localization and the ability to determine if a sound is changing in frequency are affected, and an even greater effect in macaques where, in addition to the deficits observed in cats, the ability to detect sound is affected. Although Bruce Masterton originally set out to determine the evolution of auditory cortex by studying the effect of cortical ablation in species that approximated the human evolutionary line, this work is far from complete. In addition to showing the evolutionary changes in the function of auditory cortex, knowledge of the variation of its role in different species could serve as a basis for comparative electrophysiological studies of auditory cortex to determine the neurological correlates of the differences in function.

A second line of inquiry, one that shows great promise, is the use of reversible lesions made by inactivating an area either by cooling it or by applying transmitter antagonists. One advantage of this technique is the ability to repeat the lesions in the same animals, thereby reducing variation due to individual differences. Another important advantage is that the effect of inactivating an area may be determined before any compensation by other areas can occur. However, in conducting these experiments, it is important that investigators conduct the necessary control tests to rule out alternative explanations. For example, although ablation of auditory cortex does not cause a hearing loss in rats and cats when they are tested after recovering from the surgery, it is conceivable that a hearing loss could result from sudden inactivation of auditory cortex, which would then confound the results of other auditory tests. Given the behavioral procedures currently available, it is possible to shift a well-trained animal from one auditory discrimination to another within a session, a situation that would allow the effect of reversible lesions to be determined on multiple auditory discriminations in the same animals.

15.5.3 Advances in Behavioral Procedures

Advances in behavioral procedures are the most difficult to foresee, perhaps because they tend to be conceptual in nature. For example, the improvements in the method of conditioned suppression made over the years could have been made when the procedure was first used for sensory testing. The reason they were not made then is because people tend to be conservative; after all, why change something that works? The reason we made the changes we did is because we were often the ones doing the actual testing and wanted to speed up the procedure. We were also testing non-standard species, often ones with different behavioral strategies and motor capacities, that didn't work well in tests designed for rats or monkeys. Other advances, such as the use of reversible lesions, had to await technological improvements. Nevertheless, some trends in behavioral procedures are apparent.

The latest behavioral procedure to be developed is the previously mentioned reflex modification in which the detectability of a sound can be demonstrated by showing that it reduces the amplitude of an animal's reaction to a startling stimulus. Although it has the advantage of using an animal's unconditioned startle response, and thus requiring no training of the animal, it is likely to be 10–15 dB less sensitive than tests in which an animal is trained to listen for low-intensity sounds. Where this technique could use improvement is in reducing the variability of the results. One step would be to fix an animal's head in the sound field, for example, by having it drink from a water spout, so that the amplitude of the sound at its head can be accurately measured. Another source of variation is in the startle response itself, which can vary greatly in size from one trial to the next for the same stimulus. This variation may be due to changes in an animal's muscle tension and/or its level of arousal; this technique would be improved if the variability of its results could be reduced. It would also be helpful to know how closely thresholds obtained with this technique by different laboratories agree. A second trend that can be seen is in the use of electrophysiological measures, such as the auditory brain stem response, for measuring thresholds and changes in thresholds. However, as human research has shown, electrophysiological measures are no substitute for pure-tone audiograms and there is little evidence that they can accurately indicate threshold shifts caused by either a sensorineural or a conductive hearing loss. Indeed, it is unlikely we would have discovered the comparative relationships we did had we used physiological measures of hearing. However, whereas obtaining a behavioral audiogram on an animal can take weeks or even months in some species, an electrophysiological audiogram can be obtained in less than a day, making it much cheaper to obtain. As reviewers and editors come increasingly to accept electrophysiological measures as equivalent to behavioral thresholds, then behavioral studies of hearing will become scarce, demonstrating that Gresham's Law also applies to science.

Finally, the future of this research may depend most of all on the limitations put on it by others. When pressure to restrict animal research began in earnest several decades ago, scientific organizations made two crucial decisions. One was that their organizations would not defend other users of animals, such as ranchers, fishermen, and hunters. Indeed, those opposed to animal research offered to go easy on researchers if they did not support other groups, a divide and conquer strategy. The second decision was to defend animal research by ceding that while it was bad for animals, the benefits to humans outweighed the harm done to the animals. Yet not only is this position difficult to defend, it is unnecessary. The use of animals by humans results in symbiotic relationships in which animals benefit by achieving an environment that is superior to life in the wild. However, this fact is rarely noted. Indeed, a manuscript pointing out the symbiotic nature of animal research was rejected by psychological journals before finally finding publication in a journal oriented toward biology and medicine (Heffner, 1999). The continued well-being of animal research depends on whether the nature of our interactions with animals, and the benefits to both humans and animals, are understood.

References

Dalland, J. I. (1965). Auditory thresholds in the bat: A behavioral technique. *Journal of Auditory Research*, 5, 95–108.

Heffner, H. E. (1999). The symbiotic nature of animal research. *Perspectives in Biology and Medicine*, 43, 128–139.

Heffner, H. E. (2005). The neurobehavioral study of auditory cortex. In P. Heil, R. König, E. Budinger, & H. Scheich (Eds.), *Auditory cortex: Towards a synthesis of human and animal research* (pp. 111–126). Mahwah, NJ: Lawrence Erlbaum.

Heffner, H. E. (2011). Two-choice sound-localization procedure for detecting unilateral tinnitus in animals. *Behavior Research Methods*, 43, 577–589.

Heffner, R. S., & Heffner, H. E. (1992). Visual factors in sound localization in mammals. *Journal of Comparative Neurology*, 317, 219–232.

Heffner, H. E., & Heffner, R. S. (1995). Conditioned avoidance. In G. M. Klump, R. J. Dooling, R. R. Fay, & W. C. Stebbins (Eds.), *Methods in comparative psychoacoustics* (pp. 73–87). Basel: Birkhäuser.

Heffner, H. E., & Heffner, R. S. (1998). Hearing. In G. Greenberg & M. M. Haraway (Eds.), *Comparative psychology: A handbook* (pp. 290–303). New York: Garland.

Heffner, H. E., & Heffner, R. S. (2003). Audition. In S. F. Davis (Ed.), *Handbook of research methods in experimental psychology* (pp. 413–440). Malden, MA: Blackwell.

Heffner, H. E., & Heffner, R. S. (2008). High-frequency hearing. In P. Dallos, D. Oertel, & R. Hoy (Eds.), *Handbook of the senses: Audition* (pp. 55–60). New York: Elsevier.

Heffner, H. E., & Heffner, R. S. (2012). Behavioral tests for tinnitus in animals. In J. J. Eggermont, F.-G. Zheng, A. N. Popper, & R. R. Fay (Eds.), *Tinnitus* (pp. 21–58). New York: Springer Science+Business Media.

Heffner, R. S., Koay, G., & Heffner, H. E. (2001). Audiograms of five species of rodents: Implications for the evolution of hearing and the encoding of pitch. *Hearing Research*, 157, 138–152.

Heffner, H. E., Koay, G., & Heffner, R.S. (2006). Behavioral assessment of hearing in mice: Conditioned suppression. In J. Crowley et al. (Eds.), *Current protocols in neuroscience* (pp. 8.21D.1–8.21D.15). New York: John Wiley & Sons.

Heffner, H. E., Koay, G., Hill, E. M., & Heffner, R. S. (2012). Conditioned suppression/avoidance as a method for testing hearing in birds: The domestic pigeon (*Columba livia*). *Behavior Research Methods*, 45, 383–392.

Klump, G. M., Dooling, R. J., Fay, R. R. Fay, & Stebbins, W. C., Eds. (1995). *Methods in comparative psychoacoustics*. Basel: Birkhäuser.

Masterton, R. B. (1997). Neurobehavioral studies of the central auditory system. *The Annals of Otology, Rhinology & Laryngology*, 106, 31–34.

Masterton, B., Heffner, H., & Ravizza, R. (1969). The evolution of human hearing. *Journal of the Acoustical Society of America*, 45, 966–985.

Neff, W. D., Diamond, I. T., & Casseday, J. H. (1975). Behavioral studies of auditory discrimination: Central nervous system. In W. D. Keidel & W. D. Neff (Eds.), *Handbook of sensory physiology. Auditory system*, Vol. 2 (pp. 307–400). New York: Springer-Verlag.

Simmons, A. M., & Moss, C. F. (1995). Reflex modification: A tool for assessing basic auditory function in anuran amphibians. In G. M. Klump, R. J. Dooling, R. R. Fay, & W. C. Stebbins (Eds.), *Methods in comparative psychoacoustics* (pp. 197–208). Basel: Birkhäuser.

Stebbins, W. C. (1970). Studies of hearing and hearing loss in the monkey. In W. C. Stebbins (Ed.), *Animal psychophysics: The design and conduct of sensory experiments* (pp. 41–66). New York: Appleton-Century-Crofts.

Stevens, S. S. (1951). *Handbook of experimental psychology*. New York: John Wiley & Sons.

Whitfield, I. C., & Evans, E. F. (1965). Responses of auditory cortical neurons to stimuli of changing frequency. *Journal of Neurophysiology*, 28, 655–672.

Chapter 16
Hearing in Insects: The Why, When, and How

Ronald R. Hoy

Ronald R. Hoy

R.R. Hoy (✉)
Department of Neurobiology & Behavior, Cornell University,
215 S.G. Mudd Hall, Ithaca, NY 14853, USA
e-mail: rrh3@cornell.edu

A.N. Popper and R.R. Fay (eds.), *Perspectives on Auditory Research*, Springer Handbook of Auditory Research 50, DOI 10.1007/978-1-4614-9102-6_16,

16.1 Introduction

Why are some insects so conspicuously loud? We can't avoid noticing the chorusing of loud periodical cicadas on a hot summer's day, nor the din of crickets and katydids that pierce the quiet of a summer's night. Presumably, their calls are for insects of the same kind, of course, but does this mean that insect species that don't call can't hear? What other kinds of animals inhabit the auditory scene of auditive insects? Why such insects are endowed with an acute sense of hearing and how their hearing organs work are questions that have challenged dozens of laboratories, world-wide, for more than half a century. There are good reasons why many of us have spent our entire careers and trained hundreds of students to pursue these and related questions. To the non-entomologically inclined scientist, the rewards for studying hearing in these mostly tiny animals that are so alien to our own kind may not be at all obvious, but rewards there are and it is my pleasure to recount them in this chapter.

Although naturalists have been curious about the sounds of insects for millennia, serious studies did not start until the mid-20th century when studies of insect hearing organs coincided with the emergence of comparative neurophysiology in the 1960s and neuroethology in the 1970s–'80s. It was my great, good fortune to begin my career from a privileged perch, standing upon the shoulders of the founders of insect bioacoustics: Richard Alexander, Franz Huber, and Kenneth Roeder. I have been even luckier to have recruited a succession of congenial and very talented and able graduate students and postdocs (nearly 50) into the "Hoy lab" since the 1970s. I am taking the editors at their word when they urged us to write reviews from a personal perspective. Given the space limitations it would be impossible to do otherwise. Fortunately, there are several excellent books, cited in the reference section, that do just that. Nonetheless, I apologize to all my colleagues whose work is not cited in this chapter; I hope to rectify this oversight later—but that is for another book!

The senior editors, Art Popper and Dick Fay, encouraged the authors of this volume to say a little something about why each of us wound up studying bioacoustics. I have always been interested in animals and their behavior. As a kid, growing up in eastern Washington State, I had pet hamsters and kept freshwater aquarium fish, besides being surrounded by household cats and dogs. As an undergraduate I majored in physiology and psychology, which set me up for a graduate program in neurobiology, which I pursued at Stanford. I was fortunate enough to do my PhD studies in Donald Kennedy's lab, studying the crayfish nervous system. At Stanford, I also met David Bentley, a postdoc who had worked with Franz Huber, the "father" of cricket neurobiology and bioacoustics. David joined the faculty at Berkeley, so when I finished my PhD I did my postdoc in his lab. My year postdoc with David was very productive but just short of two years. I applied to the NIH for a third year but was told that I was being turned down in spite of submitting a splendid proposal—that it was time for me to fledge and "get a job." Eventually, I brought my crickets along with me and joined the Cornell faculty in 1973, where I've remained to this day. I fell naturally into bioacoustics because of my fondness for music. It was easy for me to tell the difference between species by their song and so I began a career-long journey to study insect ears and their auditory systems in every species I could get my hands on.

16.1.1 Preliminaries

The evolutionary imperative for any organism is to survive long enough to reproduce. This means getting enough to eat without itself being eaten and then growing to sexual maturity to reproduce itself. Many animals have deployed the auditory sense to detect potential predators, potential prey, and at maturity, potential mates. Insects are no exception. I focus my review on how insects have evolved hearing organs as key adaptations that enable them to detect and localize salient auditory signals for survival.

The acoustically conspicuous mating calls of orthopteran insects such as crickets (Gryllidae), katydids (Tettigoniidae), and grasshoppers (Locustidae) have undoubtedly piqued the curiosity of humans since their African origins. Insects are far older. They have been filling the air with sound for hundreds of millions of years, dating back to at least Jurassic times (Alexander, 1962). Examination of fossil crickets reveals the presence of file and scraper structures on their forewings similar to those of their present-day counterparts. They apparently were and still are used to produce the intraspecific mating and/or territorial calls of crickets and katydids (Gu et al., 2012). These calls are behavioral adaptations for reproductive behavior—a driving force through natural and sexual selection for evolving an auditory sense. In the 1950s, portable tape recorders and microphones became commercially available, as well as sensitive neurophysiological tools for recording the activity of even single neurons. This also coincided with widespread use of the electron microscope—both transmission and scanning microscopes. These developments ushered in the golden age of insect bioacoustics. In the 1960s and '70s these tools enabled researchers to investigate how male insects produced their mating calls and how females of the same species detected and recognized them. The founding pioneers of 1950s and '60s notably include Richard Alexander in the United States, Franz Huber in Germany, and Yasuji Katsuki in Japan. By far, most of the work done in insect bioacoustics has focused on mating calls: how they are produced and emitted (mostly by males) and then detected and localized (by females, which are mostly silent). Several excellent review volumes are testaments to the published work on the role of acoustic signals in calling and mating behavior that followed (Huber et al., 1989; Gerhardt & Huber, 2002). Less well appreciated is the fact that many insects possess keen hearing organs to detect sounds that have nothing at all to do with mating and everything to do with predator detection. Because of this relative neglect in the review literature, I begin my chapter with insect hearing in relation to predator detection.

One of most remarkable discoveries in comparative auditory science dates back to the mid-20th century when Donald Griffin discovered that nocturnally flying micropteran bats navigated in the dark by means of emitting and hearing ultrasonic signals, heralding the era of animal biosonar (Griffin, 1958). It had been long known that many microchiropteran bats were obligate insectivores, especially for moths, that were hawked in mid-air. Shortly after Griffin's work, the entomologist Kenneth Roeder showed that several orders of large moth species possessed specialized hearing organs that contain only one to four auditory receptor cells, which sharply tuned to the ultrasonic frequencies that were emitted by local bats.

The detection of directional ultrasound sources enabled moths to change the direction of their flight—away from the ultrasonic sound source—a predatory escape response (Roeder, 1967). Roeder's work inspired a small army of followers, including myself.

16.2 Bugs, Bats, and Biosonar

16.2.1 Fly-by-Night Ears

In their adult stage, most insects live short, intense lives measured in weeks or at most a few months. Some species are primarily diurnal; others are mostly nocturnal. This temporal partitioning has doubtlessly been shaped by the predators that feed on insects. By day it is the birds and reptiles; by night it is insectivorous mammals and famously, microchiropteran bats.

16.2.2 Diversity of Ultrasound-Sensitive Ears in Insects

To date, ultrasound sensitivity has been reported in the following orders of insects: Orthoptera (crickets, locusts, katydids); Coleoptera (beetles); Dictyoptera (praying mantises); Lepidoptera (moths) Neuroptera (lacewings); and Diptera (flies). These are the most widely represented and speciose insect orders and they include the most commonly occurring insects.

16.2.2.1 The Biosonar Ear: Structure and Function

Insects must hear the ultrasound cries of bats from far enough away to take evasive action—flying away—at distances of a few feet to tens of meters (Roeder, 1967). This means insects must detect the pressure wave of the ultrasound signal, just as we and most terrestrial vertebrates detect the pressure waves of vocalizations and environmental sound sources. Such far-field ears of vertebrates can be deconstructed into three components: an outer ear that is an external, thin membrane (tympanum) that is set into vibration by the impinging sound wave; a middle ear consisting of set of tiny bones that converts the vibrations of the tympanum to vibrations in a fluid-filled cochlea; and a sensory organ inner ear that transduces vibrations in the cochlea into electrical signals of the sensory cells. The insect ear is "constructed" along similar lines with key differences. There is an external, thin tympanal membrane that when set into vibration excite a sensory tympanal organ but the sensory organ usually resides in an air-filled chamber—absent a fluid-filled cochlea. This is currently the standard model, based on decades of histological work—this may be due for a revision due to very recent work I'll discuss later in this chapter. There is one respect in which vertebrate hearing organs and insect ears are very different: location. All vertebrate ears are located bilaterally in the animal's head owing to a common evolutionary, embryonic origin from gill arches and jaw bones (Popper et al., 1992). However, insect ears may be located in nearly every body segment and/or appendage because there are no developmental constraints as in the vertebrates (Fullard & Yack, 1993). In particular, ultrasound-sensitive hearing organs in insects may be found in mouthparts, legs, wings, and thoracic or abdominal segments (Hoy, 1992). Insects that possess ultrasound-sensitive ears are phylogenetically more numerous than those that have ears for conspecific signaling, indicating the relentless predation pressure from insectivorous bats. In general, the number of sensory cells that service predation-related ears is less than in ears that are primarily specialized for hearing mating signals (Hoy, 1992). The number of auditory receptors in ultrasound-sensitive ears may be as low as one, in notodontid moths (Roeder, 1967), to up to a dozen or so in other insects. Contrast this to scores of auditory receptors in crickets and katydids, up to a 1000 or so in cicadas, insects in which hearing serves social communication.

16.2.2.2 Ultrasound-Triggered Startle/Escape Responses

Most insects that possess ultrasound-sensitive ears respond behaviorally when acoustically stimulated with bat-like ultrasonic stimuli (field studies: Roeder, 1967). Roeder's ground-breaking work was followed by a flurry of activity in Europe, Canada (Fullard & Yack, 1993), and the United States, resulting in the discovery of ultrasound-sensitive hearing organs in many more insects, reviewed in Hoy (1992), for example, praying mantis, beetles, and crickets. Behavioral studies are done in the laboratory using tethered insects stimulated into flapping flight by tethering and suspending them in the wind stream of flight tunnel. Lab studies also permit neural recordings from the auditory system. In one such preparation it was shown that stimulation of just one auditory interneuron, whether by acoustic stimulation of bat-like ultrasound pulses or by direct electrical stimulation in the absence of sound stimulation, was both necessary and sufficient to trigger motor responses characteristic of acoustic startle and evasive motor acts in the cricket *Teleogryllus oceanicus* (Nolen & Hoy, 1984). Ultrasound startle responses have been demonstrated in many species of nocturnal, flying insects at the level of behavior and the auditory system (reviewed in Hoy, 1992) and continue to uncover more layers to the bat-insect story (Miller & Surlykke, 2001).

16.3 Mating Calls and Species-Specific Recognition

The notable species-specificity of a male's call to attract conspecific females occurs in both insects and vertebrates like frogs and toads. The literature is huge and fortunately developments up through 2000 are covered in a single review volume: *Acoustic Communication in Insects and Anurans* by two masterful pioneers in the field (Gerhardt & Huber, 2002). I will restrict my remarks to a few salient and more recent studies.

16.3.1 Communication

The motor control of song production and perception has been advanced by the studies in England by Berthold Hedwig (Hedwig, 2000; Schoenreich & Hedwig, 2012), who has provided evidence for central command neurons in initiating cricket calls in semi-intact preparations. At the same time, Kostarakos and Hedwig (2012) make the case that the temporal pattern of chirps and trills are the key to species-specific "recognition" and phonotaxis in field crickets. Similar investigations of neural coding features that govern phonotaxis in grasshoppers, with some emphasis on the temporal features of the grasshopper songs, has been conducted by Bernhard Ronacher's group, who applied spike-coding algorithms to study information transfer between primary auditory receptor neurons and their second-order auditory interneurons, and their results seem promising (Ronacher, 2012). In terms of fulfilling the aims of 1970s and '80s project to understand the neural bases of temporal coding in the species-specific calling songs, neither definitive nor general solutions are at hand, but we are getting close.

16.3.2 The Localization Problem

Sound localization is, at first glance, a matter of simple physics. A sound wave passing across any animal can generate two kinds of difference cues detectable by its pair of ears. These are differences in interaural time (ITD) and intensity (IID)—I omit interaural phase differences from this discussion because ITD is the practical proxy for interaural phase differences in small animals. Directional hearing in very small animals is a bioacoustics challenge, because sound localization requires computation of ITD and IID. For a given wavelength of sound, both cues decrease with diminishing size of the animal (Michelsen, 1992).

The key technical development from the 1970s that enabled a breakthrough in our understanding was the Doppler laser vibrometer (DLV). It permitted measurement of minute mechanical vibrations from eardrums or other anatomical structures in the auditory chain of mechanical processing, no matter how small in amplitude or fast in frequency, from animals large and small, and as an optical method, it did not interfere with the vibrating structure in any way.

The earliest applications to DLV to insect hearing organs were in the 1970s on the field cricket (Paton et al., 1977; Larsen & Michelsen, 1978). The Danish group led by Michelsen's Odense group would tackle insect hearing for the next three decades in the cricket, katydid, and locust ears. This work would demonstrate the importance of the pressure difference principle of directional hearing and sound localization. DLV is now the standard tool for providing the gold standard measurements in investigating the response properties of hearing organs of vertebrates and invertebrates alike (Montealegre-Z et al., 2012).

By the 1990s, it was well established that ensiferan orthoptera such as crickets and katydids, possessed hearing organs that were pressure difference receivers. In the acridids, size matters. Large locusts, such as *Locusta migratoria*, the hearing organs act as pressure receivers but in small grasshoppers evidence favors pressure difference mechanisms, as in crickets and katydids (Michelsen, 1992). However, many insects are smaller than crickets and grasshoppers and some of them are known to be acoustically attracted to sounds, even specifically to cricket songs. In particular, attention focused on small parasitic flies that possess a sense of hearing (Cade, 1975; Lakes-Harlan & Heller, 1992; Robert et al., 1992). Our fascination with these *Ormia* flies was with their mysterious hearing organs (Robert et al., 1992). Previously, dipteran flies were known to possess only particle velocity sensitive hearing organs (Johnston's organs) on their antennae that are not sensitive to sound pressure waves, in the far field of sound. Robert quickly identified a pair of tympanal membranes on Ormia's prothorax, which along with Lakes-Harlan's contemporaneous finding in an emblemmasomid parasitoid, were the first tympanal hearing organs ever reported in flies (Lakes-Harlan & Heller, 1992; Robert et al., 1992). The bigger puzzle was how the tiny Ormia fly localized sound. From an acoustical physics standpoint, the fly was dealt a miserable hand of cards for sound localization. The flies detect crickets by hearing their calling song, trains of 5 kHz tone pulses. Crickets, of course, hear other crickets with pressure difference receiver auditory organs. However, Ormia's ears are smaller than a cricket's by more than an order of magnitude, effectively reducing both IID and ITD cues to near zero. Necessity is the mother of invention and Mother Nature is endlessly innovative—Ormia flies possess a mechanically coupled ear, making it a new kind of directionally sensitive hearing organ (Miles et al., 1995; Robert & Hoy, 1998). In effect, these small flies "solved" a challenging problem in acoustical physics through evolutionarily "inventing" an ear that localizes sound through a different set of design principles than pressure receivers or pressure-difference receivers, which are perfectly adequate for larger insects. Thus, a third class of directionally sensitive receivers, mechanically coupled hearing organs, was added to pressure receivers and pressure-difference receivers.

16.4 Audition in Other Insects and Spiders

In the past several decades we have come to that many more insect taxa used acoustic signals for communication, especially when the term "acoustic signal" was broadened to include substrate vibrations, in addition to air-borne pressure waves, that are transmitted and exchanged between conspecific insects. Much of this work is thoroughly reviewed in the book *Insect Sounds and Communication*, which covers the work of the 1990s and early 2000s (Drosopoulos & Claridge, 2006). This volume documents the explosion of bioacoustic data from major insect fauna such as true bugs, beetles, and bees as the application of new digital technology opened up substrate or seismic signaling in hundreds of new species.

16.4.1 Mechanical Vibrations as Communication Signals

Spiders are not insects but exhibit fascinating and compelling communication behavior that includes seismic/substrate signaling in reproductive and territorial behavior (Heberstein, 2011). Recent work in jumping spiders, which were thought to be exclusively visual communicators owing to their conspicuous semaphore-like signals using extravagantly colored front legs coupled with colored mask-like facial markings, has shown that these spiders produce species-specific and remarkably complex substrate vibratory signals that are coupled with visual displays (Elias et al., 2003). It is clear that spider acoustic signaling is an area for investigation that really opened up in the 1990s and continues to reward a growing number of investigators (Heberstein, 2011).

16.4.2 Insect That Use Mechanical Vibrations for Signals

It has long been known that beetles and true bugs (Hemiptera including the Heteroptera) signal through various substrates, plants, tree stems, and even water. A pioneering researcher who performed phylogenetically informed analyses of seismic signals is the Slovenian entomologist Matija Gogala (2006), who built an active group for hemipteran bioacoustics that flourishes today that is led by Andrej Cokl and Meta Virant-Doberlet (Cokl et al., 2006). In the United States, Rex Cocroft's (Cocroft & McNett, 2006) research into the membracid treehoppers has shown clear linkages between social behavior (including maternal care) and substrate vibratory signals.

16.5 Acoustic Signaling in *Drosophila*

The laboratory "fruitfly," *Drosophila melanogaster*, has long been a model system for studies at all levels of biological organization, and bioacoustics is no exception. Earlier studies (Ewing & Bennet-Clark, 1968) showed that acoustic signals were produced by wing movements and that these wing movements are species-specific in other *Drosophila* species (Hoikkala, 2006). Behavior aside, the nature of the particle-velocity-sensitive auditory receptor organ (Johnston's organ) was solved in the ingenious experiments of Goepfert and Robert (2001). The past decade has seen a burst of new findings about the neural basis of behavior in drosophilids because of its completed genome but more importantly, the development of optogenetic techniques that permit targeting and staining specific neuronal phenotypes.

16.6 Mosquitoes

Perhaps the most important development in terms of comparative bioacoustics as related to mechanisms common to vertebrate/mammalian hearing and insect hearing is the discovery that active signal amplification through metabolically dependent molecular mechanisms underlies hearing in several species of insects.

Active auditory amplification is the sine qua non in the mammalian cochlea, and the same appears to be true in various flies and orthopteran insects as well. A particularly convincing case for active auditory mechanics (AAM) in the Johnston's auditory organ of mosquitoes was advanced in the experiments of Goepfert and Robert (2002), whose work supported the dependence of AAM on metabolic processes, as well as presenting evidence for unbidden acoustic activity in the Johnston's organ, analogous to the spontaneous otoacoustic emissions in vertebrates.

16.7 Looking Ahead

Future progress in advioacoustics, like in any other field of science, will be driven by new technical advances, as depicted by the following example. Recent work from Daniel Robert's lab (Montealegre-Z et al., 2012) reveals an astounding example of functional and morphological convergence between mammalian and insect auditory organs. Mammals and insects phylogenetically diverged hundreds of millions of years ago. This has led, of course to vast divergences in body plans, size, and anatomy. However divergent the "design features" of an animal's anatomy, natural selection imposes functional imperatives for survival—such as feeding without being fed upon so as to survive to sexual maturity and then to reproduce. This similarity of function in the absence of common descent is the stuff of convergent evolution. It should be clear that the auditory sense has evolved in both vertebrates and insects to serve as tools for survival in predator–prey and reproductive behaviors. However, the morphological instantiation of hearing organs is grossly different, as even casual inspection between insect and vertebrate ears reveals. The textbook example of hearing organ draws from the terrestrial mammalian model, usually the human ear. It comprises three different mechanical structures for processing sound—first, a tympanic membrane (TM) that is set into vibration by impinging sound waves (changes in air pressure); second, the vibrations of the TM set the ossicles of the middle ear into corresponding vibration by means of a rigid lever system of bones; and third, one of the middle ear bones exerts pressure on the oval window of the fluid-filled cochlea, where a linear array of vibration-sensitive hair cells abutting the tectorial membrane enables frequency analysis. The need for middle ear bones is for converting vibrations in the air into a traveling wave within the fluid in the cochlea and performs impedance matching. The traveling wave enables frequency analysis performed by the tonotopically distributed hair cells. Thus, it comes as a big surprise to learn that a small South American rainforest katydid possesses an auditory organ that shares functional and structural equivalents with all three components of the mammalian ear (Montealgre-Z et al., 2012). The katydid is itself unusual. Its conspecific call is a 23-kHz tone, with some FM sweeps, which is inaudible to humans. Its 600 μm long hearing organ is among the smallest known and its 26 auditory receptors are contained within a fluid-filled organ—thus posing the same physical problem as occurs in terrestrial mammals: how to convert airborne pressure waves into corresponding vibrations in a fluid medium, the problem of impedance matching. Remarkably, this katydid "solves" the problem in the same way: through an analogous lever and fulcrum system that couples tympanal vibrations to a traveling wave within the fluid-filled vesicle that is its auditory organ. Linear tonotopy in the hearing organ of the ensiferan orthopterous insects has been known for decades (Oldfield et al., 1986) and first demonstrated in large Australian katydids. What is remarkable for this small New World katydid is that its auditory receptor cells are bathed in fluid and excited by a traveling wave by means of impedance matching mechanical coupling. The remarkable study of Montealegre-Z et al. is emblematic of bioacoustics in the 21st century. It would not have been possible without the application of the newest imaging technologies, in this case, X-ray microcomputer tomography (micro-CT), which enabled nano- and microscale imaging of the internal and external anatomy of fresh specimens of the katydid's miniscule ear. Micro-CT permits optical resolution is far beyond 20th century techniques that were limited to what could be seen through dissecting microscopes in fresh specimens and compound microscopes of histologically fixed and preserved specimens or histological sections prepared through microtomy. Micro-CT has been used in the physical sciences but its use in living organisms is relatively new. Obviously, this study invites a reexamination of the auditory organs in the much larger insects.

We can expect micro-CT to be an important tool in future studies of the biomechanics of hearing in insects. Similarly, we can expect that the spectacular application of optogenetics to the analysis of *Drosophila*'s nervous system to eventually find their way into other insects and make the job of "circuit-breaking" not only easier but also very pleasant to the eye, with images of multicolored fluorescent stained neurons. But there will always be room for intrepid young naturalist-scientists who explore the great biodiversity of insect and spider species in search of new discoveries in the realm of insect behavior. There are many more model organisms that remain to be discovered and on which to hone our imaginations and experimental inclinations. I wish them success!

References

Alexander, R. D. (1962). Evolutionary change in cricket acoustical communication. *Evolution*, 16, 443–467.

Cade, W. H. (1975). Acoustically orienting parasitoids: fly phonotaxis to cricket song. *Science*, 190, 1312–1313.

Cocroft, R., & McNett, G. D. (2006). Vibratory communication in treehoppers. In S. Drosopoulos & M. F. Claridge (Eds.), *Insect sounds and communication* (pp. 305–318). Boca Raton, FL: CRC Press.

Cokl, A., Virant-Doberlet, M., & Zorovic, M. (2006). Sense organs involved in the vibratory communication of bugs. In S. Drosopoulos & M. F. Claridge (eds.), *Insect sounds and communication* (pp. 71–80). Boca Raton, FL: CRC Press.

Drosopoulos, S., & Claridge, M. F. (2006). *Insect sounds and communication*. Boca Raton, FL: CRC Press.

Elias, D. O., Mason, A. C., Maddison, W. P., & Hoy, R. R. (2003). Seismic signals in a courting male jumping spider (Araneae: Salticidae). *Journal of Experimental Biology*, 206, 4029–4039.

Ewing, A.W., & Bennet-Clark, H. C. (1968). The courtships songs of *Drosophila. Behaviour*, 31, 288–301.
Fullard, J. H., & Yack, J. E. ((1993). The evolutionary biology of insect hearing. *Trends in Ecology and Evolution*, 8, 248–252.
Gerhardt, H., & Huber, F. (2002). *Acoustic communication in insects and aneurans*. Chicago: University of Chicago Press.
Goepfert, M. C., & Robert, D. (2001). Active auditory mechanics in mosquitoes. *Proceedings of the Royal Society of London B: Biological Sciences*, 268, 333–339.
Goepfert, M. C., & Robert, D. (2002). The mechanical basis of *Drosophila* audition. *Journal of Experimental Biology*, 205, 1199–1208.
Gogala, M. (2006). Vibratory signals produced by heteroptera. In S. Drosopoulos & M. F. Claridge (Eds.), *Insect sounds and communication* (pp. 275–296). Boca Raton, FL: CRC Press.
Griffin, D. R. (1958). *Listening in the dark*. New Haven, CT: Yale University Press.
Gu, J.-J., Montealegre-Z, F. Robert, D., Engel, M.S., Qiao, G.-X., & Ren, D. (2012). Wing stridulation in a Jurassic katydid produced low-pitched musical calls to attract females. *Proceedings of the National Academy of Sciences of the* USA, 109, 3868–3873.
Heberstein, M. E. (2011). *Spider behaviors: Flexibility and versatility*. Cambridge, UK: Cambridge University Press.
Hedwig, B. (2000). Control of cricket stridulation by a command neuron: Efficacy depends on behavioral state. *Journal of Neurophysiology*, 83, 712–722.
Hoikkala, A. (2006). Inheritance of male sound characteristics in *Drosophila* species. In S. Drosoopoulos & M. F. Claridge (Eds.), *Insect sounds and communication* (167–178). Boca Raton, FL: CRC Press.
Hoy, R. R. (1992). Tympanal hearing in insects. Annual Review of Entomology, 41, 433–450.
Huber, F., Moore, T., & Loher, W., Eds. (1989). *Cricket behavior and neurobiology*. Ithaca, NY: Cornell University Press.
Kostarakos, K., & Hedwig, B. (2012). Calling song recognition in female crickets: Temporal tuning of identified brain neurons matches behavior. *Journal of Neuroscience*, 32(28), 9601–9612.
Lakes-Harlan, R., & Heller, K.-G. (1992). Ultrasound sensitive-ears in a parasitoid fly. *Naturwissenschaften*, 79, 224–226.
Larsen, O. N., & Michelsen, A. (1978). Biophysics of the ensiferan ear. III. The cricket ear as a four-input system. *Journal of Comparative Physiology A*, 123, 217–227.
Michelsen, A. (1992). Hearing and sound communication in small animals: Evolutionary adaptations to the laws of physics. In D. M. Webster, R. R. Fay, & A. N. Popper (Eds.), *The evolutionary biology of hearing* (pp. 61–78). New York: Springer-Verlag.
Miles, R. N., Robert, D., & Hoy, R. R. (1995). Mechanically coupled ears for directional hearing in the parastoid fly *Ormia ochracea. Journal of the Acoustical Society of America*, 98, 3059–3070.
Miller, L. A., & Surlykke, A. M. (2001). How some insects detect and avoid being eaten by bats: Tactics and countertactics of prey and predator. *BioScience*, 51, 571–582.
Montealegre-Z, F., Jonsson, T., Robson-Brown, K. A., Postles, M., & Robert, D. (2012). Convergent evolution between insect and mammalian audition. *Science*, 338, 968–971.
Nolen, T. G., & Hoy, R. R. (1984). Initiation of behavior by single neurons: The role of behavioral context. *Science*, 226, 992–994.
Oldfield, B. P., Kleindienst, H.-U. & Huber, F. (1986). Physiology and tonotopic organization of auditory receptors in the cricket *Grullus bimaculatus. Journal of Comparative Physiology A*, 159, 454–464.
Paton, J. A., Capranica, R. R., Dragsten, P. R., & Webb, W. W. (1977). Physical basis for auditory frequency analysis in field crickets (Gryllidae). *Journal of Comparative Physiology A*, 119, 221–240.
Popper, A. N., Platt, C., & Edds, P. L. (1992). Evolution of the vertebrate inner ear. In D. M. Webster, R. R. Fay & A. N. Popper (Eds.), *The evolutionary biology of hearing* (pp. 49–66). New York: Springer-Verlag.

Robert, D., & Hoy, R. R. (1998). The evolutionary innovation of tympanal hearing in Diptera. In: R. R. Hoy, A. N. Popper, R. R. Fay (Eds.), *Comparative hearing: Insects* (pp. 197–227). New York: Springer-Verlag.

Robert, D., Amoroso J., & Hoy, R. R. (1992). The evolutionary convergence of hearing in a parasitoid fly and its cricket host. *Science*, 258, 1135–1137.

Roeder, K. D. (1967). *Nerve cells and insect behavior*. Cambridge MA: Harvard University Press.

Ronacher, B. (2012). Processing of species-specific signals in the auditory system of grasshoppers. In B. Hedwig (Ed.), *Topics of acoustic communication in insects*. New York: Springer Science+Business Media.

Schoenreich, S., & Hedwig, B. (2012). Cellular basis for singing motor pattern generation in the field cricket. *Brain and Behavior*, 2(6), 707–725.

Chapter 17
The Cognitive Auditory System: The Role of Learning in Shaping the Biology of the Auditory System

Nina Kraus and Trent Nicol

Nina Kraus

Trent Nicol

N. Kraus (✉)
Departments of Communication Sciences, Neurobiology & Physiology, Otolaryngology, Auditory Neuroscience Lab, Northwestern University, 2240 Campus Drive, Evanston, IL 60208, USA
e-mail: nkraus@northwestern.edu; tgn@northwestern.edu; www.brainvolts.northwestern.edu

T. Nicol
Department of Communication Sciences, Auditory Neuroscience Lab, Northwestern University, 2240 Campus Drive, Evanston, IL 60208, USA

A.N. Popper and R.R. Fay (eds.), *Perspectives on Auditory Research*, Springer Handbook of Auditory Research 50, DOI 10.1007/978-1-4614-9102-6_17,

17.1 Introduction

Since the 1980s, we have known that training will prompt a given cell in auditory cortex (AC) to alter its firing properties in response to a stimulus following training. My doctoral dissertation was among the first to demonstrate this (Kraus & Disterhoft, 1982). In the years since, we have learned a great deal about the role of the types of training, the strategies used to achieve learning, and trainee motivation on AC response plasticity. Except in rare circumstances, single-unit methodologies are unavailable to researchers interested in determining the effects of learning and experience on the human auditory system. Noninvasive electrophysiology, in the form of cortical evoked potentials (EPs), has served as an informative surrogate. Moving to suprathreshold stimulation and far-field recording methodologies, widely studied cortical responses such as the N1, mismatch negativity, P300 and "processing negativity" are valuable in characterizing neural plasticity in groups. However, they suffer from response variability, rendering them unsatisfactory as gauges of training-related plasticity *in individuals*. In addition, their slow voltage fluctuations, occurring hundreds of milliseconds after the evoking sound, offer poor renderings of the acoustics of the stimulus. Recently, there has been a resurgence of brain stem EP work. The auditory brain stem response to complex stimuli (cABR), coupled with advanced analysis methodologies, is a faithful gauge of acoustic processing, yet also reveals an auditory processing that is profoundly affected by external factors such as communication skills and training. The brainstem is a hub of sensory, cognitive and reward influences. As such, cABRs complement animal work in understanding an integrated, cohesive auditory system and the central-to-peripheral circuits that make auditory learning possible. My own scientific career path has somewhat mimicked that of evolution in the field, first exploring cortical plasticity at the single-unit level in animals, and then shifting focus to cortical evoked responses in humans and experimental animals, and now, with my focus on cABR. However, each phase of my career—though the approach and the tools have changed—has focused on *the cognitive auditory system*. And so we, along with the field as a whole, are moving toward approaching the many components of the auditory system—periphery, central structures, nonauditory cortex—as an integrated hearing circuit.

17.2 Anatomy of the Cognitive Auditory System

Information flow in the auditory system depends on a network of peripheral, subcortical and cortical nuclei with vast interconnectivity (Hackett, 2011). Auditory information processing is affected by the extensive corticofugal system, which exerts "downward" influence on the "upstream" afferent processing chain (Suga, 2008). Within the auditory pathway, corticofugal connections exist between auditory cortex and thalamic, collicular and peripheral structures (Winer, 2006). In addition, nonauditory regions, including visual, somatosensory, limbic, and association areas, enervate auditory centers (Budinger et al., 2006). Memory, attention, experience, communication skills, and learning influence the activation of the auditory system

Fig. 17.1 Schematic illustration of ascending and descending connections in the auditory system. This cross-sectional view includes the cochleae, as well as auditory brain stem and thalamus. Descending connections between auditory cortex and these lower auditory areas have been long established and studied. The investigation of the functional and physiological consequences of the reciprocal connections between cognitive centers (e.g., those responsible for executive functions like attention and memory), reward (limbic) areas, and the auditory pathway is an emerging field. (Image by Libby Karlinger Escobedo)

(Halpern & Zatorre, 1999; Tzounopoulos & Kraus, 2009). The vast interconnectedness of the auditory system, with its links to cognitive and reward centers, is represented in Fig. 17.1.

17.3 Cognitively Mediated Physiological Changes—Receptive Fields of Cortex

We now know that the auditory cortex, along with other primary sensory neocortical regions, is "a structure holding a strategic position in the interaction between bottom-up processes (dominated by the sensory input) and top-down processes (dominated by the state of the subject, including its past history, future goals and expectations, and hence mediating the meaning of stimuli, as well as states and actions associated with perception of these stimuli)" (Deliano & Ohl, 2009). Indeed, auditory cortex plasticity, both in primary (A1) (Weinberger et al., 1984) and nonprimary (Kraus & Disterhoft, 1982) regions, has been known since the 1980s.

A common visualization technique used to investigate auditory neural plasticity, both in cortex and subcortical structures, is the spectrotemporal receptive field

(STRF). Like a spectrogram, it depicts frequency on the ordinate, time on the abscissa. The color axis represents neural firing—both excitatory and inhibitory—and thus a best frequency and corresponding time delay (relative to stimulus) can be ascertained for the unit(s) under test. Used as a portrayal of characteristic frequency (CF), it enables quick visualization of changes in tuning that occur following intervention, such as noise exposure, ototoxic drugs, training, activation of other brain regions, and so forth. The exact nature and extent of the training-induced change in STRF depends on a number of *nonauditory* factors, demonstrating the role of cognitive factors such as motivation and emotion on auditory processing.

17.3.1 Type or Difficulty of Task

The sharpening, shifting, or otherwise changing of receptive fields of cortex generally is a result of a need to accomplish a task. Training establishes that a particular tone, for example, signals that a task must be performed to receive a reward or avoid punishment. Cortical tuning to this now behaviorally relevant tone is expanded and/or sharpened to accomplish the task accurately. The patterns of plasticity depend on whether the task type is avoidance or approach. Specifically, when a given tone is associated with punishment, the STRF reveals an increase in responsivity at the tone frequency; when it signals a reward there is a decrease (David et al., 2012).

17.3.2 Strategy and Attention

The strategy used to learn when to respond and when not to respond influences the tuning of auditory cortex, as does the particular element of the stimulus that bears information relevant to the task. A water reward paradigm, in rats, was designed such that shortly after a tone stops, a flashing light signals that continued licking will result in punishment. Some rats adopt a strategy to stop licking as soon as the tone ceases while others wait until the flashing light before stopping. Rats in the latter group have larger cortical reorganization after training, even when motivation level is taken into account (Berlau & Weinberger, 2008). Also in rats, a design was employed in which a particular feature of the stimulus (either frequency or intensity) was the relevant aspect for the task. The same stimulus was found to have a different reorganization effect depending on which aspect was attended to (Polley et al., 2006). Thus the attention, strategy, and motivation of an animal are crucial elements of AC reorganization.

17.3.3 Neurotransmitters

The shaping of auditory cortex is driven by neuromodulators, in particular acetylcholine. The behavioral relevance of a stimulus activates the nucleus basalis (NB), a highly cholinergic region of forebrain. Pairing tones with NB stimulation engenders changes in primary AC tuning and nonprimary AC selectivity similar to those seen with conditioning. Other neuromodulators, including dopamine and serotonin, also modulate AC learning (reviewed in Thiel, 2007).

A hot question in the auditory learning literature is whether or not a change of tuning in auditory cortex can be construed as "information storage" and thus memory. The view that AC itself meets the criteria of being a site of memory, not merely a processor of signals that lead to memories in "higher" cortical regions is held by Weinberger (2004). He argues that attributes seen in AC, such as rapid learning, consolidation, and long-term maintenance of plasticity, constitute the ingredients necessary for it to be classified as a site for storage of stimulus features of behavioral significance. Others disagree, noting that over time, AC map reorganization often normalizes to a pre-trained state; yet, the learned behavior endures (Kilgard, 2012), indicating that AC itself is not exhibiting memory. Whether AC is a storage site for learned stimulus features, whether learning is expressed in a neural code different from map reorganization, or some combination of the two, as is my view, the cognitive nature of AC is clear.

17.4 Cognitively Mediated Physiological Changes—Subcortical Regions

Auditory subcortical regions, like those in other sensory modalities, serve, in part, to propagate neural impulses from the periphery—the cochlea—to auditory cortex. Indeed, like their sensory cortical counterparts, there was a time when subcortical structures were thought to be either simple relay stations or, at most, centers of binaural processing. However, we now know that their enervation is bidirectional (Winer, 2006) and, like AC, they have plastic response properties. Of particular relevance to this chapter are the properties of the inferior colliculus (IC) of midbrain. This auditory subcortical nucleus shares some functional characteristics with primary visual cortex (Nelken, 2008), and viewed in this light, its experience-dependent plasticity is not entirely surprising. For a review of corticofugal enervation and learning-associated plasticity of IC and other subcortical regions, see Suga (2012). Here, we will mention a few studies that illustrate the functional plasticity of IC.

17.4.1 Importance of Descending Fibers in Learning

In a study of sound localization training, it was found that ferrets with ablated corticocollicular projections maintain their ability to localize sound. However, when one of their ears is plugged, resulting in the need to adjust to the new spatial cues of an altered soundscape, localization performance suffered compared to controls with intact corticocollicular fibers (Bajo et al., 2010). *Thus, facilitation of learning appears to be a chief role of the auditory efferent system, with the implication that learning can bring about subcortical physiological changes* via *top-down mechanisms*.

17.4.2 Online Implicit Learning of Sound

Inferior colliculus neurons have characteristic rate-intensity functions. That is, a given neuron, in response to sounds of varying intensities, might show increased firing up to a particular maximum intensity above which firing rates saturate—defining the dynamic range of the neuron. The dynamic range is the range over which there is a change in firing rate with increased intensity of stimulation. Above the saturation level, there is no meaningful distinction in the response to two different intensities. However, this dynamic range can be shifted in real time if the sound input is manipulated. If noise stimuli of varying intensities are presented with statistical distributions peaking, first at 39 dB and then at 63 dB, the rate-intensity functions shift such that the saturation level is higher for the 63–dB-centered stimulus set (Dean et al., 2005). These changing set points enable a much wider dynamic range than would be possible if such online modification of neural properties did not take place. Another example of online changes in response properties occurring in IC was demonstrated using an oddball paradigm in which stimulus-specific adaptation was discovered; there was a rapid increase in IC spiking if a stimulus was presented in a novel context compared to when it was presented in a train of other like-stimuli (Malmierca et al., 2009). A human analogue of online alteration of response properties was demonstrated by the differences between subcortical responses to identical tonal stimuli depending on whether they were presented in a random sequence or as components of a pseudolanguage (Skoe et al., 2013a).

17.4.3 Attention

It is also possible to see IC changes in humans with imaging and electrophysiological techniques. Recent studies have demonstrated an alteration in IC activation depending on attentional demands put on the listener (Rinne et al., 2008; Hairston et al., 2013; Lehmann & Schönwiesner, 2014). This implicates cognitive top-down control of a subcortical auditory nucleus tied into cognitive demands (Raizada & Poldrack, 2007).

The studies reviewed above largely describe online changes, coincident with the behavior that is learned. Changes brought about over longer time scales via enriched or restricted environments and past training are even more striking and have been seen in auditory cortex and subcortical regions of experimental animals (Knudsen, 1999; Engineer et al., 2004; Yu et al., 2007) and humans (Skoe & Kraus, 2012; White-Schwoch et al., 2013).

17.5 Cognitive Auditory Processing: Application to Human Communication

The importance of cognitive influence on hearing has huge implications in the realm of "training to hear." Communication skills, such as listening to conversations in background noise or parsing the sounds of language in order to learn to read are skills that can be trained via protocols that exercise attention and memory, not simply the acoustics of sound itself (Hornickel et al., 2012b). Even in the absence of explicit training, implicit learning in the form of music performance and bilingualism also affects communication skills. Bilinguals, for example, enjoy advantages in executive function over monolinguals (Carlson & Meltzoff, 2008). Likewise, musicians have superior listening in noise abilities, enhanced auditory attention, and better auditory memory skills than nonmusicians, and these skills are reflected in the auditory system's response to sound (Kraus & Chandrasekaran, 2010; Kraus et al., 2012; Strait & Kraus, 2014). Notably, these skills, though in the auditory domain, are somewhat far afield from music. Many, in fact, are in the realm of speech, a generalization which is predicted by Patel's OPERA (overlap, precision, emotion, repetition, attention) hypothesis (Patel, 2011). In the next section, we discuss a neural metric—cABR—that is accessible in humans and enables the exploration of such experiential and training effects on the functioning of the cognitive auditory system.

17.6 Accessing the Cognitive Auditory System in Humans—cABR

In the human auditory system, we must infer that auditory system anatomy and physiological mechanisms are, for the most part, similar to those verified by work in experimental animal models. However, neurophysiological and imaging techniques bring quite a bit to the table and are not limited to animal models. Investigations in musicians, for example, indicate that this specialized experience with sound impacts brain structure and function (Herholz & Zatorre, 2012). Likewise, cortical electrophysiology is malleable with short-term training (Kraus et al., 1995; van Wassenhove & Nagarajan, 2007). In addition, cortical responses in language impaired populations differ from controls (Kraus et al., 1996; Jentschke et al., 2008). The auditory brain stem is also modifiable by learning and experience and, in this section, we focus on recent work in our lab and others' on the auditory brain stem response to complex sounds (cABR) illustrating this phenomenon.

The cABR is accessible noninvasively in humans, and unlike cortical evoked responses, cABR offers a greater applicability in individuals and a more direct relevance to the evoking stimulus. Like its counterparts, the click- and tone-evoked brain stem responses, cABR is a measure of synchronous subcortical neural activity. A composite response originating largely in IC, it represents not only a gauge of afferent sound processing, but also a snapshot of the influences engendered by the corticocollicular networks. As such, *we view cABR as a metric of the entire auditory system*; *that is, we are not interested in "brainstem responses" per se, but in a measure that is accessible in humans and serves as a window into the cognitive auditory system as a whole.* In this section we review the two chief attributes of cABR—stimulus fidelity and experience dependence. In Section 7, we present some data that demonstrate the effect of training and experience on its response properties.

17.6.1 Stimulus Fidelity

Electroencephalographic recordings of sensory function are among the best tools that we have at our disposal for investigating the processing of auditory input in a noninvasive way in humans. Familiar exogenous auditory cortical evoked responses (EPs) such as P1, N1, and P2 and endogenous responses such as mismatch negativity and P300 are revealing in their ability to signal sound propagation in auditory cortex and its processing as sensory memory traces are formed and violated. However, cortical electrical activity offers a limited denotation of the features of the evoking sound, bearing only an abstract representation of the stimulus. Occurring hundreds of milliseconds after sound onset, these slow-voltage fluctuations recorded at the scalp convey little information about the complexity (or lack thereof) of the signal that evoked them. This is largely due to the lowpass filter characteristic of the auditory pathway and the convergence and overlapping of many sources that culminate at the scalp electrode. In contrast, although subcortical activity is also recorded from scalp electrodes and is subject to overlapping sources, stimulus presentation practices, selective filtering, and signal processing techniques together can largely eliminate the contribution of peripheral (i.e., cochlear microphonic) and cortical activity (Chandrasekaran & Kraus, 2010), resulting in a "brain stem" response, largely originating in IC. It is useful to point out that the nuclei responsible for the response are of secondary importance. *Although cABR features "brain stem" prominently in its name, utility as a measure of global auditory system processing should override the fact that its generation is of subcortical origin. The entire auditory system, cochlear hair cells included, is continually shaped by non-peripheral influences. Consequently, I feel that a paradigm shift away from labels such as early versus late and cortical versus subcortical is warranted.* The importance of selective filtering is that the resulting neural activity bears a remarkable resemblance to the evoking sound and exhibits experience dependence in a way that is not possible with other techniques. The response adheres to the stimulus on multiple time scales and visualization domains (Kraus, 2011; Li & Jeng, 2011). To take speech as an example (Skoe & Kraus, 2010a), at the longest timeframes such as the sentence or word level, onsets, offsets, stops, and other perturbations in speech envelope are maintained in the time-domain response as discrete transients. At the syllable level, the fundamental frequency (F_0) of the vowel is represented via the frequency-following response (FFR) that phase locks to the periodicity (voicing) of the utterance. The fine structure of the syllable—for example, the overtones that distinguish vowels or the frequency glides that characterize consonant-vowel syllables—are represented in the frequency domain by the harmonic content and phase attributes of the response and in the time domain with submillisecond timing. This high degree of correspondence between stimulus and response enables familiar signal processing routines such as autocorrelation, Fourier analysis, phase analysis, spectrograms, and so forth, to be applied to the response as well as to the stimulus in order to visualize sound processing of the auditory brain stem (Fig. 17.2).

17.6.2 Experience Dependence

Despite this high level of adherence to stimulus features, the cABR does not a represent a brain stem that is a passive conveyance of information from lower to higher auditory regions. *Subtle variations in the timing and spectrum of the response demarcate differences in individuals' auditory processing abilities based on experience*. This dual nature of the response is extremely appealing to me. Like STRFs in the cortex, it is rigorously signals-based in its elicitation and analysis approaches. Yet, this neural activity tracks with real-world experience and human communication, enabling scientists to address complex and practical questions regarding human communication with iron-clad objectivity. cABR therefore very comfortably fits my translational-educational bent, permitting a marriage of basic science and social and clinical issues. In a series of investigations our team has undertaken in the past dozen years, utilizing a number of different stimuli, cABR has been shown to track with aging, reading ability, cognitive abilities, musical experience, and bilingualism. In addition, training studies have demonstrated its short-term and online plasticity.

17.7 cABR as a Metric of Auditory System Plasticity

In this section and in Section 8, we review some work demonstrating the cognitive auditory system—comprising auditory-based skills, the factors that influence them, and experiential factors—using cABR recordings in humans, in most cases to speech sounds, as our approach. The cABR response properties include encoding of (1) the fundamental frequency (F_0), (2) pitch tracking, (3) harmonics, (4) changing formant frequencies, (5) onsets, and (6) response consistency.

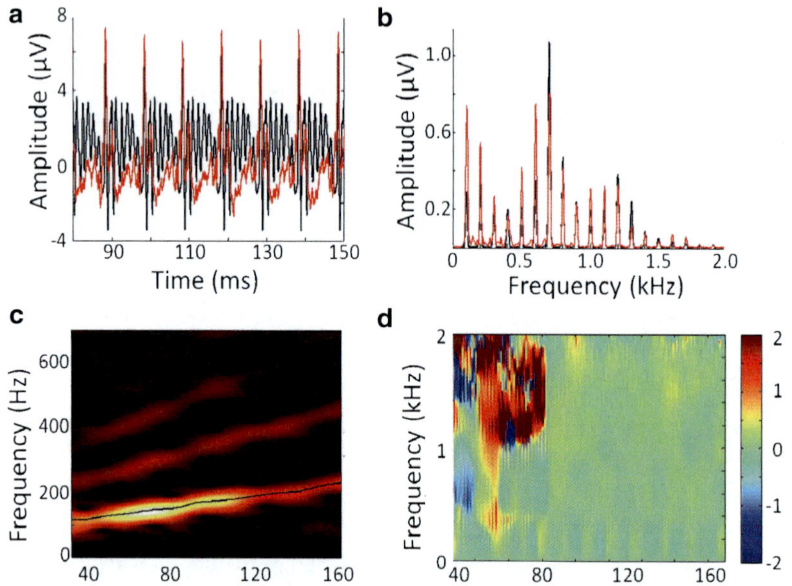

Fig. 17.2 cABR's adherence to the stimulus is demonstrated in these responses recorded from guinea pig scalp. (**a**) In the time domain, the response (red) closely follows the periodicity of this "a" syllable (black). The 70-ms portion of the stimulus waveform, which is arbitrarily scaled, has been time-shifted to visually align with the corresponding response peaks which actually occur about 9 ms later. (**b**) The same "a" syllable (black) closely matches the response (red) in the frequency domain. (**c**) Spectrogram of response to a "da" which has an upward sweeping fundamental frequency (F_0). The thin black line is an overlay of the F_0 of the stimulus. It is possible to derive an objective assessment of "pitch tracking" by comparing the frequency of the recorded F_0 to that of the stimulus over time. (**d**) A "cross phaseogram" comparing responses to "ba" and "ga." The colors depict timing differences, in radians, on a frequency-specific basis. The expected outcome, ga earlier than ba, is depicted with warm colors within the first 60 ms, corresponding to the consonant sounds. The large green field (~0 radians), from about 70 ms on, illustrates the similarity of the responses to the shared "a" sound

17.7.1 Encoding the Fundamental Frequency of a Signal

As shown in Fig. 17.2, the dominant feature in a harmonic-based signal like speech or music is the periodicity reflecting the dominant pitch. Even when the F_0 is absent, time-domain periodicity reveals the missing F_0. The F_0 in a cABR response can be viewed in either the frequency or time domain, and has been shown to be a predictor of the ability to hear speech in noise in children (Anderson et al., 2010b) and young and older adults (Anderson et al., 2011; Song et al., 2011; Anderson et al., 2012a), and in older adults it is a better predictor of hearing in noise ability than audiometric thresholds (Anderson et al., 2011). Phase locking to the F_0 also relates to the ability to selectively attend to an auditory signal amid distractors (Ruggles et al., 2011) and is better in bilinguals, and relates to their attention skills (Krizman et al., 2012). These skills use the F_0 to assist in the object grouping necessary to accomplish the task, and cABR demonstrates an objective basis of this grouping in humans.

17.7.2 Encoding the Changing F_0 of a Signal (Pitch Tracking)

Frequently, in natural stimuli, the F_0 of a signal is not flat. Questions, statements, emotional utterances, and other varieties of expression result in F_0 deviations that are perceived as pitch changes. Perception of a varying F_0 contour is especially important in tonal languages that use pitch to convey meaning. The cABR to the sweeping pitches found in Mandarin syllables has revealed differences in pitch processing between speakers of tonal versus nontonal languages. There have been demonstrations that, as with cortical EPs, the influence of language experience extends to the cABR. Specifically, to both Mandarin syllables and nonspeech analogs of Mandarin syllables (Krishnan & Gandour, 2009; Jeng et al., 2011), Mandarin speakers have more precise pitch tracking than English speakers; however, in infants born to tonal-language speaking families, this advantage is not seen at birth, confirming that the pitch-tracking advantage in tonal-language speakers is an experience-dependent effect (Jeng et al., 2011). This finding was extended in studies that saw an increase in cABR pitch tracking precision to Mandarin tones in musicians (Wong et al., 2007) and following linguistically relevant tone-syllable identification training in non-Mandarin speakers (Song et al., 2008; see also Carcagno & Plack, 2011). The cABR has also revealed group differences in F_0 encoding in children on the autism spectrum. A hallmark of autism is an inability to produce and detect prosodic elements in speech. In a study of autistic children, it was discovered that their tracking of pitch contours—a major contributor to prosody—in the brain stem was often less precise than in typically developing controls (Russo et al., 2008). Thus, the cognitive and affective meaning conveyed by F_0 fluctuations leaves its mark on brain stem processing of this sound property.

17.7.3 Encoding Speech Harmonics

Whereas the fundamental frequency can impart the percept of pitch to a speech signal, the arrangement of the harmonics is the primary information-bearing property in nontonal languages. The relative powers among the harmonics, determined by the filtering properties of the vocal tract, define the speech formants that differentiate vowels and contribute to consonant perception. Harmonics contribute to phonemic awareness and the mapping of sound to letters, which both underlie reading acquisition. Some studies from our lab support the link between speech harmonics and reading ability. Response power involving the higher harmonics of speech syllables is suppressed in cABRs of poor reading children relative to good readers (Banai et al., 2009; Hornickel et al., 2012a). The context of presentation of a longer syllable (repetitive presentation or embedded in a train of other syllables) affects the harmonic content of the response in normal readers, but the difference is suppressed in dyslexic children (Chandrasekaran et al., 2009). Indeed we have repeatedly observed cABR to relate most strongly to a complex cognitive skill such as reading rather than to basic psychophysical perception of sound properties (reviewed in Kraus & Chandrasekaran, 2010; Chandrasekaran & Kraus, 2012; Tierney & Kraus, 2013a).

17.7.3.1 Combination Tones

Nonlinearities of the auditory system also result in cABRs with spectral components not present in the evoking stimulus. Periodic activity at combination tone frequencies, that is, frequencies not present in the evoking signal, can arise in the response. The most straightforward example of this phenomenon is that demonstrated by responses to missing-F_0 stimuli (Galbraith, 1994). More complicated distortion products (e.g., cubic difference tones) also arise to two-tone interval stimulation (Krishnan, 1999), and experience with music impacts the salience with which they are represented in the cABR (Lee et al., 2009).

17.7.4 Encoding of Formant Frequency Timing

The relative timing of events in the formant transition, in stop consonants, enables the system to distinguish among them. We have studied cABR to consonant–vowel syllable pairs, for example, "ba" versus "ga." Poor reading is associated with poor phonological awareness, and this is reflected in ambiguity in cABR formant transition timing in contrastive consonants (Hornickel et al., 2009). Another related approach is the phase relationship between two responses. In the frequency domain, however, a cross-phase analysis presents a picture of the relative timing delays, such as how noise affects the entire response spectrum across time (Tierney et al., 2011). A particularly intriguing use of this cross-phase technique compares pairs of speech syllables. Stop consonant pairs, for example, t versus k or b versus g, are acoustically distinguished by differing formant trajectories. The differing formant frequencies of the syllables are represented by timing changes in far-field evoked responses and, in experimental animals, near-field responses from inferior colliculus corroborate the midbrain source (Warrier et al., 2011). The extent of the resulting inter-response phase difference shows a relationship with speech-in-noise perception ability and reading ability. People with the greatest phase differences in their responses have the best hearing in noise (Skoe et al., 2011) and in children the degree of phase difference is linked to pre-reading skills (White-Schwoch & Kraus, 2013). Musicians also have greater phase distinctions than nonmusicians (Parbery-Clark et al., 2012b). Although peak timing and phaseogram reflections of formant transition processing overlap, each also yields distinctive information (Tierney et al., 2011). By either measure—discrete timings or phase relationships—this neural reflection of critical speech–sound sensitivity represents a potent probe of a key feature of human communication.

17.7.5 Encoding the Onset of a Signal

Sound onsets, and also offsets and other transitions contributing to the envelope of a signal, evoke discrete neural events. The timings of these peaks, relative to the evoking event, are affected by nonperipheral factors. Examples of this include the ability to understand speech in noisy backgrounds (Anderson & Kraus, 2010; Hornickel et al., 2011), the aging auditory system (Anderson et al., 2012b), reading ability in children (Banai et al., 2009; Hornickel et al., 2009), and the context of the evoking sound with respect to its placement in a rhythmic musical background (Tierney 2013b). This latter finding has language implications given the importance of the processing of onsets in running speech. It should be stressed that response timing to a simple click-evoked stimulus generally does not depend on nonperipheral factors and click latencies always serve as important controls for all published cABR findings from our lab.

17.7.6 Response Consistency

Over the course of a cABR recording session, hundreds or thousands of repetitions of the signal are presented. The extent to which each evokes a similar response can be quantified via linear correlation. In practice, the signal-to-noise ratio of the response to any single stimulus event is too low to assess consistency. But, creating and correlating multiple partial averages can provide another means of assessing non-auditory influences on cABR. With techniques using first-half/last-half or odd/even pairs, response consistency was found to be lower in older adults (Anderson et al., 2012b), linguistically impoverished children (Skoe et al., 2013b), and poor readers (Hornickel & Kraus, 2013). This latter finding is consistent with increased neural variability in experimental-animal models of dyslexia (Centanni et al., 2013). Response consistency improves after auditory intervention (Hornickel et al., 2012b), is higher in bilinguals (Krizman et al, 2014) and is maintained in older adults who have musical training (Parbery-Clark et al., 2012a).

17.8 Cognitive Relationships with cABR

An observation, across studies of cABR, is that both overtly auditory skills such as the ability to hear speech in noise and general cognitive skills such as attention and memory relate to cABR metrics. An interesting pattern has emerged that the latter, more cognitive-centered skills, often show the stronger relationship to cABR. An example of this is demonstrated by the relationship of cABR to various measures of speech-in-noise perception. We use a variety of standardized measures of hearing in noise and there is a progression of strength of relationship of cABR F_0 encoding with the cognitive demand of the test (specifically, QuickSIN > HINT > Words in Noise). The general mechanism for this, we speculate, is that high-level processes such as working memory and attention are tapped in communication skills and learning, whether learning is defined as short-term, focused training, or lifelong experiential learning such as language or music. *It is the engagement of these cognitive mechanisms in the past that, in turn, shapes the nervous system's response to the acoustics of the signal in the present.*

Music, in fact, provides a superb model of the learning auditory system, in particular music training. As the OPERA hypothesis states, engaging in the study of music, for example, learning to play an instrument, is highly emotionally rewarding, relies on highly focused attention to precise acoustic sounds, and is highly repetitive, thus meeting major criteria of successful learning (Patel, 2011). With its honing of executive function and the overlap between music and language processing centers in the brain, music experience, even of limited duration (Skoe & Kraus, 2012; White-Schwoch et al., 2013), shapes communication skills as well as the nervous system's response to the acoustics of the speech signal (reviewed in Kraus & Chandrasekaran, 2010; Strait & Kraus, 2014; Kraus et al., 2012).

17.8.1 Neural Signatures of Cognitive Auditory Processing

Taken as a whole, from recent work from our lab and others, patterns have emerged associating subcomponents of cABR with communication skills, learning and experience. With cABR's capacity for application on the individual level, it is exciting to think about potential "neural signatures" that can inform the underlying mechanisms of sound-themed cognitive tasks. Toward that end, we have begun, with an admittedly limited scope, to organize some signatures that have emerged from some of the extant cABR findings. In Table 17.1, we have listed several broad categories corresponding to groups or communication activities in which we have seen particular cABR profiles.

- Successful *hearing in noise* is accomplished, in large part, by tracking the F_0 of the target speaker (Brokx & Nooteboom, 1982). Fittingly, there is a good correspondence between F_0 encoding in the brainstem and hearing in noise ability (Anderson et al., 2010b; Anderson et al., 2011, 2013; Song et al., 2011, 2012). Similarly, consonant–vowel formant transitions, as the fastest-moving and lowest-intensity components of speech, are most susceptible to noise (Nishi et al., 2010), and their representation in cABR timing aptly is associated with hearing in noise ability (Anderson et al., 2010a; Skoe et al., 2011).
- *Reading* ability, which has a strong relationship to phonological awareness (Ramus & Szenkovits, 2008) and is known to correlate with variable cognitive and sensory processing (Roach et al., 2004), aligns with processing of signal harmonics (Banai et al., 2009; Hornickel et al., 2012a), the acoustic differences between consonants (Hornickel et al., 2009; White-Schwoch & Kraus, 2013), and with response consistency (Hornickel & Kraus, 2013).
- In the *aging* system, there is a slowing of neural timing (Tremblay et al., 2002) and a decrease in inhibitory processes (Caspary et al., 2008), reflected by delayed response timing and inconsistent cABR. This is accompanied by reductions in harmonics, in response consistency, and in phase locking (Anderson et al., 2012b).
- A hallmark of *autism*, is difficulty with prosody; this is mirrored by diminished pitch tracking in the brain stem (Russo et al., 2008).

Table 17.1 Various forms of deprivation, decline and disorder (top rows) track with particular constellations of degradation in cABR measures. Experience and training (bottom rows), on the other hand, result in selective response enhancements. The cABR measures are described in Section 7

	Fundamental frequency	Pitch tracking	Harmonics	Onset timing	Formant frequency timing	Response consistency
Communication difficulty						
Hearing in noise	–				–	
Reading			–	–	–	–
Aging	–		–	–	–	–
Autism		–		–		
Linguistic impoverishment			–			–
Experience/intervention						
Musicianship	–	+	+	+	+	+
Bilingualism	+	+				+
Short term training	+	+			+	+
Online processing	+		+	+		

The bottom part of Table 17.1 lists five varieties of experience or training that impact cABR. The responses in the denoted categories are impacted by *musicianship* (Kraus & Chandrasekaran, 2010; Kraus et al., 2012; Tierney et al., 2013), *bilingualism* (Krizman et al., 2012, 2014), *linguistic impoverishment* due to poverty (Skoe et al., 2013b), *short-term training* in the form of computer-based training (Carcagno & Plack, 2011; Song et al., 2012; Anderson et al., 2013), and assistive listening devices (Hornickel et al., 2012b), or are influenced by mere *online exposure* to a stimulus (Skoe & Kraus, 2010b; Skoe et al., 2013a; Hairston et al., 2013). As can be seen by the wide variety of response property combinations, a mixing board is a better analogy for cABR than is a volume knob with respect to its ability to reflect communication ability and experience. When neural signatures of both categories of communication and experiential effects are better delineated, there exists a potential for practical remediation decisions. With more research, one might envision a cABR battery that, with the right mix of stimuli and analysis approaches, indicates, for example, an individual's difficulty in hearing in noise is best remedied by a particular intervention strategy.

17.9 Conclusions

In closing, we want to impress on readers that the auditory system can and must be viewed as a whole when any communication-related activity is to be considered. Auditory science is fractionized by disciplines—cognitive, peripheral, central. This is mirrored in the clinical realm, with discrete clinical practices targeting hearing, speech, and learning. Biology doesn't respect disciplines; and, happily, the field is moving toward a greater consideration of cross-disciplinary views. The brain works as an integrated unit, especially in learning, by which the auditory efferent system, in conjunction with nonclassical auditory brain regions, brings about fundamental changes in cortical and subcortical response properties within the classical auditory system. These cognitive mechanisms unleash the plastic properties that define and redefine "hearing" in behaving animals and human beings.

The auditory research field has been shaped, to a certain extent, by the available methodologies, especially as it pertains to human auditory function. Over the last several decades unit studies in animals, providing a cochleocentric focus on basic signal components and threshold tuning curves, have been joined by more corticocentric techniques such as functional magnetic resonance imaging (fMRI). There also has been a move toward utilizing signals at the suprathreshold levels at which we conduct the business of daily communication. The lens through which our lab and a growing number of worldwide colleagues view this cognitive auditory system happens to be the brain stem response. But the word "brain stem" must not be conflated with an outdated view of a handful of one-way signal-propagating nuclei. In fact, we have some regrets that the terminology "cABR" has caught on. Scientists using this technology—including ourselves—at times feel imprisoned by this nomenclature. Papers from my lab are guilty of propagating the problem, often including "brain stem" or "subcortical" in their titles. This automatically lessens their appeal to a segment of the scientific community concerned with auditory cortex or learning and plasticity and who equate ABR with "peripheral hearing test." We are in fact, with cABR, studying cortical influences as much as local ones. What we must do, going forward, is to carefully present cABR as a way of examining how the auditory nervous system *in its entirety* processes fundamental components of sound, and how sound processing is modulated by online, short-term, and lifelong experience and developmental life stages. The auditory system, shaped by the activation of cognitive mechanisms, is a moving target and cABR moves right along with it. *An additional appealing aspect of this technology is its broad applicability*—the same exact stimuli, recording paradigms and analysis procedures can be used in humans from infancy to old age (Skoe et al., 2014), and in animal models (Warrier et al., 2011). Owing to cABR's magnificent transparency to the evoking signal, its utility in individual humans, and its malleability in response properties, it provides an unprecedented snapshot of the inner workings of the vast, dynamic cognitive auditory system.

17.10 Future Directions

Although the plasticity of primary auditory pathway regions was confirmed 30 years ago, there is a continued hesitancy to pursue its applicability in research on learning and experience in humans. Brain stem- and early cortical-evoked EEG recordings are often dismissed as afferent-driven "obligatory" response to sound and the search for plastic response properties is largely constrained to responses such as early left anterior negativity, mismatch negativity, and others, which invoke associational processes such as sensory memory or linguistic knowledge. In the last decade, however, there has been a flurry of research in the malleability of responses formerly viewed as immutable. In particular, brain stem responses, which long have been seen as just the opposite, are proving to be an exciting window into the dynamicity of the auditory system. A huge advantage cABR has over many other evoked response measures is that it has a high level of reliability on an individual level and, in the absence of intervention, is stable and quite replicable. *As such, it has the potential to move outside the lab and into a role in the clinic, in schools, and in industry: venues where the impact of hearing on assessment and understanding of the biological bases of learning, training, and education is of vital interest.*

There are remaining questions about the manner by which auditory learning proceeds. Work remains to be done that combines cABR, cortical physiology, listening, and cognitive testing in a controlled longitudinal basis. In this way, developmental and learning-related time courses can be mapped with the goal of synthesizing the auditory cortical, subcortical, and cochlear findings into a cognitive auditory system model, each part of which is instrumental in learning and plasticity.

17.11 Summary

The auditory system comprises a vast network of interconnected peripheral, subcortical, and cortical centers. These circuits are bidirectional and extend beyond classically defined auditory pathway. Limbic and association areas have direct input to this auditory network, and there is clear evidence that cognitive processes such as attention, memory, emotion, and motivation impact the auditory processing of sound. This chapter reviews some of the anatomical and physiological underpinnings of these cognitive processes. Finally, it presents data demonstrating a means of physiologically accessing the cognitive auditory system in humans via cABR, and proffers its application in the assessment of and research into human auditory-based communications.

Acknowledgments This work was supported by NIH R01–DC010016, NSF BCS-0921275, and BCS-1057556, The Mathers Foundation, The National Academy of Recording Arts and Sciences, The National Association of Music Merchants and the Hugh Knowles Center.

References

Anderson, S., & Kraus, N. (2010). Objective neural indices of speech-in-noise perception. *Trends in Amplification*, 14(2), 73–83.

Anderson, S., Skoe, E., Chandrasekaran, B., & Kraus, N. (2010a). Neural timing is linked to speech perception in noise. *Journal of Neuroscience*, 30(14), 4922–4926.

Anderson, S., Skoe, E., Chandrasekaran, B., Zecker, S., & Kraus, N. (2010b). Brainstem correlates of speech-in-noise perception in children. *Hearing Research*, 270(1–2), 151–157.

Anderson, S., Parbery-Clark, A., Yi, H. G., & Kraus, N. (2011). A neural basis of speech-in-noise perception in older adults. *Ear and Hearing*, 32(6), 750–757.

Anderson, S., Parbery-Clark, A., White-Schwoch, T., & Kraus, N. (2012a). Auditory brainstem response to complex sounds predicts self-reported speech-in-noise performance. *Journal of Speech, Language, and Hearing Research*, 56(1), 31–43.

Anderson, S., Parbery-Clark, A., White-Schwoch, T., & Kraus, N. (2012b). Aging affects neural precision of speech encoding. *Journal of Neuroscience*, 32(41), 14156–14164.

Anderson, S., White-Schwoch, T., Parbery-Clark, A., & Kraus, N. (2013). A dynamic auditory-cognitive system supports speech-in-noise perception in older adults. *Hearing Research*, 300, 18–32.

Bajo, V. M., Nodal, F. R., Moore, D. R., & King, A. J. (2010). The descending corticocollicular pathway mediates learning-induced auditory plasticity. *Nature Neuroscience*, 13(2), 253–260.

Banai, K., Hornickel, J. M., Skoe, E., Nicol, T., Zecker, S., & Kraus, N. (2009). Reading and subcortical auditory function. *Cerebral Cortex*, 19(11), 2699–2707.

Berlau, K. M., & Weinberger, N. M. (2008). Learning strategy determines auditory cortical plasticity. *Neurobiology of Learning and Memory*, 89(2), 153–166.

Brokx, J. P. L., & Nooteboom, S. G. (1982). Intonation and the perceptual separation of simultaneous voices. *Journal of Phonetics*, 10(1), 23–36.

Budinger, E., Heil, P., Hess, A., & Scheich, H. (2006). Multisensory processing via early cortical stages: Connections of the primary auditory cortical field with other sensory systems. *Neuroscience*, 143(4), 1065–1083.

Carcagno, S., & Plack, C. J. (2011). Subcortical plasticity following perceptual learning in a pitch discrimination task. *Journal of the Association for Research in Otolaryngology*, 12(1), 89–100.

Carlson, S. M., & Meltzoff, A. N. (2008). Bilingual experience and executive functioning in young children. *Developmental Science*, 11(2), 282–298.

Caspary, D. M., Ling, L., Turner, J. G., & Hughes, L. F. (2008). Inhibitory neurotransmission, plasticity and aging in the mammalian central auditory system. *Journal of Experimental Biology*, 211(11), 1781–1791.

Centanni, T.M., Booker, A.B., Sloan, A.M., Chen, F., Maher, B.J., Carraway, R.S. Khodaparast, N., Rennaker, R., LoTurco, J.J. & Kilgard, M.P. (2013), Knockdown of the dyslexia-associated gene Kiaa0319 impairs temporal responses to speech stimuli in rat primary auditory cortex, Cerebral Cortex, 10.1093/cercor/bht028.

Chandrasekaran, B., & Kraus, N. (2010). The scalp-recorded brainstem response to speech: Neural origins and plasticity. *Psychophysiology*, 47, 236–246.

Chandrasekaran, B., & Kraus, N. (2012). Biological factors contributing to reading ability: Subcortical auditory function. In A. Benasich & R. H. Fitch (Eds.), *Developmental dyslexia: Early precursors, neurobehavioral markers and biological substrates*. Baltimore: Brookes.

Chandrasekaran, B., Hornickel, J., Skoe, E., Nicol, T., & Kraus, N. (2009). Context-dependent encoding in the human auditory brainstem. *Neuron*, 64, 311–319.

David, S. V., Fritz, J. B., & Shamma, S. A. (2012). Task reward structure shapes rapid receptive field plasticity in auditory cortex. *Proceedings of the National Academy of Sciences of the USA*, 109(6), 2144–2149.

Dean, I., Harper, N. S., & McAlpine, D. (2005). Neural population coding of sound level adapts to stimulus statistics. *Nature Neuroscience*, 8(12), 1684–1689.

Deliano, M., & Ohl, F. (2009). Neurodynamics of category learning: Towards understanding the creation of meaning in the brain. *New Mathematics and Natural Computation*, 5(1), 61–81.

Engineer, N. D., Percaccio, C. R., Pandya, P. K., Moucha, R., Rathbun, D. L., & Kilgard, M. P. (2004). Environmental enrichment improves response strength, threshold, selectivity, and latency of auditory cortex neurons. *Journal of Neurophysiology*, 92(1), 73–82.

Galbraith, G. C. (1994). Two-channel brain-stem frequency-following responses to pure tone and missing fundamental stimuli. *Electroencephalography and Clinical Neurophysiology*, 92(4), 321–330.

Hackett, T. A. (2011). Information flow in the auditory cortical network. *Hearing Research*, 271(1–2), 133–146.

Hairston, W. D., Letowski, T. R., & McDowell, K. (2013). Task-related suppression of the brainstem frequency following response. PLoS One 8(2), e55215.

Halpern, A. R., & Zatorre, R. J. (1999). When that tune runs through your head: A PET investigation of auditory imagery for familiar melodies. *Cerebral Cortex*, 9(7), 697–704.

Herholz, S. C., & Zatorre, R. J. (2012). Musical training as a framework for brain plasticity: Behavior, function, and structure. *Neuron*, 76(3), 486–502.

Hornickel, J., & Kraus, N. (2013). Unstable representation of sound: A biological marker of dyslexia. *Journal of Neuroscience*, 33(8), 3500–3504.

Hornickel, J., Skoe, E., Nicol, T., Zecker, S., & Kraus, N. (2009). Subcortical differentiation of stop consonants relates to reading and speech-in-noise perception. *Proceedings of the National Academy of Sciences of the USA*, 106(31), 13022–13027.

Hornickel, J., Chandrasekaran, B., Zecker, S., & Kraus, N. (2011). Auditory brainstem measures predict reading and speech-in-noise perception in school-aged children. *Behavioural Brain Research*, 216(2), 597–605.

Hornickel, J., Anderson, S., Skoe, E., Yi, H. G., & Kraus, N. (2012a). Subcortical representation of speech fine structure relates to reading ability. *Neuroreport*, 23(1), 6–9.

Hornickel, J., Zecker, S., Bradlow, A., & Kraus, N. (2012b). Assistive listening devices drive neuroplasticity in children with dyslexia. *Proceedings of the National Academy of Sciences of the USA*, 109(41), 16731–16736.

Jeng, F. C., Hu, J., Dickman, B., Montgomery-Reagan, K., Tong, M., Wu, G., & Lin, C. D. (2011). Cross-linguistic comparison of frequency-following responses to voice pitch in american and chinese neonates and adults. *Ear and Hearing*, 32(6), 699–707.

Jentschke, S., Koelsch, S., Sallat, S., & Friederici, A. D. (2008). Children with specific language impairment also show impairment of music-syntactic processing. *Journal of Cognitive Neuroscience*, 20(11), 1940–1951.

Kilgard, M. P. (2012). Harnessing plasticity to understand learning and treat disease. *Trends in Neurosciences*, 35(12), 715–722.

Knudsen, E. I. (1999). Mechanisms of experience-dependent plasticity in the auditory localization pathway of the barn owl. *Journal of Comparative Physiology A*, 185(4), 305–321.

Kraus, N. (2011). Listening in on the listening brain. *Physics Today*, 64(6), 40–45.

Kraus, N., & Disterhoft, J. F. (1982). Response plasticity of single neurons in rabbit auditory association cortex during tone-signalled learning. *Brain Research*, 246(2), 205–215.

Kraus, N., & Chandrasekaran, B. (2010). Music training for the development of auditory skills *Nature Reviews Neuroscience*, 11, 599–605.

Kraus, N., McGee, T., Carrell, T., King, C., & Tremblay, K. (1995). Central auditory system plasticity associated with speech discrimination training. *Journal of Cognitive Neuroscience*, 7, 27–34.

Kraus, N., McGee, T. J., Carrell, T. D., Zecker, S. G., Nicol, T. G., & Koch, D. B. (1996). Auditory neurophysiologic responses and discrimination deficits in children with learning problems. *Science*, 273(5277), 971–973.

Kraus, N., Strait, D., & Parbery-Clark, A. (2012). Cognitive factors shape brain networks for auditory skills: Spotlight on auditory working memory. *Annals of the New York Academy of Sciences*, 1252, 100–107.

Krishnan, A. (1999). Human frequency-following responses to two-tone approximations of steady-state vowels. *Audiology and Neuro-Otology*, 4(2), 95–103.

Krishnan, A., & Gandour, J. (2009). The role of the auditory brainstem in processing linguistically-relevant pitch patterns. *Brain and Language*, 110(3), 135–148.

Krizman, J., Marian, V., Shook, A., Skoe, E., & Kraus, N. (2012). Subcortical encoding of sound is enhanced in bilinguals and relates to executive function advantages. *Proceedings of the National Academy of Sciences of the USA*, 109(20), 7877–7881.

Krizman, J., Skoe, E., Marian, V., & Kraus, N. (2014). Bilingualism increases neural response consistency and attentional control: Evidence for sensory and cognitive coupling. Brain and Language, 128, 34–40.

Lee, K. M., Skoe, E., Kraus, N., & Ashley, R. (2009). Selective subcortical enhancement of musical intervals in musicians. *Journal of Neuroscience*, 29(18), 5832–5840.

Lehmann, A., & Schönwiesner, M. (2014). Selective attention modulates human auditory brainstem response: Relative contributions of frequency and spatial cues. PLoS ONE, 9(1), e85442.

Li, X., & Jeng, F. C. (2011). Noise tolerance in human frequency-following responses to voice pitch. *The Journal of the Acoustical Society of America*, 129(1), EL21–26.

Malmierca, M. S., Cristaudo, S., Pérez-González, D., & Covey, E. (2009). Stimulus-specific adaptation in the inferior colliculus of the anesthetized rat. *Journal of Neuroscience*, 29(17), 5483–5493.

Nelken, I. (2008). Processing of complex sounds in the auditory system. *Current Opinion in Neurobiology*, 18(4), 413–417.

Nishi, K., Lewis, D. E., Hoover, B. M., Choi, S., & Stelmachowicz, P. G. (2010). Children's recognition of American English consonants in noise. *Journal of the Acoustical Society of America*, 127(5), 3177–3188.

Parbery-Clark, A., Anderson, S., Hittner, E., & Kraus, N. (2012a). Musical experience strengthens the neural representation of sounds important for communication in middle-aged adults. *Frontiers in Aging Neuroscience*, 4(30), 1–12.

Parbery-Clark, A., Tierney, A., Strait, D., & Kraus, N. (2012b). Musicians have fine-tuned neural distinction of speech syllables. *Neuroscience*, 219, 111–119.

Patel, A. D. (2011). Why would musical training benefit the neural encoding of speech? The OPERA hypothesis. *Frontiers in Psychology*, 2, 142.

Polley, D. B., Steinberg, E. E., & Merzenich, M. M. (2006). Perceptual learning directs auditory cortical map reorganization through top-down influences. *Journal of Neuroscience*, 26(18), 4970–4982.

Raizada, R. D., & Poldrack, R. A. (2007). Challenge-driven attention: interacting frontal and brainstem systems. *Frontiers in Human Neuroscience*, 1, 3.

Ramus, F., & Szenkovits, G. (2008). What phonological deficit? *Quarterly Journal of Experimental Psychology*, 61(1), 129–141.

Rinne, T., Balk, M. H., Koistinen, S., Autti, T., Alho, K., & Sams, M. (2008). Auditory selective attention modulates activation of human inferior colliculus. *Journal of Neurophysiology*, 100(6), 3323–3327.

Roach, N. W., Edwards, V. T., & Hogben, J. H. (2004). The tale is in the tail: An alternative hypothesis for psychophysical performance variability in dyslexia. *Perception*, 33(7), 817–830.

Ruggles, D., Bharadwaj, H., & Shinn-Cunningham, B. G. (2011). Normal hearing is not enough to guarantee robust encoding of suprathreshold features important in everyday communication. *Proceedings of the National Academy of Sciences of the USA*, 108(37), 15516–15521.

Russo, N. M., Skoe, E., Trommer, B., Nicol, T., Zecker, S., Bradlow, A., & Kraus, N. (2008). Deficient brainstem encoding of pitch in children with autism spectrum disorders. *Clinical Neurophysiology*, 119(8), 1720–1731.

Skoe, E., & Kraus, N. (2010a). Auditory brainstem response to complex sounds: A tutorial. *Ear and Hearing*, 31(3), 302–324.

Skoe, E., & Kraus, N. (2010b). Hearing it again and again: on-line subcortical plasticity in humans. *PLoS ONE*, 5(10), e13645.

Skoe, E., & Kraus, N. (2012). A little goes a long way: how the adult brain is shaped by musical training in childhood. *Journal of Neuroscience*, 32(34), 11507–11510.

Skoe, E., Nicol, T., & Kraus, N. (2011). Cross-phaseogram: Objective neural index of speech sound differentiation. *Journal of Neuroscience Methods*, 196(2), 308–317.

Skoe, E., Krizman, J., Spitzer, E., & Kraus, N. (2013a) The auditory brainstem is a barometer of rapid auditory learning. *Neuroscience*, 243, 104–114.

Skoe, E., Krizman, J., & Kraus, N. (2013b) The impoverished brain: Disparities in maternal education affect the neural response to sound. *Journal of Neuroscience*, 33(44),17221–17231.
Skoe, E., Krizman, J., Anderson, S., & Kraus N., (2014). Stability and plasticity of auditory brainstem function across the lifespan. Cerebral Cortex, doi: 10.1093/cercor/bht311.
Song, J., Skoe, E., Wong, P. C. M., & Kraus, N. (2008). Plasticity in the adult human auditory brainstem following short-term linguistic training. *Journal of Cognitive Neuroscience*, 20(10), 1892–1902.
Song, J. H., Skoe, E., Banai, K., & Kraus, N. (2011). Perception of speech in noise: Neural correlates. *Journal of Cognitive Neuroscience*, 23(9), 2268–2279.
Song, J. H., Skoe, E., Banai, K., & Kraus, N. (2012). Training to improve hearing speech in noise: biological mechanisms. *Cerebral Cortex*, 22(5), 1180–1190.
Strait, D.L., & Kraus, N. (2014). Biological impact of auditory expertise across the life span: Musicians as a model of auditory learning. *Hearing Research*. 308, 109–121.
Suga, N. (2008). Role of corticofugal feedback in hearing. *Journal of Comparative Physiology A: Neuroethology, Sensory, Neural, and Behavioral Physiology*, 194(2), 169–183.
Suga, N. (2012). Tuning shifts of the auditory system by corticocortical and corticofugal projections and conditioning. *Neuroscience and Biobehavioral Reviews*, 36(2), 969–988.
Thiel, C. M. (2007). Pharmacological modulation of learning-induced plasticity in human auditory cortex. *Restorative Neurology and Neuroscience*, 25(3–4), 435–443.
Tierney, A., Parbery-Clark, A., Skoe, E., & Kraus, N. (2011). Frequency-dependent effects of background noise on subcortical response timing. *Hearing Research*, 282(1–2), 145–150.
Tierney, A., Krizman, J., Skoe, E., Johnston, K., & Kraus N. (2013). High school music classes enhance the neural processing of speech. Frontiers in Educational Psychology, 4, 855.
Tierney, A., & Kraus, N. (2013a) Music training for the development of reading skills. In M. Merzenich (Ed.), Changing brains: *Applying brain plasticity to advance and recover human ability*. Progress in Brain Research. Philadelphia: Elsevier.
Tierney, A., & Kraus, N. (2013b), Neural responses to sounds presented on and off the beat of ecologically valid music. Frontiers in Systems Neuroscience, 7, 14.
Tremblay, K. L., Piskosz, M., & Souza, P. (2002). Aging alters the neural representation of speech cues. *NeuroReport*, 13(15), 1865–1870.
Tzounopoulos, T., & Kraus, N. (2009). Learning to encode timing: mechanisms of plasticity in the auditory brainstem. *Neuron*, 62(4), 463–469.
van Wassenhove, V., & Nagarajan, S. S. (2007). Auditory cortical plasticity in learning to discriminate modulation rate. *Journal of Neuroscience*, 27(10), 2663–2672.
Warrier, C., Nicol, T., Abrams, D., & Kraus, N. (2011). Inferior colliculus contributions to phase encoding of stop consonants in an animal model. *Hearing Research*, 282, 108–118.
Weinberger, N. M. (2004). Specific long-term memory traces in primary auditory cortex. *Nature Reviews. Neuroscience*, 5(4), 279–290.
Weinberger, N. M., Hopkins, W., & Diamond, D. M. (1984). Physiological plasticity of single neurons in auditory cortex of the cat during acquisition of the pupillary conditioned response: I. Primary field. *Behavioral Neuroscience*, 98, 171–188.
White-Schwoch, T. & Kraus, N. (2013). Physiologic discrimination of stop consonants relates to phonological ability in pre-readers: A biomarker for subsequent reading ability? *Frontiers in Human Neuroscience*, 7, 899.
White-Schwoch, T., Woodruff Carr, K., Anderson, S., Strait, D. L., & Kraus, N. (2013). Older adults benefit from music training early in life: Biological evidence for long-term training-driven plasticity. *Journal of Neuroscience*, 33(45), 17667–17674.
Winer, J. A. (2006). Decoding the auditory corticofugal systems. *Hearing Research*, 212(1–2), 1–8.
Wong, P.C.M., Skoe, E., Russo, N.M., Dees, T., & Kraus, N. (2007). Musical experience shapes human brainstem encoding of linguistic pitch patterns. *Nature Neuroscience*, 10, 420–422
Yu, X., Sanes, D. H., Aristizabal, O., Wadghiri, Y. Z., & Turnbull, D. H. (2007). Large-scale reorganization of the tonotopic map in mouse auditory midbrain revealed by MRI. *Proceedings of the National Academy of Sciences of the USA*, 104(29), 12193–12198.

Chapter 18
Fundamentals of Hearing in Amniote Vertebrates

Geoffrey A. Manley

Geoffrey A. Manley

G.A. Manley (✉)
Department of Neuroscience, School of Medicine and Health Sciences,
and Cluster of Excellence Hearing4all, Carl Von Ossietzky Universität,
Carl von Ossietzky Strasse 9-11, 26129 Oldenburg, Germany
e-mail: geoffrey.manley@uni-oldenburg.de

A.N. Popper and R.R. Fay (eds.), *Perspectives on Auditory Research*, Springer Handbook of Auditory Research 50, DOI 10.1007/978-1-4614-9102-6_18,

18.1 The Status of Comparative Research in Auditory Science

The book *The Evolutionary Biology of Hearing* (Webster et al., 1992), which was the publication of papers stemming from a conference, appeared at the same time that the Springer Handbook in Auditory Research series was being launched. The conference provided an excellent opportunity for the first major survey of and stocktaking in the field of comparative auditory physiology. It was very refreshing to confer with colleagues who did not question the validity of working on animals, some of which were quite unrelated to humans, colleagues who took for granted the usefulness of such studies. At that time, all comparative researchers had begun to feel more strongly the pressures of granting agencies and reviewers to work on something "relevant" to the human condition, something that would clearly justify spending public money on its research.

I have at that time, and would definitely still, insist that comparative auditory research should have a firmer place in the broad picture of auditory research, for several reasons. First, there is clear evidence that the results of comparative research support and help interpretations of research on mammals and humans. Given that many nonmammalian vertebrates and invertebrates are physiologically more robust animals than mammals and are frequently the animal of choice for long in vitro experiments, there is a substantial block of basic research—for example, on transduction channels and general hair cell function—that derives from nonmammals. Our current understanding of how hair cells work would be dramatically weaker, were it not for comparative studies. Were it also not for earlier research on auditory specialists such as bats and barn owls, our current understanding of how auditory processing in the brain works would be considerably poorer. Second, the evolutionary processes that have resulted in the extraordinary variety of vertebrate hearing organs and systems have provided us with an invaluable basis for comparisons of structure and function. To ignore this would be self-defeating, but using it, we can learn how complex functions are realized in the inner ear. Third, any physiological system is prone to failure and many experimental procedures have numerous elements that can "go wrong" during an experiment. The direct comparison of results from experiments using mammalian and nonmammalian subjects offers an invaluable control for systematic, hard-to-pin-down errors. Fourth and finally, humans are a powerfully cultural species and our approach to the world is enormously influenced by knowledge. We recognize that knowledge has intrinsic value and should only secondarily expect that this knowledge will, one day, have some sort of direct or indirect economic impact or be useful, for example, for medical procedures. Arguments against the use of animals in research are short-sighted, ignoring the true nature of "Nature" and are also one-sided, ignoring the great benefits to other species with which we share our lives (e.g., Heffner, 1999). Comparative auditory research benefits humans and domestic animals for all the above reasons and does not deserve to be relegated to a position of Cinderella among more "useful" fields.

The question of course arises: To what extent can nonmammals provide useful experimental objects for understanding mammalian and human ears? Any

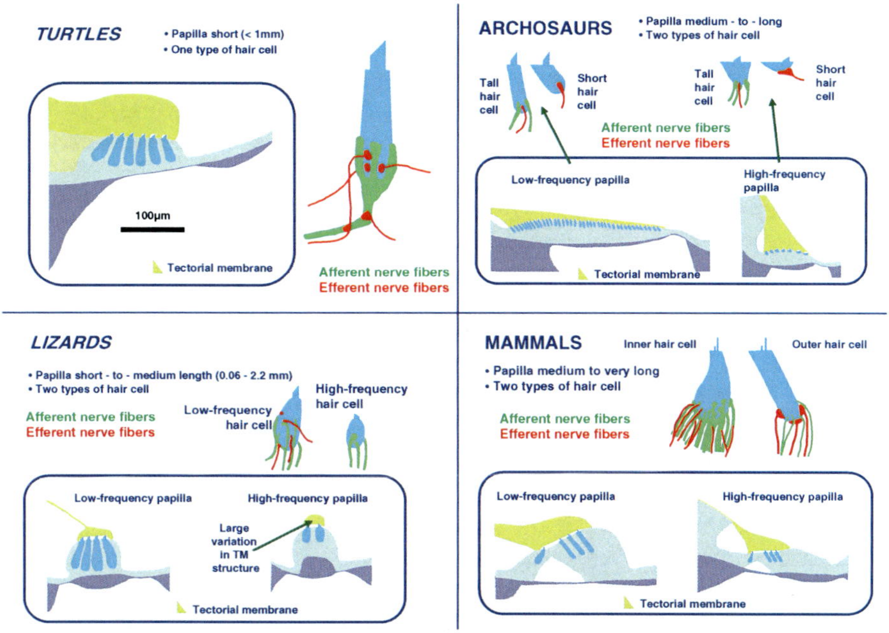

Fig. 18.1 Schematic comparison of the structure of the auditory papilla in different groups of amniotes. Each panel shows a papillar cross section and examples of hair cells. The single hair cell type with both afferent and efferent innervation in the turtle papilla is considered to be ancestral. In most lepidosaurs, such as lizards, there are specialized low- and high-frequency cells with a differentiated innervation. It is now known that in some species at least, there is some efferent innervation to high-frequency cells. Both birds and mammals have evolved two hair cell types that are present at all frequency locations and that have highly specialized innervation patterns. (Modified after Manley, 2000, Copyright (2000) National Academy of Sciences, U.S.A)

mammalian organ shows conservative features that arose early in evolution and others that arose later. Each of the modern lineages of vertebrates is a mosaic of ancestral and recent characters (Manley, 2010, 2012, 2013). Which features of mammals do we see in nonmammals? Here, we also need to differentiate between organizational levels—though, for example, hair cells themselves are definitely extremely old, a few features of mammalian hair cells are unique. This is seen, for example, in the differentiation into two unique populations that differ from the two populations seen in other groups (Fig. 18.1; Gleich et al., 2004). They also differ at the level of specific proteins and biochemical pathways. Thus, in general, the more detailed the question, the less likely it is that characteristics are the same in all vertebrate groups. Nonetheless, there are only a restricted number of evolutionary solutions to developmental and functional "problems" and the parallels between vertebrate lineages can be striking. One example is the tendency—discussed below—for hair cells to form separate populations differentiated to different functions and displaying analogous anatomical specializations, such as in their innervation. Comparative studies of hearing provide both a much broader foundation and a framework for understanding the principles of hearing. Specific cases of parallelism lead us to better-founded conclusions as to function.

For me, 1992 also marked the beginning of the increased use of a then relatively new method for studying the auditory periphery, the measurement of otoacoustic emissions—a technique that was to lead to major advances in our understanding of hearing physiology in all vertebrates. Here, I offer a brief historical review of the development of my understanding of amniote hearing and some perspectives on the future of this field.

18.2 The Background

Prior to the 1992 conference, most of the research on the peripheral auditory physiology of lizards and birds had been summarized in my—at that time relatively new—book (Manley, 1990). That volume reviewed the huge efforts of earlier years to describe the anatomy of lizard (especially by Malcolm Miller and Glen Wever) and bird (Catherine Smith and others) inner ears. It also reviewed physiological studies that had shown that "reptile" ears show an enormous but systematic variety, whereas bird inner ears were more uniform but possessed two hair cell populations that have tantalizing resemblances to those of mammals (Fig. 18.1). (The term "reptile" is placed in quotation marks here because it is a polyphyletic group whose individual lineages are not closely related to each other). While other research groups (under Tom Weiss, working with alligator lizards, Rob Fettiplace, with turtles and Jim Hudspeth, with frog sacculi) had begun to use "reptile" and frog ears to ask fundamental questions regarding hair cell physiology, I was more interested in the evolutionary aspects. How was the huge structural variety, especially in lizard papillae, to be understood? What evolutionary pressures could have led to this variety and what were its functional correlates? What are the parallels and what the differences between the different kinds of auditory papillae? What functions do the independently evolved, different hair cell populations of mammals and birds play and do these functions differ (Manley & Köppl, 1998)?

In the two decades since 1992, I have been involved in many studies to better describe the anatomy and the physiology of lizard, bird, and mammal auditory papillae. Some principles of structure–function relationships became quickly obvious. For example, there is a general evolutionary trend in all amniote groups for the basilar papilla to become longer. This trend is weak in lizards, stronger in birds, and strongest in mammals, resulting in the mammals having the greatest frequency-space constants (length of papilla devoted to one octave; Manley, 1973). Although this presumably increased the number of afferent fibers to each octave, it did not result in mammals showing the sharpest frequency tuning. Curiously, that honor belongs to the birds, followed by geckos and with mammals last (in the same frequency ranges; Manley & Köppl, 1998). In continuing these studies, I built on previous experience, including earlier microelectrode studies of the spontaneous and sound-driven activity of lizard (European wall lizards [*Podarcis*], Tokay gecko [*Gecko*], monitor lizards [*Varanus*], Australian bobtail skinks [*Tiliqua*]) and bird (chicken [*Gallus*], starling [*Sturnus*]) auditory nerve fibers. Otoacoustic emissions, which had made it possible in lizards and birds to undertake very broadly based surveys of function without the necessity of carrying out terminal experiments, of course later strongly confirmed the importance of active processes.

18.3 An Interest in Paleontology

My training as a student at Cambridge included paleontology, and I have maintained a strong interest in studies of fossils. This not only greatly influenced my anatomical and physiological work, beginning with my doctoral thesis, but it has always also served to keep my interests broad. Thus, although I later almost exclusively worked on lizards and birds, some of my work has been on mammals—including early middle ear measurements to very high frequencies in guinea pigs and bats (carried out with Brian Johnstone in W. Australia). Later Eberhard Zwicker and I reported stimulus-frequency otoacoustic emissions (SFOAEs) from guinea pigs. Anges Kronester-Frei and I developed a new technique for direct *in vivo* observation of the exact position of an electrode in the organ of Corti. Using that technique, Geerd Runhaar and I reported data on the electrophysiological profile of the guinea pig organ of Corti—the only recordings to date carried out using a technique that made it possible to know exactly where the electrode tip was placed in relation to the cellular structures. We reported that there is no endocochlear potential in the inner sulcus, a finding that, since then, has been completely ignored.

In recent years, I have returned to studying the middle and inner ears of fossil mammals by proxy, using published findings. Recent fossil finds in China and elsewhere, and especially the use of micro-CT scanning of fossils have dramatically increased our knowledge of the evolution of middle and inner ears. My main interest in these fossil data was to be a kind of mediator and collator, bringing the salient findings on the evolution of the structures underlying hearing to the attention of my colleagues in hearing research that lack a training in paleontology. This explains my recently published reviews of the evolution of mammalian middle ears and mammalian cochleae (Manley, 2010, 2012, 2013). It is remarkable to note, although not at all unusual in evolution, that both the mammalian middle and inner ears are partly the result of structural changes initially unrelated to hearing, the advantages of which for hearing only emerged much later in time. Cochlear coiling and high-frequency hearing in mammals emerged only after 100 million years of mammalian evolution and only in the therian lineage (and not, e.g., in the modern egg-laying monotreme mammals or other, now extinct, lineages, such as the Multituberculata; Fig. 18.2). Only in therians did the middle ear become light and suspended in space and did the cochlea become so elongated that coiling emerged as a means of ameliorating the problem of space. In therians also, bone merged into and stiffened cochlear soft tissue and changes in prestins occurred that were clear specializations for high frequency hearing. These three independent features gradually emerged in parallel over the last 100 million years of therian evolution and conferred some mammals—independently from birds and "reptiles"—with excellent hearing (Fig. 18.2; Manley, 2010, 2012, 2013).

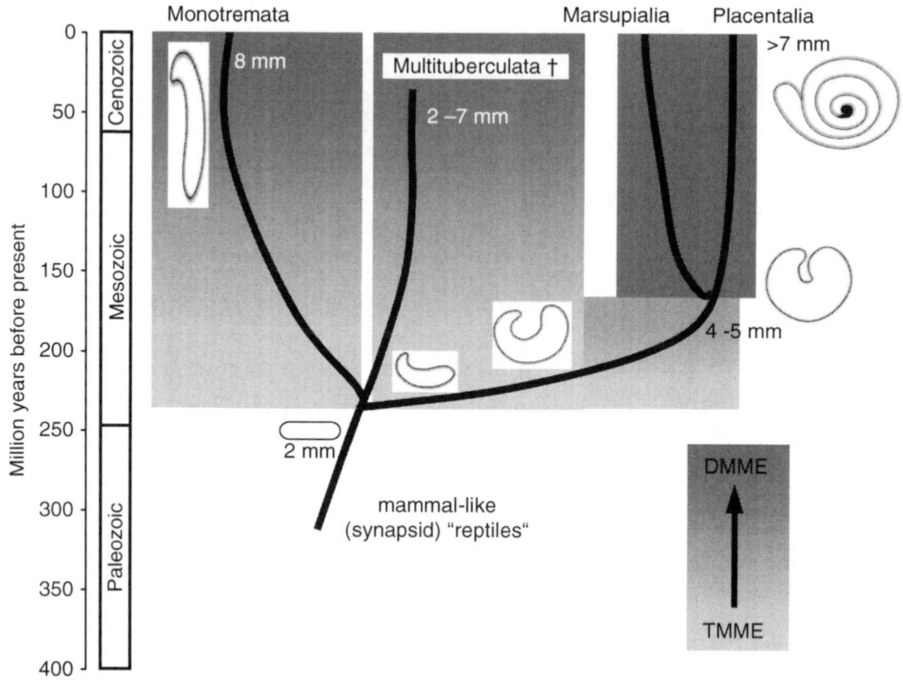

Fig. 18.2 Highly schematic diagram of the evolution of mammalian middle ears and cochleae from their common origin in the early Mesozoic, 230 million years ago, from mammal-like Synapsida. The gray shade indicates the developmental stage of the middle ear, light gray indicates a transitional mammalian middle ear (TMME), dark gray a definitive middle ear (DMME). The DMME was reached at different times in the three lineages of mammals. In addition, all lineages began with an approximately 2 mm long cochlea (outline drawing). In the extinct Multituberculata and the egg-laying Monotremate (Platypus and relatives), the cochlea remained more or less straight and never achieved a length of more than 7–8 mm (e.g., outline drawing top left). Only in the therian lineage, which gave rise to the Placentalia and Marsupialia, did the cochlea fully coil after approximately 50 million years of evolution and then continue to elongate independently in these two lineages. (From Manley, 2013, with permission)

18.4 A Huge Resource of Limited Usefulness

Glen Wever—as noted earlier—not only examined inner ear anatomy in a huge range of lizards and related species. He also carried out parallel physiological experiments on most of these species in the form of cochlear microphonic (CM) measurements. This resulted in hundreds of CM audiograms that were purported to indicate auditory thresholds in each species. These measurements would have provided a huge resource of information but for one fatal flaw, an error in thought that makes Wever's lizard CM data essentially uninterpretable (Manley, 1990). Wever did not take into account the fact that—as he and others had described—in the higher-frequency regions of all lizard species, there are two hair cell populations whose stereovillar bundles are oppositely oriented. This means that at every frequency of sound stimulation, the cm produced by the two cell populations would be out-of-phase over each cycle of a sound wave and thus electrically cancel within the inner ear. Only if one population is larger would there be a residual CM at the stimulus frequency. Thus measurements from frequency ranges processed by two populations of hair cells (that is, above 1 kHz) cannot be compared to measurements at low frequencies in lizards, where the hair cell bundles are—in most but not all species—all oriented in the same direction. Not only are low- and high-frequency CM thus not comparable, but they also cannot be compared across frequencies within one ear or across species, as the patterns of hair cell orientations also differ, both along a given papilla and also strongly between the papillae of different lizard families. An example of the cancellation effect can be seen in a comparison of Wever's CM data on the one hand and threshold measurements for single auditory nerve fibers in *Gecko gecko* as collected in my lab on the other. In Wever's data, the cm response above 1 kHz steadily loses sensitivity toward higher frequencies. By contrast, the nerve-fiber responses are sensitive and sharply tuned above 1 kHz, with a clear second sensitivity optimum near 2 kHz. The conclusion can only be drawn that very unfortunately, Wever's CM data for lizards—the only group with such orientation patterns among amniote vertebrates (Manley, 2004; Manley & Köppl, 2008)—simply cannot be used. There is, however, one exception. The exception is that in a number of species, Wever (generally while working with Yehudah Werner), compared CM measurements before and after severing the middle ear connection. This within-ear control is of course free of hair cell problems and provides very useful and interesting data on the effectiveness of lizard one-ossicle middle ears (e.g., sensitivity improvements of up to 65 dB).

18.5 Remote Sensing: Otoacoustic Emissions

By 1992, our understanding of lizard ears was best in *Tiliqua rugosa* and in the alligator lizard *Gerrhonotus*. I and Christine Köppl had recently published a comprehensive study of spontaneous and driven auditory nerve activity in *Tiliqua rugosa* and, also in cooperation with Brian Johnstone's auditory lab at the University of Western Australia, followed this up with a look at distortion-product otoacoustic emissions (DPOAEs). DPOAE suppression characteristics in bobtail skinks revealed clear resemblances to the characteristic tuning patterns in the auditory nerve, and this was one of the first indications that OAE can be directly attributed to hair cell activity (Manley & Köppl, 2008). These, and our later DPOAE studies in birds, thus clearly also supported the diagnostic use of these emissions in the medical clinic.

During DPOAE recordings, we also noticed very small peaks in the spectra that were unrelated to our stimuli. These turned out to be very stable spontaneous otoacoustic emissions (SOAE) originating in the higher-frequency area of the bobtail papilla. These SOAE provided the basis of extensive follow-up studies that I carried out with Christine Köppl and that revealed detailed parallels between the

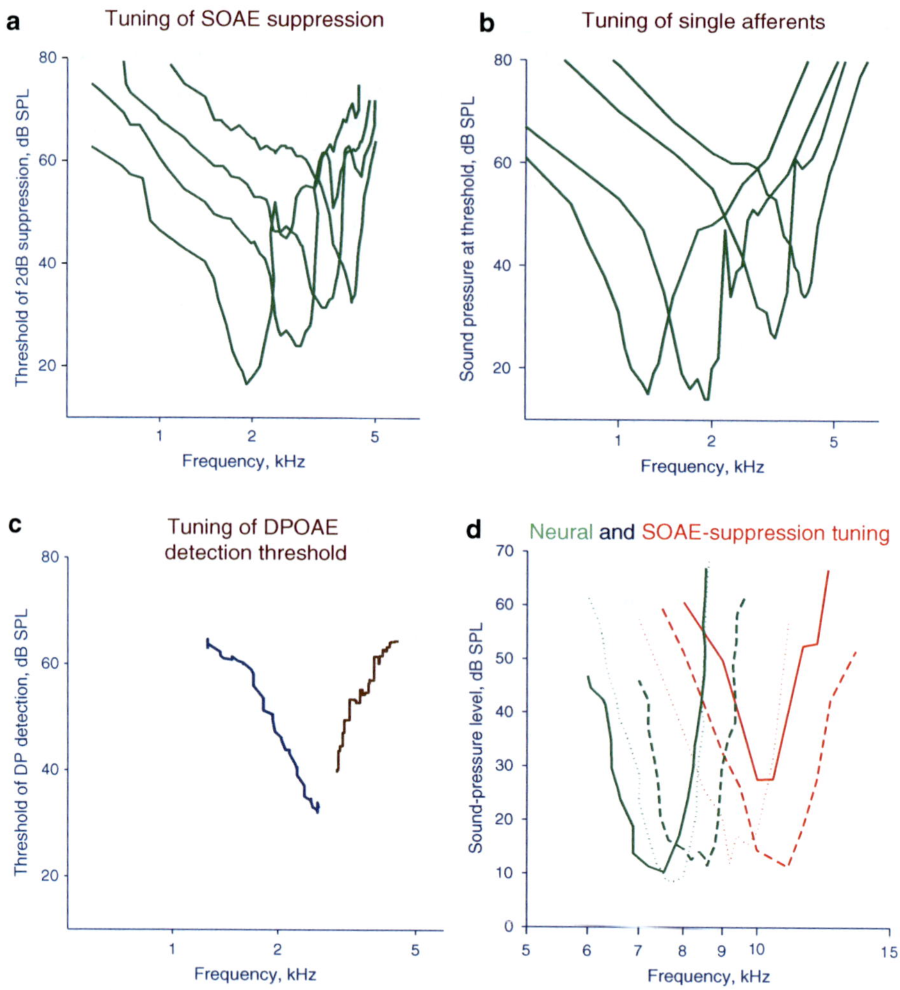

Fig. 18.3 A comparison of different measures of cochlear frequency selectivity or tuning in the bobtail skink (*Tiliqua*, **a**–**c**) and the barn owl (**d**). In (**a**), frequency tuning curves are shown for the suppression of four spontaneous otoacoustic emissions by 2 dB. In (**b**), threshold tuning curves for four single primary afferents are shown. (**c**) The lowest sound pressure levels at which distortion-product otoacoustic emissions can be detected for, in blue, the product 2f1–f2 and in brown for 2f2–f1. In (**d**), green curves are tuning curves for single primary auditory afferents, red curves are levels at which individual spontaneous otoacoustic emissions were suppressed by 2 dB. (Partially after Taschenberger and Manley, 1997, Manley and Köppl, 2008; barn owl data kindly provided by Christine Köppl)

characteristics of SOAE, DPOAE, and auditory nerve activity, including the frequency selectivity of emission suppression behavior in the presence of external tones (Fig. 18.3). To a remarkable degree, SOAE could, in spite of being a "remote sensing" technique, reveal details of peripheral sensitivity and tuning selectivity—and I later used this technique to study the auditory periphery in lizard species of a variety of families.

In concert with auditory nerve studies, these SOAE data made it possible to understand differences in activity patterns in lizard inner ears on the basis of their anatomy, in particular the papillar size and type of tectorial structures. These in turn made it possible to understand the functional consequences of the wide structural variety of the papilla that had been achieved during lizard evolution (Manley, 2011).

18.6 Simple and Complex Lizard Papillae

Basically, it can be assumed that in most lizards, the selective pressures on hearing were not great. As long as a reasonable hearing ability up to a few kilohertz was maintained, in most lizard families it was not very important how the auditory periphery achieved this. Thus even very tiny papillae, with, for example, only approximately 60 hair cells (in, e.g., some iguanid and agamid lizards) were maintained over long evolutionary time periods. All the very small papillae have lost their tectorial membrane (independently in different families; Fig. 18.4c, d). This loss reduced the coupling between hair cells arranged along the papilla and made it possible to code for several octaves of sound frequencies with a very small frequency space constant—at the cost, however, of sensitivity and frequency selectivity (Fig. 18.4e–h; Manley, 2011). The most complex lizard papillae, which presumably result from stronger selection pressures, are found in geckos, the only really vocal lizard family. In geckos, as we now know from studies in Jim Hudspeth's and Christine Köppl's laboratories, there is one population of hair cells that completely lacks an afferent innervation (in this case it is the hair cells on the inner or neural side of the papilla). This was suggested to be a parallel evolutionary development to the two hair cell populations of mammals and birds (Chiappe et al., 2007).

My recent studies with Hanna Kraus of Australian pygopod geckos produced tantalizing suggestions that there might be interactions between hair cell populations, but this is far from being understood (Manley and Kraus, 2010; Manley, 2011). In field studies of the auditory sensitivity of pygopods of the genus *Delma*, we measured compound action potentials (CAP) forward-masked by narrow-band noise. In these species, as in other geckos, Christine Köppl has shown that all auditory afferents only innervate the outer, postaxial hair cell population. CAP suppression curves derived for tones above about 3 kHz showed two sensitivity maxima, one below 8 kHz (as expected, near the probe frequency) and one above 8 kHz. It is tempting, but as yet without a mechanistic explanation, to assume that the two hair cell populations somehow interact and the noninnervated population induces the sensitivity at high frequencies. In general, geckos show the sharpest frequency tuning of all lizards and are the only lizards—indeed the only amniotes—that have a reversed tonotopic organization. As I have explained in reviews of the evolution of lizard papillae, this is a logical and thus not unexpected result of the derivation of all lizard papillae from an ancestral, tripartite papilla having two (redundant) higher-frequency areas, one at each end (Manley, 2011). A few modern lizard families maintain this organization, but in most lizard families either one or the other of the high-frequency areas was lost.

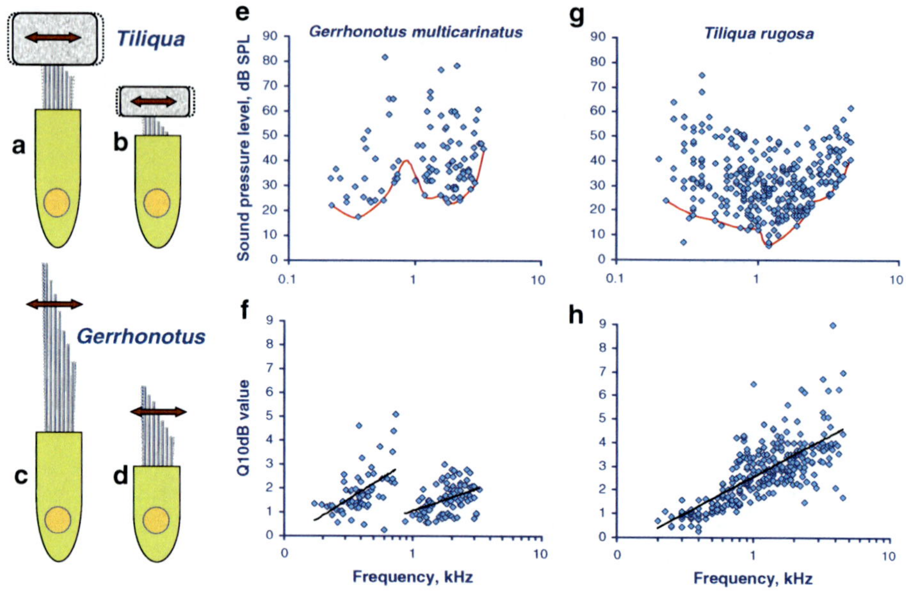

Fig. 18.4 Schematic illustration of the influence of the tectorial membrane on threshold and tuning of primary afferent fibers for high-frequency regions of lizard papillae. The hair cells of the bobtail skink (*Tiliqua*) are covered by a tectorial membrane, shown as a gray block in (**a**) and (**b**) for low- and high-frequency fibers, respectively. In the alligator lizard *Gerrhonotus* (now called *Elgaria*), there is no tectorial membrane over hair cells with best frequencies above 1 kHz and (**c**) and (**d**) illustrate hair cell structures for the 1 kHz and 4 kHz regions, respectively. Panels (**e**) and (**f**) show best thresholds and Q_{10dB} sharpness coefficients, respectively, for *Gerrhonotus* primary afferents, (**g**) and (**h**) the same for *Tiliqua*. The possession of a tectorial membrane improves threshold and raises Q_{10dB} values. (After Manley, 2000, Copyright (2000) National Academy of Sciences, U.S.A.; Manley and Köppl, 2008 and used with permission)

18.7 Active Processes and Hair Cell Specialization

Jim Hudspeth and his coworkers have suggested that in geckos, the noninnervated hair cells should be regarded as an analogue of mammalian outer hair cells, forming a subunit that has primarily a motor function (Chiappe et al., 2007). We had suggested the same for avian short hair cells (Manley and Köppl, 1998; Manley, 2000). In fact, the tendency to hair cell functional specialization—and therefore modification of their innervation patterns—is more extreme in geckos and in birds, where one hair cell population (the preaxial cells in geckos and the short hair cells in birds) completely lacks an afferent innervation. In mammals, the afferent innervation of outer hair cells is merely strongly reduced, perhaps vestigial. We can assume that ancient hair cells were all both afferently and efferently innervated. If some hair cell populations lost their afferents, it was presumably because their most important function was within the papilla. Mammalian hair cell evolution has thus not proceeded quite as far as evolution has in other lineages (Fig. 18.1). As I first pointed out in 1986, we can thus recognize an evolutionary trend towards receptor cell specialization that occurred independently in various amniote groups (Manley, 2000). This began early in the evolution of the different lineages, and certainly began in the mammal lineage before the evolution of tympanic middle ears and before the extreme specialization of prestins that we see in eutherians (Manley, 2012). Thus hair cell specialization into motor and receptor groups occurred initially at least only on the basis of the stereovillar bundle active process (Manley, 2001).

18.8 The Definitive Localization of an Active Process to the Hair Cell Bundle

In 1992, one of the most discussed issues in auditory research was the question of the mechanisms underlying active processes (Manley, 2001). Whereas on the one hand those studying mammalian ears were understandably excited by somatic motility, some much earlier studies of the fundamental properties of hair cells (Fettiplace's group in turtle papillae and Hudspeth's group in frog sacculus) had clearly shown *in vitro* that hair cell stereovillar bundles spontaneously oscillate, and had localized the motor activity to the transduction complex. The underlying active process is a phylogenetically very old mechanism that almost certainly existed in hair cells of vestibular systems that were the forerunners of auditory hair cells (Manley, 2001). Our own later studies, with Andy Forge, of hair cell membranes in geckos and barn owls, two species that we had shown to produce SOAEs, found no evidence for the presence of dense concentrations of membrane-bound particles (prestin tetrads) that are characteristic of the lateral membranes of outer hair cells and are part of the somatic motor in mammals.

Two active processes had been implicated in 1992, but could it be shown *in vivo* which one was really operating? It could—and one of our most important results from emission studies was made possible by the fact that, uniquely, in all lizard papillae, hair cell areas exist that, have their stereovillar bundles oppositely oriented. Christine Köppl and I, working together with the late Des Kirk and Graeme Yates in W. Australia, predicted that, were we able to electrically stimulate active processes in lizard hair cells and influence these with low-frequency sound, there would be two possible and clearly predictable patterns in the electrically evoked emissions (EEOAEs; Manley et al., 2001). These patterns would make it clear as to whether the active process itself was to be found in the hair cell bundle or in the cells' lateral membranes. In the first (bundle) case, EEOAEs generated by the opposite orientations of two sets of hair cell bundles would be out-of-phase and thus essentially cancel each other out within the ear. In the second case (somatic), they would add. In the second case, also, stimulation with low-frequency sound should show no sound-phase–dependent amplitude and phase fluctuations in the EEOAEs. The results of such experiments using *Tiliqua* were unambiguous (Manley et al., 2001): in the absence of an added, low-frequency sound, high-frequency electrical stimulation

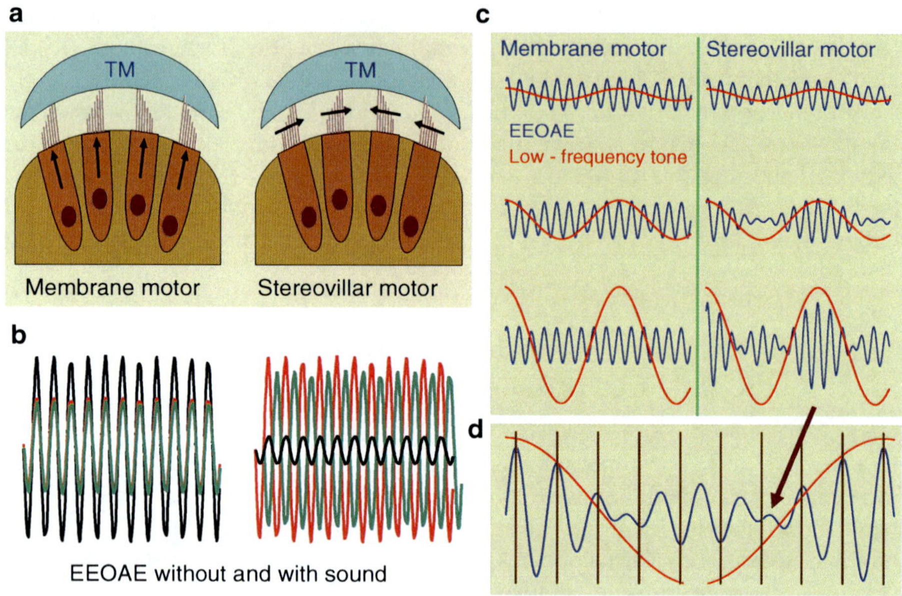

Fig. 18.5 Model predictions for the results of high-frequency current injections into the cochlea of the Bobtail skink *Tiliqua*. Panels (**a**) and (**b**) illustrate the predicted effects of putative cellular motors. In (**a**), the left figure illustrates a cellular shortening or elongation as the result of a putative motor in the lateral cell membranes. The right panel in (**a**) shows the putative motor in the stereovillar bundles and, for the same current, opposed in their effects. In (**b**), waveforms of electrically evoked otoacoustic emissions induced by current injection into scala media are shown for the anatomical situations in (**a**). Red and green curves (that are in phase and therefore almost completely overlap in the left panel but almost cancel in the left panel of **b**) are the contributions of hair cells on each side of the papilla. The black curve illustrates the resultant emission, which is large in the left panel and very small in the right panel. (**c**) Predicted emission curves (blue resultant curve) for the two motor systems, somatic membrane motor on the left and stereovillar motor on the right, when adding a very-low-frequency sound bias of increasing level from top to bottom traces (red curves) during current injection. Whereas the left panels show almost no modulation due to sound, the stereovillar motor emissions shown on the right are very sensitive to sound and, at high sound levels, change their phase 180° for every half-wave of the sound stimulus. (**d**) A sample of data from an experiment. The results are remarkably similar to the predicted results for the stereovillar motor, including the phase shift (arrow). (Partially after Manley et al., 2001, Copyright (2001) National Academy of Sciences, U.S.A)

produced only tiny EEOAEs, far smaller than predicted based on mammalian studies. In the presence of low-frequency sound, however, EEOAE size increased for each half-cycle of the sound, suggesting that the emissions were emerging from a cancellation process between the hair cell groups. The high-frequency EEOAE components were out-of-phase during the half-cycles of the sound. This was exactly as predicted for the case of an active process that resided in the hair cell stereovillar bundles (Fig. 18.5; review in Manley & Köppl, 2008). Later studies by Hudspeth's group revealed, of course, that in mammals, both active processes are to be found.

In other studies in collaboration with Pim van Dijk, we also showed that the statistical properties of lizard and bird SOAEs indicated that they were, indeed, derived from active processes and were not simply filtered noise. In *Anolis*, species that have a very small papilla, my group also demonstrated that small SOAE spectral peaks could result from the activity of only two or three hair cells (reviewed in Manley & van Dijk, 2008).

18.9 Calcium and the Evolutionary Consequences of the Loss of the Lagena Macula

Spontaneous hair-bundle activity in lizards was also examined by altering the calcium concentration in vivo in the endolymph of *Tiliqua*, work carried out in W. Australia with Des Kirk, Christine Köppl, and Ulrike Sienknecht. We either lowered calcium level using BAPTA (1,2-**b**is(*o*-**a**mino**p**henoxy)ethane-*N*,*N*,*N*′,*N*′-**t**etra**a**cetic acid, a calcium-binding molecule) or lowered or raised it by injecting various concentrations of calcium dissolved in artificial endolymph. These procedures showed that raising calcium levels above about 1 mM led to an increase in the frequency of spontaneous hair cell bundle oscillation, inducing calcium levels below 1 mM led to a fall in oscillation frequency. This was consistent with earlier *in vitro* results obtained on hair cells of the bullfrog sacculus by Pascal Martin and Jim Hudspeth. Our data also clearly suggested that in bobtail lizards, the endolymph calcium concentration is about 1 mM. Such a high concentration is presumably necessary to maintain the integrity of the otoliths of the lagena macula that lies adjacent to the auditory papilla. In birds and frogs, the concentration is lower, around 250 μM, but still at least 10 times higher than in mammalian endolymph. This interesting fact suggests that when therian mammals lost their lagena macula, it was no longer necessary to maintain such high levels of endolymphatic calcium. It still needs to be studied what—perhaps decisive—consequences this huge drop in calcium concentrations had, for example, on the tectorial membranes, the prestin active mechanism or on the transduction channel micromechanics of therian mammals (Manley, 2012).

18.10 Frequency Maps of the Papilla and the Functions of the Tectorial Membrane

Our earlier mathematical models of the frequency map and micromechanics of the papillae of *Tiliqua* (with Graeme Yates) and *Gecko* (with Stephan Authier) were based on detailed anatomical data collected by Christine Köppl. The model predictions not only correlated well with physiological results (e.g., predicting the reversed tonotopic map of geckos), but also emphasized the great importance of the tectorial membrane. Auditory papillae lacking a tectorial membrane showed poorer sensitivity and poorer frequency selectivity than those that had either a continuous tectorial structure or a chain of tectorial sallets (Fig. 18.4). Coupling hair cells via a tectorial membrane both sharpened tuning and improved sensitivity. On the other hand, the lack of hair cell coupling because of the loss of a tectorial membrane enabled species with very small papillae (e.g., <200 μm) and very few hair cells (<100) to have tuned afferents over a broad range of frequencies, but at the cost of poorer selectivity and sensitivity (Manley & Köppl, 2008; Manley, 2011). We also showed that the frequency selectivity in geckos could not be modeled using normal values for endolymph viscosity. Only the—impossible—assumption that the viscosity was about one-tenth of normal produced matching frequency selectivity. This was, of course, equivalent to assuming that an active process driving the stereovillar bundle reduced the effective viscosity, and was thus a proxy for the active process that we had helped identify in lizards. These conclusions regarding tuning and tectorial membranes have been recently confirmed by Chris Bergevin's studies of SFOAEs in various lizard and mammal species (e.g., Bergevin, 2011).

18.11 "High-Frequency" Hearing in Lizards

OAE studies in lizards from my own and from Bergevin's lab suggested that some species, including, remarkably, *Anolis*, with its tiny papilla, have a higher upper frequency limit than expected from earlier data, for example from *Gerrhonotus*, the alligator lizard. Although frequency tuning in lizards is, of course, temperature sensitive, comparisons at the same temperatures put some upper hearing limits nearer 7 or 8 kHz than the previously known rough limit near 5 kHz. This was supported by the results of my studies with Jakob Christensen-Dalsgaard of the directionality of sound-induced eardrum vibrations of lizards (review in Manley, 2011). Not only did these data indicate that lizards have (species-varying) an extremely effective pressure-gradient middle ear system that provides a strong directionality prior to the sound being detected by the inner ear. The data also showed that small gecko eardrums, for example, still show responses in the 8 kHz range. My recent detailed field study with Hanna Kraus of hearing in a group of legless geckos, the Australian pygopods, indicated that, remarkably, species of the genus *Delma* show auditory nerve activity up to 13 kHz, which is higher than the upper limit even of birds.

18.12 Hearing in Birds

After I developed the technique of recording avian auditory nerve fiber activity in the cochlear ganglion of the starling in 1977 (review in Manley, 1990), my lab continued to study physiologically avian hearing by using the chicken, especially to examine the development of the frequency map during ontogeny. Even before 1992, Jutta Brix, Alex Kaiser, and I had established that—contrary to published reports—the frequency map of hatchling chickens did not change with age. This was confirmed and extended by Jones and Jones' remarkable study of embryonic chickens and Dick Salvi's group's data on the adult chicken frequency map (Gleich et al., 2004).

In 1994, one other important fact about bird ears was established by anatomical studies carried out in my group by the late Franz Peter Fischer. Fischer demonstrated that, contrary to expectations, in all avian species, a certain—sometimes quite large—population of abneurally located hair cells totally lacked an afferent innervation (Fig. 18.1). Obviously, these hair cells must have a function restricted to the papilla, and this strengthened our concept that these "short" hair cells were in fact motor cells. Fischer proposed that these short hair cells—previously defined arbitrarily on the basis of their shape—could now be specifically defined anatomically through their lack of afferents. Their massive stereovillar bundles presumably pass energy via the tectorial membrane to the "tall" hair cells. Kuni Isawa and Christine Köppl's recent study of avian hair cell bundles supports this idea, as does recent work from Robert Fettiplace's lab. Otto Gleich had previously shown in the starling that the most sensitive afferent fibers connect to hair cells close to the neural edge of the papilla (this part of the papilla is not atop the free basilar membrane) and that the sensitivity reduced by 6 dB/ hair cell across the papilla to the middle. Modeling by Charles Steele suggested that in birds, the tectorial membrane could indeed transport motor activity from the short to the tall hair cell region (Gleich et al., 2004).

One small but interesting aspect of the activity of avian auditory nerve fibers that was later also seen in lizard data was the presence of preferred intervals in the spontaneous activity, which—although then controversial—I now see as one of the earliest signs of active processes in the hair cells of nonmammals. It was claimed that these peaks were the result of inadvertent noise stimuli. However, quite apart from our careful checks of the sound system, there were two other good reasons why this could not have been the case. First, the characteristic frequencies of the cells when driven by tonal stimuli did not always correspond to the best frequency as calculated from the preferred intervals, and second, the thresholds of some cells showing this phenomenon were too high to be even contemplated as responding to inadvertent noise. In songbird and chicken data, there is always a wide spread of thresholds (>50 dB) in ears in good condition, the result of different thresholds of the hair cells across the wide papilla. Had there been so much noise artifact, then cells with better thresholds should all have shown even stronger preferred intervals, which was not the case. Thus these data suggested a spontaneous activity at the hair cell level that is at least partly driven by an active process. The coupling of the avian tectorial membrane is strong, however, perhaps making it difficult for the activity of local hair cell groups to be transported into fluid movements, and so far SOAEs in birds have been detected only in the barn owl (Gleich et al., 2004).

18.13 Barn Owls, the Hearing Specialists

Following a research visit with Christine Köppl to Mark Konishi's lab in 1988, where we had worked on the barn owl (*Tyto alba pratincola, now known as Tyto furcata*) auditory brain stem, we established our own colony of European barn owls in order to carry out research on their auditory periphery. Much earlier studies in Johann Schwarzkopf's lab had shown that owl cochleae in general are large and the cochlea of the barn owl is about 11 mm long, far longer than in song birds (3–5 mm) and chickens (5 mm). This was confirmed and detailed anatomical data derived by Christine Köppl and Franz Peter Fischer of my lab, who showed that the barn owl papilla showed very interesting features when compared to those of other birds. Christine Köppl, Otto Gleich, and I also mapped frequency in the auditory papilla and showed that the owl hearing organ possesses a clear fovea, an area of expanded frequency representation (Fig. 18.6a). Here, the fovea extends over one octave, from 5 to 10 kHz, and occupies the entire basal half of the papilla (>5 mm). In this region, the most neural hair cells have an afferent innervation denser than yet seen in other avian species. Christine's later auditory nerve data of the barn owl showed that, remarkably, nerve fibers from this foveal region were not especially sharply tuned (Fig. 18.6b). Instead, the fovea seems to be a mechanism for producing massive parallel processing in a frequency range that is vital for the owl in sound localization and thus prey capture (Köppl, 2009).

One feature in which barn owl afferent fibers excelled was their ability to phase lock to very high frequencies. In contrast to other birds and to mammals, in which the highest phase-locking frequency is generally 3–5 kHz, higher-frequency barn owl afferents showed useful phase locking at least an octave higher, to 9 kHz (Köppl, 1997b; Fig. 18.6c). This feature is now known to be essential for the extreme ability of barn owls to compare binaural inputs and localize sound in the horizontal plane. Grit Taschenberger in my lab then found SOAEs in the barn owl, which is still to date the only bird species showing this phenomenon. Almost all SOAEs were found above 7.5 kHz, at frequencies of the foveal region. This suggested that the expanded space constant (~5 mm per octave) in the fovea coupled so many active hair cells of the same best frequency together that they were able to synchronize and drive the tectorial membrane and surrounding fluids. Suppression of these SOAEs using pure tones showed that their thresholds and their tuning sharpness was the same as that seen in single auditory nerve afferents of this species by Christine Köppl (Köppl, 1997a; Taschenberger & Manley, 1997; Fig. 18.3d). Thus within the limited frequency range of their occurrence, barn owl SOAEs reflect in detail the function of the auditory papilla (Manley & van Dijk, 2008).

Fig. 18.6 (continued) afferents of different characteristic frequency in the barn owl, illustrating that there is no increase in frequency selectivity in the foveal region of the papilla. (**c**) Data illustrating the extraordinary ability of barn owl primary auditory afferents to phase lock to high frequencies. The blue dots show vector strength of phase locking in a large number of auditory primary afferents over the hearing range. The green curve is a moving window average of the data. In comparison, the orange curve is equivalent average data from the cat auditory nerve; above about 4 kHz, barn owl afferents show equivalent phase locking an octave higher than the cat. (All data kindly supplied by C. Köppl)

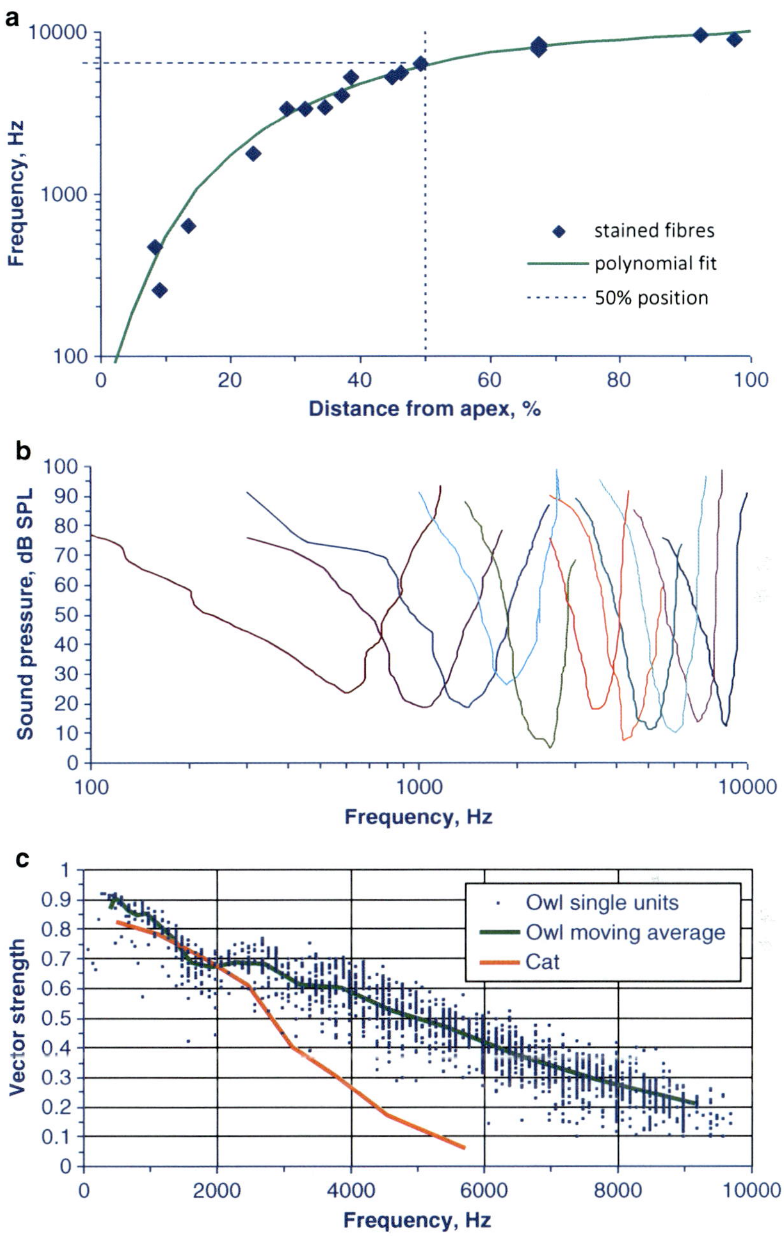

Fig. 18.6 Diagrammatic representation of the characteristics of auditory-nerve afferent fibers in the barn owl. In (**a**), the peripheral origin of characterized and stained afferents (blue diamonds) are shown as the response frequency as a function of the distance of the stain from the cochlear apex. The green curve is a fourth-order polynomial fit to the data. The dashed lines show that in the region of the auditory fovea one octave (~6 kHz–10 kHz) occupies the basal half of the auditory papilla (a length of 5.5 mm). (**b**) A sample of threshold tuning curves of auditory primary

Using contralateral sound stimuli to suppress SOAEs and DPOAEs via the efferent system, Grit Taschenberger, Horst Oeckinghaus and I also showed that in the barn owl, efferent effects can be large and are not attributable to reflex middle ear responses to the (sometimes loud) contralateral sound. As shown by Alex Kaiser in my lab, tuning of efferents in the chicken brain stem is usually relatively poor. Unlike in mammals, also, chicken efferent cells of the brain stem could show excitation or inhibition during tonal stimulation. Similarly, in the barn owl, the effects of contralateral sound stimulation on DPOAEs could be either facilitation or suppression but with a frequency tuning on average sharper than for efferent activation in the chicken.

18.14 Avian Diversity and a Unique Feature

Over the years, we studied a variety of birds. One interesting species was the emu (*Dromaius novaehollandiae*), a representative of the very basal avian group, the paleognaths. For obvious reasons, we used emu chicks (among other things, adult emus weigh more than 50 kg and can be very dangerous; one does not like to imagine the effect a loose adult emu could have on a lab full of equipment). Not unexpectedly for such a large bird, even animals just a few weeks old heard very well at low frequencies. A basal status for the ear was confirmed by the large percentage of tall hair cells and the almost perfectly logarithmic frequency map we measured for the auditory papilla. Interestingly, a recent study by Christine Köppl and Andrew Affleck of another basal bird, the New Zealand kiwi (*Apteryx*) showed clear indications of a cochlear fovea, at a position consistent with its possible use for individual call recognition in this nocturnal species.

Christine Köppl and I, in cooperation with Graeme Yates in Australia, studied the rate-intensity (RI) functions of auditory nerve fibers of emus and barn owls. Graeme had earlier made important contributions to research in mammalian hearing by providing a consistent explanation for the existence of three basic forms of RI functions in mammals. His idea was based on their thresholds in relation to the saturating rate level function of the organ of Corti–basilar membrane complex. Although bird afferents did show the same pattern of RI types, the relationship to one another differed; the data suggested that each hair cell afferent response unit in birds has its own individual threshold-response relationship and is not governed by a global response pattern as in mammals. In birds, Otto Gleich's data indicated that the most sensitive hair cells were supported not by the basilar membrane but by the solid neural limbus. Thus, unlike in mammals, hair cell activity cannot be fully integrated into a global oscillation of basilar membrane and hearing organ together (Manley & Köppl, 1998). As shown by Rainer Klinke's group in the pigeon, any reflections of hair cell activity in a traveling wave of the basilar membrane in birds are poor, at least compared to those in mammals.

One of the most useful discoveries in nonmammals in recent decades was that of Doug Cotanche, that birds are capable of quickly regenerating hair cells. Otto Gleich in my lab cooperated with Bob Dooling and others to show that in the

"Waterslager" race of canaries, a genetic defect leads to continuous hair cell degeneration, but that the hair cells also continuously regenerate. On average at any one time, enough hair cells are defective to confer birds of this race with higher auditory thresholds, which presumably explains why they sing so loudly. Such examples illustrate that—contrary to the opinion of some grant reviewers—birds can be extremely useful organisms for studying the mechanisms of hair cell regeneration.

18.15 Projecting from Birds Back to Dinosaurs

One later idea that emerged partly from data collected in my lab and collated by Otto Gleich in Regensburg was the possibility of estimating the hearing abilities of bird ancestors—the dinosaurs—from comparative studies of modern species. Gleich, Dooling, and I, and again later in cooperation with paleontologists coordinated by Stig Walsh, were able to show that in birds, hearing frequency limits correlate sufficiently well with animal size that extrapolations to extinct organisms are possible and reasonable. Using this, the upper hearing limits of early birds, of immense quadropedal dinosaurs, and of very large bipedal dinosaurs could be estimated from measurements from endocasts of fossil cochleae. Thus the largest dinosaurs were estimated to have had a best frequency response below 2 kHz. *Archaeopteryx*, an ancient bird, probably had a best hearing frequency of 3 kHz and an upper frequency limit below 7 kHz. Against this background, we can say that all the squeaks, honks, groans, and bellows of television and film animations of these animals are reasonably accurate, although vocalizations were unlikely to have been emitted as frequently as has been portrayed and probably not by all species.

18.16 What Have We Learned?

1. Evolutionary processes acted in parallel on the various lineages of amniotes and produced sensitive, frequency selective auditory papillae in all groups. Over the eons, selective pressures induced convergent and parallel effects, such as the evolution of specialized hair cell populations in concert with the utilization of active processes (Manley &Köppl, 1998; Manley, 2000, 2001).
2. Some structural changes during evolution clearly had important consequences for function. In particular, it has proven possible to understand the tectorial membrane better through the effects of its loss on sensitivity and frequency selectivity (Manley & Köppl, 2008). In addition, the frequency maps of basilar papillae can now be better understood and modeled.
3. The various hearing organs of the different lineages of amniotes show strong resemblances. Each type of papilla does, however, have unique features, and these resemblances and differences can be understood only in the context of comparative studies. The main functional difference in mammal hearing, as

compared to other amniotes, is the extension of the high-frequency range in most mammals, and this was the result of unrelated and fortuitous events in their early evolutionary history (Manley, 2010, 2012, 2013).
4. Comparative studies of hearing have provided, and still provide, a powerful and flexible tool to widen our knowledge and to understand in detail the complex mechanisms underlying the hearing of vertebrate organisms, including humans.

18.17 Perspectives for the Future

The research fields covered above are so diverse that it is difficult to select areas for special attention in the future. Here, I touch briefly on four potentially fruitful fields.

1. Obviously, the interactions between hair cell populations are of huge general interest and here, the birds and the geckos certainly deserve more attention. Very recent data from the chicken from Fettiplace's lab is already providing fascinating insights into bird hearing.
2. The remarkable ability of barn owl auditory afferents to phase lock one octave higher than any other species needs an explanation at the cellular and biochemical levels.
3. The huge fall in the calcium concentration in the endolymph of therian mammals during evolution likely had profound consequences for various biochemical processes. It presumably led to the changed constitution of the tectorial membrane (which is highly sensitive to the ionic medium) and affected the further evolution of prestins and the transduction machinery. But what were these effects and how did they influence hearing in mammals?
4. Both lizards and birds can show obvious effects of anesthesia, up to a total loss of responses in the ear. We showed, for example, that DPOAE amplitudes in barn owls drift over time during anesthesia. It is possible that this sensitivity to anesthetics, which has not been reported in mammals, is related to another effect not obvious in mammals, the effects of temperature on frequency responses. The latter can be quite large in lizards and birds. What biochemical mechanisms underlie these differences from mammals?

References

Bergevin, C. (2011). Comparison of otoacoustic emissions within gecko subfamilies: Morphological implications for auditory function in lizards. JARO, 12, 203–217.

Chiappe, M. E., Kozlov, A. S., & Hudspeth, A. J. (2007). The structural and functional differentiation of hair cells in a lizard's basilar papilla suggests an operational principle of amniote cochleas. *Journal of Neuroscience*, 27, 11978–11985.

Gleich, O., Fischer, F. P., Köppl, C., & Manley, G. A. (2004). Hearing organ evolution and specialization: Archosaurs. In G. A. Manley, A. Popper, and R. R. Fay (Eds.), *Evolution of the vertebrate auditory system* (pp. 224–255). New York: Springer Science+Business Media.

Heffner, H. E. (1999). The symbiotic nature of animal research. *Perspectives in Biology and Medicine*, 43, 128–139.

Köppl, C. (1997a). Frequency tuning and spontaneous activity in the auditory nerve and cochlear nucleus magnocellularis of the barn owl *Tyto alba*. *Journal of Neurophysiology*, 77, 364–377.

Köppl, C. (1997b). Phase locking to high frequencies in the auditory nerve and cochlear nucleus magnocellularis of the barn owl, *Tyto alba*. *Journal of Neuroscience*, 17, 3312–3321.

Köppl, C. (2009). Evolution of sound localization in land vertebrates. *Current Biology*, 19, R635–R639.

Manley, G. A. (1973). A review of some current concepts of the functional evolution of the ear in terrestrial vertebrates. *Evolution*, 26, 608–621.

Manley, G. A. (1990). *Peripheral hearing mechanisms in reptiles and birds*. New York: Springer-Verlag.

Manley, G. A. (2000). Cochlear mechanisms from a phylogenetic viewpoint. *Proceedings of the National Academy of Sciences of the USA*, 97, 11736–11743.

Manley, G. A. (2001). Evidence for an active process and a cochlear amplifier in non-mammals. *Journal of Neurophysiology*, 86, 541–549.

Manley, G. A. (2004). The lizard basilar papilla and its evolution. In G. A. Manley, A. Popper, & R. R. Fay (Eds.), *Evolution of the vertebrate auditory system* (pp. 200–223). New York: Springer Science+Business Media.

Manley, G. A. (2010). An evolutionary perspective on middle ears. *Hearing Research*, 263, 3–8.

Manley, G. A. (2011). Lizard auditory papillae: An evolutionary kaleidoscope. *Hearing Research*, 273, 5–64.

Manley, G. A. (2012). Evolutionary paths to mammalian cochleae. *JARO*, 13, 733–743.

Manley, G.A. (2013). *Mosaic evolution of the mammalian auditory periphery*. Advances in Experimental Medicine and Biology. 01/2013; 787, 3–9.

Manley, G. A., & Köppl, C. (1998). Phylogenetic development of the cochlea and its innervation. *Current Opinion in Neurobiology*, 8, 468–474.

Manley, G. A., & Köppl, C. (2008). What have lizard ears taught us about auditory physiology? *Hearing Research*, 238, 3–11.

Manley, G. A., & van Dijk, P. (2008). Otoacoustic emissions in amphibians, lepidosaurs and archosaurs. In G. A. Manley, R. R. Fay, & A. Popper (Eds.), *Active processes and otoacoustic emissions in hearing* (pp. 211–260). New York: Springer Science+Business Media.

Manley, G. A., & Kraus, J. E. M. (2010). Exceptional high-frequency hearing and matched vocalizations in Australian pygopod geckos. *Journal of Experimental Biology*, 213, 1876–1885.

Manley, G. A., Kirk, D., Köppl, C., & Yates, G. K. (2001). *In-vivo* evidence for a cochlear amplifier in the hair-cell bundle of lizards. *Proceedings of the National Academy of Sciences of the USA*, 98, 2826–2831.

Taschenberger, G., & Manley, G. A. (1997). Spontaneous otoacoustic emissions in the barn owl. *Hearing Research*, 110, 61–76.

Webster, D. B., Fay, R. R., & Popper, A. N. (1992). *The evolutionary biology of hearing*. New York: Springer-Verlag.

Chapter 19
Directional Hearing in Insects and Other Small Animals: The Physics of Pressure-Difference Receiving Ears

Axel Michelsen and Ole Næsbye Larsen

Axel Michelsen and Ole Næsbye Larsen

A. Michelsen (✉) • O.N. Larsen
Department of Biology, University of Southern Denmark, DK-5230 Odense M, Denmark
e-mail: A.Michelsen@biology.sdu.dk; onl@biology.sdu.dk

A.N. Popper and R.R. Fay (eds.), *Perspectives on Auditory Research*, Springer Handbook of Auditory Research 50, DOI 10.1007/978-1-4614-9102-6_19,

19.1 Introduction

Animals may use several strategies for determining the direction to sound sources. Animals that are large relative to the wavelengths of the sounds of interest may base their directional hearing on variations of sound pressure at the ears caused by diffraction of sound by the body. Most mammals and birds can also exploit the differences in the time of arrival of the sounds at the ears. This is not so, however, for the majority of the hearing animals. Most insects live in dense vegetation, which filters away high-frequency sounds, and their bodies are often smaller than the wavelengths of the sounds of interest. In addition, the brains of insects cannot exploit the minute differences in the time of arrival of sound at their ears. For many years it has been suspected that their directional hearing could be based on a directional sensitivity of the ears, caused by sound reception at both the external and internal surfaces of the eardrums. We now know that this is actually so, and that pressure-difference reception also is common in several groups of vertebrates.

This chapter outlines the methods used in this branch of auditory research: Very accurate measurements of ears in animals situated in very homogeneous sounds fields and interpretation of the data by means of mathematical models of the physics of sound transmission from ear to ear. We have chosen a few examples from our own research (bushcrickets, grasshoppers, birds, and crickets) for illustrating the methods and some of the variations in the animals' strategies.

This research has been made possible by the contributions during 200 years from scientists and engineers in four different areas: anatomy, acoustics, instrumentation, and computing. During the 19th century, a few scientists such as Müller (1826) described the detailed anatomy of most of the insect ears that are known today. Some of these pioneers had an admirable capacity for hard work and published up to one printed page per day, year after year.

In the last part of the 19th century the field of acoustics was transformed into an exact science by Lord Rayleigh and others. However, the experimental biologists had to wait until the middle of the 20th century before they could exploit the methods of electrophysiology for measuring the output of ears. Simultaneously, a few scientists had speculated about the physics of hearing in insects. In 1940 two very different ideas were published by R. J. Pumphrey (1940) and by H. Autrum (1940), who wrote in English and German, respectively. The war and the following years witnessed a shift of the balance between English and German as the dominant language of science. It was therefore many years before the views of Autrum (that the directionality of insect ears could be accounted for if the ears worked like the pressure gradient receivers studied by Harry Olson in the 1930s) were accepted. Pumphrey's view (that the delicate insect ears follow the movements of the air particles) is now known to be correct only for the sensory hairs on the bodies of insects.

A major reason for the progress made in biological acoustics during the careers of the two of us is the technological development. In the beginning of the 1960s, nerve impulses were displayed on an oscilloscope screen and filmed. The film was developed, and the spikes were counted by the investigator. One of us (A. M.)

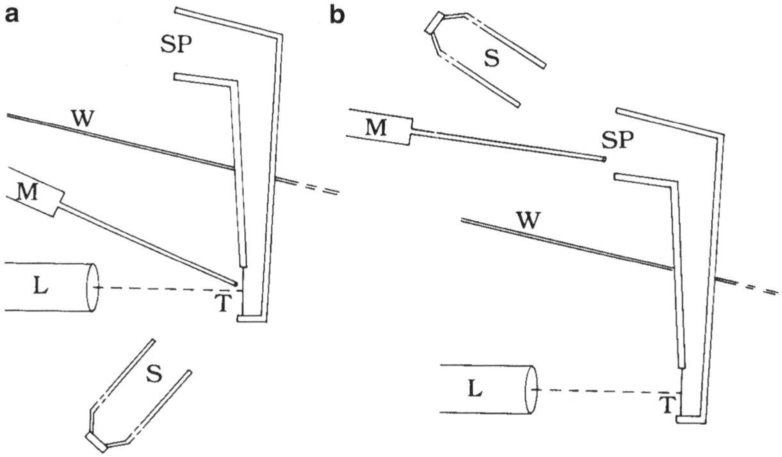

Fig. 19.1 Method for measuring the gain of a transmission path to the inner surface of an eardrum (tympanum T, here in a bushcricket). (**a**) The eardrum is calibrated with sound acting on its outer surface. (**b**) The calibrated eardrum is used for measuring the sound acting on the inner surface of the eardrum. Further explanations in the text. (From Michelsen et al., 1994a)

actually counted more than one million spikes when earning his doctorate! During the following decades, computers took over most of the trivial work and also made complex calculations like Fourier transformations a practical tool. Work on hearing and sound emission in small animals was eased by the invention of precision microphones with excellent long-term stability and diameters from 3 mm and up. An even higher spatial resolution of sound fields could now be obtained by means of probe microphones with probe diameters of only 1 mm and a long distance (e.g., 20 cm) from the tip to the bulky microphone.

Several very productive research groups took advantage of the new possibilities. In 1989, three leaders in the study of cricket biology published a book, in which the major themes of the behavior and neurobiology were covered by specialists. We contributed a chapter about sound reception (Larsen et al., 1989), in which the part about directional hearing was more descriptive than analytical. It was based on several attempts from 1978 to 1984 to understand the reasons for the directivity of the ears of crickets and bushcrickets. Such attempts showed, for instance, that the eardrum does indeed respond to the vectorial difference between external and internal sound pressures and that eardrum motion is a necessary part of the sensory transduction process and that it correlates strongly with ear directionality (Kleindienst et al., 1983; Larsen et al., 1984). However, we gave up further analysis when we realized that in order to arrive at a *real* understanding of the system, it was necessary for us to measure the sound pressure acting on the inner surface of the eardrum but no microphone could possibly do the job.

Approximately 10 years later during a shower early in the morning it suddenly occurred to A. M. what we should do. Figure 19.1 shows how one can measure the gain of a horn-shaped trachea connecting the surface of the thorax with the inner surface of an eardrum in a foreleg of a bushcricket (*Poecilimon laevissimus*, Tettigoniidae). The gain of the horn-shaped trachea is the change of the amplitude and phase angle of the sound from the spiracular entrance of the tracheal system (SP) to the inner surface of the eardrum. The laser vibrometer (L) measures the vibrations of a small hollow, silver coated glass sphere (weight 0.5 ng, diameter 10 μm) placed on the eardrum (tympanum, T), which is set in motion by sound from a loudspeaker (S), while a probe microphone (M) records the local sound pressure at the outer surface of the eardrum (Fig. 19.1a). The next step is to measure the vibration spectrum when the eardrum is driven by sound reaching mainly its inner surface as a wall of beeswax (W) attenuates the sound between the two inputs (Fig. 19.1b). The ratio between the two recorded vibrations is then a measure of the gain of the horn-shaped spiracle and trachea guiding sound to the inner surface of the eardrum.

Why do the bushcrickets need a horn-shaped trachea to guide sound to their ears? Most ears seem to have evolved from existing sense organs, which were not always located at the most favorable position from an acoustical point of view. The ears of bushcrickets and crickets are located at the middle of their long and slender forelegs. This is not an ideal position for directional hearing, as the ears are located too far away to exploit the diffraction of sound by the main body. The "hearing trumpets" open at the lateral sides of the bushcricket body and guide the sounds to the inner surface of the eardrums in the legs. In many bushcrickets the hearing trumpets are horn-shaped and have a gain of 5–10 times. The sound acting on the outer surface of the eardrum therefore is much less intense than that acting on the inner surface and not a significant input to the ear. With respect to directional hearing, the ears are therefore in the same situation as if they had been located at the lateral surfaces of the body. One could say that the bushcrickets have built-in hearing aids!

19.2 Pressure-Difference Receivers

Most small animals cannot process the tiny differences in time, which are one of our cues for determining the direction to sound emitters. However, the ears may become sensitive to direction if the sound waves can reach both surfaces of the eardrums. Such sound transmitting pathways inside the heads or bodies are known in several insect groups and in frogs, reptiles, birds, and even some mammals, potentially coupling the two ears acting as reciprocal pressure-difference receivers (Köppl, 2009). In most cases, some experimental evidence supports the notion that the ears function as pressure-difference receivers and provide information about the direction of sound waves. Most studies in vertebrates have been performed with recordings of neural responses. A physical analysis has so far been carried out in some lizards (Christensen-Dalsgaard & Manley, 2008), bushcrickets, grasshoppers, crickets, budgerigars, and barn owls.

The existence of an anatomical air space leading to the inner surface of an eardrum from openings on the body surface or from the contralateral eardrum and middle ear is a necessary prerequisite, but it does not automatically create a pressure-difference receiver with a useful directionality. The sound has to arrive at the inner surface of the eardrum with a proper amplitude and phase relative to that acting on the outer surface. In addition, the sound propagating through a sound guide inside the animal (an interaural canal) should be affected in a suitable manner by the direction from which the sound reaches the outer surface of the animal. The mere presence of an air-filled interaural pathway does not necessary mean that the ear functions as a directional pressure-difference receiver in the relevant frequency range. This is seen, for instance in the barn owl, which possesses a substantial interaural canal and can locate sound emitting prey in total darkness with remarkable precision, yet the ears are functionally uncoupled in the frequency range relevant for prey localization (Moiseff & Konishi, 1981). One complicating factor in the analysis is that the sound arriving at the inner surface may have entered the body through several auditory inputs (e.g., through the other ear and through two spiracles in crickets; through the other ear and through the lungs in frogs).

The air-filled cavities leading to the inner surface of the eardrum are often a part of (or connected to) the respiratory pathways. This may have undesirable consequences since the large pressure fluctuations during respiration may affect the mechanics of the eardrums. In grasshoppers the eardrums (tympana) may be displaced outside their linear range (so that Hooke's law is no longer obeyed). This may affect the threshold for hearing and distort the frequency analysis (Michelsen et al., 1990). Large displacements coupled with the respiration can also be observed in frogs (in which the middle ear cavity and the mouth are connected through a wide Eustachian tube). Obviously, a reduction of such effects (at the expense of the directionality) may have been an important factor in the evolution of pressure-difference receivers. The air-filled spongy bone connecting the middle ears in birds and moles, and the middle-ear systems that are open to the buccal cavity in reptiles may be examples of this.

In the following, we present a few examples of pressure-difference receiving ears that we have analyzed. We start with two examples of ears with only two acoustical inputs (grasshoppers and budgerigars), and then consider the more complicated situation in crickets, where each ear receives sounds from four acoustical inputs.

19.2.1 Grasshoppers

Grasshoppers (Acrididae) have an ear at each side of the first abdominal segment. A sclerotized ring encircles an eardrum, to which 60–80 receptor cells attach in four groups, each having a different frequency preference. Between the ears are air-filled tracheal sacs, through which sound can propagate from one ear to the other. The physics of the pressure-difference receiver has been examined in locusts (*Schistocerca gregaria*) and in a three to four times smaller grasshopper (*Chorthippus biguttulus*) (Michelsen & Rohrseitz, 1995).

The locusts are 5–6 cm long and about 1 cm wide at the position of the ears. Above 8 kHz most of their directional hearing can be based on diffraction of sound

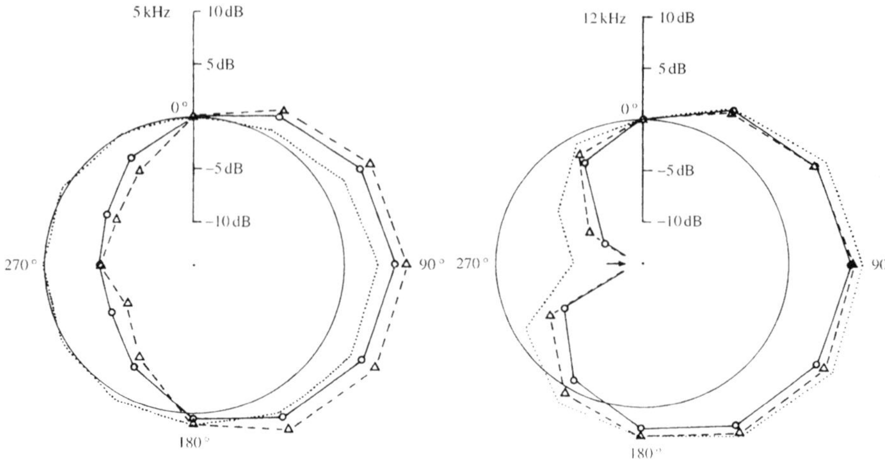

Fig. 19.2 Directional patterns at 5 kHz and 12 kHz for the right ear of the locust. Dotted lines: amplitude of sound pressure at the external surface of the eardrum. Solid lines with circles and dashed lines with triangles: Observed and calculated vibration velocity of the eardrum, respectively. (From Michelsen & Rohrseitz, 1995)

by the body, as the sound arriving from the other ear and acting on the inside of the eardrum is only 20–30% of the sound pressure acting at the outside of the eardrum. In contrast, below 8 kHz the amplitude of the sound from the other ear is about 50% of the sound acting on the outer surface of the eardrum. The duration of the sound propagation from one eardrum to the other can be estimated from the change of phase. The results suggest that the propagation velocity through the air sacs is less than in free space. A similar trend has also been found in other insects.

The directional diagrams for the locust at 5 and 12 kHz are shown in Fig. 19.2. The agreement between the calculated directional dependence and the actual values (measured with laser vibrometry) suggest that a two-input model is a valid description of the acoustics of the single ear, both at 5 kHz and at 12 kHz. There is therefore no reason to believe that sounds arriving at the internal surface of the eardrum through other routes should play a significant role.

From Fig. 19.2 it is obvious that the sound transmission from ear to ear is essential at 5 kHz, whereas at 12 kHz it only slightly improves the left–right gradient in the forward direction (when the animal turns so that the direction to the sound source changes from 330° to 30° or vice versa). The small grasshopper is much more dependent on the transmission of sound from ear to ear, and from 3 to 18 kHz the amplitude of the sound transmitted to the inner surface of the eardrum is 60–80% of the amplitude acting on the outer surface. In theory, the locust should be able to move directly toward targets singing either at low or high frequencies. In contrast, the small grasshopper has a clear gradient in eardrum vibration only at high frequencies. In ideal sound fields the directional hearing of small grasshoppers should thus improve with frequency. This prediction may not be true in natural habitats, where the presence of soil and vegetation may cause a substantial degradation of the directional cues.

Very careful studies of the strategies for grasshopper phonotaxis performed by the late Dagmar von Helversen (review 1997) demonstrated that the small grasshopper makes use of a very specific strategy for locating a conspecific sound source. Males searching for females sing at regular intervals, and the females respond, but it is up to the male to take the risk of approaching. When it hears a female, the male turns abruptly toward the side from which her signal arrives, moves forward, and sings again. The turning angle is usually larger than that needed to bring the female into the male's frontal auditory field, and the behavior thus results in a zigzag course. The female sings for about 1 s, and the male may therefore obtain closed-loop directional information during the turn. If the turn brings the female into his frontal auditory field, he is likely to jump forward. It is interesting that he will also jump forward, if the sound comes from behind, so he is probably not able to distinguish sounds from the front from sounds from the back (as suggested by the front–back symmetrical directional patterns (Fig. 19.2)).

19.2.2 Birds

The middle ears of birds are connected through an air-filled interaural canal located below the brain. So, it is reasonable to hypothesize that birds also take advantage of the pressure-difference receiver principle for directional hearing. Despite the superficial similarity to the situation in grasshoppers, however, it has been much more difficult to test this hypothesis in birds. For more than 50 years the role of the interaural canal in birds has therefore remained an open question. The main reason is that during ketamine–xylazine anesthesia (and perhaps also other types of anesthesia) birds do not regularly open their Eustachian tubes to equalize the intracranial air pressure with the ambient pressure. The resulting decrease in intracranial air pressure displaces the eardrums inward, increasing their tension. We can experience a similar situation when on board a passenger airplane descending to land. If we do not equalize the pressure in our middle ears, the low-frequency noise from the jet engines seems to disappear or at least reduce substantially but once we open our Eustachian tubes the enervating noise immediately returns!

This insight did not come easy to us. For a long time we just noticed that laser vibrometry recordings of vibrations in eardrums of birds were "unstable" and that recordings had to be performed very quickly to "keep stability" as the eardrum seemed to move (Klump & Larsen, 1992). It was only when one of us (O. N. L.) one day was very clumsy (probably too much coffee) when routinely trying to place a glass microsphere on the eardrum. The needle with the microsphere slipped and ripped a small hole in the eardrum, which responded by immediately moving much further out into the ear canal. Then he finally understood the obvious causation.

The increased tension in the eardrum during anesthesia substantially reduces the eardrum vibrations at frequencies below 3–4 kHz and hence the interaural coupling leading to a significant decrease in ear directionality in this frequency range (Larsen et al., 1996). We avoided this problem by ventilating the middle ears by means of a

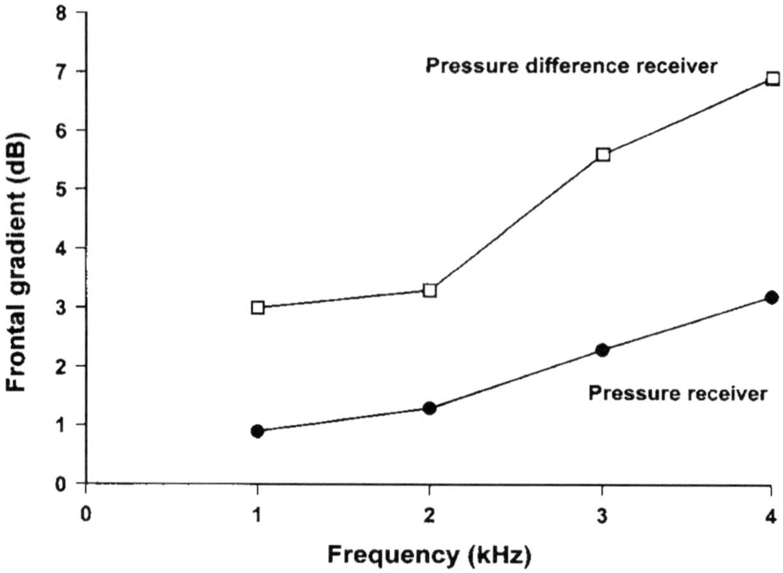

Fig. 19.3 Calculated effect of sound transmission through the interaural canal on the difference in the vibration velocities of the two eardrums of a budgerigar. The direction of sound incidence differs by 30° from the forward direction. In the lower curve (filled symbols), the eardrums are activated only by sound at their outer surfaces. In the upper curve (open symbols), the eardrums also receive sound at their inner surfaces through the interaural canal. (From Larsen et al., 2006)

thin injection needle when we studied the directional hearing of budgerigars (Larsen et al., 2006). We found that sound transmission through the interaural pathway considerably improves the directional hearing in the horizontal (azimuth) plane of the bird for two reasons: The frontal gradients of eardrum vibration become larger (Fig. 19.3) and the vibrations of the eardrums differ more in time. The latter effect is not relevant in grasshoppers, because the brains of insects cannot exploit so small time cues.

The methods were similar to those used with the grasshoppers, except that the anaesthetized bird had to be supported in such a manner that we kept a free sound field around the body of the bird. This was achieved with steel rods with a diameter, which was smaller than one tenths of the wavelength of the highest frequency investigated (4 kHz).

Although the bodies of most birds are larger than those of insects, bird skulls are often small relative to the body. One would therefore expect that the diffraction of sound by the head would be best suited for directional hearing at high frequencies, which unfortunately are easily absorbed by the vegetation. In the budgerigar, the skull has a diameter of about 16 mm at the position of the ears. At a point facing the sound source on the surface of a hard sphere with a diameter of 16 mm one expects a surplus pressure of 1.3 dB at 4 kHz. However, we found a surplus pressure of 3.3 dB, which is approximately the value expected for a sphere with a diameter of 28 mm. The reason for this difference is probably the presence of soft feathers on the head, but the close proximity of the body may also play a role. This phenomenon deserves further study.

The physical theories for calculating the difference in time of arrival at opposite positions on a sphere reflect the complicated kinds of waves that a thought to exist at the surface. Contrary to intuitive expectation, the speed of sound close to the skull may be much lower than the ambient speed. In other words, the time difference depends on the frequency of sound (Kuhn, 1977). For a sound source facing one of two opposite positions on a sphere, the expected difference in the time of arrival of sound is $3a/c$ at low frequencies, but $2a/c$ at high frequencies (where a is the radius of the sphere, and c the ambient speed of sound). "Low" means that $2\pi a/\lambda << 1$ (where λ is the wavelength of sound), and "high" that $2\pi a/\lambda >> 1$.

The measured values for the difference in the time of arrival (when one ear was facing the sound source) were 118 and 91 μs at 1 and 4 kHz, respectively. For a sphere of the size of the skull ($a = 8$ mm) the expected differences are 70 μs at low frequencies and 47 μs at high frequencies, that is, much lower values than those observed. For $a = 14$ mm (the size of the head determined from the surplus pressure), the expected time differences are 122 μs and 81 μs at low and high frequencies, respectively. The measured value at 1 kHz is thus close to the expected value at low frequencies, and the observed value at 4 kHz is a transition value toward the 81 μs expected at higher frequencies.

The presence of the air-filled interaural canal not only allows the eardrums to operate as coupled pressure-difference receivers, but also creates substantial interaural delays at low frequencies. These delays may be much larger than the delays caused by the path lengths around the head, and they increase with the amplitude of the sound transmitted through the interaural canal. However, the price for a large interaural transmission is a decrease of the sensitivity to sound in the forward direction. The actual amplitude of the sound in the interaural canal thus seems to be a compromise between sensitivity, forward gradients, and interaural time cues (Michelsen & Larsen, 2008).

The conclusions drawn from these biophysical studies have recently been supported and extended by behavioral experiments on budgerigars equipped with headphones, through which interaural time differences (ITD) and interaural level differences (ILD) were independently manipulated (Welch & Dent, 2011). These experiments confirmed that budgerigars can lateralize sounds behaviorally based on ITD and ILD cues with thresholds of 18–47 μs (0.5–4 kHz) and 2.3–3.4 dB, respectively, that is, well below the maximum interaural differences predicted from the biophysical analysis of the coupled pressure-difference receiving ears.

19.2.3 The Tuned Cricket

In contrast to many bushcrickets, the crickets (Gryllidae) generally communicate at sonic frequencies, and their calling song is often a pure tone. Like the bushcrickets, the crickets have their ears located in the thin front leg tibiae, where diffraction does

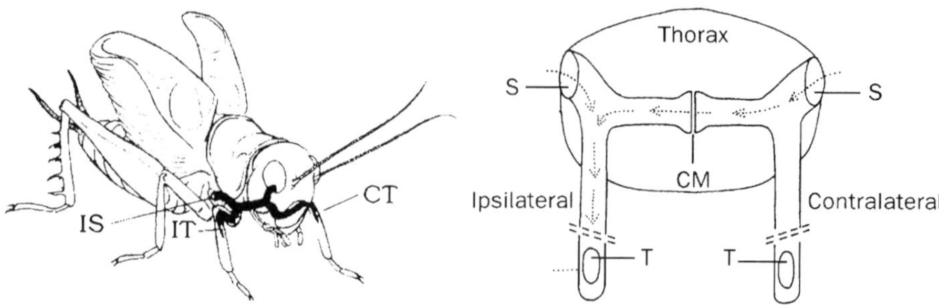

Fig. 19.4 (Left) The two ears in a cricket share four acoustic inputs: two eardrums (T, tympanum) and two spiracles (S). Each ear receives sound at the external surface of its eardrum (IT, ipsilateral tympanum), but also at the internal surface of its eardrum from the eardrum of the other ear (CT, contralateral tympanum), from the ipsilateral spiracle (IS), and from the contralateral spiracle (CS). Right: Sounds from the contralateral inputs pass the central membranes (CM), which act as a mechanical phase shifter. (From Michelsen et al., 1994b, and Michelsen & Löhe, 1995)

not provide useful directional cues. However, the crickets have solved the problem in a very different way. A pressure-difference receiver mechanism provides the ear with an excellent directionality within a narrow frequency band around the calling song.

A horn-shaped tracheal tube known as the acoustic trachea connects the inner surface of the eardrum with an ipsilateral acoustic spiracle (IS) on the thorax (Fig. 19.4). The acoustic trachea is also linked through a connecting trachea with the acoustic trachea on the other side of the body (Fig. 19.4, right). Sounds can therefore propagate from an ear and/or acoustic spiracle across the midline to the inner surface of the contralateral eardrum. The connecting trachea from the one side ends in a central membrane, which is a close neighbor to the central membrane from the other side.

For many years, very different opinions were held of the importance of the contralateral inputs to creating the directional characteristics of the ear (review: Weber & Thorson, 1989). It had been found that at 5 kHz, the frequency of the song, body screening effects could only account for a few dB of the directionality, but greater directionalities had been measured in receptor axons and thoracic neurons. Some investigators thought that the ear was mainly responding to sound from the contralateral ear, while others favored the sound from the contralateral spiracle. Finally, the observation that disrupting the central tracheal connection does not hinder sound localization in very homogeneous sound fields was regarded as evidence to "toll the death of cross-body-theories." Obviously, in order to settle these disputes, it was necessary to determine the transmission gains of the three internal sound pathways. The transmission gain is the change of the amplitude and phase angle of the sound from the entrance of the tracheal system to the inner surface of the eardrum.

Such experiments were performed a few years later in the field cricket *Gryllus bimaculatus* (Michelsen et al., 1994b). The transmission gain from the ipsilateral spiracle to the inner surface of an ear was fairly simple and close to that expected for a delay line. In contrast, the transmission gains from the contralateral ear or spiracle, through the midline and to the ear were far from simple, both with respect

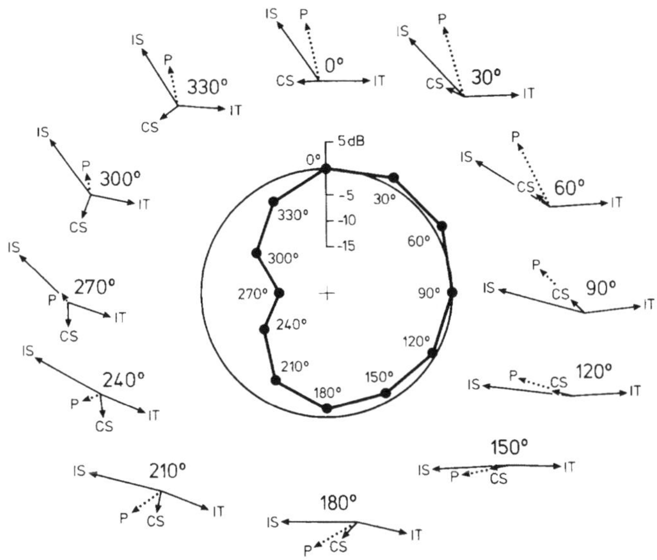

Fig. 19.5 Calculation of the directional hearing at 4.5 kHz in the right ear of the cricket, *Gryllus bimaculatus*. Three vectors (sounds from three sound inputs) add at each direction of sound incidence to produce the vector P, which is the net force acting on the eardrum. CS: sound from the contralateral spiracle. IS: sound from the ipsilateral spiracle. IT: sound acting directly on the eardrum. (From Michelsen et al., 1994b)

to amplitude and to phase. Apparently, the central membrane connection between the two acoustic tracheae behaves like an eight-pole filter. This finding was much more complicated than the ideas discussed during the previous decade, so the heated debate had no winners.

In addition to sound transmission, we measured the frequency spectra and time of arrival of sounds at the outer surface of the eardrum and from the 3 entrances at 12 directions of sound incidence. By combining these data we calculated how the total driving force at the eardrum depends on the direction of sound (Fig. 19.5). The results are in excellent agreement with the dependence on sound direction of the eardrum vibrations.

When measuring the amplitude and phase angle of the sounds from each of the four auditory inputs at various angles of sound incidence, our reference values were the amplitude and phase at the outer surface of the right eardrum when sound arrived from the frontal direction. In this manner we obtained values, which were not true values for the effects of diffraction, but those needed in the calculations of directionality. As we shall see, this is especially important for the phase angles of the sounds from the contralateral inputs.

From these data one can make some predictions about the mechanism of directional hearing. It is obvious that the amplitude of the sound pressure at the outer surface of the eardrum changes only little with the direction of sound incidence. In the frontal directions (around 0°), which are of prime interest with respect to how a cricket localizes a sound source, a change in sound direction would cause the forces driving the two eardrums to differ by up to 1.3 dB. A pressure-difference receiver is obviously needed for providing more directionality, but from which input(s) should the sound at the inner surface of the eardrum originate? By testing all combinations of inputs, we found that more than two inputs were needed in order to account for the observed directionality, and that the two contralateral inputs are better potential contributors of directional cues than the ipsilateral spiracle. Both the amplitude and the phase of the contralateral sounds change in opposite directions to the values for the ipsilateral sounds when the sound source moves from one frontal direction to another. The change of phase is especially prominent and thus the most likely contributor to the directionality of the ear.

The transmission gains from the contralateral eardrum and the two thoracic spiracles were measured in the following manner. A small local sound source was used for delivering sound at one of the auditory inputs, while walls of beeswax between the auditory inputs ensured that the sound levels at the other inputs were at least 20 dB down (cf. Fig. 19.1). We first determined the transfer function of the eardrum by applying sound at its outer surface. The transfer function is the drum velocity divided by the sound pressure; it has an amplitude and a phase part. We then determined the transfer function for each route to the inner surface of the eardrum, but this time the sound pressure was measured at the input in question. These transfer functions consist of the gain of the transmission path times the transfer function of the eardrum. The gains of the transmission paths could then be obtained by dividing these transfer functions with the transfer function of the eardrum.

For the transmission of sound from the ipsilateral spiracle (IS), the amplitude gain is close to 1 at low frequencies and increases to a maximum around 6–8 kHz and again at 17–19 kHz. At low frequencies, the phase at the inner surface of the eardrum is close to that at the outer surface. With increasing frequency the sound at the inner surface becomes progressively delayed, as one would expect in a transmission line where the propagation of sound takes a certain time. The phase changes approximately 360° between 1 and 22 kHz. At a temperature of 21 °C (and a sound propagation velocity in free space of 344 m/s), the length of the tracheal tube was calculated to be 15.6 mm, which is significantly larger than the anatomical length (about 12 mm).

This means that the sound propagates with a lower velocity inside the tube than in the air outside the animal. The propagation velocity estimated from our data is 264 m/s. This value is in excellent agreement with that determined by Larsen (1981), who found an average value of 263 m/s by measuring the delays of very short impulse sounds. He pointed out that this value is close to that expected for isothermal wave propagation in air (245 m/s), and he suggested that an exchange of heat may occur at the tracheal walls.

The transmission of sound from the contralateral spiracle (CS) differs very much from this simple pattern. The amplitude is at a maximum at 5–7 kHz and again around 18 kHz (much like the sound from IS), but it is almost zero below 3.5 kHz. In the frequency range 4–5 kHz (around the frequency of the calling song, which is at 4.6–4.7 kHz in *G. bimaculatus*), the amplitude of the sound from CS varies drastically with frequency. Between 4.0 and 4.6 kHz the amplitude increases by a factor of 4. The average increase between 4.6 kHz and 5.0 kHz is only 10%. The strong frequency dependence of the amplitude is accompanied by a large change of phase. From 4 to 10 kHz the phase angle of the sound from CS changes by approximately 560° (for comparison: the phase of the sound from the ipsilateral spiracle (IS) changes by only 100° from 4 to 10 kHz). From 10 to 20 kHz the changes of phase in the sounds from IS and CS have approximately the same magnitude. The transmission from the contralateral tympanum (CT) follows the pattern observed in the transmission from CS. However, the amplitude of the sound arriving at the ipsilateral eardrum is considerably smaller.

The measured diffraction, time delays, and transmission gains of the four sounds acting on the tympanum were now combined in an attempt to account for the dependence of the tympanal vibrations on the direction of sound incidence. We decided to use averaged data for the diffraction and time delays (these data show only moderate scatter). For the transmission gains we have chosen "typical" values for the amplitudes and varied the phase values within the ranges observed in the experiments. In this way we have obtained an impression of how robust the directional patterns are.

For frequencies up to 5 kHz, the amplitude gain of the transmission of sound from the contralateral tympanum (CT) is so small (below 0.1) that it does not have much effect on the directional pattern. In the following example at 4.5 kHz we ignore the sound from CT. The problem is then reduced to considering three vectors: the sound at the outer surface of the tympanum (IT) and the sounds at the inner surface arriving from the ipsilateral and contralateral spiracles, respectively (IS and CS).

The amplitude and phase of IT for sounds arriving from the frontal direction are, as a matter of definition, 1 and 0°. At 4.5 kHz, the amplitude gain and the change of phase of the sounds transmitted from IS and CS are 1.5 and 154° and 0.44 and 208°, respectively. The first step in the calculation is to multiply the gain and the diffraction/time-of-arrival factor for each of the sound components and at each angle of sound incidence. In Fig. 19.5, the results are represented as three vectors, which are then added to produce a sound pressure (P), which is proportional to the force that causes the eardrum to vibrate. It should be noted that the phase angles for the transmission gains include a 180° phase shift, which means that in adding the three vectors, sound components acting on the inner surface (IS and CS) are subtracted from the sound component acting at the outer surface (IT).

The solid curve in Fig. 19.5 shows the pressure driving the eardrum. The polar plot is surrounded by the 12 vector diagrams. The calculated directional pattern has the most important of the features seen in the measured patterns: The driving force is at a maximum at the ipsilateral directions 30° and 60°; the force decreases by approximately 6 dB from 30°, through 0° to 330°, and the force is at a minimum at 270° (the "contralateral null").

When examining Fig. 19.5 one may start by looking at the vector diagram for the 270° direction. Obviously, the null is caused by the fact that the sum of CS and IT has approximately the same amplitude, but the opposite direction of IS. A deeper minimum (more perfect null) would require only a slight reduction in the amplitude of the sound from IS. In the vector diagrams for 300°, 330°, 0°, and 30°, the amplitudes of the three vectors change only little, and only little variation is seen in the phase angles for IS and IT. The only major change is in the phase angle for CS. At 30° and 60°, CS has almost the same phase as IS, and the amplitude of P is now at a maximum. The slope of the driving force in the forward direction is therefore caused almost entirely by changes in the phase angle of the sound from CS.

Several directional diagrams have been calculated by selecting other values for IS and CS within the ranges observed during the measurements. The most conspicuous difference between the diagrams is in the magnitude and direction of the contralateral null, which is not always in the 270° direction. A closer examination of the data revealed that a change of sound frequency from 4.5 kHz toward 4 kHz causes the null to move backwards toward 240° or even 210°, whereas an increase of frequency from 4.5 toward 5 kHz causes the null to move forward toward 300°. This trend was, in fact, observed in directional diagrams of the tympanal vibrations measured with laser vibrometry during the collection of the data on diffraction and time-of-arrival.

In most of these crickets, the vibration velocity of the eardrum is a maximum at 4.6 kHz where the difference between the eardrum velocities for sound arriving from the 30° and 330° directions is about 10 dB (Michelsen & Löhe, 1995). The difference is lower than 5 dB below 4.3 kHz and above 4.8 kHz. The tuning to 4.6 kHz is destroyed if the central membranes are perforated by pushing a human hair through the spiracle. The difference between the eardrum velocities for sound arriving from the 30° and 330° directions is then only 1–2 dB.

In summary, despite the large variations in the shapes of the measured and calculated directional patterns, the biologically important forward slope was very consistent. Furthermore, the same tendency for the position of the minimum to move with frequency was observed in both measured and calculated patterns. We conclude that the measured data on transmission, diffraction, and time delays can account for the most prominent features of the directivity. It is obvious from Fig. 19.5 that the slope of sensitivity in the forward directions depends upon a change in the relative phase angles of the three vectors. The sound from the contralateral spiracle (CS) plays a prominent role in creating this directionality. The sound from the ipsilateral spiracle (IS) is necessary, however, for producing the cardioid pattern shown in Fig. 19.5.

Several investigators have studied the accuracy of the phonotactic steering in crickets. Some investigators have studied crickets walking on a closed-loop trackball system that compensated the animals' walking movements and found that the crickets meandered by 30°– 60° around the frontal midline (e.g., Weber & Thorson, 1989). Others have studied directional orientation in a Y-maze and observed similar uncertainties. However, a recent study of *G. bimaculatus* females walking on an open-loop trackball system found that for angles of sound incidence between 1° and 6° the animals precisely walked towards the sound source (Schöneich & Hedwig, 2010). These results reveal hyperacute directional hearing and place the cricket at the same level of directional hearing as the fly *Ormia ochracea* (Mason et al., 2001), barn owls, and some mammals (humans, bats, elephants) and it is surpassed only by dolphins (Renaud & Popper, 1975).

19.3 Perspectives for Future Research

This chapter is focused on three examples of the physical mechanisms in pressure-difference receiving ears. Similar mechanisms probably occur in several other groups of insects and vertebrates, but the physical mechanisms have not been studied in detail. In some groups it is known that sound is entering the body, but that the sound inputs are too numerous and/or diffuse to allow a quantitative study. For example, large parts of the surface of cicadas are set in motion by sound, and cicadas may therefore behave as almost omnidirectional sound emitters and sound receivers. Although sound emitting tymbals and sound receiving eardrums are known anatomically, cicadas may radiate sound through their eardrums and receive sound partly through their tymbals. A similar confusing situation also seems to exist in some frogs.

In other animal groups, mainly lizards, there is evidence for pressure-difference reception (review: Christensen-Dalsgaard, 2011), but more comprehensive physical studies are needed in order to understand the exact mode of operation. The vital importance of the phase shifting filter for the directional hearing in crickets is one example of a discovery that was made possible by the quantitative physical approach.

From the findings described here one may guess about suitable strategies for behaviorally locating sound emitters in the three animals studied. The predictions were close to the actual strategies observed in grasshoppers and crickets, but not in small birds. We calculated the difference in eardrum vibration amplitude at 330° and 30° direction of sound incidence in birds with normal interaural sound transmission and facing sound sources. We guessed that exploiting this gradient in the forward direction would be a good strategy for a small bird to localize a sound source. Nevertheless, field experiments with small passerine birds (Nelson & Suthers, 2004) showed that prior to flying toward a sound source the birds (eastern towhees, *Pipilo erythroophthalmus*) turned their heads so that the angles between the beaks and the sound sources were 30° to 50°. This observation deserves further study in other species both in the field and in the laboratory. This example demonstrates the importance of confronting laboratory observations with field studies. In addition, comparative studies in a larger number of species and environments will give a more comprehensive picture than using just one species and one habitat as is usually seen.

We suggest that a major future theme is the study of pressure-difference receivers operating in natural habitats. In a study of sound localization of grasshoppers (Michelsen & Rohrseitz, 1997) we found that amplitude cues degrade much faster with distance than phase (time) cues. Animals exploiting phase cues may therefore maintain a reasonable directional hearing when the amplitude cues no longer make sense. The pressure-difference receiver type of ears responds to phase-differences, and these ears may be particularly suited to overcoming the degradation of directional cues. This suggests that the possession of such ears may be an adaptation not only to small body size relative to wavelength but also to the acoustic properties of the complicated natural habitat.

A number of authors still question the role of the avian interaural canal and find the pressure-difference receiver hypothesis controversial. We find it highly likely that many other small birds make use of this mechanism. The pressure-difference receiver properties of avian ears should therefore be studied in more species and with careful biophysical methods as those described here to arrive at a more comprehensive understanding of directional hearing in birds. We especially encourage careful measurements of the transmission through the interaural canal before designing experiments on awake and behaving birds. In addition, we urge experimenters to take into consideration the potential effects of anesthesia when designing physiological experiments on pressure-difference receivers.

Although we are very satisfied with the results of the investigations of the mechanisms for the directional hearing in crickets, there is a possible flaw. It is very difficult to see whether a tracheal spiracle is open or closed, because the opening is covered by a lid. During the preparations for the experiments the lid was fastened with beeswax, either in an open or a closed position. The actual state was then controlled by observing whether a local sound source had an effect on an eardrum. In theory, it is possible that the animals may control the degree of open/closed and thus be able to vary the properties of the directional receiver system. We will continue to think of a possible solution to this problem.

Acknowledgments Original research for this chapter was supported by grants from the Danish Natural Science Research Council and from the Danish National Research Foundation. We thank Jakob Christensen-Dalsgaard for comments on the manuscript.

References

Autrum, H. (1940). Über Lautäußerungen und Schallwahrnehmung bei Arthropoden. II. Das Richtungshören von *Locusta* und Versuche einer Hörtheorie für Tympanalorgane vom Locustidentyp. *Zeitschrift für vergleichende Physiologie*, 28, 326–352.

Christensen-Dalsgaard, J. (2011). Vertebrate pressure-gradient receivers. *Hearing Research*, 273, 37–45.

Christensen-Dalsgaard, J. & Manley, G. A. (2008). Acoustic coupling of lizard eardrums. *JARO*, 9, 407–416.

Helversen, D. von (1997). Acoustic communication and orientation in grasshoppers. In M. Lehrer (Ed.), *Orientation and communication in arthropods* (pp. 301–341). Basel: Birkhäuser Verlag.

Kleindienst, H.-U., Wohlers, D. W., & Larsen, O. N. (1983). Tympanal membrane motion is necessary for hearing in crickets. *Journal of Comparative Physiology A*, 15, 397–400.

Klump, G. M., & Larsen, O. N. (1992) Azimuthal sound localization in the European starling (*Sturnus vulgaris*): I. Physical binaural cues. *Journal of Comparative Physiology A*, 170, 243–251.

Köppl, C. (2009). Evolution of sound localization in land vertebrates. *Current Biology*, 19(15), R635–R639.

Kuhn, G. F. (1977). Model for the interaural time differences in the azimuthal plane. *Journal of the Acoustical Society of America*, 62, 157–167.

Larsen, O. N. (1981). Mechanical time resolution in some insect ears. II. Impulse sound transmission in acoustic tracheal tubes. *Journal of Comparative Physiology*, 143, 297–304.

Larsen, O. N., Surlykke, A., & Michelsen, A. (1984). Directionality of the cricket ear: A property of the tympanal membrane. *Naturwissenschaften*, 71, 538–540.
Larsen, O. N., Kleindienst, H.-U., & Michelsen, A. (1989). Biophysical aspects of sound reception. In F. Huber, T. E. Moore & W. Loher (Eds.), *Cricket behavior and neurobiology* (pp. 364–390). Ithaca, NY: Cornell University Press.
Larsen, O. N., Dooling, R. J., & Ryals, B. M. (1996). Roles of intracranial air pressure in bird audition. In E. R. Lewis, G. R. Long, R. F. Lyon, P. M. Narins, C. R. Steele, & E. Hecht-Poinar (Eds.), *Diversity in auditory mechanics* (pp. 11–17). Singapore: World Scientific.
Larsen, O. N., Dooling, R. J., & Michelsen, A. (2006). The role of pressure difference reception in the directional hearing of budgerigars (*Melopsittacus undulatus*). *Journal of Comparative Physiology A*, 192, 1063–1072.
Mason, A. C., Oshinsky, M. L., & Hoy, R. R. (2001). Hyperacute directional hearing in a microscale auditory system. *Nature*, 410, 686–690.
Michelsen, A., & Larsen, O. N. (2008). Pressure difference receiving ears. *Bioinspiration & Biomimetics*, 3, 011001. doi: 10.1088/1748-3182/3/1/011001
Michelsen, A., & Löhe, G. (1995). Tuned directionality in cricket ears. *Nature*, 375, 639.
Michelsen, A., & Rohrseitz, K. (1995). Directional sound processing and interaural sound transmission in a small and a large grasshopper. *Journal of Experimental Biology*, 198, 1817–1827.
Michelsen, A., & Rohrseitz, K. (1997). Sound localization in a habitat: An analytical approach to quantifying the degradation of directional cues. *Bioacoustics*, 7, 291–313.
Michelsen, A., Hedwig, B., & Elsner, N. (1990). Biophysical and neurophysiological effects of respiration on sound reception in the migratory locust *Locusta migratoria*. In F. G. Gribalin, K. Wiese, & A. V. Popov (Eds.), *Sensory systems and communication in arthropods* (pp. 199–203). Basel: Birkhäuser Verlag.
Michelsen, A., Heller. K.-G., Stumpner, A., & Rohrseitz, K. (1994a). Directional hearing and the gain of the acoustic trachea in bushcrickets. *Journal of Comparative Physiology A*, 175, 145–151.
Michelsen, A., Popov, A. V., & Lewis, B. (1994b). Physics of directional hearing in the cricket *Gryllus bimaculatus*. *Journal of Comparative Physiology A*, 175, 153–164.
Moiseff, A., & Konishi, M. (1981). The owl's interaural pathway is not involved in sound localization. *Journal of Comparative Physiology*, 144, 299–304.
Müller, J. (1826). Zur vergleichenden Physiologie des Gesichtssinnes des Menschen und der Tiere. Leipzig.
Nelson, B. S., & Suthers, R. A. (2004). Sound localization in a small passerine bird: Discrimination of azimuth as a function of head orientation and sound frequency. *Journal of Experimental Biology*, 207, 4121–4133.
Pumphrey, R. J. (1940). Hearing in insects. *Biological Reviews*, 15, 107–132.
Renaud, D. L., & Popper, A. N. (1975). Sound localization in the bottle nose porpoise *Tursiops truncatus*. *Journal of Experimental Biology*, 63, 569–585.
Schöneich, S., & Hedwig, B. (2010). Hyperacute directional hearing and phonotactic steering in the cricket (*Gryllus bimaculatus* de Geer). *PloS ONE*, 5, e15141. doi: 10.1371/journal.pone.0015141
Weber, T., & Thorson, J. (1989). Phonotactic behavior of walking crickets. In F. Huber, T. E. Moore, & W. Loher (Eds.), *Cricket behavior and neurobiology* (pp. 310–339). Ithaca, NY: Cornell University Press.
Welch, T. E., & Dent, M. L. (2011). Lateralization of acoustic signals by dichotically listening budgerigars (*Melopsittacus undulatus*). *Journal of the Acoustical Society of America*, 130(4), 2293–2301.

Chapter 20
Distributed Cortical Representation of Sound Locations

John C. Middlebrooks

John C. Middlebrooks

J.C. Middlebrooks (✉)
Departments of Otolaryngology, Neurobiology & Behavior, Cognitive Sciences, and Biomedical Engineering, University of California at Irvine, Irvine, CA 92697-5310, USA
e-mail: j.midd@uci.edu

A.N. Popper and R.R. Fay (eds.), *Perspectives on Auditory Research*, Springer Handbook of Auditory Research 50, DOI 10.1007/978-1-4614-9102-6_20,

20.1 Introduction and Overview

"Location" is a prominent feature of a stimulus, be it visual, somatic, or auditory. In the visual and somatosensory systems, stimulus locations are coded directly by activity on the retina or body surface, respectively, and cortical representations of location amount to representations of the respective sensory surfaces. It is well known that frequency, rather than location, is represented on the auditory sensory epithelium and that the cortical representation of the cochlea amounts to a tonotopic representation of frequency. The locations of sounds somehow must be computed from the interaction of sounds with the head and external ears, and the central representation of sound location must be something other than a point-to-point map of the ear.

Despite the clear functional differences of the auditory system from visual and somatosensory systems, physiologists long have sought a cortical representation of auditory space in a form analogous to that of the visual and somatic representations; such a representation could be called "topographic," "point-to-point," or simply a "map." An idealized auditory space map might consist of cortical neurons exhibiting narrow spatial receptive fields, with the receptive fields of such neurons shifting systematically as a function of location in the cortex. The activity of any particular neuron would signal the presence or absence of a sound within its spatial receptive field. The spatial location of a sound source would be represented by a restricted focus of neural activity within such a space map. An existence proof for an auditory space map is provided by the superior colliculus of the midbrain, where auditory, visual, and somatic maps align with maps of motor error.

Early studies of the auditory cortex were designed to trace the borders of auditory spatial receptive fields and to plot their locations as a function of cortical location, much as one would plot visual or somatic receptive fields. None of those studies were successful in discovering an auditory space map, including our first systematic study, which produced results that clearly were contrary to the presence of a map of auditory space in the cat's primary auditory cortex (area A1; Middlebrooks & Pettigrew, 1981). One might argue that we were simply looking in the wrong place or that failure to find a map might have been due to the use of general anesthesia. Results from multiple cortical areas in both anesthetized and awake conditions, however, also were inconsistent with space maps.

The frustration of finding a preponderance of very large receptive fields eventually led us to look at how patterns of neural activity might vary as a function of stimulus location within those large fields. We realized that temporal firing patterns of neurons vary systematically such that a given neuron can signal sound locations across as much as 360° of auditory space. Such neurons could be said to be "panoramic" (Middlebrooks et al., 1994). Subsequent work by our group and others have examined how sound locations might be represented by firing patterns of broadly tuned neurons and by populations of such neurons.

In this chapter, I look at how our views of the cortical representation of auditory space have evolved from a hypothetical model analogous to the space-mapped organization of the visual system to an appreciation that the representation of any particular point in space is distributed across large populations of neurons within multiple cortical fields. I review the importance of temporal features of neural response patterns in transmitting information about sound locations, and I show examples of neural spatial selectivity that correlate with some oddities of auditory spatial perception. I present recent results demonstrating dynamics of spatial selectivity observed when animals are engaged in listening tasks and present some striking differences among cortical fields seen in those awake/behaving conditions. Finally, I consider the implications from studies of location representation for understanding of hearing in complex auditory scenes. This chapter is based primarily on results obtained from domestic cats because cats have been the species of choice for the majority of systematic experiments and because we can tell a more coherent story by focusing on one species.

20.2 Spatial Receptive Fields and (the Absence of) Spatial Topography

I began my graduate work in the laboratory of Michael Merzenich at UCSF. Merzenich had recently made a big splash by demonstrating a precise tonotopic map of frequency in the auditory cortex, and he was soon to make bigger splashes by describing highly dynamic maps in the somatosensory cortex. That lab was all about maps. Also about that time, Knudsen and Konishi had demonstrated a precise map of auditory space in the midbrain of the barn owl. My mission was clear—discover the space map in the mammalian auditory cortex. That effort was aided by Mark Konishi's generous offer to permit me to use his anechoic chamber at Caltech and by the opportunity to collaborate with Konishi's colleague at Caltech, Jack Pettigrew.

Pettigrew and I optimistically set out to discover the space map in cats by presenting sounds from a small loudspeaker that could be positioned anywhere on the surface of an imaginary sphere (Middlebrooks & Pettigrew, 1981). Stimuli were characteristic-frequency (CF) tones played at levels ≥10 dB above each neuron's threshold at its most sensitive location. We estimated borders of spatial receptive fields by varying the position of the sound source and inspecting neural responses on the oscilloscope.

We encountered three types of spatial receptive field. About half of the units, which we called "omnidirectional," responded on at least a subset of trials to tones presented from anywhere in the front half of space; locations in the rear half of space were not tested consistently in that study. Although omnidirectional units did not exhibit any receptive field borders, we noted that many omnidirectional units tended to respond most vigorously to sounds near the frontal midline. The other half of the sampled unit population exhibited one of two types of spatial receptive field that were bounded within the front half of space. "Hemifield" units (Fig. 20.1a) responded to sounds through most locations contralateral to the recording site and exhibited a vertically oriented boundary typically within 20° of the frontal midline; most of those units had CFs below 12 kHz. "Axial" units (Fig. 20.1b) had spatial

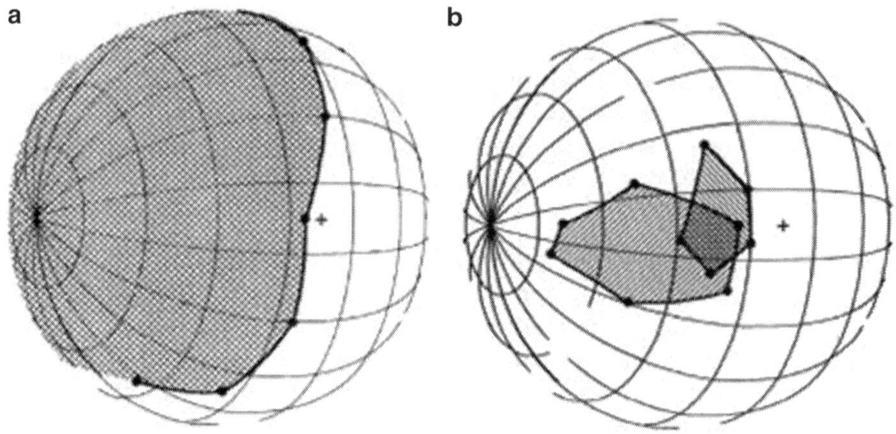

Fig. 20.1 Spatial receptive fields of single neurons in cat primary auditory cortex (A1). Hatched areas indicate regions of space in which characteristic-frequency (CF) tone bursts ≥10 dB above threshold elicited responses on a majority of trials. The plus sign indicates the point in front of the cat (i.e., 0° azimuth and elevation). (**a**) A hemifield receptive field. (**b**) Two axial receptive fields. (From Middlebrooks & Pettigrew, 1981)

receptive fields that were completely circumscribed within the frontal contralateral quadrant of space; most of those units had CFs above 12 kHz. The receptive fields of axial units tended to align with the acoustical axis of the contralateral pinna. Tangential electrode penetrations oriented parallel to isofrequency laminae encountered sequences of units all of one receptive-field class segregated from sequences of units of a different class, consistent with a modular columnar organization.

At the time, it was standard practice to excise the pinnae of experimental animals so as to permit delivery of calibrated sound levels to the tympanic membranes. We were criticized by some of our peers for our shoddy practice of having left the pinnae intact, and were accused of studying what was merely a "pinna effect." Eventually, however, it became obvious that animals normally hear sounds that have been filtered by their pinnae and that the pinnae provide essential cues for the shaping of neural spatial sensitivity. We showed in our 1981 paper that spatial receptive fields shifted in location when we mechanically deflected the contralateral pinna, and that dynamic pinna directionality must contribute to spatial hearing in the many mammals that have movable pinnae. Admittedly, our study unintentionally emphasized the effects of pinna directionality by using pure-tone stimuli that, in most cases, were little more than 10 dB above neural thresholds. At such low sound levels, sounds at many locations likely were below the threshold to activate one or the other ear and, therefore, did not activate binaural pathways from all locations. Nevertheless, it is now clear that the directionality of the pinnae is essential for spatial hearing.

In retrospect, one can see that, in our first study, we should have used higher sound levels and broadband sounds. Near-threshold sound levels can give the impression of smaller spatial receptive fields. Listeners generally localize sounds better at levels well above threshold, however, so any reasonable model of spatial representation must account for performance at moderate sound levels. Pure tones probably were a poor choice for the stimuli because they lack the full variety of localization cues that are available from most natural sounds. In particular, mammals cannot localize pure tones in elevation due to the absence of spectral-shape cues—the top and bottom borders that we measured for axial units must have been due to pinna directionality and would have vanished at higher sound levels. We note, however, that a later comparison of area-A1 responses to pure-tone versus broad-band stimuli showed no consistent difference in spatial sensitivity (Rajan et al., 1990a).

The most significant, and disappointing, result of our early study of area A1 was that there was no evidence of a systematic map of auditory space. The only spatially tuned low-CF neurons had receptive fields that spanned approximately 180° and that were centered near the contralateral pole of the sound field. The only spatially tuned high-CF neurons had receptive fields that were centered on the axis of the contralateral pinna, thereby representing only a restricted region of sound locations. Contrary to the requirements for a topographical representation of auditory space: (1) Receptive fields were large, providing a very low-precision location signal if one thinks only in terms of a neuron representing only the presence or absence of a stimulus within its receptive field. (2) Receptive fields did not cover space uniformly—all of the hemifield receptive fields were centered near contralateral 90°, and most of the axial receptive fields were centered between 20° and 40° (varying between cats). (3) Nearby units could have widely disparate spatial tuning, and widely separated units could have similar spatial tuning, both of which are inconsistent with a systematic progression of spatial tuning as a function of neural location. These properties, which essentially refute the hypothesis that auditory space is represented topographically, have been encountered in subsequent studies employing broadband sounds at moderate levels in cat areas A2 and anterior ectosylvian sulcus area (AES; Middlebrooks et al., 1994, 1998; Las et al., 2008), the posterior auditory field (PAF; Stecker et al., 2003), the dorsal zone (DZ; Stecker et al., 2005b), and the anterior auditory field (AAF; Harrington et al., 2008).

Three papers published in 1990 (Imig et al., 1990; Rajan et al., 1990a, b) confirmed many of our earlier observations and extended them by testing broader ranges of sound level, by using broadband sounds, and by quantifying spike counts as a function of sound-source azimuth. Both groups used a more restrictive definition of omnidirectional or "low-directionality" tuning than we did and, not surprisingly, found a substantially lower percentage of omnidirectional units. Both groups also found ipsi- or near-midline-preferring units, which probably would have fallen within our omnidirectional class. Both groups observed a variety of changes in spatial tuning resulting from increasing stimulus level, which we had shown only anecdotally in 1981. Both groups replicated our observation of a modular columnar segregation of units showing differing classes of spatial tuning. Again, there was no consistent evidence for a topographical auditory space map. Many of the characteristics of spatial receptive fields observed with free-field stimulation by our group and by the Imig and Rajan groups have been replicated and expanded on by Brugge and colleagues (1994) using virtual–auditory–space stimuli.

20.3 Panoramic Neurons, Distributed Representations, and Population Codes

After some years of plotting spatial receptive fields, and finding generally broad spatial tuning, we began to look more closely at the spatial sensitivity of neural spike patterns. An example of such spatial sensitivity is given in Fig. 20.2 for a unit in area AES; raster plots are shown as a function of sound location in azimuth. This unit would have been termed variously as omnidirectional, contralateral preferring, or contrafield by the Middlebrooks, Imig, or Rajan groups. I began to feel as if I was discarding useful information by collapsing the rich variety of temporal firing patterns down to one-dimensional spike-count-versus azimuth functions and then by summarizing the azimuth function by just a single preferred azimuth. By that time, I was an assistant professor at the University of Florida, and I had the opportunity to work with my senior colleague, David Green, who introduced me to some quantitative tools with which to evaluate my data. We hypothesized that the temporal firing patterns of single neurons might carry location-related information and might signal the location of a sound source throughout a broad range of locations.

We initially addressed this hypothesis empirically for neurons recorded in areas A2 and AES (Middlebrooks et al., 1994, 1998). For each neuron, we used spike patterns from a subset of trials to train a computerized pattern classifier (an artificial neural network [ANN] in that study), and then we used the trained classifier to estimate the locations of sounds based on spike patterns from a different subset of trials from that neuron. Figure 20.3 displays the performance of the ANN in estimating locations based on spike patterns from the unit shown in Fig. 20.2. The location estimates tend to cluster around the positive diagonal line that indicates perfect performance, and the mean directions of estimates (the solid line) tend to follow the positive diagonal. We referred to neurons that could signal locations throughout up to 360° of azimuth as "panoramic."

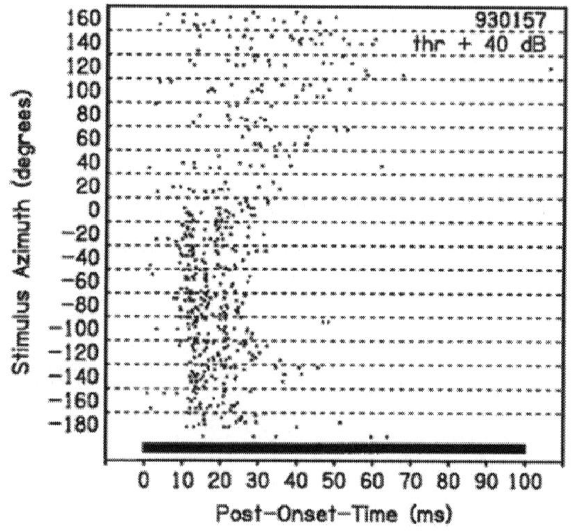

Fig. 20.2 Temporal firing pattern of a unit in area AES. Each dot indicates one action potential. Stimulus locations are indicated on the abscissa. Responses to 10 noise bursts are each location are shown. The bar at the bottom of the figure indicates the duration of the sound. (From Middlebrooks et al., 1998)

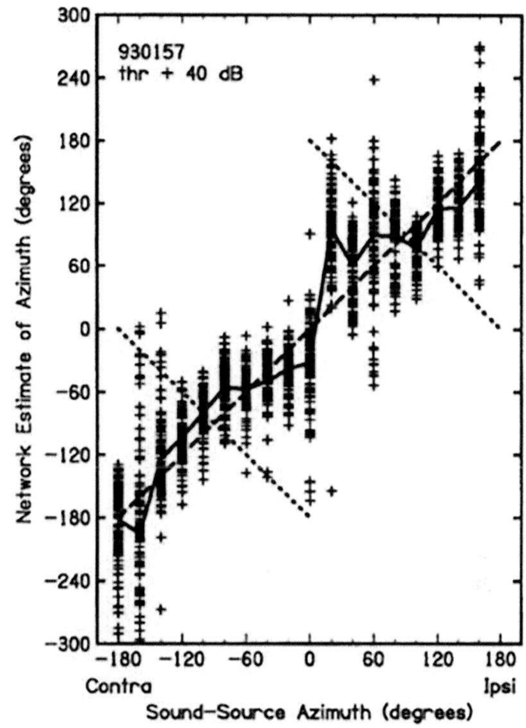

Fig. 20.3 Artificial neural network (ANN) estimates of sound-source locations. Each plus sign indicates one estimate based on a boot-strapped spike pattern from the unit shown in Fig. 20.2. The actual sound location is plotted on the abscissa, and the estimates are plotted on the ordinate. The diagonal line with positive slope indicates the loci of perfect estimates. The two diagonals with negative slope indicate the loci of perfect front/back confusions. (From Middlebrooks et al., 1998)

Of course, there was considerable variation across our sample in the accuracy with which single neurons could signal sound locations panoramically, but location estimates by every neuron showed errors smaller than those predicted by random chance, and more than half of the sample showed errors smaller than half of the random-chance level. That a single neuron can signal individual sound-source locations more or less panoramically and that such panoramic neurons are widely distributed through the auditory cortex (or at least areas A2 and AES in the 1994 and 1998 papers) indicate that information about any particular sound-source location is represented throughout the auditory cortex. This is a highly "distributed" representation.

A reassuring aspect of the neural panoramic coding that we have observed is that it performs analogously to some special cases of human psychophysical performance. I give two examples. First is the phenomenon of the precedence effect. When a listener is exposed to a pair of sounds separated by a brief time interval, the localization judgment varies with the duration of that interval (Litovsky et al., 1999). When the interval is <1 ms, the listener tends to report a location somewhere between the two sources; this is "summing localization." When the time interval is longer, approximately 1–5 ms, listeners tend to report a location corresponding to that of the first sound source; this is "localization dominance." Correlates of both summing localization and localization dominance were evident in neural spike patterns (Mickey & Middlebrooks, 2001). The second phenomenon is that of illusory vertical localization of band-passed sounds. When presented with 1/6th-octave

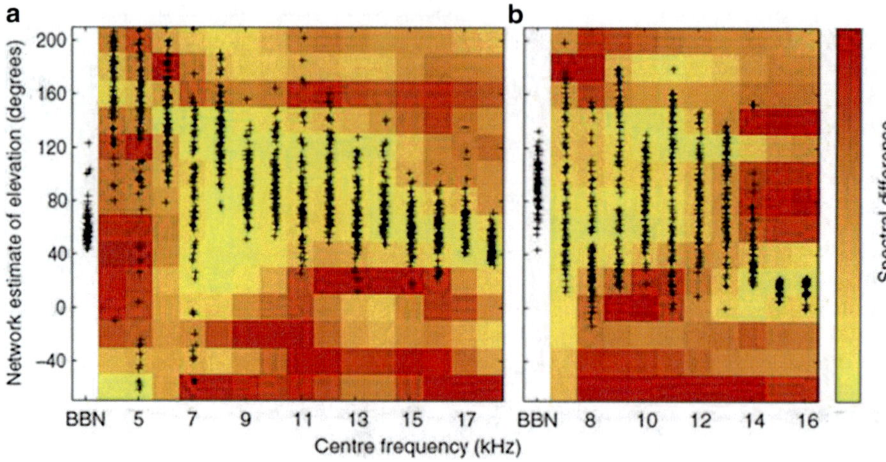

Fig. 20.4 Estimates of sound-source elevation based on temporal spike patterns of two neurons recorded in area A2. One-sixth-octave noise bands were presented from +80° elevation; the center frequencies of the noise bands are indicated on the abscissa. Neural spike patterns, and the corresponding elevation estimates, varied systematically with stimulus frequency. The colors indicate predictions of responses based on the difference of stimulus spectra and the head-related-transfer functions of the cats' ears at various elevations. Light colors indicate a small difference and, thus, a high probability of response. Elevation estimates based on neural responses tend to lie at elevations at which spectral differences were small. The estimates for the broadband noise (BBN) fell near the actual stimulus location of +80°. (From Xu et al., 1999)

band-passed sounds, human listeners tend to make localization judgments that are determined by the center frequency of the sound, not by its location. We were able to predict such erroneous judgments in humans based on correspondence of the stimulus spectrum and the location-specific filter properties of each listener's ears (Middlebrooks, 1992). We adapted that model to incorporate the filter functions of cats' ears, and we made predictions of judgments of the vertical location of narrowband sounds (Xu et al., 1999). Those predictions corresponded nicely with localization judgments computed from our ANN analysis of spike patterns elicited by corresponding narrowband stimuli. Two examples are shown in Fig. 20.4.

I wish to emphasize two key points from our 1994 and 1998 panoramic coding studies. First, the responses of many neurons could effectively point to sound sources across a broad range of locations, not just indicating the presence or absence of a sound within a limited "best area" or "preferred location." We have duplicated that finding in subsequent studies in other cortical fields and other experimental conditions. Panoramic coding by single neurons has not been pursued much by other research groups, although a number of studies have quantified the spatial information carried by populations of cortical neurons (Reale et al., 2003) or have formed estimates of sound location based on responses of neural populations (in primates; Miller & Recanzone, 2009). Second, substantial stimulus-related information is transmitted by the timing of neural action potentials. Neurons could signal sound locations with substantially greater accuracy when their responses were represented by temporal spike patterns than when the responses were reduced to simple spike counts (Middlebrooks et al., 1994, 1998). We subsequently demonstrated that, at least in anesthetized conditions, first-spike latencies carried more stimulus-related information than did spike counts, nearly as much as did complete spike patterns (Furukawa & Middlebrooks, 2002). Coding of sound location and other sound parameters by spike timing has been explored further by other research groups. For example, Reale and colleagues (2003) have examined the spatial and sound-level sensitivity of first-spike latencies in area A1 and quantified the stimulus-related information carried by just the latencies of neuronal populations. Nelken and colleagues (2005) showed that essentially all the sound-location-related information in a (ferret auditory cortex) spike train can be captured by just the mean spike count and the first spike latency, but that more than half of that information is lost if latency information is eliminated.

20.4 Population Codes

The accuracy of location signaling by any single neuron is far worse than that of behavioral performance. One assumes that the high quality of behavioral sound localization in some way reflects the integrated activity of multiple neurons. I describe here two studies that explored such integration. The first considers a parallel classification of spike patterns of small ensembles of neurons. The second recognizes the inhomogeneity of auditory spatial tuning and tests a simple population rate code based on a small number of opponent neural population.

We examined location coding by ensembles of neurons that were assembled off line from neurons studied at multiple multichannel probe placements in multiple cats (Furukawa et al., 2000). Temporal spike patterns were combined across neurons by concatenating their spike patterns. Stimulus locations were estimated by classification of those concatenated patterns with ANNs. We compared a condition in which spike times were expressed relative to stimulus onset ("absolute timing") with a condition in which spike times were expressed relative to the first spike in the ensemble ("between-unit timing"). In many cases, location estimates were nearly as accurate in the biologically plausible between-unit condition as in the rather artificial condition in which spike times were measured relative to stimulus onset. Not surprisingly, location estimation improved with increasing ensemble size. Performance by ensembles of 128 neurons (the largest that we tested) approached that reported in a cat behavioral study by May and Huang (1996). I note that the location estimates from neural ensembles in that study relied on a simplistic pattern recognition of concatenated spike patterns and did not explicitly exploit any information that might have been carried by details of relative timing among neurons.

In the second study, we looked at the inhomogeneity in single-neuron spatial tuning and attempted to relate it to published psychophysical localization results (Stecker et al., 2005a). A traditional view is that neurons primarily represent locations near the peaks of their spike-rate-versus-azimuth functions (i.e., the neurons' "best areas"). In the case of auditory cortex, however, most best areas tend to lie well away from the frontal midline, i.e., away from where behavioral localization accuracy is greatest (e.g., May & Huang, 1996). We explored an alternative hypothesis, that location signaling is most accurate near the steepest portions of neural rate-azimuth functions, which tend to lie near the midline; a similar hypothesis was advanced previously in regard to spatial coding by brain stem neurons (Harper & McAlpine, 2004). We tested location discrimination by a cortical spike-pattern classifier and confirmed that, like psychophysical subjects, most neurons show their best spatial acuity for sound locations near the frontal midline.

Given that most neurons discriminate near-midline locations accurately and that most response much more vigorously to sounds in one sound hemifield than the other, we tested a simple population model that would localize sound based on the relative activity of neurons tuned to the right or left sound hemifield. There has been some discussion of models that would compare activity in right-versus-left cortical hemispheres. Such models, however, predict that a unilateral cortical lesion would result in a bilateral deficit in sound localization, which conflicts with observations that unilateral auditory cortical lesions (e.g., Jenkins & Masterton, 1982) or inactivation (e.g., Malhotra et al., 2004) produce strictly unilateral deficits. Instead of an inter-hemisphere comparison, we tested a localization model that compared normalized spike rates between populations of right- or left-favoring neurons in one hemisphere. That model disregarded all the fine details of location-specific temporal fine structure, but it demonstrated reasonable accuracy in identifying the locations of sounds that were presented in 20° increments of azimuth and that varied in level across a 20-dB range. Like psychophysical localization performance, the model showed greatest accuracy around the frontal midline with accuracy declining for more eccentric targets.

We assume that location signaling by neural populations could only improve with more sophisticated models that would exploit the full variety of spatial tuning that we have observed and that has been reported by others. I would like to point out one simple extension of our contra-versus-ipsilateral opponent process model. That is, in awake animals (described in Section 6) we see a sizable population of neurons tuned to locations near the midline. Similarly, the Phillips group (Dingle et al., 2010) has presented psychophysical evidence for a third, frontally tuned, channel in human location discrimination. It remains to be tested whether localization is simply a process of comparison among a small number of differently tuned populations (e.g., contralateral, ipsilateral, and frontal) or whether the psychophysical evidence for a small number of channels just indicates local expansions in an otherwise widely distributed representation of space.

20.5 Specialization Among Cortical Fields

The auditory cortex has long been known to be essential for sound localization in that unilateral lesions result in contralesional localization deficits (e.g., Jenkins & Masterton, 1982). The Lomber group more recently has introduced an elegant technique that permits identification of specific cortical areas that are necessary for normal localization. This involves training an animal to perform a sound-localization or other auditory task and then to use small cooling coils to inactivate specific cortical areas. Lomber and colleagues found sound-localization deficits associated with inactivation of any of four auditory areas: areas A1, DZ, PAF, and AES (Malhotra et al., 2004, 2008). The deficits varied among areas. Inactivation of area AES or PAF resulted in a profound contralateral deficit, with performance dropping to around chance levels (Malhotra et al., 2004). In contrast, inactivation of A1 alone resulted in a large number of relatively small localization errors, whereas localization DZ alone resulted in a smaller number of errors, which were relatively large (Malhotra et al., 2008). Inactivation of A1 and DZ together produced profound contralateral deficits, like those produced by AES or PAF inactivation. An interesting "double dissociation" was demonstrated in areas AAF and PAF (Lomber & Malhotra, 2008). Cats were trained to perform both a localization task and a discrimination of temporal patterns. Bilateral inactivation of area PAF disrupted localization but had no effect on the pattern discrimination. Conversely, bilateral AAF inactivation disrupted pattern discrimination while preserving localization.

More or less in parallel with the behavior/inactivation work by the Lomber group, we characterized spatial sensitivity of neurons in α-chloralose-anesthetized cats in cortical areas AES, A2, A1, PAF, DZ, and AAF. In each of these areas, neurons showed at least some degree of spatial sensitivity, although that sensitivity varied considerably among areas. In each of these areas, we observed features of neural sensitivity and functional architecture that were consistent with a distributed representation of sound location and that were contrary to a topographical representation. Despite their similarities, these cortical areas showed some interesting differences, some that accord to some degree with expectations from the behavioral/inactivation studies by the Lomber group and others that conflict with those expectations. We consider some of those characteristics here.

Of the cortical areas that we have studied in anesthetized cats, areas PAF and DZ exhibit the strongest spatial sensitivity in terms of narrowness of spatial receptive fields, sharpness of receptive field cutoffs, and depth of modulation of spike counts by variation in sound location (Stecker et al., 2003, 2005b). Those observations accord with Lomber's demonstrations that inactivation of PAF produces a profound contralateral localization deficit and that inactivation of DZ can produce large localization errors. Imig and colleagues (1990) noted that it was the neurons in area A1 having nonmonotonic spike-count-versus-level functions that showed the strongest, most level-invariant, spatial tuning. Consistent with that observation, areas PAF and DZ contain larger numbers of neurons showing non-monotonic level functions and greater level invariance of spatial sensitivity than is seen in other areas. Area DZ is distinguished from PAF in anesthetized conditions by having more neurons tuned to frontal ipsilateral locations. Both areas PAF and DZ have markedly longer first-spike latencies than other auditory fields that we have studied as well as greater modulation of latency by sound location.

Neurons in areas AES showed slightly but significantly sharper location sensitivity than did A2 neurons (Middlebrooks et al., 1998). Areas AES and A2 showed no significant difference, however, in the distribution of median errors of location estimates.

The minimal differences in neural spatial coding between anesthetized areas A2 and AES conflicts with the observations from the Lomber group (Malhotra et al., 2004) that inactivation of area A2 produces no localization deficit whereas inactivation of area AES produces a profound deficit. I note that Las and colleagues (2008) have demonstrated an increase in the proportion of spatially selective neurons and an increase in the proportion of neurons tuned to midline locations as they shifted recording sites toward the posterior pole of auditory area AES, which lies under areas A1 and A2. We primarily studied the portion of area AES that is on the bank of the sulcus, and we probably missed the posterior pole. I imagine, however, that the Lomber cooling procedure likely missed that area as well. Another conflict between our results and those of the Lomber group is seen in the contrast between areas A1 and AAF. Neurons in A1 show only slightly greater spatial sensitivity than do neurons in AAF, and there is no significant difference in the amount of location-related information carried by spike patterns in those two areas (Harrington et al., 2008). Nevertheless, behavior/inactivation results show substantial localization deficits following inactivation of A1 and no such deficits following AAF inactivation (Malhotra et al., 2004).

Based on our survey of spatial sensitivity in six cortical fields, I could hazard a ranking of increasing sensitivity, from AAF to A1 to A2 to AES to DZ to PAF. The choice of AAF and PAF as least and most spatially sensitive, respectively, fits well with the Lomber inactivation results, but the Lomber results probably would call for moving A1 higher in the ranking than A2 and AES higher than DZ. What I find most significant is that the same basic patterns of spatial sensitivity (i.e., generally broad, generally contralaterally biased) are found in every area that we have examined. Any consistent differences are only quantitative. Walker and colleagues (2011) similarly have found spatial sensitivity distributed among multiple auditory cortical fields in the ferret. My interpretation is that the role (or absence of a role) of any cortical area in localization behavior is determined by its connections with other substrates of that behavior, not by the details of its spatial sensitivity. The widespread finding of spatial sensitivity in the cortex reminds us that identification of the source location is but one of many key auditory functions that are enabled or enhanced by spatial sensitivity. Probably the greatest contribution of spatial hearing, more than localization per se, is in detecting and recognizing communication sounds (including speech) in a complex auditory scene. Any cortical area primarily involved in discrimination or recognition of such signals, then, would be likely to show some spatial sensitivity.

20.6 Dynamic Spatial Sensitivity in Awake Animals

The final place in which we had not looked for an auditory space map was the auditory cortex of unanesthetized animals. Could it be that our general anesthetics were somehow disrupting an orderly point-to-point map? That hypothesis was refuted by the results of our first recordings from area A1 in awake cats (Mickey & Middlebrooks, 2003). Those recordings demonstrated that, as in anesthetized conditions, spatial tuning curves were much larger than the scale of behavioral localization errors or spatial acuity and that the organization of spatial tuning was inconsistent with the presence of a map. Nevertheless, several characteristics of responses in the awake cats were quite different than in anesthetized conditions. First, there was a greater variety of spike patterns. In anesthetized A1 we usually saw only an onset response, occasionally followed by a second transient burst of spikes. In contrast, the awake cortex exhibited examples of robust tonic responses, transient bursts at stimulus onset and offset, spontaneous activity suppressed by sounds, and other patterns. Often, the various onset, offset, and tonic components of the spike patterns of single units differed in their spatial sensitivity. Second, the spatial sensitivity of units was sharper, primarily in the sense that the magnitude of modulation by changes in sound-source location was deeper. That is, the modulation depth of responses could be larger than 100 % because responses were elevated well above the spontaneous firing rate for some stimulus locations and suppressed below the spontaneous rate for others. Third, increases in sound level produced much less broadening of spatial tuning curves than is seen in anesthetized conditions. I imagine that general anesthetic disturbs the fine balance of excitation and inhibition that is needed to maintain stable spatial sensitivity across changing sound levels and that use of an awake preparation eliminates that artifact.

We compared spatial sensitivity in awake conditions among neurons in areas A1, DZ, and PAF (Lee & Middlebrooks, 2013), all of which have been shown to be necessary for normal sound-localization behavior (Malhotra et al., 2004, 2008). Among the sampled neuron populations in those cortical areas there was a wide variety of temporal firing patterns and of spatial sensitivity, but one can make some generalizations. A small majority of neurons sampled in A1 and DZ and about one third of the neurons in PAF responded only to the onset of sounds, with the onset response typically showing relatively broad spatial sensitivity. Many neurons showed an additional tonic response following the onset response. Tonic responses typically showed equal or, often, sharper spatial sensitivity than did the onset responses. Area DZ showed sharper spatial sensitivity than A1 and PAF in that DZ's average tuning width was narrower and that a larger percentage of units showed more that 50 % modulation of spike rates by sound-source location. Areas DZ and PAF differed clearly in their spatial preferences. Area DZ had a large population of neurons with best areas around the frontal midline, whereas area PAF neurons showed a broader distribution of best-area locations, showing a fairly uniform representation of the entire contralateral sound hemifield including rear locations; spatial preferences of A1 neurons were intermediate between those of DZ and PAF.

We explored the possibility that cortical responses might adapt to the demands of a behavioral task (Lee & Middlebrooks, 2011, 2013). We tested cortical neuronal responses in three task conditions: off- task ("Idle") and while the animal was engaged in one of two tasks. The "Periodicity Detection" task required the animal to attend to the sound and to act in response to a sound change, but the location of the stimulus was irrelevant to reward. The "Localization" condition, in contrast, required the animal to evaluate the location of every stimulus and to act when there was a change in the vertical location.

A sizeable minority of units in areas A1, DZ, and PAF sharpened their spatial sensitivity significantly in either of the task conditions compared to the Idle condition, and many units showed further significant sharpening in the Localization compared to Periodicity Detection conditions. The sharpening primarily affected the onset portion of responses and was most conspicuous among units showing the broadest tuning in the Idle condition. Area A1 showed a higher proportion of broadly tuned onset-only neurons than did DZ and PAF, so the proportion of units showing significant sharpening was somewhat larger in A1 than in DZ and PAF.

In all three cortical areas that we studied, sharpening of spatial sensitivity of onset responses was accomplished more by increased suppression of responses to non-favored locations than by enhancement of responses to favored locations. In contrast, the tonic portion of responses tended to show relatively little task-dependent change in spatial sensitivity, but the magnitudes of tonic responses tended to increase in the on-task conditions. Based on those observations, one can infer that when an animal is not actively listening for a target, a sound burst elicits a transient response in a relatively large population of neurons. When an animal is actively listening for a target, however, the cortical representation of a sound source is sparser (i.e., involving fewer active neurons) and the responses of the active neurons are more sustained.

The differences in spatial sensitivity between neurons in areas DZ and PAF suggested that they might differ in the accuracy with which neurons, or small ensembles of neurons, could estimate the locations of sound sources. We tested that hypothesis, using a procedure that classified onset, tonic, and offset portions of the responses of single neurons and of small randomly selected ensembles of neurons. In both areas, performance improved (i.e., errors decreased) as the sizes of ensembles were increased from 1 to 4 to 16 units (Fig. 20.5). Ensembles of neurons in areas DZ and PAF showed orthogonal patterns of performance. Area-DZ ensembles showed greatest accuracy within about 30° contralateral to the frontal midline. In contrast, area-PAF ensembles performed relatively poorly for midline locations but showed accurate performance at the contra- and ipsilateral extremes of the sound field. Both areas showed an increase in the accuracy of location estimates when animals were on task.

20.7 How Far Have We Come, and Where Do We Need to Go?

We have come a long way toward a first-order description of spatial representation, at least in the cat. We have abundant evidence that there is no point-to-point space map in the cortex, and we can infer that the representation of any point in space is distributed throughout the majority of neurons in multiple cortical areas. We can state that neurons vary in both the magnitude and timing of their responses as a function of sound location across receptive fields as wide as 360° of azimuth. Many

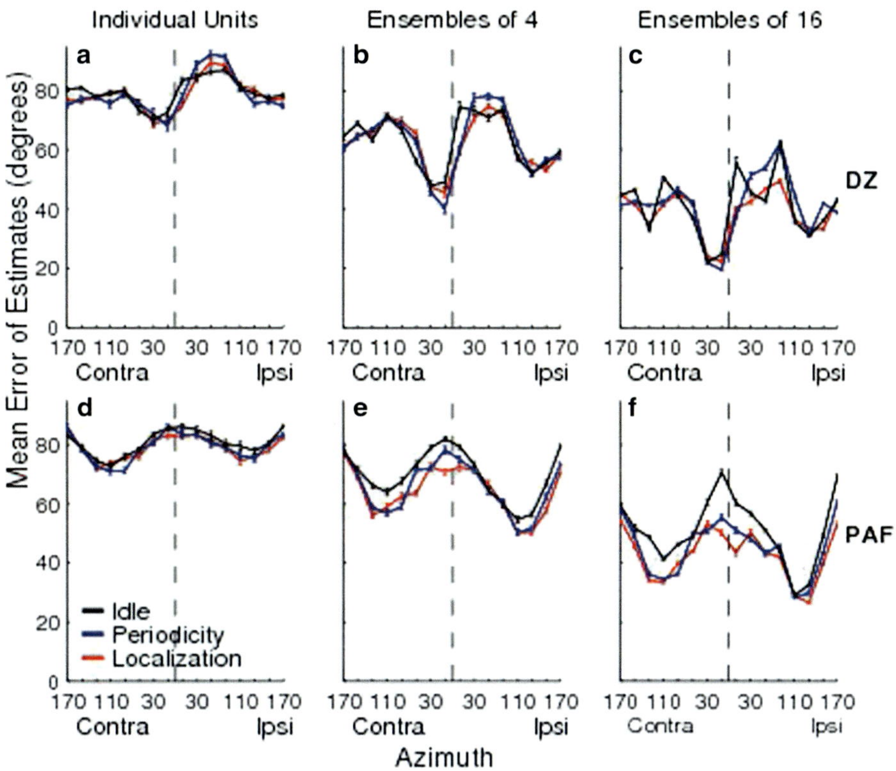

Fig. 20.5 Mean errors of azimuth locations based on classification of spike patterns of ensembles of 1, 4, or 16 units in area DZ (upper row) and area PAF (lower row) in awake, behaving cats. Line colors indicate the three task conditions. Error bars are standard errors of the mean. (From Lee and Middlebrooks, 2013)

neurons show the greatest modulation of response magnitude for changes in sound location across the frontal midline, which corresponds well with the region of greatest behavioral localization accuracy. Two cortical areas in the cat, demonstrably necessary for localization behavior, show somewhat complementary representations of space, with many area DZ neurons responding best to frontal targets and area PAF neurons exhibiting widespread spatial preferences including locations outside of the visual field. Area AES also has been shown to be necessary for normal localization behavior, but I am inclined to think that its role is more sensorimotor in nature, involving modulation of gaze-control circuits, than in localization per se. We can infer that the cortical representation of space is broad and transient when an animal is inactive and that it becomes sharper and more sustained when an animal is engaged in a listening task.

On the other hand, if our goal is to understand how cortical activity produces sound-localization behavior and perception, we are nowhere near the destination. Had we found a space map in the cortex, we might have been able to model feed-forward projections into systems for spatial behavior and perception, but lacking such a map it is difficult to know what to do with a highly distributed representation. We find that the cortical representation becomes sparser in on-task conditions, and one might hope to find even greater sharpening of location tuning if one were to create a more demanding task, but there still is no evidence to suggest that a map will emerge. There is some evidence in humans and other primates that spatial information from the auditory cortex projects to posterior parietal cortex and, there, is integrated with spatial representations of other sensory modalities. As is the case in the superior colliculus of the midbrain, the space-mapped visual and somatic representations might enforce a topographic representation of auditory space, but such a multimodal map has not been demonstrated, nor is it clear how the distributed representation in auditory cortex would project to a topographic representation in multimodal parietal cortex.

A cat uses localization to catch a mouse, and humans use localization to assign locations to sounds in our perceptual space. Spatial hearing, however, gives us much more than localization. Perhaps the most important contributions of spatial hearing are to enhance detection and recognition of sounds of interest in a background of other competing sounds. These contributions of spatial hearing can include spatial release from informational masking and spatial segregation of multiple temporally interleaved sequences of sounds. In ongoing experiments, we already have found some surprising contrasts between representation of the locations of single sounds sources and spatial segregation of multiple sequences of sounds (Middlebrooks & Bremen, 2013). There are ongoing efforts by our research group and others to understand the brain mechanisms by which spatial hearing facilitates hearing in complex auditory scenes. We hope that our present understanding of sound location representation will provide a useful foundation for those efforts.

Acknowledgments I am grateful to my talented collaborators in this research, especially to my mentors, Michael Merzenich, Eric Knudsen, and David Green. My research is funded by the National Institute on Deafness and Other Communication Disorders.

References

Brugge, J. F., Reale, R. A., Hind, J. E., Chan, J. C., Musicant, A. D., & Poon, P. W. (1994). Simulation of free-field sound sources and its application to studies of cortical mechanisms of sound localization in the cat. *Hearing Research*, 73, 67–84.

Dingle, R. N., Hall, S. E., & Phillips, D. P. (2010). A midline azimuthal channel in human spatial hearing. *Hearing Research*, 268, 67–74.

Furukawa, S., & Middlebrooks, J. C. (2002). Cortical representation of auditory space: Information-bearing features of spike patterns. *Journal of Neurophysiology*, 87, 1749–1762.

Furukawa, S., Xu, L., & Middlebrooks, J. C. (2000). Coding of sound-source location by ensembles of cortical neurons. *Journal of Neuroscience*, 20, 1216–1228.

Harper, N. S., & McAlpine, D. (2004). Optimal neural population coding of an auditory spatial cue. *Nature*, 430, 682–686.

Harrington, I. A., Stecker, G. C., Macpherson, E. A., & Middlebrooks, J. C. (2008). Spatial sensitivity of neurons in the anterior, posterior, and primary fields of cat auditory cortex. *Hearing Research*, 240, 22–41.

Imig, T. J., Irons, W. A., & Samson, F. R. (1990). Single-unit selectivity to azimuthal direction and sound pressure level of noise bursts in cat high-frequency primary auditory cortex. *Journal of Neurophysiology*, 63, 1448–1466.

Jenkins, W. M., & Masterton, R. B. (1982). Sound localization: Effects of unilateral lesions in central auditory system. *Journal of Neurophysiology*, 47, 987–1016.

Las, L., Shapira, A. H., & Nelken, I. (2008). Functional gradients of auditory sensitivity along the anterior ectosylvian sulcus of the cat. *Journal of Neuroscience*, 28, 3657–3667.

Lee, C.-C., & Middlebrooks, J C. (2011). Auditory cortex spatial sensitivity sharpens during task performance. *Nature Neuroscience*, 14, 108–114.

Lee, C.-C., & Middlebrooks, J. C. (2013). Specialization for sound localization in fields A1, DZ, and PAF of cat auditory cortex. *Journal of the Association for Research in Otolaryngology*, 14, 61–82.

Litovsky, R. Y., Colburn, H. S., Yost, W. A., & Guzman, S. J. (1999). The precedence effect. *Journal of the Acoustical Society of America*, 106, 1633–1654.

Lomber, S. G., & Malhotra, S. (2008). Double dissociation of 'what' and 'where' processing in auditory cortex. *Nature Neuroscience*, 11, 609–616.

Malhotra, S., Hall, A. J., Lomber, S. G. (2004). Cortical control of sound localization in the cat: Unilateral cooling deactivation of 19 cerebral areas. *Journal of Neurophysiology*, 92, 1625–1643.

Malhotra, S., Stecker, G. C., Middlebrooks, J. C., & Lomber, S. (2008). Sound localization deficits during reversible deactivation of primary auditory cortex and/or the dorsal zone. *Journal of Neurophysiology*, 99, 1628–1642.

May, B. J., & Huang, A. Y. (1996). Sound orientation behavior in cats. I. Localization of broadband noise. *Journal of the Acoustical Society of America*, 100, 1059–1069.

Mickey, B. J., & Middlebrooks, J. C. (2001). Responses of auditory cortical neurons to pairs of sounds: Correlates of fusion and localization. *Journal of Neurophysiology*, 86, 1333–1350.

Mickey, B. J., & Middlebrooks, J. C. (2003). Representation of auditory space by cortical neurons in awake cats. *Journal of Neuroscience*, 23, 8649–8663.

Middlebrooks, J. C. (1992). Narrowband sound localization related to external ear acoustics. *Journal of the Acoustical Society of America*, 92, 2607–2624.

Middlebrooks, J. C., & Pettigrew, J. D. (1981). Functional classes of neurons in primary auditory cortex (A1) of the cat distinguished by sensitivity to sound location. *Journal of Neuroscience*, 1, 107–120.

Middlebrooks, J. C., & Bremen, P. (2013). Spatial stream segregation by auditory cortical neurons. *Journal of Neuroscience*, 33, 10986–11001.

Middlebrooks, J. C., Clock, A. E., Xu, L., & Green, D. M. (1994). A panoramic code for sound location by cortical neurons. *Science*, 264, 842–844.

Middlebrooks, J. C., Xu, L., Eddins, A. C., & Green, D. M. (1998). Codes for sound-source location in nontonotopic auditory cortex. *Journal of Neurophysiology*, 80, 863–881.

Miller, L. M., & Recanzone, G. H. (2009). Populations of auditory cortical neurons can accurately encode acoustic space across stimulus intensity. *Proceedings of the National Academy of Sciences of the USA*, 106, 5931–5935.

Nelken, I., Chechik, G., Mrsic-Flogel, T. D., King, A. J., & Schnupp, J. W. (2005). Encoding stimulus information by spike numbers and mean response time in primary auditory cortex. *Journal of Computational Neuroscience*, 19, 199–221.

Rajan, R., Aitkin, L. M., Irvine, D. R., & McKay, J. (1990a). Azimuthal sensitivity of neurons in primary auditory cortex of cats. I. Types of sensitivity and the effects of variations in stimulus parameters. *Journal of Neurophysiology*, 64, 872–887.

Rajan, R., Aitkin, L. M., & Irvine, D. R. (1990b). Azimuthal sensitivity of neurons in primary auditory cortex of cats. II. Organization along frequency-band strips. *Journal of Neurophysiology*, 64, 888–902.

Reale, R. A., Jenison, R. L., & Brugge, J. F. (2003). Directional sensitivity of neurons in the primary auditory (AI) cortex: Effects of sound-source intensity level. *Journal of Neurophysiology*, 89, 1024–1038.

Stecker, G. C., Mickey, B. J., Macpherson, E. A., & Middlebrooks, J. C. (2003). Spatial sensitivity in field PAF of cat auditory cortex. *Journal of Neurophysiology*, 89, 2889–2903.

Stecker, G. C., Harrington, I. A., & Middlebrooks, J. C. (2005a). Location coding by opponent neural populations in the auditory cortex. *PLoS Biology*, 3, 520–528.

Stecker, G. C., Harrington, I. A., Macpherson, E. A., & Middlebrooks, J. C. (2005b). Spatial sensitivity in the dorsal zone (area DZ) of cat auditory cortex. *Journal of Neurophysiology*, 94, 1267–1280.

Walker, K. M., Bizley, J. K., King, A. J., & Schnupp, J. W. (2011). Multiplexed and robust representations of sound features in auditory cortex. *Journal of Neuroscience*, 31, 14565–14576.

Xu, L., Furukawa, S., & Middlebrooks, J. C. (1999). Auditory cortical responses in the cat to sounds that produce spatial illusions. *Nature*, 399, 688–691.

Chapter 21
Pitch: Mechanisms Underlying the Pitch of Pure and Complex Tones

Brian C.J. Moore

Brain C.J. Moore

B.C.J. Moore (✉)
Department of Experimental Psychology, University of Cambridge,
Downing Street, Cambridge CB2 3EB, UK
e-mail: bcjm@cam.ac.uk

A.N. Popper and R.R. Fay (eds.), *Perspectives on Auditory Research*, Springer Handbook of Auditory Research 50, DOI 10.1007/978-1-4614-9102-6_21,

21.1 Introduction

The perception of pitch is one of the oldest topics in hearing research. Despite this, there is still controversy about the underlying mechanisms, both for pure tones and for complex tones. This controversy prompted me to focus on pitch for my PhD thesis, which I started in 1968. Evidently, I did not resolve the controversy. In this chapter I consider the current situation with emphasis on the lines of experimental evidence that seem to me to be the most critical. The chapter is divided into two major sections, the first dealing with the pitch of pure tones and the second with the pitch of complex tones.

21.2 The Pitch of Pure Tones

21.2.1 Mechanisms Underlying Pitch

Two mechanisms have been proposed to account for the ability to estimate the pitch of a pure tone and to detect changes in frequency of a pure tone. The first, called the place mechanism, is based on the filtering that occurs on the basilar membrane within the cochlea. A pure tone produces a pattern of vibration with a distinct peak at a specific place. Von Békésy (1960) proposed that the pitch value corresponds to the position of the peak vibration. However, physiological measurements show that the position of this peak shifts markedly with level, by an amount equivalent to a half-octave shift in frequency, or even more (Sellick et al., 1982). Psychophysical data based on forward masking provide evidence for similar effects in humans, at least for high frequencies (Moore et al., 2002). In contrast, the pitch perceived by human listeners shifts only slightly with level (Terhardt, 1974b; Verschuure & van Meeteren, 1975). To account for this discrepancy, Zwislocki and Nguyen (1999) proposed that the pitch value corresponds to the position of the apical edge of the vibration pattern. This changes in position much less than the peak when the level of the sound is altered.

A problem with the argument of Zwislocki and Nguyen (1999) is that the pitch of a pure tone is little affected by the presence of a background noise that would mask the apical edge of the vibration pattern (Webster & Muerdter, 1965; Houtsma, 1981). An alternative possibility is that pitch is based on the position of the peak of the vibration pattern, as proposed by von Békésy (1960), but that the brain has stored knowledge about the way that the position of the peak changes with level, and that it uses this knowledge to compensate for the changes so as to produce a pitch percept that is approximately invariant with level.

The other mechanism for frequency discrimination is called the temporal mechanism. It is based on the assumption that the frequency of a pure tone is estimated from the patterns of phase locking in the auditory nerve to the "temporal fine structure" (TFS) of the tone (Siebert, 1970; Srulovicz & Goldstein, 1983; Heinz et al., 2001).

It should be emphasized that it is the time intervals between peaks in the nerve spikes that are important, not the firing rate (the number of nerve spikes per second). The firing rate increases with increasing level until neural saturation occurs, but over a wide range of sound levels, the nerve spikes occur at roughly a constant phase within the cycle of the stimulus, so the time intervals between spikes are approximately integer multiples of the period of the sound (Rose et al., 1967).

In most mammals, phase locking is relatively precise for frequencies up to about 1 kHz, but the precision weakens at high frequencies, and in most mammals phase locking becomes difficult to measure at all for frequencies above 4–5 kHz (Johnson, 1980; Palmer & Russell, 1986). However, weak phase locking does occur in mammals at frequencies above 5 kHz (Heinz et al., 2001; Recio-Spinoso et al., 2005), and the upper limit at which it might be usable in humans is still debated.

It is possible that perception of the pitch of pure tones depends on there being a correspondence between place information and temporal information (Evans, 1978). Evidence relevant to this idea is presented later in this chapter.

21.2.2 Evidence for Different Mechanisms at Low and High Frequencies

Several lines of evidence are consistent with the idea that estimation of the pitch of pure tones depends on different mechanisms at low and high frequencies. For example, the sense of musical interval, as assessed using judgments of musical intervals such as the octave (Ward, 1954) or by the recognition of familiar melodies (Attneave & Olson, 1971), tends to be weak or absent for tones with frequencies above about 5 kHz. Also, the ability of people with absolute pitch to label tones worsens for stimulus frequencies above 5 kHz (Ohgushi & Hatoh, 1991).

Another aspect of pitch that changes with frequency is the ability to "hear out" partials in complex tones or chords containing multiple inharmonically spaced partials, all with the same amplitude (Plomp, 1964; Plomp & Mimpen, 1968). For medium and low frequencies, a partial that is "surrounded" by other partials can be heard out with 75% accuracy (in a two-alternative forced-choice task) when it is separated from neighboring partials by about 1.25ERB$_N$ (Moore & Ohgushi, 1993; Moore et al., 2006b). ERB$_N$ stands for the equivalent rectangular bandwidth of the auditory filter as measured using subjects with normal hearing at moderate sound levels (Glasberg & Moore, 1990; Moore, 2012). Because each one-ERB$_N$ step in frequency corresponds to a distance along the basilar membrane of about 0.89 mm (Moore, 1986), this means that a partial can be heard out when the peak in its vibration pattern is separated from that of neighboring partials by about 1.1 mm. However, for partials with frequencies above about 5 kHz, a separation much greater than 1.25ERB$_N$ is needed to achieve 75% accuracy (Hartmann et al., 1990; Hartmann & Doty, 1996). Indeed, 75% accuracy may not be reached even for a separation of 3ERB$_N$ (Moore et al., 2006b).

These findings have led to the idea that a temporal mechanism dominates at low frequencies and a place mechanism dominates at high frequencies (Moore, 2012). It is assumed that the place mechanism does not lead to a clear sense of musical interval and it does not allow the pitches of partials in complex tones or chords to be heard out.

21.2.3 Discrimination of Changes in Frequency of Pure Tones

A change in frequency of a pure tone produces a shift in the vibration pattern on the basilar membrane along the tonotopic axis, and this shift may provide the basis for detection of the change in frequency. Models based on this approach usually involve the concept of the excitation pattern, which may be considered as a psychoacoustic analog of the vibration pattern on the basilar membrane. For the purpose of this chapter, the excitation pattern is defined as the output of the auditory filters plotted as a function of filter center frequency (Moore & Glasberg, 1983a; Glasberg & Moore, 1990). The shapes of the excitation patterns are based on experiments involving notched-noise or rippled-noise simultaneous masking (Patterson, 1976). However, it may be the case that a closer correspondence with the tuning on the basilar membrane is obtained when nonsimultaneous masking is used (Oxenham & Shera, 2003; Oxenham & Simonson, 2006).

For a place mechanism, the smallest detectable difference in frequency between successive steady tones, the difference limen for frequency (DLF), may depend on detecting changes in excitation level at a single point on the excitation pattern (Zwicker, 1956; Zwislocki & Nguyen, 1999) or on the combination of information from changes in excitation level over the whole audible part of the excitation pattern (Moore & Sek, 1994). In either case, the DLF is predicted to depend on the slope of the excitation pattern and on the smallest detectable change in excitation level. For example, Zwicker (1956) proposed that a change in frequency can be detected whenever the excitation level at some point on the excitation pattern changes by more than about 1 dB. The change in excitation level is greatest on the steeply sloping low-frequency side of the excitation pattern. The steepness of the low-frequency side is roughly constant when expressed in units of ERB_N; the slope is about 27 dB/ERB_N. Thus, Zwicker's model predicts that the smallest detectable change in frequency in Hertz at any given center frequency should be about $ERB_N/27$. Zwicker intended his model to apply to thresholds for the detection of frequency modulation, called here FMDLs, but the model has also been applied to the DLF.

Data from many studies show that DLFs are not consistent with this prediction (Moore, 1973; Moore, 1974; Sek & Moore, 1995). If the DLF is expressed as a proportion of ERB_N, the result is not invariant with center frequency, but tends to increase markedly at high frequencies (Sek & Moore, 1995). An illustration of this is given by the solid line in Fig. 21.1 (the dashed line is described in Section 2.4). This is consistent with the idea that DLFs are determined by a temporal mechanism at low frequencies. The increase in the DLF at high frequencies is assumed to reflect the decreased precision of phase locking, as I argued in my PhD thesis (Moore,

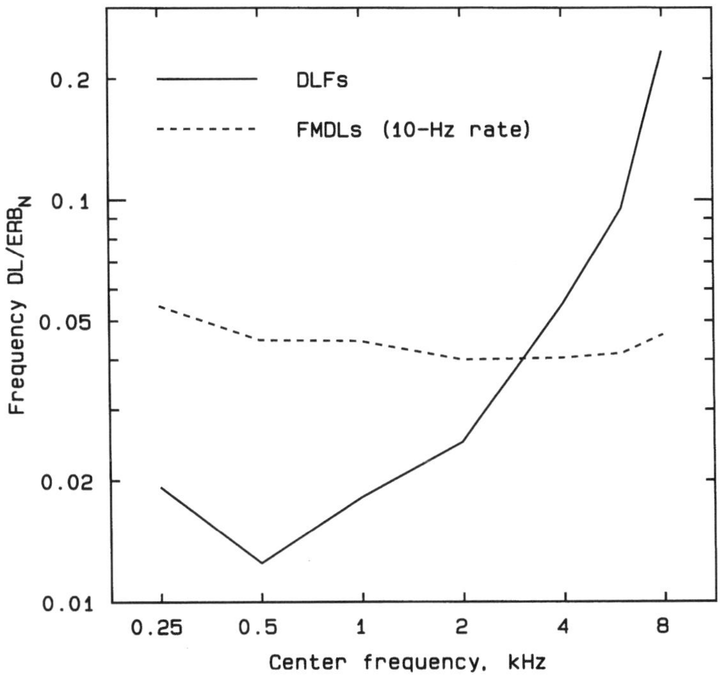

Fig. 21.1 The solid and dashed lines show the ratios DLF/ERB$_N$ and FMDL/ERB$_N$, respectively, plotted as a function of center frequency. The FM rate was 10 Hz. The data are from Sek and Moore (1995)

1973). Presumably, at some sufficiently high frequency, the temporal information carried by phase locking becomes so weak that place information dominates. I consider next the pattern of results that would be expected over the frequency range where a place mechanism dominates.

At high center frequencies, auditory filters and excitation patterns have a sharpness that is roughly invariant with center frequency, or even increases slightly with increasing center frequency, when plotted on a log-frequency scale (Oxenham & Shera, 2003; Oxenham & Simonson, 2006; Moore, 2012). Also, the ability to detect changes in excitation level is roughly invariant with center frequency, although it does worsen slightly at very high frequencies, especially for medium levels (Carlyon & Moore, 1984; Florentine et al., 1987). If DLFs at high frequencies are based purely on a place mechanism, then, over the range where the place mechanism is dominant, DLFs expressed as a proportion of center frequency, as Δf/f, should be approximately constant. If there is a transition from a temporal to a place mechanism, then Δf/f should increase with increasing center frequency above 1–2 kHz (as phase locking weakens), but eventually should reach an approximate plateau when the place mechanism has taken over fully from the temporal mechanism.

Until recently, published data did not show evidence for such a “breakpoint.” For most studies, Δf/f increased with increasing center frequency above 2 kHz up to the

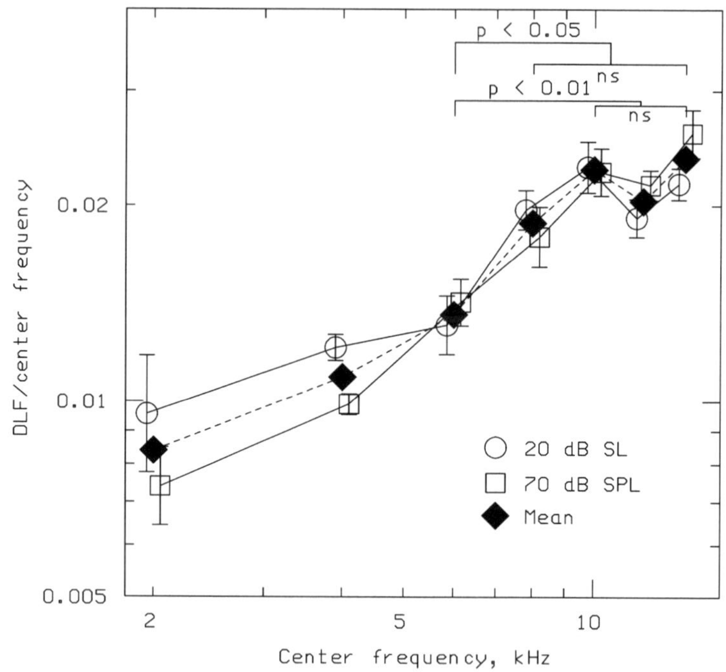

Fig. 21.2 Data from Moore and Ernst (2012) showing the ratio DLF/center frequency as a function of center frequency. The mean stimulus level was 20 dB SL (circles) or 70 dB SPL (squares). Filled diamonds show geometric mean values for the two levels. The outcomes of pairwise comparisons of the ratios for different center frequencies are shown at the top

highest frequency tested, which has usually been 8 kHz; for a review, see Micheyl et al. (2012). To account for the failure to find the expected breakpoint, Stephan Ernst and I (Moore & Ernst, 2012) suggested that usable phase locking may occur even for frequencies above 5 kHz. We proposed that a breakpoint might be observed, but at a higher center frequency than has typically been used in previous studies. To test this idea, we measured DLFs for center frequencies from 2 to 14 kHz, using earphones designed to produce a flat response at the eardrum. To make it difficult for subjects to use loudness cues to detect the frequency changes, the level of each tone was varied randomly over a level range of ±4 dB (uniform distribution) around the nominal level. This range was chosen to be large enough to disrupt the use of loudness cues but not so large that there would be substantial changes in pitch with level (Terhardt, 1974b; Verschuure & van Meeteren, 1975). For each frequency, the DLF was estimated for a mean level of 70 dB SPL and for a mean sensation level (SL) of 20 dB. The nine subjects were selected to have relatively low absolute thresholds for frequencies up to 14 kHz.

Geometric mean DLFs across subjects, expressed as $\Delta f/f$, are plotted as a function of frequency in Fig. 21.2. Open circles show mean DLFs obtained at 20 dB SL, and open squares show DLFs obtained at 70 dB SPL. Filled diamonds show geometric means for the two levels. The values of Δf/f increased progressively with increasing frequency up to 8 or 10 kHz and then tended to flatten off. An analysis of variance showed a significant effect of frequency, but no significant effect of level and no interaction. Pairwise comparisons showed that there was no significant difference between the values of Δf/f for center frequencies from 8 to 14 kHz. The values of Δf/f for all these center frequencies were significantly greater than the value of Δf/f at 6 kHz.

These results are consistent with the idea that DLFs depend on a transition from a temporal mechanism at low frequencies to a place mechanism at high frequencies. However, the transition appears to occur at 8–10 kHz rather than at 4–5 kHz. It remains to be explained why the transition suggested by the data of Moore and Ernst (2012) occurs at a frequency well above the frequency of 4–5 kHz at which other changes in pitch perception occur, as discussed in Section 2.2. Possibly, the weak phase locking that occurs for frequencies between 4 and 8 kHz is sufficient to allow reasonably small DLFs, but not sufficient to provide a strong sense of musical interval.

21.2.4 Detection of Frequency Modulation

As predicted by Zwicker's (1956) model, the ratio FMDL/ERB_N is approximately invariant with center frequency if the frequency modulation rate is 10 Hz or above (Sek & Moore, 1995), as illustrated by the dashed line in Fig. 21.1. This is consistent with the idea that FMDLs for high modulation rates are determined by a place mechanism for all carrier frequencies. Why should a place mechanism determine FMDLs but not DLFs? Aleksander Sek and I (Moore & Sek, 1996) proposed that the temporal mechanism may be "sluggish" and unable to track rapid changes in frequency. However, the temporal mechanism may be usable for very low FM rates. Figure 21.3 shows data from Sek and Moore (1995); the ratio FMDL/ERB_N is plotted as a function of center frequency for FM rates of 2, 5, and 10 Hz. The ratios for the 10-Hz rate are the same as those plotted in Fig. 21.1; they are approximately invariant with center frequency. For the two lower rates, FMDLs are smaller than for the 10-Hz rate for low center frequencies, but are higher than for the 10-Hz rate for high center frequencies. The pattern of results for the 2-Hz rate is similar to that for DLFs, as shown in Fig. 21.1, except that the changes in the ratio with frequency are somewhat smaller.

These results can be understood in terms of the variation of the effectiveness of the temporal and place mechanisms with carrier frequency and FM rate. For the highest carrier frequency (8 kHz), a place mechanism probably dominates for all FM rates. Performance worsens with decreasing FM rate because the ability to detect fluctuations in excitation level, as measured by thresholds for the detection of amplitude modulation, worsens with decreasing rate over the range 10–2 Hz (Moore & Sek, 1995; Ernst & Moore, 2010; Ernst & Moore, 2012). For lower carrier frequencies, a temporal mechanism probably plays a role when the FM rate is sufficiently slow. Hence performance for carrier frequencies below 2 kHz improves with decreasing FM rate.

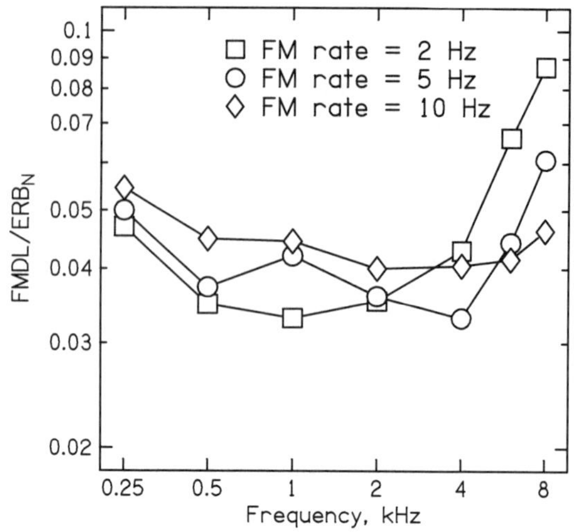

Fig. 21.3 The ratio FMDL/ERB$_N$, plotted as a function of carrier frequency for FM rates of 2 Hz (squares), 5 Hz (circles), and 10 Hz (diamonds). The data are from Sek and Moore (1995)

Another way of assessing the role of place cues in FM detection is to add amplitude modulation to the stimulus in all intervals of a forced-choice trial. For example, in a two-interval forced-choice task, the FM is present in only one randomly chosen interval, while the AM is present in both intervals. The amount of AM is chosen to be large relative to the changes in excitation level induced by the FM, hence making it more difficult to detect FM using a place mechanism. The AM can be sinusoidal with the same rate as the FM, or it can be noise-like, but with fluctuations in the same range as the FM. Experiments of this type have shown that, for low and medium carrier frequencies, the disruptive effect of the added AM tends to be greater for high FM rates than for low FM rates, consistent with the place mechanism being dominant for high rates (Moore & Sek, 1996; Lacher-Fougère & Demany, 1998; Moore & Skrodzka, 2002; Ernst & Moore, 2010). For high carrier frequencies, the disruptive effect of the added AM is similar across different FM rates (Moore & Sek, 1996; Moore & Skrodzka, 2002; Ernst & Moore, 2010), consistent with a dominant role for a place mechanism for all FM rates.

In summary, FM detection for FM rates of 10 Hz and above is probably dominated by a place mechanism for all carrier frequencies. For FM rates below 10 Hz, a temporal mechanism may dominate FM detection for low and medium carrier frequencies. The temporal mechanism may be sluggish and unable to track rapid changes in frequency.

21.2.5 *Evidence that Pitch Perception Depends on Both Place and Temporal Information*

Evidence that temporal information alone or place information alone are not sufficient for the perception of a clear pitch comes from studies of people with "dead regions" in the cochlea; these are regions with no or very few functioning inner hair cells (IHCs) and/or neurons (Moore et al., 2000; Moore, 2004). I coined the phrase "dead region," and have received complaints from audiologists that the phrase is too blunt; they prefer to tell their clients that they have "holes" in their hearing (Shannon et al., 2002) or to use some other more gentle term. However, for better or worse, the phrase dead regions has become commonly used.

The extent of a dead region can be defined in terms of the characteristic frequencies (CFs) of the functioning IHCs and neurons adjacent to the dead region. One way of demonstrating the existence of a dead region is by the measurement of psychophysical tuning curves (PTCs). For a subject without a dead region, the tip of the PTC (i.e., the frequency at which the masker level is lowest) lies close to the signal frequency (Moore, 1978). However, if the signal frequency falls within a dead region, the tip of the PTC may be above the signal frequency (for a low-frequency dead region) or below the signal frequency (for a high-frequency dead region) (Moore et al., 2000; Moore & Alcántara, 2001; Kluk & Moore, 2005, 2006).

When a tone has a frequency falling within a dead region, the peak in the neural excitation pattern may occur at a place very different from that normally associated with that frequency. The place theory predicts that the perceived pitch of the tone in such a case should be very different from normal. Consider as an example a person with an apical dead region, extending downwards from 1 kHz. Any tone with a frequency below 1 kHz, if it is made sufficiently intense to be detectable, would produce a peak in the neural excitation pattern at the same place, a place with CF just above 1 kHz. Also, the apical edge of the excitation pattern would always fall at the same place. Hence, according to the place theory, all tones with frequencies below 1 kHz should have the same pitch. However, the neural spikes evoked by a low-frequency tone would still be synchronized to a specific phase of the tone, that is, phase locking would occur. Hence, if temporal information alone were sufficient for pitch perception, the perceived pitch should correspond to the frequency of the stimulus.

Florentine and Houtsma (1983) studied a subject with a moderate to severe unilateral low-frequency hearing loss of cochlear origin. For a 1-kHz signal, the tip of the PTC for the impaired ear fell between 2.2 and 2.85 kHz, depending on the exact level of the signal, suggesting a low-frequency dead region with an edge frequency between 2.2 and 2.85 kHz. Florentine and Houtsma obtained pitch matches between the two ears of their subject. Pitch shifts between the two ears were small, even for tones whose frequencies fell within the dead region in the impaired ear. Thus, the tone in the impaired ear was perceived with a roughly "normal" pitch, despite being detected at the "wrong" place in the cochlea. However, the variability of the pitch matches was rather large, indicating that the pitch in the impaired ear was not clear. Turner et al. (1983) studied six subjects with low-frequency cochlear hearing losses. Three subjects gave PTCs with tips close to the signal frequency; they did not have dead regions. The other three subjects showed PTCs with tips well above the signal frequency, indicating that they had low-frequency dead regions. Pitch perception was studied either by pitch matching between the two ears (for subjects with unilateral losses) or by octave matching (for subjects with bilateral losses, but with some musical ability). The subjects whose PTCs had tips above the signal frequency gave results similar to the subjects whose PTCs had tips close to the signal frequency; no distinct pitch anomalies were observed. These results are not consistent with the place theory.

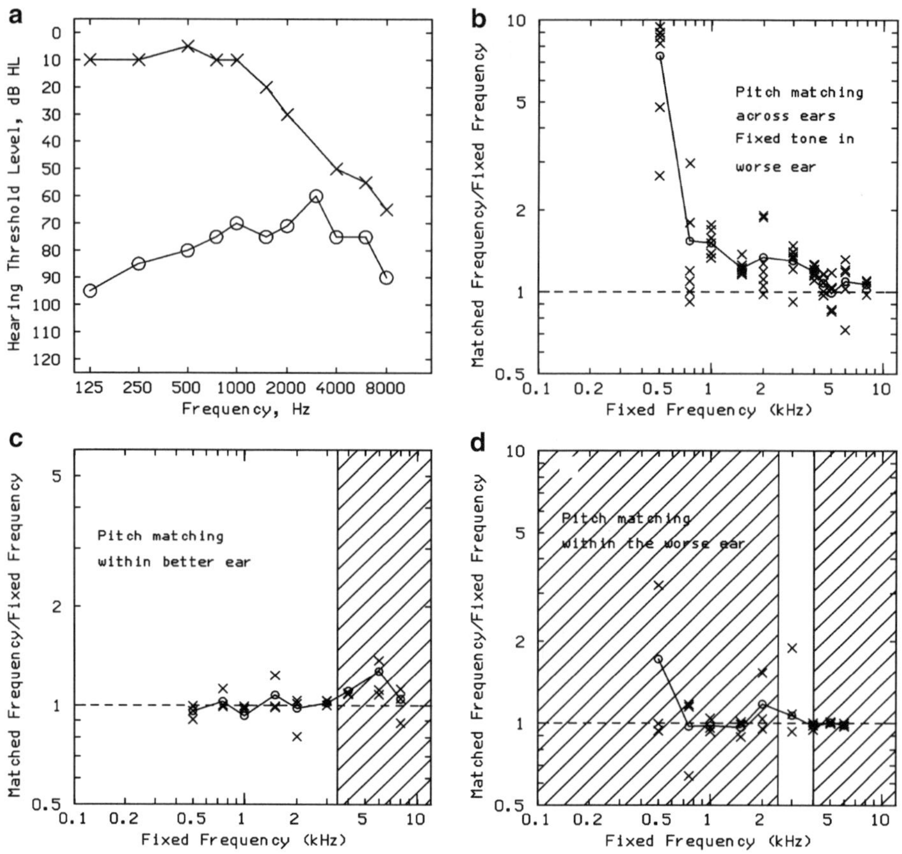

Fig. 21.4 Pitch-matching results for a subject with highly asymmetric hearing loss. The audiogram for the left ear (crosses) and right ear (circles) is shown in (**a**). The bottom panels show the results of pitch matching within the better ear (**c**) and worse ear (**d**). Crosses show individual matches and circles show geometric means. The scatter of the points around the dashed lines indicates the inherent variability of the matches. Shaded areas indicate dead regions. (**b**) Results of pitch matching across ears, with the fixed-frequency tone in the worse ear. The dashed line indicates where the matches would lie if a given frequency evoked the same pitch in the two ears. Data are from Huss and Moore (2005)

A similar study was conducted by Martina Huss and I (Huss & Moore, 2005). Both pitch-matching and octave-matching tasks were used. The level for each frequency was chosen using a loudness model (Moore & Glasberg, 1997), so as to give a calculated loudness level of either 50 or 60 phons. Results of the pitch-matching task for a subject with severe hearing loss in the right ear and a moderate high-frequency loss in the left ear are shown in Fig. 21.4. The audiogram of the subject is shown in panel A (circles—right ear, crosses—left ear). On the basis of the "threshold equalizing noise" test described by Moore et al. (2000), and on the basis of PTCs, this subject was diagnosed as having extensive low-frequency and high-frequency dead regions in the right ear, with an "island" of functioning IHCs around

3.5 kHz. The left ear had a dead region above 3.4 kHz. In the remaining panels, each × denotes one match, and means are shown by open circles. Shaded areas indicate dead regions. Matches within the better ear (Fig. 21.4c) were reasonably accurate at low frequencies, but became less accurate at high frequencies. Matches within the worse ear (Fig. 21.4d), were more erratic, indicating a less clear pitch percept. Matches across ears, with the fixed tone in the worse ear (Fig. 21.4b), showed considerable variability, but also some consistent deviations. A fixed tone with a frequency of 0.5 kHz in the worse ear was matched with a tone of about 3.5 kHz in the better ear. Generally, the matched frequency lay above the fixed frequency, for all fixed frequencies up to about 4 kHz, indicating upward pitch shifts in the worse ear.

Huss and Moore (2005) also obtained pitch matches and octave matches for subjects with high-frequency dead regions. Results for a subject with an extensive high-frequency dead region are shown in Fig. 21.5 (results for one ear only; the other ear had no usable hearing). The dead region was estimated to start at about 1.2 kHz. Pitch matches within one ear (Fig. 21.5b) were reasonably accurate for frequencies up to 1.25 kHz, and then became much more erratic, indicating that a clear pitch percept was not obtained for frequencies falling well within the dead region. Octave matches with the lower tone fixed in frequency (Fig. 21.5d) resulted in frequency ratios around 2 (the "expected" value) for fixed frequencies up to 0.5 kHz. For a fixed frequency of 1 kHz, the upper tone was adjusted to about 1.4 kHz; when the upper tone fell within the dead region its pitch was higher than "normal". Octave matches with the upper tone fixed in frequency (Fig. 21.5c) resulted in frequency ratios around (but a little above) 0.5 (the "expected" value) for fixed frequencies up to 1 kHz. For fixed frequencies of 1.76 and 2 kHz, octave matches clearly deviated from a ratio of 0.5. For tones whose frequencies fell well within the dead region, the perceived pitch was shifted upwards, although it was also unclear.

Taken together, the results of studies of pitch perception using people with dead regions indicate the following:

1. Pitch matches are often erratic for tones with frequencies falling well inside a dead region. This indicates that such tones do not evoke a clear pitch sensation.
2. Pitch matches across the ears of subjects with asymmetric hearing loss, and octave matches within ears, indicate that tones falling within a dead region sometimes are perceived with a near-"normal" pitch and sometimes are perceived with a pitch distinctly different from "normal."
3. The pitch does not generally correspond to the CF of the place that is excited most. This is not consistent with the place theory.
4. The shifted pitches found for some subjects indicate that the pitch of low-frequency tones is not represented solely by a temporal code. Possibly, there needs to be a correspondence between place and temporal information for a "normal" pitch to be perceived (Evans, 1978; Loeb et al., 1983; Srulovicz & Goldstein, 1983). Alternatively, temporal information may be "decoded" by a network of coincidence detectors or cross-correlators whose operation depends on the phase response at adjacent points along the basilar membrane (Loeb et al., 1983; Shamma & Klein, 2000; Carlyon et al., 2011). Detection of a tone at the "wrong" place in the cochlea may prevent effective use of temporal information because

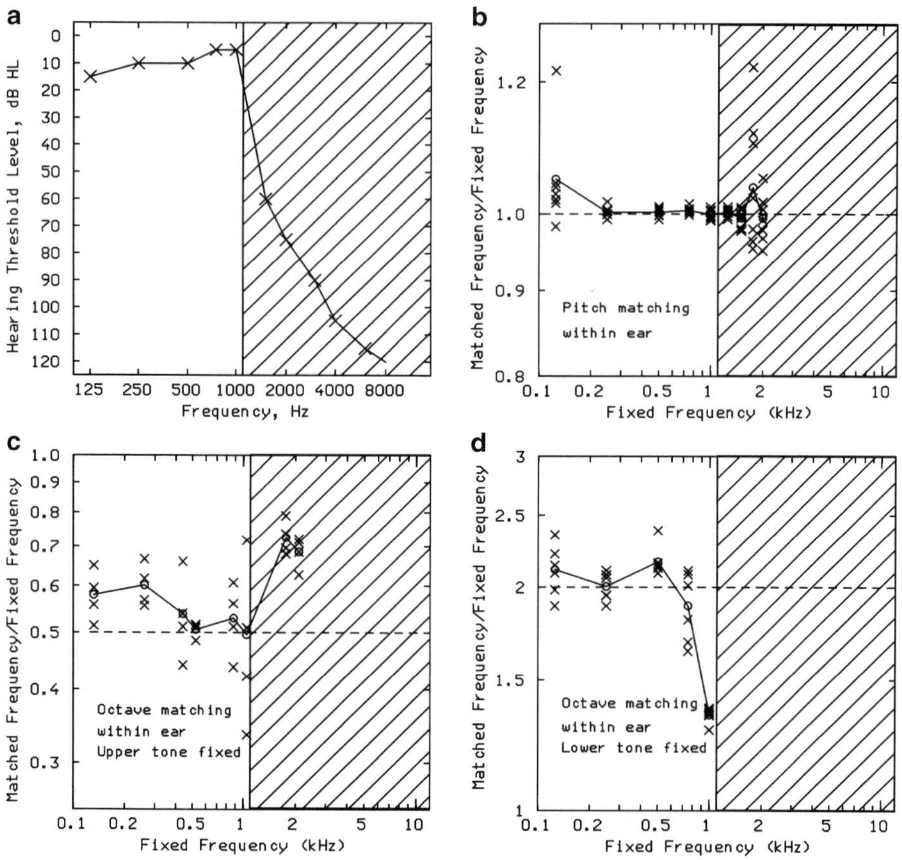

Fig. 21.5 Pitch and octave-matching results for a subject with a precipitous hearing loss in one ear; the other ear had no usable hearing. The audiogram for the better ear is shown in (**a**). The shaded area indicates a dead region. (**b**) Results of pitch matching within the better ear. Crosses show individual matches and circles show geometric means. The scatter of the points around the dashed line indicates the inherent variability of the matches. The bottom panels show the results of octave-matching with the upper tone fixed in frequency (**c**) and the lower tone fixed in frequency (**d**). The dashed lines indicate where the matches would lie if the octave match corresponded to a frequency ratio of 2. Data are from Huss and Moore (2005)

the relative phase responses around this place are very different from the relative phase responses around the "correct" place.

21.2.6 Conclusions on the Pitch of Pure Tones

1. Several aspects of the perception of the pitch of pure tones, including the ability to identify musical intervals and the ability to hear out partials from complex tones or chords, show a worsening in performance at high frequencies (above 4–5 kHz), consistent with a loss of temporal information (reduced precision of phase locking).

2. DLFs expressed as a proportion of center frequency worsen with increasing center frequency up to about 8 kHz, and then reach an approximate plateau. This is consistent with a transition from a temporal to a place mechanism, but the transition occurs at a higher frequency than is usually assumed.
3. FMDLs for FM rates of 10 Hz and above seem to be mainly determined by a place mechanisms for all carrier frequencies. FMDLs for lower rates are probably partly determined by a temporal mechanism for frequencies below about 4 kHz. The temporal mechanism may be sluggish, and unable to track rapid changes in frequency.
4. Data obtained from subjects with dead regions in the cochlea suggest that the pitch of pure tones is not determined by the place of maximum excitation in the cochlea. The pitch of tones whose frequency falls within a dead region is usually indistinct. The pitch value sometimes corresponds approximately with what would be expected from the use of temporal information, but the pitch can also differ from the expected value. Possibly there needs to be a correspondence between place and temporal information for a clear pitch to be perceived. Alternatively, temporal information may be "decoded" on a place-specific basis.

21.3 The Pitch of Complex Tones

21.3.1 Pitch Mechanisms

Within the cochlea, each spectral component in a complex tone gives rise to maximum vibration at a specific place along the basilar membrane. The harmonics in a periodic complex tone have a spacing equal to the fundamental frequency, F0. However, the bandwidths of the filters on the basilar membrane increase with increasing center frequency (von Békésy, 1960; Robles & Ruggero, 2001). Similarly, the bandwidths of the auditory filters as measured in masking experiments increase with increasing center frequency (Glasberg & Moore, 1990), from about 30 Hz at very low frequencies (Jurado & Moore, 2010) to about 1600 Hz at very high frequencies (Zhou, 1995). As a result, the lower harmonics in a complex tone are resolved on the basilar membrane; each gives rise to a distinct peak along the tonotopic axis. In contrast, the high harmonics do not give rise to distinct separate peaks; each place responds to several harmonics. Such harmonics are said to be unresolved.

Many researchers have proposed that there are different pitch mechanisms for low and for high harmonics (de Boer, 1956; Houtsma & Smurzynski, 1990; Shackleton & Carlyon, 1994). For tones with low harmonics, the pitch may be derived from temporal information and/or place information about the frequencies of individual harmonics (Goldstein, 1973; Terhardt, 1974a). For tones with only very high harmonics (above about the 14th), the pitch may be derived from the temporal envelope of the waveform evoked on the basilar membrane by the interference of several harmonics. Tones containing only very high harmonics have a less distinct pitch than tones with low harmonics, and changes in their F0 are less well discriminated (Hoekstra & Ritsma, 1977; Moore & Rosen, 1979; Moore & Glasberg,

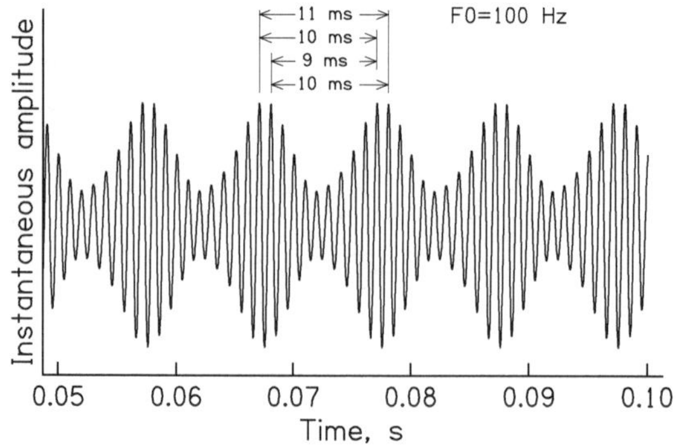

Fig. 21.6 Waveform at the output of a simulated auditory filter centered at 1000 Hz in response to a complex tone containing many harmonics of an F0 of 100 Hz. Nerve spikes are evoked at times close to the largest peaks in the TFS (close to envelope maxima). The most prominent time intervals between nerve spikes are shown at the top

1988b; Moore & Peters, 1992; Houtsma & Smurzynski, 1990; Bernstein & Oxenham, 2003, 2006). For tones for which the number of the lowest harmonic, *N*, is in the range 6–14, the pitch may be derived either from estimates of the frequencies of partially resolved harmonics or from the TFS of the waveform evoked on the basilar membrane by the interference of two or more harmonics. Nerve spikes tend to occur at peaks in the TFS close to envelope maxima, and pitch is assumed to correspond to the time interval between peaks in the TFS close to adjacent envelope maxima (Schouten, 1940; Schouten et al., 1962; Moore et al., 2009). This concept is illustrated in Fig. 21.6, which shows the output of a simulated auditory filter centered at 1000 Hz in response to a complex tone containing many equal-amplitude harmonics of a 100-Hz F0. For this example, the time interval that occurs most often is 10 ms (=1/F0), but other intervals such as 9 and 11 ms also occur. Such intervals are represented in the temporal patterns of discharge in the auditory nerve (Javel, 1980; Moore, 1980; Cariani & Delgutte, 1996a, b).

Discrimination of F0 worsens when *N* increases above about 5–7, reaching a plateau when *N* is about 14–15 (Hoekstra & Ritsma, 1977; Houtsma & Smurzynski, 1990; Bernstein & Oxenham, 2003; Moore et al., 2006a). The worsening as *N* is increased from 7 to about 14 has been interpreted by some researchers as resulting from a progressive reduction of the ability to resolve the components in the complex tone (Houtsma & Smurzynski, 1990; Shackleton & Carlyon, 1994). This is referred to as the "resolvability hypothesis." Other researchers have interpreted the worsening as resulting from a progressive loss of the ability to use TFS information (Moore et al., 2006a; Hopkins & Moore, 2007; Ives & Patterson, 2008; Moore & Sek, 2009a, b). This is referred to as the "TFS hypothesis."

The decision as to whether the resolvability hypothesis or the TFS hypothesis is more nearly correct depends on the extent to which harmonics with numbers in the range 7–14 are resolved, which is still a matter of debate; for a review, see Moore and Gockel (2011).

21.3.2 The Limits of Resolvability

One of the least controversial ways of estimating the extent to which components in a complex tone are resolved is to measure masking patterns in forward masking. The use of nonsimultaneous masking avoids effects resulting from interactions between the signal and masker. Plomp (1964) used as a masker a complex tone with an F0 of 500 Hz. Each harmonic had the same loudness level. The amount of forward masking of a 20-ms pure tone signal was measured as a function of signal frequency. There were clear peaks in the masking pattern corresponding to the first five harmonics. However, no such peaks were apparent for the sixth or higher harmonics.

A similar method was used by myself and Brian Glasberg (Moore & Glasberg, 1983b). We used as forward maskers three complex tones with F0s of 100, 200, and 400 Hz, all with equal-level components (61 dB SPL per component). The masking patterns showed distinct peaks only for the first three or four harmonics. We argued that the measurements should have been sufficiently accurate to measure ripples in the excitation pattern with a depth of 3 dB or more. Thus, the results suggest that ripples in the excitation pattern are smaller than 3 dB for the fifth and higher harmonics.

An alternative method that involves nonsimultaneous presentation of the signal and masker is the pulsation-threshold method (Houtgast, 1972, 1974). The sinusoidal signal is presented in alternation with the complex tone in a repeating sequence. The level of the signal is adjusted to find the value at which the signal changes from appearing to pulsate to appearing to be continuous. This is called the "pulsation threshold." Houtgast (1972) argued that, at the pulsation threshold, the excitation evoked by the signal approximately matched the excitation evoked by the masker in the frequency region of the signal.

Houtgast (1974) used as a masker a complex tone containing the first 10 harmonics of an F0 of 250 Hz. Each harmonic had the same level and two overall levels were used, separated by 30 dB. The pulsation-threshold patterns showed clear peaks for harmonics up to the sixth. The ripples in the pulsation-threshold patterns were smaller at the higher level, consistent with the idea that frequency selectivity decreases with increasing level (Moore & Glasberg, 1987). There was a very small peak corresponding to the seventh harmonic for both levels used, and no peaks corresponding to higher harmonics for either level. These results are broadly consistent with those obtained using forward-masking patterns, and suggest that only harmonics up to the sixth or seventh are resolved. Thus, the results support the TFS hypothesis rather than the resolvability hypothesis.

21.3.3 The Dominant Region

For complex tones containing many harmonics, the pitch is determined mainly by the harmonics in a limited frequency region, called the "dominant region." Two approaches have been used for determining the dominant region. In one approach, a group of harmonics (Plomp, 1967; Ritsma, 1967) or a single harmonic (Moore et al., 1985;

Gockel et al., 2005) are shifted in frequency (mistuned) and the effect of this on the pitch of the whole sound is determined. It is assumed that the harmonics whose shift has the greatest effect on the overall pitch of the complex are the dominant harmonics. In the other approach, thresholds for discriminating the frequency of a group of harmonics (Miyazono et al., 2009, 2010) or a single harmonic (Moore et al., 1984; Gockel et al., 2007) are determined. It is assumed that the harmonics giving the lowest discrimination thresholds are the dominant ones (Moore & Glasberg, 1986). Consistent with this assumption, difference limens for the discrimination of F0 (F0DLs) can be predicted from the thresholds for frequency discrimination of single harmonics within the complex tones, based on the optimal combination of information across harmonics (Moore et al., 1984; Gockel et al., 2007).

For complex tones with F0s in the range 100–400 Hz, the dominant region corresponds to harmonics in the range 2–5 (Plomp, 1967; Ritsma, 1967; Moore et al., 1984; 1985), although some researchers have argued that the dominant region corresponds to a fixed frequency region around 500–600 Hz, rather than a fixed range of harmonic numbers (Terhardt, 1979; Dai, 2000). The finding that the dominant harmonics are generally reasonably well resolved is consistent with the idea that resolution of harmonics is important for a clear pitch to be perceived. However, data for tones with very low F0s are not consistent with this view.

For low center frequencies, the auditory filter bandwidth, expressed as a proportion of center frequency, increases (Jurado & Moore, 2010; Jurado et al., 2011). This means that fewer low harmonics are resolved (Plomp, 1964; Plomp & Mimpen, 1968). As described earlier, for a low-frequency harmonic to be resolved, it should be separated from neighboring harmonics by 1.25 ERB_N or more. For an F0 of 50 Hz, only the first two harmonics would be resolved (at 100 Hz, $ERB_N \approx 35$ Hz, while at 150 Hz $ERB_N \approx 41$ Hz, and 41 Hz × 1.25 is greater than the spacing of 50 Hz between harmonics). If resolvability is the key factor determining the dominant region, then for an F0 of 50 Hz, the first and second harmonics should be dominant.

The data do not support this prediction. Moore and Glasberg (1988a, b) showed that for an F0 of 50 Hz, F0DLs were larger for complex tones that contained harmonics 1–5 than for complex tones that contained harmonics 6–12. This suggests that the dominant region falls in a range where the harmonics are not resolved. Helen Jackson and I (Jackson & Moore, 2013) investigated the dominant region for an F0 of 50 Hz more directly. F0DLs were measured for a group of harmonics (group B) embedded in a group of fixed (nonoverlapping) harmonics (group A) with the same mean F0 of 50 Hz. It was assumed that F0DLs would be lowest when the harmonics in group B fell in the dominant region. The rank of the lowest harmonic in group B, N, was varied from 1 to 15. The components in the two groups started either in random or in cosine phase. I consider here only the results for the random-phase condition. F0DLs increased with increasing N, but the increase did not occur until N was above 5. At the center frequency of the 5th harmonic, 250 Hz, the value of ERB_N is about 52 Hz (Glasberg & Moore, 1990), which is greater than the spacing between harmonics. It seems likely that the 5th harmonic was barely if at all resolved. The fact that performance did not worsen when N was increased from 1 to 5 suggests that resolvability is not the critical factor determining the dominant region.

21.3.4 Further Evidence that Resolvability Is Not Critical

Bernstein and Oxenham (2003) measured F0DLs for complex tones with successive harmonics of a 100- or 200-Hz F0 that were presented dichotically (even harmonics to one ear and odd harmonics to the other ear) or diotically (all harmonics presented to both ears). For dichotic presentation, the ear receiving the even harmonics was determined randomly for every stimulus. Under these conditions, the harmonics were perceptually fused across ears, and a single pitch was heard. The rank of the lowest and highest harmonics in each stimulus was roved, to encourage performance based on comparisons of the (missing) F0 rather than comparisons of the pitch of individual harmonics. With dichotic presentation, the harmonics in each ear were more widely separated in frequency than for diotic presentation, making the harmonics more resolvable. Performance was measured as a function of the (mean) rank of the lowest harmonic, *N*. If F0DLs for diotic stimuli with high *N* are large because the harmonics are unresolved, then dichotic presentation should lead to markedly smaller F0DLs. Bernstein and Oxenham (2003) found that F0DLs increased when *N* was above 9, but the pattern of the results was similar for dichotic and diotic presentation. Thus, subjects could not take advantage of the greater resolvability of the harmonics in the dichotic condition to improve F0 discrimination for high *N*.

Bernstein and Oxenham (2008) later found somewhat different results. In this study, for dichotic presentation it was determined randomly on each trial which ear received the even harmonics and which received the odd harmonics, but within a trial the even harmonics were always presented to the same ear. Also, the rank of the lowest and highest harmonics was not roved from one stimulus to the next. Under these conditions, F0DLs were smaller (performance was better) for dichotic than for diotic presentation. However, the changes in F0DL with *N* were still similar for dichotic and diotic presentation. Thus, improving the resolvability of the harmonics through dichotic presentation did not change the range of values of *N* over which performance worsened. Again, this leads to the conclusion that the worsening in performance with increasing *N* is not primarily caused by reduced resolution of harmonics.

Bernstein and Oxenham (2008) also explored the effect of increasing the frequencies of the odd harmonics by 3%. This led to perceptual segregation of the odd and even harmonics, even when all harmonics were presented diotically. The mistuning led to smaller F0DLs for both dichotic and diotic presentation for high *N*. However, the mistuning did not lead to an improvement in the ability to hear out individual partials in the complex sound, as measured in a separate experiment.

21.3.5 Modeling the Effects of N

Bernstein and Oxenham (2005, 2008) presented a modified auto-correlation model to account for their results. In this model, the stimulus was passed through an array of simulated auditory filters, and simulated neural spike trains were generated for each filter. An autocorrelation function (ACF) was calculated for the spike trains in each channel.

A weighting function was applied to limit the range of lags that could be measured for each center frequency, which effectively introduces a dependence on harmonic number. This is based on the idea that the range of interspike intervals that can be measured at a given center frequency is limited to about 15 divided by the center frequency, as proposed in the second edition of my book *An Introduction to the Psychology of Hearing* (Moore, 1982). The ACFs were then combined across channels to give a summary ACF (SACF). It was assumed that F0 discrimination was based on changes in the SACF. The model was able to account for their data reasonably well. Note that in this model, as in earlier models based on autocorrelation (Meddis & O'Mard, 1997), use is made of TFS information, but no distinction is made as to whether that information comes from resolved harmonics, partially resolved harmonics, or unresolved harmonics.

The model of Bernstein and Oxenham (2005) accounts for the worsening of F0 discrimination with increasing N using a limit in the range of autocorrelation lags that can be measured at each center frequency. A possible origin of this limit comes from the work of de Cheveigné and Pressnitzer (2006). They proposed a model in which the delays that are required to perform operations such as autocorrelation or measurement of interspike intervals are synthesized from cross-channel phase interaction. According to this model, the delays that can be measured are limited by the impulse response times of the auditory filters. Because these impulse response times decrease with increasing center frequency (and correspondingly with increasing filter bandwidth), a center-frequency-dependent limit to the range of delays that can be measured is automatically produced.

Another mechanism that might limit the range of delays that can be measured was proposed by Brian Glasberg and I (Moore & Glasberg, 2010). We suggested that the auditory system has difficulty distinguishing interspike intervals that are very close to one another, because of jitter in the timing of nerve spikes (Johnson, 1980) and limitations in the precision with which central mechanisms can measure the time intervals. Consider the outputs of auditory filters centered at 1000 and 2000 Hz in response to a complex tone with many harmonics of an F0 of 100 Hz. For the center frequency of 1000 Hz, the most prominent time interval between nerve spikes would be 10 ms (=1/F0), but other intervals such as 9 ms and 11 ms would also occur, as illustrated in Fig. 21.6. Hence, to extract the F0 from the TFS, the auditory system must distinguish the intervals of 9, 10 and 11 ms. For the center frequency of 2000 Hz, the interspike intervals would cluster around 9, 9.5, 10, 10.5, and 11 ms. These intervals may be harder to distinguish from one another than the intervals for the center frequency of 1000 Hz, because they are closer together. Hence, for an F0 of 100 Hz, useful TFS information may be extracted for a center frequency of 1000 Hz, but not for a center frequency of 2000 Hz. As F0 increases, the highest CFs from which useful TFS information can be extracted would also increase, because the limit is related to the relative magnitude of the intervals that have to be discriminated.

It should be noted that the limitations discussed above apply to the extraction of TFS information from individual neurons or small groups of neurons. Once time-interval information has been extracted, it may be combined across neurons and over time to yield F0DLs that are markedly smaller than the 5% limitation that might apply to the precision of interval extraction within individual neurons.

21.3.6 Conclusions on the Pitch of Complex Tones

The clarity of pitch decreases and F0DLs increase when the rank of the lowest harmonic, *N*, is increased from about 6 or 7 to about 14. It has been proposed that this reflects decreasing resolution of harmonics with increasing *N* (the resolvability hypothesis) or a decrease in the ability to use TFS information with increasing *N* (the TFS hypothesis). The following aspects of the results reviewed in Sections 3.2 to 3.5 are more consistent with the TFS hypothesis than with the resolvability hypothesis:

1. Data obtained using forward masking or the pulsation threshold method indicate that harmonics above the seventh in complex tones are not resolved.
2. For tones with a low F0 of 50 Hz, the dominant harmonics are not the lowest (resolved) harmonics; rather the dominant harmonics are not resolved.
3. Dichotic presentation (even harmonics to one ear and odd harmonics to the other ear), which would improve the resolution of harmonics, does not change the way that F0DLs vary with *N*.
4. Mistuning the odd harmonics by 3% improves F0 discrimination, but does not improve the resolution of harmonics.

The results are compatible with models in which TFS information from both resolved and unresolved harmonics is used, but with a limitation to the range of time intervals that can be measured at each center frequency.

Acknowledgments My thinking about pitch was strongly influenced by a visit to the Netherlands while I was working on my PhD. I had especially useful discussions with Reinier Plomp, Guido Smoorenburg, Tammo Houtgast, Roel Ritsma, Jan Schouten, Ben Lopez-Cardozo, and Egbert de Boer. I am grateful to all of them for the time that they graciously spent with a then-unknown PhD student.

References

Attneave, F., & Olson, R. K. (1971). Pitch as a medium: A new approach to psychophysical scaling. *American Journal of Psychology*, 84, 147–166.

Bernstein, J. G., & Oxenham, A. J. (2003). Pitch discrimination of diotic and dichotic tone complexes: Harmonic resolvability or harmonic number? *Journal of the Acoustical Society of America*, 113, 3323–3334.

Bernstein, J. G., & Oxenham, A. J. (2005). An autocorrelation model with place dependence to account for the effect of harmonic number on fundamental frequency discrimination. *Journal of the Acoustical Society of America*, 117, 3816–3831.

Bernstein, J. G., & Oxenham, A. J. (2006). The relationship between frequency selectivity and pitch discrimination: Effects of stimulus level. *Journal of the Acoustical Society of America*, 120, 3916–3928.

Bernstein, J. G., & Oxenham, A. J. (2008). Harmonic segregation through mistuning can improve fundamental frequency discrimination. *Journal of the Acoustical Society of America*, 124, 1653–1667.

Cariani, P. A., & Delgutte, B. (1996a). Neural correlates of the pitch of complex tones. I. Pitch and pitch salience. *Journal of Neurophysiology*, 76, 1698–1716.

Cariani, P. A., & Delgutte, B. (1996b). Neural correlates of the pitch of complex tones. II. Pitch shift, pitch ambiguity, phase invariance, pitch circularity, rate pitch and the dominance region for pitch. *Journal of Neurophysiology*, 76, 1717–1734.

Carlyon, R. P., & Moore, B. C. J. (1984). Intensity discrimination: A severe departure from Weber's Law. *Journal of the Acoustical Society of America*, 76, 1369–1376.

Carlyon, R. P., Long, C. J., & Micheyl, C. (2011). Across-channel timing differences as a potential code for the frequency of pure tones. *Journal of the Association for Research in Otolaryngology*, 13, 159–171.

Dai, H. (2000). On the relative influence of individual harmonics on pitch judgment. *Journal of the Acoustical Society of America*, 107, 953–959.

de Boer, E. (1956). Pitch of inharmonic signals. *Nature*, 178, 535–536.

de Cheveigné, A., & Pressnitzer, D. (2006). The case of the missing delay lines: Synthetic delays obtained by cross-channel phase interaction. *Journal of the Acoustical Society of America*, 119, 3908–3918.

Ernst, S. M., & Moore, B. C. J. (2010). Mechanisms underlying the detection of frequency modulation. *Journal of the Acoustical Society of America*, 128, 3642–3648.

Ernst, S. M., & Moore, B. C. J. (2012). The role of time and place cues in the detection of frequency modulation by hearing-impaired listeners. *Journal of the Acoustical Society of America*, 131, 4722–4731.

Evans, E. F. (1978). Place and time coding of frequency in the peripheral auditory system: Some physiological pros and cons. *Audiology*, 17, 369–420.

Florentine, M., & Houtsma, A. J. M. (1983). Tuning curves and pitch matches in a listener with a unilateral, low-frequency hearing loss. *Journal of the Acoustical Society of America*, 73, 961–965.

Florentine, M., Buus, S., & Mason, C. R. (1987). Level discrimination as a function of level for tones from 0.25 to 16 kHz. *Journal of the Acoustical Society of America*, 81, 1528–1541.

Glasberg, B. R., & Moore, B. C. J. (1990). Derivation of auditory filter shapes from notched-noise data. *Hearing Research*, 47, 103–138.

Gockel, H., Plack, C. J., & Carlyon, R. P. (2005). Reduced contribution of a nonsimultaneous mistuned harmonic to residue pitch. *Journal of the Acoustical Society of America*, 118, 3783–3793.

Gockel, H., Moore, B. C. J., Carlyon, R. P., & Plack, C. J. (2007). Effect of duration on the frequency discrimination of individual partials in a complex tone and on the discrimination of fundamental frequency. *Journal of the Acoustical Society of America*, 121, 373–382.

Goldstein, J. L. (1973). An optimum processor theory for the central formation of the pitch of complex tones. *Journal of the Acoustical Society of America*, 54, 1496–1516.

Hartmann, W. M., & Doty, S. L. (1996). On the pitches of the components of a complex tone. *Journal of the Acoustical Society of America*, 99, 567–578.

Hartmann, W. M., McAdams, S., & Smith, B. K. (1990). Hearing a mistuned harmonic in an otherwise periodic complex tone. *Journal of the Acoustical Society of America*, 88, 1712–1724.

Heinz, M. G., Colburn, H. S., & Carney, L. H. (2001). Evaluating auditory performance limits: I. One-parameter discrimination using a computational model for the auditory nerve. *Neural Computation*, 13, 2273–2316.

Hoekstra, A., & Ritsma, R. J. (1977). Perceptive hearing loss and frequency selectivity. In E. F. Evans & J. P. Wilson (Eds.), *Psychophysics and physiology of hearing* (pp. 263–271). London: Academic Press.

Hopkins, K., & Moore, B. C. J. (2007). Moderate cochlear hearing loss leads to a reduced ability to use temporal fine structure information. *Journal of the Acoustical Society of America*, 122, 1055–1068.

Houtgast, T. (1972). Psychophysical evidence for lateral inhibition in hearing. *Journal of the Acoustical Society of America*, 51, 1885–1894.

Houtgast, T. (1974). *Lateral suppression in hearing*. Ph.D. thesis, Free University of Amsterdam.

Houtsma, A. J. (1981). Noise-induced shifts in the pitch of pure and complex tones. *Journal of the Acoustical Society of America*, 70, 1661–1668.
Houtsma, A. J. M., & Smurzynski, J. (1990). Pitch identification and discrimination for complex tones with many harmonics. *Journal of the Acoustical Society of America*, 87, 304–310.
Huss, M., & Moore, B. C. J. (2005). Dead regions and pitch perception. *Journal of the Acoustical Society of America*, 117, 3841–3852.
Ives, D. T., & Patterson, R. D. (2008). Pitch strength decreases as F0 and harmonic resolution increase in complex tones composed exclusively of high harmonics. *Journal of the Acoustical Society of America*, 123, 2670–2679.
Jackson, H. M., & Moore, B. C. J. (2013). The dominant region for the pitch of complex tones with low fundamental frequencies. *Journal of the Acoustical Society of America*, 134, 1193–1204.
Javel, E. (1980). Coding of AM tones in the chinchilla auditory nerve: Implications for the pitch of complex tones. *Journal of the Acoustical Society of America*, 68, 133–146.
Johnson, D. H. (1980). The relationship between spike rate and synchrony in responses of auditory-nerve fibers to single tones. *Journal of the Acoustical Society of America*, 68, 1115–1122.
Jurado, C., & Moore, B. C. J. (2010). Frequency selectivity for frequencies below 100 Hz: comparisons with mid-frequencies. *Journal of the Acoustical Society of America*, 128, 3585–3596.
Jurado, C., Pedersen, C. S., & Moore, B. C. J. (2011). Psychophysical tuning curves for frequencies below 100 Hz. *Journal of the Acoustical Society of America*, 129, 3166–3180.
Kluk, K., & Moore, B. C. J. (2005). Factors affecting psychophysical tuning curves for hearing-impaired subjects. *Hearing Research*, 200, 115–131.
Kluk, K., & Moore, B. C. J. (2006). Detecting dead regions using psychophysical tuning curves: A comparison of simultaneous and forward masking. *International Journal of Audiology*, 45, 463–476.
Lacher-Fougère, S., & Demany, L. (1998). Modulation detection by normal and hearing-impaired listeners. *Audiology*, 37, 109–121.
Loeb, G. E., White, M. W., & Merzenich, M. M. (1983). Spatial cross correlation: A proposed mechanism for acoustic pitch perception. *Biological Cybernetics*, 47, 149–163.
Meddis, R., & O'Mard, L. (1997). A unitary model of pitch perception. *Journal of the Acoustical Society of America*, 102, 1811–1820.
Micheyl, C., Xiao, L., & Oxenham, A. J. (2012). Characterizing the dependence of pure-tone frequency difference limens on frequency, duration, and level. *Hearing Research*, 292, 1–13.
Miyazono, H., Glasberg, B. R., & Moore, B. C. J. (2009). Dominant region for pitch at low fundamental frequencies (F0): The effect of fundamental frequency, phase and temporal structure. *Acoustical Science and Technology*, 30, 161–169.
Miyazono, H., Glasberg, B. R., & Moore, B. C. J. (2010). Perceptual learning of fundamental frequency (F0) discrimination: Effects of F0, harmonic number, and component phase. *Journal of the Acoustical Society of America*, 128, 3649–3657.
Moore, B. C. J. (1973). Frequency difference limens for short-duration tones. *Journal of the Acoustical Society of America*, 54, 610–619.
Moore, B. C. J. (1974). Relation between the critical bandwidth and the frequency-difference limen. *Journal of the Acoustical Society of America*, 55, 359.
Moore, B. C. J. (1978). Psychophysical tuning curves measured in simultaneous and forward masking. *Journal of the Acoustical Society of America*, 63, 524–532.
Moore, B. C. J. (1980). Neural interspike intervals and pitch. *Audiology*, 19, 363–365.
Moore, B. C. J. (1982). *An introduction to the psychology of hearing*, 2nd ed. London: Academic Press.
Moore, B. C. J. (1986). Parallels between frequency selectivity measured psychophysically and in cochlear mechanics. *Scandinavian Audiology*, Supplement 25, 139–152.
Moore, B. C. J. (2004). Dead regions in the cochlea: Conceptual foundations, diagnosis and clinical applications. *Ear and Hearing*, 25, 98–116.
Moore, B. C. J. (2012). *An introduction to the psychology of hearing*, 6th ed. Leiden, Netherlands: Brill.

Moore, B. C. J., & Rosen, S. M. (1979). Tune recognition with reduced pitch and interval information. *Quarterly Journal of Experimental Psychology*, 31, 229–240.

Moore, B. C. J., & Glasberg, B. R. (1983a). Suggested formulae for calculating auditory-filter bandwidths and excitation patterns. *Journal of the Acoustical Society of America*, 74, 750–753.

Moore, B. C. J., & Glasberg, B. R. (1983b). Forward masking patterns for harmonic complex tones. *Journal of the Acoustical Society of America*, 73, 1682–1685.

Moore, B. C. J., & Glasberg, B. R. (1986). The role of frequency selectivity in the perception of loudness, pitch and time. In B. C. J. Moore (Ed.), *Frequency selectivity in hearing* (pp. 251–308). London: Academic Press.

Moore, B. C. J., & Glasberg, B. R. (1987). Formulae describing frequency selectivity as a function of frequency and level and their use in calculating excitation patterns. *Hearing Research*, 28, 209–225.

Moore, B. C. J., & Glasberg, B. R. (1988a). Pitch perception and phase sensitivity for subjects with unilateral and bilateral cochlear hearing impairments. In A. Quaranta (Ed.), *Clinical audiology* (pp. 104–109). Bari, Italy: Laterza.

Moore, B. C. J., & Glasberg, B. R. (1988b). Effects of the relative phase of the components on the pitch discrimination of complex tones by subjects with unilateral cochlear impairments. In H. Duifhuis, H. Wit, & J. Horst (Eds.), *Basic issues in hearing* (pp. 421–430). London: Academic Press.

Moore, B. C. J., & Peters, R. W. (1992). Pitch discrimination and phase sensitivity in young and elderly subjects and its relationship to frequency selectivity. *Journal of the Acoustical Society of America*, 91, 2881–2893.

Moore, B. C. J., & Ohgushi, K. (1993). Audibility of partials in inharmonic complex tones. *Journal of the Acoustical Society of America*, 93, 452–461.

Moore, B. C. J., & Sek, A. (1994). Effects of carrier frequency and background noise on the detection of mixed modulation. *Journal of the Acoustical Society of America*, 96, 741–751.

Moore, B. C. J., & Sek, A. (1995). Effects of carrier frequency, modulation rate and modulation waveform on the detection of modulation and the discrimination of modulation type (AM vs FM). *Journal of the Acoustical Society of America*, 97, 2468–2478.

Moore, B. C. J., & Sek, A. (1996). Detection of frequency modulation at low modulation rates: Evidence for a mechanism based on phase locking. *Journal of the Acoustical Society of America*, 100, 2320–2331.

Moore, B. C. J., & Glasberg, B. R. (1997). A model of loudness perception applied to cochlear hearing loss. *Auditory Neuroscience*, 3, 289–311.

Moore, B. C. J., & Alcántara, J. I. (2001). The use of psychophysical tuning curves to explore dead regions in the cochlea. *Ear and Hearing*, 22, 268–278.

Moore, B. C. J., & Skrodzka, E. (2002). Detection of frequency modulation by hearing-impaired listeners: Effects of carrier frequency, modulation rate, and added amplitude modulation. *Journal of the Acoustical Society of America*, 111, 327–335.

Moore, B. C. J., & Sek, A. (2009a). Development of a fast method for determining sensitivity to temporal fine structure. *International Journal of Audiology*, 48, 161–171.

Moore, B. C. J., & Sek, A. (2009b). Sensitivity of the human auditory system to temporal fine structure at high frequencies. *Journal of the Acoustical Society of America*, 125, 3186–3193.

Moore, B. C. J., & Glasberg, B. R. (2010). The role of temporal fine structure in harmonic segregation through mistuning. *Journal of the Acoustical Society of America*, 127, 5–8.

Moore, B. C. J., & Gockel, H. (2011). Resolvability of components in complex tones and implications for theories of pitch perception. *Hearing Research*, 276, 88–97.

Moore, B. C. J., & Ernst, S. M. (2012). Frequency difference limens at high frequencies: Evidence for a transition from a temporal to a place code. *Journal of the Acoustical Society of America*, 132, 1542–1547.

Moore, B. C. J., Glasberg, B. R., & Shailer, M. J. (1984). Frequency and intensity difference limens for harmonics within complex tones. *Journal of the Acoustical Society of America*, 75, 550–561.

Moore, B. C. J., Glasberg, B. R., & Peters, R. W. (1985). Relative dominance of individual partials in determining the pitch of complex tones. *Journal of the Acoustical Society of America*, 77, 1853–1860.

Moore, B. C. J., Huss, M., Vickers, D. A., Glasberg, B. R., & Alcántara, J. I. (2000). A test for the diagnosis of dead regions in the cochlea. *British Journal of Audiology*, 34, 205–224.

Moore, B. C. J., Alcántara, J. I., & Glasberg, B. R. (2002). Behavioural measurement of level-dependent shifts in the vibration pattern on the basilar membrane. *Hearing Research*, 163, 101–110.

Moore, B. C. J., Glasberg, B. R., Flanagan, H. J., & Adams, J. (2006a). Frequency discrimination of complex tones; assessing the role of component resolvability and temporal fine structure. *Journal of the Acoustical Society of America*, 119, 480–490.

Moore, B. C. J., Glasberg, B. R., Low, K.-E., Cope, T., & Cope, W. (2006b). Effects of level and frequency on the audibility of partials in inharmonic complex tones. *Journal of the Acoustical Society of America*, 120, 934–944.

Moore, B. C. J., Hopkins, K., & Cuthbertson, S. J. (2009). Discrimination of complex tones with unresolved components using temporal fine structure information. *Journal of the Acoustical Society of America*, 125, 3214–3222.

Ohgushi, K., & Hatoh, T. (1991). Perception of the musical pitch of high frequency tones. In Y. Cazals, L. Demany, & K. Horner (Eds.), *Ninth International Symposium on Hearing: Auditory Physiology and Perception* (pp. 207–212). Oxford: Pergamon.

Oxenham, A. J., & Shera, C. A. (2003). Estimates of human cochlear tuning at low levels using forward and simultaneous masking. *Journal of the Association for Research in Otolaryngology*, 4, 541–554.

Oxenham, A. J., & Simonson, A. M. (2006). Level dependence of auditory filters in nonsimultaneous masking as a function of frequency. *Journal of the Acoustical Society of America*, 119, 444–453.

Palmer, A. R., & Russell, I. J. (1986). Phase-locking in the cochlear nerve of the guinea-pig and its relation to the receptor potential of inner hair-cells. *Hearing Research*, 24, 1–15.

Patterson, R. D. (1976). Auditory filter shapes derived with noise stimuli. *Journal of the Acoustical Society of America*, 59, 640–654.

Plomp, R. (1964). The ear as a frequency analyzer. *Journal of the Acoustical Society of America*, 36, 1628–1636.

Plomp, R. (1967). Pitch of complex tones. *Journal of the Acoustical Society of America*, 41, 1526–1533.

Plomp, R., & Mimpen, A. M. (1968). The ear as a frequency analyzer II. *Journal of the Acoustical Society of America*, 43, 764–767.

Recio-Spinoso, A., Temchin, A. N., van Dijk, P., Fan, Y. H., & Ruggero, M. A. (2005). Wiener-kernel analysis of responses to noise of chinchilla auditory-nerve fibers. *Journal of Neurophysiology*, 93, 3615–3634.

Ritsma, R. J. (1967). Frequencies dominant in the perception of the pitch of complex sounds. *Journal of the Acoustical Society of America*, 42, 191–198.

Robles, L., & Ruggero, M. A. (2001). Mechanics of the mammalian cochlea. *Physiological Reviews*, 81, 1305–1352.

Rose, J. E., Brugge, J. F., Anderson, D. J., & Hind, J. E. (1967). Phase-locked response to low-frequency tones in single auditory nerve fibers of the squirrel monkey. *Journal of Neurophysiology*, 30, 769–793.

Schouten, J. F. (1940). The residue and the mechanism of hearing. *Proceedings of the Koninklijke Nederlands Akademie van Wetenschap*, 43, 991–999.

Schouten, J. F., Ritsma, R. J., & Cardozo, B. L. (1962). Pitch of the residue. *Journal of the Acoustical Society of America*, 34, 1418–1424.

Sek, A., & Moore, B. C. J. (1995). Frequency discrimination as a function of frequency, measured in several ways. *Journal of the Acoustical Society of America*, 97, 2479–2486.

Sellick, P. M., Patuzzi, R., & Johnstone, B. M. (1982). Measurement of basilar membrane motion in the guinea pig using the Mössbauer technique. *Journal of the Acoustical Society of America*, 72, 131–141.

Shackleton, T. M., & Carlyon, R. P. (1994). The role of resolved and unresolved harmonics in pitch perception and frequency modulation discrimination. *Journal of the Acoustical Society of America*, 95, 3529–3540.

Shamma, S., & Klein, D. (2000). The case of the missing pitch templates: How harmonic templates emerge in the early auditory system. *Journal of the Acoustical Society of America*, 107, 2631–2644.

Shannon, R. V., Galvin, J. J., 3rd, & Baskent, D. (2002). Holes in hearing. *Journal of the Association for Research in Otolaryngology*, 3, 185–199.

Siebert, W. M. (1970). Frequency discrimination in the auditory system: Place or periodicity mechanisms. *Proceedings of the IEEE*, 58, 723–730.

Srulovicz, P., & Goldstein, J. L. (1983). A central spectrum model: A synthesis of auditory-nerve timing and place cues in monaural communication of frequency spectrum. *Journal of the Acoustical Society of America*, 73, 1266–1276.

Terhardt, E. (1974a). Pitch, consonance, and harmony. *Journal of the Acoustical Society of America*, 55, 1061–1069.

Terhardt, E. (1974b). Pitch of pure tones: Its relation to intensity. In E. Zwicker & E. Terhardt (Eds.), *Facts and models in hearing* (pp. 350–357). Berlin: Springer-Verlag.

Terhardt, E. (1979). Calculating virtual pitch. *Hearing Research*, 1, 155–182.

Turner, C. W., Burns, E. M., & Nelson, D. A. (1983). Pure tone pitch perception and low-frequency hearing loss. *Journal of the Acoustical Society of America*, 73, 966–975.

Verschuure, J., & van Meeteren, A. A. (1975). The effect of intensity on pitch. *Acustica*, 32, 33–44.

von Békésy, G. (1960). *Experiments in hearing*. New York: McGraw-Hill.

Ward, W. D. (1954). Subjective musical pitch. *Journal of the Acoustical Society of America*, 26, 369–380.

Webster, J., & Muerdter, D. (1965). Pitch shifts due to low-pass and high-pass noise bands. *Journal of the Acoustical Society of America*, 37, 382–383.

Zhou, B. (1995). Auditory filter shapes at high frequencies. *Journal of the Acoustical Society of America*, 98, 1935–1942.

Zwicker, E. (1956). Die elementaren Grundlagen zur Bestimmung der Informationskapazität des Gehörs (The foundations for determining the information capacity of the auditory system). *Acustica*, 6, 356–381.

Zwislocki, J. J., & Nguyen, N. (1999). Place code for pitch: A necessary revision. *Acta Oto-Laryngologica*, 119, 140–145.

Chapter 22
Unavoidably Delayed: A Personal Perspective of Twenty Years of Research on a Sound Localization Cue

Alan R. Palmer

Alan R. Palmer

A.R. Palmer (✉)
Medical Research Council Institute of Hearing Research,
Science Road, University Park, Nottingham NG7 2RD, UK
e-mail: arp@ihr.mrc.ac.uk

A.N. Popper and R.R. Fay (eds.), *Perspectives on Auditory Research*, Springer Handbook of Auditory Research 50, DOI 10.1007/978-1-4614-9102-6_22,

22.1 The World Before 1992

I came to binaural hearing rather late. This is perhaps not surprising, as I was an accidental hearing scientist anyway. My undergraduate degree was in Biological Sciences in a largely invertebrate department and I managed to graduate pretty much a blank canvas as far as hearing and ears were concerned. I wanted to work on vision. Attempting to follow this career path, I attended an interview for a PhD studentship to study vertebrate vision, only to be told it was no longer available, so would I consider working on hearing? OK I was in. The idea of all-night experiments working on hearing brain science in a relevant vertebrate animal model was seductive. Little did I realize at the time how much I would not like being in the lab in the wee small hours of the night, tired, stale, and hearing voices, with only an anesthetized animal for company (the voices were *mostly* real—picked up from local radio stations or taxis).

At the time (1972) I was under the supervision of Ted Evans, recording in the cochlear nerve and cochlear nucleus, which are generally (but not correctly) considered to be strictly monaural parts of the auditory pathway. My introduction to binaural processing was only really in the way of my general education in all things auditory, necessary to convince examiners that I was worthy of a doctorate. Indeed, most of what I took from one of the absolutely undisputed landmark binaural papers (Goldberg & Brown, 1969) was the vector strength analysis that I applied to my monaural recordings. When many years later, after working on binaural processing, I met Paul Brown, in Morgantown West Virginia, I rather embarrassingly (given my age at the time) had a few seconds acting like a stage door groupie (you really are THE Paul Brown?). I have to say he appeared flattered and well pleased that I held his early work in such high esteem. He is a jovial raconteur and we enjoyed an excellent and entertaining evening.

In 1978 at the age of 28 I moved on to my own laboratory at the National Institute for Medical Research in London. I arrived to be confronted by an empty rectangular box (my laboratory) into which was shortly to be delivered another rectangular box, an anechoic chamber that had been ordered, but that the commissioner no longer wanted (!). This turned out to be extremely fortuitous. I was recruited by Mike Keating, a developmental biologist, who was then head of neurophysiology. Mike was one of the most impressive intellects that I have had the pleasure to meet and, better still, work with. He was also a warm and caring human being and superb mentor, and my time in his department was probably the most enjoyable and productive of my career. Sadly, he died far too prematurely. Possibly most important in the present context, Mike encouraged me to consider looking for auditory responses in the superior colliculus (SC). Anyone with knowledge of the visual literature at the time would have known that such responses existed and were probably topographically organized in some way, but I this was news to me.

Andy King was a PhD student with Mike and Shin-Ho Chung at the time and had been working on current source density measurements in SC, so he joined me in investigating auditory responses in the SC to stimuli presented from a speaker ring in anechoic space. After three fruitless months we were on the verge of giving up, having decided that the auditory inputs to deep SC shown in other species were very small or nonexistent in the guinea pig. And then it started to work! We found responses to auditory stimulation and were able demonstrate the presence of an auditory space map in the deep layers of the SC that was aligned with the maps of visual space (Palmer & King, 1982). My one regret is that, in the impetuosity of youth, we didn't invite Mike Keating to co-author the paper. That aside, this was the time when I really began to read and try to understand the binaural literature, such as it was. The space mapped SC cells were all relatively high frequency and hence the spatial cue had to be interaural level differences, and this was demonstrated convincingly by a closed-field study by Lisa Wise and Dexter Irvine (1985). Note that up to this point, while interaural level differences and head-related transfer functions had played a part in my work, interaural time differences (ITDs) were notable by their absence. Obviously, our work was largely prompted by the barn owl story, which is one of the most successful accounts in neurobiology and all started (at least for me) with the 1978 demonstration of the space map (Knudsen & Konishi, 1978). As the years rolled on, the barn owl story has been elaborated in ways that brook no argument. It is quite clear that in this species evolution has worked to make the very best of the available binaural cues, and the manner of their processing is apparently by a mechanism first proposed by Lloyd Jeffress in 1948 (Jeffress, 1948).

But hold fast, I am clearly getting ahead of myself here. The fact is that psychophysical experimentation and thinking was in the early 1970s way ahead of the available empirical physiological observations. Some of the most important observations of sensitivity to cues for sound localization, laying the foundation for the duplex theory, had been made by Lord Rayleigh around 1900 in his garden with the stalwart assistance of his wife and his butler (Rayleigh, 1907). However, it took until the 1940s for Lloyd Jeffress to proposed a simple and elegant neural circuit that could provide a means by which the incredibly small interaural time differences, that humans and other animals are able to use to localize low-frequency sounds, could be converted into a firing rate code and further that this might provide the basis for a topographical map of azimuthal space. It was almost too simple: a network of cells acting as coincidence detectors whose connections from the two ears were in the form of a series of delay lines. Each cell was characterized by a specific interaural delay at which it fired maximally and this was considered to be the delay at which internal conduction delays, across the delay line network, exactly compensated for the interaural delay due to the source location. Hence the delay that characterized the cell could be taken as a surrogate for the azimuthal position and in the crudest approximation the cells could be considered to be responding only to defined azimuthal positions. Beautiful!

Steve Colburn's (1973) paper evaluating information content in the auditory nerve ended up with something very close to a Jeffress model, and full instantiations of this model have proliferated with subsequent embellishments by a number of people including Richard Stern. This in turn led to a cottage industry in binaural psychoacoustics that examined all aspects of ITD sensitivity, that for the most part depended on Jeffress type networks (embodied in the correlogram type display) for their interpretation and that appeared to give a remarkably full account of psychophysical observations.

So, now I can say that in the barn owl the anatomical and physiological observations were consistent with exactly the type of circuitry that Jeffress proposed. Like many others I took this *de facto* to be indicative that this was likely to be the same mechanism in mammals in general and the early physiological measurements were apparently consistent with such a conclusion. I have already mentioned the Goldberg and Brown paper. This seemed to me to be an excellent study although I always found it surprising that it was conducted in the dog (the only auditory physiology that I know of), especially because many or even most of the recent data have been obtained very much smaller mammals such as the gerbil and guinea pig. They demonstrated that the response of cells in auditory brain stem [the medial superior olive (MSO) although, as with other studies, being sure of the recording location within the MSO was and has always been an issue] depended on the interaural time delay and responded in a manner consistent with coincidence detection. Further, each cell was characterized by the presence of a characteristic delay (ergo the essential elements of the Jeffress network).

It is unfortunate that recording in MSO is somewhat more technically demanding that in other binaural nuclei, as it seems to me this has been an issue that has held us back until recently when braver people than I have deemed the extraordinary effort worthwhile, and indeed so it has proved. Early work in the inferior colliculus by Jerzy Rose and his colleagues (Rose et al., 1966) added further evidence, and although the population samples were not large in any of these early studies, as proof of principle they were more than adequate. The Rose studies introduced the idea of the characteristic delay (CD): the delay at which the cell's output is at the same relative level irrespective of the frequency of the stimulus. This has proved a powerful concept, and the combination of CD with characteristic phase (CP) defines the response of an interaural time sensitive cell and gives real insight into the nature of the neural interactions that underlie the sensitivity.

And then came Tom Yin and Shig Kuwada. Just about the time their spectacular series of papers were emerging from Madison, I was moving to the University of Sussex and, under the influence of Darwin and Russell, had been seduced back to the dark side (monaural processing in the periphery and wait for it.... speech encoding). For some time at least, my interests in binaural processing took a back seat and, even when I moved to the Institute of Hearing Research in 1986, I continued to work monaurally, though Adrian Rees and I had ventured as far up the pathway as the inferior colliculus (IC). Indeed, it was a reviewer of one of our monaural IC papers that made the rather salutary comment that, though this was still state-of-the-art at this time, soon monaural studies of hearing anywhere above the cochlear nucleus would be unacceptable. I thought this an extremely fair comment and soon moved to entirely diotic/dichotic presentation when recording more centrally. Interestingly, decades later, monaural studies are still being published. However, I digress again.

The papers by Yin and Kuwada recording interaural time difference sensitivity in the IC were elegant, comprehensive, and left very few gaps (Kuwada & Yin, 1983; Yin & Kuwada, 1983; Yin et al., 1986). They confirmed the principle of characteristic delay; showed distributions of the various binaural sensitivities; introduced the binaural beat stimulus to physiology; and showed that the responses at the IC level were remarkably linear, showing superposition: adding up the responses to many tonal stimuli of different frequencies gave basically the same interaural delay sensitivity as using a wideband noise stimulus. When I worked with speech signals I kept finding that, after making any kind of observation, going back to the classic papers by Murray Sachs and Eric Young (Sachs & Young, 1979; Young & Sachs, 1979) it was already there. When I started working on binaural processing the experience was *déjà vu*: find something really interesting and then go and see what Tom and Shig had already written about it. Yes, I know I should have read their papers more carefully the first time, but it was a little depressing at times, to say the least. However, thankfully they sometimes only mentioned things in passing, without really elaborating, and I have published a few papers that have basically been that elaboration, for example, the interaction of ITD and ILD (Palmer et al., 2007). Thanks, chaps!

It was at about this time in the mid to late 1980s that I started working again on binaural processing. The main impetus for this was a collaboration with David Caird (then in Rainer Klinke's laboratory) and Adrian Rees, who was with me in Nottingham. Dave had been trying and failing to find a physiological basis for the binaural masking level difference (the BMLD: an unmasking effect first discovered during the Second World War which is taken as a surrogate for the improvement in detectability when a target sound and a masker are from different directions: not really a good analogy, but used nevertheless). Because his work in Germany had failed to find any such basis in recordings from the IC (Caird et al., 1989), we thought maybe that combining our expertise with speech signals [as in the original psychophysical observation by Licklider (1948)] and his binaural expertise we might be more successful. Thus began a long series of studies that to some extent reflected our growing understanding of the mechanisms underlying what seemed at first to be a simple phenomenon. While Dave, Adrian, and I published papers that appeared to provide good evidence for the basis of the BMLD (Caird et al., 1991) it wasn't until later work with Dan Jiang and Dave McAlpine (McAlpine et al., 1996b; Jiang et al., 1997; Palmer et al., 2000) that I think we really nailed it. But again I am jumping ahead. If you want to understand the physiological mechanisms of the BMLD it proved absolutely essential to understand the underlying encoding of ITD by the cells. So, our routine first measurements on any cell we encountered, from that point on, were a detailed frequency response area, the sensitivity to the ITD of characteristic frequency (CF) tones and, ideally, also the sensitivity to interaural time differences of noise. At this point, I accepted entirely that the basis of interaural delay sensitivity in mammals could be nothing other than an almost literal embodiment of the Jeffress model.

Having obtained interaural time sensitivity in a species (the guinea pig) not previously reported, we wrote the data up for publication. In doing so, we (Dave, Adrian, and I) noticed that the distribution of the best interaural time delays (the position of the peak of the ITD function) was almost exactly the same as Tom and Shig had published for the cat (Palmer et al., 1990). From the perspective of a traditional biologist this seemed surprising. Evolution being such a spectacularly successful process, I expected the range of ITD sensitivities to be tailored to match the ecologically relevant range of the animal. Guinea pigs with smaller heads than cats should have had a narrower distribution. However, I still manage to rationalize this finding with a delay line network: maybe the network is hard wired genetically and different animals are able to make use of greater or lesser parts of this range, dependent on their head size. To some extent this presaged the way we came to view things later.

In the 1990s two different groups (Smith et al., 1993; Beckius et al., 1999) provided anatomical evidence, in the cat, that they considered to demonstrate the presence of axonal delay line systems making up the Jeffress network. It appeared that the regular arrangement of axons required was present on the contralateral inputs to the medial superior olive, but not on the ipsilateral inputs. No matter, this variant of the model would work just as well. Taken a face value, these data were compulsive and further constrained all explanations to operate within the framework of a Jeffress network. It was only much later that these data were re-evaluated and found not to be quite as convincing as they first appeared. About this time Tom Yin and Joe Chan published a careful study of responses to ITD in the MSO (Yin & Chan, 1990). Like those before, the sample was not large, but it did hint (but didn't convincingly demonstrate) at the possibility that there might be a topographical organization of ITD sensitivity in this nucleus (again pure Jeffress). Similar hints, but no definitive demonstration, were also provided at the IC level.

22.2 The World After 1992

Our early forays into searching for the physiological basis for the BMLD, though by no means a failure, were intellectually unsatisfactory. We hadn't got a complete story and we didn't fully understand the underlying mechanisms. It was about this time in the early 1990s that I took on a couple of new postdocs. Normally I only had one postdoc, but for reasons not relevant here, I doubled up. Eventually both of them (Dave McAlpine and Dan Jiang) moved on to take up where we had left off on the BMLD project. This was an exhilarating time in my lab. Dave and Dan were young, enthusiastic, and hard working and almost competed with each other to see who could gather the best data. Really good for me: all I had to do was point them at the science and referee the resulting intellectual fisticuffs. We continued to accrue ITD functions to noise and tones and eventually wrote up a second paper on the guinea pig. In this one we demonstrated that there was a very strong dependence of best ITD on CF (McAlpine et al., 1996a). In our data at least, we could find very few cells at very low CF that had short best ITDs. Better still, in this paper with data available at different tonal frequencies, we showed that the relationship between CF and delay held when we computed the characteristic delay. We seemed to be showing that a requirement of the Jeffress model, that all ITDs are represented in all frequency regions, wasn't coming through in our data. I am not sure at that point that we had fully thought through the possible implications of this result and, I have to say, that particular finding sunk almost without causing a ripple in any pond.

A few other interesting results were emerging that didn't quite fit with a hard-wired series of delay lines. For example, Mal Semple and Matt Spitzer demonstrated that the position of the best ITD was not invariant, but varied with stimulation context (Spitzer & Semple, 1991, 1993). This in itself, for me at least, didn't represent a complete heresy, as at least some of these effects were readily explicable by allowing for neural adaptation, which is an almost invariant property of sensory neurones.

Eventually, after some years working in the IC, we ended up with several hundred noise ITD functions. I had begun to prefer this measure to pure tone ITD functions, as following Tom and Shig's work it seemed to provide a convenient summary of a cell's ITD sensitivity. With a rather large sample of carefully measured functions we were able to reconfirm the dependence of the best delay on CF and to take it a few steps further. The peaks in our data were displaced from zero by an ITD equivalent to approximately one eighth of a cycle of the CF. The net effect of this was that the maximum slopes of the curves, whatever the CF, passed through the midline (zero ITD) (McAlpine et al., 2001). It had been shown many times psychophysically that the ITD resolution was best at this region and our result seemed entirely congruent with this. The fact that all the slopes irrespective of CF passed through midline suggested that the old model of von Békésy updated by van Bergeijk involving some kind of comparison of activities on the two sides of the brain might be more appropriate than the Jeffress formulation of a topographical organization of ITD sensitivity (von Békésy, 1960; van Bergeijk, 1962). It also had the benefit of bringing ITD processing and ILD processing more into line. Probably because, at the insistence of Dave McAlpine, we submitted this to, and were successful in publishing in, a high impact non-auditory journal, this result garnered rather more interest.

I have to say that, though no aspersions were actually cast, I had a distinct impression that our data suggesting that the full range of best ITDs were not available at all frequencies were not well received. As far as I was concerned, the McAlpine et al. (2001) paper was the third time we had found such a relationship in our guinea pig data and I for one could not see a flaw in our methodology. Though the guinea pig might be rather different from other mammals in this respect, my biological background still suggested to me that while convergent evolution sometimes came up with the same result, more often, when a solution is found that works, it is retained (i.e., it seemed likely that these data were representative of other mammals, although the possibility still remains that larger mammals might need a different solution).

Soon afterward Dave McAlpine and Dan Jiang left the lab. Dave, as we all know, has continued to actively pursue research in binaural processing, while Dan went back to medicine and is now a very successful consultant surgeon who continues to do excellent research. For our part, Trevor Shackleton and I were prompted by Brett Skottun to evaluate whether the slopes of the ITD functions and the distributions of spike counts actually provided enough resolvability for changes in ITD to account for the astonishing abilities shown psychophysically. Slightly to my surprise (given that discharge rates in the IC were never very high, and often seemed a little too noisy and unreliable), we demonstrated that the spike trains did provide sufficient information to sustain the psychophysical thresholds (Shackleton et al., 2003).

Several groups have subsequently re-examined the distribution of best ITD with CF, either by new measurements or by reanalyzing existing data sets, and similarities with what we had found have emerged. However, to the best of my knowledge there still seems to be a greater representation of best ITDs near zero in low-CF neurons in the cat and possibly rabbit than was the case in our guinea pig data (Hancock & Delgutte, 2004; Joris et al., 2006) or in the gerbil.

Probably the most controversial paper dealing with ITD sensitivity in recent years is the work of Antje Brand from Benedikt Grothe's laboratory (Brand et al., 2002). Benedikt came out of the rich German tradition of bat biology spearheaded originally in Frankfurt and later in Munich by Gerhard Neuweiler. The bats that they studied had, for the most part, very small heads and thus the maximum ITD they could experience naturally was extremely small, making this an unlikely cue to be useful for localization and bats generally did not hear so well at low frequency anyway. This being so, it was surprising that bats appeared to have an extremely well developed MSO (which, as we all know, in mammals is generally associated with initial processing of ITDs). What was also clear was that in the bat and other mammals there was a strong inhibitory input to the MSO cells. It appeared that this inhibition, in concert with the afferent excitatory input, imbued bat MSO cells with excellent sensitivity to amplitude modulation (Grothe, 2000). This, in the case of bats, is very useful, because, for example, such modulation is imposed on returning echoes by the wing beats of moths. So how does this relate at all to ITD sensitivity in other mammals? In the first place there is also a profuse inhibitory input to the mammalian MSO, the function of which at that time was completely unknown and largely overlooked in a world in which it was considered that MSO cells acted only as coincidence detectors between excitatory inputs. With the knowledge and data from the bat, Grothe's laboratory addressed the question of what this inhibition was actually doing in the MSO of other mammals.

The difficulties of recording from the MSO have already been alluded to and it hasn't got any easier, despite several notable publications from, for example, Yin and Chan (1990), Spitzer and Semple (1995), Batra et al. (1997a), and so forth. At the end of a seminar I gave in Munich in 2003, in which I questioned the sanity of those working for so little reward in MSO, I met Antje Brand, who confided that it took huge dedication and a very large number of experiments to obtain what, in the final paper, was only just an adequate sample.

This, for me, was so reminiscent of Bill Rhode's (1971) paper: the hugely important result of nonlinearity in the basilar membrane motion was based on a relatively small data set that had been obtained from a large set of experiments. So, what was so controversial in the Brand *et al.* study of MSO? First, though possibly by then not quite so controversial, is that in the gerbil MSO they showed that the best ITD varied with CF in the same way as in the guinea pig: two small rodenty-type animals looking quite similar! The big result, however, was that when the inhibitory inputs to the MSO were blocked by local iontophoretic application of strychnine (a blocker of glycinergic inputs) the best ITD moved toward zero ITD and the variation in discharge rate across the animal's physiological range was considerably reduced (Brand et al., 2002). Taken at face value, this result suggested that rather than a hard-wired delay line (e.g., made up of axons of different length) that the delay that characterized the response of the MSO neuron was generated by phase-locked inhibitory inputs.

However, the results in the Brand paper were not as unequivocal as one might have hoped. In a later review it was pointed out that the behavior of the side lobes of the ITD function suggested that what was reported might be limited only to the responses to the stimulus onset (Joris & Yin, 2007). Acknowledging that this might be the case, and despite the huge effort involved, Grothe's group made more recordings to ascertain that the shifts also applied to the sustained response. Indeed, in a series of studies this group has constructed quite an edifice of circumstantial evidence to support the role of inhibition in sculpting the shape of the ITD sensitivity, and in particular in shifting the peak away from zero to position the maximum slope through zero (Kapfer et al., 2002; Seidl & Grothe, 2003, 2005). When measured in naive animals and in animals raised to adulthood in omnidirectional noise, to eliminate directional signals, the peaks of the ITD functions were found to cluster around zero ITD. During development, inhibitory synapses start out uniformly distributed across soma and dendrites and end up only on the soma, but this does not occur when the animals are raised in omnidirectional noise. Taken altogether these data present, to me at least, a convincing case for an involvement of precisely timed inhibition in the shaping of ITD sensitivity in the gerbil.

Actually, delay lines were not the only show in town even before this introduction of a possible role for inhibition. An original idea by Manfred Schroeder extended and elaborated by Shihab Shamma, who coined the term "stereausis" (Schroeder, 1977; Shamma, 1989), suggested that the traveling wave vibration in the cochlea provided a ready source of delays. Exact matching of the positions in the two cochleas (and by implication the matching of frequency) would result in zero internal delay. Mismatching of the positions of the peak displacements on the basilar membrane could generate a delay between the two ears that could perform the same function as the Jeffress delay line. Computer models of this mismatching by Ben Bonham and Ed Lewis and by ourselves (Bonham & Lewis, 1999; Shackleton et al., 2000) suggested that it was indeed feasible and to generate appropriate delay values would require mismatches of only a few tens of Hertz, which would be almost undetectable given the resolution that is typically used to measure frequency response areas. However, a much more compulsive empirical test has recently been provided by the Joris laboratory. They cross correlated the activity of auditory nerve fibres in the cat with very close CFs and found that the functions resembled ITD sensitivity functions (Joris et al., 2006). All of this is again circumstantial, but provides evidence that such a model is a real practical possibility.

Given the apparent anatomical evidence of an arrangement in the cat that looked like a workable delay line, I had been thinking in terms of hybrid methods to develop the delays. For example, the anatomical Jeffress network, though not looking very precise, might provide delays in the right ball park which might then be fine tuned by inhibition or mismatched CFs. Certainly, an excellent paper in the chick (Seidl et al., 2010), from Ed Rubel's laboratory, convincingly demonstrated that the delay lines present in that species (when taking into account axon diameter, nodal distances, branching, etc.) seemed to be capable alone of generating delays of appropriate magnitudes. Interestingly, even in the chick, inhibitory circuits play an important role in shaping the inputs to the coincidence detectors and there are strong GABAergic inputs to the nucleus laminaris (the avian equivalent of the MSO), which appear capable of modulating ITD sensitivity in a frequency dependent manner (Burger et al., 2011; Tang & Lu, 2012). More recently, a careful reanalysis of the cat data, by one of the original authors, cast doubt as to whether these anatomical arrangements really could provide the necessary delays (Karino et al., 2011). One point that emerged for me, that I had certainly not taken on board from the original 1993 paper, was that the fiber that showed the most Jeffress-like delay-line structure had a CF of around 7 kHz!

Should we really be looking for one size fits all? Is it likely that a single solution to processing small ITDs should be universally applicable across species with different ecological constraints? No. Or at least a modeling study by Harper and McAlpine (2004) suggested that the strict Jeffress model applicable in birds (in particular barn owl, small head, high-frequency ITD sensitivity) is not optimal for all species and that the comparison of two hemispheres is more appropriate in the human (large head, only low-frequency ITD sensitivity). I am well aware that there are issues concerning a simple model of hemispheric comparisons, but would argue that information enough is present in terms of monaural activity levels to allow even a single-sided localization metric.

What is also quite clear is that we do not fully understand even the processing that takes place in the MSO. Early reports from IC showed ITD sensitivities that were not simply accounted for by a coincidence detection mechanism: plots of best phase against stimulus frequency that did not produce a characteristic phase near zero (pure peak type coincidence of excitatory inputs to the MSO) or 0.5 cycles (pure trough type coincidence of excitation and inhibitory input in the low-frequency part of the LSO). We called these non-simple coincidence detectors intermediate types, but I honestly prefer the endearing term coined by Shig Kuwada "tweeners" which bizarrely reminds me of sausages. Because these were described at the level of the IC, we and probably others initially assumed that they were created by convergence of cells with different types of ITD sensitivity onto IC cells. Indeed, one of the more ingenious experiments that we devised was a way of switching off some of these converging inputs to reveal the simpler coincidence detector inputs (McAlpine et al., 1998). However, though we demonstrated that some of the tweeners at the IC resulted from convergence, others actually showed that such sensitivities could be also demonstrated at the level of the MSO (Spitzer & Semple, 1995; Batra et al., 1997b), so something else (apart from strictly coincidence between excitatory inputs) is taking place at this level in the generation of the ITD sensitivity.

22.3 Where Are We Now?

The Jeffress model of the way in which ITDs are initially processed was prescient, elegant, and relatively simple. Astonishingly, given the state of neurophysiological knowledge existent when it was first proposed it has stood the test of time remarkably well. There seems no doubt whatsoever that in chicks and barn owls a more or less pure form of this model appears to operate. The model has several key features: (1) Each cell acts as coincidence detector, so that the relative timing of its inputs determines its output, and the cell fires maximally at a characteristic interaural delay; (2) the selectivity for interaural delay is generated by a delay line system; (3) all delays within the ecological range are represented at all frequencies; and (4) there is a topographical map of delay in which it is the position of the maximum activity in the map that represents the spatial position. The first, and most fundamental, of these has been shown repeatedly to be true for cells in the MSO and at stations above this level, to which it projects. Recent work in mammals has, however, questioned whether the mechanism by which characteristic delays are created is a delay line or one of several alternatives. The completeness of the representation of ecological delays at all frequencies has also been questioned by our work and that of others. The topographical arrangement has yet to be convincingly demonstrated in any mammal. Finally, whether it is the peak of activity or some other characteristic, such as maximum slope, that is important in encoding spatial position has been questioned by recent theoretical work.

We still do not know what generates the internal delay in larger mammals. What modeling and recent work by Phillip Joris appears to have established is that stereausis is a possible source of the delay and that the anatomical basis for delay lines in larger mammals is not so secure. However, even stereausis implies as its decoding mechanism a relatively straightforward coincidence detection across excitatory inputs and let's not forget that there is a good deal of inhibitory input to the MSO. Clearly, because there is or has been considerable resistance to the idea that the eighth cycle offset of best phase from zero is facilitated by precisely timed inhibition, another laboratory needs to try to replicate the effects of switching off this inhibition. To give face validity, this would have to be done probably in the cat, because, for completely arbitrary and undefined reasons, this animal seems to have assumed the *de facto* status of the gold standard mammalian model of hearing.

The implication of the optimal coding model of Harper and McAlpine is that the disposition and distribution of best delays might well be species dependent, but this has not yet been fully explored.

I have concentrated here on a single localization cue in isolation, which is the way we and others have sought to simplify things, in the hope of getting a more or less complete understanding at this level. This is not how it works in any real-world situation. Any realistic sound source, except one emitting a single low-frequency tone (pretty rare in my world), will be subject to a number of other factors that will make it more or less localizable. There will of course be ITDs of the carrier and envelope, there will also be interaural level differences and spectral cues generated by the pinna, all of which correspond with the position of the sound source in space. All of these cues will be dynamically changing with alterations in the position of the head with respect to the source and modified by the acoustic environment (echoes, etc.). It is perhaps unsurprising, therefore, that Chase and Young (2005), among others, using more realistic real-world signals generated using head-related transfer functions, have been able to show that cells in the IC are often sensitive to more than one localization cue. How the various cues for a single source are combined to give a single-source location and stabilized against the dynamic variation of movement through the environment or just head normal movements is completely unknown at present and has to be a fruitful and important area of future research.

For me, working at in this field has been enormously rewarding. Binaural hearing research has been challenging, and with the various minor or major controversies has remained fascinating, at least for those of us at the coal face. Most of all though, it has been a joy to be able to associate and work with many interesting and clever people, only some of whom I have been able to mention here, and to whom I say a sincere thank you.

Acknowledgments I am really grateful to Trevor Shackleton and Adrian Rees for casting a critical eye over this piece of writing which, to be honest, is a little outside my usual comfort zone.

References

Batra, R., Kuwada, S., & Fitzpatrick, D. C. (1997a). Sensitivity to interaural temporal disparities of low- and high-frequency neurons in the superior olivary complex. I. Heterogeneity of responses. *Journal of Neurophysiology*, 78, 1222–1236.

Batra, R., Kuwada, S., & Fitzpatrick, D. C. (1997b). Sensitivity to interaural temporal disparities of low- and high- frequency neurons in the superior olivary complex. I. Heterogeneity of responses. *Journal of Neurophysiology*, 73, 1222–1236.

Beckius, G. E., Batra, R., & Oliver, D. L. (1999). Axons from anterventral cochlear nucleus that terminate in medial superior olive of the cat: Observations related to delay lines. *Journal of Neuroscience*, 19, 3146–3161.

Bonham, B. H., & Lewis, E. R. (1999). Localization by interaural time difference (ITD): Effects of interaural frequency mismatch. *Journal of the Acoustical Society of America*, 106, 281–290.

Brand, A., Behrend, O., Marquardt, T., McAlpine, D., & Grothe, B. (2002). Precise inhibition is essential for microsecond interaural time difference coding. *Nature*, 417, 543–547.

Burger, R. M., Fukui, I., Ohmori, H., & Rubel, E. W. (2011). Inhibition in the balance: Binaurally coupled inhibitory feedback in sound localization circuitry. *Journal of Neurophysiology*, 106, 4–14.

Caird, D. M., Pillman, F., & Klinke, R. (1989). Processing of binaural masking level difference signals in the cat inferior colliculus. *Hearing Research*, 43, 1–24.

Caird, D. M., Palmer, A. R., & Rees, A. (1991). Binaural masking level difference effects in single units of the guinea pig inferior colliculus. *Hearing Research*, 57, 91–106.

Chase, S. M., & Young, E. D. (2005). Limited segregation of different types of sound localization information among classes of units in the inferior colliculus. *Journal of Neuroscience*, 25, 7575–7585.

Colburn, H. S. (1973). Theory of binaural interaction based on auditory-nerve data. I. General strategy and preliminary results on interaural discrimination. *Journal of the Acoustical Society of America*, 54, 1458–1470.

Goldberg, J. M., & Brown, P. B. (1969). Response of binaural neurons of dog superior olivary complex to dichotic tonal stimuli: Some physiological mechanisms of sound localization. *Journal of Neurophysiology*, 32, 613–636.

Grothe, B. (2000). The evolution of temporal processing in the medial superior olive, an auditory brainstem structure. *Progress in Neurobiology*, 61, 581–610.

Hancock, K. E., & Delgutte, B. (2004). A physiologically based model of interaural time difference discrimination. *Journal of Neuroscience*, 24, 7110–7117.

Harper, N. S., & McAlpine, D. (2004). Optimal neural population coding of an auditory spatial cue. *Nature*, 430, 682–686.

Jeffress, L. A. (1948). A place code theory of sound localization. *Journal of Comparative and Physiological Psychology*, 44, 35–39.

Jiang, D., McAlpine, D., & Palmer, A. R. (1997). Detectability index measures of binaural masking level difference across populations of inferior colliculus neurones. *Journal of Neuroscience*, 17, 9331–9339.

Joris, P. X., & Yin, T. C. T. (2007). A matter of time: Internal delays in binaural processing. *Trends in Neuroscience*, 30, 70–78.

Joris, P. X., Van de Sande, B., Louage, D. H., & van der Heijden, M. (2006). Binaural and cochlear disparities. *Proceedings of the National Academy of Sciences of the USA*, 103, 12917–12922.

Kapfer, C., Seidl, A. H., Schweizer, H., & Grothe, B. (2002). Experience-dependent refinement of inhibitory inputs to auditory coincidence-detector neurons. *Nature Neuroscience*, 5, 247–253.

Karino, S., Smith, P. H., Yin, T. C., & Joris, P. X. (2011). Axonal branching patterns as sources of delay in the mammalian auditory brainstem: A re-examination. *Journal of Neuroscience*, 31, 3016–3031.

Knudsen, E. I., & Konishi, M. (1978). A neural map of auditory space in the owl. *Science*, 200, 795–797.

Kuwada, S., & Yin, T. C. (1983). Binaural interaction in low-frequency neurons in inferior colliculus of the cat. I. Effects of long interaural delays, intensity, and repetition rate on interaural delay function. *Journal of Neurophysiology*, 50, 981–999.

Licklider, J. C. R. (1948). The influence of interaural phase relations upon the masking of speech by white noise. *Journal of the Acoustical Society of America*, 20, 150–159.

McAlpine, D., Jiang, D., & Palmer, A. R. (1996a). Monaural and binaural responses of low-best frequency neurones in the inferior colliculus of the guinea pig. *Hearing Research*, 97, 136–152.

McAlpine, D., Jiang, D., & Palmer, A. R. (1996b). Binaural masking level differences in the inferior colliculus of the guinea pig. *Journal of the Acoustical Society of America*, 100, 490–503.

McAlpine, D., Jiang, D., & Palmer, A. R. (1998). Convergent input from brainstem coincidence detectors onto delay-sensitive neurones in the inferior colliculus. *Journal of Neuroscience*, 18, 6026–6039.

McAlpine, D., Jiang, D., & Palmer, A. R. (2001). A neural code for low-frequency sound localisation in mammals. *Nature Neuroscience*, 4, 396–401.

Palmer, A. R., & King, A. J. (1982). The representation of auditory space in the mammalian superior colliculus. *Nature*, 299, 248–249.

Palmer, A. R., Rees, A., & Caird, D. (1990). Interaural delay sensitivity to tones and broad-band signals in the guinea-pig inferior colliculus. *Hearing Research*, 50, 71–86.

Palmer, A. R., Jiang, D., & McAlpine, D. (2000). Responses of neurones in the inferior colliculus to binaural masking level differences created by inverting the noise in one ear (NoSo vs NpSo). *Journal of Neurophysiology*, 84, 844–852.

Palmer, A. R., Lui, L., & Shackleton, T. M. (2007). Changes in interaural time sensitivity with interaural level differences in the inferior colliculus. *Hearing Research*, 223, 105–113.

Rayleigh, L. (1907). Our perception of sound direction. *Philosophical Magazine*, 13, 214–232.

Rhode, W. S. (1971). Observations of the vibration of the basilar membrane in squirrel monkeys using the Mössbauer technique. *Journal of the Acoustical Society of America*, 49, 1218–1231.

Rose, J., Gross, N., Geisler, C., & Hind, J. (1966). Some neural mechanisms in the inferior colliculus of the cat which may be relevant to localization of a sound source. *Journal of Neurophysiology*, 2, 288–314.

Sachs, M. B., & Young, E. D. (1979). Encoding of steady-state vowels in the auditory nerve: representation in terms of discharge rate. *Journal of the Acoustical Society of America*, 66, 470–479.

Schroeder, M. R. (1977). New viewpoints in binaural interactions. In E. F. Evans & J. P. Wilson (Eds.), *Psychophysics and physiology of hearing* (pp. 455–467.). New York: Academic Press.

Seidl, A. H., & Grothe, B. (2003). Acoustic experience is essential for the development of normal ITD processing. *Association for Research in Otolaryngology*, 26, 930.

Seidl, A. H., & Grothe, B. (2005). Development of sound localization mechanisms in the mongolian gerbil is shaped by early acoustic experience. *Journal of Neurophysiology*, 94, 1028–1036.

Seidl, A. H., Rubel, E. W., & Harris, D. M. (2010). Mechanisms for adjusting interaural time differences to achieve binaural coincidence detection. *Journal of Neuroscience*, 30, 70–80.

Shackleton, T. M., McAlpine, D., & Palmer, A. R. (2000). Modelling convergent input onto interaural-delay sensitive inferior colliculus neurones. *Hearing Research*, 149, 199–215.

Shackleton, T. M., Skottun, B. C., Arnott, R. H., & Palmer, A. R. (2003). Interaural time difference discrimination thresholds for single neurons in the inferior colliculus of Guinea pigs. *Journal of Neuroscience*, 23, 716–724.

Shamma, S. A. (1989). Stereausis: Binaural processing without neural delays. *Journal of the Acoustical Society of America*, 86, 989–1006.

Smith, P. H., Joris, P. X., & Yin, T. C. T. (1993). Projections of physiologically characterized spherical bushy cell axons from the cochlear nucleus of the cat: Evidence for delay lines to the medial superior olive. *Journal of Comparative Neurology*, 331, 245–260.

Spitzer, M. W., & Semple, M. N. (1991). Interaural phase coding in auditory midbrain: Influence of dynamic stimulus features. *Science*, 254, 721–724.

Spitzer, M. W., & Semple, M. N. (1993). Responses of inferior colliculus neurones to time-varying interaural phase disparity: Effects of shifting the locus of virtual motion. *Journal of Neurophysiology*, 69, 1245–1263.

Spitzer, M. W., & Semple, M. N. (1995). Neurons sensitive to interaural phase disparity in gerbil superior olive: Diverse monaural and temporal response properties. *Journal of Neurophysiology*, 73, 1668–1690.

Tang, Z.-Q., & Lu, Y. (2012). Two $GABA_A$ responses with distinct kinetics in a sound localization circuit. *Journal of Physiology*, 590, 3787–3805.

van Bergeijk, W. A. (1962). Variation on a theme of Bekesy: A model of binaural interaction. *Journal of the Acoustical Society of America*, 34, 1431–1437.

von Békésy, G. (1960). *Experiments in hearing*. New-York: McGraw-Hill.

Wise, L. Z., & Irvine, D. R. (1985). Topographic organization of interaural intensity difference sensitivity in deep layers of cat superior colliculus: Implications for auditory spatial representation. *Journal of Neurophysiology*, 54, 185–211.

Yin, T., & Kuwada, S. (1983). Binaural interaction in low-frequency neurons in inferior colliculus of the cat. III. Effects of changing frequency. *Journal of Neurophysiology*, 50, 1020–1042.

Yin, T., Chan, J., & Irvine, D. (1986). Effects of interaural time delays of noise stimuli on low-frequency cells in the cat's inferior colliculus. I. Responses to wideband noise. *Journal of Neurophysiology*, 55, 280–300.

Yin, T. C., & Chan, J. C. (1990). Interaural time sensitivity in medial superior olive of cat. *Journal of Neurophysiology*, 64, 465–488.

Young, E. D., & Sachs, M. B. (1979). Representation of steady-state vowels in the temporal aspects of the discharge patterns of populations of auditory-nerve fibers. *Journal of the Acoustical Society of America*, 66, 1381–1403.

Chapter 23
Size Matters in Hearing: How the Auditory System Normalizes the Sounds of Speech and Music for Source Size

Roy D. Patterson and Toshio Irino

Roy D. Patterson

Toshio Irino

R.D. Patterson (✉)
Centre for the Neural Basis of Hearing, PDN, University of Cambridge,
Downing Site, Cambridge CB2 3EG, UK
e-mail: rdp1@cam.ac.uk

T. Irino
Faculty of Systems Engineering, Wakayama University,
930 Sakaedani, Wakayama 640-8510, Japan
e-mail: irino@sys.wakayama-u.ac.jp

A.N. Popper and R.R. Fay (eds.), *Perspectives on Auditory Research*, Springer Handbook of Auditory Research 50, DOI 10.1007/978-1-4614-9102-6_23,

23.1 Introduction to Pulse-Resonance Sounds and Size Information

The sounds that animals use to communicate at a distance, to declare their territories and attract mates, are typically *pulse-resonance* sounds (Fig. 23.1). They are the basis of the calls produced by most vertebrates (mammals, birds, reptiles, frogs, and fish). Pulse-resonance sounds are produced by what are referred to as source-filter systems. In mammals for example, the larynx produces abrupt pulses of acoustic energy and the pulses excite resonant cavities in the vocal tract above the larynx (Fig. 23.1a). The pulses mark the start of each cycle of the communication while the resonances provide the "message," that is, distinctive information about the shape and structure of the resonators in the sender's body (Patterson et al., 2008). In the magnitude spectrum of the sound (Fig. 23.1b), the pulse rate of the source appears as the spacing between the spectral components of the sound (vertical lines), while the vocal resonances appear as spectral prominences, or formants, which collectively define the spectral envelope of the sound (the thick gray line).

In the majority of animals, nature has adapted existing body parts to create the structures that animals use to produce their sounds. In mammals, the vocal tract is part of the tube that carries food from the entrance of the mouth to the stomach, while the larynx serves to close off the bronchial tubes during the ingestion of food. As a young animal grows, the digestive track and the larynx have to grow, and this causes dilation of both the vocal resonance and the pulse stream, simply because larger/heavier things vibrate more slowly than smaller/lighter things. The rate of oscillation of a vocal resonance is a physical property of the sound; it is referred to as the "acoustic *s*cale of the *f*ilter element of the sound," S_f. Similarly, the pulse rate is the "acoustic *s*cale of the *s*ource element of the sound," S_s. The values of these acoustic variables describe the size of acoustic features in the waveform, and it is primarily these properties of the sound that inform the listener about the size of the sender (Smith & Patterson, 2005). In Fig. 23.1b, the magnitude spectrum is plotted on a logarithmic frequency axis to emphasize the form of acoustic scale information and the changes effected by dilation. On this axis, as a child grows up, the formants move, *as a unit*, toward the origin without changing their (log) spacing or their (log) shape; the position of the envelope, S_f, can be specified in terms of the geometric, mean formant frequency (MFF). Similarly, as a child grows up, the set of harmonics that constitute the fine structure of the spectrum moves, *as a unit*, toward the origin without any change in the (log) spacing of the components; the position of the set of harmonics, S_s, is typically specified in terms of the fundamental of the harmonic series, F0.

The basic structure and physiology of the sounder in an animal do not change as the individual grows; it is primarily the size of the source and the size of the filter that change. As a result, the calls produced by individuals of a species have the same "message" and vary primarily in terms of S_f and S_s, which are often referred to collectively as "vocal characteristics" to distinguish the information they carry from the message of the communication. Within a species, there is a correlation between S_s and S_f

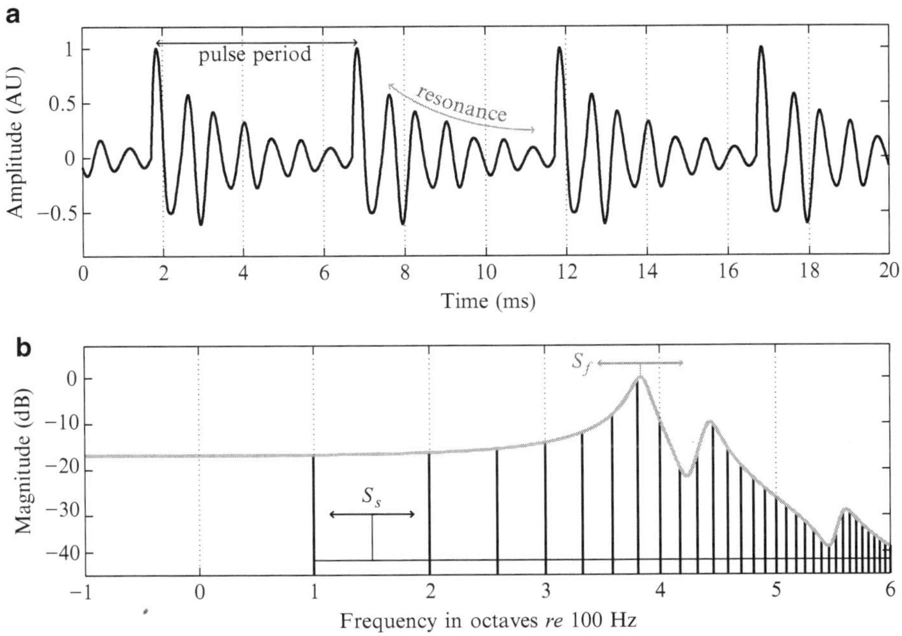

Fig. 23.1 The waveform (**a**) and magnitude spectrum (**b**) of the vowel /a/, as in the musical note name "la"

inasmuch as both the source and the filter grow as the individual grows. However, the correlation is far from perfect because most animals have a degree of voluntary control over the pulse rate of the source, and some animals (e.g., deer and dogs) have a degree of voluntary control over the length of the pharynx (when roaring or barking). Thus, in the natural world, the species message is conveyed by sounds that commonly vary in two aspects of acoustic scale (Patterson et al., 2008). Finally, note that humans have the ability to vary the shape of their vocal track rapidly and reliably to produce different vowel sounds, and in the current context, each vowel type is considered to be a distinct message.

The variation of S_s and S_f within a population of animal calls presents the perceptual system of the listener with a classic categorization problem. The problem is how to discriminate correctly when two individuals are from different species and, at the same time, correctly generalize across the calls of individuals of a single species. Distinguishing two sounds is not difficult, especially when they vary in pulse rate and resonance rate. The problem is to recognize that sounds that differ may, nevertheless, represent animals of the same species saying the same thing. Correct generalization involves isolating the shape of the resonance and recognizing that the calls of large and small members of a species have the same resonance shape, despite all the variability introduced by differences in S_s and S_f.

Moreover, in natural communication, the discrimination/generalization problem is exacerbated by the fact that orderly variation of pulse rate during a sequence of sounds is sometimes used to add information to the communication. For example, pulse rate is varied in human speech to convey prosodic information. For purposes of the current discussion, however, the variability introduced by voluntary manipulation of pulse rate can be grouped with the variability associated with size because it produces a change in one of the acoustic scale variables, S_s. It is also the case that the pulse-resonance sounds of cold-blooded animals (e.g., frogs) change their acoustic scale with temperature. But again, this leads primarily to changes in the S_s and/or S_f of the sound, which affects the discrimination/generalization problem in the same way. Thus, in a more general sense, the problem is to segregate the voice characteristics of the sender from the message information.

The sustained notes of many orchestral instruments are also pulse-resonance sounds produced by source-filter systems. The perception of instrument family is associated with the shape of the resonance; the acoustic scale of the source and filter components of the sound identify the instrument within its family, largely through the perception of its register, or size (Patterson et al., 2010). So pulse-resonance sounds are very common in the everyday world.

Despite the apparent conflict between generalization and discrimination, recent experiments have shown that auditory perception is singularly robust to changes in source size. People can recognize vowel type in sounds with virtually any combination of S_s and S_f, and they can discriminate speaker size from sequences of randomly chosen vowels (e.g., Smith et al., 2005). This suggests that the internal, auditory representation of sound is size invariant (Irino & Patterson, 2002), or, to be more precise, that the auditory representation of communication sounds is "scale-shift covariant" (Patterson et al., 2007) with respect to both S_s and S_f. It is clear that automatic, peripheral normalization of pulse-resonance sounds would greatly benefit communication by adults and the development of communication skills by the young. Central mechanisms of learning, memory and recognition would be insulated from the variability of S_s and S_f found in the natural environment, leaving them free to process phonemes in a size-invariant form. This would help explain how children learn speech from a small number of people with widely differing sizes and vocal characteristics. The question, then, is the form of this internal, automatically normalized representation of sound and how the auditory system might construct it.

23.2 The Auditory Image and the Normalization of the Auditory Image

In the years just before 1992, computer power increased to the point where it was possible to model cochlear processing with auditory filter banks, and this made it possible to simulate the neural activity patterns (NAPs) produced in the auditory nerve by everyday sounds such as vowels and musical tones. Figure 23.2 shows the NAP of an /ae/ vowel, like that in the word "hat," produced with a gammatone

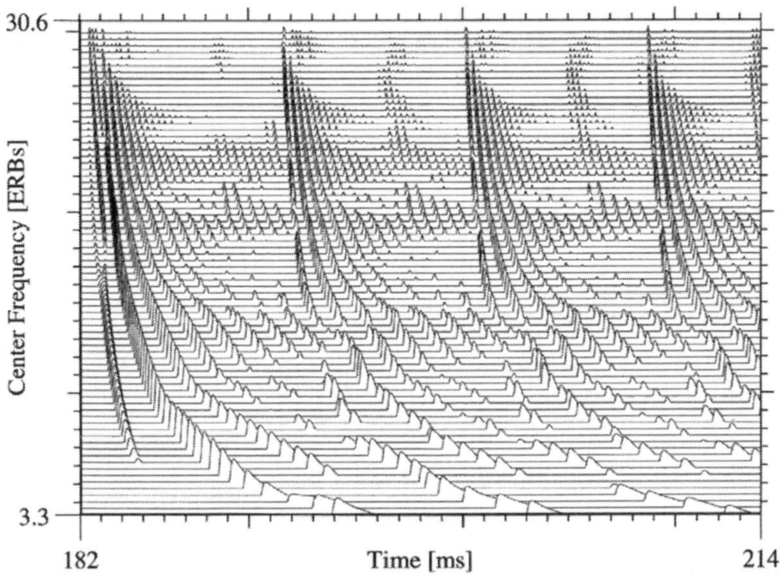

Fig. 23.2 The NAP of an /ae/ vowel, as in the word "hat"

auditory filter bank (Patterson et al., 1995). These NAPs of vowels revealed distinctive, phase-locked firing patterns that repeated once per cycle of the sound, because each glottal pulse excites a wide range of auditory filters, and the filters in the region of a formant ring longer than those in regions between formants. The NAPs were used to study monaural phase perception in complex tones which revealed that small, low-level changes in the details of the NAP were often heard as timbre changes. This was a puzzle because these sounds produce exceptionally stable perceptions, and if the auditory system used a sliding temporal window to smooth NAP fluctuations, the small differences that affect perception would be obliterated by the averaging long before the window duration was sufficient to produce a stable perception of level.

To explain monaural phase perception, we proposed a form of *strobed* temporal integration (STI) with a bank of integration units, one for each channel of the NAP. Each unit monitors the activity in its channel looking for the peak that marks the start of a new cycle, and when such a peak is found, the time-interval information in the NAP is copied into a decaying image buffer. The multichannel result of this STI process has two dimensions (quasi-log) frequency and (linear) time interval; it was referred to as the *auditory image*. It was argued that the stabilized auditory image (SAI) can explain our initial experience of tonal sounds better than the NAP. The SAI derived from the /ae/ NAP of Fig. 23.2 is shown in Fig. 23.3. STI synchronizes temporal integration to the period of the sound so that successive copies of the vowel pattern are averaged without destroying the fine-grain temporal information within the glottal cycle. With STI, the voiced parts of speech (and all other periodic

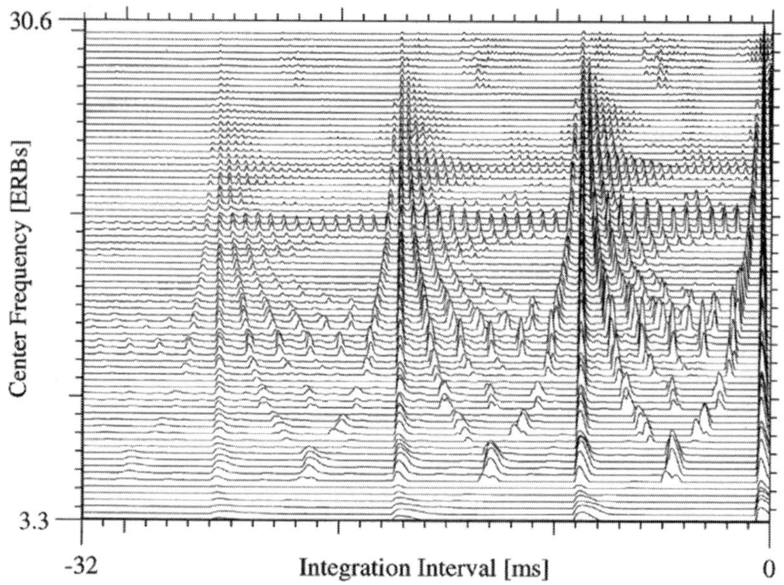

Fig. 23.3 The auditory figure produced by an /ae/ vowel, as in the word "hat"

complex tones) produce static patterns with distinctive shapes, referred to as *auditory figures*, and they provide important information about the sender. For example:

1. The vertical ridges of activity above time intervals of 8.5 and 17 ms show that the period of the vowel was 8.5 ms. This is the internal representation of S_s. We hear this feature as the pitch of the voice. A period of 8.5 ms corresponds to a frequency of 118 Hz. When the pitch of the voice rises, the ridge shifts to the right, and when the pitch falls, it shifts to the left.
2. The set of four rightward pointing triangular shapes hanging on a vertical ridge tells us that the sound is probably from a human vocal tract. These asymmetric, triangular features are the formants of speech as they appear in auditory figures.
3. The relative positions of the formants within the auditory figure indicate that the human is saying, or singing, the vowel /ae/ at this moment in time. When the vowel type changes, the triangles move up and down the vertical ridge. For example, in the /i/ vowel of "beet," the first formant shifts down and the second formant shifts up.
4. The vertical position of the auditory figure is the internal representation of the acoustic scale of the filter, S_f, and in this case, the position indicates that the person speaking has a relatively long vocal tract and is probably an adult. If a child with a relatively short vocal tract spoke the same vowel, the auditory figure would have a similar shape but it would move up the ordinate and the width of the figure would shrink.

5. The combination of a relatively low pitch with relatively low formants indicates that the person who spoke this vowel was probably a man rather than a woman.

These five properties of the sender's voice are specified by each and every vowel as features of the auditory figures that appear in the auditory image as the speech progresses. Many musical tones and animal calls also produce distinctive auditory figures in the auditory image, and the positions of those figures provide information about the size of the instrument or the animal. Collectively, the set of processes that produce the SAI are referred to as the Auditory Image Model (AIM) of auditory perception.

The decay rate of the auditory image is on the order of 30 ms and so STI averages four to eight cycles of vowel sounds. As a result, *auditory figures* form and dissolve at the rate we hear change in the sound. When the input is a sentence or a melody, the auditory figures appear as brief *auditory events* synchronized to the rhythm of the syllables of the sentence or the notes of the melody.

Videos of the auditory images produced by speech and music appeared at about the same time as videos of the autocorrelogram representation of speech and music (~1990). The dynamics of these two representations of sound are quite similar because they have similar dimensions and time constants. The main difference is in the details of the auditory figures; autocorrelation distorts the glottal patterns observed in the NAP and imposes an unwarranted degree of symmetry on auditory figures. These differences prompted research into the perception of temporal fine structure and the perception of temporal asymmetry. The experiments and the implications for models of temporal processing are discussed in Section 4, along with the form of the time-interval dimension in the auditory image.

At the same time, models of auditory filtering were being extended to new data showing that the auditory filter becomes broader and more asymmetric as sound level increases. This meant that the carrier of the filter's impulse response must chirp at higher levels. This led us to consider what the optimal form of auditory filter is in a system where the bandwidth of the filter is proportional to its centre frequency and filter asymmetry changes with level. The concept of optimality in time-frequency trading was introduced by Gabor (1946), who proved that a symmetric Gaussian window is optimal for the construction of spectrograms where channel bandwidth is fixed. When the optimality criterion is used to derive an auditory filter with proportional bandwidth and level-dependent asymmetry, the result is a filter with a gamma*chirp* impulse response. The derivation is briefly described in Section 5. It led to the discovery of the Mellin transform and the realization that it was possible to modify the dimensions of the auditory image to produce scale-shift-invariant versions of auditory figures and auditory events. This prompted behavioral experiments on the robustness of auditory perception to variation in acoustic scale, and these experiments are described in Section 3.

The results place rather strong constraints on the form of the internal, auditory representation of sound. It appears that the auditory figure needs to be *scale-shift* covariant to explain the robustness of perception to changes in speaker size while auditory events need to be time-shift invariant to explain why we hear the same thing each time a specific recorded sound is played. Scale-shift covariance and time-shift invariance are compatible only in very particular spaces, and that is the topic of the remainder of this section.

23.2.1 The Normalized Auditory Image

The architecture of the normalized version of AIM is very similar to that of the original AIM; an auditory filter bank simulates the operation of the cochlea and strobed temporal integration converts the repeating neural patterns produced by pulse-resonance sounds into stabilized auditory figures that evolve dynamically as auditory events.

In the current version of AIM, the operation of the cochlea is simulated with a dynamic, compressive GammaChirp (dcGC) auditory filter bank (dcGC-AFB); as before, filter center frequency is distributed evenly along a tonotopic frequency axis, and the bandwidth of the auditory filter is proportional to filter center frequency, but now the filter shape varies with sound level becoming more asymmetric as level increases. Neural transduction is simulated by half-wave rectification. Figure 23.4 shows the response of the dcGC-AFB to a synthetic, two-formant vowel. The formants are exponentially damped sinusoids and the damping parameter is proportional to center frequency, so it is like a vowel produced by a vocal tract in which the mouth cavity is a scaled down version of the throat cavity. The envelope of the impulse response is a gamma function whose duration decreases as frequency increases. This means that, mathematically, the operation of the cochlea is more like a wavelet transform than a windowed Fourier transform, where the window duration is the same in all channels. Formally, this suggests that the auditory system is transforming the time waveform of sound into a {time, scale} representation, rather than a {time, frequency} representation. By its nature, the time dimension of the NAP is linear; that is, the NAP exists in a {linear time, quasi-log scale} planar space.

In Fig. 23.4, the change in resonator size from large to small causes a change in the shape of the distribution of formant information; it moves up the scale dimension and shrinks in the time dimension. Similarly, if the vocal tract length (VTL) of this hypothetical speaker were reduced, the combined distribution of activation would move up in scale and shrink in time. To achieve scale-*shift* covariance, we need to expand the distribution within each channel as the size of the resonator decreases. This can be accomplished by multiplying time as it exists in each channel of the NAP by the center frequency of the channel in question. This changes the unit of the abscissa from time to (cycles/second) × time (second), which reduces to cycles of the impulse response of the filter that produces a specific channel of the NAP. The transformation is motivated by a consideration of the operators that can transform {time, scale} space into a scale-shift covariant space—operators that are, at the same time, *unitary*. The derivation of the logarithmic operator is described in Patterson et al. (2007). The dimensions of the new space are $\log_2$(cycles), or $\log_2 c$, and $\log_2$(scale), or $\log_2 s$. The operator transforms the {time, scale} NAP of Fig. 23.4 into the $\{\log_2 c, \log_2 s\}$ NAP shown in Fig. 23.5.

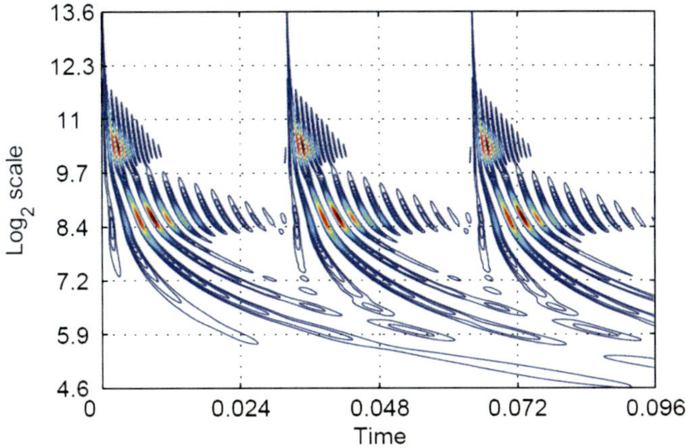

Fig. 23.4 The response of a gammachirp cochlea simulation to a synthetic, two-formant vowel. This version of the NAP is time-shift covariant but not scale-shift covariant because formant duration decreases as scale increases

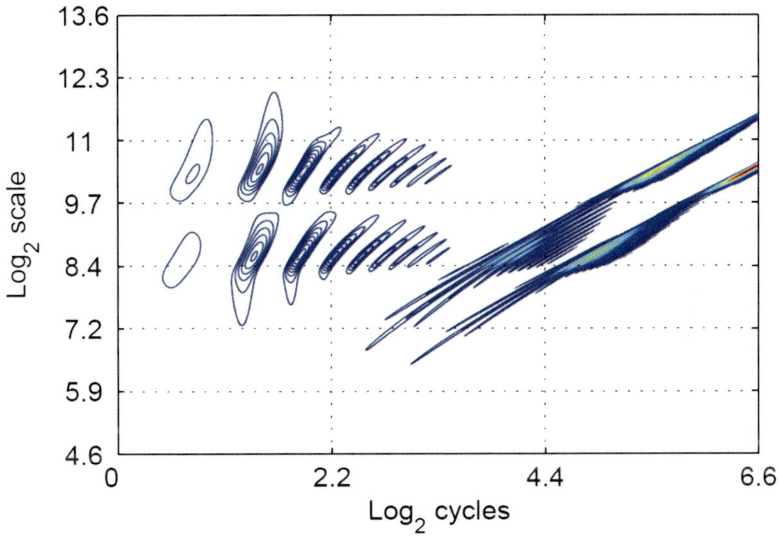

Fig. 23.5 The response of a gammachirp cochlea simulation to a synthetic, two-formant vowel. This version of the NAP is scale-shift covariant but not time-shift covariant because the formant distribution is time-compressed in successive periods of the sound

Consider the activity of the first cycle of the NAP; the progressive stretching of the time base by frequency expands the distribution that represents the upper formant more than the distribution that represents the lower formant. As a result, in this representation, the two distributions of resonator activity are the same size and have effectively the same shape. This is a concrete demonstration of scale-shift covariance; the smaller resonator is a scaled version of the larger resonator. It is also the case that the overall pattern of activity in the first cycle of the NAP does not change with variation in S_f, like that produced by a change in vocal tract length. Changes in S_f cause the auditory figure formed by the pair of formants to shift vertically, as a unit without deformation. Similarly, changes in S_s do not affect the shape of the auditory figure; indeed, they do not even affect the vertical position of the auditory figure; rather, the diagonal that marks the start of the next pulse period, and the detail attached to it, move back and forth without any change in either the shape of the tilted pattern *or* its angle. The operator shows that, when we use logarithmic scale units, the shift of the neural pattern with a change in S_f will be restricted to the vertical dimension; that is, the shift will be orthogonal to the log cycles dimension.

The onset of the second cycle of the pattern occurs along a diagonal with a 45° angle whose position is determined by S_s. The start of the third cycle of activity is a parallel diagonal that is shifted along the cycle axis by one more period of the sound. When we use logarithmic cycle units, a change in S_s produces a vertical shift of the period-terminating diagonal, and the change is orthogonal to the cycles dimension. In summary, the auditory figure associated with a given vowel type is a fixed pattern of activity in the first cycle of the $\{\log_2 c, \log_2 s\}$ NAP, independent of voice pitch (S_s) or vocal tract length (S_f). This $\{\log_2 c, \log_2 s\}$ NAP would not, however, make a good model of the internal, auditory representation of sound because it is not *time-shift* invariant. The pattern of time intervals in the second, and successive, periods of the sound is skewed and progressively compressed. Auditory perception is time-shift invariant, at least at the level of auditory events; we perceive the same sequence of events when a sentence or melody is played at two different times. This suggests that auditory figures occupy a separate plane that is orthogonal to the temporal dimension of auditory perception. This in turn means that the representation of the space of auditory perception has to be three-dimensional; the fine-grain, time-interval information responsible for the shape and detail of auditory figures exists in a normalized plane that is orthogonal to the progression of time as we perceive it in auditory events.

In AIM, the extra dimension is created by strobed temporal integration as described in the preceding text. In the normalized version of AIM, however, the buffer dimensions are $\log_2 c$ and $\log_2 s$ rather than linear time-interval and log frequency. The auditory figure of the two-formant vowel that forms in the $\{\log_2 c, \log_2 s\}$ plane of the auditory image is shown in Fig. 23.6. In the normalized version of AIM, the activity of auditory figures is limited in the cycles dimension to one period of the sound; that is, when a new period of the sound is detected by the temporal integration unit in a given channel, the transfer of information from the previous period of the sound is terminated and the information transfer for the new period begins again at the left-hand edge of the $\{\log_2 c, \log_2 s\}$ plane of the auditory image. In this algorithm, each element of the activity within a period of the sounds is used once and only once, which matches the magnitude of activity in the auditory image more closely to that in the NAP than in previous versions of AIM.

The auditory figures of pulse-resonance sounds have a fixed form in the $\{\log_2 c, \log_2 s\}$ plane. When there is a change in VTL from small to large, the activity just moves vertically down as a unit without deformation, and the extent of the shift is the logarithm of the ratio of the S_f values for the two individuals. When there is an increase in pitch, the auditory figure does not change shape *and it does not move*

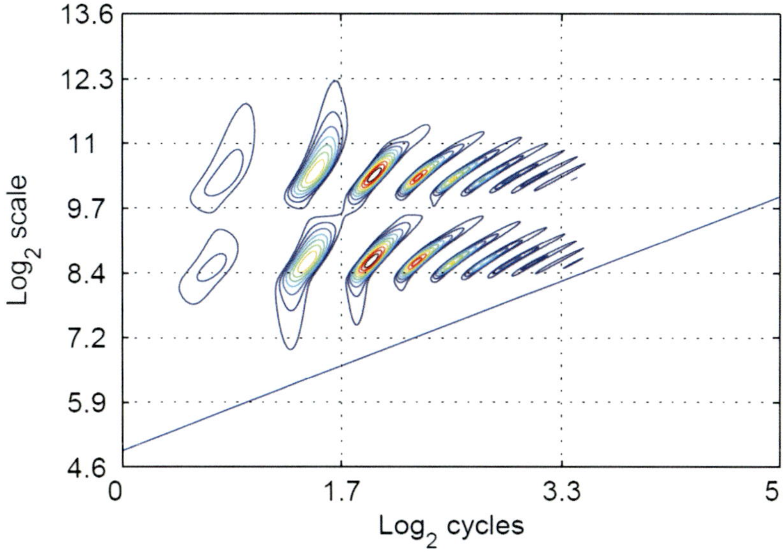

Fig. 23.6 The auditory figure of the two-formant vowel that forms in the {$\log_2 c$, $\log_2 s$} plane of the normalized auditory image. This version of the auditory figure is scale-shift covariant and so it is largely independent of voice pitch. An auditory event showing how the vowel evolves over time would be time-shift invariant

vertically; rather, the diagonal that marks the extent of the period of the sound, moves vertically without changing either its shape *or* its angle. This changes the size of the area available for the auditory figure produced by the sound, but it does not change the shape of the figure, other than to cut off the tail of the lowest resonance when the glottal-pulse period is short relative to the duration of the resonance. The extent of the vertical shift of the diagonal is the logarithm of the ratio of the S_s values associated with the old and new pulse periods.

In AIM, snapshots of the {$\log_2 c$, $\log_2 s$} plane of the auditory image can be recorded at a regular rate and used to construct a video of the auditory figure as it develops, evolves, and decays. The auditory event is *time-shift* invariant insofar as, when the vowel is repeated, it produces the same video event. In this way, the normalized version of AIM can explain why auditory perception is robust to changes in acoustic scale and why the perception of auditory events is, nevertheless, time-shift invariant.

23.3 Size Invariance in Auditory Perception

In the paper in which we initially illustrated how the peripheral auditory system could use a form of "stabilized wavelet–Mellin transform" to normalize vowel sounds for vocal tract length (VTL) (Irino & Patterson, 2002), we cited several sources of circumstantial evidence from the literature to indicate that the auditory system did, indeed, normalize speech sounds. At that time, however, there had been little research on the topic because it was not possible to produce balanced sets of speech sounds in which S_f and S_s could be controlled systematically. Fortunately, at about this time, a colleague (Hideki Kawahara) developed a sophisticated speech processing package (STRAIGHT) that made it possible to manipulate the S_f and S_s of natural speech sounds, so that utterances recorded by a man could be transformed to sound like those of a women or a child and vice versa. STRAIGHT made it possible for us to record the speech sounds of one individual and then synthesise a population of speakers with widely varying vocal characteristics who were all speaking the same set of utterances with the same accent and with the same prosodic details. The version of STRAIGHT used in these experiments is described in Appendix A of Irino et al. (2012).

With carefully tailored populations of speakers, we performed a sequence of experiments to provide quantitative evidence for what everyone intuitively knows, namely that, auditory perception is singularly robust to changes in the S_f and S_s values of communication sounds. We have no difficulty whatsoever understanding when a child and an adult have spoken the same speech sounds (vowels, syllables, or words), despite substantial differences in the S_f and S_s values of their waves. Indeed, we can recognize speech sounds when the S_f and S_s of the voice have been scaled well beyond the range of normal experience, and when the combination of S_f and S_s values is not one that occurs in the normal population. The experiments also support the hypothesis that λ_f, the internal version of S_f, functions as an independent variable in the auditory system, much like λ_s, the internal version of S_s. When presented with a pair of speakers, listeners know not only which one has the higher pitch, but also which speaker has the longer vocal tract, although they perceive it as a change in speaker size rather than a change in vocal tract length. In short, the experiments support the hypothesis that the auditory system automatically normalizes communication sounds for S_f and S_s to segregate the message from the vocal characteristics.

In this section of the chapter, we briefly describe three experiments designed to document the robustness of speech perception to changes in acoustic scale, and we present a model of how information produced by a vowel in the cochlea might be transformed into the perception of a vowel spoken by a person with a specific size and sex.

23.3.1 The Role of S_f in Speech Perception

A schematic of some of the variables involved in speech perception is presented in Fig. 23.7. The sound in this example is a voiced vowel and the bottom panel presents the spectral envelope of the vowel with its three formant peaks. In a psychoacoustic experiment, the peaks might be manipulated individually to determine, for example, the minimal frequency shift of the second formant required to detect a change

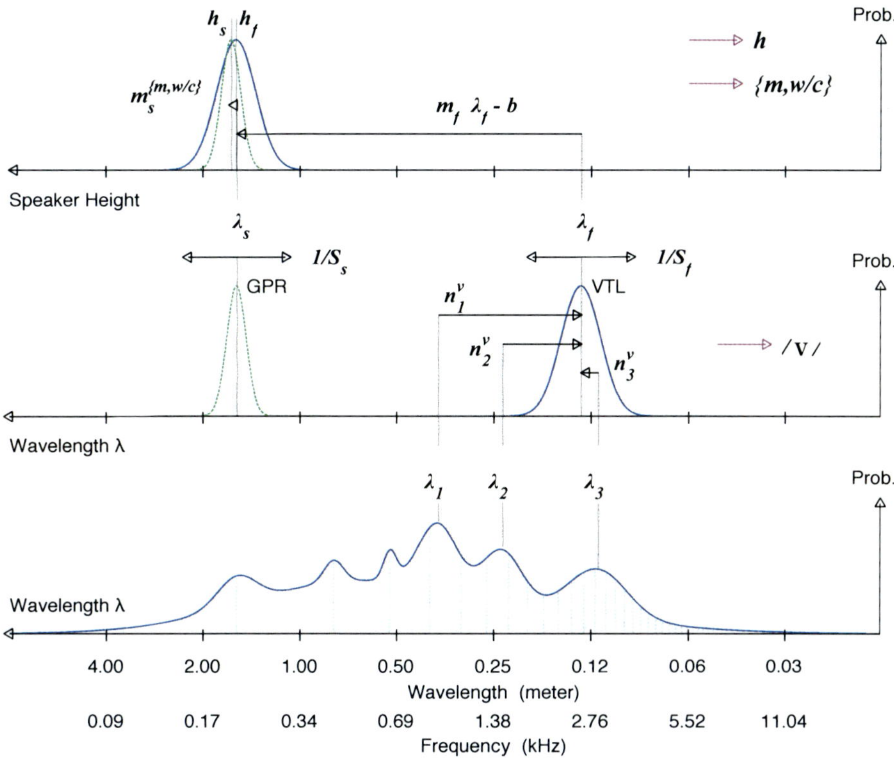

Fig. 23.7 Auditory variables and mechanisms involved in estimating speaker size and sex. Bottom: Spectral profile of the NAP of a vowel, showing three resolved harmonics and three formant peaks with their formant wavelengths. Middle: (left) voice pitch; (right) conversion of formant wavelengths to vowel type, /V/, and the associated VTL. Top: Conversion of the wavelengths associated with voice pitch and VTL into a common code for height estimation

in the sound. These physical variables and psychoacoustic discriminations were not, however, the focus of our experiments; rather, they were designed to encourage people to listen to the sounds as speech—either to identify the vowel being spoken or to make a judgment about the size and/or sex of the speaker. For vocal resonances, formant wavelength is directly related to the length of the section of the vocal tract that produces it, and for any given vowel, these lengths are highly correlated with body size. Accordingly, the formants are designated by their wavelengths on a quasi-logarithmic scale similar to the tonotopic dimension of the cochlea; wavelength (in meters/cycle) is velocity/frequency and the velocity of sound in air is 345 m/s. The pitch information of a vowel is distributed across the spectrum so it does not appear in the bottom row. The extraction of voice pitch is described in Section 4 and it appears in the schematic as the pitch of the vowel, λ_s, in the middle row.

The transformation which leads to the sound being perceived as a vowel, rather than three pitches corresponding to the formant peaks, is illustrated in the middle row. The formant wavelengths of the bottom row, λ_1, λ_2, λ_3, are transformed into a different measure of size, namely, λ_f, by three scalar constants, n_1, n_2, n_3, that specify formant wavelength as a proportion of VTL. If the speaker is a woman saying the vowel /a/ with an average VTL of 0.134 m (Fitch & Giedd, 1999), the formant frequencies would be about 0.930, 1.470, and 2.910 kHz (Lee et al., 1999, table III); the wavelengths would be 0.368, 0.233, and 0.118 m; and the scalars, n_1, n_2, n_3, would be 0.036, 0.058, and 0.114, respectively. The scalars are like formant ratios but they specify formant wavelength with regard to one common standard wavelength, λ_f, rather than with regard to the other formant wavelengths. There is a different set of scale factors for each vowel type in a given language and these scale factors represent the phonological information that a child acquires about vowels when learning to speak a specific language.

Within this framework, once a person has learned to speak, vowel recognition and VTL estimation could proceed as follows: The learned scale factors (n_1^v, n_2^v, n_3^v) for each vowel type, v, are applied to an incoming vector of formant wavelengths (λ_1, λ_2, λ_3) to provide three estimates of λ_f, and the differences between the estimates are summed to provide an error term. The process is performed for each vowel type in the language to see if one vowel type produces three coincident estimates of λ_f, as shown in the middle row of Fig. 23.7; if so, that is the system's candidate vowel type for that segment of speech. The estimate of speaker VTL is the average of the three λ_f values for that vowel type. For a sequence of vowels or syllables from a given speaker, VTL is effectively fixed, so the sequence of VTL estimates will converge on speaker VTL rapidly as an utterance proceeds. In speech production, λ_1 is less reliable than λ_2 or λ_3, so the estimation of λ_f is probably performed statistically with reference to knowledge about formant variability, as illustrated in Section III of Turner et al. (2009).

Our initial experiments in this field were designed to demonstrate how the perception of speech sounds might be based on variables such as those in the middle row of Fig. 23.7 and to test the hypothesis that the auditory system automatically normalizes the message of a pulse-resonance sound. The first experiment was performed by Smith et al. (2005) using a single-interval recognition experiment with the vowels (/a/, /e/, /i/, /o/, /u/). The values of glottal pulse rate (GPR) ranged from 10 to 640 Hz in doublings; that is from more than two octaves below the lower limit of voice pitch to about an octave above the pitch of young children. The VTL values ranged from 5 to 32 cm in seven equal steps on a logarithmic scale. If we extrapolate from the known relationship between VTL and body height, these shortest and longest VTLs correspond, respectively, to a small baby 0.3 m in length to a giant adult 3.5 m tall. Performance is largely independent of GPR throughout the range from 10 to 640 Hz; it remains near 100 % even when the GPR is so low that the sound does not produce a perception of pitch. Performance is near ceiling for VTLs within the normal range (9–18 cm); it drops to threshold only just before VTL reaches the extremes of 5 cm at the short end and 32 cm at the long end. The studies confirm that performance is excellent for voices with combinations of S_s and S_f throughout the normal range and for voices with combinations that do not exist naturally, thus supporting the hypothesis that the auditory system automatically normalizes speech sounds for both S_s and S_f.

If the normalization process concentrates VTL information in a perceptual variable, λ_f, as suggested in the middle row of Fig. 23.7, listeners should be able to use λ_f to discriminate the relative size of speakers. This led us to perform a series of S_f discrimination experiments, with vowels, syllables, and eventually words. The experiments were performed with "phrases" of four vowels or four syllables to engage the brain in speech mode and to ensure that listeners could not perform the task simply by focusing on one particular spectral peak (like those in the bottom row of Fig. 23.7). Ives et al. (2005) used a database of syllables to measure the JND for speaker size; the entire set of syllables was scaled with STRAIGHT to produce a wide range of natural sounding voices or speakers. There were two syllable formats, consonant–vowel (cv) and vowel–consonant (vc), for each of three consonant types: plosive, sonorant, and fricative. Within each syllable type, six examples of each consonant type were paired with each of the five canonical vowels, giving a total of $2\times3\times6\times5$, or 180, syllables. Psychometric functions were produced for five "standard" voices with combinations of GPR and VTL spanning the range from small children to large adult males and somewhat beyond. The just noticeable difference (JND) was measured by contrasting a (randomly chosen) syllable phrase from a standard speaker with a (randomly chosen) syllable phrase from a test speaker having a somewhat smaller or larger S_f, and asking which speaker was smaller. (The pitch of the voice was varied from syllable to syllable in order to eliminate simple spectral cues.) Over trials, the S_f difference was varied to determine the JND, that is, the change in S_f required to support 76 % correct choice of the smaller speaker. The results were surprisingly simple: the JND for VTL was close to 5 % independent of syllable type (cv or vc), consonant type (plosive, sonorant, or fricative), or the position of the standard in the GPR-VTL plane. For comparison, the JNDs for noise level (loudness), light level (brightness), and chemical density (odour) are about 10 %, 15 %, and 25 % respectively.

Subsequently, we replicated the identification and discrimination experiments using Japanese words, and recently Irino et al. (2012) compared identification and discrimination performance for voiced words with that for whispered words. The discrimination experiment is interesting inasmuch as whispered speech is missing one of perceptual cues used to estimate speaker size, namely, S_s. The recognition experiments showed that performance with whispered words is almost as robust to variation in S_f as performance with voiced words, and so long as the whispered words are clearly audible, performance is constrained only by the audibility of features at the extremes of the spectrum in these scaled speech sounds. Whispered word recognition is a little worse than voiced word recognition at all values of S_f when the GPR of the voiced words is in the middle of the normal range. But as pitch decreases below 80 Hz, voiced-word performance decreases and eventually becomes *worse* than whispered-word performance. The internal representation of voiced speech is not entirely stable at very low GPRs and this may well explain why performance falls off at low GPRs. The discrimination experiments showed that the average JND for the S_f of whispered words was effectively the same as for voiced words (~5 %), and both were the same as the JND observed for English subjects listening to with English syllables. In summary, these results show listeners can make fine judgments about S_f from whispered words, just as they can from voiced words, when the whispered words are clearly audible.

23.3.2 The Interaction of S_s and S_f in the Estimation of Speaker Height

The discrimination experiments with vowels, syllables, and words were designed to demonstrate the role of VTL, or S_f, in speaker perception; S_f was fixed for the set of vowels or syllables within a specific phrase and it was varied between intervals to prompt the perception that the speaker had changed. Recognizing the speech of an isolated speaker does not depend on pitch perception; the S_s distribution in the middle row of Fig. 23.7 is not involved in the process of vowel identification illustrated in the middle row. Accordingly, in those experiments, the pitch of the voice was manipulated to neutralize its role in the perception of speaker size. In everyday listening, however, S_s does play a role in the perception of speaker size. Before the advent of STRAIGHT, studies of size perception were limited to databases of sounds recorded from homogeneous groups of adult males or adult females—typically undergraduates at a specific university. Many of these studies concluded that you cannot predict height differences, within sex, from segments of speech. However, the range of GPR and VTL values in those studies was necessarily rather limited because the standard deviation for height in adult populations of men and women is only about 4 %. This is a little less than the JND for S_f with syllables and less than half the JND for isolated vowels. STRAIGHT made it possible to manipulate the S_s and S_f of vowels accurately over a huge range and investigate their interaction in speaker size discrimination.

In one study, listeners were presented two intervals with four-vowel phrases on each trial, one spoken by a "standard" speaker with a fixed combination of S_s and S_f, and the other spoken by a "test" speaker whose S_s and S_f values both varied from those of the standard speaker, and by enough to support size discrimination. The JND for S_f is roughly four times the JND for S_s, so on log S_s versus log S_f coordinates, the locus of test speakers that are equally discriminable from the standard is an ellipse about the point occupied by the standard speaker. The details are presented in Patterson et al. (2008). For each standard voice, there were eight test voices spaced around the ellipse and for each we measured the probability that the listener would choose the test speaker as smaller. A plane was fitted to the eight probabilities and the angle of the line of steepest descent on the plane revealed the trade-off between S_f and S_s in the perception of speaker size. The experiment was replicated for 16 standard speakers with a wide range of combinations of S_s and S_f, and the gradient vector was effectively the same across the entire plane, with $\log S_s$ having about twice the effect of $\log S_f$ in the discrimination of speaker size.

The middle and upper rows of Fig. 23.7 illustrate how the wavelengths, λ_f and λ_s, might be converted into estimates of speaker height, h_f and h_s, by two more scale factors, m_f and m_s, and how the interaction of S_f and S_s might arise. Consider first the conversion of the VTL information, λ_f, into the filter estimate of height, h_f, as it is the less complicated of the two. The fMRI data of Fitch and Giedd (1999) show that height is about 14.8 times VTL minus 41 cm from age five onwards (Turner et al., 2009, Fig. 23.4). So, in Fig. 23.7, the scalar m_f is 14.8 and it converts λ_f into a measure of height, h_f+41 cm. For example, in the United States of America, adult women have an average VTL of 13.8 cm and an average height of 163 cm.

Now consider the conversion of the voicing information, λ_s, into a source estimate of height, h_s. As children grow up, the locus of the voice in the logGPR–logVTL plane is initially a line that is the same for boys and girls. For girls, pitch decreases along the line without deflection until they reach their full height as women. For boys, pitch decreases along the line until they reach puberty when there is an abrupt decrease in pitch (an increase in λ_s), with the result that there is a largely separate region of the logGPR–logVTL plane for men's voices (see Turner et al., 2009, Fig. 23.13). This means that the scalar that converts λ_s into h_s has one of two values, m_s^m for men and $m_s^{w/c}$ for women/children; the parameter is designated $m_s^{\{m,w/c\}}$ in the figure to indicate that there are two separate categories. For women and children, the wavelength of their average voice pitch is roughly equal to their height. For example, adult American women have an average GPR of about 220 Hz (Lee et al., 1999, table III) which is equivalent to a λ_s of 156 cm. The average height of adult women is 163 cm, so $m_s^{w/c}$ is 1.04λ_s. In contrast, the average height of American men is 1.77 m and their average λ_s is 2.75 m (corresponding to an average GPR of 125 Hz). So m_s^m is 0.65 and the source estimate of speaker height, h_s^m, is 0.65λ_s.

The fact that there are two scalars suggests that the decoding of speaker size may occur jointly with the decoding of speaker type (man or woman/child), in a manner somewhat similar to that proposed for the joint decoding of vowel type and VTL. Specifically, we imagine that both of the source scalars, m_s^m and $m_s^{w/c}$, are applied to λ_s without attempting to determine speaker type, and the resulting height estimates (h_s^m and $h_s^{w/c}$) from the source information, λ_s, are compared with the height estimate, h_f, from the filter information, λ_f, to determine which source-height estimate produces the better match to the filter-height estimate. The value that produces the better match determines our perception of speaker type (man or woman/child) and speaker height. Specifically, h is a weighted average of h_f and the version of h_s that produced the better match to h_f, and the h_s value is given greater weight in the averaging because λ_s is more precise than λ_f.

In summary, the mechanism illustrated in Fig. 23.7 can convert the compressed, spectral-magnitude information produced by a vowel in the cochlea into the perception of a specific vowel spoken by a man, woman, or child with a specific size. In natural speech, as a sentence proceeds, vowel type changes frequently whereas the vocal characteristics remain fixed, and it is the stability of the sex and size information over syllables which tells us that we are hearing one specific person speaking even when we do not understand the language being spoken.

23.4 Time-Interval Information in Auditory Perception and Auditory Models

The majority of the information in speech and music is carried by frequencies below about 3 kHz where the inner hair cells phase lock to basilar membrane motion. As a result, the pulse-resonance cycles of communication tones produce NAPs with distinctive, repeating time-interval patterns (e.g., Fig. 23.3). In AIM, STI creates a separate dimension in auditory space for the time-interval information and it generates a stabilized version of the repeating NAP pattern in the auditory image (Patterson, 1994). Thus, for pulse-resonance tones, STI performs temporal integration over cycles without destroying the details of the pattern within the cycle (Patterson et al., 1992). The auditory image decays exponentially with a time constant of 30 ms and, as a result, the NAP of a sentence or a melody is converted into a dynamic auditory image in which detailed auditory figures emerge, evolve, and dissolve rapidly, forming a sequence of auditory events synchronized to the syllables of the sentence or the notes of the melody. These intriguing figures and events led us to run behavioral experiments to investigate the robustness of STI and to demonstrate the role of temporal asymmetry in auditory perception.

The first series of experiments was concerned with the robustness of STI and the salience of the pitch of iterated rippled noise (IRN). Rippled noise (RN) is constructed by delaying a random noise and adding it back to the original. Iterated rippled noise (IRN) is constructed by repeating the delay-and-add process. IRN produces a two-component perception: a buzzy tone with a pitch equal to the reciprocal of the delay and a background noise that sounds like the original random noise. The perceived tone/noise ratio increases with the number of iterations. The temporal regularity of IRN stimuli is not apparent in the NAPs flowing from an auditory filter bank; they look like the NAPs of noise (see Patterson et al., 2002, Fig. 23.1). The experiments were performed with Bill Yost and Stan Sheft, who were using short-term autocorrelation (STAC) rather than STI to convert NAPs into autocorrelograms (ACGs). There is a vertical ridge in the ACG at the IRN delay and the height of the ridge relative to the background activity increases with the number of iterations; there is a similar ridge in the auditory image. Autocorrelation is an optimal process for extracting periodicity information from temporal waveforms, and Yost and Sheft expected that STAC and the ACG would provide a better explanation of the pitch strength of IRN than STI and the auditory image. STI has the advantage that its computational load is about 1/10th that of STAC. The friendly rivalry that arose from discussions of optimality and efficiency led to a long and productive collaboration.

We measured the effective tone/noise ratio of IRN sounds in a discrimination matching experiment with Steve Handel and Jay Datta. Then we measured the perceptual tone-to-noise ratios of merged IRNs. With Lutz Wiegrebe, we investigated the role of envelope modulation in the perception of spectrally unresolved IRN, and this led to experiments on the temporal dynamics of pitch strength with Lauren Demany, Bob Carlyon, Stefan Hirsch, and Hugo Fastl. Lutz Wiegrebe also recorded single unit responses to IRN with Ian Winter to show that the firing patterns of cochlear nucleus cells did, indeed, contain the fine-grain time-interval information of IRN NAPs. We also extended the research to auditory masking with Katrin Krumbholz, Andrea Nobbe, and Hugo Fastl; we showed that threshold for IRN masked by noise was about 10 dB greater than threshold for noise masked by IRN. In most of these papers, we argue that the results cannot be explained with spectrographic models of auditory perception and many of them present computational models of pitch based on the relative height of the pitch ridge as it appears in the ACG and/or the auditory image. Detailed comparison of the models led to the conclusion that STAC and STI produce effectively the same estimates of pitch value and pitch salience for IRN and other pitch-producing stimuli.

The rivalry between STAC and STI then turned to pulse-resonance sounds like vowels. They are highly *asymmetric* in time and the NAPs of such sounds are highly *asymmetric* in time. STI preserves the majority of the temporal asymmetry in the NAP and so the auditory figures that appear in the auditory image are asymmetric in the time-interval dimension. Autocorrelation is *symmetric* in time and it converts *asymmetric* NAP patterns into virtually *symmetric* ACG patterns; a direct comparison of the ACG and auditory image of a vowel is presented in Patterson et al. (1995). This suggested that STI would prove to be a better model than STAC of those aspects of timbre perception associated with temporal asymmetry, and it prompted us to develop experiments on the perception of "damped" and "ramped" sounds with Michael Akeroyd. The damped version of the sound has a repeating, exponentially decaying envelope; the ramped version has a repeating, exponentially rising envelope. The carrier was either a sinusoid or a broadband noise. As the half-life of the exponential increases from 1 to 200 ms, the relative loudness of the carrier component of the sound increases. Pairs with the same half-life have identical power spectra, but they are, nevertheless, discriminable over a wide range of half-lives (2–32 ms) and envelope periods (10–160 ms). In the NAPs of damped and ramped sounds, the width of the spectral peak of the damped sound is narrower than that of the ramped equivalent. Nevertheless, it is the ramped version of the sound that produces the stronger perception of the carrier, and the ramped sound produces more time intervals at the carrier frequency in the auditory image.

The importance of temporal asymmetry in natural sounds prompted us to refine the damped/ramped discrimination experiment to provide a direct measure of auditory temporal asymmetry (Irino & Patterson, 1996). In a two-alternative, forced-choice experiment, listeners were presented a ramped sound in one interval and a damped version of the same sound having the same or greater half life in the other interval and they were asked to choose the interval containing the sound with the louder carrier component. Between trials the half-life of the damped version was varied to determine how much longer it needed to be to make the carrier sound equally loud to the carrier in the ramped sound. Broadly speaking, the half-lives of damped tones and noises have to be 4 and 2.5 times larger than those of ramped tones and noises, respectively, when the half-life is in the range 2–32 ms. With Christian Lorenzi we showed that cochlear implantees can discriminate damped and ramped sinusoids when the stimuli are presented on a single electrode, and that their *performance was far superior* to that of normals at longer half-lives. With Dick Fay, we showed that goldfish, which have no basilar membrane, can nevertheless discriminate damped and ramped sinusoids. With Daniel Pressnitzer, Ian Winter, and

Veronica Neuert, we showed that the firing rate of cochlear nucleus cells and inferior colliculus cells was different for damped and ramped sinusoids. With Stefan Uppenkamp and Sandra Fobel, we showed that short frequency sweeps (chirps) have clearly different timbres when time reversed. Because damped and ramped sounds have the same long-term magnitude spectra, these experiments pose a serious problem for traditional, spectrographic models of auditory perception.

The cochlea simulations of ACG models are conceptually similar to the gammachirp filterbank in the original version of AIM, and they all produce NAPs with strong temporal asymmetry when presented with pulse-resonance sounds. The main difference between an ACG model and AIM is the use of STAC rather than STI in the final stage. Accordingly, we implemented a version of AIM in which STI could be replaced with STAC (Patterson & Irino, 1998). Whereas STI increases the temporal asymmetry of the auditory image over that observed in the NAP, STAC reduces the temporal asymmetry in the ACG to the point where it is not sufficient to explain the perceived temporal asymmetry.

In summary, detailed comparison indicates that time-domain auditory models based on STI and STAC can both explain the perception of pitch in minute detail, but the same is not true for the perception of timbre. STI is an *asymmetric* process that can explain the timbre differences associated with the discrimination of temporal asymmetry; STAC is a *symmetric* process and it cannot explain the perception of temporal asymmetry.

23.4.1 STI, Autocorrelation, and Scale-Shift Covariance

In Section 2, we noted that the normalized version of the NAP with dimensions $\{\log_2 c, \log_2 s\}$ would not make a good model of the internal representation of sound that underlies auditory perception because it is not time-shift invariant, whereas auditory perception is. We concluded that the normalized auditory figures of pulse-resonance sounds exist in a $\{\log_2 c, \log_2 s\}$ plane that is orthogonal to the progression of time as we perceive it, and that it is auditory *events* that appear and evolve in the auditory image plane that require the property of time-shift invariance. The version of STI that was used to create the original {time-interval, log-frequency} representation of the auditory image can also be used to create a normalized auditory image in a $\{\log_2 c, \log_2 s\}$ plane, and the normalized image will have the same dynamic properties as the original auditory image, including the property of time-shift invariance for auditory events. The only difference is that the STI of the normalized version of AIM has to limit the representation of the auditory figure to a single copy in the $\{\log_2 c, \log_2 s\}$ plane. A thorough review of STI is presented in Walters (2011), including the constraints imposed on the process when sounds pass through an auditory filter with a gamma envelope.

Finally, it is important to note that it is not possible to produce a scale-shift covariant ACG because the lag dimension of autocorrelation does not preserve time-interval order within the period of a pulse-resonance sound. Multiplying the lag dimension in each channel of the ACG by the frequency of the channel does not align the periods of the impulse response across channels as it does in a scale representation. Autocorrelation averages periodicity information across the analysis window without regard to the period of the sound and this distorts the neural patterns of pulse-resonance NAPs.

23.5 The Gammachirp Auditory Filter and Joint Time-Frequency Representations of Sound

In the latter half of the 20th century, psychophysical and physiological data repeatedly demonstrated that the operation of the cochlea was much more like a wavelet transform (WT) than a Fourier transform (FT) (e.g., Irino & Kawahara, 1993). In a WT, as center frequency increases, the bandwidth of the channels (or filters) increases, and the duration of the kernel (or impulse response) decreases. The gammatone auditory filter bank (GT-AFB) is one common example of a WT, and it was successfully used to explain human masking data and revcor functions *at moderate stimulus levels* in the years before 1992. The success of the GT-AFB, and the exquisite detail in the stabilized auditory images of vowel sounds, led us to argue that the speech recognition community should stop using the short-term Fourier transform (STFT) as a "front end" for automatic speech recognition (ASR) and switch to the auditory solution—a stabilized WT—as human speech recognition is so much superior to ASR. The advice was largely ignored and one of the reasons cited for the continued use of the STFT was that Gabor (1946) had shown that the STFT was mathematically "optimal." Specifically, Gabor had shown that if the window that restricts the FT in time to produce the STFT is Gaussian in shape, then the STFT satisfies an important criterion for transforms that produce joint, time-frequency representations of sound; specifically, it minimizes uncertainty in the joint representation. The implication was that the auditory system is nonoptimal because it has to satisfy some mechanical or physiological constraint that is not compatible with minimal uncertainty. Although this discussion may seem arcane, it led us to consider an important alternative argument, which is that the auditory system *is optimal*, but it is optimal for a different representation of sound—one based on linear time and *logarithmic* frequency rather than *linear* frequency.

Gabor (1946) used operator methods to derive the window shape that produced minimum uncertainty for the spectrogram. We used very similar operator methods to derive the function that provides minimal uncertainty in a joint representation that involves linear time and logarithmic frequency (Irino & Patterson, 1997). The result was the gammachirp auditory filter—a generalized version of the gammatone with the same gamma envelope but a somewhat different carrier that glides (or chirps) in to the center frequency of the filter over the course of the impulse response. Impulse responses that chirp have spectral magnitude functions that are asymmetric in frequency. The auditory filter exhibits level-dependent asymmetry, with the low-frequency skirt becoming progressively shallower as level increases. Accordingly, we developed a gammachirp auditory filter bank (GC-AFB) that could explain the level dependent asymmetry of auditory masking *and* the level-dependent gain and compression observed physiologically around the peak frequency of the filter (Irino & Patterson, 2006). Moreover, in the ultimate version, the chirp rate does not vary with level as required by physiological revcor data. Patterson et al. (2003) provides a short history of the gammatone/gammachirp filter family and their uses.

In the derivation of the gammachirp function as the optimal auditory filter, the role of the Fourier transform (FT) is played by the Mellin transform (MT). The MT is invariant to the dilations produced by changes in vocal tract length, and this prompted us to propose a stabilized, wavelet–Mellin transform (SWMT) as a mechanism for the automatic normalization of pulse-resonance sounds in the auditory system (Irino & Patterson, 2002). From the mathematical perspective, a wavelet transform is the optimal filter bank to precede the MT because it is transparent to dilation, and the gammachirp is an ideal kernel for the WT because it satisfies the minimum uncertainty criterion. For example, it minimizes distortion of the resonant features of pulse-resonance sounds (Fig. 23.6). Thus, the SWMT provides the mathematical underpinnings for auditory models that simulate cochlear processing with a level-dependent auditory filter bank—models that include provision for stabilization and normalization by more central stages of the auditory system. The details are presented in Irino and Patterson (2002) and Patterson et al. (2007).

23.6 Summary

We have argued that the interesting sounds in everyday life, such as speech, music, and animal calls, have a special "pulse resonance" form that is not well represented by the spectrogram, and that our perception of these sounds is better represented by what is referred to as a stabilized auditory image (SAI). Over the past 20 years, we have pursued three related streams of research into the form of the SAI and the associated model of auditory perception (AIM): The first stream on level-dependent auditory filtering is described in Section 5; the second stream on the perception of the time-interval patterns in the SAI is described in Section 4. The chapter is concerned mainly with the third stream involving the size information in pulse-resonance sounds and our attempts to develop size-invariant and size-covariant versions of the SAI that can explain the size-invariant properties of auditory perception.

In Section 1, we explained that pulse-resonance sounds are produced by source-filter systems. The source and the filter both grow as animals mature, and the acoustic scale of the source and filter components of these sounds, S_f and S_s, change in fairly simple and predictable ways as the animal matures. In Section 2, we showed how the auditory system might construct a scale covariant version of the SAI that separates the size information of the source, S_s, and the size information of the filter, S_f, from the message of the communication. In Section 3, we described a series of experiments intended to demonstrate the value of scale-shift covariance in the perception of speech and music, and we showed how the acoustic scale information in the speech sounds might be converted into an estimate of speaker size with the inclusion of a small amount of contextual information.

Acknowledgments We thank Jess Monaghan and Etienne Gaudrain for their contributions to the model of speaker-size estimation. We thank Tom Walters for assistance with Fig. 23.1 and Ralph van Dinther for assistance with Figs. 23.4, 23.5, and 23.6.

References

Fitch, W. T., & Giedd, J. (1999). Morphology and development of the human vocal tract: A study using magnetic resonance imaging. *Journal of the Acoustical Society of America*, 106, 1511–1522.

Gabor, D. (1946). Theory of communication. *Journal of the Institute of Electronic Engineers* (London), 93, 429–457.

Irino, T., & Kawahara, H. (1993). Signal reconstruction from modified auditory wavelet transform. *IEEE Transactions of Signal Processing*, 41, 3549–3554.

Irino, T., & Patterson, R. D. (1996). Temporal asymmetry in the auditory system. *Journal of the Acoustical Society of America*, 99, 2316–2331.

Irino, T., & Patterson, R. D. (1997). A time-domain level-dependent auditory filter: The gammachirp. *Journal of the Acoustical Society of America*, 101, 412–419.

Irino, T., & Patterson, R. D. (2002). Segregating information about the size and shape of the vocal tract using a time-domain auditory model: The stabilised wavelet-Mellin transform. *Speech Communication*, 36, 181–203.

Irino, T., & Patterson, R. D. (2006). A dynamic compressive gammachirp auditory filterbank. IEEE *Transactions of Audio Speech & Language Processing*, 14, 2222–2232.

Irino, T., Aoki, Y., Kawahara, H., & Patterson, R. D. (2012). Comparison of performance with voiced and whispered speech in word recognition and mean-formant-frequency discrimination. *Speech Communication*, 54, 998–1013.

Ives, D. T., Smith, D. R. R., & Patterson, R. D. (2005). Discrimination of speaker size from syllable phrases. *Journal of the Acoustical Society of America*, 118, 3186–3822.

Lee, S., Potamianos, A., & Narayanan, S. (1999). Acoustics of children's speech: Developmental changes of temporal and spectral parameters. *Journal of the Acoustical Society of America*, 105, 1455–1468.

Patterson, R. D. (1994). The sound of a sinusoid: Time-interval models. *Journal of the Acoustical Society of America*, 96, 1419–1428.

Patterson, R. D., & Irino, T. (1998). Modeling temporal asymmetry in the auditory system. *Journal of the Acoustical Society of America*, 104, 2967–2979.

Patterson, R. D., Robinson, K., Holdsworth, J., McKeown, D., Zhang, C., & Allerhand, M. (1992). Complex sounds and auditory images. In Y. Cazals, L. Demany, & K. Horner (Eds.), *Auditory physiology and perception* (pp. 429–446). Oxford: Pergamon Press.

Patterson, R. D., Allerhand, M. H., & Giguère, C. (1995). Time-domain modeling of peripheral auditory processing: A modular architecture and a software platform. *Journal of the Acoustical Society of America*, 98, 1890–1894.

Patterson, R. D., Uppenkamp, S., Johnsrude, I., & Griffiths, T. D. (2002). The processing of temporal pitch and melody information in auditory cortex. *Neuron*, 36, 767–776.

Patterson, R. D., Unoki, M., & Irino, T. (2003). Extending the domain of center frequencies for the compressive gammachirp auditory filter. *Journal of the Acoustical Society of America*, 114, 1529–1542.

Patterson, R. D., van Dinther, R., & Irino, T. (2007). The robustness of bio-acoustic communication and the role of normalization. In *Proceedings of the 19th International Congress on Acoustics* (Madrid), pp. a-07–011.

Patterson, R. D., Smith, D. R. R., van Dinther, R., & Walters, T. C. (2008). Size information in the production and perception of communication sounds. In W. A. Yost, A. N. Popper, & R. R. Fay (Eds.), *Auditory perception of sound sources* (pp. 43–75). New York: Springer Science + Business Media.

Patterson, R. D., Gaudrain, E. & Walters, T. C. (2010). The perception of family and register in musical tones. In M. R. Jones, R. R. Fay, & A. N. Popper (Eds.), *Music perception* (pp. 13–50). New York: Springer Science + Business Media.

Smith, D. R. R., & Patterson, R. D. (2005). The interaction of glottal-pulse rate and vocal-tract length in judgements of speaker size, sex and age. *Journal of the Acoustical Society of America*, 118, 3177–3186.

Smith, D. R. R., Patterson, R. D., Turner, R. E., Kawahara, H., & Irino, T. (2005). The processing and perception of size information in speech sounds. *Journal of the Acoustical Society of America*, 117, 305–318.

Turner, R. E., Walters, T. C., Monaghan, J. J. M., & Patterson, R. D. (2009). A statistical formant-pattern model for estimating vocal-tract length from formant frequency data. *Journal of the Acoustical Society of America*, 125, 2374–2386.

Walters, T. C. (2011). *Auditory-based processing of communication sounds*. Ph.D. dissertation, University of Cambridge.

Chapter 24
A Changing View of the Auditory System Obtained from the Ears of Bats

George D. Pollak

George D. Pollak

G.D. Pollak (✉)
Section of Neurobiology and Center for Perceptual Systems,
The University of Texas at Austin, Austin, TX 78712, USA
e-mail: gpollak@austin.utexas.edu

A.N. Popper and R.R. Fay (eds.), *Perspectives on Auditory Research*, Springer Handbook of Auditory Research 50, DOI 10.1007/978-1-4614-9102-6_24,

24.1 Introduction

I began my career in 1970 when I was a postdoctoral fellow working on bats at Yale University with Bill Henson and Alvin Novick. Neuroethology was all the rage back then, and bats were one of the "hot" model systems. The guiding principle of neuroethology is based on the Krogh Principle, which states in essence that organisms that exhibit extremes of adaptation may reveal general principles not readily observable in less extreme species (Krebs, 1975). This idea resonated with me, and bats, with their high reliance on hearing for survival, were the animals of choice for studies of the auditory system. Actually it was not the auditory system per se that I wanted to study, but rather my goal was to discover the specialized mechanisms that could explain echolocation. And I got off to a blazing start. My first publication was a cover picture in *Science* (Pollak et al., 1972). The study revealed that the cochlea in the mustache bat (*Pteronotus parnellii*) has a specialized cochlear resonance that imparts remarkably sharply tuning at 60 kHz. There was nothing like it in the animal kingdom. The resonance at 60 kHz corresponds exactly to the frequencies adjusted during echolocation in this species. It was a striking correlation between a highly specialized physiological adaptation and behavior, and went a long way toward explaining how faint echoes were perceived during a unique form of echolocation called Doppler shift compensation. I was delighted by the wide attention the study attracted, and it was a big boost to my job prospects.

In 1973 I accepted a faculty appointment at the University of Texas at Austin and off I went to start my own laboratory. I was full of energy and optimism. If the cochlea so readily yielded its secrets, the special adaptations in the central auditory system should be just as yielding and just as dramatic. All I had to do was stick microelectrodes into an auditory nucleus, present stimuli that simulated the sounds the bats hear during echolocation, and the special adaptations that enable these animals to "see their world through their ears" would be revealed. It was just a matter of effort and time, or so I thought. It is now 43 years and more than 100 publications later, and I have no idea how echolocation works or how the auditory systems of these animals form images of the world via the echoes they receive.

My studies at Texas investigated the processing in brainstem auditory nuclei with particular attention directed at the inferior colliculus (IC). Try as I might, and I tried hard, I could find no unique mechanisms in the auditory systems of these animals that could explain echolocation. What I found after my first 10 years of study were the same physiological and anatomical features that my colleagues were finding in cats, rats, gerbils, and the other mammals auditory neuroscientists use in their studies. There were no highly specialized mechanisms or unique structural features of the sort that I found in the cochlea of mustache bats in 1972. However, many of the features I found were far more pronounced and more abundant in bats than in other mammals. What I began to see was a fundamentally mammalian auditory system in highly magnified form. I then began a transformation, changing my orientation and view of the auditory system from that of a "bat neurophysiologist" to an auditory neurophysiologist. I was reminded of the Krogh Principle and I decided that I would exploit the magnified features in bats' auditory systems to study basic processing mechanisms in the mammalian auditory system.

During the period when the basic but magnified features of the bat's auditory system were emerging, I also settled on a particular question for investigation. I was, as were all of my colleagues, captivated by the space map in the owl's auditory system that Eric Knudsen and Mark Konishi (Knudsen & Konishi, 1978) discovered. The ability to localize a sound source in space is common to all animals that hear, but because bats earn their livings by catching flying insects in the night sky, they must have pronounced mechanisms for sound localization. It was for these reasons that, in 1983, I began to focus my research on the binaural processing of interaural intensity disparities (IIDs), the cues animals use to localize high-frequency sounds.

The studies of binaural processing that my colleagues and I conducted over the next 10 years changed my thinking and appreciation of the auditory system. I gradually learned that the processing in the auditory system is far more complex, diverse, and expansive than I had previously thought, or even imagined. It is full of surprises because nothing is quite as it seems. Nuance prevails, and the nuance profoundly impacts the functioning of the circuitry in the ascending auditory pathway, changing it from something that at first appeared simple into something far more complex, but also more interesting and conceptually important. I also came to see the dominant role that inhibition plays in shaping the response properties of IC cells. Indeed, I have come to view inhibition as the primary "sculptor" in the auditory system, which shapes the highly selective properties in IC neurons out of less selective excitatory inputs.

And then I came full circle. Although I view the auditory systems of bats as basically mammalian in design, I have always been troubled by the fact that bats can echolocate with such great precision while most other mammals cannot. Why is this? Is there, in fact, something fundamentally different and special about bats that had eluded me? Recent events suggest that I did not misinterpret the data. In the final section I will argue, and present evidence to support the argument, that whatever enables bats to echolocate is also part and parcel of the human auditory system. In the following sections I recount how the discoveries my colleagues and I made at each stage of my career changed my understanding of the mammalian auditory system.

24.2 The Early Years and My First Visit to Germany

When I began my career in the early 1970s, the study of the bat's auditory system was in its very early stages. A seminal event occurred in 1976 when a series of studies were published by the group headed by Gerhard Neuweiler, who was then in Frankfurt (Bruns, 1976; Schnitzler et al. 1976; Suga et al. 1976). The studies were stunning and represented a quantum leap in both the conceptual and technical quality than anything that had come before. All the major players in the field, Nobuo Suga, Alan Grinnell, and Jim Simmons, were flocking to Frankfurt. So I too had to make the pilgrimage and find out what was going on.

Fig. 24.1 Gerhard Neuweiler

This wasn't easy for me, as I am the son of Viennese immigrants. Indeed, I wasn't sure that I could stay in Germany even for a few days, but things turned out well. I was greeted in Frankfurt by Gerhard Neuweiler (Fig. 24.1), who opened his home and heart to me and my family. Gerhard, whose father had been a Nazi, and I, whose parents were Viennese Jews forcibly torn from their homes and lives by the Nazis, became close friends, and I will never forget the profound impact Gerhard had on both my personal and professional life.

Gerhard was a gifted visionary. He understood the important questions in science and sent his students throughout the world to learn the techniques required to answer those questions. He took talented people and placed them in positions where their talents could be exploited to best advantage. He was a genius at this. And Gerhard was an incurable romantic. For Gerhard, biology was not just a bunch of facts and data; biology was beauty. His lectures, even in English, were simply galvanizing and were like poetry. It was from Gerhard that I derived a far more integrated and coherent view of biology, and of bats in all their forms, than I had before. I owe him, and will always owe him, a great debt of gratitude.

24.3 Returning to Austin

I returned to Austin invigorated, and began a series of studies of the mustache bat's IC. Around this time I met John Zook, who had conducted the first connectional studies of the auditory system of mustache bats when he was a graduate student with Pete Casseday at Duke (Zook & Casseday, 1982a, b). They were truly majestic studies and showed that mustache bats possess the same complement of lower auditory nuclei as other mammals and that the projections from each of the lower nuclei are basically the same as those found in other, less specialized mammals. John was continuing his postdoctoral work with Mike Merzenich at the University of California at San Francisco, but would come to my lab during the summer and we, together with Robert Bodenhamer, who was a graduate student in my lab, mapped out the tonotopic organization of the mustache bats' IC.

What we found was stunning. One isofrequency contour was massively over-represented and occupied about a third of the entire volume of the IC. All of the neurons in that contour were tuned to 60 kHz, exactly the same frequency as the resonance in the cochlea that Bill Henson and I had discovered in 1972. John then introduced me to Jeff Winer at Berkeley, who was a master with the Golgi technique. Together we not only mapped the tonotopic arrangement of the mustache bat IC, but also evaluated its cytoarchitecture from Nissl stained material and its neuronal architecture from Golgi impregnated material (Zook et al., 1985).

Both Jeff and John were first class comparative neuroanatomists, and they convinced me that the anatomical features that we found in the mustache bat's IC were basically mammalian in design. There was nothing special about the cells that distinguished the bat's IC from the IC of other animals. The greatly magnified isofrequency contour was anatomically just like any of the other isofrequency contours in the mammalian IC, but much larger. I was impressed, but not fully convinced. If the 60-kHz contour was really just a typical frequency contour in magnified form, then it should also have the same connections with lower nuclei and its cells should expressed the same response features as those found in the IC of other mammals.

24.4 Mapping the Functional Organization of the Mustache Bat's IC

The next obvious step was to map the functional organization of the magnified contour. Because the IC is a surface structure in bats, this was not a difficult experiment and I had a talented group of students and postdocs eager to take up the question. Jeff Wenstrup, Nick Fuzessery, and Linda Ross had just joined my lab; Jeff and Nick were postdocs and Linda was a graduate student. Jeff and Linda conducted a mapping study and showed that monaural and binaural neurons were spatially segregated in the magnified 60 kHz contour (Wenstrup et al., 1985, 1986). Moreover, the EI cells, the cells that code for interaural intensity disparities, were also segregated and confined to the ventrolateral sector of the contour. The map of aural types was not only distinct, but was also reproducible from animal to animal. Indeed, the arrangement of aural types in this isofrequency contour was in principle similar to the arrangement of aural types that Mal Semple and Lindsey Aitkin had found along is frequency contours the cat's IC (Semple & Aitkin, 1979).

But Jeff and Linda also discovered something else that caused great excitement (Wenstrup et al., 1986). They found the sensitivities for IIDs were arranged in an orderly manner in the EI region of the 60-kHz contour. The exciting part of this discovery is that the topography of IID sensitivities provided a functional substrate for coding the particular IID that reached the ears. It revealed that the cues for localizing sound were represented as a population code, where a locus was generated by a particular IID that separated active from inactive EI cells. The locus that separated active from inactive cells then shifted with IID and thus with spatial location.

At the same time, Nick Fuzessery was conducting studies on the spatial receptive fields of IC neurons, where he could correlate the binaural properties of each neuron recorded with speakers inserted into the ear canals, with the spatial receptive fields obtained when the speakers were removed and sounds were presented in the free field from various azimuthal locations (Fuzessery & Pollak, 1984, 1985). Because these bats have three dominant harmonics in their echolocation calls, it was relatively simple to also obtain the head-related transfer functions (HRTFs) for the three frequencies by using the ear as a microphone, that is, by recording cochlear microphonic potentials evoked by the three frequencies from various regions of space.

Nick and Jeff then teamed up to conduct two, very elegant experiments (Wenstrup et al., 1988a, b). What they did is to use earphones to record both the monaural and binaural responses from the EI neurons in the 60-kHz contour. They then removed the speakers and recorded each neuron's spatial receptive field, just as Nick had done previously. They then used the monaural and binaural responses, scaled them with the HRTF at 60 kHz, and obtained a predicted spatial receptive field. They then compared the predicted spatial receptive field with the actual spatial receptive fields recorded in the free field. The agreement was striking, and showed that spatial receptive fields could be accurately predicted by just knowing the HRTF and the binaural and monaural response properties of the neurons. It also showed that for each EI neuron, the IID at which discharges were suppressed gave an accurate picture of the locations in space that could drive the neuron and the spatial location at which the neuron was inhibited and stopped responding. It was a striking complement to our previous proposal of that the IID received at the ear is represented by a locus in the EI region that separates a population of excited from inhibited neurons.

24.5 Mapping the Projections to the Aural Regions in the 60-kHz Contour

These were exciting times but there were also a number of questions that just begged to be studied. Because the various aural types were topographically segregated in the EI region, the question of which lower nuclei provided innervation to each aural subregion could be evaluated by making small injections of a retrograde tracer into each of the subregions. This question was taken up by Linda Ross and John Zook, who continued to be a regular visitor to my lab. Linda and John, as it turned out, had a budding romance and ultimately married, but at this time they teamed up to conduct the connectional study. They first made large injections of tracers confined to the 60-kHz contour and found that the contour receives the full complement of projections from lower nuclei, as occurs in other mammals (Ross et al., 1988). This finding convinced me that the mustache bat's auditory system is not special but rather is typically mammalian, and crystallized my transformation from a "bat neurophysiologist" to a mammalian neurophysiologist.

In a follow-up study, Linda then made small injections confined each of the aural subregions in the 60-kHz contour and found that each subregion receives a unique complement of projections (Ross & Pollak, 1989). The projections to the EI region of the 60-kHz contour were of particular interest because we had so much information about the response properties of those neurons. Linda's results showed that the EI region received bilateral projections from the lateral superior olives (LSOs), bilateral projections from the dorsal nuclei of the lateral lemniscus (DNLLs), and a host of projections from monaural nuclei. Projections from the LSO were not surprising, as the LSO is the nucleus were EI properties are first formed (Boudreau & Tsuchitani, 1968). Nor were projections from the two DNLLs surprising, as previous studies in cats showed that high-frequency DNLL neurons were also EI (Brugge et al., 1970) and that the DNLL sends strong bilateral projections to the inferior colliculus (Adams, 1979).

Nevertheless, knowing the complement of projections to the EI region raised a profoundly important question; if the EI properties are already formed in the LSO, why are the other projections even needed and what, exactly, are they doing? Projections from the DNLL were especially puzzling since it was already known that the DNLL was a purely GABAergic nucleus (Adams & Mugnaini, 1984) and thus was providing inhibitory innervation to the EI cells in the IC. But how did this inhibitory innervation impact the EI cells in the IC and how could one determine what that impact is by recording discharges with extracellular electrodes?

24.6 A Return Trip to Germany and Where I Met Benedikt Grothe

I struggled with this problem of how to evaluate the role of inhibition at the IC, but without resolution. During this period, my old friend Gerhard Neuweiler had moved to Munich and invited me back for another visit to his new lab and group. So off I went for a most memorable visit. When I settled in, I met a young graduate student, Benedikt Grothe, and we hit it off from the start. Although Benedikt was working on his dissertation research, he sort of latched onto me and we collaborated on a routine project that we later published together (Grothe et al., 1994). This was the beginning of a friendship and collaboration that continues to this day.

I was very impressed with Benedikt, and I was especially impressed by his dissertation project and the rationale for choosing it. Benedikt was working on the medial superior olive (MSO) in mustache bats (yes, bats have an MSO), but in this species the inputs from the ipsilateral ear are greatly attenuated. His reasoning was that excitatory inputs from the contralateral cochlear nucleus respond with phase-locked discharges to a wide range of amplitude modulations rates, where the neurons fire to the amplitude modulations imposed on a high-frequency carrier. Because the MSO also receives inhibitory inputs from the medial nucleus of the trapezoid body (MNTB), the phase-locked inhibition from the MNTB should be just slightly delayed relative to the excitation at the MSO. That delay of inhibition should, in turn, cause the inhibition to interlace with the excitation, and at higher modulation rates, the excitation and inhibition should coincide or interlace with periods too brief for the inhibition to decay. In short, the inhibition should convert an all pass amplitude modulation rate to a low pass rate in the MSO cell. I'll never forget the day that Benedikt was recording from an MSO cell and dragged me into the lab to watch the experiment. The MSO cell did indeed fire only to low modulation rates, but when he iontophoretically applied strychnine, which blocked the glycinergic inhibition from the MNTB, the cell was magically transformed, and now responded to both low and high modulation rates.

When I saw this transformation, I was absolutely speechless! It was, for me at least, an epiphany. From previous studies, Benedikt knew the excitatory and inhibitory innervation that played upon the MSO and proposed an explicit hypothesis to explain how the inhibitory innervation could shape an emergent property. The important thing was that he could actually test and confirm the hypothesis (Grothe et al., 1992; Grothe, 1994). I had never seen or heard of anything like it in the auditory system. When I saw the response transformation myself, at that moment I knew that I had to learn iontophoresis, because it was the technique that would allow me to evaluate the roles of inhibition in the IC.

24.7 Back to Austin and the Tom Park Years

After Jeff, Nick and Linda started their own labs, Tom Park joined my lab as a postdoc. Tom is an exceptional man and an exceptional scientist. Tom is not only great fun to work with, but he also somehow makes everything work, regardless of the level of difficulty. What we wanted to know is how does the inhibitory innervation shape the IID functions of the EI cells in the IC. Although Benedikt tried to teach me how to make the multibarrel electrodes required for iontophoresis, I was not particularly good with such manipulations and too klutzy to learn. So we asked Don Caspary, the master of this methodology in the auditory system, to give us some lessons. Don is not only a fine scientist, but he is also an exceptionally generous man. He graciously devoted several days to teaching us how to make the multibarrel electrodes and some of the nuances of microiontophoresis.

Tom and I returned to Austin and eagerly conducted a series of experiments that utilized microiontophoretic application of drugs that block inhibition. On conducting these studies we found three principal changes in EI properties when inhibition was blocked and we proposed that each type of change was due to a different set of inputs received by each type of EI cell (Park & Pollak, 1993, 1994). The first type is the simplest (Fig. 24.2a). In these cells blocking inhibition at the IC did not change the spike suppression caused by ipsilateral stimulation. Because these cells received no direct inhibitory innervation from the ipsilateral ear, we proposed that they inherited their EI properties entirely from the LSO, the lower nucleus in which the EI property is first formed (Fig. 24.2a). In the second type of cell, the spike suppression due to ipsilateral stimulation was largely abolished when inhibition was blocked (Fig. 24.2b). We proposed that the EI property in these cells is formed de novo in the IC from an excitatory monaural projection driven by the contralateral (excitatory) ear and an inhibitory projection from the contralateral DNLL driven by the ipsilateral (inhibitory) ear. Thus, blocking GABAergic inhibition in these cells transformed a strongly suppressed EI cell into a very weakly suppressed or even

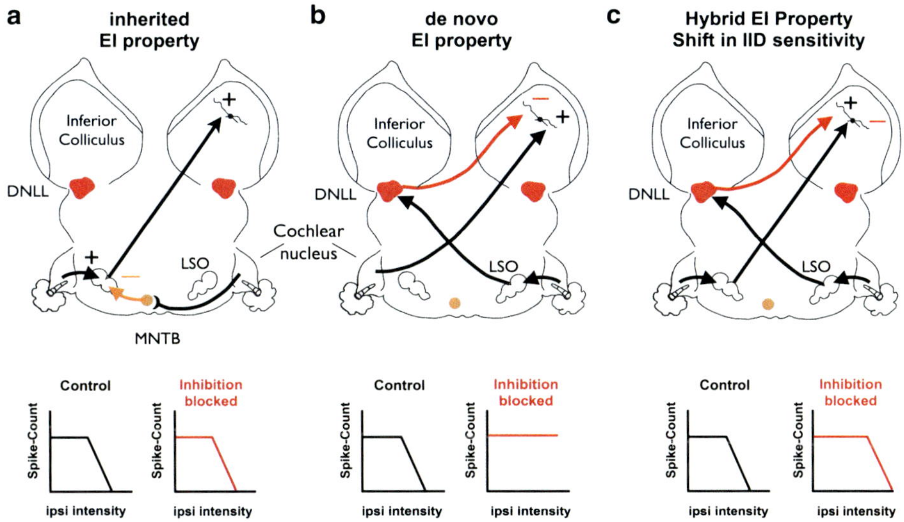

Fig. 24.2 Three principal ways that EI cells are constructed in the IC from the complement of inputs that innervate each EI type. The circuits were proposed from neurophysiological studies in which IID functions were obtained before and when inhibition was blocked at the IC. Schematic IID functions and the way they changed or did not change when inhibition was blocked are shown below each circuit

into a monaural cell, in which ipsilateral stimulation caused little or no spike suppression. In the third EI type, blocking inhibition did not change the degree of spike suppression, but rather shifted the IID at which the cell was suppressed (Fig. 24.2c). We proposed that these cells were "hybrids," in that we thought that they are innervated by both the LSO and the contralateral DNLL. In this scenario, the EI feature is formed in the LSO, but due to the input from the DNLL, a lower intensity at the ipsilateral ear now generates the criterion degree of inhibition. When the inhibition from the DNLL is blocked at the IC, a stronger ipsilateral signal is required to generate the criterion degree of spike suppression, thereby shifting their IID functions.

The circuits we proposed received support from studies conducted in rats by Li and Kelly (Li & Kelly, 1992), and by Carl Faingold (Faingold et al., 1993). Both of those studies found that pharmacological inactivation of the DNLL transformed the IID functions of EI cells in ways that were in agreement with the circuits proposed for the de novo and "hybrid" EI cells that we proposed based on the microiontophoretic studies in the bat's IC.

The results of these experiments told me that things are often not what they seem in the auditory system. Even garden-variety EI cells, all of which seem to be the same binaural type, comprise a diverse population. The diversity is due to the unique complement of projections each type of EI neuron receives, and the properties of at least two types are strongly shaped by the GABAergic inhibition from the contralateral DNLL. The question raised by these results is what functional significance should we attribute to the different complement of inputs. The answers to these questions are not at all apparent. I will them take up later, but first, another visit to Munich.

24.8 A Third Visit to Munich, This Time with Tom Park, and the First Montana Meeting

Tom finished his postdoctoral work in my lab and secured a faculty appointment at the University of Illinois at Chicago. Before Tom moved to Chicago, Gerhard Neuweiler invited us back to Munich to work with Benedikt, who at that time held the rank of assistant professor in Neuweiler's group. So off we went. I was in Munich for the semester but Tom spent the entire year working with Benedikt. We focused on the LSO and the roles of interaural timing and interaural intensity shaped each neuron's IID function (Park et al., 1996). It was a remarkably productive period for both Tom and Benedikt (Park & Grothe, 1996; Grothe et al., 1997; Park et al., 1998) and they formed a lifelong friendship that is very special. Indeed, in subsequent visits to Munich, Tom played a pivotal role in the experiments that made Benedikt one of the icons in auditory neuroscience.

But I digress. The close friendship Tom and Benedikt formed on Tom's initial visit to Munich somehow prompted them to organize a meeting that summer. The meeting they had in mind would be small, only 20 or so people, and would be devoted to binaural processing, and would be held at a dude ranch that Tom knew of in Montana. When they told me about their plans, I told them to forget it; they were both young investigators just beginning their careers, and my admonition was that they could not afford either the time or effort required to organize even a small international meeting. Fortunately, they did not heed my sage advice. The meeting was a raving success. It was so successful that it is now held on a biannual basis and invitations to the meeting are highly sought after. It is now known as the Wyoming Meeting, as the original dude ranch closed and a new ranch in Wyoming had to be found.

24.9 Up Next, the DNLL

While the iontophoretic experiments showed that projections from the DNLL shape IID functions in the IC, exactly why their inhibitory projections should create EI cells de novo in some IC cells and shift the IID functions in others was unclear. Stated differently, because the full range of IID sensitivities is already established in the LSO (Park et al., 1996, 1997), why should these features be created again in the IC by DNLL inhibitory projections? No one had the foggiest idea. Something was missing, and finding the missing features drove research in my laboratory for the next several years.

The missing features emerged when a graduate student in my lab, Lichuan Yang, a.k.a. Lenny, asked how EI properties are formed in the DNLL (Yang & Pollak, 1994a, b). Each feature he discovered answered one question but raised even more puzzling questions, until the puzzle was finally resolved. The sequence of discoveries is interesting and is recounted below to show how the mystery was solved.

It was well known from numerous studies that the DNLL receives strong projections not only from the contralateral LSO, but also from the opposite DNLL via the commissure of Probst (Ross et al., 1988; Oliver & Shneiderman, 1989; Merchan et al., 1994; Yang et al., 1996; Chen et al., 1999). It was, therefore, thought that in DNLL cells the ipsilaterally evoked spike suppression was caused by the inhibitory projections from the opposite DNLL. However, when Lenny blocked inhibition at the DNLL, there was no reduction in the ipsilaterally evoked suppression, and the IID functions of DNLL cells were virtually unchanged (Yang & Pollak, 1994b). When Lenny first told me this I did not believe him. So I sat down at the rig with him and blasted the DNLL cells with as much bicuculline and strychnine as our instruments could eject. But no matter what we did, we could not transform any of the EI cells in the DNLL into monaural cells, as Tom and I did in the IC. We concluded that the DNLL must inherit its EI properties from the LSO. But where was the inhibition from the opposite DNLL? It had to be there because the commissure of Probst is a large, GABAergic projection, but how it was expressing its inhibitory influences on the opposite DNLL was not apparent.

Lenny's experiments found the answer and the answer was not only surprising but also highly significant. The answer lay in the operation of the circuits that connect one DNLL with the LSOs on both sides and with the opposite DNLL (Yang & Pollak, 1994a, 1998; Pollak et al., 2003). In essence the DNLL inherits its EI property via an excitatory projections from the opposite LSO. But whether the contralateral LSO provides excitatory input to the DNLL or whether the ipsilateral LSO provides inhibitory input to the DNLL, depends on the IID.

To illustrate how this would work, let's consider a cell in the left DNLL, which inherits its EI property from the right LSO (Fig. 24.3). A monaural stimulus presented only to the right ear drives the DNLL via the crossed excitatory projection from the right (contralateral) LSO (Fig. 24.3a) while a monaural stimulus presented to the left ear inhibits the DNLL via the pathways activated by the left (ipsilateral) LSO (Fig. 24.3b). As shown in Fig. 24.3c, binaural stimuli with IIDs that are louder in the right ear, drive the right LSO and thus drive the left DNLL. At an IID of about 0 dB, a complete spike suppression is produced in both LSOs and thus a complete spike suppression in the left DNLL as well. That is why blocking inhibition at the DNLL has no effect on the DNLL cell's IID function. But as the IID changes and becomes more intense in the left ear (Fig. 24.3e), activation shifts from the right LSO, which is now inhibited, to the left LSO, which is now excited. The left LSO sends a glycinergic projection directly to the left DNLL. The left LSO also excites the right DNLL, which then sends a strong GABAergic input to the opposite (left) DNLL via the commissure of Probst. Hence, blocking inhibition in the left DNLL has no effect EI properties of its cells because the EI property is generated entirely in the LSO and projected on the DNLL. Moreover, the DNLL receives both GABAergic and glycinergic inhibition, but only *after* its discharges are completely suppressed. The inhibition plays no role in suppressing the excitation from the LSO.

This scenario explains why the EI properties were unaffected when inhibition was blocked, and it showed that DNLL cells do indeed receive inhibitory input from the opposite DNLL. But it was also a very puzzling result; what is the point of inhibiting DNLL cells that are already completely suppressed? There was still something missing.

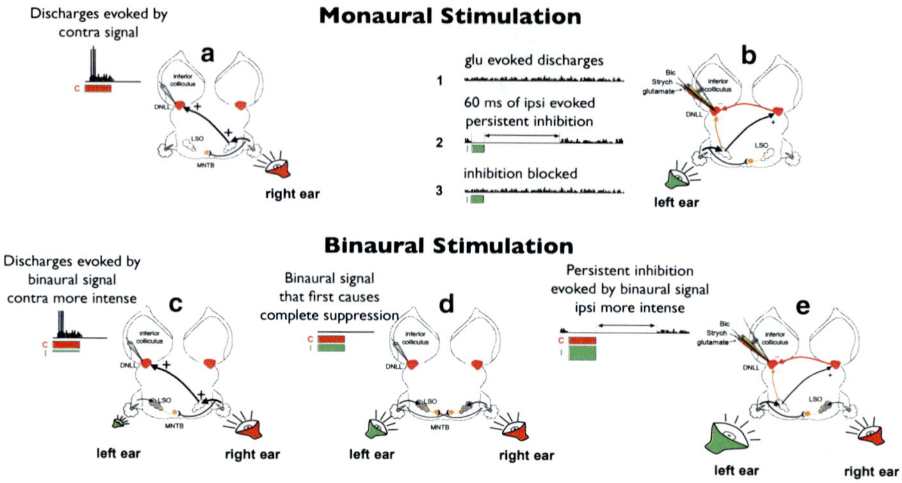

Fig. 24.3 The circuits that regulate the activity of DNLL cells evoked by monaural (**a, b**) and binaural (**c–e**) stimulation. (**a**) Discharges are evoked by monaural stimulation at the contralateral ear. (**b**) Stimulation of ipsilateral ear. (**b1**) Background activity evoked by iontophoretic application of glutamate. (**b2**) Monaural stimulation of ipsilateral ear evokes inhibition that generates the gap in the carpet of glutamate evoked background activity. (**b3**) Ipsilaterally evoked inhibition is blocked by iontophoretic application of bicuculline and strychnine. The gray LSOs in the binaural stimulation panels represent LSO cells that are completely suppressed. See text for further explanation

24.9.1 The Discovery of Persistent Inhibition

Lenny then discovered that the missing feature of the DNLL that brings all of these features into a functional coherence is persistent inhibition (Yang & Pollak, 1994b, 1998). Stimuli presented to the ear ipsilateral to the DNLL evoke an inhibition in the DNLL that outlasts the duration of the signal that evoked the inhibition for periods of 10–80 ms (Fig. 24.3b2, e). This was shown by iontophoretically applying glutamate to the DNLL cell, which generated a carpet of background discharges (Fig. 24.3b1). When a signal was then presented to the ipsilateral ear, the DNLL was inhibited as shown by the gap in the background discharges, but the gap was much longer than the signal that generated the inhibition (Fig. 24.3b2). We called the inhibition that lasts for a period beyond the duration of the stimulus that evoked it "persistent inhibition."

Persistent inhibition, of course, plays no role in generating the basic EI property of the DNLL cell or in generating the IID function with a single binaural signal because the EI property is generated in the LSO. However, and this is the point, the persistent inhibition should influence the processing of multiple signals that follow each other in time and that emanate from different regions of space, and thus have different IIDs. Specifically, an initial signal in the ipsilateral sound field, which is more intense in the ear ipsilateral to the DNLL, will evoke a persistent inhibition in the DNLL. The persistent inhibition, in turn, should prevent that DNLL cell from responding to trailing signals from other regions of space, which have IIDs that normally evoke discharges in the DNLL cell. In addition, the suppression of responses to the trailing signals should occur only during the period of persistent inhibition in the DNLL generated by the first signal.

24.9.2 *The Functional Circuits That Innervate the DNLL Were Experimentally Tested and the Hypothesis Confirmed*

The hypothesis is not only consistent with the circuitry, but most importantly, it also makes specific predictions about how the system should respond to multiple IIDs, and it predicts exactly how the responses to the same IIDs should change when inhibition is blocked. We were excited and immediately had the computer programmed to present the multiple stimuli required for the experiment. We were almost incredulous when the experiment worked on the first DNLL cell we recorded, and subsequently comforted when the experiment worked on all the other DNLL cells we tested (Fig. 24.4) (Yang & Pollak, 1994a, b, 1998; Pollak, 1997). The circuit worked exactly as predicted.

I was reminded of the afternoon in Benedikt's lab when he confirmed his hypothesis about the role of glycinergic inhibition for shaping amplitude modulation rate selectivity in the mustache bat's MSO. These experiments also confirmed a hypothesis about the roles of inhibition, and to me, they were and will always be among the most important and most satisfying experiments that were ever conducted in my laboratory. They were satisfying because they showed how the complex interactions among several brainstem nuclei act cooperatively to shape the selective responses to multiple binaural stimuli, and that inhibition was the mechanism that shaped the selective responses. They were important because of the insights they provided about the functional significance of the projections from the DNLL to the IC.

24.9.3 *The Role of the DNLL in the Processing of Multiple Sound Sources in the IC*

The feature that links two of the three types of EI neurons presented previously is innervation from the DNLLs, which play a special role in the differential responsiveness to multiple sounds that emanate from different regions of space (that have different IIDs). Thus by presenting an initial binaural signal that is more intense in the inhibitory ear than the excitatory ear, the DNLL on that side is persistently inhibited and thereby deprives its targets in the IC of its inhibitory innervation for the duration of the persistence. The effects of the DNLL then must be expressed in the IC but in an inverted way; when the responses of trailing signals are suppressed in the DNLL, the persistent inhibition that caused the suppression removes inhibition from the IC, allowing the IC to respond to trailing signals whose discharges were suppressed by inhibition from the DNLL when presented alone.

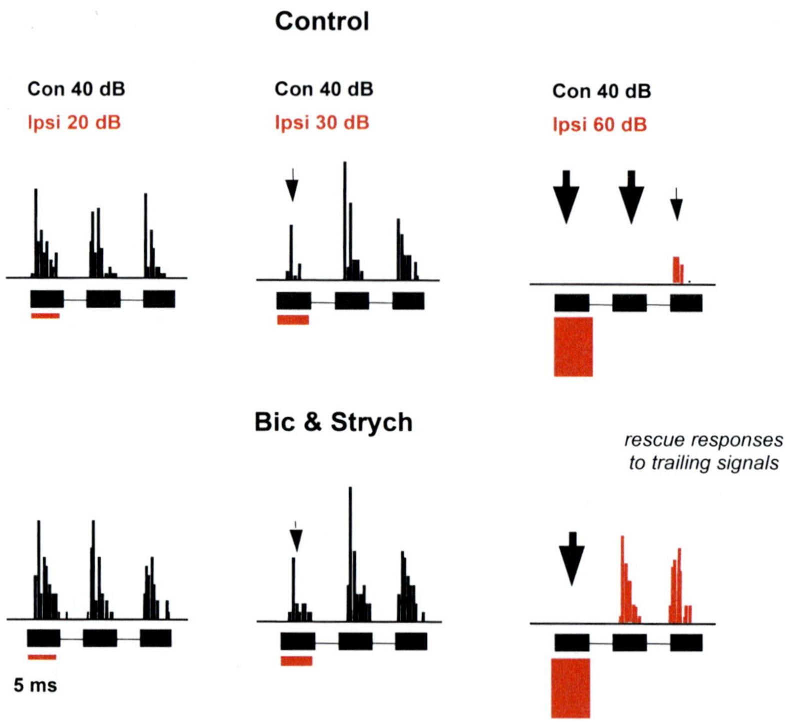

Fig. 24.4 Persistent inhibition generated by an initial binaural signal suppresses responses to trailing sounds (T_1 and T_2) in DNLL neurons. (Top panel) Control responses before inhibition was blocked. The first signal is binaural while the two trailing sounds are monaural and presented only to the contralateral (excitatory) ear. The contralateral intensity of the first, binaural sound is held constant while the ipsilateral intensity (red) is at the inhibitory ear progressively increased. The increasing ipsilateral intensity at first only inhibits the responses to the first, binaural sound. The inhibition occurs at the LSO. Increasing ipsilateral intensities then generate a persistent inhibition in the DNLL that completely suppresses the responses to the first trailing sound (T_1) and partially suppresses the responses to the second trailing sound (T_2). (Lower panel) Blocking inhibition eliminates persistent inhibition and rescues responses to trailing sounds but does not rescue responses to the initial, binaural sound, because the inhibition of responses evoked by the binaural sound occurred in the LSO

The influence of the DNLL on the responses to multiple signals in the IC was also tested and confirmed (Burger & Pollak, 2001; Pollak et al., 2003) (Fig. 24.5). I was again fortunate to have a superb graduate student, Mike Burger, to conduct the experiments. Mike recorded from EI cells in the IC while reversibly inactivating the contralateral DNLL with kynurenic acid, which blocks glutamatergic receptors (not shown). He first presented a binaural tone that was louder in ear ipsilateral to the IC than the contralateral ear, which evoked no discharges (Fig. 24.5a). He then inactivated the contralateral DNLL with kynurenic acid, and the same binaural signal now evoked discharges (Fig. 24.5b). As shown in Fig. 24.5, the effects of inactivating the DNLL pharmacologically were then reproduced in the same neuron by first presenting a binaural signal that was more

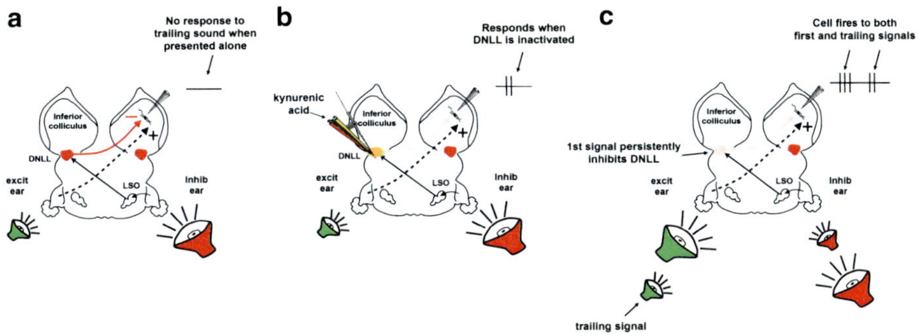

Fig. 24.5 Responses of EI neurons innervated by the contralateral DNLL to multiple sounds. (**a**) No spikes are evoked by a binaural signal that is more intense at the ipsi- than the contralateral ear, as occurs in the ipsilateral sound field. (**b**) When the DNLL is inactivated by kynurenic acid, the neuron now fires to the binaural signal that is more intense at the ipsilateral ear, the same signal that was presented in **a**. (**c**) Responses to multiple sounds. When the first sound is louder in the contralateral ear, as occurs with sounds in the contralateral sound field, it evokes discharges in the IC and persistently inhibits the DNLL. A trailing sound from the ipsilateral sound field, which evokes no spikes when presented alone, now evokes discharges. See text for further explanation

intense at the contralateral ear followed by a trailing signal that was louder in the ipsilateral ear, a signal that previously evoked no discharges. The cell now fired to both the first and the trailing signals. The rescue of the responses to the trailing sound occurred because the persistent inhibition in the DNLL, evoked by the first signal, now deprived the IC of its inhibitory projection from the DNLL. In the absence of inhibition, the IC was free to respond to the weaker signal that the trailing signal produced at the contralateral ear. The rescue of trailing responses only occurred if the trailing sound followed the first sound within the period of DNLL persistent inhibition induced by the first binaural signal. Thus he showed that an initial binaural signal could reversibly inactivate the DNLL that affected the binaural responses of trailing signals as effectively as reversibly inactivating the DNLL pharmacologically. Importantly, he also showed the initial signals only affected EI properties in cells that receive innervation from the DNLL and had no effects on the EI properties of cells that did not receive innervation from the DNLL.

24.9.4 DNLL Innervation of the IC Contributes to Precedence Effect

The paradigm whereby temporally separated binaural signals that emanated from different regions of space were presented, is a stimulus configuration used in psychophysical studies of the precedence effect (Zurek, 1987). The precedence effect was discovered in human psychophysical studies and is caused by a mechanism that suppresses the *directional information* carried by echoes (trailing sounds). When two sounds are presented from different locations in close temporal sequence, listeners hear a single composite sound and perceive the composite sound as originating from the location of the leading sound. The second sound is heard but is not perceived as a separate sound, nor does it influence the perceived location of the first sound.

Mike Burger and I suggested that the differential processing of trailing sounds might be a mechanism that underlies the precedence effect (Burger & Pollak, 2001). Many others, however, felt that the features of the DNLL that we demonstrated must be unique to bats. They could not accept the argument that bats experience precedence, as these animals must distinguish and localize sequential echoes. Although the role of such processing in echolocation is unclear, we pointed out that these animals also spend more than half of their lives in dark caves where they use a rich repertoire of vocal communication calls in a wide range of social interactions (Bohn et al., 2008, 2009). The caves are reverberant and they must, like all other animals, use precedence to suppress the localization of reverberant echoes. Our arguments were met with deafening silence. Most of my colleagues felt that we had demonstrated some esoteric features unique to bats.

24.10 Back to Munich Again and Confirmation That DNLL Innervation of the IC Contributes to Precedence

I visited Munich again in 2006 and met Michael Pecka, who was already a budding star in auditory neuroscience. We immediately became friends and my affection for Michael was enhanced when he showed me experiments that he and Ida Kolmar (now married and known as Ida Siveke) conducted on the DNLL of gerbils. Michael and Ida found that persistent inhibition is a pronounced feature of the high-frequency neurons in the gerbil DNLL, and that the gerbil DNLL behaves in exactly the same way to multiple sounds as does the DNLL in bats. They also showed that the influence of persistent inhibition on its targets in the IC contributes significantly to precedence.

I was almost beyond myself with joy on learning of their results. Mike Burger, Lenny Yang, and I were vindicated! Michael had not yet written the manuscript for publication and I offered to help in any way possible. In discussions with the Grothe group, a number of additional experiments were required and conducted. Michael and Benedikt were most gracious, and included my name on the list of authors (Pecka et al., 2007).

Munich has become a second home to me, and I have now worked with two generations of scientists from the group, something that is personally satisfying. But so were Michael's results because they again illustrate both the importance of inhibition and that the circuits and mechanisms found in the auditory systems of bats are not unique to bats, but rather are features present in the auditory systems of other, if not all mammals.

24.11 Using In Vivo Whole Cell Recordings: The Deeper You Look, the More You See

Ellen Covey was the first to use in vivo whole cell recordings to evaluate how the interplay of excitation and inhibition generated monaurally evoked responses in the IC of bats (Covey et al., 1996). In 2005 my friend and colleague at Texas, Nace Golding, convinced me to use in vivo whole cell recordings in our experiments, and I will be eternally grateful to Nace for his support.

I had a talented graduate student, Ruili Xi, who took on the project. Although Nace helped us in our initial attempts, it was clear that in vivo whole cell recordings required an entirely different skill set than either Nace, Ruili, or I had. Fortunately, my friend Gary Rose, who uses in vivo whole cell recordings in his work on electric fish and frogs, came to the rescue. Gary invited us out to his lab at the University of Utah and showed us how it was done. When we returned to Austin, Ruili launched into experiments with patch electrodes. But even after Gray's instructions and Nace's help, working out the techniques for in-vivo whole cell recordings in awake bats was not trivial. Ruili was highly skilled with extracellular recordings and iontophoresis, and it took him three frustrating months to finally get a respectable patch recording. He then conducted the initial series of in vivo whole cell recordings studies in the IC of awake bats (Xie et al., 2007, 2008; Pollak et al., 2011a, b).

My lab at the time had switched to another question: How is the rich repertoire of vocal communication calls the bats use for social communication processed and represented in the IC (Bauer et al., 2002; Klug et al., 2002; Xie et al., 2005)? Josh Gittelman joined the lab as a postdoc in 2007. Josh is also a very talented scientist, and he adopted a procedure pioneered by Nick Priebe (whose lab is now down the hall from mine) and David Ferster in which the excitatory and inhibitory conductances could be computed from current clamp recordings (Priebe & Ferster, 2005).

This was a revelation to me. Now one could derive the relative timing, the relative magnitudes, and the shapes of the conductance waveforms of both the excitation and inhibition that were evoked by an acoustic signal. Moreover, Josh then showed that he could work backwards, and compute a predicted response from the conductances, and that the predicted responses were in remarkably close agreement with the responses that were evoked by that sound. Because the conductances predicted the actual responses, Josh then manipulated one or the other conductance to evaluate the roles of the excitatory and inhibitory inputs in shaping FM directional selectivity (Gittelman et al., 2009, 2012; Gittelman & Pollak, 2011). It is still stunning to me that one can actually manipulate the timing or the magnitude of an excitatory or inhibitory input to determine its contribution to the response. Josh used these techniques with great imagination and skill, and revealed aspects of processing that could never have been seen, much less determined, with extracellular recordings.

24.11.1 In Vivo Recordings of EI Neurons in the IC Show That the Cells Are Even More Complex Than Previously Thought

But back to binaural processing in the IC and EI cells. Based on the previous experiments with microiontophoresis that Tom Park, Mike Burger, and I had conducted, we proposed specific sets of excitatory and inhibitory inputs to each of the various types of EI types in the IC. Na Li, who was a graduate student in my lab at the time, decided to use in vivo whole cell recordings to test those predictions directly (Li et al., 2010; Pollak, 2012). The results of those experiments showed that circuitry that innervates most, but not all EI cells in the IC is even more complex and more diverse than we had previously suggested from the extracellular studies.

A few EI cells had no responses evoked by ipsilateral stimulation, and thus their EI property must have been inherited via an excitatory projection from the LSO. These were the cells that were in agreement with the EI cells whose binaural properties were not changed when inhibition was blocked in the iontophoretic studies. However, patch electrodes recorded both postsynaptic potentials (PSPs) and spikes, and we found that virtually all the other types EI cells displayed subthreshold EPSPs evoked by ipsilateral stimulation. These ipsilateral projections were invisible to extracellular recordings because they were subthreshold.

Even more surprises were revealed when we computed the excitatory and inhibitory conductances that underlie each response. The analyses of the synaptic conductances suggest that the projections that innervate EI neurons are far more diverse and even more complex than previously inferred either from extracellular or intracellular studies that recorded only PSPs and spikes. An example of just one type is shown in Fig. 24.6. The point here is that with extracellular recordings this cell would have been classified as one that inherited its properties from the LSO. Intracellular recordings showed that there was also an ipsilaterally evoked, subthreshold EPSP. The circuit obtained from conductances, however, revealed an even more complex set of inputs that were not detectable from recordings of spikes or PSPs alone. In a nutshell, those studies suggest that almost all EI cells receive excitatory projections from the LSO, and inhibitory projections from the ipsilateral or contralateral DNLL, or both DNLLs, in addition to other projections, some of which exert such subtle influences that they could not have been detected with extracellular records or even from intracellular recordings of PSPs. It is not sufficient to obtain only spikes and subthreshold PSPs to obtain a view of the circuitry that innervates the EI cell from which recordings are made. Rather, it is necessary also to evaluate the conductances underlying the responses.

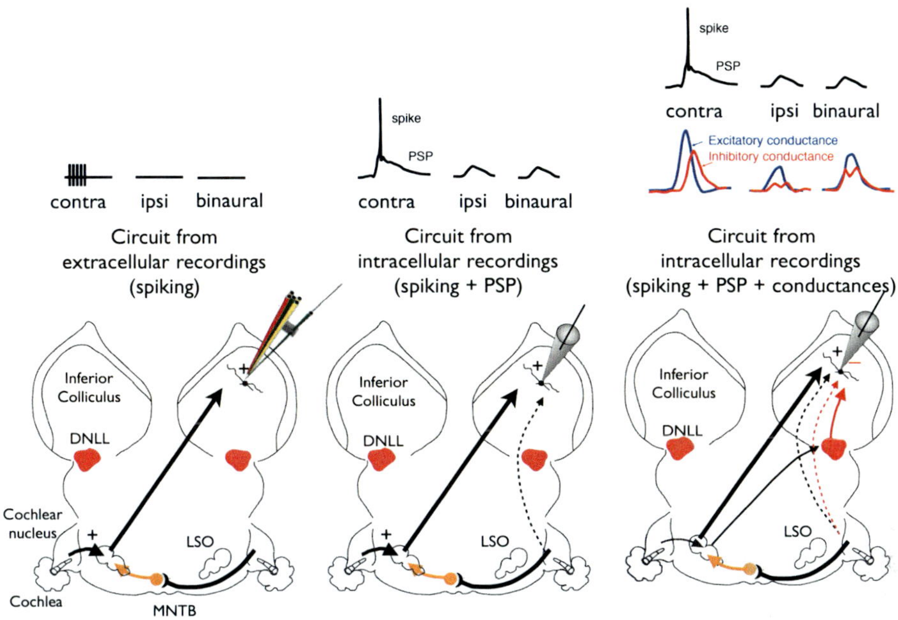

Fig. 24.6 Progressively more complex circuits that innervate an EI cell are seen with in vivo whole cell recordings from which conductances were computed. Left panel shows spikes recorded with monaural and binaural stimulation and the circuit that could explain the spiking responses. Middle panel shows intracellular responses, both spikes and PSPs, from the same cell. Ipsilateral stimulation evoked a subthreshold EPSP that was invisible to extracellular recordings because it was subthreshold. Thus blocking inhibition in this cell would not have changed its EI property, and thus from spikes alone it would appear that the cell inherited its EI property from the LSO. Right panel shows the excitatory and inhibitory conductances evoked by monaural and binaural stimulation in same EI cell. Both the excitatory and inhibitory conductances evoked by binaural stimulation are smaller than the conductances evoked by contralateral stimulation, showing that both the excitatory and inhibitory innervation must have originated in lower nuclei that were themselves binaural. The circuit suggested by the conductances is more complex than the circuit suggested by either the spike or the PSPs evoked by monaural and binaural stimulation

24.12 Summary, Some Conclusions, and Questions for the Future

Our understanding of the circuits involved in the binaural processing of high frequencies has come a long way over the past 30 or so years. Studies have revealed the importance of inhibition of the IC and knowing the circuitry that innervates each EI type, as the circuitry explains how the cell derives it properties. It has also become apparent that the EI cells in the IC cells are designed not only to process an isolated sound source in space, as would occur in an anechoic environment. Rather they seem also to be designed for processing more complex binaural stimuli, such as dynamic IIDs, signals composed of multiple IIDs that change over time. We live in a complex and very "noisy" world, where we are constantly bombarded by a multitude of sounds, and the system has to be adept at handling complex stimuli with IIDs that change over time. One example of such differential processing of dynamic IIDs is the IC cells that are innervated by the contralateral DNLL. As Mike Burger and I showed (2001) many years ago, the EI properties of these cells change when multiple sounds are presented that follow in time. But other studies that utilized different binaural configurations also showed that with stimuli composed of IIDs that change over time, that is, dynamic IIDs, there is in many IC cells, a dramatic shift from the way IC neurons responded to static IIDs that were previously presented (there is a change in the IID that evokes the criterion inhibition) (Sanes et al., 1998; Dahmen et al., 2010). What all of this suggests to me is that changes in IID sensitivities with dynamic stimuli are a universal feature of the IC.

Although considerable progress has been made in understanding the circuits that play upon different types of EI neurons, I am still left with a host of questions. The circuits, and how those circuits might generate differential responsiveness to dynamic IIDs, were only suggestions based on the response properties of lower nuclei and how the impact of the lower nuclei might be expressed in the IC. Future studies should be directed at providing more definitive proof of these suggestions as well as extending our understanding of the ways that IC neurons process and code for multiple stimulus features. Below I list five of the many questions that I wonder about, and that, I believe, need to be and will be answered in future studies.

1. The first question concerns the circuitry that underlies the responses to dynamic IIDs. The various EI types have a unique complement of excitatory and inhibitory inputs that must respond differently to dynamic IIDs. But what exactly are those differences? Stated differently, how does circuitry suggested by the conductances and PSPs recorded from monaural and binaural stimulation translate into differential responses to particular forms of dynamic IIDs?
2. The second question concerns circuitry. The innervation of each type of cell in the IC has to be more thoroughly understood. Techniques that can actually identify which of the lower nuclei innervate the IC cell being studied and the impact of those innervations are critical. Reversible inactivation of lower nuclei, as was done with the DNLL, or with inputs from the dorsal cochlear nucleus that Davis (Davis, 2002) accomplished in cats, are just a first step. In this regard, the development of viruses that can selectively transfect cells in a lower nucleus and cause the expression of channel and/or halorhodopsins is a major advance (Bamann et al., 2010; Zhang et al., 2010). With these techniques, investigators will be able to optically activate and inactivate specific lower nuclei rapidly and repeatedly, and thereby assess the impact of one or another projection on its target in the IC in far more neurons and in greater detail than has been possible previously.
3. The third question also concerns circuitry, but in this case it is not the functional impact of ascending projections but rather the impacts of intrinsic connections (Malmierca et al., 1995) and those of the commissure of the IC (Oliver, 1984; Saldana & Merchan, 1992; Oliver et al., 1994; Malmierca et al., 1995), both of which are poorly understood, to say the least. The commissure provides one of

the largest, if not the largest, projection to the IC and yet there have been only a few studies on its putative role on processing in general (Moore et al., 1998; Malmierca et al., 2003) and none on its influence on binaural processing. In this regard, the channel and halorhodopsins will be of exceptional value. The possibility of recording from neurons in one IC while reversibly inactivating the opposite IC with light holds enormous promise for elucidating the roles that the activation of one IC has on the processing in the other IC.

4. The fourth question concerns how IC cells simultaneously encode multiple stimulus attributes. This chapter only addressed how IC cells respond to IIDs of tones. However, animals do not listen to tones but rather to sounds composed of various combinations of frequency and amplitude modulations (FM and AM) that have broad spectral contents. It is well established that many if not most IC neurons respond selectively to the direction of FMs (Poon et al., 1991; Fuzessery et al., 2006; Gittelman et al., 2009; Andoni & Pollak, 2011; Kuo & Wu, 2012) and to restricted rates of AMs (Langner & Schreiner, 1988; Burger & Pollak, 1998; Caspary et al., 2002). A question about which little is known is how the circuits that innervate each type of IC cell operate cooperatively to create selectivity, or nonselectivity, for AM and FM as well as their selectivities for static and dynamic IIDs?
5. And the fifth question is the degree to which attention and behavior modify response properties. Do the EI cells that we think we know so well respond to the same way when the animal is listening passively as the same cells do while the animal is attempting to actually locate the sound source? I would be surprised if behavior had little or no affects on their response properties.

24.13 Humans Can Echolocate Like a Bat

It is generally assumed, implicitly as well as explicitly, that echolocation is such a unique perceptual ability that it requires circuits and mechanisms so specialized as to render the processing of acoustic information in auditory systems of bats different from the processing that occurs in other mammals. However, as far as I can determine, the adaptations in the brainstem auditory nuclei of bats are primarily, although not exclusively, a matter of quantity, where a species expresses certain features that are shared by other species but to a greater degree or in a more pronounced form, rather than expressing wholesale qualitative changes in the mode of processing. This is well illustrated by the greatly enlarged 60-kHz isofrequency contour in the mustache bat's IC; it is massive relative to other contours yet it receives the same set of projections and expresses the same response properties as the neurons in the isofrequency contours in the IC of other mammals. I have pointed out in previous sections of this chapter the common features of binaural processing in the bat's auditory brainstem nuclei. If space permitted, I could list the entire litany of features that apply to monaural processing as well. In addition, higher order features that other investigators first reported in the IC of bats, duration tuning (Casseday et al., 1994) and combination sensitivity (Suga et al., 1983; Mittmann & Wenstrup, 1995; Portfors & Wenstrup, 1999), were also subsequently seen in the auditory systems of other mammals (Adams, 1979; Brand et al., 2000; Portfors & Felix, 2005), once investigators looked for them. The evidence indicates that the principal mechanisms for processing acoustic information are conserved among mammals.

If the brainstem auditory systems of bats are so similar to those of other mammals, why can bats echolocate while most other mammals cannot? But maybe they can, though they just don't know how. Echolocation, though exotic, is not as unique a perceptual ability as many believe. It evolved several times, and is present in two species of birds; in cetaceans; in one species of megachiropteran bat, the tomb bat, *Rosettus aegyptiacus*; as well as in all microchiropteran bats (Griffin, 1986). And now comes the real surprise: Humans can also use echolocation with remarkable precision.

There have always been anecdotal stories about one or another blind person who displayed echolocation abilities that were so good that he or she appeared not to be blind at all. Recently, one person, Daniel Kish, has received particular attention. Kish has been sightless since he was a year old. Yet he can mountain bike, navigate the wilderness alone, and recognize a building at hundreds of feet away, all with echolocation. He echolocates by emitting clicks with his tongue, the same way birds and tomb bats do, and can form remarkably precise images from the echoes he receives.

One might suspect that Kish is an anomaly, some sort of savant with abilities that almost no one else possesses. Apparently, he is not anomalous. Kish has founded an organization, *World Access for the Blind*, dedicated to teaching the blind how to echolocate, which is now being done successfully on a large scale. Blind children as well as adults are learning to echolocate almost as well as, if not even better than, Kish! You can see Daniel Kish explain his echolocation ability, how he is teaching other blind people to use echolocation, and then watch demonstrations of bike riding, playing soccer, skateboarding, and even shooting baskets by his students at http://www.youtube.com/watch?v=CRA-asTuP_Y and at http://www.youtube.com/watch?v=xATIyq3uZM4&feature=related.Check them out because you will be absolutely amazed. Whatever the circuits and mechanisms are that enable bats to form images of objects in their environment by listening to echoes, they are also present in humans.

Acknowledgments I would like to express my sincerest appreciation to the National Institutes of Health for their generous support of my research over the past 40 years. I also thank Carl Resler for his expertise with computers and his invaluable technical support throughout my career. Finally, I thank my graduate students, postdoctoral fellows, and all of my colleagues who have been so supportive and have taught me so much.

References

Adams, J. C. (1979). Ascending projections to the inferior colliculus. *Journal of Comparative Neurology*,183, 519–538.

Adams, J. C., & Mugnaini, E. (1984). Dorsal nucleus of the lateral lemniscus: A nucleus of GABAergic projection neurons. *Brain Research Bulletin*,14, 585–590.

Andoni, S., & Pollak, G. D. (2011). Selectivity for spectral motion as a neural computation for encoding natural communication signals in bat inferior colliculus. *Journal of Neuroscience*,31(46), 16529–16540.

Bamann, C., Nagel, G., & Bamberg, E. (2010). Microbial rhodopsins in the spotlight. *Current Opinion in Neurobiology*,20, 610–616.

Bauer, E. E., Klug, A., & Pollak, G. D. (2002). Spectral determination of responses to species-specific calls in the dorsal nucleus of the lateral lemniscus. *Journal of Neurophysiology*,88(4), 1955–1967.

Bohn, K. M., Schmidt-French, B., Ma, S. T., & Pollak, G. D. (2008). Syllable acoustics, temporal patterns, and call composition vary with behavioral context in Mexican free-tailed bats. *Journal of the Acoustical Society of America*,124(3), 1838–1848.

Bohn, K. M., Schmidt-French, B., Schwartz, C., Smotherman, M., & Pollak, G. D. (2009). Versatility and stereotypy of free-tailed bat songs. *PLoS One*,4(8), e6746.

Boudreau, J. C., & Tsuchitani, C. (1968). Binaural interaction in the cat superior olive S segment. *Journal of Neurophysiology*,31(3), 442–454.

Brand, A., Urban, R., & Grothe, B. (2000). Duration tuning in the mouse auditory midbrain. *Journal of Neurophysiology*,84(4), 1790–1799.

Brugge, J. F., Anderson, D. J., & Aitkin, L. M. (1970). Responses of neurons in the dorsal nucleus of the lateral lemniscus of cat to binaural tonal stimulation. *Journal of Neurophysiology*,33(3), 441–458.

Bruns, V. (1976). Peripheral auditory tuning for fine frequency analysis by the CF-FM bat, *Rhinolophus ferrumequinum. Journal of Comparative Physiology*,106, 77–86.

Burger, R. M., & Pollak, G. D. (1998). Analysis of the role of inhibition in shaping responses to sinusoidally amplitude-modulated signals in the inferior colliculus. *Journal of Neurophysiology*,80(4), 1686–1701.

Burger, R. M., & Pollak, G. D. (2001). Reversible inactivation of the dorsal nucleus of the lateral lemniscus reveals its role in the processing of multiple sound sources in the inferior colliculus of bats. *Journal of Neuroscience*,21(13), 4830–4843.

Caspary, D. M., Palombi, P. S., & Hughes, L. F. (2002). GABAergic inputs shape responses to amplitude modulated stimuli in the inferior colliculus. *Hearing Research*,168(1–2), 163–173.

Casseday, J. H., Ehrlich, D., & Covey, E. (1994). Neural tuning for sound duration: Role of inhibitory mechanisms in the inferior colliculus. *Science*,264, 847–850.

Chen, L., Kelly, J. B., & Wu, S. H. (1999). The commissure of probst as a source of GABAergic inhibition. *Hearing Research*,138(1–2), 106–114.

Covey, E., Kauer, J. A., & Casseday, J. H. (1996). Whole-cell patch-clamp recording reveals subthreshold sound-evoked postsynaptic currents in the inferior colliculus of awake bats. *Journal of Neuroscience*,16(9), 3009–3018.

Dahmen, J. C., Keating, P., Nodal, F. R., Schulz, A. L., & King, A. J. (2010). Adaptation to stimulus statistics in the perception and neural representation of auditory space. *Neuron*,66(6), 937–948.

Davis, K. A. (2002). Evidence of a functionally segregated pathway from dorsal cochlear nucleus to inferior colliculus. *Journal of Neurophysiology*,87(4), 1824–1835.

Faingold, C. L., Anderson, C. A., & Randall, M. E. (1993). Stimulation or blockade of the dorsal nucleus of the lateral lemniscus alters binaural and tonic inhibition in contralateral inferior colliculus neurons. *Hearing Research*,69, 98–106.

Fuzessery, Z. M., & Pollak, G. D. (1984). Neural mechanisms of sound localization in an echolocating bat. *Science*,225(4663), 725–728.

Fuzessery, Z. M., & Pollak, G. D. (1985). Determinants of sound location selectivity in bat inferior colliculus: A combined dichotic and free-field stimulation study. *Journal of Neurophysiology*,54(4), 757–781.

Fuzessery, Z. M., Richardson, M. D., & Coburn, M. S. (2006). Neural mechanisms underlying selectivity for the rate and direction of frequency-modulated sweeps in the inferior colliculus of the pallid bat. *Journal of Neurophysiology*,96(3), 1320–1336.

Gittelman, J. X., & Pollak, G. D. (2011). It's about time: How input timing is used and not used to create emergent properties in the auditory system. *Journal of Neuroscience*,31(7), 2576–2583.

Gittelman, J. X., Li, N., & Pollak, G.D. (2009). Mechanisms underlying directional selectivity for frequency modulated sweeps in the inferior colliculus revealed by in vivo whole-cell recordings. *Journal of Neuroscience*,29, 13030–13041.

Gittelman, J. X., Wang, L., Colburn, H. S., & Pollak, G. D. (2012). Inhibition shapes response selectivity in the inferior colliculus by gain modulation. *Frontal Neural Circuits*,6(article 67), 67.

Griffin, D. R. (1986). *Listening in the dark*. Ithaca, NY: Cornell University Press.

Grothe, B. (1994). Interaction of excitation and inhibition in processing of pure tone and amplitude-modulated stimuli in the medial superior olive of the mustached bat. *Journal of Neurophysiology*,71(2), 706–721.

Grothe, B., Vater, M., Casseday, J. H., & Covey, E. (1992). Monaural interaction of excitation and inhibition in the medial superior olive of the mustached bat: An adaptation for biosonar. *Proceedings of the National Academy of Sciences of the USA*,89(11), 5108–5112.

Grothe, B., Schweizer, H., Pollak, G. D., Schuller, G., & Rosemann, C. (1994). Anatomy and projection patterns of the superior olivary complex in the Mexican free-tailed bat, *Tadarida brasiliensis mexicana. Journal of Comparative Neurology*,343(4), 630–646.

Grothe, B., Park, T. J., & Schuller, G. (1997). Medial superior olive in the free-tailed bat: response to pure tones and amplitude-modulated tones. *Journal of Neurophysiology*,77(3), 1553–1565.

Klug, A., Bauer, E. E., Hanson, J. T., Hurley, L., Meitzen, J., & Pollak, G. D. (2002). Response selectivity for species-specific calls in the inferior colliculus of Mexican free-tailed bats is generated by inhibition. *Journal of Neurophysiology*,88(4), 1941–1954.

Knudsen, E. I., & Konishi, M. (1978). A neural map of auditory space in the owl. *Science*, 200(4343), 795–797.

Krebs, H. A. (1975). The August Krogh principle: "For many problems there is an animal on which it can be most conveniently studied." *Journal of Experimental Zoology*,194(1), 221–226.

Kuo, R. I., & Wu, G. K. (2012). Generation of direction selectivity in the auditory system. *Neuron*,73, 1016–1027.

Langner, G., & Schreiner, C. E. (1988). Periodicity coding in the inferior colliculus of the cat. I. Neuronal mechanisms. *Journal of Neurophysiology*,60(6), 1799–1822.

Li, L., & Kelly, J. B. (1992). Inhibitory influence of the dorsal nucleus of the lateral lemniscus on binaural responses in the rat's inferior colliculus. *Journal of Neuroscience*,12(11), 4530–4539.

Li, N., Gittelman, J. X., & Pollak, G. D. (2010). Intracellular recordings reveal novel features of neurons that code interaural intensity disparities in the inferior colliculus. *Journal of Neuroscience*,30(43), 14573–14584.

Malmierca, M. S., Rees, A., Le Beau, F. E., & Bjaalie, J. G. (1995). Laminar organization of frequency-defined local axons within and between the inferior colliculi of the guinea pig. *Journal of Comparative Neurology*,357(1), 124–144.

Malmierca, M. S., Hernandez, O., Falconi, A., Lopez-Poveda, E. A., Merchan, M., & Rees, A. (2003). The commissure of the inferior colliculus shapes frequency response areas in rat: An in vivo study using reversible blockade with microinjection of kynurenic acid. *Experimental Brain Research*,153(4), 522–529.

Merchan, M. A., Saldana, E., & Plaza, I. (1994). Dorsal nucleus of the lateral lemniscus in the rat: concentric organization and tonotopic projection to the inferior colliculus. *Journal of Comparative Neurology*,342(2), 259–278.

Mittmann, D. H., & Wenstrup, J. J. (1995). Combination-sensitive neurons in the inferior colliculus. *Hearing Research*,90(1–2), 185–191.

Moore, D. R., Kotak, V. C., & Sanes, D. H. (1998). Commissural and lemniscal synaptic input to the gerbil inferior colliculus. *Journal of Neurophysiology*,80(5), 2229–2236.

Oliver, D. L. (1984). Neuron types in the central nucleus of the inferior colliculus that project to the medial geniculate body. *Neuroscience*,11(2), 409–424.

Oliver, D. L., & Shneiderman, A. (1989). An EM study of the dorsal nucleus of the lateral lemniscus: Inhibitory, commissural, synaptic connections between ascending auditory pathways. *Journal of Neuroscience*,9(3), 967–982.

Oliver, D. L., Winer, J. A., Beckius, G. E., & Saint Marie, R. L. (1994). Morphology of GABAergic neurons in the inferior colliculus of the cat. *Journal of Comparative Neurology*,340(1), 27–42.

Park, T. J., & Pollak, G. D. (1993). GABA shapes sensitivity to interaural intensity disparities in the mustache bat's inferior colliculus: Implications for encoding sound location. *Journal of Neuroscience*,13(5), 2050–2067.

Park, T. J., & Pollak, G. D. (1994). Azimuthal receptive fields are shaped by GABAergic inhibition in the inferior colliculus of the mustache bat. *Journal of Neurophysiology*,72(3), 1080–1102.

Park, T. J., & Grothe, B. (1996). From pattern recognition to sound localization: A by-product of growing larger during evolution. *Naturwissenschaften*,83(1), 30–32.

Park, T. J., Grothe, B., Pollak, G. D., Schuller, G., & Koch, U. (1996). Neural delays shape selectivity to interaural intensity differences in the lateral superior olive. *Journal of Neuroscience*,16(20), 6554–6566.

Park, T. J., Monsivais, P., & Pollak, G. D. (1997). Processing of interaural intensity differences in the LSO: Role of interaural threshold differences. *Journal of Neurophysiology*,77(6), 2863–2878.

Park, T. J., Klug, A., Oswald, J. P., & Grothe, B. (1998). A novel circuit in the bat's midbrain recruits neurons into sound localization processing. *Naturwissenschaften*,85, 176–179.

Pecka, M., Zahn, T. P., Saunier-Rebori, B., Siveke, I., Felmy, F., Wiegrebe, L., Klug, A., Pollak, G. D., & Grothe, B. (2007). Inhibiting the inhibition: A neuronal network for sound localization in reverberant environments. *Journal of Neuroscience*,27(7), 1782–1790.

Pollak, G., Henson, O. W., Jr., & Novick, A. (1972). Cochlear microphonic audiograms in the "pure tone" bat *Chilonycteris parnellii parnellii. Science*,176(4030), 66–68.

Pollak, G. D. (1997). Roles of GABAergic inhibition for the binaural processing of multiple sound sources in the inferior colliculus. *Annals of Otology, Rhinology, and Laryngology Supplement*,168, 44–54.

Pollak, G. D. (2012). Circuits for processing dynamic interaural intensity disparities in the inferior colliculus. *Hearing Research*,288, 47–57.

Pollak, G. D., Burger, R. M., & Klug, A. (2003). Dissecting the circuitry of the auditory system. *Trends in Neuroscience*,26(1), 33–39.

Pollak, G. D., Gittelman, J. X., Li, N., & Xie, R. (2011a). Inhibitory projections from the ventral nucleus of the lateral lemniscus and superior paraolivary nucleus create directional selectivity of frequency modulations in the inferior colliculus: A comparison of bats with other mammals. *Hearing Research*,273(1–2), 134–144.

Pollak, G. D., Xie, R., Gittelman, J. X., Andoni, S., & Li, N. (2011b). The dominance of inhibition in the inferior colliculus. *Hearing Research*,274(1–2), 27–39.

Poon, P. W., Chen, X., & Hwang, J. C. (1991). Basic determinants for FM responses in the inferior colliculus of rats. *Experimental Brain Research*,83(3), 598–606.

Portfors, C. V., & Wenstrup, J. J. (1999). Delay-tuned neurons in the inferior colliculus of the mustached bat: Implications for analyses of target distance. *Journal of Neurophysiology*,82(3), 1326–1338.

Portfors, C. V., & Felix, R. A., 2nd (2005). Spectral integration in the inferior colliculus of the CBA/CaJ mouse. *Neuroscience*,136(4), 1159–1170.

Priebe, N. J., & Ferster, D. (2005). Direction selectivity of excitation and inhibition in simple cells of the cat primary visual cortex. *Neuron*,45(1), 133–145.

Ross, L. S., & Pollak, G. D. (1989). Differential ascending projections to aural regions in the 60 kHz contour of the mustache bat's inferior colliculus. *Journal of Neuroscience*,9(8), 2819–2834.

Ross, L. S., Pollak, G. D., & Zook, J. M. (1988). Origin of ascending projections to an isofrequency region of the mustache bat's inferior colliculus. *Journal of Comparative Neurology*,270(4), 488–505.

Saldana, E., & Merchan, M. A. (1992). Intrinsic and commissural connections of the rat inferior colliculus. *Journal of Comparative Neurology*,319(3), 417–437.

Sanes, D. H., Malone, B. J., & Semple, M. N. (1998). Role of synaptic inhibition in processing of dynamic binaural level stimuli. *Journal of Neuroscience*,18(2), 794–803.

Schnitzler, H.-U., Suga, N., & Simmons, J.A. (1976). Peripheral auditory tuning for fine frequency analysis by the CF-FM bat, *Rhinolophus ferrumequinum. Journal of Comparative Physiology*,106, 99–110.

Semple, M. N., & Aitkin, L. M. (1979). Representation of sound frequency and laterality by units in central nucleus of cat inferior colliculus. *Journal of Neurophysiology*,42(6), 1626–1639.

Suga, N., Neuweiler, G., & Moller, J. (1976). Peripheral auditory tuning for fine frequency analysis by the CF-FM bat, *Rhinolophus ferrumequinum. Journal of Comparative Physiology*,106, 111–125.

Suga, N., O'Neill, W. E., Kujirai, K., & Manabe, T. (1983). Specificity of combination-sensitive neurons for processing of complex biosonar signals in auditory cortex of the mustached bat. *Journal of Neurophysiology*,49(6), 1573–1626.

Wenstrup, J. J., Ross, L. S., & Pollak, G. D. (1985). A functional organization of binaural responses in the inferior colliculus. *Hearing Research*,17(2), 191–195.

Wenstrup, J. J., Ross, L. S., & Pollak, G. D. (1986). Binaural response organization within a frequency-band representation of the inferior colliculus: Implications for sound localization. *Journal of Neuroscience*,6(4), 962–973.

Wenstrup, J. J., Fuzessery, Z. M., & Pollak, G. D. (1988a). Binaural neurons in the mustache bat's inferior colliculus. I. Responses of 60-kHz EI units to dichotic sound stimulation. *Journal of Neurophysiology*,60(4), 1369–1383.

Wenstrup, J. J., Fuzessery, Z. M., & Pollak, G. D. (1988b). Binaural neurons in the mustache bat's inferior colliculus. II. Determinants of spatial responses among 60–kHz EI units. *Journal of Neurophysiology*,60(4), 1384–1404.

Xie, R., Meitzen, J., & Pollak, G. D. (2005). Differing roles of inhibition in hierarchical processing of species-specific calls in auditory brainstem nuclei. *Journal of Neurophysiology*,94(6), 4019–4037.

Xie, R., Gittelman, J. X., & Pollak, G. D. (2007). Rethinking tuning: In vivo whole-cell recordings of the inferior colliculus in awake bats. *Journal of Neuroscience*,27(35), 9469–9481.

Xie, R., Gittelman, J. X., Li, N., & Pollak, G. D. (2008). Whole cell recordings of intrinsic properties and sound-evoked responses from the inferior colliculus. *Neuroscience*,154(1), 245–256.

Yang, L., & Pollak, G. D. (1994a). Binaural inhibition in the dorsal nucleus of the lateral lemniscus of the mustache bat affects responses for multiple signals. *Auditory Neuroscience*,1, 1–17.

Yang, L., & Pollak, G. D. (1994b). The roles of GABAergic and glycinergic inhibition on binaural processing in the dorsal nucleus of the lateral lemniscus of the mustache bat. *Journal of Neurophysiology*,71(6), 1999–2013.

Yang, L., Liu, Q., & Pollak, G. D. (1996). Afferent connections to the dorsal nucleus of the lateral lemniscus of the mustache bat: Evidence for two functional subdivisions. *Journal of Comparative Neurology*,373(4), 575–592.

Yang, L., & Pollak, G. D. (1998). Features of ipsilaterally evoked inhibition in the dorsal nucleus of the lateral lemniscus. *Hearing Research*,122(1–2), 125–141.

Zhang, F., Gradinaru, V., Adamantides, A. A., Durand, R., Airan, R. D., de Lecea, L., et al. (2010). Optogenetic interrogation of neural circuits: Technology for probing mammalian brain. *Nature Protocols*,5(3), 439–456.

Zook, J. M., & Casseday, J. H. (1982a). Cytoarchitecture of auditory system in lower brainstem of the mustache bat, *Pteronotus parnellii. Journal of Comparative Neurology*,207(1), 1–13.

Zook, J. M., & Casseday, J. H. (1982b). Origin of ascending projections to inferior colliculus in the mustache bat, *Pteronotus parnellii. Journal of Comparative Neurology*,207(1), 14–28.

Zook, J. M., Winer, J. A., Pollak, G. D., & Bodenhamer, R. D. (1985). Topology of the central nucleus of the mustache bat's inferior colliculus: Correlation of single unit properties and neuronal architecture. *Journal of Comparative Neurology*,231(4), 530–546.

Zurek, P. M. (1987). The precedence effect. In W. A. Yost & G. Gourevitch (Eds.), *Directional hearing* (pp. 85–105). New York: Springer Verlag.

Chapter 25
From Cave Fish to Pile Driving: A Tail of Fish Bioacoustics

Arthur N. Popper

Arthur N. Popper

> Fishes have neither organs of hearing, nor yet the exterior orifice. And yet, it is quite certain that they do hear; for it is a well-known fact that in some fish-ponds they are in the habit of being assembled to be fed by the clapping of hands.
>
> —Pliny the Elder (about 50 CE), 1890 edition, p. 547[1]

[1] Or, if one wants the original Latin: "*Pisces quidem auditus nec membra habent nec foramina, audire tamen eos palam est, utpote cum plausu congregari feros ad cibum adsuetudine in quibusdam vivariis spectetur et in piscinis Caesaris genera piscium ad nomen venire quosdamve singulos. itaque produntur etiam clarissime audire mugil, lupus, salpa, chromis et ideo in vado vivere.*" (From http://penelope.uchicago.edu/Thayer/L/Roman/Texts/Pliny_the_Elder/10*.html – see part LXXXIX).

A.N. Popper (✉)
Department of Biology and Center for Comparative and Evolutionary Biology of Hearing, University of Maryland, College Park, MD 20742, USA
e-mail: apopper@umd.edu

A.N. Popper and R.R. Fay (eds.), *Perspectives on Auditory Research*, Springer Handbook of Auditory Research 50, DOI 10.1007/978-1-4614-9102-6_25,

25.1 A Bit of History

Interest in fish sounds and hearing goes back about 2000 years to its first mention by the ancient Roman Pliny (the Elder). However, the first experimental studies of fish hearing did not come until the early 20th century with work by such prominent investigators as G. H. Parker (1902), Karl von Frisch (1923), and von Frisch's student Sven Dijkgraaf (e.g., 1932).

The "modern era" in studying fish hearing can be clearly linked to the classic and pioneering study by Tavolga and Wodinsky (1963). In this study, Bill Tavolga (Fig. 25.1) and Jerry Wodinsky, working at the Lerner Marine Laboratory in Bimini, the Bahamas, used psychophysics to measure the hearing sensitivity in nine species of marine fishes. Tavolga and Wodinsky trained fish to swim over a barrier in a "shuttlebox" to avoid a mild electric shock (something that might not get through an animal study regulatory committee today!) (Fig. 25.2a). The results of the study were the first comparative psychophysical hearing data for fishes, and the first to give accurate and repeatable thresholds and fish hearing ranges (Fig. 25.2b).

Although anyone can read this classic paper (available at http://digitallibrary.amnh.org/dspace/handle/2246/1122), there are two stories associated with it that Bill told me and that had a significant impact on how I do science. I pass these along to all of my students.

First, Bill talks often about how Wodinsky insisted on taking notes in pencil. Bill thought this was a bit ridiculous, but he went along with Jerry. On his return trip home (New York), Bill took advantage of being able to buy very good gin at low prices in Bimini, and packed two bottles in his luggage along with the notebooks from the summer work. As one might guess, both bottles broke on the way home. But thanks to Jerry's insistence that the notes be taken in pencil (which is resistant to alcohol), the data were preserved even though the ink-drawn lincs in the books

Fig. 25.1 Eugenie Clark and Bill Tavolga (about 2003). Dr. Clark, who is known internationally as the "Shark Lady," and Bill were graduate students together (along with Bill's late wife Margaret) at NYU, and both were students of Dr. Breder. I first met Genie in about 1967 when she was on the faculty of City College of New York. I later became Genie's "boss" when I served as Chair of Zoology at the University of Maryland and she was a professor in the department

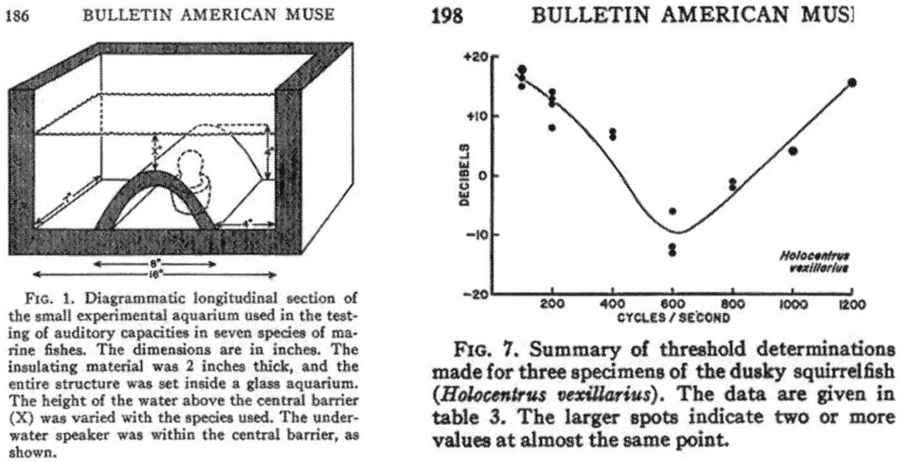

Fig. 25.2 Figure from Tavolga and Wodinksy (1963). The left picture shows a shuttlebox used in the experiments. On the right is an audiogram they determined for the dusky squirrelfish. (Courtesy of The American Museum of Natural History)

were badly smudged. Our whole field would have been different had Jerry not insisted on pencil!

The second story Bill tells is about the actual study. The setup involved having a fish in a shuttlebox inside a chamber that was opaque on all sides but the top. A mirror was placed over the tank and reflected the view of the fish to another mirror across the room where the experimenter could monitor the response (this was before Bill and Jerry adopted photocells and 1960s modern electronics) and control the sound and the shock that would follow if the fish did not cross the barrier. Bill had trained a dusky squirrelfish (now *Sargocentron vexillarium*) to cross the barrier whenever it heard a sound and was testing its hearing by using the staircase (or tracking) method to raise and lower the sound level depending on the response of the fish in the previous trial (e.g., Fig. 25.2b). After some time, Bill found that the fish was detecting every sound played, no matter how low it was. Bill was quite befuddled until he realized that while he was watching the fish, the fish had learned to watch him and had figured out that when Bill moved his hand to turn on the sound, this would be followed by a shock. Thus, the fish learned not to respond to the sound per se but to the hand movement! The moral of this story is that one has to think very hard about the consequences of anything one does in experimental design and also work very hard to make sure that controls really are controls!

25.2 Blind Cave Fish and Georg von Békésy

I can trace my interest and excitement in biology to a single person—my sixth grade teacher, Thomas Vinci. Mr. Vinci was the only male teacher in my school (which went from kindergarten to sixth grade and had perhaps 40 teachers). He was (and is) a truly gifted teacher. Although Mr. Vinci taught us many other things besides science, science was his first love, and he passed that love on to a number of his students, including me. Mr. Vinci encouraged science, fostered curiosity, and profoundly impacted the lives of his students.

Some years later, I was an undergraduate at the now-defunct Bronx campus (called the Heights) of New York University (NYU). On my walk to school one day I noticed a new pet store along the way. I walked in, looked at the tropical fishes, and "discovered" a tank with fish that did not have eyes, Mexican blind cave fish (then called *Anoptichthys jordani* but now *Astyanax mexicanus*). These fish totally intrigued me, and so when I got to campus, I went straight to my comparative anatomy professor, Douglas B. Webster (another amazing teacher), and asked him about these fish. Doug, being a consummate teacher, did not answer my question, but instead encouraged me to do research on the fish. This started as a library exercise but wound up in a two-year undergraduate research project on the general morphology of this species and its eyed ancestor. And because Doug's own research was on hearing and the ear (then working on desert rodents; e.g., Webster, 1962), I was also exposed to auditory neuroscience.

In learning about the fish, I read papers by the great ichthyologist Charles M. Breder, Jr. (e.g., Breder, 1943) and by one of his students Phyllis Cahn (Cahn, 1958). I mention this because both Drs. Breder and Cahn were on the faculty at the American Museum of Natural History (AMNH) in New York City, then, as now (in my view), the finest natural history museum in the world. Coincidentally, Doug Webster (who joined Dick Fay and me later to organize a meeting on the evolution of hearing; Webster et al., 1992) arranged for me to work at the AMNH for Dr. Donn E. Rosen (another student of Dr. Breder), chair of the Ichthyology Department. Here I met Phyllis Cahn, whose office was next to my work area.

Demonstrating serendipity in science, one day I was working on the skeleton of a large parrotfish (Family Scaridae) and in walked one of the other faculty in the Ichthyology Department, Dr. C. Lavett Smith. I cannot recall why Dr. Smith came in, but I do vividly recall his picking up the skeleton, taking out a structure he called an otolith, and telling me that it was part of the ear. While in Doug's lab, I was introduced to ears; I did not even know that fishes had ears, much less ear bones, until Dr. Smith came into my lab.

A few years later I was a graduate student at Queen's College of the City University of New York (CUNY) and looking around for a doctoral mentor. I heard about Dr. William Tavolga, a faculty member at another CUNY campus, City College of New York (CCNY). I suspect that one of the things that interested me about Tavolga was that he was one of the few among the faculty in CUNY working on fishes. It may also have been because Bill's lab was in the fabled and historic Department of Animal Behavior at the AMNH (the first animal behavior department in the United States) and I wanted to continue working at the AMNH.

On my first visit to Bill, he agreed that he might take me on as a student, but I had to come up with my own project. He gave me some papers and told me to come back in a week or so. I pondered the topic and then I found a comment in one of Bill's papers that said that no one yet knew if fishes can localize sound.

I suggested to Bill that I study sound localization, and his immediate response was that it would take 17 years to complete a study, and I signed on (although I did not believe it would take 17 years). Bill's point was that sound localization was one of the really difficult questions about fish hearing, and, in fact, Willem van Bergeijk (1964) had just argued very clearly that fishes cannot localize sound (see Chapter 7 by Fay and Chapter 14 by Hawkins for further discussions of fish sound source localization).

So how does one study something so enigmatic? I devised a plan to use the shuttlebox to train fishes to respond when they heard sounds shifting sides, with the idea that they would be able to localize a sound if they could detect shifts in the sound source. Then, Bill made the suggestion that I use a fish without eyes to avoid any visual cue and I came back to my blind cave fish. I won't continue the story with localization other than to say that Bill's 17 years were way off; even today, we really don't have a clear understanding of the mechanisms and capabilities of fishes for sound localization, although there is enough evidence to say that at least some species can do it (e.g., Fay & Popper, 2012). Because localization studies, at least in a lab on the fifth floor of the AMNH, were not possible (all kinds of issues with tank acoustics), I finally switched my dissertation to a study of comparative hearing in the Mexican blind cave fish and its eyed ancestors. This proved to be very tractable and was the first study to show, using psychophysics, that fish could hear to over 4 kHz (Popper, 1970). The study also showed that hearing in both the blind and eyed forms was about the same and that the two groups (they are now considered to be morphs of the same species) had similar hearing structures (Popper, 1971).

As an aside, I was immensely fortunate in meeting Bill Tavolga. He was (and is to this day) an amazing and caring mentor and exceptional scholar (and musician!). He not only facilitated my doctoral work, but he also provided a role model for mentorship and doing science for which I am deeply grateful. I also think it is fair to say that Bill "invented" the field of marine bioacoustics (Tavolga, 1964, 1967).

It also turns out that Dr. Breder, in whose work I read about blind cave fish, was doctoral advisor to Bill Tavolga (and Phyllis Cahn and Eugenie Clark; Fig. 25.1), and so Dr. Breder is my academic grandfather! I should also add that, quite independently, Bill had a major impact on my wife Helen as well. Helen met Bill in her first semester biology course at CCNY[2] when Bill was her lab instructor [CCNY had no teaching assistants in those days and so labs were run by professors] and then took other classes with him; Helen remembers Bill as an exceptionally good undergraduate instructor.

While doing my dissertation, I was invited by Phyllis Cahn, by then a major figure in lateral line research, to attend a 1966 meeting on lateral line in New York City (Cahn, 1967). The keynote speaker at the meeting was Georg von Békésy, winner of the 1960 Nobel Prize in Medicine and Physiology for his work on hearing (von

[2] Until the mid-1960s CCNY was strictly an undergraduate institution. It cost perhaps $50/year to attend, and was considered one of the finest colleges in the United States. Indeed, CCNY has produced more graduates who went on to win Nobel Prizes than any other college or university in the world –something like 10 or 11 to date.

Békésy, 1967). Von Békésy had just left Harvard and moved to the University of Hawai'i (where, unbeknownst to me then, I would move a few years later). I managed a few moments to speak with von Békésy and asked him what kind of research he would be doing in Hawai'i. His answer was that he was going to study fish hearing! I vividly recall a sudden feeling of depression when I realized that my whole field had disappeared; what could I contribute once this Nobel Prize winner got into the topic?

Several years later, when I was an assistant professor at the University of Hawai'i, I met von Békésy at the garbage dump of our apartment building; we both lived in faculty housing and he was (quite literally) our upstairs neighbor. During a later meeting, I asked him why he was no longer working on fish hearing, and his answer was that "it was too hard"! A feeling of depression again!

So, although von Békésy kept giving me doubts about my field, he also did me a great favor. For the first time in his career, von Békésy decided to invite a postdoc to join him, and he turned to his friend Professor E. G. Wever at Princeton. Wever sent a recent doctoral student of his to work with von Békésy; his name was (is) Richard Fay (see Chapter 7 by Fay). Dick and I met for the first time on December 26, 1971 when he, his wife Cathy, and their son Chris came to our house for a barbeque.

Indeed, in thinking back on it, although von Békésy never contributed to studies of fish hearing per se, perhaps his "arranging" for Dick and me to meet was his major contribution to the field. While I cannot begin to imagine how Dick's and my careers might have progressed separately, I think that we would both agree that as a team we've accomplished more than two individuals separately and we have had opportunities to contribute in ways that neither of us would ever have alone. Most certainly there would be no Springer Handbook of Auditory Research.

25.3 Fish Hearing

25.3.1 Early Comparative Studies

One of the most interesting questions with regard to fish hearing focuses on comparative issues, something that I was introduced to by Bill Tavolga. Indeed, one "difference" in approach for Dick Fay and myself is that I come from a strong comparative perspective, whereas Dick has focused on a wide range of studies of the hearing capabilities of one species, the goldfish (*Carassius auratus*). As a result of my comparative interests and the work done in a number of labs as well as my own, it is now clear that among the 32,000 or more species of fishes there are substantial differences in hearing capabilities and mechanisms (reviewed by Ladich & Fay, 2013).

The initial understanding of the variation in ear structure and the potential in capabilities came from the anatomical descriptions of Ernst H. Weber (1820), who described ears in a number of species and first described a series of bones, now known as the Weberian ossicles, that connect the swim bladder to the inner ear in the otophysan fishes (goldfish, catfishes, etc.; Fig. 25.3 next page). This was followed

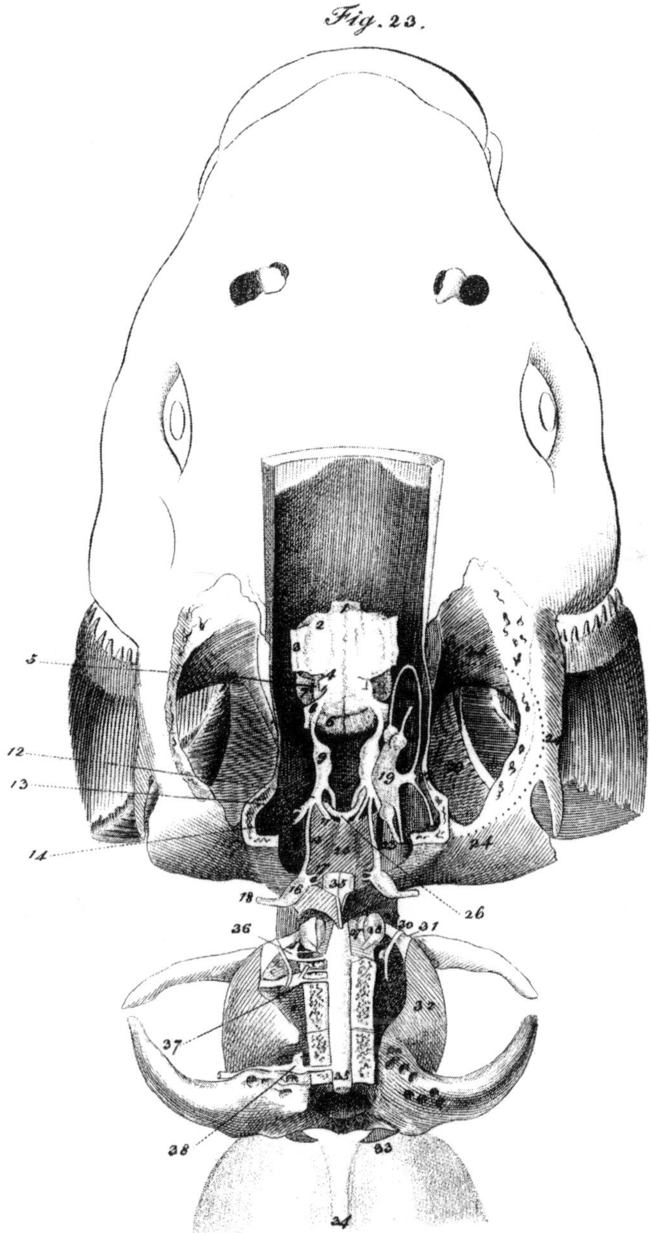

Fig. 25.3 (Above) Drawing of the head of a carp by Weber (1820, plate IV, figure 23) showing the head of a carp. The skull is opened in this picture and the brain cut away to show the ears. The right ear is labeled no. 19. The Weberian ossicles are the bones to the left and right of the vertebral column just behind the skull. (Next Page) Drawings of the ears of the salmon (*Salmo*) from Retzius (1881, plate XIV). I thank David Corey for providing the very high resolution image from Retzius shown here

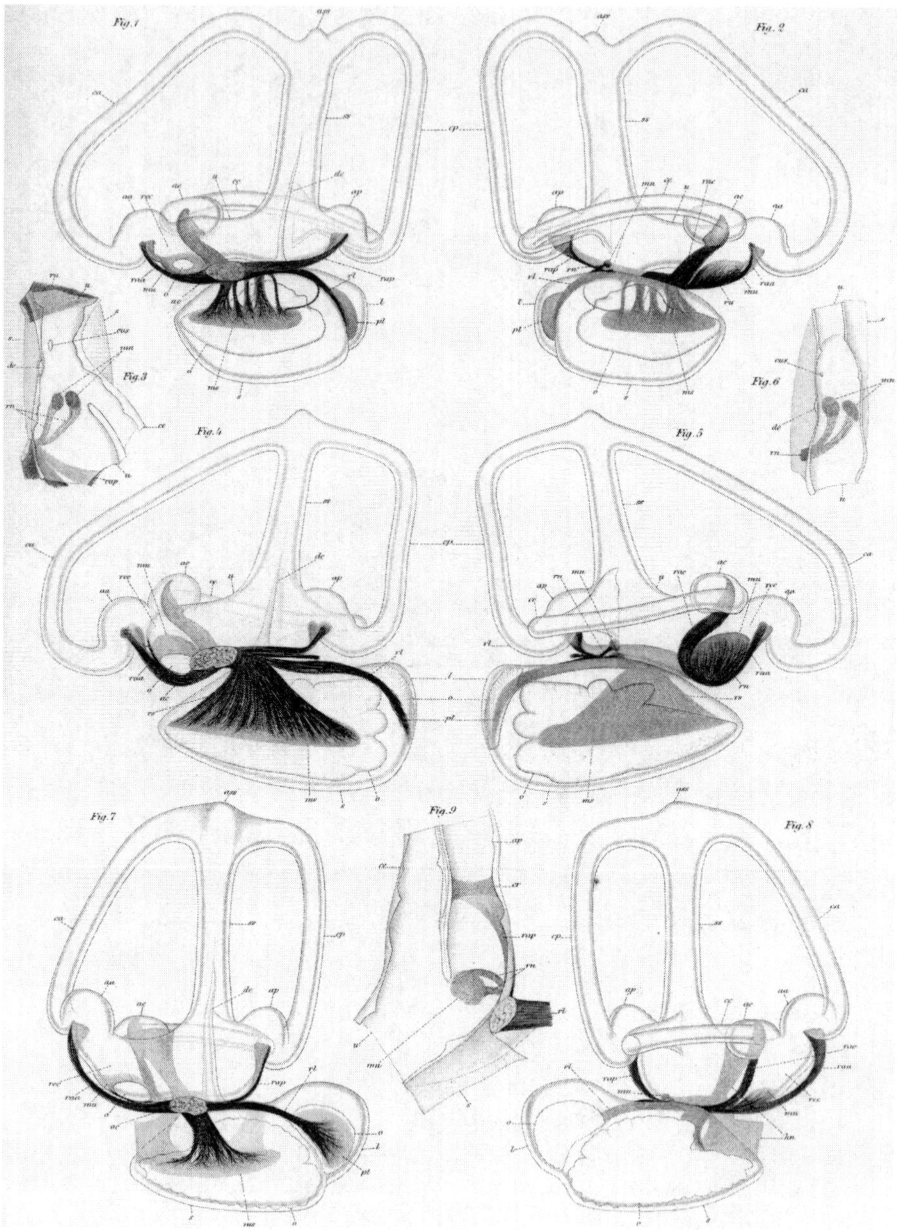

Fig. 25.3 (continued)

by Gustav Retzius (1881), who described the ears, and its variations, in dozens of fish species (Fig. 25.3 above). Indeed, Retzius' anatomical drawings are still invaluable today in helping us understand the variations in the ears of fishes and they have been instrumental in helping decide which species to study for measures of hearing capabilities. One conclusion coming from the body of work by Retzius is that there is probably far more diversity in ear structure in fishes than in all the other vertebrates combined.

25.3.2 *Comparative Hearing*

Fish hearing capabilities vary in several dimensions. Most notably, different species have different hearing bandwidths. The narrowest hearing ranges, from below 50 Hz to perhaps 500 Hz, are found in species that do not have a swim bladder (an air chamber in the abdominal cavity that likely evolved for buoyancy control but that later evolved for use in hearing and sound production in many species). Other species may hear up to 1000–1500 Hz, and these often involve the swim bladder to a greater or lesser degree (reviewed in Popper et al., 2003).

A third "group" of fishes hears sounds up to 3000–4000 Hz. These fishes have a specialized connection between the inner ear and the swim bladder that enhances the hearing range and hearing sensitivity (Jacobs & Tavolga, 1967; Popper, 1971). The goldfish, for example, has a series of bones, the aforementioned Weberian ossicles (Fig. 25.3), that serve as a direct path for sound from the swim bladder to the inner ear, whereas some squirrelfish (genus *Myripristis*) have anterior projections from the swim bladder that directly contact the inner ear (Coombs & Popper, 1979).

We now have hearing data on more than 100 species of fish (see Ladich & Fay, 2013), and what is striking is the wide variation in hearing bandwidth and thresholds for various species. However, as Fritz Ladich and Dick Fay (2013) point out, much of these data have to be considered with care because thresholds were often measured in terms of sound pressure and not in terms of particle motion, the major auditory stimulus in most species (e.g., Popper & Fay, 2011; see also Chapter 14 by Hawkins). Moreover, with few exceptions, most studies of hearing have been done in small tanks where, as my late friend Antares Parvulescu (1964) clearly pointed out, it is virtually impossible to calibrate the sound field. As a consequence, though many of the studies to date are important and reveal a good deal about interspecific variation in fish hearing, we are still a reasonably long way from knowing as much as we need to know about auditory sensitivity in fishes, and particularly as it relates to detection and use of both pressure and particle motion.

25.4 Comparative Ears

There is extraordinary diversity in the structure of fish ears. As noted by my close friend Christopher Platt, from the great lithographs of Retzius on vertebrate ears (Retzius, 1881), much variation is found in the semicircular canals (Platt, 1983). Though some of this variation may be related to the shape of the skull bones in which the canals are embedded, some variation in these tubular shapes may determine sensitivity to certain frequencies or amplitudes of head movements (see Platt, 1983).

The variation that has most intrigued me is in the three otolithic end organs: the saccule, lagena, and utricle. Although it was first proposed by von Frisch (1923) that the saccule, and perhaps the lagena, are the main hearing organs in fishes, recent evidence strongly supports the idea that the utricle may be involved in hearing as well (and see Section 5 on ultrasonic hearing).

In 1975 I had the opportunity to learn scanning electron microscopy (SEM) and apply it to fish ears. I first examined the saccular epithelium of the lake whitefish (*Coregonus clupeaformis*) and recall one of the "eureka" moments in my career, the realization that rather than having hair cells oriented in two opposing directions as had been described for virtually all other vertebrates analyzed to date (e.g., Wersäll et al., 1965), this salmon relative had saccular hair cells oriented in four directions, two dorsoventrally (as in all other saccules) and two rostrocaudally, something that had never been observed (Fig. 25.4). This work was published in *Science* (Popper, 1976). To be fair, just as my paper was published, another paper came out in Europe by Tor Dale (1976) on the ear of the Atlantic cod (*Gadus morhua*) that also showed hair cells oriented in four directions, confirming that this "plan" is not unique to one fish group.

Following this study, I was involved in a series of investigations that explored the structure and ultrastructure of the ears in widely diverse species (e.g., Popper, 1977, 1978, 1980; Popper & Platt, 1979; Popper & Northcutt, 1983). These investigations revealed that there is substantial variation in the hair cell orientation patterns in fishes, with particular focus on the saccule. We found that most fishes have saccular hair cells oriented in at least four distinct directions (Figs. 25.4 and 25.5). However, the plan was often not nearly as "simple" as found in lake whitefish or Atlantic cod, but instead, the orientation patterns on the rostral end of the epithelium were often complex and highly specialized (Figs. 25.5 and 25.6).

Several questions arose. First, why do fishes have variation in saccular hair cells (with there being much less variation in lagenar hair cells and even less in the utricular hair cells; see Section 5)? Second, why are their hair cells oriented in multiple directions? Both questions are still open, but we think we have some basic suggestions for both.

25.4.1 Variation

For her doctoral dissertation in my lab, Sheryl Coombs examined hearing in several different species of squirrelfish (Holocentridae). She found that a species of *Adioryx* could hear to about 1500 Hz, whereas the closely related species *Myripristis* was able to detect sounds up to 4000 Hz (Fig. 25.5) (Coombs & Popper, 1979). Putting this together with work from Tavolga and Wodinsky (1963) on a third species (Fig. 25.2), we came to the conclusion that perhaps the variation in hearing could be correlated with hearing structures. This idea was supported by a morphological

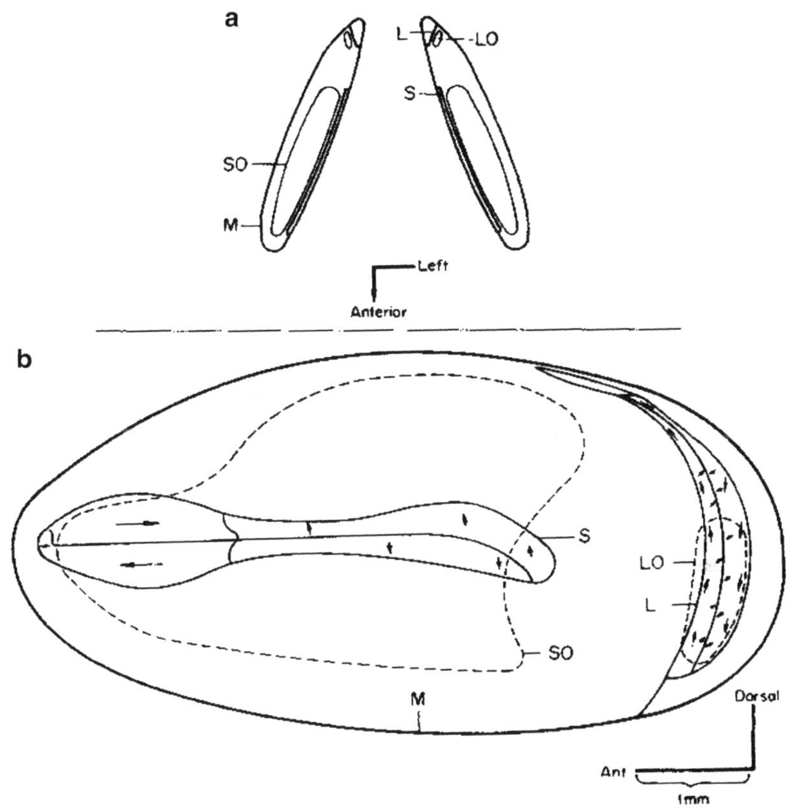

Fig. 25.4 Hair cell orientation patterns from lake whitefish (Popper, 1976). (**a**) Dorsal view of the two ears showing the lagena (L) and its otolith (LO) and the saccule (S) and its otolith (SO). The otolith lies in close proximity to the sensory epithelia in each end organ. (**b**) A lateral view of the saccule and lagena showing the otoliths (dashed lines) and the sensory epithelia. The hair cells are divided into "orientation groups" based on the position of the kinocilium in each ciliary bundle relative to the rest of the bundle. The tip of the arrow indicates the direction of orientation (toward the kinocilium) in each region. Regions are divided by solid lines

study by O'Connell (1955), who showed that the swim bladder is farthest from the ear in *Adioryx*, progressively closer in *Holocentrus*, and intimate to the ear in *Myripristis*. Moreover, my studies of the saccular epithelium in the two species that Sheryl studied showed that the epithelium in *Myripristis* is far larger and more complex in terms of orientation patterns than that in *Adioryx* (Fig. 25.5) (Popper, 1977).

Sheryl and I then started to examine what we knew about hair cell orientation patterns and hearing in a wide range of species. We proposed the hypothesis that fishes with the most highly elaborate (in terms of orientation pattern) saccules inevitably are species that (a) have a wider bandwidth of hearing and (b) have specializations that somehow mechanically "connect" the swim bladder to the inner ear (Popper & Coombs, 1982). Indeed, we then predicted that we could make

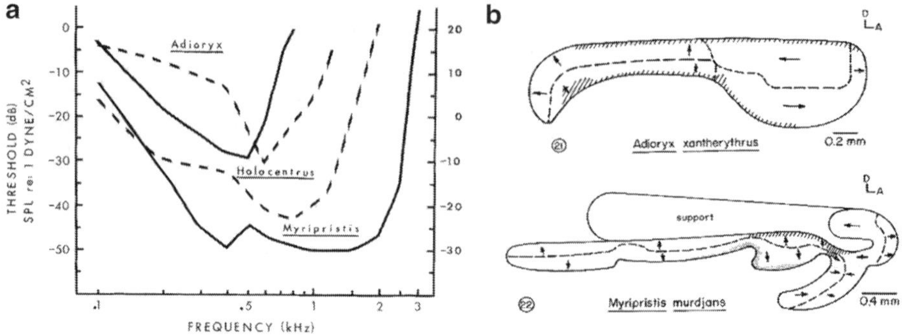

Fig. 25.5 Hearing capabilities and hair cell orientation patterns in squirrelfish. (**a**) Hearing thresholds for various species of squirrelfish. Solid lines are for a species of *Adioryx* and *Myripristis* measured by Coombs and Popper (1979). The dashed lines are for another *Adioryx* species and a species in the genus *Holocentrus* as determined by Tavolga and Wodinsky (1963). (Figure from Coombs and Popper, 1979.) (**b**) Saccular hair cell orientation patterns for the species used by Coombs and Popper (from Popper, 1977). The figure illustrates that the species with the most elaborate hair cell orientation pattern also has the widest hearing bandwidth. Interestingly, *Holocentrus* has hearing capabilities between those of the other two species, and there is evidence from O'Connell (1955) that the adaptation for enhanced hearing lies between the two other species. We do not, however, have ultrastructural data for any species of *Holocentrus*

suggestions about the bandwidth of hearing in fish just from seeing the ultrastructure and hair cell orientation pattern on the saccule.

Since then, we have tested this idea a number of times. For example, John Ramcharitar, when a graduate student in my lab, examined the ears and hearing in a number of species of Sciaenidae (croakers), a commercially very important group of fish (e.g., Ramcharitar et al., 2006). John found a very close correlation between hearing bandwidth and complexity of saccular structure, with fishes having the most complex structure having the widest bandwidth.

To my thinking, the ultimate test for this hypothesis would come from work that I did in the 1980s on the ears of species from very great depths (several thousand meters) and from work recently completed by my last graduate student Xiaohong Deng and an undergraduate working with Xiaohong, Bradley Buran (Popper, 1980; Buran et al., 2005; Deng et al., 2011, 2013). In these studies, we found that most deep-sea species we examined have highly specialized saccules (e.g., Fig. 25.6a), often with hair cells having exceptionally long ciliary bundles. Moreover, Xiaohong has shown that similarly distinct orientation patterns and other ultrastructure features show up time and again in taxonomically unrelated species. Thus, although we predict that many deep-sea species (living in areas without light) are likely to have excellent hearing based on the structure of their ears, it is virtually impossible to do hearing tests on them because they cannot be kept alive when brought to the surface.

At the same time, I don't want to leave the impression that fishes must have four hair cell orientation patterns in the saccule to detect higher frequencies (3000–4000 Hz). For example, goldfish are known to hear to 3000 Hz (Jacobs & Tavolga, 1967; see Chapter 7 by Fay), yet have a relatively simple bidirectional saccular

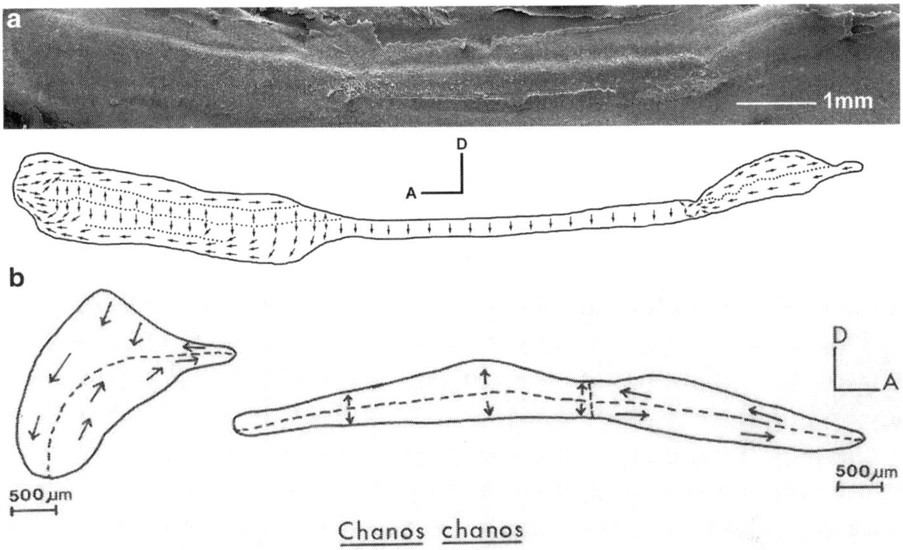

Fig. 25.6 (**a**) Scanning electron microscopic view (top) and drawing of hair cell orientation patters on the saccule of the deep-sea fish *Antimora rostrata* (blue antimora). This species, typical of many other deep-sea species, has a very complex hair cell orientation patter on the rostral end of the epithelium (left). In addition, there are rostral–caudal oriented cells and the caudal end of the epithelium as well, something found in some, but far from all, taxonomically diverse species. (Image from Deng et al., 2011. *Deep-Sea Research I*. Reprinted with permission). (**b**) Hair cell orientation patterns from the saccule (right) and lagena (left) of the milkfish *Chanos chanos*. Note that the rostral end of the saccule (right) has hair cells oriented rostral-caudally, whereas its descendants, the otophysan fishes, have an equally elongate saccular epithelium but only hair cells oriented dorsally and ventrally (like the caudal end of the milkfish saccule). (From Popper and Platt, 1983, *Journal of Morphology, reprinted with permission*)

pattern. Platt (1977) did an SEM analysis of the hair cell orientation patterns in all the end organs in the ears of goldfish (and later zebrafish [*Danio rerio*]; Platt, 1993) and showed that the whole saccule has only dorsally and ventrally oriented hair cells. This pattern has been confirmed for all members of the Otophysi (fishes with Weberian ossicles) (Platt & Popper, 1984). The ability to hear well in the Otophysi is no doubt related to the presence of the aforementioned the Weberian ossicles. It may be that their relatively simple saccular pattern is related to oscillation vectors produced by this specialized direct coupling of the gas bladder to the sensory macula. In any case, this finding of Platt and others argues against hearing specialization always being correlated with four hair cell orientation patterns.

However, Platt and I obtained samples of the milkfish *Chanos chanos*, a species representing the ancestor of all of the otophysan fishes. We discovered that *Chanos* has a saccule that is intermediate between the modern Otophysi and fishes with four hair cell patterns; the shapes of the saccule and lagena are very similar to those in Otophysi, but the saccule has hair cells oriented in four directions (Fig. 25.6b) (Popper & Platt, 1983). *Chanos* also has bones that are clearly on the way to becoming Weberian ossicles. The conclusion we reached was that the presence of only two hair cell orientation directions in the Otophysi is a derived characteristic and one that is an extreme specialization for a broad bandwidth of underwater hearing!

In fact, this idea is supported by work done by Cathy McCormick when she was a postdoc with me. Cathy studied hearing in mormyrid fishes (*Gnathonemus* sp.), elephant-nosed fishes that use electroreception for communication. She found that *Gnathonemus* has hearing sensitivity that rivals that of the Otophysi (McCormick & Popper, 1984), although they are not at all closely related taxonomically, and we found that the saccular hair cell orientation pattern is bidirectional (Popper, 1981) as in Otophysi. However, instead of having Weberian ossicles, there is a large air bubble attached to the saccule in this species, thereby enhancing hearing capabilities.

Indeed, having bidirectional hair cells in the saccule appears to be the "norm" for other vertebrates, whereas four directions are found in many bony fishes. This idea is supported by findings that show that sharks and rays have saccular hair cells oriented in two directions (Corwin, 1981a), and we have found the same thing in a whole range of more primitive fish that representing the origins of the species that have hair cells in four directions (Popper, 1978; Popper & Northcutt, 1983; Mathiesen & Popper, 1987). Moreover, a similar pattern is found in lungfishes, a group that is thought to have given rise to terrestrial vertebrates (Platt et al., 2004).

25.4.2 Why Multiple Hair Cell Patterns?

Of course, a singularly interesting question is why so many fish species have hair cells oriented in multiple directions. Related to this is the question of why fishes with particularly wide hearing bandwidths often have even more elaborate orientation patterns. And, although not being discussed here, what is the functional significance for fishes of having, on a single epithelium, hair cells with different length ciliary bundles (Popper, 1977; Platt, 1983; Popper & Platt, 1983)?

One very reasonable suggestion for having hair cells oriented in different directions may reflect back to the question I initially wanted to ask for my dissertation, sound source localization. Fishes do not have available to them the same acoustic cues for localization used by terrestrial vertebrates (e.g., interaural differences) owing to the close proximity of fish ears to one another and the much higher speed of sound in water than in air (e.g., van Bergeijk, 1964). At the same time, the basic mechanism by which the sensory hair cells of the ear are stimulated results from the relative motion between the dense overlying otolith and the sensory cells, and this motion, which is in response to the particle motion component of the underwater sound field, is directional (Popper et al., 2003; Fay & Popper, 2012).

Thus, one can imagine that if the relative motion between the epithelium and otolith changes with the direction of the impinging sound field, and knowing that the physiological response of the hair cells is directional, it follows that by having hair cells oriented in different directions, each would respond maximally to sound from different directions. We therefore speculated that by combining inputs from hair cells oriented in multiple directions, it should be possible to localize a sound source (Rogers et al., 1988). Although the actual process is somewhat more complex owing to the presence of pressure as well as particle motion fields and many other factors (e.g., Rogers & Zeddies, 2008), it is likely that one important role of hair cells oriented in different directions is to aid in determining the direction of a sound source. At the same time, when one considers that each of the three end organs in most fishes lie on different planes, this means that not only can fishes use directional responses from the two saccules, but also that the other end organs can potentially contribute to directional responses as well, thereby refining the information about direction.

Still, sound localization by fishes remains an enigmatic problem. Although it is clear that the hair cell orientation patterns of the ear are involved in determining direction, there are few data showing how well fishes can localize (e.g., Rogers & Zeddies, 2008). This remains an area that, even 40+ years since my doctorate, has not been solved, making that Tavolga's 17-year prediction way off!

25.4.3 So Why Variation?

But the question still remains as to why some fishes have more complex orientation patterns than others and why many of the species with the most complex patterns have a wider bandwidth of hearing than fishes without complex patterns. Perhaps this has to do with refinements in systems for sound localization. Or perhaps this has to do with other aspects of hearing.

There are two basic hypotheses (Popper et al., 2003). One is that in the evolution of numerous species, fishes have "experimented" in widely different ways to extract the same information from sound. That is, as we pointed out years ago (Fay & Popper, 2000), for fishes to glean the maximum amount of information from the acoustic scene, they need to be able to detect sound, discriminate between sounds, localize sound, and detect signals in the presence of noise. If we assume that every fish needs to be able to do these functions, then it is possible that the different inner ear patterns are all different ways to help accomplish the same tasks (also see Chapter 7 by Fay). Alternatively, if all fishes do not have to do the same basic things in terms of hearing, it is possible that the different patterns have evolved to do different auditory tasks. There is no ready way to resolve which hypothesis is correct, but this is a question worth asking in the future.

25.5 Ultrasound Detection

In the late 1990s, I read several papers that suggested that some fishes in the herring family (Clupeidae) could be kept from entering the water intakes of nuclear power plants by projecting ultrasound into the water around the intakes (e.g., Dunning et al., 1992). This was my first introduction to the idea that sound could potentially be used to control fish behavior, and I found the whole idea of fish detecting ultrasound somewhat "ridiculous." But, because clupeids are some of the most commercially important fishes in the world, we submitted a proposal to the National Science

Foundation to examine hearing, and ultrasonic hearing, in a clupeid fish, the American shad (*Alosa sapidissima*). The grant was funded (those were the days when "wild" ideas could still get funding) and David Mann, then a postdoc in the lab, led a project to explore hearing in this species. Another postdoc (and former Fay doctoral student) Zhongmin Lu collaborated with David on the project.

We discovered that American shad and their relatives in the subfamily Alosinae (shads, menhaden, and some freshwater herring) are able to detect sounds to at least 180 kHz (Mann et al., 1997). Thus, these fishes may have the widest hearing bandwidth of any known vertebrate including echolocating dolphins and bats.

Once we realized that Alosids detect ultrasound, we struggled to figure out why they have such an extraordinary hearing range. We finally realized that perhaps these fishes are detecting the high-frequency echolocation sounds of dolphins and avoiding predation. Although the idea seems far-fetched, it, in part, arose from our familiarity with the work on moths and other insects that have evolved high-frequency hearing to detect echolocating bats and avoid being eaten (e.g., Roeder & Treat, 1961).

Although it was not possible for us to expose American shad to actual dolphins, my postdoc Dennis Plachta developed a behavioral paradigm that exposed American shad to ultrasonic signals and enabled us to observe the response (Plachta & Popper, 2003). We found that the American shad did not show a behavioral response to low-frequency sounds (e.g., 500 or 1000 Hz) but that they would react to ultrasonic signals. We also found that responses of American shad to ultrasound, like those in moths evading bats, were "graded." At lower received intensity signals, the American shad would swim away from the source (demonstrating, by the way, sound source localization). As sound levels got higher, the responses became more rapid, and at the highest sound levels, the fish showed highly random and "chaotic" behavior. The conclusion we reached is that when an echolocation click is just audible, the American shad may not pay attention, "thinking" that the dolphin was far away. But as the sound gets louder, and potentially the dolphin closer, the fish start to actively swim away from the sound source. Finally, if the dolphin is very close and the sound loud, the fish school (often tens of thousands of fish) go into random rapid motion so that the predator can no longer echolocate on a single animal, thereby lowering the likelihood of predation.

Along with trying to understand why American shad detect ultrasound, it is equally of interest to know *how* ultrasound detection operates in a fish. This turned out to be a far more difficult problem and one that has yet to be solved.

The mechanism we proposed for ultrasound hearing involves use of the very highly specialized utricle that is known in clupeids to be in close contact with a small gas bubble in the head (Denton & Gray, 1979). This utricle, unlike those found in any other species of vertebrate, is tripartite (Popper & Platt, 1979). Moreover, the central epithelial region in ultrasound-detecting species is very thin, as demonstrated in a project led by my postdoc Dennis Higgs (Higgs et al., 2004). Thus, we hypothesized that the bubble resonates at ultrasonic frequencies and stimulates the middle utricular epithelium.

The problem in testing this hypothesis is that all clupeids are very fragile, with the lab "joke" being that just looking at a clupeid can result in its dying. Although American shad are perhaps the heartiest of the clupeids, even these fish could not tolerate electrodes being placed near the utricle because the only possible surgical approach would be through the air bubble by the utricle. This would destroy the bubble and, of course, eliminate its being tested for ultrasonic hearing. However, Plachta, working with postdoc Michele Halvorsen and our colleague Jiakun Song, devised an approach that allowed us to test ultrasonic responses at various brain levels (Plachta et al., 2004). These studies showed that there are, indeed, neurons in the brain that respond best to ultrasound and there was good evidence that these units arose in the utricle.

Of course, things may not be as simple as we thought. Recent work by Wilson et al. (2009) suggests that connections between the air chamber in the head and the lateral line receptors on the body may also play a role in ultrasound detection. Although we are still convinced that the ultimate receptor is the utricle and that the air bubble is involved, the whole mechanism may be rather more complex and something that one would hope would be studied by future investigators.

Another question that arose was *how* fishes could have evolved ultrasonic hearing. Although there is no fossil record to give an answer, we do think we have a reasonable answer. The answer comes out of a paper by two friends and colleagues, Peter Rogers and Mardi Hastings (Rogers & Cox, 1988).

Rogers and Cox pointed out to that sound propagation is very different in shallow versus deep water. In shallow water (e.g., streams, shallow rivers), low frequencies propagate very poorly, and only higher frequencies propagate greater distances—the shallower the water the poorer the propagation of low frequencies. We thus proposed that fishes that have evolved sound detection above 1 kHz or so probably arose in shallow water, and, indeed, most fishes that hear above about 1500 Hz are in shallow water or evolved from species there. (As an aside, the same observation was made by Pliny the Elder [1890, p. 547], who noted that "…the mullet, the wolf-fish, the salpa, and the chromis, have very exquisite sense of hearing, and that it is for this reason that they frequent shallow water.") We therefore proposed that the ancestors of clupeid fishes evolved in shallow water. And, indeed, American shad and many other clupeid species are anadromous; they breed in shallow rivers and streams and then move out to the ocean to grow, as described by the Pulitzer Prize winning author John McPhee (2003).

Exactly why American shad and other *Alosa* evolved ultrasound detection is not clear, but perhaps selective pressures placed on them by echolocating dolphins resulted in the increased bandwidth, using the same utricular structures that are in all other clupeids. Again, this is a really interesting problem for future study.

25.6 Addition of Sensory Hair Cells in the Ear

In the early 1980s, Jeff Corwin (who completed his MS with Albert Tester, Ian Cooke, and me in Hawai'i) did a series of studies showing that sharks and rays had very large numbers of sensory hair cells in their ears and also that there was a continuous addition of such cells over the life of these animals (Corwin, 1981b). We started to wonder if the same phenomenon occurs in bony fishes. Thus, Becky Hoxter and I examined the number of sensory cells in the ears of different-sized

Oscars (*Astronotus oscellatus*) and found a substantial proliferation of cells in the saccule as the fishes grew (Popper & Hoxter, 1984). This has been confirmed for other species but perhaps most dramatically in a study done by my postdoc Antoni Lombarte. Toni, a fisheries biologist from Barcelona, obtained specimens of the Mediterranean hake (*Merluccius merluccius*), a relative of the Atlantic cod. We demonstrated a very substantial addition of sensory hair cells in each of the otolithic end organs for at least the first nine years of life and that this increase was several hundred cells a day (Lombarte & Popper, 1994). Nine-year-old animals had more than 2 million hair cells in each ear. Although speculative, we also noted that the largest "spurt" of hair cell addition came at about an age when *Merluccius* switched its feeding habits and started to feed on myctophid fishes, one of the most widely distributed fish groups and one thought to make sounds. Our very tentative suggestion was that *Merluccius* start to hear their prey when they switch to myctophids as a diet, and so hearing becomes more important in their lives.

Still, there is nothing known about why fishes increase hair cells. One possibility is that as fishes add cells, they increase hearing sensitivity. However, this makes little sense because it would mean that large fishes would have different hearing capabilities than smaller brethren, affecting communication and the detection of the acoustic scene. Alternatively, because fishes continue to grow through most of their lives, the structures associated with hearing change relative positions and the additional sensory cells are needed to maintain hearing at a certain level. Two lines of evidence support the latter hypothesis. First, in a study of walleye pollock (*Theragra chalcogramma*), we showed that the hearing sensitivity of small fish to both pressure and particle motion was virtually the same as that in fish 3 years older (Mann et al., 2009). Second, in a study on zebrafish hearing that he did while a postdoc, Dennis Higgs showed that hair cell proliferation stopped when zebrafish (which do not seem to grow substantially after they reach a certain adult size) stopped growing (Higgs et al., 2002).

Although there is no space to go into it here, one other point about the ability of fishes to proliferate new hair cells for most (if not all) of their lives is worth noting. In the early 1990s my postdoc Hong Young Yan demonstrated that treatment with the ototoxic drug gentamicin would destroy sensory cells in some parts of each sensory epithelium (Yan et al., 1991). This was followed by a study led by Lombarte, who showed regeneration of the cells over about 10–15 days post exposure to gentamicin (Lombarte et al., 1993). More recently, my postdoc Michael Smith and graduate student Allison Coffin, studying temporary threshold shift (TTS) in fishes, showed that exposure to loud sounds results in damage to sensory cells as well as TTS in some species, and that hearing recovers as hair cells return (Smith et al., 2006).

25.7 Bridge Construction and Other Applied Issues

As mentioned in Section 5, the study of ultrasound detection arose because of the use of high-frequency sounds to control fish behavior. Before hearing about this, I had never given thought to an applied use for our work on fish hearing, but over

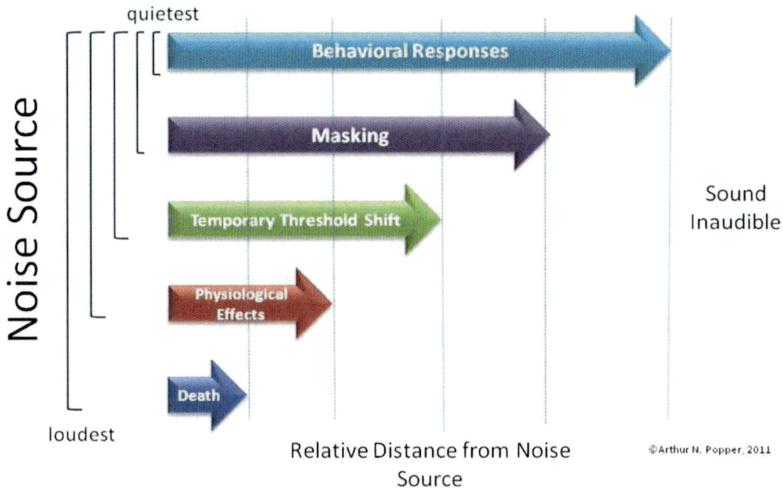

Fig. 25.7 Relationship between noise levels, relative distance, and potential effects

the past 10 years or so, the work in my lab has been heavily focused on asking questions that directly relate to the use of sound to control fish behavior and, far more importantly, on the effects of man-made sound (also called anthropogenic sound) on fishes. Indeed, Dick Fay and I looked back at a paper we published 20 years earlier (Popper & Fay, 1993; Fay & Popper, 2012) and came to the conclusion that one of the major driving forces now and for the foreseeable future for studies of fish hearing lies with these, and related, more applied questions.

The issue of effects of man-made sound on fishes is worldwide. Human activities are increasingly adding sound to the aquatic environment from a variety of sources. Commercial ships (e.g., oil tankers) are very noisy and, along with other kinds of boating, tend to increase ambient noise levels (Fig. 25.7). This may be particularly the case in places like shipping lanes and harbors, where increased ambient noise has the potential to mask sounds that are of biological relevance to fishes, including their own communication sounds and sound of the acoustic scene.

Perhaps more dramatic, however, are sounds produced in the construction of structures such as bridges and wind farms and from geological exploration for undersea gas and oil. These sounds tend to be very intense (in some cases over 220 dB re 1 μPa rms) and have the potential to kill or dramatically affect the behavior of fishes.

One of the real problems in considering the effect of man-made sounds on fishes is that the number of well-controlled and peer-reviewed studies is very limited (Popper & Hastings, 2009). Investigations of the effects of intense sounds on fishes are very difficult because the sound sources of concern cannot easily be brought into a lab because of their sizes. And studies in the field are very difficult because the sound-producing devices used in construction and/or exploration are very expensive. As a consequence, investigators have no control over the sources during their experiments.

For example, one of the sources of concern is the seismic air guns used in oil and gas exploration. The specialized vessels and sources used cost hundreds of thousands of dollars a day to operate, so no vessel operator will "donate" time for a research project. Similarly, the major source of sounds during construction is from impact driving of piles used to support structures such as bridges and wind turbines. Pile drivers are very large machines (and very expensive to operate) and so they cannot be brought into the lab or purchased for use in an experiment.

Despite these limitations we have been able to make a number of contributions over the past few years that not only help to understand the effects of these intense sources on fishes, but, perhaps more importantly, to contribute directly to developing regulatory practices that both protect animals and allow needed construction and exploration.

25.7.1 Seismic Air Guns in the Arctic Circle

There is major international concern over the potential effects of seismic air guns on fishes (see references in Popper & Hawkins, 2012). These devices use release of highly compressed air to project sounds into the substrate. The reflected signals are picked up by long arrays of hydrophones and the data are analyzed to determine geologic formations that indicate the presence of gas and oil. In 2004 David Mann and I were invited to work with Bruce Hanna and Peter Cott of the Canadian Department of Fisheries and Oceans in Inuvik, Canada, to determine if the sounds from a small air gun array used in the Mackenzie River (the second longest river in North America) was harming local fishes. This was a great concern of the local Inuit population because they are subsistence fishers.

We exposed several species to air guns and measured hearing post exposure to determine if there was hearing loss and any damage to the sensory cells of the ear. This was the first study to measure hearing in fishes exposed to air guns. (Earlier, working with Rob McCauley in Perth, Australia, we showed that exposure to a seismic air gun could damage sensory hair cells in the ears of one species of fish; McCauley et al., 2003.) The fish were exposed to sounds up to received sound levels of around 198 dB re 1 μPa rms and then tested to ascertain whether there was TTS. We found that several of the species showed TTS, but that complete recovery took place within about 18 hours. Exposure to this intense sound did not kill any animals nor was there any evidence of damage to internal tissues (Popper et al., 2005). Subsequent analysis of the sensory epithelia of the inner ear also showed no effect (Song et al., 2008). This study was the first to examine potential hearing loss in any fish exposed to very high-intensity impulsive sounds, and it succeeded in showing that fish exposed to high levels of such sounds may not be killed or permanently harmed and that not every species shows the same effects from such exposure as others.

Fig. 25.8 A picture of the HICI-FT used in studies of effects of pile driving on fish. The large gray chamber on the left and the steel cylinder on the right are the shakers used to produce the sounds (the one on the left looks different because it is enclosed in plastic to keep out water when fish are put into the device. The central steel region is a "fish tank" in which the animals are exposed to sound. See Halvorsen et al. (2012b) for a description of how the device works. The hoses and plastic tubes are for temperature control and water flow

25.7.2 Pile Driving

During construction of the east span of the San Francisco Bay Bridge in the mid-2000s, concern arose that the very extensive pile driving may kill or harm migrating endangered salmon that pass the construction site. There was then very little useful scientific data to actually help understand the effects of the pile-driving sounds on fishes since it was not easy to take a pile-driving apparatus into the lab to do controlled studies.

One morning over breakfast, however, my colleague Tom Carlson and I came up with an idea to bring pile-driving sounds into the lab with the same sound levels and acoustic characteristics as found during actual pile driving in the field. The idea involved use of a device invented by Pete Rogers, something he called a "ratabrator." Tom and I managed to get funding and then worked with Pete and his group to redesign the "ratabrator" so that it could generate very high intensity pile-driving signals in my lab at the University of Maryland. The device was named the HICI-FT (pronounced hissy fit) (Fig. 25.8) by my postdoc (and project leader) Michele Halvorsen because of its initial fickleness. Once we figured out how to isolate the HICI-FT from the building structure (its first uses shook our five-story steel-framed building!) and cool it so that the very powerful shakers needed to generate the sound field would not heat the water and cook the fish, we were able to do a range of experiments on the effects of pile-driving sounds on fish.

In initial studies on young Chinook salmon (*Oncorhynchus tshawytscha*), we showed that the onset of physiological effects (damage) occurred only when fish were exposed to accumulated sound energy (cumulative sound exposure level [SEL_{cum}]) of 210 dB re 1 μPa^2·s, a signal level that was equivalent to giving the fish 1920 pile strikes (at 1.2-s intervals) of 177 dB re 1 μPa^2·s total sound energy in each strike (Halvorsen et al., 2012b). These provide the experimentally derived sound exposure levels that can be used in developing regulations to protect the fish. At the same time, these proposed levels are substantially higher than those currently used by regulators and that are not science based, meaning that although protective of fish, the levels from our studies also mean that construction is less likely to harm the animals than previously thought.

Subsequently, with the added collaboration from my last postdoc Brandon Casper, additional studies have shown that the levels that result in the onset of physiological effects are about the same for a morphologically diverse group of species (Halvorsen et al., 2012a; Casper et al., 2013), and so the levels we proposed with Chinook salmon may be broadly applicable to other species, although more studies are needed because there is so much variability in fish ear morphology.

25.7.3 The Significance of Applied Studies

Although the work described starts to contribute to understanding of the effects of man-made sounds on fishes, this is an area ripe for further study that has worldwide implications as shipping and the exploration for traditional and alternative energy sources increase (see Popper & Hawkins, 2012). Indeed, the interest in this area is shown by the very high participation in international meetings that my friend and colleague Tony Hawkins and I have been organizing over the past several years (e.g., Popper & Hawkins, 2012).

From the perspective of someone who has spent more than 45 years studying fish hearing and the evolution of hearing in vertebrates, moving into the applied arena is a big change. However, what I have learned (and what Dick Fay and I expressed in our 2012 paper) is that there are unique opportunities afforded by delving into applied questions. Not only does one get access to nontraditional funding sources, but there are also creative ways to use the applied research to continue to explore important basic science questions (such as hearing capabilities and sound source localization). Moreover, one gets a chance to use what one has learned over decades to help solve problems that are important for the environment and for humans.

25.8 Final Thoughts

In fact, the past years have seen increases in our understanding of fish hearing in the evolution of vertebrate hearing. Still, there are a myriad of open questions (also see Chapter 14 by Hawkins), starting with my favorite, sound source localization, but including numerous comparative questions, some of which were raised in this chapter. It will be exciting to see what the next generation of investigators finds. But, at the same time, I anticipate that many of these discoveries will be made in the context of solving problems that have a "translational" or applied aspect.

Acknowledgments I have been truly fortunate over the last 40+ years to have a large number of great collaborators, including a wonderful group of students and postdocs. I would very much have liked to mention the work of every one of these people, but because Dick and I have limited the lengths of the chapters in this volume, I had to focus on just a few of the many questions that my lab has dealt with over the years. I do want all those students and collaborators not mentioned here to know, however, that I value and appreciate working with every one of them and that perhaps, in the future, I can discuss their work. I also thank the many funding agencies who have, for 43 years without interruption, supported the work in my lab. I also want to thank Allison Coffin, Sheryl Coombs, Brandon Casper, and Christopher Platt for reviewing the manuscript, and my oldest friend, Steve Weinberg (from Mr. Vinci's sixth grade class), for asking penetrating questions that helped me improve the manuscript substantially. Finally, I am most grateful to my wife Helen for editing this manuscript—she has not only totally supported me in my work over the years, but she is also the very best editor, and I am grateful for her sharp eye and pen.

References

Breder, C. M. Jr. (1943). Problems in the behavior and evolution of a species of blind cave fish. *Transactions of the New York Academy of Sciences*, 5, 168–176.

Buran, B. N., Deng, X., & Popper, A. N. (2005). Structural variation in the inner ears of four deep-sea elopomorph fishes. *Journal of Morphology*, 265, 215–225.

Cahn, P. H. (1958). Comparative optic development in *Astyanax mexicanus* and in two of its blind cave derivatives. *Bulletin of the American Museum of Natural History*, 115, 75–112.

Cahn, P. H., Ed. (1967). *Lateral line detectors*. Bloomington: Indiana University Press.

Casper, B. M., Smith, M. E., Halvorsen, M. B., Sun, H., Carlson, T. J., & Popper, A. N. (2013). Effects of exposure to pile driving sounds on fish inner ear tissues. *Comparative Biochemistry and Physiology A*, 166, 352–360.

Coombs, S., & Popper, A. N. (1979). Hearing differences among Hawaiian squirrelfishes (Family Holocentridae) related to differences in the peripheral auditory system. *Journal of Comparative Physiology*, 132, 203–207.

Corwin, J. T. (1981a). Audition in elasmobranchs. In W. N. Tavolga, A. N. Popper, & R. R. Fay (Eds.), *Hearing and sound communication in fishes*, (pp. 81–105). New York: Springer-Verlag.

Corwin, J. T. (1981b). Postembryonic production and aging in inner ear hair cells in sharks. *Journal of Comparative Neurology*, 201, 541–553.

Dale, T. (1976). The labyrinthine mechanoreceptor organs of the cod *Gadus morhua* L. (Teleostei, Gadidae). *Norwegian Journal of Zoology*, 24, 85–128.

Deng, X., Wagner, H.-J., & Popper, A. N. (2011). The inner ear and its coupling to the swim bladder in the deep-sea fish *Antimora rostrata* (Teleostei: Moridae). *Deep Sea Research, Part I*, 58, 27–37.

Deng, X., Wagner, H.-H., & Popper, A. N. (2013). Interspecific variations of inner ear structure in the seep-sea fish family Melamphaidae. *Anatomical Record*, 296, 1064–1082. doi: 10.1002/ar.22703.

Denton, E. J., & Gray, J. A. (1979). The analysis of sound by the sprat ear. *Nature*, 282, 406–407.

Dijkgraaf, S. (1932). Untersuchunger über die Funktion der Seitenorgane an Fischen. *Zeitschrift für vergleichende Physiologie*, 20, 162–214.

Dunning, D. J., Ross, Q. E., Geoghegan, P., Reichle, J. J., Menezes, J. K., & Watson, J. K. (1992). Alewives avoid high-frequency sound. *North American Journal of Fisheries Management*, 12(3), 407–416.

Fay, R. R., & Popper, A. N. (2000). Evolution of hearing in vertebrates: The inner ears and processing. *Hearing Research*, 149, 1–10.

Fay, R. R. & Popper, A. N. (2012). Fish hearing: New perspectives from two "senior" bioacousticians. *Brain, Behaviour and Evolution*, 792, 215–217.

Halvorsen, M. B., Casper, B. C., Matthews, F., Carlson, T. J., & Popper, A. N. (2012a). Effects of exposure to pile driving sounds on the lake sturgeon, Nile tilapia, and hogchoker. *Proceedings of the Royal Society of London B: Biological Sciences*, 279, 4705–4714.

Halvorsen, M. B., Casper, B. M., Woodley, C. M., Carlson, T. J., & Popper, A. N. (2012b). Threshold for onset of injury in Chinook salmon from exposure to impulsive pile driving sounds. *PLoS ONE*, 7(6) e38968.

Higgs, D. M., Souza, M. J., Wilkins, H. R., Presson, J. C., & Popper, A. N. (2002). Age- and size-related changes in the inner ear and hearing ability of the adult zebrafish (*Danio rerio*). *Journal of the Association for Research in Otolaryngology*, 3, 174–184.

Higgs, D. M., Plachta, D. T. T., Rollo, A. K., Singheiser, M., Hastings, M. C., & Popper, A. N. (2004). Development of ultrasound detection in of American shad (*Alosa sapidissima*). *Journal of Experimental Biology*, 207, 155–163.

Jacobs, D. W., & Tavolga, W. N. (1967). Acoustic intensity limens in the goldfish. *Animal Behaviour*, 15, 324–335.

Ladich, F., & Fay, R. R. (2013). Auditory evoked potential audiometry in fish. *Reviews in Fish Biology and Fisheries* 23, 317–364. doi:10.1007/s11160-012-9297-z

Lombarte, A., & Popper, A. N. (1994). Quantitative analyses of postembryonic hair cell addition in the otolithic endorgans of the inner ear of the European hake, *Merluccius merluccius* (Gadiformes, Teleostei). *Journal of Comparative Neurol*ogy, 345, 419–428.

Lombarte, A. Yan, H. Y., Popper, A. N., Chang, J. C., & Platt, C. (1993). Damage and regeneration of hair cell ciliary bundles in a fish ear following treatment with gentamicin. *Hearing Research*, 66, 166–174.

Mann, D. A., Lu, Z., & Popper, A. N. (1997). A clupeid fish can detect ultrasound. *Nature*, 389, 341.

Mann, D. A., Wilson, C. D., Song, J., & Popper, A. N. (2009). Hearing sensitivity of the Walleye Pollock, *Theragra chalcogramma*. *Transactions of the American Fisheries Society*, 138, 1000–1008.

Mathiesen, C., & Popper, A. N. (1987). The ultrastructure and innervation of the ear of the gar, *Lepisosteus osseus*. *Journal of Morphology*, 194, 129–142.

McCauley, R. D., Fewtrell, J., & Popper, A. N. (2003). High intensity anthropogenic sound damages fish ears. *Journal of the Acoustical Society of America*, 113, 638–642.

McCormick, C. A., & Popper, A. N. (1984). Auditory sensitivity and psychophysical tuning curves in the elephant nose fish, *Gnathonemus petersii*. *Journal of Comparative Physiology*, 155, 753–761.

McPhee, J. (2003). *The founding fish*. New York: Farrar, Straus and Giroux.

O'Connell, C. P. (1955). The gas bladder and its relation to the inner ear in *Sardinops caerulea* and *Engraulis mordax*. *Fishery Bulletin*, 56, 505–533.

Parker, G. H. (1902). Hearing and allied senses in fishes. *Bulletin of the U.S. Fisheries Commission*, 22, 45–64.

Parvulescu, A. (1964). Problems of propagation and processing. In W. N. Tavolga (Ed.), *Marine bio-acoustics* (pp. 87–100). Oxford: Pergamon Press.

Plachta, D. T. T., & Popper, A. N. (2003). Evasive responses of American shad (*Alosa sapidissima*) to ultrasonic stimuli. *Acoustical Research Letters Online*, 4, 25–30.

Plachta, D. T. T., Song, J., Halvorsen, M. B., & Popper, A. N. (2004). Neuronal encoding of ultrasonic sound by a fish. *Journal of Neurophysiology*, 91, 2590–2597.

Platt, C. (1977). Hair cell distribution and orientation in goldfish otolith organs. *Journal of Comparative Neurology*, 172, 283–297.

Platt, C. (1983). The peripheral vestibular system in fishes. In R. G. Northcutt & R. E. Davis (Eds.), *Fish neurobiology* (pp. 89–124). Ann Arbor: University of Michigan Press.

Platt, C. (1993). Zebrafish inner ear sensory surfaces are similar to those in goldfish. *Hearing Research*, 63, 133–140.

Platt, C., & Popper, A. N. (1984). Variation in lengths of ciliary bundles on hair cells along the macula of the sacculus in two species of teleost fishes. *Scanning Electron Microscopy*, 1984, 1915–1924.
Platt, C., Jørgensen, J. M., & Popper, A. N. (2004). The inner ear of the lungfish *Protopterus*. *Journal of Comparative Neurology*, 471, 277–278.
Pliny (The Elder) (1890). *The natural history of Pliny*, Vol. II. Translated by J. Bostock and H. T. Riley, London: George Bell & Sons. Available at http://books.google.com/books?id=BUEMAAAAIAAJ&printsec=frontcover&source=gbs_ge_summary_r&cad=0#v=onepage&q=clapping&f=false
Popper, A. N. (1970). Auditory capacities of the Mexican blind cave fish *Astyanax jordani* and its eyed ancestor *Astyanax mexicanus*. *Animal Behaviour*, 18, 52–562.
Popper, A. N. (1971). The morphology of the Weberian ossicles in two species of *Astyanax* (Ostariophysi: Characidae). *Journal of Morphology*, 133, 179–188.
Popper, A. N. (1976). Ultrastructure of the auditory regions in the inner ear of the lake whitefish. *Science*, 192, 1020–1023.
Popper, A. N. (1977). A scanning electron microscopic study of the sacculus and lagena in the ears of fifteen species of teleost fishes. *Journal of Morphology*, 153, 397–418.
Popper, A. N. (1978). Scanning electron microscopic study of the otolithic organs in the bichir (*Polypterus bichir*) and shovel-nose sturgeon (*Scaphirhynchus platorynchus*). *Journal of Comparative Neurology*, 18, 117–128.
Popper, A. N. (1980). Scanning electron microscopic studies of the sacculus and lagena in several deep-sea fishes. *American Journal of Anatomy*, 157, 115–136.
Popper, A. N. (1981). Comparative scanning electron microscopic investigations of the sensory epithelia in the teleost sacculus and lagena. *Journal of Comparative Neurology*, 200, 357–374.
Popper, A. N., & Coombs, S. (1982). The morphology and evolution of the ear in Actinopterygian fishes. *American Zoologist*, 22, 311–328.
Popper, A. N., & Fay, R. R. (1993). Sound detection and processing by fish: Critical review and major research questions. *Brain Behaviour and Evolution*, 41, 14–38.
Popper, A. N., & Fay, R. R. (2011). Rethinking sound detection by fishes. *Hearing Research*, 273, 25–36.
Popper, A. N., & Hastings, M. C. (2009). Effects of anthropogenic sources of sound on fishes. *Journal of Fish Biology*, 75, 455–498.
Popper, A. N., & Hawkins, A. D., Eds. (2012). *The effects of noise on aquatic life*. New York: Springer Science+Business Media.
Popper, A. N., & Hoxter, B. (1984). Growth of a fish ear: 1. Quantitative analysis of sensory hair cell and ganglion cell proliferation. *Hearing Research*, 15, 133–142.
Popper, A. N., & Northcutt, R. G. (1983). Structure and innervation of the inner ear of the bowfin, *Amia calva*. *Journal of Comparative Neurology*, 21, 279–286.
Popper, A. N., & Platt, C. (1979). The herring ear has a unique receptor pattern. *Nature*, 280, 832–833.
Popper, A. N., & Platt, C. (1983). Sensory surface of the saccule and lagena in the ears of ostariophysan fishes. *Journal of Morphology*, 176, 121–129.
Popper, A. N., Fay, R. R., Platt, C., & Sand, O. (2003). Sound detection mechanisms and capabilities of teleost fishes. In S. P. Collin & N. J. Marshall (Eds.), *Sensory processing in aquatic environments* (pp. 3–38). New York: Springer Science+Business Media.
Popper, A. N., Smith, M. E., Cott, P. A., Hanna, B. W., MacGillivray, A. O., Austin, M. E., & Mann, D. A. (2005). Effects of exposure to seismic airgun use on hearing of three fish species. *Journal of the Acoustical Society of America*, 117, 3958–3971.
Ramcharitar, J., Gannon, D. P., & Popper, A. N. (2006). Bioacoustics of the family Sciaenidae (croakers and drumfishes). *Transactions of the American Fisheries Society*, 135, 1409–1431.
Retzius, G. (1881). *Das Gehörorgan der Wirbelthiere*. Vol. I. Stockholm: Samson and Wallin.
Roeder, K. D., & Treat, A. E. (1961). The detection and evasion of bats by moths. *Readings in the Psychology of Perception*, 49, 35.
Rogers, P. H., & Cox, M. (1988). Underwater sound as a biological stimulus. In J. Atema, R. R. Fay, A. N. Popper & W. N. Tavolga (Eds.), *Sensory biology of aquatic animals* (pp. 131–149). New York: Springer-Verlag.

Rogers, P. H., & Zeddies, D. G. (2008). Multipole mechanisms for directional hearing in fish. In J. F. Webb, A. N. Popper, & R. R. Fay (Eds.), *Fish bioacoustics* (pp. 233–252). New York: Springer Science+Business Media.

Rogers, P. H., Popper, A. N., Cox, M., & Saidel, W. M. (1988). Processing of acoustic signals in the auditory system of bony fish. *Journal of the Acoustical Society of America*, 83, 338–349.

Smith, M. E., Coffin, A. B., Miller, D. L., & Popper, A. N. (2006). Anatomical and functional recovery of the goldfish (*Carassius auratus*) ear following noise exposure. *Journal of Experimental Biology*, 209, 4193–4202.

Song, J., Mann, D. A., Cott, P. A., Hanna, B. W., & Popper, A. N. (2008). The inner ears of northern Canadian freshwater fishes following exposure to seismic air gun sounds. *Journal of the Acoustical Society of America*, 124, 1360–1366.

Tavolga, W. N., Ed. (1964). *Marine bio-acoustics*. Oxford: Pergamon Press.

Tavolga, W. N., Ed. (1967). *Marine bio-acoustics*, II. Oxford: Pergamon Press.

Tavolga, W. N., & Wodinsky, J. (1963). Auditory capacities in fishes. Pure tone thresholds in nine species of marine teleosts. *Bulletin of the American Museum of Natural History*, 126, 177–240.

van Bergeijk, W. A. (1964). Directional and nondirectional hearing in fish. In W. N. Tavolga (Ed.). *Marine bio-acoustics* (pp. 281–289). Oxford: Pergamon Press.

von Békésy, G. (1967). Some similarities in sensory perception of fish and man. In P. H. Cahan (Ed.). *Lateral line detectors* (pp. 417–435). Bloomington: Indiana University Press.

von Frisch, K. (1923). Ein Zwergwels der kommt, wenn man ihm pfeift. *Biology Zentralblatt Leipzig*, 43, 439–446.

Weber, E. H. (1820). *De Aure et Auditu Hominis et Animalium. Pars I. De Aure Animalium Aquatilium.* Leipzig: Gerhard Fleischer.

Webster, D. B. (1962). A function of the enlarged middle-ear cavities of the kangaroo rat, *Dipodomys. Physiological Zoology*, *35*(3), 248–255.

Webster, D. B., Fay, R. R., & Popper, A. N., Eds. (1992). *Evolutionary biology of hearing*, New York: Springer-Verlag.

Wersäll, J., Flock, Å., & Lundquist, P. G. (1965). Structural basis for directional sensitivity in cochlear and vestibular sensory receptors. *Cold Spring Harbor Symposium on Quantitative. Biology*, 30, 115–132.

Wilson, M., Montie, E. W., Mann, K. A., & Mann, D. A. (2009). Ultrasound detection in the Gulf menhaden requires gas-filled bullae and an intact lateral line. *Journal of Experimental Biology*, 212(21), 3422–3427.

Yan, H. Y., Saidel, W. M., Chang, J., Presson, J. C., & Popper, A. N. (1991). Sensory hair cells of the fish ear: evidence of multiple types based on ototoxicity sensitivity. *Proceedings of the Royal Society of London B: Biological Sciences*, 245, 133–138.

Chapter 26
Current Topics in the Study of Sound Conduction to the Inner Ear

John J. Rosowski, Hideko H. Nakajima, and Jeffrey T. Cheng

John J. Rosowski

Hideko H. Nakajima

Jeffrey T. Cheng

J.J. Rosowski (✉) • H.H. Nakajima • J.T. Cheng
Department of Otology and Laryngology, Harvard Medical School, Boston, MA 02114, USA

Massachusetts Eye and Ear Infirmary, Boston, MA 02114, USA
e-mail: John_Rosowski@meei.harvard.edu; Heidi_Nakajima@meei.harvard.edu; Tao_Cheng@meei.harvard.edu

A.N. Popper and R.R. Fay (eds.), *Perspectives on Auditory Research*, Springer Handbook of Auditory Research 50, DOI 10.1007/978-1-4614-9102-6_26,

26.1 Introduction

Nearly 20 years ago one of us wrote a chapter for SHAR 4, *Comparative Hearing: Mammals*, which reviewed what was then known regarding sound conduction through the middle ears in mammals (Rosowski, 1994). Since that time there have been multiple new ideas regarding sound transmission to the inner ear, some of which were anticipated in that chapter and others not. We have defined eight such areas: (1) new high-resolution descriptions of ossicular structure; (2) factors that control the bandwidth of the middle ear's response to sound; (3) how the tympanic membrane (TM) couples sound to the ossicular chain; (4) the contribution of complex ossicular motions to sound transmission; (5) the reverse coupling of sound generated in the inner ear to the external ear; (6) the question of inner ear "third windows"; (7) new ways to diagnose conductive hearing loss; and (8) new data and models of how bone conduction stimulates the inner ear.

Some of these topics are related, for example, investigations of middle ear bandwidth are related to the function of the TM, and to the significance of different modes of motion within the ossicular chain. Also the question of cochlear third windows is tied to a previously undescribed inner ear pathology associated with conductive hearing loss, and is of some importance in our interpretation of new data on how bone conduction stimuli reach the inner ear. The rest of this chapter looks into each of the eight areas in more detail, though not exhaustively. Readers interested in more detailed information on the middle ear are directed toward the cited papers or recent reviews for more details (Puria et al., 2013).

26.2 An Overview of Middle Ear Sound Conduction

Although middle ear function was described in many previous reviews (e.g., Rosowski, 1994; Puria et al., 2013), let's spend some time discussing the basics. The middle ear transfers sound energy from the external ear to the auditory inner ear (Fig. 26.1). We usually describe this transmission by comparing the middle ear output with some input to enable analyses of transfer functions. A common middle ear input is the sound pressure P_{EC} within the ear canal just lateral to the TM. ($P_{EC}(f)$ is a complex variable that describes the magnitude and phase of the sinusoidal sound pressure measured at a particular frequency f.) Commonly measured middle ear outputs are the sound pressure within the vestibule of the inner ear P_V or the sound-induced velocity of the stapes V_S. ($P_V(f)$ and $V_S(f)$ are also complex numbers that depend on the frequency of the sound.) The ratio of the frequency-dependent complex amplitudes of an output and input measured at many frequencies defines a transfer function. The ratio of the magnitude of the output and input sinusoids is the magnitude of the transfer function; the difference between the phase of the output and input is the phase of the transfer function. A measurement of the transfer function describing sound pressure transmission through the middle ear of the chinchilla is plotted in several formats in Fig. 26.2.

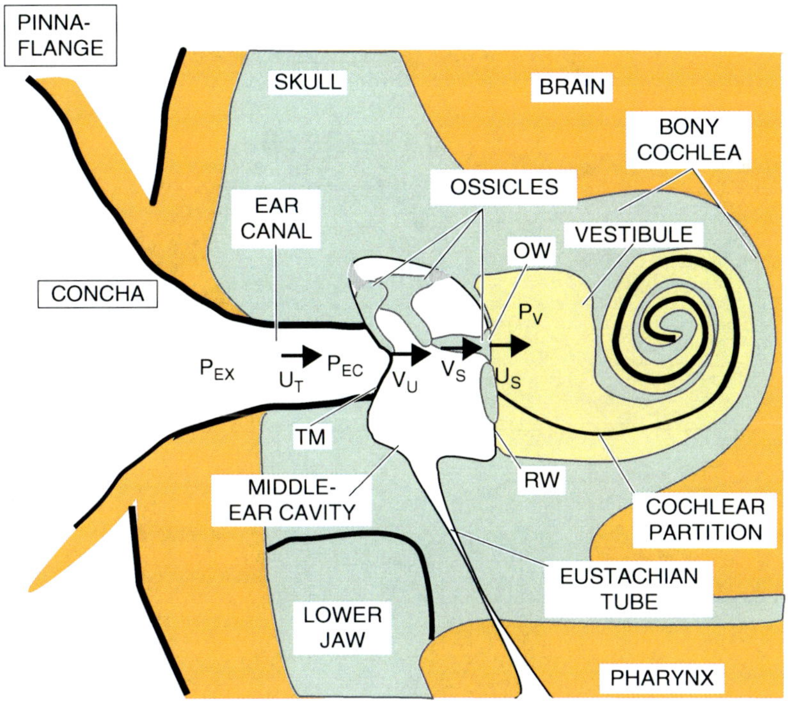

Fig. 26.1 A schematic of the mammalian auditory periphery. From left to right, the three ossicles are the malleus, incus and stapes. Superimposed on the anatomical figure are sound pressures measured in three locations: At the entrance to the ear canal, P_{EX}; at the termination of the ear canal near the TM, P_{EC}; and within the fluids of the vestibule of the inner ear, P_V. Two volume velocities are noted: the volume velocity of the tympanic membrane, U_T, and of the stapes footplate, U_S. The two labeled mechanical velocities are V_U, the sound-induced velocity of the umbo—the tip of malleus embedded near the center of the TM, and V_S the sound-induced velocity of the stapes. The figure is based on another in Rosowski (1994)

26.3 New Developments in Middle Ear Function

26.3.1 New High-Resolution Descriptions of Ossicular Structure

Since 1994 there have been significant improvements in imaging technology that have produced much more accurate descriptions of the individual ossicles from a number of species. More traditional imaging methods such as serial histological sections have been used to construct three-dimensional computer images of the ossicles of a few ears of a number of species (Funnell et al., 1992; Koike et al., 2002; Sun et al., 2002). More recently high-resolution computed tomography (CT) imaging, with and without the use of dyes to help label the soft connecting tissues, have been used to describe ossicular structure for comparative as well as modeling purposes (Sim & Puria, 2008; Lee et al., 2010; Lavender et al., 2011; Buytaert et al., 2012; Salih et al., 2012).

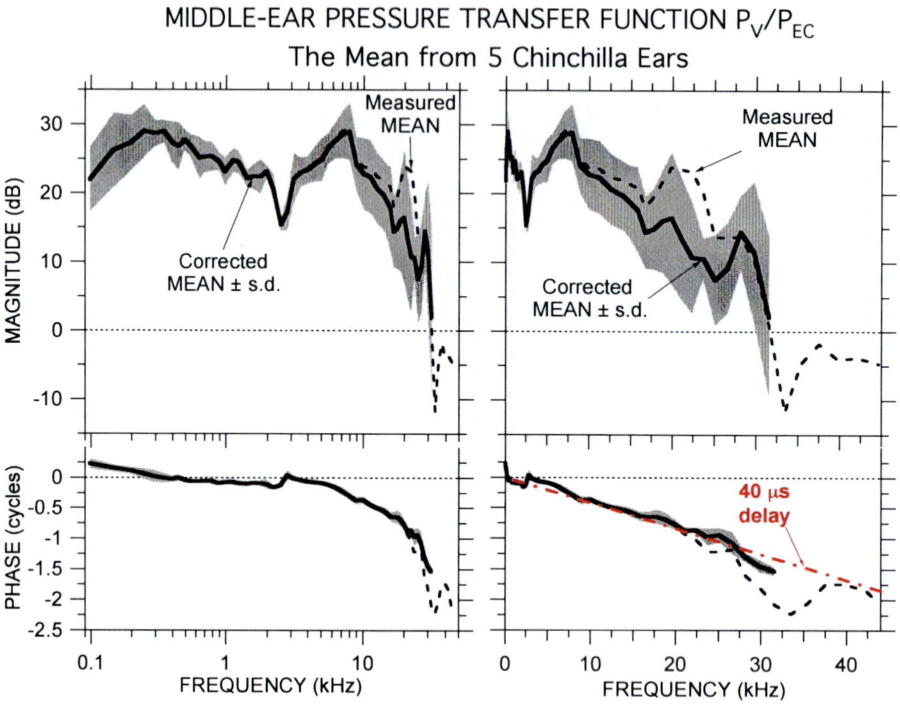

Fig. 26.2 An example of a middle ear transfer function, the ratio of sound pressure in the cochlear vestibule, P_V, to sound pressure at the TM, P_{EC}, measured in 5 chinchillas (data from Ravicz & Rosowski, 2013). The left side shows the magnitude (top) and phase angle (bottom) of the transfer function plotted against a logarithmic frequency scale, which is roughly consistent with the coding of sound along the basilar membrane. The right side shows the same magnitude and phase plotted against a linear frequency scale, which emphasizes the high-frequency response of the ear at the expense of the response to sound frequencies less than 2 kHz. The linear frequency scale does allow estimation of middle ear delay from the slope of the phase vs. frequency function. The prominent magnitude notch seen near 2 kHz on both the right and left is caused by an open hole placed in the middle ear wall, which was needed to place the pressure transducers within the ear. The linear phase gradient observed at frequencies between 5 and 25 kHz is consistent with a 40-μs transmission delay through the middle ear

Such methods have also been used to help "anatomically register" physiological measurements in individual specimens to better describe the three-dimensional motions of the ossicles (Decraemer et al., 2002; Decraemer & Khanna, 2004).

26.3.2 Factors That Control the Bandwidth of the Middle Ear's Response to Sound

In the past 20 years new middle ear transfer-function measurements (e.g., Olson, 1998; Overstreet & Ruggero, 2002; Ravicz et al., 2008, 2010) and a reinterpretation of older data (Ruggero & Temchin, 2002) have led to conflicting views on the frequency range (or bandwidth) over which the magnitude of the middle ear transfer function remains at a high level. The generally accepted viewpoint before 1994 was that the transfer function of the middle ear was band limited (allowing efficient transfer of sound signals in the mid-frequencies, but limiting transfer at low and high frequencies) and was a major contributor to the bandwidth of the hearing response in different mammals. One form of evidence for this view is that animals such as toads are sensitive to sounds of frequencies that are greatly different from the frequencies of sounds heard by mice, and that measurements of the bandwidth of middle ear transfer functions in these different species covaried with the species' range of "audible" frequencies (e.g., Zwislocki, 1965; Dallos, 1973; Rosowski, 2003). Beginning in the late 1990s, newer measurements of middle ear sound transfer in gerbil (*Meriones unguiculatus*: Olson, 1998; Overstreet & Ruggero, 2002) suggested that although there was some correlation between the frequency dependence of the middle ear and hearing sensitivity at frequencies less than a few kilohertz, the response of the middle ear at higher frequencies surpassed that suggested by the audibility curve. This point was generalized to other species by Ruggero and Temchin (2002), who reanalyzed data available in the literature. To complicate the argument further, some of the newest measurements of middle ear transfer functions in gerbil (Ravicz et al., 2008) and chinchilla (*Chinchilla laniger*: Ravicz et al., 2010) show a roll-off in at higher frequencies that is in the same range as the roll-off of hearing sensitivity in these animals, whereas others do not show such a clear high-frequency limit in mouse (Dong et al., 2013).

The controversy has pointed out several significant factors that need consideration in future discussions of this topic.

1. As pointed out by Ruggero and Temchin (2002), one cannot ignore the restrictions placed by the frequency dependence of the inner ear on the high-frequency hearing sensitivity. Just as there are significant correlations between middle ear structure and the limits of the hearing range (Rosowski, 1992, 1994), there is a significant correlation between the high end of the tonotopic cochlear map and high-frequency hearing limits.
2. A significant complication in using measurements of ossicular velocity to define middle ear sound transfer is that most of these measurements only quantify ossicular motion along one direction, and that direction is often not in line with the direction of stapes piston-like motion. Such one-dimensional measurements can be affected by any motion that gives rise to a motion component in the measurement direction, as do many of the complex ossicular motions that occur in response to sound frequencies above 2–5 kHz (Decraemer & Khanna, 1995, 2004; Heiland et al., 1999; Hato et al., 2003); therefore, the presence of such complex motions greatly increases the difficulty of attempts to quantify the piston-like translation of the stapes in the middle- and high-frequency range from one-dimensional measurements (Voss et al., 2000; Sim et al., 2010; Lauxmann et al., 2012). A possible solution is the use of inner ear sound pressure measurements, which are thought to quantify the stimulus to the inner ear regardless of the complexities in three dimensional ossicular motion, such as done by Decraemer et al. (2007) and de La Rochefoucauld et al. (2008). However, see Section 3.4 on the possible contribution of complex ossicular motions to sound transmission.

3. Another complication in describing the middle ear transfer function at frequencies above 10 kHz is the quantification of the stimulus sound pressure in the ear canal at the tympanic membrane (Pearce et al., 2001; Overstreet & Ruggero, 2002; Ravicz et al., 2007; Ravicz & Rosowski, 2012). The small dimensions of the ear canal, the possible effect of the measurement device itself, and the small wavelengths of sound at the high-frequency end of the hearing range in many animals makes quantifying the input sound pressure a challenge, and indeed some of the differences in high-frequency response reported in the literature are directly related to differences in the estimation of the sound input (Overstreet & Ruggero, 2002; Ravicz & Rosowski, 2012; Dong et al., 2013). To investigate this issue and decrease the uncertainty in estimates of the stimulus presented to the middle ear, Neely and co-workers have proposed using measurements of sound power flow into the external ear to describe the sound input to the middle ear (Neely & Gorga, 1998; Lewis et al., 2009; McCreery et al., 2009).

Returning to the topic of factors that influence the bandwidth of the middle ear, recent work has concentrated on the influence of the ossicular joints and the distribution of mass within the ossicular system. The three ossicles (the malleus, incus, and stapes) are usually joined by two ligamentous joints—the induco-malleolar (IM) and the incudo-stapedial (IS). In many mammals these joints appear as synovial joints made up of a fluid-filled capsule surrounded by cartilage and other fibrous tissues (Henson, 1974; Mason & Farr, 2012). There is a wide variety in the flexibility and relative size of the joints: In humans the IM joint is relatively large and highly flexible (Willi et al., 2002; Nakajima et al., 2005); in chinchillas the IM joint is nearly vestigial and the incus and malleus are rigidly connected (Puria & Steele 2010; Mason & Farr, 2012). There are several measurements that suggest that the slippage within the IM joint limits the bandwidth of middle ear sound transfer (e.g., Guinan & Peake, 1967). One of the more recent of these studies, performed in human temporal bones, demonstrates that flexibility in the IM joint introduces a decrease in sound-induced stapes and incus motion at frequencies above 2 kHz, and also reduces middle ear output at lower frequencies (Willi et al., 2002). The reduction in response to low-frequency sound stimuli may be a byproduct of the IM joint's role in protecting the human inner ear from large variations in static pressure between the middle ear and the environment (Hüttenbrink, 1988), where humans, with their good low-frequency hearing, may be especially susceptible to such low-frequency pressure variations.

Another view of the possible effect of flexible (or compliant) ossicular joints was espoused by Puria and Allen (1998), who noted that the joints separated the ossicular chain into a series of elements whose mechanics depended on the mass of the bone and the compliance of the joints, where each mass element is separated from the others by the spring-like compliant joints. Such an arrangement is equivalent to a mechanical transmission line, where the transfer of stimulus energy along the lines depends on the magnitudes of the springs and masses. Further, it is theoretically possible to select combinations of masses and springs to "match" the impedances distributed along the line, where matching leads to an optimized transfer of energy over a broad frequency range. Such matching is associated with a wide-band frequency response and a frequency-independent transmission delay. Measurements of middle ear sound transfer at high frequencies that demonstrate good high-frequency response support this point of view, as do the presence of middle ear transmission delays (Fig. 26.2) that dominate the phase of the transfer function at frequencies above a few kilohertz (Puria et al. 1997; Olson, 1998; Overstreet & Ruggero, 2002; O'Connor & Puria, 2008; Ravicz et al., 2008; Nakajima et al., 2009) although some of these measurements do describe a high-frequency limit above which the middle ear transfer function seems to decline in magnitude (Fig. 26.2). The presence of a high-frequency roll-off in middle ear function has been observed repeatedly in cadaveric human middle ears, and Ruggero and Temchin (2003) suggested the roll-off was a result of postmortem artifact. However, comparisons of stapes-velocity measurements in cadaveric and live human ears undergoing cochlear implant surgery show similar frequency response after taking into account significant differences in the angulation of the laser to the stapes that results from differences in the surgical approach to the stapes in live and cadaveric middle ears (Chien et al., 2009). The methodological corrections (Chien et al., 2006) also take into account the effect of the three-dimensional motion of the stapes on the one-dimensional measurements.

26.3.3 How the Tympanic Membrane Couples Sound to the Ossicular Chain

New ideas and data have been used to describe how the tympanic membrane (TM) is set in motion by sound and how those motions are coupled to the ossicular chain. Some of these ideas are strongly related to the matched transmission-line and delay models of Puria and Allen (1998) and have led to new transmission line–based models of TM function that incorporate transverse surface waves (much like ocean waves) that travel on the TM from the periphery to the center of the TM as a means of coupling sound energy in air to the ossicular chain (Parent & Allen, 2007, 2010; Goll & Dalhoff, 2011). These transmission line ideas ignore the more traditional mechanics-based plate and membrane models of how two-dimensional surfaces are set into motion by sound stimuli, which dominated the field since the 1970s (Tonndorf & Khanna, 1970; Shaw & Stinson, 1983; Fletcher, 1992; Fay et al., 2006). Both the transmission line models and the plate and membrane models can predict the presence of standing waves—with their regions of large motion separated by regions of little motion—on the TM surface, where such waves have been observed (Khanna & Tonndorf, 1972; Tonndorf & Khanna, 1972; Fay et al., 2005) or intuited from other measurements (Puria & Allen, 1998). However, there is a significant difference in how the standing waves arise in the two models.

The transmission line models, by their very nature, assume that sound energy is delivered to one end of the line and travels to the other end of the line as a traveling wave; partial reflections from the other end of the line produce a wave of motion that travels in the opposite direction, and the two traveling waves interact to produce a standing wave-like component of motion. The same types of wave motion are predicted by the string model of Goll and Dalhoff (2011), though it assumes a graded stimulus to each membrane string. In the plate/membrane models, it is assumed that the entire TM surface is excited uniformly and simultaneously by the sound stimulus in the ear canal. Such stimulation produces "standing-wave-like" two-dimensional modal motion patterns without the generation of traveling waves; however, because of damping (energy losses) within the TM these patterns can exhibit spatial phase gradients similar to those of traveling-waves (Fletcher, 1992; Rosowski et al., 2011; Cheng et al., 2013). An important distinction between the models is that transmission line models generally support only one motion wavelength for any given frequency of stimulation, whereas modal responses are best described by the sum of motions of varied wavelengths (Fletcher, 1992).

New measurements of the sound-induced motion of the TM performed with high spatial density (Decraemer et al., 1999; Cheng et al., 2010; de La Rochefoucauld & Olson, 2010; Rosowski et al., 2011; Cheng et al., 2013), as in Fig. 26.3, are consistent with the presence of dominant low-order (large-wavelength) standing-wave–like motions in which much of the membrane moves back and forth in phase in combination with smaller amplitude higher order (small-wavelength) motions that may result from damped higher order modal motions or smaller wavelength waves that travel on the TM surface in response to sound frequencies above 1–2 kHz. There are also data consistent with the presence of longitudinal (in-plane) waves that travel within the TM (Jackson et al., 2012; Rosowski et al., 2013). What is still largely unknown is how each of the observed motion patterns (modal motions, transverse waves traveling on the TM surface, and longitudinal motions traveling within the membrane) induces motions of the ossicular chain. One extreme hypothesis is that the low-order large wavelength modal motions approximate piston-like motions of the surface that directly drive the malleus (de La Rochefoucauld & Olson, 2010; Rosowski et al., 2011); however, this hypothesis does not readily account for the observations of delay between sound pressure and ossicular motion (Puria & Allen, 1998). At the other extreme, the transmission line and string models are not consistent with the relatively small magnitudes of the traveling waves observed on the TM surface at most frequencies (Rosowski et al., 2011; Cheng et al., 2013).

26.3.4 The Contribution of Complex Ossicular Motions to Sound Transmission

As mentioned in Section 3.1, multiple investigations have measured the three-dimensional motion of the ossicular chain of various mammals in attempts to describe more precisely the role of individual ossicles and of different motion components in middle ear function (Decraemer & Khanna, 1995, 2004; Heiland et al., 1999; Hato et al., 2003). These studies point out that the simple model of rigid ossicles rotating about a fixed axis defined by the ossicular ligaments does not account for observations of more complex ossicular motions such as bending and

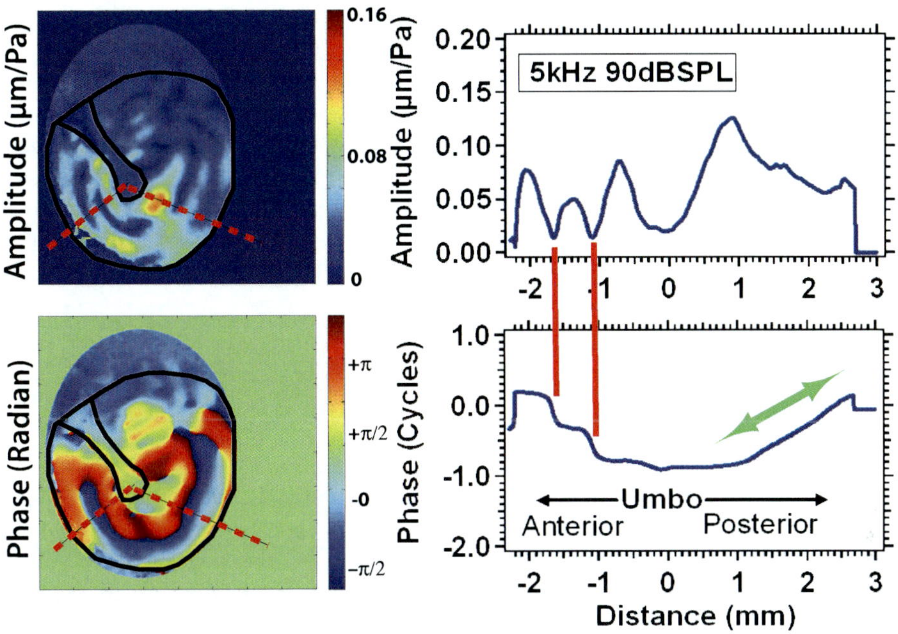

Fig. 26.3 Magnitude and phase of the motion on the surface of the TM measured in a human temporal bone. The stimulus is a 90 dB SPL 5 kHz tone. The measurements were made using computer-aided opto-electronic holography (Cheng et al., 2013). On the left are two surface maps that illustrate the magnitude (upper) and phase (lower) of TM surface motion at each of more than 200,000 points on the TM. The black outlines show the limits of the TM area and the position of the manubrium of the malleus within the TM. The dashed red lines on the map show two radii that radiate from the umbo (the tip of the manubrium) toward the edge of the TM: The radius on the left is directed anteriorly from the umbo; the radius on the right is directed posteriorly and inferiorly from the umbo. In the right two panels, we see the measured magnitude (upper plot) and angle (lower plot) of TM displacement along the two radial lines illustrated on the left. The umbo is at position 0. Negative numbers code the distance from the umbo along the anterior radius. Positive numbers code the distance from the umbo along the posterior–inferior radius. The vertical red lines point to locations consistent with modal nodes in the motion pattern where the magnitude is at a minimum and where the phase of the displacement changes rapidly by half a cycle. The green arrow labels a region where the phase is changing gradually, which is suggestive of traveling waves, but could also be a result of losses within the TM (Funnell et al., 1987; Aernouts, 2012)

twisting of the ossicles (Decraemer et al., 1991; Decraemer & Khanna, 2004; Homma et al., 2009; Puria & Steele, 2010). Although several studies have demonstrated sound-induced bending of the manubrium of the malleus in different mammalian species (e.g., de La Rochefoucauld & Olson [2010] in gerbils), the significance of manubrium bending is unclear. Simple lumped models of the ear predict bending leads to a decrease in middle ear output (e.g., Zwislocki, 1962); ossicular transmission line advocates could argue that such bending adds another set of distributed ossicular masses and stiffness to the ossicular chain that could help regulate the flatness of the middle ear response (e.g., Puria & Allen, 1998).

Another area of disagreement concerns the significance of the complex components of stapes motion: Some data suggest the complex twisting motions of the stapes have little effect on the relatively smooth frequency dependence of the middle ear (Decraemer et al., 2007), and other data indicates that more complex motions such as rocking of the stapes footplate can actually stimulate the cochlea (Huber et al., 2008; Sim et al., 2010; Eiber et al., 2012). The resolution of this conflict is still to come, though a significant complication is how one precisely quantifies three-dimensional motions in structures with irregular shapes such as the mammalian stapes (Lauxmann et al., 2012).

26.3.5 The Reverse Coupling of Sound Generated in the Inner Ear to the External Ear

A significant area of research in the last 20 years concerns the transmission of sound produced within the cochlea to the external ear, where this "reverse" middle ear function is an integral part of the process that generates measureable oto-acoustic emissions within the ear canal of patients and subjects. As of now, reverse middle ear transmission has been quantified in guinea pig (*Cavia procellus*: Magnan et al., 1997), cat (*Felis catus*: Voss & Shera, 2004), gerbil (Dong & Olson, 2006; Dong et al., 2012), and human temporal bone (Puria, 2003), and has been estimated in live humans (Keefe, 2002). One of the findings of these studies includes the dependence of the reverse transfer function on whether the entrance of the external ear is opened or plugged. Specifically, the impedance looking out the ear canal from the tympanic membrane influences the magnitude of the oto-acoustic emissions measured there, where larger magnitude emissions are observed with higher terminating impedance magnitudes. Another finding is that the frequency dependence of reverse sound transmission through the middle ear can vary significantly from the frequency dependence of forward transmission. This difference in transmission is consistent with the property of "reciprocity" that constrains sound transfer through passive, linear, and time invariant acoustical and mechanical systems (Shera & Zweig, 1992a), and reflects the differences in the acoustic impedance that terminates the acoustical-mechanical middle ear in forward (where the impedance of the inner ear is the termination) and reverse transmission (where the impedance looking out the external ear terminates the middle ear).

26.3.6 The Question of Inner Ear "Third Windows"

An area of study that is relevant to basic understanding of hearing mechanisms and is also highly clinically relevant is the accuracy of the two-window noncompressible cochlea model of sound stimulation of the inner ear. That model builds on the concept that stimulation of the inner ear depends on the production of a difference in sound pressure across the cochlear partition, and notes that given the assumption of incompressible cochlear fluid and incompressible walls the trans-partition pressure difference is simply related to a difference in sound pressure external to the two cochlear windows. This model is consistent with comparisons that suggest the sound-induced volume displacements of the stapes footplate and round window membrane are equal and opposite in phase (an inward stapes displacement is accompanied by a simultaneous outward round-window displacement of near equal volume) during normal ossicular stimulation of the inner ear (e.g., Kringlebotn, 1995; Stenfelt et al., 2004). However, the strongest support for the incompressible model comes from data demonstrating great reductions in the response of the inner when it is presented simultaneously with sound pressures of equal magnitude and opposite phase at the oval and round windows (Wever & Lawrence, 1950; Voss et al., 1996).

Newer clinical data suggesting direct mechanical stimulation of the round window is useful in the treatment of conductive pathologies that include stapes fixation (e.g., Colletti et al., 2006, 2010; Beltrame et al., 2009) go against the two-window model. True stapes fixation with an entirely incompressible inner ear would lead to a nonexistent sound pressure difference across the cochlear partition (and little stimulation of the organ of Corti) in response to round-window stimulation. The new clinical data, therefore, suggest the presence of either a compressible cochlea (Shera & Zweig, 1992b) or additional "windows" by which sound can leave the inner ear (Ranke et al., 1951; Tonndorf & Tabor, 1962). Whether these newer observations can be explained in terms of the older data is a new point of study (e.g., Lupo et al., 2012).

Some direct evidence for cochlear compressibility or an additional window into the normal ear exists. Stenfelt et al. (2004) demonstrated that the volume displaced by motion of the round and oval window in a temporal bone were not equal with bone conduction stimulation, and Stieger et al. (2013) presented data, obtained by a combination of stapes velocity and intracochlear pressure measurements, consistent with a normal "third" window on the vestibular side of the inner ear. The presence of such windows might still be consistent with the stimulus cancellation data of Wever and Lawrence (1950) and Voss and colleagues (1996). The critical factor in this argument is the relative impedance looking "out" the third window from the inner ear and looking "into" the cochlea from the vestibule. If the "third" window impedance is high enough, it would have a small effect on the pressure difference during normal ossicular stimulation, but could produce a large enough cochlear compressibility to benefit stimulation of the inner ear from the round window in cases of stapes fixation.

Although the evidence for normal inner ear "third" windows is still in the evaluation stage, there is significant clinical and basic science evidence that abnormal inner ear third windows can result from pathology of the bone surrounding the inner ear, and that these pathologic windows can produce a large (30–50 dB) conductive hearing loss (Minor, 2000; Mikulec et al., 2004; Limb et al., 2006; Songer & Rosowski, 2010; Pisano et al., 2012).

26.3.7 New Ways to Diagnose Conductive Hearing Loss

Differentiating the cause of conductive hearing loss can be problematic, especially in cases of ossicular disorders where the TM is intact and the middle ear air spaces are aerated (Rosowski et al., 2008). Therefore, the application of acoustic and mechanical measurements of middle ear function to clinical diagnosis of different disorders continues to be a point of study. Earlier efforts of using single-frequency impedance measurements at atmospheric or varied static ear canal pressures (tympanometry; Jerger, 1975; Margolis & Hunter, 1999) are being supplemented by newer "wideband" measurements of ossicular velocity (Rosowski et al., 2008) or ear canal impedance and reflectance that have shown increased sensitivity and selectivity to various middle ear disorders (Feeney et al., 2003, 2009; Allen et al., 2005; Shahnaz et al., 2009; Nakajima et al., 2012). Wideband ear canal reflectance/admittance measurements have also shown promise in the screening of infants and young children for the conductive hearing loss associated with middle ear effusions (Keefe et al., 2000; Hunter et al., 2010; Prieve et al., 2013). Significantly, although these tests show promise at detecting specific pathologies, none can quantify the degree of conductive hearing loss. This lack of quantitative ability stems from the basic limitation that all of these tests measure the motion of the tympanic membrane, and it is the sound-induced motion of the stapes that is most related to the efficiency of sound conduction to the ear. Several groups (Subhash et al., 2012; Chang et al., 2013) are working on a technology based on optical coherence tomography (OCT) that will allow measurements of the motion of the ossicles through an intact TM, and thereby allow better objective estimation of the magnitude of conductive hearing loss with a single test.

26.3.8 New Data and Models of Bone Conduction Stimulation of the Inner Ear

The final area we discuss are investigations of the mechanisms that couple acoustic-frequency vibrations into the inner ear, or what is generally called bone conduction. The rather intense interest in bone conduction mechanisms is motivated by a number of practical and theoretical questions: (1) What do clinical bone conduction measurements tell us about the working of the middle and inner ear, in particular: Why do pathological third windows often lead to an increased sensitivity to vibration? and Why does fixation of the stapes lead to a decrease in sensitivity? (2) How can we increase the benefit of bone conduction hearing aids to the hearing impaired? (3) How can we limit the intense vibratory stimulation of the inner ear caused by modern heavy equipment and weaponry? These questions have been addressed in several different manners. One research group has investigated how mammals specialized for the reception of bone-conducted stimuli use substrate vibrations to help them communicate and feed (Narins et al., 1997; Mason & Narins, 2002; Mason, 2003).

Another set of workers, primarily Stenfelt and his co-workers, has been investigating how clinical bone conduction stimuli stimulate the inner ear, and have refined our knowledge of the contributions of the multiple previously described vibratory paths to the hearing sensation produced by direct vibration of the skull (Stenfelt & Goode, 2005; Stenfelt, 2011). A third group (Sohmer and colleagues) has investigated the possibility that vibrations at normal acoustic frequencies reach the inner ear via fluid-filled passages (normal third windows) that connect the soft tissues and fluids of the brain to the inner ear (Sohmer et al., 2000; Sohmer & Freeman, 2004; Perez et al., 2011). A fourth area of work has been to develop accurate models of the conduction of bone vibrations to the inner ear, for example, Kim et al. (2011), which include body structures in an attempt to understand how vibration of the whole body affects the sensory structures within the inner ear.

26.4 The Next Twenty Years

Based on the recent work summarized above, the last 20 years have seen continued intense study of the processes by which sound is conducted to the inner ear. This activity is evidence that the common perception that these processes are well described is not accurate. One of the factors driving these studies is that middle ear surgery continues to be performed at high rates and that the treatment of conductive hearing loss is still a clinical challenge. New ideas regarding the basic function of the middle ear (e.g., Ruggero & Temchin, 2002) have arisen and are still contending with the older ideas for acceptance. We have learned more about the important features of ossicular form and how they move in response to sound, though the consequences of much of that knowledge is yet to be determined. In particular, how the middle ear's motion transduces high-frequency sound to the inner ear continues to be a challenging question. Better understanding of how sound is transduced to the inner ear by various means is paramount in understanding the consequences of the wide range in middle ear structure among animals, and in helping the many patients that require efficient and accurate diagnosis and treatment of conductive and mixed hearing losses.

The increased application of more precise imaging techniques (high-intensity CT and OCT) to quantify middle ear structure and function in live animals and patients will be one of the major thrusts of the next 20 years. Another area of major study will be the melding of measurements of wide-area TM and ossicular geometry and sound-induced motions with comprehensive mechanical models. This combination will elucidate the factors that influence the coupling of sound from the ear canal to the inner ear and will lead to improvements in the relatively poor surgical results observed after more complex middle ear reconstructions (Merchant et al., 2003). A third area of future achievement will be to improve the understanding of the mechanisms of sound conduction within the inner ear, both in terms of how bone-conducted sound stimulates the inner ear and the mechanics of, and treatment for, "inner ear" conductive disorders such as fistulas and canal dehiscences.

Technological advances in imaging and measurements of motion and pressure, combined with increased knowledge of middle and inner ear mechanics, promise major advances in our understanding of sound transmission to the inner ear.

Acknowledgments It is necessary for us to acknowledge the great debt we owe to our late co-worker Saumil N. Merchant, MD. Saumil was an important part of our lab group during the 20 years covered by this perspective. He (along with Mike Ravicz) performed some of the significant experiments that came out of our laboratory during that time. He was also our fountainhead of knowledge regarding things clinical and pathological. Saumil's leadership, scientific and clinical acumen, and wonderful personality will be sorely missed by all of us and by those who work in the fields to which he contributed. We also thank Bill Peake, Mike Ravicz, Melissa Wood McKinnon, and all of the others who have contributed to our thinking on sound conduction over the years, and acknowledge the continuous funding from the National Institute on Deafness and other Communication Disorders (NIDCD) that made our work possible.

References

Aernouts, J. (2012). *Mechanical properties of the tympanic membrane: Measurement and modeling.*, Ph. D. thesis, Department of Physics, University of Antwerp, Belgium.

Allen, J. B., Jeng, P. S., & Levitt, H. (2005). Evaluation of human middle ear function via an acoustic power assessment. *Journal of Rehabilitation Research and Development*, 42, 63–78.

Beltrame, A. M., Martini, A., Prosser, S., Giarbini, N., & Streitberger, C. (2009). Coupling the Vibrant Soundbridge to cochlea round window: Auditory results in patients with mixed hearing loss. *Otology & Neurotology*, 30, 194–201.

Buytaert, J. A. N., Salih, W. H. M., Dierick, M., Jacobs, P., & Dirckx, J. J. J. (2012). Realistic 3D computer model of the gerbil middle ear, featuring accurate morphology of bone and soft tissue structures. *Journal of the Association for Research in Otolaryngology*, 12, 681–696.

Chang, E.W., Cheng, J.T., Röösli, C., Kobler, J.B., Rosowski, J.J., & Yun, S.H. (2013). Simultaneous 3D imaging of sound-induced motions of the tympanic membrane and middle ear ossicles. *Hearing Research*, 304, 49–56.

Cheng, J. T., Aarnisalo, A. A., Harrington, E., Hernandez-Montes, M. D. S., Furlong, C., Merchant, S. N., & Rosowski, J. J. (2010). Motion of the surface of the human tympanic membrane measured with stroboscopic holography. *Hearing Research*, 263, 66–77.

Cheng, J. T., Hamade, M., Harrington, E., Furlong, C., Merchant, S. N., & Rosowski, J. J. (2013). Wave motion on the surface of the human tympanic membrane: Holographic measurement and modeling analysis. *Journal of the Acoustical Society of America*, 133, 918–937.

Chien, W., Ravicz, M. E, Merchant, S. N., & Rosowski, J. J. (2006). The effect of methodological differences in the measurement of stapes motion in live and cadaver ears. *Audiology & Neurotology*, 11, 183–197.

Chien, W., Rosowski, J. J., Ravicz, M. E., Rauch, S. D., Smullen, J., & Merchant, S. N. (2009). Measurements of stapes velocity in live human ears. *Hearing Research*, 249, 54–61.

Colletti, V., Soli, S. D., Carner, M., & Colletti, L. (2006). Treatment of mixed hearing loss via implantation of a vibratory transducer on the round window. *International Journal of Audiology*, 45, 600–608.

Colletti, L., Carner, M., Mandalà, M., Veronese, S., & Colletti, V. (2010). The floating mass transducer for external auditory canal and middle ear malformations. *Otology & Neurotology*, 32, 108–115.

Dallos, P. (1973). *The auditory periphery*. New York: Academic Press.

Decraemer, W., & Khanna S. (1995). Malleus vibration modeled as rigid body motion. *Acta Oto-Rhino-Laryngologica Belgica*, 49, 139–145.

Decraemer, W. F, & Khanna, S. M. (2004). Measurement, visualization and quantitative analysis of complete three-dimensional kinematical data sets of human and cat middle ear. In K. Gyo, H. Wada, N. Hato, & T. Koike (Eds.), *Middle ear mechanics in research and otology* (pp. 3–10). Singapore: World Scientific.

Decraemer, W. F., Khanna, S. M, & Funnell, W. R. J. (1991). Malleus vibration mode changes with frequency. *Hearing Research*, 54, 305–318.

Decraemer, W. F., Khanna, S. M., & Funnell, W. R. J. (1999). Vibrations at a fine grid of points on the cat tympanic membrane measured with a heterodyne interferometer. Presented at *EOS/Spie International Symposia on Industrial Lasers and Inspection, Conference on Biomedical Laser and Metrology and Applications*, Munich, Germany.

Decraemer, W. F. S., Khanna, S. M., & Dirckx, J. J. J. (2002). The integration of detailed 3-dimensional anatomical data for the quantitative description of 3-dimensional vibration of a biological structure: An illustration from the middle ear. *SPIE Proceedings*, 4827, 148–158.

Decraemer, W. F., de La Rochefoucauld, O., Dong, W., Khanna, S. M., Dirckx, J. J., & Olson, E. S. (2007). Scala vestibuli pressure and three-dimensional stapes velocity measured in direct succession in gerbil. *Journal of the Acoustical Society of America*, 121, 2774–2791.

de La Rochefoucauld, O., & Olson, E. S. (2010). A sum of simple and complex motions on the eardrum and manubrium in gerbil. *Hearing Research*, 263, 9–15.

de La Rochefoucauld, O., Decraemer, W. F., Khanna, S. M., & Olson, E. S. (2008). Simultaneous measurements of ossicular velocity and intracochlear pressure leading to the cochlear input impedance in gerbil. *Journal of the Association for Research in Otolaryngology*, 9, 161–177.

Dong, W., & Olson, E. S. (2006). Middle ear forward and reverse transfer function. *Journal of Neurophysiology*, 95, 2951–2961.

Dong, W., Decraemer, W. F., & Olson, E. S. (2012). Reverse transmission along the ossicular chain in gerbil. *Journal of the Association for Research in Otolaryngology*, 13, 447–459.

Dong, W., Varavva, P., & Olson, E. S. (2013). Sound transmission along the ossicular chain in common wild-type laboratory mice. *Hearing Research*, 301, 27–34

Eiber, A., Huber, A., Lauxmann, M., Chatzimichalis, M., Sequeira, D., & Sim, J. H. (2012). Contribution of complex stapes motion to cochlea activation. *Hearing Research*, 284, 82–92.

Fay, J., Puria, S., Decraemer, W. F., & Steele, C. (2005). Three approaches for estimating the elastic modulus of the tympanic membrane. *Journal of Biomechanics*, 38, 1807–1815.

Fay, J. P., Puria, S., & Steele, C. R. (2006). The discordant eardrum. *PNAS*, 103, 19743–19748.

Feeney, M. P., Grant, I. L., & Marryott, L. P. (2003). Wideband energy reflectance in adults with middle-ear disorders. *Journal of Speech Language and Hearing Research*, 46, 901–911.

Feeney, M. P, Grant, I. L., & Mills, D. M. (2009). Wideband energy reflectance measurements of ossicular chain discontinuity and repair in human temporal bone. *Ear and Hearing*, 30, 391–400.

Fletcher. N. H (1992). *Acoustic systems in biology*. New York: Oxford University Press.

Funnell, W. R., Decraemer, W. F., & Khanna, S. M. (1987). On the damped frequency response of a finite-element model of the cat eardrum. *Journal of the Acoustical Society of America*, 81, 1851–1859.

Funnell, W. R., Khanna, S. M., & Decraemer, W. F. (1992). On the degree of rigidity of the manubrium in a finite-element model of the cat eardrum. *Journal of the Acoustical Society of America*, 91, 2082–2090.

Goll, E., & Dalhoff, E. (2011). Modeling the eardrum as a string with distributed force. *Journal of the Acoustical Society of America*, 103, 1452–1462.

Guinan, J. J., Jr., & Peake, W. T. (1967). Middle-ear characteristics of anesthetized cats. *Journal of the Acoustical Society of America*, 41, 1237–1261.

Hato, N., Stenfelt, S., & Goode, R. L. (2003). Three-dimensional stapes footplate motion in human temporal bones. *Audiology & Neuro-Otology*, 8, 140–152.

Heiland, K. E, Goode, R. L., Asai, M., & Huber, A. M. (1999). A human temporal bone study of stapes footplate movement. *American Journal of Otology*, 20, 81–86.

Henson, O. W., Jr. (1974). Comparative anatomy of the middle ear. In W. D. Keidel & W. D. Neff (Eds.), *Handbook of sensory physiology: The auditory system* (pp. 39–110). New York: Springer-Verlag.

Homma, K., Du, Y., Shimizu, Y., & Puria, S. (2009). Ossicular resonance modes of the human middle ear for bone and air conduction. *Journal of the Acoustical Society of America*, 125, 968–979.

Huber, A. M., Sequeira, D., Breuninger, C., & Eiber, A. (2008). The effects of complex stapes motion on the response of the cochlea. *Otology Neurotology*, 29, 893–899.

Hunter, L. L., Feeney, M. P., Miller, J. A., Jeng, P. S., & Bohning, S. (2010). Wideband reflectance in newborns: Normative regions and relationship to hearing screening results. *Ear and Hearing*, 31, 599–610.

Hüttenbrink, K. B. (1988). The mechanics of the middle-ear at static air pressures. *Acta Oto-Laryngologica Supplementum*, 451, 1–35.

Jackson, R., Cai, H., & Puria, S. (2012). In search of in-surface modes of sound transmission on the eardrum using three dimensional laser doppler vibrometry. In S.–H. Lee & J.–H. Cho (Eds.), *MEMRO* 2012*: The 6th International Symposium on Middle-Ear Mechanics in Research and Otology*, Daegu, Korea, pp. 24–25.

Jerger, J. (1975). Diagnostic use of impedance measures. In J. Jerger (Ed.), *Clinical impedance audiometry* (pp. 149–174). New York: American Electromedics Corporation.

Keefe, D. H. (2002). Spectral shape of forward and reverse transfer functions between ear canal and cochlea estimated using DPOAE input/output functions. *Journal of the Acoustical Society of America*, 111, 249–260.

Keefe, D. H., Folsom, R. C., Gorga, M. P., Vohr, B. B., Bulen, J. C., & Norton, S. J. (2000). Identification of neonatal hearing impairment: Ear canal measurements of acoustic admittance and reflectance in neonates. *Ear and Hearing*, 21, 443–461.

Khanna, S. M., & Tonndorf, J. (1972). Tympanic membrane vibrations in cats studied by time-averaged holography. *Journal of the Acoustical Society of America*, 51, 1904–1920.

Kim, N., Homma, K., & Puria, S. (2011). Inertial bone conduction: Symmetric and anti-symmetric components. *Journal of the Association for Research in Otolaryngology*, 12, 261–279.

Koike, T., Wada, H., & Kobayashi, T. (2002). Modeling of the human middle ear using the finite-element method. *Journal of the Acoustical Society of America*, 111, 1306–1317.

Kringlebotn, M. (1995). The equality of volume displacements in the inner ear windows. *Journal of the Acoustical Society of America*, 98, 192–196.

Lauxmann, M., Eiber, A., Heckler, C., Ihrle, S., Chatzmichalis, M., Huber, A., & Sim, J. H. (2012). In-plane motions of the stapes in human ears. *Journal of the Acoustical Society of America*, 132, 3280–3291.

Lavender, D., Taraskin, S. N, & Mason, M. J. (2011). Mass distribution and rotational inertia of "microtype" and "freely mobile" middle ear ossicles in rodents. *Hearing Research*, 282, 97–107.

Lee, D. H., Chan, S., Salisbury, C., Kim, N., Salisbury, K., Puria, S., & Blevins, N. H. (2010). Reconstruction and exploration of virtual middle-ear models derived from micro-CT datasets. *Hearing Research*, 263, 198–203.

Lewis, J. D., McCreery, R. W., Neely, S. T., & Stelmachowicz, P. G. (2009). Comparison of in-situ calibration methods for quantifying input to the middle ear. *Journal of the Acoustical Society of America*, 126, 3114–3124.

Limb, C. J., Carey, J. P., Srireddy, S., & Minor, L. B. (2006). Auditory function in patients with surgically treated superior semicircular canal dehiscence. *Otology & Neurotology*, 27, 969–980.

Lupo, J. E., Koka, K., Jenkins, H. A., & Tollin, D. J. (2012). Third-window vibroplasty with an active middle ear implant: Assessment of physiological responses in a model of stapes fixation in *Chinchilla lanigera. Otology & Neurotology*, 33, 425–431.

Magnan, P., Avan, P., Dancer, A., Smurzynski, J., & Probst, R. (1997). Reverse middle-ear transfer function in the guinea pig measured with cubic difference tones. *Hearing Research*, 107, 41–45.

Margolis, R. H., & Hunter, L. L. (1999). Tympanometry. In F. Musiek & W. F. Rintlemann (Eds.), *Contemporary perspectives on hearing assessment* (pp. 89–130). Boston: Allyn & Bacon.

Mason, M. J. (2003). Bone conduction and seismic sensitivity in golden moles (Chrysochloridae). *Journal of Zoology London*, 260, 405–413.

Mason, M. J., & Narins, P. M. (2002). Seismic sensitivity in the desert golden mole (*Eremtalpa granti*): A review. *Journal of Comparative Psychology*, 116, 159–163.

Mason, M. J., & Farr, M. R. B. (2012). Flexibilty within the middle ears of vertebrates. *Journal of Laryngology & Otology*, doi: 10.1017/S0022215112002496.

McCreery, R. W., Pittman, A., Lewis, J., Neely, S. T., & Stelmachowicz, P. G. (2009). Use of forward pressure level to minimize the influence of acoustic standing waves during probe-microphone hearing-aid verification. *Journal of the Acoustical Society of America*, 126, 15–24.

Merchant, S. N., Rosowski, J. J., & McKenna, M. J. (2003). Tympanoplasty. *Operative Techniques in Otolaryngology-Head and Neck Surgery*, 14, 224–236.

Mikulec, A. A., McKenna, M. J., Ramsey, M. J., Rosowski, J. J., Herrmann, B.S., Rauch, S. D., Curtin, H. D., & Merchant, S. N. (2004). Superior semicircular canal dehiscence presenting as conductive hearing loss without vertigo. *Otology & Neurotology*, 25, 121–129.

Minor, L. B. (2000). Superior canal dehiscence syndrome. *American Journal of Otology*, 21, 9–19.

Nakajima, H. H., Ravicz, M. E., Merchant, S. N., Peake, W. T., & Rosowski, J. J. (2005). Experimental ossicular fixations and the middle-ear's response to sound: Evidence for a flexible ossicular chain. *Hearing Research*, 204, 60–77.

Nakajima, H. H., Dong, W., Olson, E. S., Merchant, S. N, Ravicz, M. E., & Rosowski, J. J. (2009). Differential intracochlear sound pressure measurements in normal human temporal bones. *Journal of the Association for Research in Otolaryngology*, 10, 23–36.

Nakajima, H. H., Pisano, D. V., Röösli, C., Hamade, M. A., Merchant, G., Mafoud, L., Halpin, C. F., Rosowski, J. J., & Merchant, S. N. (2012). Comparison of ear-canal reflectance and umbo velocity in patients with conductive hearing loss: A preliminary study. *Ear and Hearing*, 33, 35–43.

Narins, P. M., Lewis, E. R., Jarvis, J. J. U. M., & O'Riain, J. (1997). The use of seismic signals by fossorial southern African mammals: A neuroethological gold mine. *Brain Research Bulletin*, 44, 641–646.

Neely, S. T., & Gorga, M. P. (1998). Comparison between intensity and pressure as measures of sound level in the ear canal. *Journal of the Acoustical Society of America*, 104, 2925–2934.

O'Connor, K. N., & Puria, S. (2008). Middle-ear circuit model parameters based on a population of human ears. *Journal of the Acoustical Society of America*, 123, 197–211.

Olson, E. S. (1998). Observing middle and inner ear mechanics with novel intracochlear pressure sensors. *Journal of the Acoustical Society of America*, 103, 3445–3463.

Overstreet, E. H. III, & Ruggero, M. A. (2002). Development of wide-band middle-ear transmission in the Mongolian gerbil. *Journal of the Acoustical Society of America*, 111, 261–270.

Parent, P., & Allen, J. B. (2007). Wave model of the cat tympanic membrane. *Journal of the Acoustical Society of America*, 122, 918–931.

Parent, P., & Allen, J. B. (2010). Time-domain "wave" model of the human tympanic membrane. *Hearing Research*, 263, 152–167.

Pearce, M., Richter, C. P., & Cheatham, M. A. (2001). A reconsideration of sound calibration in the mouse. *Journal of Neuroscience Methods*, 106, 57–67.

Perez, R., Adelman, C., & Sohmer, H. (2011). Bone conduction activation through soft tissues following complete immobilization of the ossicular chain, stapes footplate and round window. *Hearing Research*, 280, 82–85.

Pisano, D. V., Niesten, M. E., Merchant, S. N., & Nakajima, H. H. (2012). The effect of superior semicircular canal dehiscence on intracochlear sound pressures. *Audiology & Neuro-Otology*, 17, 338–348.

Prieve, B. A., Vander Werff, K., Preston, J. L., & Georgantas, L. (2013). Identification of conductive hearing loss in young infants using tympanometry and wideband reflectance. *Ear and Hearing*, 34, 168–178.

Puria, S. (2003). Measurements of human middle ear forward and reverse acoustics: Implications for otoacoustic emissions. *Journal of the Acoustical Society of America*, 113, 2773–2789.

Puria, S., & Allen, J. B. (1998). Measurements and model of the cat middle ear: Evidence of tympanic membrane acoustic delay. *Journal of the Acoustical Society of America*, 104, 3463–3481.

Puria, S., & Steele, C. (2010). Tympanic-membrane and malleus-incus-complex co-adaptations for high-frequency hearing in mammals. *Hearing Research*, 263, 183–190.

Puria, S., Peake, W. T., & Rosowski, J. J. (1997). Sound-pressure measurements in the cochlear vestibule of human cadavers. *Journal of the Acoustical Society of America*, 101, 2745–2770.

Puria, S., Fay, R. R., & Popper, A. N. (2013). *The middle ear: Science, otosurgery and technology*. New York: Springer Science+Business Media.

Ranke, O. F, Keidel, W. D., & Weschke, H. G. (1951). [Hearing with closure of the fenestra ovalis.]. *Zeitschrift Laryngologie Rhinologie Otologie*, 31, 467–475.

Ravicz, M. E., & Rosowski, J. J. (2012). Chinchilla middle-ear admittance and sound power: High-frequency estimates and effects of inner-ear modifications. *Journal of the Acoustical Society of America*, 132, 2437–2454.

Ravicz, M. E., & Rosowski, J. J. (2013). Inner-ear sound pressures near the base of the cochlea in chinchilla: Further investigation. *Journal of the Acoustical Society of America*, 133, 2208–23.

Ravicz, M. E., Olson, E. S., & Rosowski, J. J. (2007). Sound pressure distribution and energy flow within the gerbil ear canal from 100 Hz to 80 kHz. *Journal of the Acoustical Society of America*, 122, 2154–2173.

Ravicz, M. E., Cooper, N. P., & Rosowski, J. J. (2008). Gerbil middle-ear sound transmission from 100 Hz to 60 kHz. *Journal of the Acoustical Society of America*, 124, 363–380.

Ravicz, M. E., Slama, M. C. C., & Rosowski, J. J. (2010). Middle-ear pressure gain and basilar membrane differential pressure in the chinchilla. *Hearing Research*, 263, 16–25.

Rosowski, J. J. (1992). Hearing in transitional mammals: Predictions from the middle-ear anatomy and hearing capabilities of extant mammals. In D. B. Webster, A. N. Popper, & R. R. Fay (Eds.), *The evolutionary biology of hearing* (pp. 615–631). New York: Springer-Verlag.

Rosowski, J. J. (1994). Outer and middle ear. In A. N. Popper & R. R. Fay (Eds.), *Comparative hearing: Mammals* (pp. 172–247). New York: Springer-Verlag.

Rosowski, J. J. (2003). The middle and external ears of terrestrial vertebrates as mechanical and acoustic transducers. In: F. G. Barth, J. A. C. Humphrey, & T. W. Secomb (Eds.), *Sensors and sensing in biology and engineering* (pp. 59–69). New York: Springer-Verlag.

Rosowski, J. J., Dobrev, I., Khaleghi, M., Lu, W., Cheng, J. T., Harrington, E., & Furlong, C. (2013). Measurements of three-dimensional shape and sound-induced motion of the chinchilla tympanic membrane. *Hearing Research*, 2013, 44–52.

Rosowski, J. J., Nakajima, H. H., & Merchant, S. N. (2008). Clinical utility of laser-Doppler vibrometer measurements in live normal and pathologic human ears. *Ear and Hearing*, 29, 3–19.

Rosowski, J. J., Cheng, J. T., Merchant, S. N., Harrington, E. & Furlong, C. (2011). New data on the motion of the normal and reconstructed tympanic membrane. *Otology Neurotology*, 32, 1559–1567.

Ruggero, M. A., & Temchin, A. N. (2002). The roles of the external, middle and inner ears in determining the bandwidth of hearing. *PNAS*, 99, 13206–13210.

Ruggero, M. A., & Temchin, A. N. (2003). Middle-ear transmission in humans: Wide-band, not frequency-tuned? *Acoustics Research Letters Online*, 4, 53–58.

Salih, W. H., Buytaert, J. A., Aerts, J. R., Vanderniepen, P., Dierick, M., & Dirckx, J. J. (2012). Open access high-resolution 3D morphology models of cat, gerbil, rabbit, rat and human ossicular chains. *Hearing Research*, 284, 1–5.

Shahnaz, N., Bork, K., Polka, L., Longridge, N., Bell, D., & Westerberg, B. D. (2009). Energy reflectance and tympanometry in normal and otosclerotic ears. *Ear and Hearing*, 30, 219–233.

Shaw, E. G., & Stinson, M. R. (1983). The human external and middle ear: Models and concepts. In E. deBoer & M. A. Viergever (Eds.), *Mechanics of hearing* (pp. 3–10). Delft University Press.

Shera, C. A., & Zweig, G. (1992a). Middle-ear phenomenology: The view from the three windows. *Journal of the Acoustical Society of America*, 92, 1356–1370.

Shera, C. A., & Zweig, G. (1992b). An empirical bound on the compressibility of the cochlea. *Journal of the Acoustical Society of America*, 92, 1382–1388.

Sim, J. H., & Puria, S. (2008). Soft tissue morphometry of the malleus-incus complex from micro-CT imaging. *Journal of the Association for Research in Otolaryngology*, 9, 5–21.

Sim, J. H., Chatzimichalis, M., Lauxmann, M., Röösli, C., Eiber, A., & Huber, A. M. (2010). Complex stapes motions in human ears. *Journal of the Association for Research in Otolaryngology*, 11, 329–341.

Sohmer, H., & Freeman, S. (2004). Further evidence for a fluid pathway during bone conduction auditory stimulation. *Hearing Research*, 193, 105–110.

Sohmer, H., Freeman, S., Geal-Dor, M., Adelman, C., & Savion, I. (2000). Bone conduction experiments in humans—a fluid pathway from bone to ear. *Hearing Research*, 146, 81–88.

Songer, J. E., & Rosowski, J. J. (2010). Superior semicircular canal dehiscence and air-bone gap in chinchilla: Measurements and models. *Hearing Research*, 269, 70–80.

Stenfelt, S. (2011). Acoustic and physiologic aspects of bone conduction hearing. *Advances in Oto-rhino-laryngology*, 71, 10–21.

Stenfelt, S., & Goode R. L. (2005). Bone-conducted sound: physiological and clinical aspects. *Otology & Neurotology*, 26, 1245–1261.

Stenfelt, S., Hato, N., & Goode, R. L. (2004). Fluid volume displacement at the oval and round windows with air and bone conduction stimulation. *Journal of the Acoustical Society of America*, 115, 797–812.

Stieger, C., Rosowski, J. J., & Nakajima, H. H. (2013). Comparison of forward (ear-canal) and reverse (round-window) sound stimulation of the cochlea. *Hearing Research*, 301,105–114.

Subhash, H. M., Nguyen-Huynh, A., Wang, R. K., Jacques, S. L., Choudhury, N., & Nuttall, A. L. (2012). Feasibility of spectral-domain phase-sensitive optical coherence tomography for middle ear vibrometry. *Journal of Biomedical Optics*, 17, 060505.

Sun, Q., Gan, R. Z., Chang, K. H., & Dormer, K. J. (2002). Computer-integrated finite element modeling of human middle ear. *Biomechanics and Modeling in Mechanobiology*, 1, 109–122.

Tonndorf, J., & Tabor, J. R. (1962). Closure of the cochlear windows: Its effects upon air and bone-conduction. *Annals of Otology, Rhinology & Laryngology*, 71, 5–29.

Tonndorf, J., & Khanna, S. M. (1970). The role of the tympanic membrane in middle ear transmission. *Annals of Otology, Rhinology and Laryngology*, 79, 743–753.

Tonndorf, J., & Khanna, S. M. (1972). Tympanic-membrane vibrations in human cadaver ears studied by time-averaged holography. *Journal of the Acoustical Society of America*, 52, 1221–1233.

Voss, S. E., & Shera, C. A. (2004). Simultaneous measurement of middle-ear input impedance and forward/reverse transmission in cat. *Journal of the Acoustical Society of America*, 116, 2187–2198.

Voss, S. E., Rosowski, J. J., & Peake, W. T. (1996). Is the pressure difference between the oval and round windows the effective acoustic stimulus for the cochlea? *Journal of the Acoustical Society of America*, 100, 1602–1616.

Voss, S. E., Rosowski, J. J., Merchant, S. N., & Peake, W. T. (2000). Acoustic response of the human middle ear. *Hearing Research*, 150, 43–69.

Wever, E. G., & Lawrence, M. (1950). The acoustic pathway to the cochlea. *Journal of the Acoustical Society of America*, 22, 460–467.

Willi, U. B., Ferrazzini, M. A., & Huber, A. M. (2002). The incudo-malleolar joint and sound transmission loss. *Hearing Research*, 174, 32–44.

Zwislocki, J. (1962). Analysis of the middle ear function. Part I: Input impedance. *Journal of the Acoustical Society of America*, 34, 1514–1523.

Zwislocki, J. (1965). Analysis of some auditory characteristics. In R. D. Luce, R. R. Bush, & E. Galanter (Eds.) *Handbook of mathematical psychology* (pp. 3–97). New York: John Wiley & Sons.

Chapter 27
From Degenerative Debris to Neuronal Tracing: An Anterograde View of Auditory Circuits

David K. Ryugo

David K. Ryugo

D.K. Ryugo (✉)
Garvan Institute of Medical Research, Sydney, NSW 2010, Australia
e-mail: d.ryugo@garvan.org.au

A.N. Popper and R.R. Fay (eds.), *Perspectives on Auditory Research*, Springer Handbook of Auditory Research 50, DOI 10.1007/978-1-4614-9102-6_27,

27.1 Introduction

There is an inescapable relationship between structure and function in biological systems—the composition, design, and engineering of structures are key to revealing how the interrelated parts are assembled, which in turn, inform us how they might function together. In this context, knowledge about axon pathways, their course, and their linkage to other brain structures yields clues about the processing of neural information. The configuration of pathways can suggest sequential versus parallel action, signal amplification, feedback and/or lateral inhibition, coincidence detection, and so on. In short, structure is one foundation of function.

When one thinks about understanding how the brain works, which was my goal when I started graduate school, the daunting nature of the question quickly emerges. How to begin? Where to begin? In the liver, 70–80 % of the cells are hepatocytes and hepatocytes are more or less the same; in contrast, in the brain no two neurons are the same. This bold comparison is made because each neuron, by virtue of its unique three-dimensional position in the brain, receives its own unique complement of synaptic inputs from other neurons.

The importance of how neurons are connected to other neurons becomes more evident because of this relationship. The auditory system becomes defined as those sets of neurons and axons that are directly or indirectly connected to the inner ear. Likewise, the visual system is composed of those neurons that have direct or indirect connections to the photoreceptors of the eye. And so on. How the brain processes sound or sight or smell, for example, will depend to a great extent on the nature of the connections of specific neurons to different receptors, and how various specific populations of neurons are connected to other specific populations of neurons throughout the central nervous system. Topographic maps representing the sensory epithelium yield a spatial organization for signal processing. Computational maps are calculated and generated by temporal processes many of which are yet to be unraveled.

The "wiring" pattern that underlies the different maps can inform us about the neural codes. The degree of myelination of connecting fibers will determine the speed with which signals are sent. How large are the synapses? Do they contact the cell body, dendrite, or initial segment? Are there inhibitory circuits that enter in the synaptic organization? Different kinds of connectivity patterns will influence how one thinks about response patterns in the brain when recording with a microelectrode. Over time, it became clear that functional neuroanatomy—the combined application of electrophysiological and anatomical methods—was essential in establishing a foundation for understanding brain function.

As an aside, I was taught as a graduate student that Ralph Waldo Gerard invented the microelectrode. I harbored this myth until recently. While researching for this essay, I discovered that an American woman, Dr. Ida Henrietta Hyde, was not only the first female doctoral graduate of Heidelberg University, the first female member of the American Physiology Society, and first woman researcher at Harvard Medical School, but also the actual inventor of the microelectrode (Hyde, 1921). She developed her devices while at the Physiology Laboratory at the University of Kansas as a consequence of studies begun at Harvard concerning respiratory mechanisms in skates (Hyde, 1904a) and changes in electrical potentials that accompanied maturation of developing eggs (Hyde, 1904b). The stimulating micropipette and recording microelectrode were simply tools Dr. Hyde fabricated to facilitate her research; little did she know the extent to which her invention would advance biological research. For the record, I stand in awe.

27.2 History

27.2.1 *Early Dissections*

Humans have long known about the relationship between structure and function. Early neuroanatomy involved dissections of the eye, optic nerve fibers, and brain structures of animals including humans to see how things were connected. These first pioneers of pathway tracing were Egyptians, Sumerians, Assyrians, Hindus, and Chinese who through blunt dissection found that fibers peeled away along the trajectory of travel like string cheese. Fortunately, when I was first learning anatomical methods as a graduate student, the science of pathway tracing had advanced considerably by use of microscopes. Pathways could be studied by three basic methods: chromatolysis caused by retrograde degeneration of damaged axons, the Golgi method, and silver stains that relied on neurofibril staining. I didn't have to trudge two miles barefoot through the snow to get to the lab, but self-correcting typewriters were still a thing of the future.

27.2.2 *Microscopy and Retrograde Degeneration*

When neurons are damaged and undergo cell death, the reaction as witnessed under a light microscope is the following: There is a redistribution of Nissl substance—free ribosomes and Nissl bodies represented by stacks of rough endoplasmic reticulum—to the perimeter of the cell body, the dislocation of the nucleus to an eccentric position, and a swelling of the cell body as revealed by vacuoles within the cytoplasm (Fig. 27.1). The idea in the middle of the 20th century was that experimental lesions in cerebral cortex would damage the rising axons from the thalamus, and that after some period of weeks to months, this damage would cause chromatolysis of thalamic neurons whose axons were damaged (Rose & Woolsey, 1949). The result of the lesion would be localized degeneration in the thalamic nuclei that projected to the specific cortical damage with the inference that a specific connection between thalamus and cortex could be made.

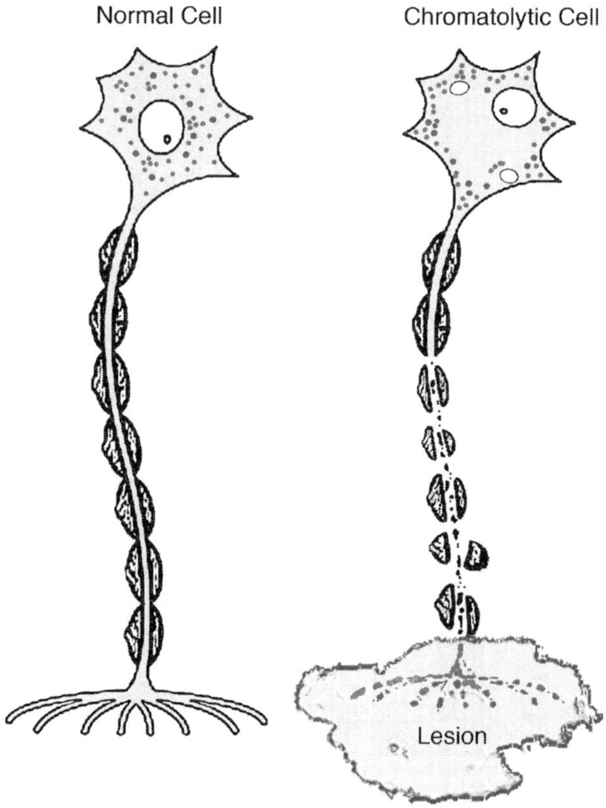

Fig. 27.1 Schematic illustration of chromatolysis. A normal cell is on the left; the cell on the right is undergoing retrograde degeneration (chromatolysis) following damage to its axon terminal. A lesion in cerebral cortex would produce degenerative changes of neurons in the thalamus where the axon originated

This method of retrograde degeneration was suggested to me by Norm Weinberger, my PhD supervisor at the University of California, Irvine. Norm had my scientific interests at heart but he was not a neuroanatomist. A difficult situation was created because there was not a trained expert on microscopic anatomy for consultation. Anatomy is more than looking through a microscope. There is an art to proper fixation of tissue. An underlying premise of anatomy is that one is studying structure as it would appear in the living condition. I had to be vigilant for artifacts created by anoxia, postmortem autolysis, and tissue bruising. Other artifacts can be introduced during the histological processing. Humason's (1967) textbook was crucial for an introduction to the science of tissue preparation. Then, I needed to learn how to use a microscope, which included aligning for Köhler illumination (align and focus the condenser, adjust the field diaphragm, center the beam). These basic steps before starting ensured that the image would be uniformly illuminated with optimal contrast.

These concerns were only the basic technical aspects of the art; the crucial component was the interpretation of what was observed.

A difficult first step was to recognize and differentiate regions of the brain. There were no labels so I had to get comfortable with uncertainty. Microscopy was demanding because when I started looking through the eyepieces, all cells looked alike—blue and small—and they tended to hide behind a veil of eyelashes. More frustrating was that I didn't necessarily know where in the thalamus to begin looking for chromatolytic neurons. I had to know how normal neurons in the various regions appeared so that identification could be made of the pathologic neurons. I mostly didn't realize how large the cat thalamus was until I had to examine it using a light microscope with a 40× objective.

This first graduate school experience resulted in frustration. The project was too difficult for a first-year graduate student. I didn't know which region of cortex I was examining. I didn't have enough knowledge at any level to make it work. Out of this struggle rose an epiphany: call it a collateral benefit. I realized that a good research question was mandatory. The question had to be clear with definable answers; in other words, it had to articulate a testable hypothesis. Moreover, there had to be methods for attacking the question that were feasible and that would provide data that were interpretable. Contemplation of possible answers prompted a thorough consideration of techniques. Good techniques make results possible and reliable. I also learned some basics about the microscope. If you're going to look at tissue, you needed a good microscope. Microscopes need to have glass that doesn't distort the image—no color or spherical aberrations. I learned to ignore my eyelashes that were always between the eyepieces and me. The focus on eyepieces had to be independently adjustable. By focusing the eyepiece for each eye, I could get binocular fusion; otherwise, my dominant eye suppressed input from my weak eye, rendering me monocular (and gave me splitting headaches). I had to learn to "scan" using low-magnification objectives and to "study" using high-magnification objectives. Scanning with a 40× objective often gave me motion sickness. Listening to music while looking was great—both eyes and ears were stimulated. Lastly, I got a comfortable chair.

27.2.3 Golgi Methods

Ramón y Cajal (1909) mastered the use of the mysterious Golgi method, which caused silver chromate or mercuric chromate to precipitate within a small number of random neurons. His advice to future hodologists was to use the immature brains of small animals to maximize the chances of finding isolated neurons whose axonal trajectory could be traced the short distance to their termination in other brain regions. But then, by looking at the figures in Cajal's (1909) classic tome, it seemed that he had already figured out everything so why would anyone want to study neuronal connections? Not only that but there were prominent researchers already studying the subjects I thought I were interesting. In fact, when I was about to complete my Ph.D., I went to a symposium talk by George Pollak at one of the Society for Neuroscience conferences. He basically reported on everything I was thinking about; talk about a bummer. That same meeting, when asking a very senior auditory neuroanatomist about cytoarchitecture, he chided me that if I had to actually measure and quantify data, I didn't have "the eye of a morphologist" and so should look for a new field of study. So how does a young scientist keep from being discouraged? How does one decide what to study? Georg von Békésy (1960) commented that one could read the literature and try to find something that hadn't been done, or ignore the literature and make your own way. In the end, the key to survival is finding good mentors.

The Golgi method was demanding and capricious. The dogma was that the method revealed 5–10 % of neurons in some random fashion. The other part of the dogma was that this method stained neurons in their entirety *sans* axons, a conclusion that may or may not be true. The rapid Golgi method was used for unfixed neonatal brains and relied on the precipitation of silver chromate within the neurons (Ryugo & Fekete, 1982); this method stained everything in moderation and was especially good for revealing axons and their terminal ramifications. The key to applying various stains, however, was not in the literature but came from the advice of colleagues. Nell Cant, Sonal Jhaveri, Tom Parks, and the late Sandy Palay were founts of information and help with this procedure. Their recipes and tricks were crucial to getting my own work going.

The Golgi–Cox method used unfixed fresh tissue of young or mature animals, mercuric chromate precipitate, celloidin embedding, and sliding microtome for preparation (Van der Loos, 1956). This method was excellent for staining cell bodies, dendrites, and spines, but wasn't very successful in staining axons. A modification of the rapid Golgi method used a perfusion technique of the Golgi reagents, and this method yielded better preservation of the tissue, stained cells in mature animals, and often revealed axons and their terminal ramifications (Colonnier, 1964; Adams, 1979). One used different methods for different objectives. Even with only 5 % of the neurons and processes stained, there are too many intersecting axons and dendrites to isolate individual neurons unambiguously (Fig. 27.2).

The Golgi method, despite the modifications for use in adult animals, was still applied mostly to immature brains. The idea was that myelin interfered with the intracellular penetration of reagents so staining in adult animals was poor. Most of our anatomical knowledge about neurons and their structure up until the 1980s was therefore based on immature brains. And yet, because most electrophysiology was performed in adult animals, there were those who doubted that Golgi descriptions were true for mature systems (Kiang et al., 1982).

27.2.4 Silver Stains and Pathway Tracing

An alternative method that had been around for years involved the use of silver to stain neurofibrils (Fig. 27.3). Neurofibrils as observed in a light microscope decades ago, are now recognized to be the condensation of neurofilaments by silver.

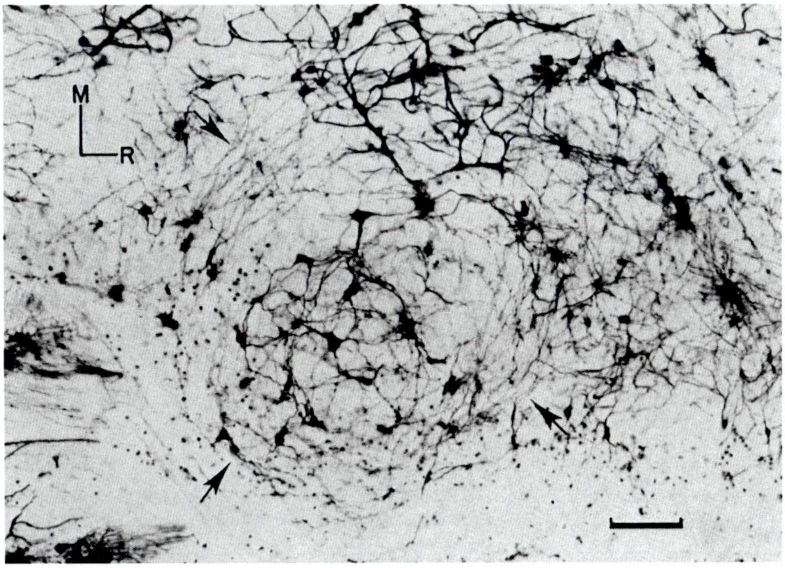

Fig. 27.2 Golgi–Cox stained preparation through the dorsal nucleus of the lateral lemniscus. In the horizontal plane, the core of the nucleus contains a reticulum of neurons and their dendrites. The surrounding fibers are arranged in a spiral pattern. The arrows indicate the nuclear boundary. The tangled overlap of dendrites and axons make separation of individual neurons impossible (Willard & Ryugo, 1983). Scale bar equals 100 μm

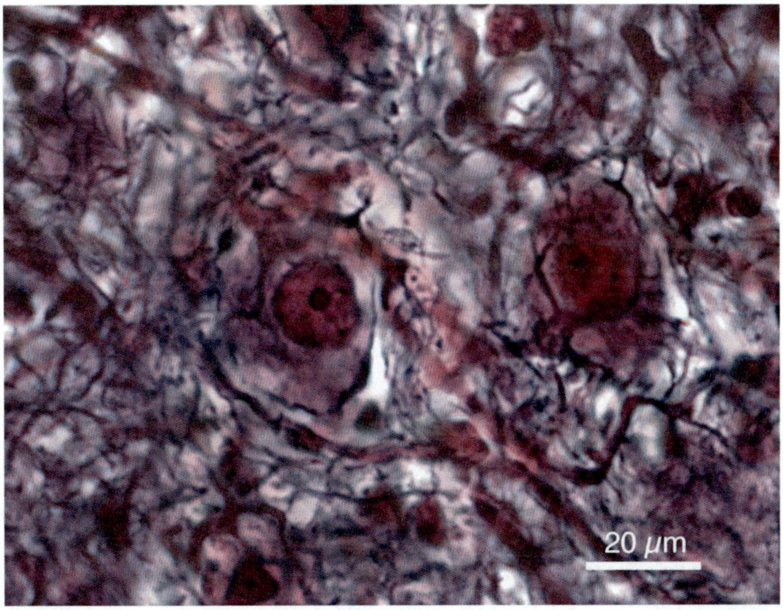

Fig. 27.3 Bodian's protargol–silver stain of the cat cochlear nucleus. Neurofibrils are stained darkly, and the claw-like structures that appear to clasp the magenta-colored cell bodies represent endbulbs. This image of the endbulb cytoskeleton was created by the condensation of neurofilaments

This method, attributed to Max Bielschowsky (1904), used silver nitrate and ammonium hydroxide to impregnate nerve fibers. Modification of this method (Bodian, 1936; Glees, 1946) used a double impregnation procedure by doing a second silver reduction using dilute formaldehyde (Fig. 27.3). These reduced silver techniques were used to trace normal fiber trajectories. Other recipes used pyridine, perhaps one of the worst smelling reagents around, so I abandoned them.

The intrepid pathfinder could use variations of the silver method (Marchi & Algeri, 1885; Bodian, 1936; Glees, 1946; Nauta & Gygax, 1951). Modified methods were developed to map pathways by combining lesion and degeneration methods with staining but one had to master the art of discrete lesions with survival surgery and near mystical procedures to obtain tissue sections with interpretable results. These modified methods revealed the affinity of degenerating axons for silver (agrygophilia). Electron microscopy confirmed that degenerating axons and terminals contained silver grains (Guillery, 1970). When successfully used, silver stains would leave a "silver" trail from the lesion to the axons' terminal field. The downside was that the morphology of the endings was not revealed, the target (dendrite or cell body) was a guess, and the presence of synapses could only be inferred.

Quirks of the methods, undocumented and often apocryphal, accompanied the various methods. For example, I was told that the Glees method only worked in certain regions of England because of idiosyncrasies of the different well waters. Post-lesion survival time, fixation methods, species studied, and specific neuronal systems all played into the calculation for how to conduct the research. Often, stained tissue yielded a relatively dense shroud of silver grains (3–10 μm in diameter) distributed everywhere over the tissue, attributed to dirty solutions causing silver to precipitate. Other times, the tissue revealed axons and nuclei of cells but not degenerating axons and terminals. There seemed to be two schools of neuroanatomists using two related but slightly different methods to trace pathways: these methods were the Nauta method and the Fink–Heimer method. There were politics involved in the choice, but as a graduate student and then young faculty member, I was not privy to the arguments. My selection of the Fink–Heimer (1967) method was completely arbitrary, and once it worked on my tissue, I simply stuck with it. The Fink–Heimer method introduced hydroquinone and oxalic acid and/or uranyl nitrate treatment to suppress the staining of normal axons. The addition of citric acid was said to be helpful so I kept fresh lemons in the lab.

The cleaning of all glassware with dilute (5 %) nitric acid was encouraged to prevent "dirt" from precipitating silver from the staining solutions. This meticulous cleaning step seemed to help—perhaps because once I started being obsessively compulsive in cleanliness, the habit transferred to the preparation of solutions as well. One couldn't be too careful, however, or your clothes tended to acquire acid holes that wouldn't appear until after washing. The goal of silver staining was to suppress normal axons but reveal degenerating axons as line fragments (resembling "dashes") and terminals as a localized cloud of "pepper" throughout the termination zone (Fig. 27.4). Professor Ann Graybiel offered one important piece of advice to me when I was still a graduate student, and that was to examine the entire brain for the "pepper" or else you would only find what you expected. Professor Graybiel

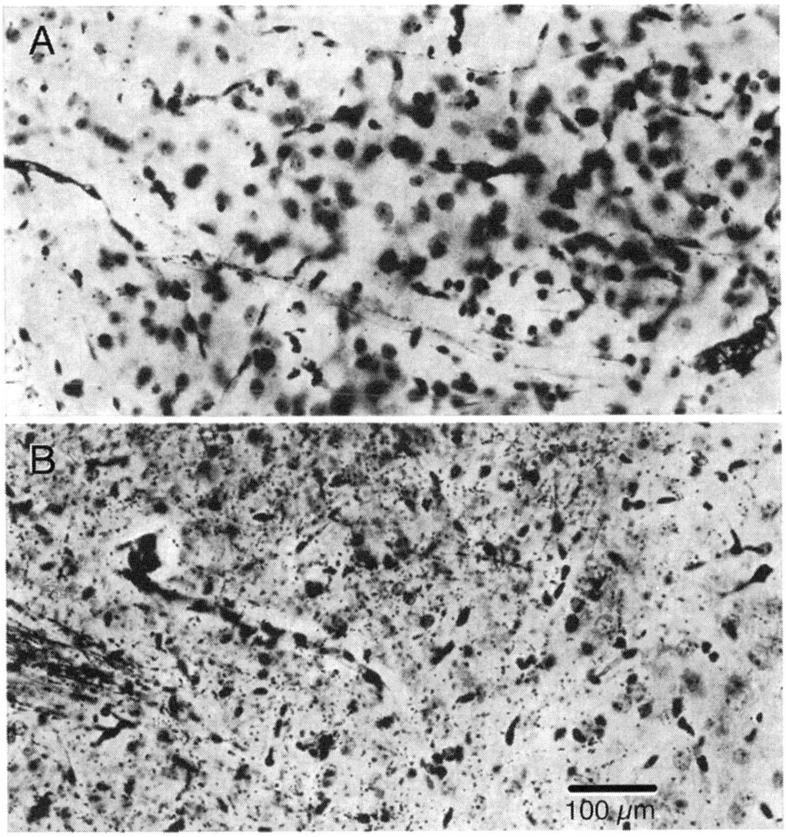

Fig. 27.4 Fink–Heimer stained tissue through the caudoputamen of the rat, 7 days after a lesion to the medial geniculate nucleus. In (**a**), the lesion was in the ventral division of the MGN, and there is a general lack of silver grains. In (**b**), however, the lesion was in the medial division of the MGN, and there are plentiful numbers of silver grains, indicating a dense projection to this area (Ryugo & Killacky, 1974). It is still unclear why the auditory system has connections to the striatum

was known for her ability to discover new pathways because of her meticulous studies of her experimental tissue. Her advice led me to discover a projection from the medial division of the medial geniculate nucleus to the caudoputamen in rats (Ryugo & Killackey, 1974).

27.2.5 Autoradiography

An anterograde method for studying pathways was also being explored using radioactive amino acids. Autoradiography was popularized by the documented observation that radioactive amino acids, when injected in the vicinity of neuronal cell bodies, would be taken up by the cells, incorporated into macromolecules, and transported down the axons to accumulate in the terminals (Rogers, 1979). Tritiated leucine and/or proline were common agents used for this purpose but leucine was preferred due to its more reliable incorporation into protein. Once incorporated within the cell body, it would be transported via fast axonal transport (100–400 mm per day) to the terminals. To work with radioactive compounds, the licensing procedure required the completion of an extensive course and a passing score on a rigorous exam. The exam experience brought back unpleasant memories of sitting for final exams in college. As a result of this test, I became much more sympathetic to the anxiety of medical students before finals, and I became a better teacher.

The post-injection survival period (1–10 days) was important because the aim was to get the radioactivity to accumulate in the terminal ending; otherwise, the radioactivity in axons confounded the terminal field. The tissue had to be fixed with buffered formaldehyde, cut into thin sections (thick sections meant that extra "noise" would be in the signal from the presence of a greater signal), dipped in photographic emulsion while working in a completely lightless darkroom, and stored in the cold for up to 6 months. The decay of tritium would leave a signal in the emulsion that could only be seen via photographic development, and the cold would lower background emissions. Mechanical stress, heat, and light would produce noise in the signal. Even when the experiment was a total success, interpretation of the results still faced several problems. First, there was the identification of the "effective" injection site. There was the problem of necrosis at the core of the injection site caused by expulsion of the isotope; finally, did the injection site "grow" with longer exposure and/or development times?

A small injection might not provide enough signal to be adequately visible for scanning at low magnifications and bright field conditions. If you couldn't scan tissue sections relatively quickly, microscopy became a very tedious process. To my good fortune, dark field microscopy emerged, which is an illumination technique that excludes the unscattered beam from the image, producing the dark field. The light scattered off the developed autoradiographic silver grants was collected, yielding an image that resembled the night sky with a globular cluster. Thus, instead of looking for small black grains against a sea of blue-stained neurons and white background, the signal became small beacons of bright light against a black background. Easy.

There is, however, the difficult issue of interpreting the pattern of transported label. The presence of the developed silver grains is inferred to result from the emitted radioactivity at the location of the terminals labeled from the injection site. The number of developed "silver grains" cannot indicate the absolute number of terminals in light microscopic preparations because terminals are themselves not labeled in this method. Even with electron microscopy, the number of grains was not a reliable indicator of the proportionate number of labeled terminals. As such, counts of grain density could provide an indication of signal versus background but otherwise didn't have much value in absolute terms. Evidence of emitted radioactivity could also be varied by varying the concentration of the labeled amino acid in the injection solution, increasing or decreasing the volume of the injection, varying the thickness of the emulsion covering the tissue, and changing the length of time during which emulsion was exposed to the radioactivity. The long exposure time meant that "feedback" on the experiment and associated methods was delayed.

My first collaborations as a faculty member involved autoradiography as a means to describe the development of projections up and down the neuraxis in the rat. These experiments were conducted with Tom Parks and Zaid Smith in Ed Rubel's lab at Yale. Ed wasn't in town, so we had our run of the lab, ordered everything we needed (which was probably rather expensive in those days), and set about injecting a cocktail of tritiated leucine and proline into the inferior colliculus of a series of age-graded neonatal mice. We spent hours working, doing surgery, cutting tissue, and dipping sections in emulsion in the pitch dark. And we had a great time, optimistically speculating on the results. Too bad we got no results. Alas, there is nothing quite like doing an experiment, waiting 3–6 months for the results, and discovering that the procedures failed. It was near impossible to retrace our steps after that much time to try to figure out what went wrong. I don't think we ever told Ed of our lab caper. To Ed's credit, I'm certain that he would not have cared. The fact that we were excited to do experiments would have been enough for him. Ed, incidentally, is also one of the great boosters of research on the planet.

27.2.6 Neuronal Labeling

Intracellular staining methods were first revealed when Procion yellow, one of a new class of textile dyes, was used to reveal individual, living nerve cells (Stretton & Kravitz, 1968). At the same time, horseradish peroxidase (HRP) was being used in cell biology laboratories to trace the recycling of intracellular vesicles (Graham & Karnovsky, 1965; Karnovsky, 1967). The LeVails took advantage of the endocytosis of HRP by axon terminals and intra-axonal retrograde transport to label cell bodies and study brain connections (LaVail & LeVail, 1972). This important study demonstrated that an extracellular injection of HRP into a localized location would result in axon terminals within the injection site taking up the enzyme and transporting it intra-axonally back to the cell body (and dendrites) where the HRP could be visualized using a histochemical procedure. The presence of HRP labeled somata was interpreted to mean that those neurons projected axons into the injection site, thus establishing a monosynaptic pathway. The beauty of this method is that there was virtually no background staining to confound the signal because only neurons with axons in the injection site would get stained. One could use a standard light microscope to follow the labeled axons continuously away from the injection site, along its path, and to the somata of origin. For the first time, pathway tracing could be accomplished using direct visual methods. Just as importantly, the method is relatively uncomplicated so that there was an explosion of labs that began pathway tracing.

My introduction to the HRP method came in Gerald Schneider's laboratory at MIT in 1976. I went to Boston for a few days from the University of Vermont to learn the technique, and Jerry assigned me to his postdoctoral fellow, Susan Udin.

Susan was using HRP to study retino-tectal plasticity in the developing hamster, and graciously allowed me to sit in on her experiments. In her encouraging and modest way, she simply said "if I can do it, so can you." I took her positivity to heart, restated her advice, and turned it into my mantra "if it can be done, I can do it." I use this mantra with my own students when they need encouragement.

This HRP method was timely because it provided a means of trying to resolve a long-standing question of how cochlear sensory receptor cells were innervated. Lorente de Nó (1937) had illustrated individual spiral ganglion cells, stained by the Golgi method, sending peripheral processes to inner hair cells, to outer hair cells, or to both. This observation, however, was made in neonatal animals. In contrast, Spoendlin (1978) had compared ganglion cell counts and fiber counts at the habenula perforata and the tunnel of Corti in normal and pathologic adult cochleae to infer that type I ganglion cells innervated inner hair cells, whereas type II ganglion cells innervated outer hair cells. This conclusion, however, was simply a strong inference. More direct observations of sensory receptor innervation was needed because how we thought about mechanisms of hearing depended on whether there was segregated input from the two classes of sensory receptor cells or whether the inputs were mixed.

The HRP method was ideal for trying to answer the question. A team effort was initiated under the direction of Nelson Kiang at the Eaton Peabody Laboratory where the still relatively new method of HRP had to be pioneered for fiber tracing in the inner ear. Our concern was whether the enzymatic activity of HRP could be retained and histochemically revealed after decalcifying the bony cochleae with acid. Fortunately, HRP was not inactivated by the tissue processing. We observed a diffuse and continuous filling of the primary neurons from the injection site, through the cell body, and out to the endings under the cochlear hair cells, illustrating a segregated innervation of the separate sensory receptors by the two types of primary neurons (Fig. 27.5; Kiang et al., 1982).

Further experimentation with HRP revealed that it could also be transported in an anterograde direction, that is, toward the axonal terminals (Dietrichs et al., 1981). HRP could also be conjugated to other agents, such as wheat germ agglutinin, and used as efficient, intracellular anterograde markers (Robertson et al., 1983). Phaseolus Vulgaris Agglutinin emerged as an efficient anterograde tracer that was prized because of its sensitivity and clarity, made so by the use of antibodies directed against the tracer (Wouterlood & Groenewegen, 1985; Rouiller & Welker, 1991; Wright & Ryugo, 1996). Biotinylated dextran amine (BDA) soon became a favorite pathway tracer because of its anterograde and retrograde properties; it was efficiently transported and methods for visualization were reliable. High molecular weight BDA (10 kDa) tended to label axons and terminals, whereas low molecular weight BDA (3 kDa) was a retrograde tracer.

The use of BDA was facilitated by the histochemical reaction using of avidin–biotin–HRP. The avidin–biotin complex was also used with fluorescent dyes, and when different dyes were injected into separate locations in the same experimental animal, interactions between fiber systems could be observed. Fast Blue and

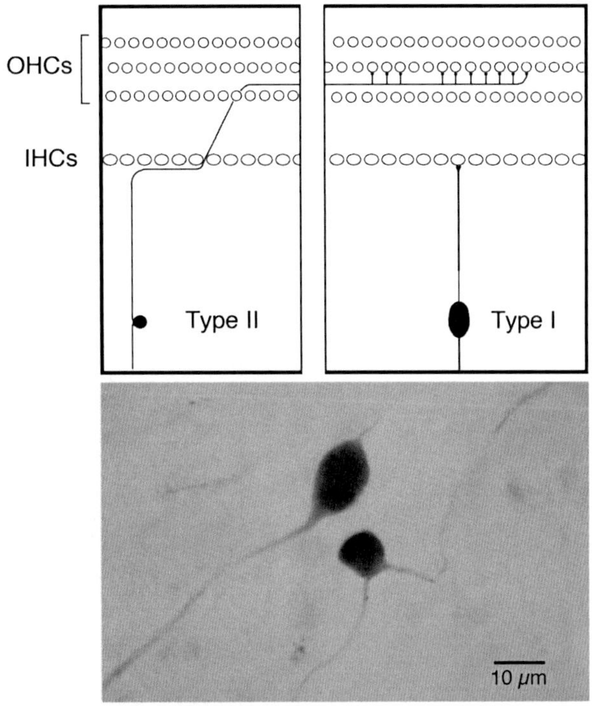

Fig. 27.5 Representative type I and type II spiral ganglion neurons in the cat labeled by HRP. The top panels illustrate their respective peripheral course and innervation. The bottom panel is a photomicrograph of the neurons (Kiang et al., 1982)

diamidino yellow are fluorescent retrograde tracers that require no histochemical tissue processing to visualize the results (Haase & Payne, 1990; Weedman & Ryugo, 1996a, b). Fast Blue is especially sensitive as a retrograde tracer (Fig. 27.6). And the beauty of these methods is that tracing experiments could be combined with immunocytochemical procedures to double and triple label neuronal systems—projection patterns could be analyzed and labeled terminals could also be directly assessed for neurotransmitter content (Wright & Ryugo, 1996; Doucet et al., 1999).

New tracers continue to be developed but HRP remained popular for several reasons. First, it could travel throughout a cell's axon trajectory in 24 hours, whereas most of the others required more time. This attribute was especially attractive when the animal's survival after the dye injection was difficult, such as following a long electrophysiology recording session or if the dye was put into a part of the brain where postsurgical recovery of the animal would be difficult. Second, HRP histochemistry was direct and had few steps. These fewer processing steps expedited tissue preparation for electron microscopy. Preservation of cell membranes for ultrastructural analysis required fixation of lipids with osmium, which was generally one of the last steps because osmium turned tissue opaque. The sooner the tissue was fixed with osmium, the better the membranes were preserved.

HRP could be loaded into micropipettes and injected axonally into single neurons after first recording their physiological response properties, in order to establish

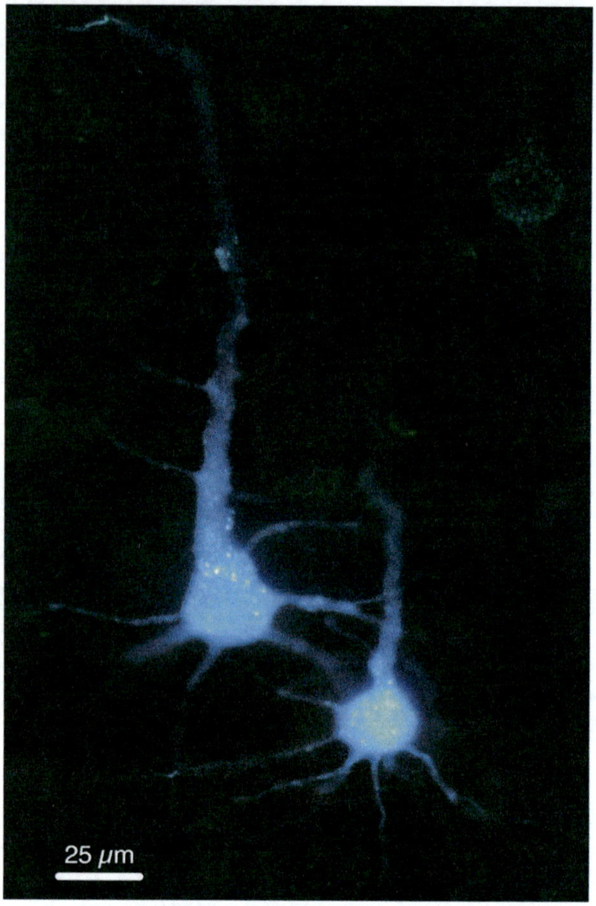

Fig. 27.6 Fast Blue labeled pyramidal cells in layer V of primary auditory cortex of the rat. These cells were labeled following a Fast Blue injection in the ipsilateral dorsal cochlear nucleus. This pathway indicates that auditory cortex can influence ascending auditory information at the earliest stages of the central auditory system (Weedman & Ryugo, 1996a, b)

structure–function relationships (e.g., Liberman, 1982a, b; Fekete et al., 1984; Sento & Ryugo, 1989). Charlie Liberman developed the single fiber recording and staining procedure in the Eaton–Peabody Lab and used HRP to create an unambiguous frequency map for the cat cochlea (Liberman, 1982a). He used the method to also show that single myelinated fiber threshold was determined by the caliber of the peripheral process and location of the terminal around the inner hair cell (Liberman, 1982b) and correlated to the morphology of the synapse (Merchan-Perez & Liberman, 1996). Charlie generously shared his technique and my lab benefited greatly (Fig. 27.7).

The number of dyes for labeling neurons increased to include Lucifer Yellow, biocytin (biotin amide), neurobiotin (aminoethyl biotinamide), and subunit B of cholera toxin. These dyes can be revealed using the fluorescent form or through histochemical reactions involving antibodies or avidin–biotin. These methods are fabulous because of the long distance labeling, the excellent signal-to-noise levels, and the compatibility for both light and electron microscopic analyses.

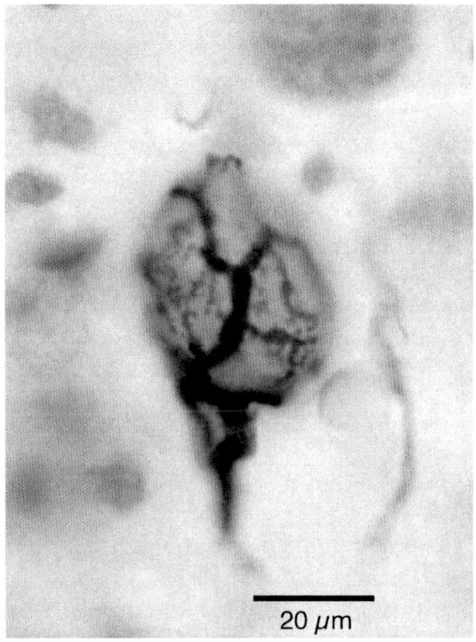

Fig. 27.7 The use of HRP marked the start of an era of selective staining of neurons. This popular marker reveals much of the structural details of the endbulb (compare to silver stained-endbulbs of Fig. 27.3) and the reaction product is electron dense for the benefit of electron microscopy (Ryugo & Fekete, 1982)

27.2.7 Pathway Tracing and the Future

Neuronal tracing methods continue to evolve. The incorporation of molecular biology techniques into research strategies has brought an entirely new level of analysis for pathway tracing. Of particular interest to our research are the transgenic mice that contain additional genetic material that labels the transporter for the amino acid neurotransmitters, glutamate, glycine, and γ-aminobutyric acid (GABA). In the past, labeling these transmitters involved conjugating the amino acid to bovine serum albumin (BSA) and making antibodies against the complex (Wenthold et al., 1986, 1987). The method was fussy and the antibody was assumed to recognize the site where BSA was bound to the amino acid. With antibodies directly labeling the transporter that packages the amino acid into synaptic vesicles, the terminals containing the particular transmitter are immunolabeled (Tirko & Ryugo, 2012). The situation with transgenic mice means that in the living animal, all neurons belonging to a specific circuit are already labeled by, for example, *enhanced green fluorescent protein* (Tamamaki et al., 2003; Zeilhofer et al., 2005; Borgius et al., 2010) or *yellow fluorescent protein* (Feng et al., 2000), thereby bypassing the need to perform immunocytochemistry on the tissue. These new methods will permit population studies of single neurons unified by a common neurotransmitter. One can imagine following the degenerative time course of dopamine-containing neurons in Parkinson's disease models when new drugs are tried.

Advances in technology continue to propel the field further. The development and growth of two-photon laser scanning microscopy have provided surprising data.

Two-photon imaging emerged from a desire to study living cells, peer deeper into tissue, and have improved resolution. The single photon confocal technique, while brilliant in using a narrow plane of focus and a pinhole to reject "out-of-focus" background light, was limited because of photobleaching and phototoxicity. Two separate photons beams, however, would activate the fluorochrome but not damage cells from heat. Single neurons, even individual spines on identifiable dendrites, could be monitored over time and observe to rapidly "come and go" (Bonhoeffer & Yuste, 2002); similar patterns of axonal sprouting and pruning accompanied functional reorganization in primary visual cortex (Yamahachi et al., 2009). This method revealed a novel and dichotomous organization pattern in mouse auditory cortex where frequency tuning was independent from intensity tuning (Bandyopadhyay et al., 2010).

The next generation of tissue processing is already here. One advance involves transgenic mice where unified cell populations are labeled. The technique, called *Brainbow*, allows the genetically labeling of individual neurons with multiple, distinct colors (Livet et al., 2007). A genetic construct was generated that could be recombined in various arrangements to produce different colors depending on the fluorescent proteins being implemented. Multiple copies of the same transgenic construct can be inserted into members of a target strain. The result would be a random expression of different fluorescent protein ratios in individual cells of a system so that each cell exhibited a different color. Now, scientists could identify, distinguish, and study the axons and dendrites of neighboring neurons to gain detailed information on connectivity and patterns. This method would be especially useful to determine if there are, for example, true gradients in dendritic morphology when comparing spherical bushy cells with respect to frequency sensitivity or for trying to delineate the structural boundary between spherical and globular bushy cells.

The significance of neuronal connections in terms of the computational power of the brain is underscored by ever-increasing descriptions of connectivity architecture. Several initiatives have been started that acknowledge structure–function mappings as representing an indispensable foundation for the interpretation of dynamic brain data, encompassing single cell marking studies to functional whole brain neuroimaging. One effort, called the *Open Connectome Project*, is an online, open forum for sharing state-of-the-art neuroscience data (http://openconnecto.me/). The other effort, supported by the NIH, has been coined the *Human Connectome Project* and it proposes to attack the question of brain networks with the same vigor as with the human genome project. The goals are ambitious, seeking to map neural connections from the subcellular level of synapses to the network level of neuronal circuits using standard pathway tracing techniques as well as functional imaging through magnetic resonance data. Here, then, lies the emerging future of neuroanatomy.

There is a vast armament of tools to approach questions of brain connections and function in the auditory system. The techniques have improved, there are lots of data to digest but many questions still remain: what are the circuits, what do they do, and how are they wired? The combination of anatomical, physiological, biochemical, molecular, quantitative, and behavioral methods will provide the broadest perspective of the problem. At the end of the day, however, answers will still depend on the cleverness, care, and integrity of the investigators.

Acknowledgments The author gratefully acknowledge the support from NIH grants DC000232 and DC004395; NMHRC grant no. 1009482; and grants from the NSW Office of Science and Medical Research, The Garnett Passe and Rodney Williams Memorial Foundation, The Curran Foundation, and The Fairfax Foundation. I am grateful to my mentors—Jay Braun, Norm Weinberger, Herb Killackey, Nelson Kiang, Sandy Palay, and Ed Rubel—as well as my many colleagues and lab mates who made science more fun than work.

References

Adams, J. C. (1979). A fast, reliable silver-chromate Golgi method for perfusion-fixed tissue. *Stain Technology*, 54(4), 225–226.

Bandyopadhyay, S., Shamma, S. A., & Kanold, P. O. (2010). Dichotomy of functional organization in the mouse auditory cortex. *Nature Neuroscience*, 13(3), 361–368.

Bielschowsky, M. (1904). Die Silberimprägnation der Neurofibrillen. *Journal of Psychology and Neurology (Lpz.)*, 3, 169–188.

Bodian, D. (1936). A new method for staining nerve fibers and nerve endings in mounted paraffin sections. *Anatomical Record*, 65, 89–97.

Bonhoeffer, T., & Yuste, R. (2002). Spine motility: Phenomenology, mechanisms, and function. *Neuron*, 35(6), 1019–1027.

Borgius, L., Restrepo, C. E., Leao, R. N., Saleh, N., & Kiehn, O. (2010). A transgenic mouse line for molecular genetic analysis of excitatory glutamatergic neurons. *Molecular and Cellular Neurosciences*, 45(3), 245–257.

Colonnier, M. (1964). The tangential organization of the visual cortex. *Journal of Anatomy*, 98, 327–344.

Dietrichs, E., Walberg, F., & Nordby, T. (1981). On retro- and anterograde transport of horseradish peroxidase in the pontocerebellar fibers as studied with the Mesulam TMB technique. *Brain Research*, 204(1), 179–183.

Doucet, J. R., Ross, A. T., Gillespie, M. B., & Ryugo, D. K. (1999). Glycine immunoreactivity of multipolar neurons in the ventral cochlear nucleus which project to the dorsal cochlear nucleus. *Journal of Comparative Neurology*, 408(4), 515–531.

Fekete, D. M., Rouiller, E. M., Liberman, M. C., & Ryugo, D. K. (1984). The central projections of intracellularly labeled auditory nerve fibers in cats. *Journal of Comparative Neurology*, 229(3), 432–450.

Feng, G., Mellor, R. H., Bernstein, M., Keller-Peck, C., Nguyen, Q. T., Wallace, M., Nerbonne, J. M., Lichtman, J. W., & Sanes, J. R. (2000). Imaging neuronal subsets in transgenic mice expressing multiple spectral variants of GFP. *Neuron*, 28(1), 41–51.

Fink, R. P., & Heimer, L. (1967). Two methods for selective silver impregnation of degenerating axons and their synaptic endings in the central nervous system. [Comparative Study]. *Brain Research*, 4(4), 369–374.

Glees, P. (1946). Terminal degeneration within the central nervous system as studied by a new silver method. *Journal of Neuropathology and Experimental Neurology*, 5, 54–59.

Graham, R. C., Jr., & Karnovsky, M. J. (1965). The histochemical demonstration of monoamine oxidase activity by coupled peroxidatic oxidation. *The Journal of Histochemistry and Cytochemistry*, 13(7), 604–605.

Guillery, R. W. (1970). Light and electron-microscopical studies of normal and degenerating axons. In W. J. H. Nauta & S. O. E. Ebbesson (Eds.), *Contemporary research methods in neuroanatomy* (pp. 77–105). New York: Springer-Verlag.

Haase, P., & Payne, J. N. (1990). Comparison of the efficiencies of true blue and diamidino yellow as retrograde tracers in the peripheral motor system. *Journal of Neuroscience Methods*, 35(2), 175–183.

Humason, G. L. (1967) *Animal tissue techniques*, 2nd ed. San Francisco: W. H. Freeman.

Hyde, I. H. (1904a) Localization of the respiratory center in the skate. *American Journal of Physiology*, 10, 236–258.

Hyde, I. H. (1904b) Differences in electrical potential in developing eggs. *American Journal of Physiology*, 12, 241–275.

Hyde, I. H. (1921) A micro-electrode and unicellular stimulation. *Biological Bulletin*, 40(3), 130–133.

Karnovsky, M. J. (1967). The ultrastructural basis of capillary permeability studied with peroxidase as a tracer. *The Journal of Cell Biology*, 35(1), 213–236.

Kiang, N. Y., Rho, J. M., Northrop, C. C., Liberman, M. C., & Ryugo, D. K. (1982). Hair-cell innervation by spiral ganglion cells in adult cats. *Science*, 217(4555), 175–177.

LaVail, J. H., & LaVail, M. M. (1972). Retrograde axonal transport in the central nervous system. *Science*, 176(4042), 1416–1417.

Liberman, M. C. (1982a). The cochlear frequency map for the cat: Labeling auditory-nerve fibers of known characteristic frequency. *Journal of the Acoustical Society of America*, 75, 1441–1449.

Liberman, M. C. (1982b). Single neuron labelling in the cat auditory nerve. *Science*, 216, 1239–1241.

Livet, J., Weissman, T. A., Kang, H., Draft, R. W., Lu, J., Bennis, R. A., Sanes, J. R., & Lichtman, J. W. (2007). Transgenic strategies for combinatorial expression of fluorescent proteins in the nervous system. *Nature*, 450(7166), 56–62.

Lorente de Nó, R. (1937). The neural mechanisms of hearing. I. Anatomy and Physiology. (b) The sensory endings in the cochlea. *Laryngoscope*, 47, 373–377.

Marchi, V., & Algeri, G. (1885). Sulle degenerazioni discendenti consecutive a lesioni sperimentale in diverse zone della corteccia cerebrale. *Rivista Sperimentale di Freniatria e Medicina Legale delle Alienazioni Mental*, 11, 492–494.

Merchan-Perez, A., & Liberman, M. C. (1996). Ultrastructural differences among afferent synapses on cochlear hair cells: Correlations with spontaneous discharge rate. *Journal of Comparative Neurology*, 371, 208–221.

Nauta, W. J., & Gygax, P. A. (1951). Silver impregnation of degenerating axon terminals in the central nervous system: (1) Technic. (2) Chemical notes. *Stain Technology*, 26(1), 5–11.

Ramón y Cajal, R. (1909). *Histologie du système nerveux de l'homme et des vertébrés* (Vol. 1). Madrid: Instituto Ramón y Cajal.

Robertson, B., Grant, G., & Bjorkeland, M. (1983). Demonstration of spinocerebellar projections in cat using anterograde transport of WGA-HRP, with some observations on spinomesencephalic and spinothalamic projections. *Experimental Brain Research*, 52(1), 99–104.

Rogers, A. W. (1979). *Techniques of autoradiography*. New York: Elsevier Science.

Rose, J. E., & Woolsey, C. N. (1949). The relations of thalamic connections, cellular structure and evocable electrical activity in the auditory region of the cat. *Journal of Comparative Neurology*, 91(3), 441–466.

Rouiller, E. M., & Welker, E. (1991). Morphology of corticothalamic terminals arising from the auditory cortex of the rat: A *Phaseolus vulgaris*-leucoagglutinin (PHA-L) tracing study. *Hearing Research*, 56(1–2), 179–190.

Ryugo, D. K., & Killackey, H. P. (1974). Differential telencephalic projections of the medial and ventral divisions of the medial geniculate body of the rat. *Brain Research*, 82(1), 173–177.

Ryugo, D. K., & Fekete, D. M. (1982). Morphology of primary axosomatic endings in the anteroventral cochlear nucleus of the cat: A study of the endbulbs of Held. *Journal of Comparative Neurology*, 210, 239–257.

Sento, S., & Ryugo, D. K. (1989). Endbulbs of held and spherical bushy cells in cats: morphological correlates with physiological properties. *Journal of Comparative Neurology*, 280(4), 553–562.

Spoendlin, H. (1978). The afferent innervation of the cochlea. In R. F. Naunton & C. Fernandez (Eds.), *Evoked electrical activity in the auditory nervous system* (pp. 21–39). London: Academic Press.

Stretton, A. O., & Kravitz, E. A. (1968). Neuronal geometry: Determination with a technique of intracellular dye injection. *Science*, 162(3849), 132–134.

Tamamaki, N., Yanagawa, Y., Tomioka, R., Miyazaki, J., Obata, K., & Kaneko, T. (2003). Green fluorescent protein expression and colocalization with calretinin, parvalbumin, and somatostatin in the GAD67-GFP knock-in mouse. *Journal of Comparative Neurology*, 467(1), 60–79.

Tirko, N. N., & Ryugo, D. K. (2012). Synaptic plasticity in the medial superior olive of hearing, deaf, and cochlear-implanted cats. *Journal of Comparative Neurology*, 520(10), 2202–2217.

Van der Loos, H. (1956). Une combinaison de deux vieilles methodes histologiques pour le systeme nerveus central. *Monatsschrift für Psychiatrie und Neurologie*, 132, 330–334.

von Békésy, G. (1960). *Experiments in hearing*. New York: McGraw-Hill.

Weedman, D. L., & Ryugo, D. K. (1996a). Projections from auditory cortex to the cochlear nucleus in rats: Synapses on granule cell dendrites. *Journal of Comparative Neurology*, 371(2), 311–324.

Weedman, D. L., & Ryugo, D. K. (1996b). Pyramidal cells in primary auditory cortex project to cochlear nucleus in rat. *Brain Research*, 706(1), 97–102.

Wenthold, R. J., Zempel, J. M., Parakkal, M. H., & Reeks, K. A. (1986). Immunocytochemical localization of GABA in the cochlear nucleus of the guinea pig. *Brain Research*, 380, 7–18.

Wenthold, R. J., Huie, D., Altschuler, R. A., & Reeks, K. A. (1987). Glycine immunoreactivity localized in the cochlear nucleus and superior olivary complex. *Neuroscience*, 22, 897–912.

Willard, F. H., & Ryugo, D. K. (1983). Anatomy of the central auditory system. In J. F. Willott (Ed.), *The Auditory Psychobiology of the mouse* (pp. 201–304). Springfield, IL: Charles C. Thomas.

Wouterlood, F. G., & Groenewegen, H. J. (1985). Neuroanatomical tracing by use of Phaseolus vulgaris-leucoagglutinin (PHA-L): Electron microscopy of PHA-L-filled neuronal somata, dendrites, axons and axon terminals. *Brain Research*, 326(1), 188–191.

Wright, D. D., & Ryugo, D. K. (1996). Mossy fiber projections from the cuneate nucleus to the cochlear nucleus in the rat. *Journal of Comparative Neurology*, 365, 159–172.

Yamahachi, H., Marik, S. A., McManus, J. N., Denk, W., & Gilbert, C. D. (2009). Rapid axonal sprouting and pruning accompany functional reorganization in primary visual cortex. *Neuron*, 64(5), 719–729.

Zeilhofer, H. U., Studler, B., Arabadzisz, D., Schweizer, C., Ahmadi, S., Layh, B., Bösl, M. R., & Fritschy, J. M. (2005). Glycinergic neurons expressing enhanced green fluorescent protein in bacterial artificial chromosome transgenic mice. *Journal of Comparative Neurology*, 482(2), 123–141.

Chapter 28
Adventures in Bionic Hearing

Robert V. Shannon

Robert V. Shannon

R.V. Shannon (✉)
Department of Otolaryngology, University of Southern California,
806 W. Adams Blvd., Los Angeles, CA 90007, USA
e-mail: rshannon@usc.edu

A.N. Popper and R.R. Fay (eds.), *Perspectives on Auditory Research*, Springer Handbook of Auditory Research 50, DOI 10.1007/978-1-4614-9102-6_28,

28.1 Introduction

In 1978 NIH funding levels were low and jobs were few. I had just finished my second postdoctoral position and was looking for a job and having little success. One job involved working on a new and controversial device called a cochlear implant (CI) that electrically stimulated the auditory nerve. At that time there were only a few patients in the world with the experimental device and no commercial products. The CI was intriguing to me because I had been trying to connect auditory psychophysics to physiology. Although I was trained in psychophysics I was interested in auditory physiology and was trying to develop models relating aspects of nerve firing to perception. CIs provided a great opportunity for testing such models. We could now activate highly unnatural patterns of neural activity (in both space and time) and see if we could predict the perceptual outcome. How would loudness and pitch relate to the stimulus parameters of electric stimulation? How could we translate acoustic sound into an electric pattern that would produce the desired loudness or pitch or speech percept?

But many colleagues warned me not to take this position. They were concerned that putting electronic stimulators in human volunteers was borderline unethical. The long-term consequences of electrical stimulation were not well understood. Animal experiments appeared to show safety, but they were mostly short-term experiments. What might happen to the cochlea and nerve after decades of electrical stimulation? Might the electronics leak poisonous chemicals into the body? How could it possibly work? The very thought of replacing the highly complex micromechanical system of the cochlea with a handful of electrodes seemed preposterous. How could we ever expect to get any useful hearing by replacing the 3000–4000 hair cells and 30,000 independent nerve fibers with 12 electrodes activating broad areas of neurons all at once? But I took the position at UC San Francisco anyway, working on early CIs. My part in the project was psychophysics—I guess you should call it psychoelectrics in the case of implants because we were quantitatively relating electric current to perceptual magnitudes. Now, more than 30 years later, CIs, a project that some felt was ethically questionable, have turned out to be the most successful prosthesis ever developed. More than 200,000 people worldwide have received the CI and most recipients are now young children.

CIs provide far better speech understanding than most auditory neuroscientists ever expected. In the 1970s, when cochlear implants were first developed, the prediction was mostly that implants would provide sound awareness and help with lip-reading, and few researchers thought they would ever allow people to converse on the telephone. Most people thought that the relatively crude representation across a small number of electrodes would not be sufficient to convey the complex spectral-temporal patterns of speech. Auditory research greatly underestimated the power of the brain's pattern recognition. Today, most people with cochlear implants can understand more than 80 % of words in sentences using only the sound from their implant. Many can understand sentences at 100 % correct. And when that level of sound-only performance is combined with lip-reading and the predictability of speech in normal conversations, many CI recipients can function at a high level in the hearing world. Children with CIs are mostly able to attend normal schools and function almost normally in a hearing world.

Auditory brain stem implants (ABIs) are similar to CIs in that they stimulate auditory neurons to restore hearing sensations. However, ABIs target the cochlear nucleus in the brain stem in patients who have no remaining auditory nerve and so cannot use a CI. When ABIs were first introduced auditory researchers again underestimated their potential. The initial ABI patients received sound awareness and a significant boost in lip-reading, but little open set speech understanding. Recent results in adults and children show that high levels of open set speech recognition are possible with ABIs, even though they bypass the auditory nerve and directly stimulating the cochlear nucleus in a non-tonotopic way.

Why have auditory researchers consistently underestimated the potential of auditory prostheses? How can auditory prostheses work so well? This chapter briefly reviews the past and present of CIs and ABIs and speculates about physiological mechanisms that might underlie the pattern of results.

28.2 Cochlear Implants

Early CIs stimulated auditory nerves with a single electrode placed on the round window or into the scala tympani. These early devices were well accepted by patients in spite of their limited ability to convey tonotopic information. The fundamental frequency of the voice was conveyed through periodicity and the overall speech envelope was conveyed by the low-frequency (<20 Hz) modulation in the electrical stimulus. These temporal patterns were sufficient to provide sound awareness and discrimination of some common environmental sounds based on their temporal patterns (e.g., telephone ringing vs. dog barking). In addition, the periodicity, when combined with lip reading in face-to-face communication, allowed much more fluid conversation than lip-reading alone. In general, patients obtained a 30–50 percentage point improvement in speech understanding with their implant compared to lip-reading alone. Although the prosthetic information was highly limited, the early CIs were enthusiastically received by postlingually deaf adults. Even these early CIs broke down the social isolation many deaf people feel in the absence of sound. Early CIs gave them awareness and some discrimination and identification of sounds, and a significant boost in the ease of face-to-face conversation.

Multichannel CIs provided additional electrodes distributed along the tonotopic axis of the scala tympani to provide more information about spectral shape and transitions. With early multichannel CIs performance immediately improved compared to single electrode CIs. These early multichannel CI users were able to recognize about 20 % of words in sentences without lip-reading. The additional spectral shape and spectral transition cues were crude compared to the detailed spectral information in a normal cochlea, but it was sufficient to allow CI users to recognize words. With this additional level of auditory quality, CI patients were now able to follow face-to-face conversations with ease.

Signal processing algorithms improved over the 1980s and early 1990s and CI performance continued to improve. By the mid-1990s most CI recipients were able to understand about 50 % of the words in sentences. This level of word recognition is enough to allow conversations on the telephone. Owing to the redundant and predictable nature of normal conversation, 50 % recognition, combined with knowledge of the personal speaking style of the talker and knowledge of the conversation topic, was enough to allow relatively good conversation over the telephone (or with a person whose lips are not clearly visible). The signal processing advance that allowed this improvement was a change in philosophy. Initial signal processing strategies assumed that there was only a limited amount of information that the implants could convey and so algorithms were developed to extract the most important features of speech from the running speech stream and code only those aspects into the CI. Such a strategy can work reasonably well in quiet listening conditions but break down badly in noisy listening conditions. Computer algorithms are not very good at reliably extracting speech cues in noisy conditions. In addition, the coded representation of the key speech features was presented in an unnatural manner across the electrodes and so constituted a new pattern of information that must be at least partially learned by the listener. A large improvement in performance came with the introduction of the continuous interleaved sampling (CIS) and the similar Advanced Combination Encoder (ACE) strategy. Electric pulses were interleaved in time across electrodes to avoid the complex interaction of simultaneously presented electric fields. These strategies simply filtered the sound stream into multiple-frequency bands and then presented an electric pulse train on each electrode representing the time-varying energy from each frequency band to the electrode assigned to that band. The pattern of stimulation produced was still very crude compared to a normal cochlea, but it was unselected in the sense that speech features were not explicitly extracted and presented. Instead, the brain's own speech feature extraction was allowed to work on the CI stimulation pattern. Although this pattern was probably shifted and distorted in frequency relative to the normal cochlea's tonotopic representation, and was a very coarse representation of spectral and temporal fine structure, most CI listeners were able to adapt their pattern recognition to the shifted pattern after a few months. The tonotopic patterns, though coarse, were sufficient to identify 40–50 % of random words presented without lip-reading cues. Further improvements in signal processing have led to improvements in speech recognition so that modern multichannel CIs provide more than 80 % recognition of words in sentences (Spahr et al., 2007). This result shows that the fine structure, both spectrally and temporally, is not necessary for speech recognition in quiet.

Research has shown that performance increases as the number of spectral bands increases and that as few as four bands were sufficient for high levels of speech recognition (Shannon et al., 1995). More bands are necessary for speech understanding in a variety of other conditions: As the complexity of the speech increases (Shannon et al., 2004); as noise interference increases (Fu et al., 1998); or when language familiarity is underdeveloped, such as in young children (Eisenberg et al., 2000), or people listening in a second language (Padilla, 2003). It appears that the brain is highly over-trained in recognition and categorization of speech patterns in our native language from millions of repetitions over our hearing lifetime. Under-optimal listening conditions speech can be recognized with surprisingly little spectro-temporal detail. As listening conditions deteriorate more fine structure is necessary.

One of the remaining puzzles about CI results is the wide variation in individual performance. It was once thought that a large part of the individual variability in outcomes was due to poor device parameter fitting. The standard clinical device fitting procedure may produce a good fit for some patients but not for others. Unlike the fitting of prescription lenses for vision problems, the fitting of CIs is not yet well developed in terms of individual fitting. It was hoped that improvements in individual fitting would convert poor CI users into good users, while already good users may get little of no benefit from fine tuning of implant parameters. This has not turned out to be the case. When individual customization of fitting parameters has been applied, the scores of all patients improve. Although poorly performing patients do show improvement with better parameter adjustments, rarely has a patient with a poor outcome been converted into one with a good outcome (Wilson et al., 1993).

Individual differences in implant performance have also been resistant to training. It was thought that brain plasticity could overcome some of the deficiencies in individual CI parameter fits and that training on speech materials would shape the brain's experience to improve performance (Wilson et al., 1993). Training, like individual parameter adjustments, improves performance for all patients (Fu & Galvin, 2008; Zhang et al., 2012); it does not have a differentially larger improvement in patients with poor outcomes.

This pattern of outcomes presents a puzzle: what is the source of the large individual variability in outcomes. If it is not fine adjustments in customizing the device to the individual patient and it is not something that can be learned, what is it? The differences in outcomes might be due to differences in the underlying pathology of the deafness, possibly related to the degree and uniformity of the surviving nerve population. For another perspective on individual variability in outcomes we next look at outcomes with the ABI.

28.3 Auditory Brain Stem Implant

The ABI is similar to the CI but is intended to stimulate the cochlear nucleus in the brain stem. It was originally designed for patients with neurofibromatosis type 2 (NF2), a genetic disorder that produces bilateral tumors on the vestibular portion of the eighth cranial nerve (VIIIn). These patients are deafened after tumor removal severs both auditory and vestibular branches of VIIIn. Such patients are deafened in a way that cannot be helped by a CI because they have no remaining auditory nerve.

The first ABI was done in 1979 by Bill House and Bill Hitselberger at the House Research Institute (Hitselberger et al., 1984) and that first patient has used the device every waking hour since that time. The device evolved over time to have multiple electrodes (Brackmann et al., 1993; Shannon et al., 1993) and was first commercialized by Cochlear Corporation in 1992. FDA approval was received in 2000 and now there is also an ABI device available from MedEl Corporation. As of 2012 there are more than 1100 ABI patients worldwide and most of them lost their auditory nerves from bilateral tumors (NF2). Overall the ABI provides sound awareness and environmental sound discrimination and some minor recognition of words (Lenarz et al., 2001; Nevison et al., 2002; Otto et al., 2002). Although the psychophysical measures of ABI performance were similar to those seen in CIs (Shannon & Otto, 1990), speech recognition was significantly poorer.

The difference between CI and ABI outcomes may provide some insight into the function of the auditory system. This early result suggested that auditory implants might have reached the point of diminished returns in terms of implant function; activation of the cochlear nucleus may produce more complex and less tonotopically organized patterns of auditory activation that didn't allow speech recognition. Stimulation at the level of the cochlear nucleus might bypass too much critical intrinsic processing so that the more central auditory structures do not have the fundamental information they need. Another possibility was that the surface array was not sufficiently tonotopically selective. Electric stimulation on the surface of the cochlear nucleus produces mostly low pitch sensations because high-frequency neurons are below the surface. ABI electrodes can interfere with each other because there is considerable overlap in the nerve populations activated by adjacent electrodes.

28.4 PABI: Penetrating Electrode ABI

It was thought that the limiting factor in ABI performance was that the surface electrode array was not making good contact with the tonotopic dimension of the human cochlear nucleus because it does not project to the surface of the nucleus. Surface electrodes primarily access low-frequency neurons and most ABI patients commented that the sound quality was low pitch and sounded "muffled." To gain access to higher frequency neurons lying below the surface of the nucleus, the ABI device was modified to include an array of 10 penetrating microelectrodes, with the goal of providing selective activation of high pitch tonotopic layers of the posterior ventral cochlear nucleus (PVCN) beneath the surface. The PABI was developed over a period of 15 years, including electrode design, materials selection and biocompatibility, and long-term safety of insertion and stimulation in animal experiments. Animal studies showed that insertion and stimulation of microelectrodes in the PVCN was safe, and that stimulation of electrodes at different depths could activate different tonotopic regions as measured in the inferior colliculus (IC; McCreery et al., 1998). Clinical trials in humans were initiated in the fall of 2003, and 10 patients were implanted with the PABI device.

The penetrating array produced auditory sensations in eight of the ten patients, with threshold levels less than 1 nanoCoulomb (nC), indicating good positioning into the PVCN. Classical psychophysical measures from the penetrating electrodes were quantitatively similar to those measured with CIs and surface-electrode ABIs. No interaction could be measured between penetrating electrodes, confirming good spatial selectivity and small area of excitation. Patients commented that the perception elicited by the penetrating electrodes was "clean and sharp and high pitch." In one case the patient still had temporary acoustic hearing in the contralateral ear and it was possible to match the pitch of each PABI electrode with acoustic tones in the nonimplanted ear, so the mapping of acoustic frequency information to electrode place was correct. In spite of successful implantation and the achievement of targeted psychophysical goals, speech performance with the PABI has been no better than with the surface electrode ABI (Otto et al., 2008). Even highly selective microstimulation of the cochlear nucleus was not sufficient to allow good speech recognition. Again, it appeared that stimulation of the cochlear nucleus, even with selective microstimulation, might have bypassed too much important neural processing, so that more central auditory nuclei didn't have sufficient information. However, the picture changed dramatically in the 2000s.

28.5 ABI in Nontumor Adults

Vittorio Colletti, a surgeon in Verona, Italy provided the ABI to patients who lost their VIIIn from causes other than NF2—such as from head trauma, severe ossification that obliterated the nerve, neurodegenerative diseases and several other causes (Colletti et al., 2002, 2004). These patients did not have tumors, but still had no auditory nerve and so were not candidates for a CI. Colletti's initial results showed excellent open set speech recognition in some of these nontumor (NT) patients. His results were met with considerable skepticism because ABI results in NF2 patients had never led to high levels of open set speech recognition. Independent testing verified Colletti's claims and showed that these patients also had better ability to detect small sinusoidal modulations in electric stimuli (Colletti & Shannon, 2005). Several of Colletti's NT ABI patients were able to achieve speech recognition scores near 100 % correct for simple sentences presented in quiet. Several could converse on the telephone as well as CI patients. One used a cell phone as his primary business contact as an independent contractor.

This exciting result showed that electric stimulation of the human cochlear nucleus could provide functional hearing comparable with CIs, even though the ABI had less access to the tonotopic gradient of the auditory system. The difference between ABI performance of NF2 and NT patients suggested that the difference in performance was related to the difference in etiology. Surgical removal of the NF2 tumor may damage some neural structure that is important for speech perception. Most psychophysical measures were similar between NF2 and NT ABI patients, but modulation detection was clearly better in the NT ABI patients and significantly correlated with speech recognition. What physiological structure that might be related to speech recognition is in a position to be damaged by tumor removal and plays a role in modulation detection? We will return to this question after reviewing two more patient populations and their results with the ABI.

28.6 ABI in Children

Based on his success with ABIs in NT adults, Colletti began a program to implant the ABI in children born without an auditory nerve. These children have a developmental or genetic abnormality in which the cochlear and/or auditory nerve fails to develop. Sometimes a full cochlea develops without an auditory nerve, and sometimes a nerve is present with a badly malformed cochlea. If a nerve is present the child may be suitable for a CI if the electrode array can be positioned near the nerve in the abnormal cochlea. Other children may have had hearing at birth but developed severe ossification following meningitis. In some cases the ossification is so severe that not only is the cochlea filled with new bone, but also the growth continues to invade the modiolus and internal auditory meatus, obliterating the auditory nerve. In such cases the children would have had some experience with hearing early in life but lost the hearing as the cochlea and then cochlear nerve were damaged by bone growth. In some cases these children may have received a CI, but it is now well known that children with this etiology do poorly or even obtain no benefit from a CI (Buchman et al., 2011). It was initially controversial to place an ABI in these children because ABI placement requires a transdural craniotomy to reach the brain stem. However, several cases showed excellent auditory development with the ABI (Colletti & Zoccante, 2008; Colletti et al., 2012)—even comparable developmental trajectories to that of congenitally deaf children with CIs (Eisenberg et al., 2008). Complications from surgery were minimal (Colletti et al., 2010). Several children developed open set sound recognition sufficient to attend mainstream schools. In 2012 there are more than 100 children with ABIs in the world and the number is growing rapidly. Again, these results show that the information delivered by the ABI to the brain stem is not only sufficient for experienced adult brains to recognize speech patterns, but it is also sufficient to allow a completely naïve child's brain to learn these patterns from the beginning.

28.7 New ABI Outcomes in NF2

One more twist in the ABI story is important before considering the potential physiological basis for these good results. Early ABI results showed useful but limited speech performance in NF2 patients. Following the excellent results in NT adults and children it was thought that the NF2 tumor removal must damage some critical structure or pathway that remains intact in these NT patient populations. However, some surgeons started seeing CI-like auditory performance even in NF2 ABI patients (Skarzynski et al., 2000; Behr et al., 2007). Some of these patients could understand sentences at 100 % correct, and could even understand 50 % of the words in sentences at a speech to noise ratio of +3 dB—a level that is rarely achieved even in CI patients. Many of these patients could converse on the phone without difficulty. If the original interpretation was correct that the tumor removal was damaging some structure in the brain stem, how were these surgeons able to remove similar tumors without such damage? Or was there another explanation for the good outcomes of these patients?

28.8 The Auditory Midbrain Implant: Electrical Stimulation of the Inferior Colliculus

The difference in outcomes between NT and NF2 ABI patients suggests that NF2 tumors and/or their removal can cause damage to auditory pathways that interfere with speech recognition. Assuming that this damage is local to the CN, it may be possible to bypass the CN region and produce better speech recognition by stimulating higher neural centers of the brain stem and midbrain. The inferior colliculus (IC) is a prime candidate for such stimulation because it has a regular and well-documented tonotopic structure and it is surgically accessible. If good speech recognition can be achieved from nontonotopic activation of the cochlear nucleus, then it might be possible to achieve good speech recognition with stimulation of a higher nucleus in the auditory pathway. If speech pattern recognition was flexible enough that top-down processing could utilize highly unnatural patterns of activation in auditory nerve and cochlear nucleus, then it might be possible to achieve similar success from stimulating the IC. At present, two implants are under development to provide electrical stimulation of the IC: (1) the inferior colliculus implant (ICI), which uses an ABI 12-electrode array placed on the surface if the IC, and (2) the auditory midbrain implant (AMI), which uses a penetrating 21-electrode array. The first patient to receive the ICI was implanted in December 2005 (Colletti et al., 2007). Five patients have now received the AMI (Lim et al., 2009). Most ICI and AMI patients receive sound sensations from stimulation and many hear different pitch sensations across the electrode array, indicating that the arrays do access different tonotopic regions of the IC. However, no significant speech recognition has been observed from electric stimulation of the IC. Although it is possible to place electrodes in or on the IC and achieve tonotopic activation, speech recognition has not been achieved. Patients receive useful auditory information from these devices but they are not receiving open-set speech recognition.

28.9 Auditory Neuropathy

Another group of patients of interest are those diagnosed with auditory neuropathy (AN) (Starr et al., 1991). Although AN may represent more than one pathology, results suggest that, like poorer-performing CI and ABI users, AN patients exhibit poor modulation sensitivity and poor speech recognition (Zeng et al. 1999, 2005). It is possible that the pathology that causes at least some types of AN is rooted in the hair cell-neuron synapse, or in the VIII nerve itself. If so, a CI might not provide benefit, but an ABI might. However, if the pathology causing AN is rooted in damage to the putative neural subsystem in the CN that is critical for speech recognition (similar to NF2 patients), then an ABI may provide limited benefit.

28.10 Possible Physiological Substrates of Speech Recognition

Now that there is more than 30 years of experience with electrical stimulation of the auditory system it may be possible to look at possible neural underpinnings of the pattern of results observed. There is a large variation in performance across CIs, but most patients can achieve high levels of open set speech recognition, including the ability to converse easily on the telephone. Similar excellent speech recognition in some ABI patients shows that it is possible to achieve excellent open set speech recognition from an ABI stimulating the cochlear nucleus, even after NF2 tumor removal. The lack of good speech recognition from stimulation of the IC suggests that we may have reached a point of diminishing returns. It is possible that activation of the IC bypasses too much intrinsic processing in the auditory brain stem. Now I consider possible physiological mechanisms that might underlie the pattern of results observed.

The dichotomy in ABI patient outcomes provides considerable leverage on a key question in auditory processing: Is there a specialized physiological pathway for speech recognition? The differences between these patient groups appear to be subtle—both groups have no functioning auditory nerve, no known central pathology; both groups are implanted with same ABI device and both groups use the same stimulation strategy. Preliminary psychophysical results show that both groups have similar threshold levels, similar degrees of electrode selectivity, and similar pitch and loudness ranges (Shannon & Otto, 1990; Shannon & Colletti, 2005). The most significant performance difference (besides speech recognition) is for modulation detection; ABI patients with good speech recognition have significantly better modulation detection thresholds (MDTs) than those of ABI patients with poor speech recognition, regardless of etiology (Colletti & Shannon, 2005). MDTs were significantly correlated with speech vowel recognition and sentence recognition in both ABIs and CIs (Fu, 2002). Thus, whatever physiological differences exist seems to impact both speech recognition and modulation detection, but not other perceptual measures.

One possible explanation for the difference between good and poor ABI patient outcomes is that the NF2 tumor and/or its removal causes some sort of damage to a neural system that is critical for speech recognition. The most likely causes of damage during tumor removal are: (1) physical damage to the brain stem neurons,

(2) anoxia related to venous bleeding, or (3) excitotoxicity from electrocautery to stop bleeding. If the presence and/or removal of tumors indeed damages a specific cell type or region of the CN, and that damage decreases speech understanding, this would be an important new finding and advance our understanding of the role of peripheral physiology in complex perception. Recent results of excellent speech recognition in NF2 patients suggest that surgical removal of NF2 tumors may not always result in damage. One common element in the NF2 ABI patients who achieved high levels of speech recognition is that most had surgery in the semi-sitting position, which lowers the venous pressure in the tumor region so that little or no cautery was used during tumor removal. Local anoxia or excitotoxicity would likely affect cells near the surface of the cochlear nucleus. If damage to the surface of the CN is causing a large difference in outcomes, what type of cells might be damaged?

NF2 tumors are benign schwannomas that originate near the myal/glial junction on the vestibular branch of the VIIIn. The myal/glial junction is near the medial opening of the internal auditory meatus. As they grow, vestibular schwannomas balloon into the cerebello-pontine angle and tumors larger than 2 cm typically contact the surface of the brain stem. Although benign, NF2 tumors produce an angiogenesis factor on their surface that attracts vascular blood supply from the surface of the brain stem, in this case the surface of the CN. The vascular supply of the CN in this region branches off of the posterior-inferior cerebellar artery (PICA). CN vessels travel along the surface and then dive into the interior of the nucleus. The tumor angiogenesis entangles the tumor's blood supply with the blood supply to the surface of the cochlear nucleus. The mere existence of the tumor and the shared vasculature may not impair the functioning of the CN because some patients with 4- to 5-cm tumors retain normal hearing and speech understanding prior to surgical removal. Tumor removal and surgical cautery may damage CN cells that share blood supply with the tumor, either through anoxia or excitotoxicity.

The small cell cap (SCC) of the cochlear nucleus is a candidate for such vascular/excitotoxic damage or direct mechanical damage, due to its physical location on the surface of the CN. Physiologically, the SCC predominantly receives input from primary auditory neurons with high thresholds and low spontaneous rates (SRs). According to Liberman (1978, 1991): "The small cell cap was almost exclusively innervated by low- and medium-SR fibers, i.e., those with the highest acoustic thresholds." Although the SCC is not well characterized, it is thought to project to the medial olivary complex (MOC; Ye et al., 2000) and to possibly have a role in intensity coding because of the wide dynamic range (DR) of its neurons (Ghoshal & Kim, 1996, 1997). The low spontaneous rate auditory neurons that project to the SCC also show little saturation with level and exhibit wide DRs (Sachs & Abbas, 1974; Winter et al., 1990).

High-threshold, low-SR neurons may provide the basis for rate coding of spectral profiles, because they are able to preserve spectral profiles at moderate loudness levels without saturating (Sachs & Young, 1979). Low spontaneous rate (LSR) neurons and cells in the SCC are also known to code modulation well because of their large dynamic range. Phylogenetically, the SCC is small region of the cochlear nucleus with an unknown function. The small cells probably cannot project to remote target sites

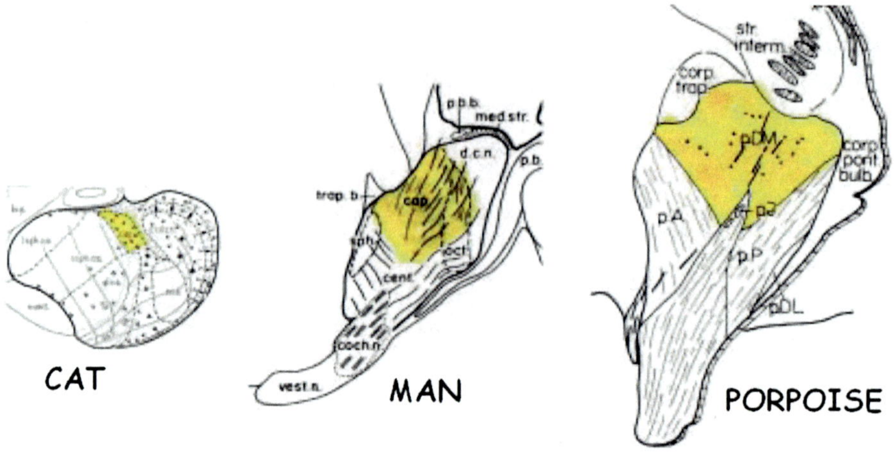

Fig. 28.1 Comparison of the SCC in cats, humans and porpoise (figure represents a composite based on figures from Osen & Jansen, 1965 and Moore, 1987). The SCC is indicated in yellow (coloring of SCC courtesy of Jean Moore). Note the large difference in relative size across species

because they are too small to metabolically support a long axon. In most mammals the SCC is quite small and it is hypertrophied only in humans and porpoises.

Figure 28.1 shows a comparison of the SCC in cats, humans, and porpoise (figure represents a composite from Moore, 1987; Osen & Jansen, 1965, coloring of SCC courtesy of Jean Moore). Note the large difference in relative size across species. Even other primates have a relatively small SCC compared to humans and cetaceans (Moore & Osen, 1979). Is the SCC a recent evolutionary structure specialized for complex pattern perception? Central mechanisms that may be selectively attentive to these neurons might provide a specialized pathway for coding complex pattern information (modulation of firing rate vs. tonotopic place). The SCC is the primary target for the initial synapses of the LSR neural population and so could provide a physiological subsystem specialized for spectral pattern processing. Damage to the SCC as a consequence of tumor removal might explain the difference in speech understanding between ABI patients who can understand speech and those who cannot. Another convergent piece of evidence is the fact that modulation detection is correlated with speech pattern recognition and the SCC neurons are also good at coding modulation because of their large dynamic range (Ghoshal & Kim, 1996, 1997). Loss of low spontaneous rate VIIIn fibers or SCC fibers may contribute to the loss of speech recognition, even when electrical thresholds and dynamic ranges and other psychophysical measures appear to be normal. If non-SCC fibers are still intact and stimulated by the ABI then they could still produce auditory sensations but may not contribute to speech recognition.

Alternatively, onset-chopper cells (OCCs) may also be candidates for structural damage in the CN. OCCs in the CN are known to enhance modulation relative to VIII nerve (Rhode & Greenberg, 1994; Frisina, 2001), and therefore could well be the physiological substrate important to both modulation detection and speech recognition. OCCs are metabolically more labile because their size makes them more susceptible to transient anoxia, which almost certainly occurs during surgical cautery. However, OCCs are so broadly distributed in the VCN that they would not be any more susceptible to mechanical surgical injury than other large cells.

Whether the physiological difference between good and poor ABI outcomes is due to damage to the SCC, OCCs, or some other cell type is not crucial for the concept of a specialized speech system. A simple physiological difference may or may not explain the large difference in performance between the two groups. All ABI patients have presumably intact central auditory processing; all had normal speech recognition before the loss of hearing, and all use the same electrode and similar speech processing strategies. Some ABI recipients have audiologically normal hearing right up to the tumor removal surgery. And in a few cases, temporary normal acoustic hearing on the contralateral ear allowed balancing of acoustic and electric stimulation in pitch and loudness, so the assignment of acoustic frequencies to tonotopically appropriate neural populations was possible. In spite of all this, most of these patients are not able to recognize speech with the ABI even though they had only a short period of deafness. But some (as many as 35 % in some clinics) can understand simple sentences at nearly 100 % correct and can converse on the telephone. The large difference in performance combined with the seemingly minor differences in etiology/pathology suggests that damage to a specific physiological mechanism may be at the root of this dichotomy in ABI patient outcomes. Whatever the physiological underpinnings of the perceptual differences between these patient groups, it is important to comprehensively characterize the perceptual capabilities of these patients. If the SCC/LSR hypothesis is correct, research with these NF2 and NT ABI patients may illuminate underlying physiological substrates/pathways for speech pattern recognition that may be independent of other auditory processing.

28.11 An Acoustic Fovea?

Consider an analogy between auditory and visual systems. Let us assume for a moment that the low spontaneous rate (LSR)/high spontaneous rate (HSR) system is analogous to the differential contribution of rods and cones to vision. Cones make up only 5 % of the retinal epithelium but perform a large part of visual pattern recognition. LSR neurons only make up 5–10 % of auditory neurons. Rods are specialized for high sensitivity and low thresholds, as are high spontaneous rate auditory neurons. Rods/HSR neurons are highly important evolutionarily because they allow early detection of predators and/or the ability to detect prey at low sound/light levels. In contrast, retinal cones and LSR auditory neurons are less sensitive but have a larger dynamic range of responsiveness. These systems are evolutionarily younger and may represent a more recent adaptation for processing more complex patterns of sensory information. It is known that complex pattern recognition like reading and face recognition is poor in the visual periphery where the receptors are mostly rods. People who experience a loss of foveal cones have great difficulty reading or recognizing people. We hypothesize that a loss of LSR neurons (or SCC neurons to which they project) may result in a loss of speech pattern recognition. Whether it is an intrinsic difference or due to experience, rods and cones have difference functions in visual processing. It is possible that the LSR auditory neurons represent an "auditory fovea"; specialized for complex pattern processing rather than sensitivity. Differences in speech recognition across implant users may reflect differences in the health of this LSR/SCC system. It may even explain some aspects of auditory neuropathy; loss of speech recognition and poor modulation sensitivity even with good threshold sensitivity. We should consider there might be multiple processing pathways in the auditory system as early as the auditory nerve and brain stem. Maybe there is an auditory fovea.

28.12 Summary

At present, there remains great variability in CI patient outcomes. Although most CI recipients show high levels of speech recognition, some do not. It has been assumed that this variability in outcomes relates to patients' neural survival or to nonoptimized speech processor settings. However, studies in which speech processor parameters were varied have shown that relative performance levels across CI patients were preserved across parameter manipulations. No matter what processing parameters were tested, the top-performing patients always performed best and poorest-performing patients performed worst (Wilson et al., 1993). Thus, optimized speech processing for individual patients did not reduce the variability in patient outcomes. This result suggests that there is a physiological basis for the differences in performance across patients.

Consider the possibility that there may be two different sources of variability in CI patient outcomes: damage to the VIIIth nerve and/or damage to the putative speech-specific pathway. If poor performance in CI patients is due to damage to the VIIIth nerve, the ABI may provide some benefit. The good speech recognition performance by NT ABI patients (who have no functioning auditory nerve) suggests that the ABI may provide a new option for patients who receive little benefit from the CI. Indeed, greatly improved speech recognition was observed in several NT ABI patients who previously received little benefit from their CI (Colletti et al., 2002, 2004). However, if poor performance in CI patients is due to the loss of a more central speech pathway, the ABI may not provide any more benefit than the CI.

One hypothesis is that the SSC of the cochlear nucleus is a possible physiological substrate for a speech pathway. It is known that the SCC primarily receives input from the LSR auditory neurons. Recent results (Kujawa & Liberman, 2009; Lin et al., 2011) show that LSR neurons are more susceptible to acoustic overstimulation than other neurons. It is possible that the LSR–SCC system is essential for speech pattern processing. People with damage to either LSR neurons or SCC neurons may still have normal auditory thresholds and normal psychophysics mediated by high spontaneous rate neurons. But if they have damage to the SSC–LSR system they might have poor modulation detection and poor speech pattern recognition —a pattern exhibited in AN patients and poor users of CI and ABI devices.

Auditory research has traditionally underestimated the role of the brain in processing complex patterns of information from the cochlea. In the 1960s, auditory neuroscientists were convinced that CIs would not work because the complexity of cochlear processing could not be replaced with a few stimulating electrodes. At that time, most auditory researchers were fixated on the complexities of cochlear processing, and thought that the highly unnatural patterns of neural activation provided by electrical stimulation would only allow only rudimentary auditory sensations. Now it is clear that central processing of complex patterns of sensory information allow high levels of speech recognition, even though the peripheral pattern of activation is spectrally impoverished and highly unnatural. High levels of speech recognition have now been documented even from stimulation of the cochlear nucleus with a pattern of electric activation that is far more unnatural than that produced by a CI. Some factor seems to be limiting NF2 ABI patients' ability to synthesize speech from the stimulation patterns provided by the ABI. Since there is no known central manifestation of NF2, the problem is most likely localized to the CN. This suggests that some physiological mechanism/structure/pathway in the CN may be damaged during NF2 tumor removal. Without this pathway, speech understanding and modulation detection are poor even in the presence of relatively normal psychophysical abilities.

This chapter proposed a hypothetical processing pathway that may be essential for speech recognition—the low spontaneous rate auditory neurons connecting to the small cell cap of the cochlear nucleus. Damage to such a putative pathway could potentially underlie the pattern of poor speech recognition and poor modulation detection documented in patients with auditory neuropathy, poor-performing CI patients, and ABI patients.

Whether or not the specific mechanisms proposed are correct is of little importance. There two principal messages of this chapter. Message one is that patient outcomes can provide important leverage on understanding the neuroscience of auditory processing. Quantitative study of pathologies and functional differences can suggest underlying mechanisms. We suggest that the linkage between patient pathology and auditory neuroscience is underutilized and can provide leverage on scientific questions. Message two is that, in spite of widespread acceptance that auditory processing is "massively parallel," most theories of speech processing are serial/sequential. As a field we need to develop better insights and models of parallel processing in the auditory system.

References

Behr, R., Müller, J., Shehata-Dieler, W., Schlake, H. P., Helms, J., Roosen, K., Klug, N., Hölper, B., & Lorens, A. (2007). The high rate CIS auditory brainstem implant for restoration of hearing in NF-2 patients. *Skull Base*, 17(2), 91–107.

Brackmann, D. E., Hitselberger, W. E., Nelson, R. A., Moore, J. K., Waring, M., Portillo, F., Shannon, R. V., & Telischi, F. (1993). Auditory brainstem implant. I: Issues in surgical implantation. *Otolaryngology, Head and Neck Surgery*, 108, 624–634.

Buchman, C. A., Teagle, H. F., Roush, P. A., Park, L. R., Hatch, D., Woodard, J., Zdanski, C., & Adunka, O. F. (2011). Cochlear implantation in children with labyrinthine anomalies and cochlear nerve deficiency: Implications for auditory brainstem implantation. *Laryngoscope*, 121(9), 1979–1988.

Colletti, V., & Shannon, R. V. (2005). Open set speech perception with auditory brainstem implant? *The Laryngoscope*, 115, 1974–1978.

Colletti, L., & Zoccante, L. (2008). Nonverbal cognitive abilities and auditory performance in children fitted with auditory brainstem implants: Preliminary report. *Laryngoscope*, 118(8), 1443–1448.

Colletti, V., Fiorino, F., Carner, M., Sacchetto, L., Miorelli, V., & Orsi, A. (2002). Auditory brainstem implantation: The University of Verona experience. *Otolaryngology Head & Neck Surgery*, 127, 84–96.

Colletti, V., Carner, M., Miorelli, V., Colletti, L., Guida, M., & Fiorino, F. (2004). Auditory brainstem implant in posttraumatic cochlear nerve avulsion, *Audiology and Neurotology*, 9, 247–255.

Colletti, V., Shannon, R. V., Carner, M., Colletti, L., Sacchetto, L., Turazzi, S., & Masotto, B. (2007). The first successful case of hearing produced by electrical stimulation of the human midbrain, *Otology and Neurotology*, 28(1), 39–43.

Colletti, V., Shannon, R. V., Carner, M., Veronese, S., & Colletti, L. (2010). Complications in auditory brainstem implant surgery in adults and children. *Otology and Neurotology*, 31(4), 558–564.

Colletti, L., Shannon, R. V., & Colletti, V. (2012). The development of auditory function in children following auditory brainstem implantation. *Audiology and Neurotology*.

Eisenberg, L., Shannon, R. V., Martinez, A. S., Wygonski, J., & Boothroyd, A. (2000). Speech recognition with reduced spectral cues as a function of age, *Journal of the Acoustical Society of America*, 107(5), 2704–2710.

Eisenberg, L. S., Johnson, K. C., Martinez, A. S., DesJardin, J. L., Stika, C. J., Dzubak, D., Mahalak, M. L., & Rector, E. P. (2008). Comprehensive evaluation of a child with an auditory brainstem implant. *Otology & Neurotology*, 29(2), 251–257.

Frisina, R. (2001). Subcortical neural coding mechanisms for auditory temporal processing, *Hearing Research*, 158, 1–27.

Fu, Q.-J. (2002). Temporal processing and speech recognition in cochlear implant users, *NeuroReport*, 13, 1635–1639.

Fu, Q.-J., & Galvin, J. J. 3rd. (2008). Maximizing cochlear implant patients' performance with advanced speech training procedures. *Hearing Research*, 242(1–2), 198–208.

Fu, Q.-J., Shannon, R. V., & Wang, X. (1998). Effects of noise and spectral resolution on vowel and consonant recognition: Acoustic and electric hearing. *Journal of the Acoustical Society of America*, 104(6), 3586–3596.

Ghoshal, S., & Kim, D. O. (1996). Marginal shell of the anteroventral cochlear nucleus: Intensity coding in single units of the unanesthetized, decerebrate cat. *Neuroscience Letters*, 205, 71–74.

Ghoshal, S., & Kim, D. O. (1997). Marginal shell of the anteroventral cochlear nucleus: single-unit response properties in the unanesthetized decerebrate cat. *Journal of Neurophysiology*, 77(4), 2083–2097.

Hitselberger, W. E., House, W. F., Edgerton, B. J., & Whitaker, S. (1984). Cochlear nucleus implants. *Otolaryngology Head & Neck Surgery*, 92(1), 52–54.

Kujawa, S. G. & Liberman, M. C. (2009). Adding insult to injury: Cochlear nerve degeneration after "temporary" noise-induced hearing loss. *Journal of Neuroscience*, 29(45), 14077–14085.

Lenarz, T., Moshrefi, M., Matthies, C., Frohne, C., Lesinski-Schiedat, A., Illg, A., Rost, U., Battmer, R. D., & Samii, M. (2001). Auditory brainstem implant: part I. Auditory performance and its evolution over time. *Otology & Neurotology*, 22(6), 823–833.

Liberman, M. C. (1978). Auditory-nerve responses from cats raised in a low-noise chamber, *Journal of the Acoustical Society of America*, 63, 442–455.

Liberman, M. C. (1991). Central projections of auditory nerve fibers of differing spontaneous rate: I. Anteroventral cochlear nucleus, *Journal of Comparative Neurology*, 313, 240–258.

Lim, H. H., Lenarz, M., & Lenarz, T. (2009). Auditory midbrain implant: A review. *Trends in Amplification*, 13(3), 149–180.
Lin, H. W., Furman, A. C., Kujawa, S. G., & Liberman, M. C. (2011). Primary neural degeneration in the Guinea pig cochlea after reversible noise-induced threshold shift. *Journal of the Association for Research in Otolaryngology*, 12(5), 605–616.
McCreery, D. G., Shannon, R. V., Moore, J. K., Chatterjee, M., & Agnew, W. F. (1998). Accessing the tonotopic organization of the ventral cochlear nucleus by intranuclear microstimulation, *IEEE Transactions on Rehabilitation Engineering*, 6(4), 391–399.
Moore, J. K. (1987). The human auditory brainstem as a generator of auditory evoked potentials. *Hearing Research*, 29, 33–43.
Moore, J. K., & Osen, K. K. (1979). The cochlear nuclei in man. *American Journal of Anatomy*, 154, 393–417.
Nevison, B., Laszig, R., Sollmann, W. P., Lenarz, T., Sterkers, O., Ramsden, R., Fraysse, B., Manrique, M., Rask-Andersen, H., Garcia-Ibanez, E., Colletti, V., & von Wallenberg, E. (2002). Results from a European clinical investigation of the nucleus multichannel auditory brainstem implant. *Ear & Hearing*, 23(3), 170–183.
Osen, K. K., & Jansen, J. (1965). The cochlear nuclei in the common porpoise, Phocaena Phocaena. *Journal of Comparative Neurology*, 125, 223–257.
Otto, S. A., Brackmann, D. E., Hitselberger, W. E., Shannon, R. V., & Kuchta, J. (2002). The multichannel auditory brainstem implant update: Performance in 60 patients. *Journal of Neurosurgery*, 96, 1063–1071.
Otto, S. R., Shannon, R. V., Brackmann, D. E., Hitselberger, W. E., McCreery, D. B., Moore, J. K., & Wilkinson, E. P. (2008). Audiological outcomes with the penetrating electrode auditory brainstem implant. *Otology & Neurotology*, 29, 1147–1154.
Padilla, M. (2003). *English phoneme and word recognition by nonnative English speakers as a function of spectral resolution and English experience*. PhD dissertation, USC, Los Angeles.
Rhode, W. S., & Greenberg, S. (1994). Encoding of amplitude modulation in the cochlear nucleus of the cat. *Journal of Neurophysiology*, 71, 1797–1825.
Sachs, M. B., & Abbas, P. J. (1974). Rate versus level functions for auditory-nerve fibers in cats: Tone-burst stimuli. *Journal of the Acoustical Society of America*, 56, 1835–1847.
Sachs, M. B., & Young, E. D. (1979). Encoding of steady-state vowels in the auditory nerve: Representation in terms of discharge rate. *Journal of the Acoustical Society of America*, 66, 470–479.
Shannon, R. V., & Otto, S. R. (1990). Psychophysical measures from electrical stimulation of the human cochlear nucleus. *Hearing Research*, 47, 159–168.
Shannon, R. V., & Colletti, V. (2005). Evidence from auditory brainstem implants of a modulation specific auditory pathway that is critical for speech recognition. *Association for Research in Otolaryngology Abstracts*, p. 184.
Shannon, R. V., Fayad, J., Moore, J. K., Lo, W., O'Leary, M., Otto, S., & Nelson, R. A. (1993). Auditory brainstem implant. II: Post-surgical issues and performance. *Otolaryngology, Head and Neck Surgery*, 108, 635–643.
Shannon, R. V., Zeng, F.-G., Wygonski, J., Kamath, V., & Ekelid, M. (1995). Speech recognition with primarily temporal cues. *Science*, 270, 303–304.
Shannon, R. V., Fu, Q.-J., & Galvin, J. (2004). The number of spectral channels required for speech recognition depends on the difficulty of the listening situation. *Acta Oto-Laryngologica Supplementum*, 552, 50–54.
Skarzynski, H., Szuchnik, J., Lorenz, A., & Zawadzki, R. (2000). First auditory brainstem implant in Poland: Auditory perception results over 12 months. *Journal of Laryngology and Otology Supplement*, 27, 44–45.
Spahr, A. J., Dorman, M. F., & Loiselle, L. H. (2007). Performance of patients using different cochlear implant systems: Effects of input dynamic range. *Ear & Hearing*, 28(2), 260–75.

Starr, A., McPherson, D., Patterson, J., Don, M., Luxford, W, Shannon, R., Sininger, Y., Tonakawa, L., & Waring, M. (1991). Absence of both auditory evoked potentials and auditory percepts dependent on timing cues. *Brain*, 114, 1157–1180.

Wilson, B. S., Lawson, D. T., Finley, C. C., & Wolford, R. D. (1993). Importance of patient and processor variables in determining outcomes with cochlear implants. *Journal of Speech & Hearing Research*, 36, 373–379.

Winter, I. M., Robertson, D., & Yates, G. K. (1990). Diversity of characteristic frequency rate-intensity functions in guinea pig auditory nerve fibers. *Hearing Research*, 45, 191–202.

Ye, Y., Machado, D. G., & Kim, D. O. (2000). Projection of the marginal shell of the anteroventral cochlear nucleus to the olivocochlear neurons in the cat. *Journal of Comparative Neurology*, 420, 127–138.

Zeng, F. G., Oba, S., Garde, S., Sininger, Y., & Starr, A. (1999). Temporal and speech processing deficits in auditory neuropathy. *NeuroReport*, 10(16), 3429–3435.

Zeng, F. G., Kong, Y. Y., Michalewski, H. J., & Starr, A. (2005). Perceptual consequences of disrupted auditory nerve activity. *Journal of Neurophysiology*, 93(6), 3050–3063.

Zhang, T., Dorman, M. F., Fu, Q. J., & Spahr, A. J. (2012). Auditory training in patients with unilateral cochlear implant and contralateral acoustic stimulation. *Ear & Hearing*, 33(6), 133–141.

Chapter 29
My Dull Deaf Ears: Four Millennia of Acquired Hearing Loss

Jochen Schacht

Jochen Schacht

J. Schacht (✉)
Kresge Hearing Research Institute, The University of Michigan,
1150 West Medical Center Drive, Ann Arbor, MI 48109-5616, USA
e-mail: schacht@umich.edu

A.N. Popper and R.R. Fay (eds.), *Perspectives on Auditory Research*, Springer Handbook of Auditory Research 50, DOI 10.1007/978-1-4614-9102-6_29,

29.1 Where We Are Coming from, Where We Are Going

My wasting lamps some fading glimmer left,
My dull deaf ears a little use to hear

Shakespeare's aging Aegeon might have phrased the problem most poetically (Comedy of Errors, Act V, Scene 1), but he was not the first one to complain about the deterioration of his senses. Hippocrates (c. 400 BCE) already listed "dullness of hearing" as one of the ailments afflicting elderly people, and many generations earlier (c. 2200 BCE) Ptahhotep of Egypt bemoaned old age with its infirmities including deaf ears (Adams, 1886; Grajetzki & Quirke, 2002). Likewise, diseases of the ear, although we do not know of which origin, were mentioned in ancient (2200–1122 BCE) Chinese medical texts (Kong et al., 2006). Drug-induced and noise-induced hearing loss do not have such a venerable history as age-related hearing impairment, but their first mention goes back at least to Avicenna in the 10th century, noting the auditory toxicity of mercury vapors, and to Ambroise Paré in the 16th century, who diagnosed gunners losing their hearing due to "great thunderous noise, large bells and artillery" (Hawkins & Schacht, 2008).

Four millennia later we are facing the same problems. Beginning at about 40 years of age, and in men earlier than in women, we start losing our hearing acuity to a noticeable extent so that by age 70 (now considered the "young old") one half of the population experiences presbycusis, age-related hearing loss. The outlook seems even worse for today's young generation. The self-inflicted sonic pollution of our environment and the use of personal music players or the unabashed enjoyment of live concerts or clubs will result in a much greater level of age-related (and partially noise-induced) hearing loss. Fortunately, the Guinness Book no longer publishes world records in loudness of rock concerts; the last entry was a whopping 126 dB SPL by *The Who* in 1976, and other bands subsequently proudly proclaimed up to 137 dB SPL. These levels are now being dwarfed in national and international "dB drag racing" competitions where 150–160 dB SPL are routinely reached and where the current record stands at an incredible 181.6 dB SPL (dBDRA, 2012). Unfortunately for our children and grandchildren, safety standards for recreational sound exposure have not yet been set.

The majority of noise-induced hearing loss is, however, associated with occupational settings (Masterson et al., 2012). More than 20 million workers in the United States are regularly exposed to potentially damaging noise, and the World Health Organization estimates that 10 % of the world's population is at risk for hearing loss. Specifically, military personnel are affected in all countries. Tinnitus was the most prevalent disability for military veterans in the United States in 2010, followed by hearing loss, for a total of more than 40 % of all claims for compensation (Yankaskas, 2013).

Yet another bane of acquired hearing loss is precisely those drugs that help us survive life-threatening infections. There are many potentially ototoxic drugs, but I will not consider in this essay the anticancer agent cisplatin that affects the ears of an estimated 75–100 % of patients and the numerous other agents that might cause sporadic or reversible auditory effects. I will focus on aminoglycoside antibiotics, which are still essential as a treatment option for Gram-negative pathogens almost seven decades after their discovery and used by millions of people. They can cause hearing loss in 10–20 % of patients when given in a short course of a week or so; the incidence reaches 100 % in long-term treatment for tuberculosis.

What behooves us, then, is to reverse four millennia of history and provide the guidance on how to preserve our hearing. I did not start my scientific career with such a goal in mind. Rather, I was happily deciphering enzymatic mechanisms in anaerobic *E. coli* and, later, labeling polyphosphoinositides with ^{32}P in goldfish brains. Hearing research was not on my horizon but by serendipity (and an enticing job offer from Merle Lawrence and Joe Hawkins) became my fascination for the last 40 years, particularly the question of why aminoglycosides kill our precious hair cells. I knew nothing of these drugs when Joe Hawkins initiated me to them in 1972. In our long professional association and friendship he not only introduced me to those drugs but to everything I know about the auditory system. Lucky is the scientist who had a mentor like him.

And after my 40 years in research there are still many unanswered questions; unwritten grant applications; and stacks of notes about what we must do, should do, and might do. And what we should not do. The following essay is a walk through my scrapbooks.

29.2 On the Road to the Cure

If our ultimate goal is to eradicate acquired hearing loss, then we must first elucidate the underlying mechanisms that lead to the demise of hair cells or the degeneration of their associated nerve fibers. Such knowledge would take protection out of a largely empirical realm into rational and targeted interventions. I will focus on hair cells, a subject close to my heart and the work of my laboratory.

29.2.1 High Hopes

For a short while there were high hopes to cure all acquired hearing loss. The "silver bullet" of pharmacological protection seemed close at hand when the first evidence emerged that the formation of reactive oxygen species (ROS) might be a common denominator of the triad of drug-induced, noise-generated, and age-related hearing pathologies (Yamane et al., 1995; Priuska & Schacht, 1995; Clerici et al., 1996; Kopke et al., 1999). This notion is, to some extent, still correct, but reality soon reared its ugly head showing that the nature of the ROS and the mechanisms of their generation were quite varied and that the downstream molecular responses were considerably more complex than first assumed. Although not the universal remedy, antioxidant treatment to neutralize ROS has nevertheless been highly successful in animal models of drug-induced hearing loss (see Xie et al., 2011). Further, our clinical trial of co-administration of aspirin with gentamicin demonstrated a 75 % reduction of the incidence of hearing loss, providing a first and essential proof of principle that animal experiments in this field can be translated to the clinic (Sha et al., 2006). Antioxidants are also suggestively successful in averting noise damage in animals and are being explored for their efficacy in clinical trials (see Oishi & Schacht, 2011). Age-related hearing loss remains an enigma: although accompanied by ROS formation (Jiang et al., 2007), a causal relationship as a basis for protection has remained rather tentative. After some initial encouraging results (Seidman et al., 2000), recent studies have failed to attenuate presbycusis by boosting mitochondrial energy supplies or by long-term feeding with antioxidant supplements (Bielefeld et al., 2008; Sha et al., 2012).

We should not be surprised by such setbacks. The complexity of the mechanisms underlying acquired hearing loss reflects the drastically divergent nature of the noxious stimuli that target the inner ear: an acute, high-impact sound; a week-long drug treatment that allows the affected tissues time to muster intermediate defense strategies; or a life-long aging process into which we have little insight. A compromised physiological state is also prone to influence a patient's treatment success; undernourishment or disease, for example, might lower endogenous antioxidant defense mechanisms and hence render a person more susceptible to ototoxic insults. We learned this first hand when glutathione supplementation attenuated gentamicin-induced hearing loss in guinea pigs in one of our studies (Garetz et al., 1994) but not in an attempt to repeat it (Lautermann et al., 1995). The explanation of this dilemma was the fact that the animals in the first study carried an infection while those in the second study were healthy and did not require an artificial boost to their resistance mechanisms.

Another point that we have to consider: Acquired (and, therefore, by definition not genetic) hearing loss is modulated by genetic factors. All three pathologies that I am discussing here reside at an intersection of environmental influences and genetic predispositions, confounding the interpretation of potential mechanisms or the establishment of prospective therapies.

29.2.2 Bespoke Interventions

Today's trend toward "personalized medicine" takes into account such genetically influenced individual responses to disease-causing factors, as well as to medications ("pharmacogenetics"). Dealing with acquired hearing loss would benefit from a similar approach. Some examples might help to illustrate my point.

We must suspect genetic factors, mostly as yet unknown, to be at work in determining the incidence and severity of aminoglycoside ototoxicity because only a fraction of patients undergoing chemotherapy sustain hearing loss, at least from a short course of treatment. One of the more striking examples of genetically influenced susceptibility is the A1555G mitochondrial mutation which confers extreme sensitivity to even a single injection of aminoglycosides (Fischel-Ghodsian, 2005).

Screening for as yet unknown "susceptibility" genes or alleles would steer these patients away from aminoglycosides or at least would allow us to seek protective measures for carriers of such genes or mutations. This targeted approach would circumvent the common reluctance in the medical community—which I have repeatedly encountered—to treat all patients preventively with antidotes when only a few might need them.

Individual susceptibility and variable responses to protective measures mar our current inroads into noise-induced hearing loss as well. Here, again, both genetic and environmental factors collude. For example, based on the results from animal experimentation, the antioxidant *N*-acetylcysteine should be expected to afford protection. However, only a subset of workers employed at a steel manufacturing company benefited from *N*-acetylcysteine supplementation. Those were individuals deficient in glutathione *S*-transferase, one of the enzymes of the cellular antioxidant system; supplementation in others made no difference (Lin et al., 2010). Similarly, dietary magnesium seemed to limit permanent threshold shifts in army recruits undergoing basic military training. However, regardless of treatment or placebo, the degree of threshold shift was low in subjects with high serum Mg^{2+} levels and higher in subjects with low serum Mg^{2+} levels (Attias et al., 1994). Clearly, individual genetics and physiology (in these cases, the expression of antioxidant enzymes or the tendency for hypomagnesemia) determine the vulnerability to trauma and the efficacy of interventions.

Even if ongoing clinical trials arrive at a suitable pharmacological intervention, and even if we tailor to individuals, it will still be necessary to alter our current approaches to protection. Most interventions begin before or commensurate with the exposure to drugs or noise. Such timing is fine for scheduled chemotherapy but does not reflect the dire situation, say, on a battlefield. Two modifications are essential in this scenario. First, daily blanket protection of an entire battalion (or, for that matter, of an entire factory) is logistically prohibitive and could be compromised by noncompliance. Targeting susceptible individuals (who ought not be in the endangering environment to begin with) might help. But it is more imperative to design a posttraumatic rescue for those individuals for whom noise exposure has indeed reached a dangerous level, information that can now be gained with personal pressure sensors. Intervention post factum should be possible, as animal experimentation suggests (Yamashita et al., 2005), but the temporal "window of rescue" for human patients and the most appropriate medication need further exploration.

A similar challenge exists for presbycusis. A life-long dietary or pharmacological regimen will have problems of compliance, let alone the unanswered question of adverse health effects of long-term administration, even of nutritional supplements. The fact that potentially beneficial vitamins and antioxidants may adversely affect certain parts of the population (β-carotene may enhance the risk for lung cancer in smokers; Goralczyk, 2009) exemplify the need of thorough scrutiny. A just-in-time treatment and then only for individuals with risk factors is not yet on the horizon but seems inescapable.

The sum of our current knowledge demonstrates that no single treatment can fit all ototoxic traumata and all individuals. That does not mean that I am pessimistic about achieving success: in addition to establishing more precise molecular mechanisms, the characterization of individual susceptibility will aid in our quest for better treatments. Success is not out of reach; it might just be harder to come by than an experiment in an inbred mouse strain suggests.

29.3 The Scientist and Her Models

The search for answers in the laboratory is intimately linked to the search for the most appropriate model in which to ask the questions. Even in the relatively narrow field of acquired hearing loss a variety of approaches have been used that have given us useful bits and pieces towards solving the puzzles but also sometimes conflicting information. I will take a brief look at some popular models for in vivo and in vitro studies.

29.3.1 Alive and Well (?)

The morphological and physiological similarities of the cochlea between mammalian species seems to suggest in vivo models as most appropriate. However, differences between animals and humans can exist in precisely the pathways that might be involved in acquired hearing loss and protection. Let's take the involvement of ROS in auditory pathologies and antioxidant intervention as an example. Humans and guinea pigs require dietary antioxidant vitamin C but rats and mice synthesize their own. Perhaps mice and rats have an advantage over us in maintaining redox homeostasis thereby skewing the experimentally observed responses in these species?

The various C57 mouse strains, frequently employed in presbycusis research, carry a mutation in cadherin 23 that predisposes to accelerated hearing loss. Is this genetic defect a welcome aid in elucidating mechanisms of age-related hearing loss or more of a confounding factor? In humans, Cdh23 mutations are present in type I Usher syndrome, but a connection of this mutation with presbycusis is tenuous. To the contrary, a recent population study in 1175 subjects found no association between Cdh23 mutations and age-related hearing impairment (Hwang et al., 2012). This fact challenges the extrapolation of results from C57 mice to presbycusis and renders a translation of any ameliorating treatments to the clinic rather questionable. I prefer to see research efforts focus on animals or strains that do not carry a disease-causing mutation and better approximate the slow progression of age-related hearing loss in humans.

Finally, I want to make the point that we should embrace diversity in our models. Given the heterogeneity of the human population, inbred animal strains may not be appropriate models. Outbred and hence genetically heterogeneous strains provide more robust results, that is, results more likely to be reproducible in other populations of experimental animals and perhaps more realistic for clinical translation. The National Institute on Aging set an example when it adopted a four-way cross mouse population as its standard stock for aging research (Miller et al., 1999). Each mouse in the progeny is genetically unique, but each shares a random 50 % of its genetic heritage with each other mouse. The advantages of this breed include robustness and genetic tractability, features that we exploited in a recent study on alleles modulating late-life hearing (Schacht et al., 2012) and that deserve to be exploited further.

Animal models are also essential to test new medications before they enter translational research or the market. Assessment of hair cell loss or functional deficits in auditory performance are the current criteria for determining the detrimental or the protective potential of a compound. Surely, in vivo studies of ototoxicity will remain indispensable as a step toward clinical trials, but drug screening could get a boost from recent developments in pharmacology that have yet eluded the auditory field.

29.3.2 Fishing for Drugs: Screening Tests

Advanced drug searches in modern medicine use transcriptome matching or target identification by molecular fit computations of drugs with human protein structures (Dakshanamurthy et al., 2012). This is particularly effective for "repurposing" approved drugs for new therapies and thousands of compounds can be assessed in a short time. Novel interactions can be quickly confirmed by direct binding studies and then extended to biochemical assays and in vivo experiments. As fruitful as this method is, we are still far from being able to apply it to acquired hearing loss. We do not know which specific proteins are involved in the cellular response to noise trauma or aging and which, hence, would need to be stimulated, inhibited, or in any other way modified in order to achieve a protective response. Moreover, chances are that both noise and age elicit complex metabolic changes that are not easily amenable to a single targeted intervention.

In contrast to noise and age, however, we should suspect specific drug-binding sites for aminoglycosides that might allow for molecular fit computations. At the moment the field suffers from an overabundance of suggested proteins and enzymatic reactions influenced by these drugs and a dearth of information as to which of these interactions (if any single one) is causally related to ototoxicity. Furthermore, not all drug targets might be proteins. There is growing evidence that the affinity of aminoglycosides to mitochondrial RNA is a key feature of ototoxicity (Matt et al., 2012). But even in this case, proteomics or metabolomics can still provide us with information and assess drug effects on expression profiles of potential markers of toxicity independent of knowledge of the target. This methodology is currently suitable for tissues such as kidney or liver (Collins et al., 2012) but needs to be further developed in order to apply to the inner ear.

No current screening system is perfect to the human auditory system and I find it even surprising that the lateral line organ of zebra fish larvae (related to the vestibular system, not the cochlea) and the postnatal murine organ of Corti (immature) both yield results that appear somewhat capable of predicting drug ototoxicity.

Within limitations, though; there exist bothersome exceptions that call for multiple screens for added robustness of the prediction (Chiu et al., 2008; Brand et al., 2011). Dose dependency and timing might be confounding factors in any one model and the use of more than a single condition or system appears mandatory. For such reasons we might be tempted to consider cochlear cell lines as additional screening systems, but those have also elicited strong criticism (Chen et al., 2011).

Until we know more about the molecular targets of ototoxic agents, the screening for otoprotective pharmaceuticals will continue in the tedious fashion of morphological observation of hair cell death and its functional consequences. But I prefer alternative solutions for a future chemotherapy without the risk of ototoxicity.

29.4 Let's Stop Cleaning Up the Mess: Development of Non-ototoxic Drugs

Although my laboratory has spent years working on this topic, I consider protection essentially a crutch, an exhausting process of cleaning up after damage has been done through the introduction of ototoxic drugs. Historically, physicians and scientists were unexpectedly confronted with adverse effects of new medications and then had to adjust therapeutic regimens or look for alternatives. Streptomycin and cisplatin are cases in point where ototoxicity only became manifest during their first clinical applications. Aggravating the problem, protection is only a short-term solution that ignores today's major challenge in chemotherapy with aminoglycosides; microbial resistance proceeds at an unprecedented pace. As of 2012, multi-drug-resistant tuberculosis is present in virtually all countries surveyed by the World Health Organization and 9 % of these cases constitute extensively drug-resistant tuberculosis (WHO, 2012). The trend is accelerating and the prospect of untreatable diseases is threatening. The long-term solution for future chemotherapy, therefore, is the development of new drugs that overcome current bacterial resistance yet are intrinsically free of ototoxic potential. This process requires a multidisciplinary approach, and toxicity testing must guide the drug development process rather than being tacked on at the end of it.

Examples of this new approach are "designer aminoglycosides." These novel derivatives fall into two major categories: drugs for the traditional role as broad-spectrum antibacterials and for the more recently developed therapeutic applications to mitigate genetic disorders by their ability to suppress disease-causing stop codons. Tired of cleaning up the ototoxic mess, we have recently teamed up with two groups in order to eliminate ototoxicity from the start. The tactic works in a highly promising fashion. Based on a mechanistic concept that postulates a key role for the mitochondria in aminoglycoside ototoxicity, we evaluated the correlation between affinity of drugs to the mitochondrial ribosome (mitoribosome) and ototoxicity. Surprisingly, we rediscovered an old drug (Matt et al., 2012): Apramycin, a structurally unique aminoglycoside antibiotic in veterinary use since the 1970s, shows little ototoxicity while exhibiting strong antibacterial activity even against strains resistant to currently clinically used aminoglycosides. Based on this proof-of-concept that antibacterial activity can be dissected from ototoxicity, the door is open for the development of further aminoglycoside derivatives with even lower ototoxic potential, eventually eliminating the need for supplemental treatment with protective agents.

The second team effort explored the hypothesis that it is possible to separate the structural elements of aminoglycosides that cause toxicity from those that are required for inducing nonsense suppression. Nonsense suppression is the therapeutic application of aminoglycosides to alleviate the adverse consequences of certain genetic disorders. Many human genetic diseases and numerous types of cancer are caused by single-point alterations in DNA, creating incorrect stop codons in coding regions and leading to the premature termination of translation and to nonfunctional proteins. Such nonsense mutations represent about 12 % of all mutations reported, including allelic variants of cystic fibrosis, Duchenne muscular dystrophy, Usher syndrome, and Hurler syndrome. Aminoglycosides can promote a selective translational read-through of premature stop codons, restoring (partial) expression of a full-length protein. Preliminary clinical trials have shown the potential efficacy of aminoglycosides in patients with genetic disorders but hearing loss through the life-long application of these drugs is a threat. We are on the way to nontoxic derivatives that will minimize such risks, again obviating the need for additional protective procedures (Nudelman et al., 2009).

Nevertheless, the search for better protective agents can still be useful if we find drugs that are already approved in other contexts and could immediately put to use. However, if we have to resort to clinical trials in order to test novel compounds, I would rather see our efforts (and money) go toward establishing a safe chemotherapy with non-ototoxic medications.

29.5 A Closer Look at Death and Dying

In my discourse on the translational aspects of protection against acquired hearing loss I have frequently mentioned the need for more basic information on mechanisms of otopathology. My guess is that investigations into hair cell pathology will continue as a mainstay of research and that we will collect more information on details of cell death and survival pathways. Such work will round out the emerging picture of a glut of molecular responses that—not surprisingly—largely follow canonical pathways already established in other systems.

So, where will really novel information come from? Perhaps we should take the road less traveled in auditory research and venture into regions ranging from neglected cochlear structures all the way into our gut. Here are some hopefully stimulating ideas.

29.5.1 Secrets of Shiny Tiny Droplets

Supporting cells of the mammalian cochlea are intriguing creatures. While the term "supporting" was originally coined for structural anatomical reasons (without Deiters' cells the outer hair cells would be blowing in the wind) we have plenty of evidence now that they also support both life and death in the cochlear neuroepithelium. Exploration of their role has been somewhat neglected because one of the hallmarks of otopathology is the loss of hair cells, which have therefore garnered most of our attention. On the other hand, supporting cells take up aminoglycosides, develop reactive oxygen species, express death-promoting signals, and eventually engulf and dispose of dying hair cells. They may be facilitators of hair cell death through the activation of trauma-signaling pathways (Lahne & Gale, 2008) and, conversely, may be promoters of cell survival. They respond to homeostatic signaling by ATP and acetylcholine and also might be involved in protection by, for example, glucocorticoids such as dexamethasone. Annexin A1 is stored inside Hensen cells within lipid droplets from which glucocorticoids drive it into the external milieu as an anti-inflammatory mediator (Kalinec et al., 2009).

The shiny "lipid droplets" have long been observed prominently in the cochlea in Hensen cells, although any cell type can contain these structures. They have mostly been ignored as inert storage depots but that notion seems to be a huge underestimation of their function. Cytoplasmic lipid droplets are well preserved evolutionarily from bacteria to yeast, to plants, to invertebrates, and to humans and are beginning to be recognized as dynamic organelles with complex functions. True, lipid droplets can store excess fatty acids as an energy source or to safeguard against apoptosis. However, proteomic analysis has revealed hundreds of proteins belonging (not surprisingly) to lipid metabolism but also to membrane trafficking, regulatory signaling, and protein degradation (Hodges & Wu, 2010). Abnormal metabolism in these multifunctional organelles has been linked to a variety of metabolic diseases, including diabetes, atherosclerosis, obesity, and cancer (Greenberg et al., 2011).

The involvement of lipid droplets in inner ear physiology or pathology remains entirely speculative. There is, however, a tantalizing hint of a link: mutations in the gene *C2ORF43* might be associated with hearing loss (Currall et al., 2012) and its protein product UPF0554 has been found in cytosolic lipid droplets, albeit from enterocytes (Bouchoux et al., 2011). In any case, the emerging information on the importance of this organelle in the development of many diseases should prompt a closer look at lipid droplets and at supporting cells in general. Interestingly, and perhaps relevant for a cochlear connection, endoplasmic reticulum stress (ER stress) promotes the formation of lipid droplets. ER stress can be caused by a variety of biochemical and pharmacological stimuli and might accompany cochlear pathologies because ER stress can lead to oxidative stress and vice versa. Lipid droplets can also be formed following mitochondrial dysfunction, a potential consequence of drug treatment, noise trauma, or aging in the cochlea. The connection of lipid droplets to ER stress and inflammatory mediators may just be the tip of an iceberg that could sink our sensory cells.

29.5.2 *Modifying the Message: Epigenetics*

I find it surprising that despite the explosive awareness of epigenetics in shaping health and disease of cells and organs, this topic has only made few inroads into cochlear physiology and pathology. While the genome holds the information for every cell's potential, modifications to the DNA itself or to the transcriptional machinery determine the expression of the information and the differentiation and fate of individual cells. DNA methylations or histone modifications are major functionally relevant mechanisms to steer differentiation. Very importantly, these mechanisms are fluent and able to respond to external stimuli in order to modulate the phenotype. Epigenetic changes not only continually reprogram gene expression during the life time of an individual, they might also be inheritable, passing "experiences" to later generations (Jablonka, 2012).

Possible lifetime influences on the epigenome are not limited to obvious noxious environmental conditions, although chemical exposure and drugs (and not just ototoxic drugs) loom large. Epigenetic changes occur with aging and play a role in changing cell physiology into pathology in diseases, among them cancer, obesity, diabetes, and nervous system disorders. Epigenetic mechanisms have been discerned in auditory development where they are part of the expected machinery of differentiation and in sensory regeneration regulating cell proliferation (Slattery et al., 2009). In addition, we must suspect ototoxic drugs, age, noise trauma and noise conditioning as modifiers of the cochlear epigenome. As a case in point, aminoglycoside antibiotics alter histone deacetylation in the cochlea, and histone deacetylase inhibitors have a profound mitigating influence on ototoxicity (Chen et al., 2009); histone modifications may also occur in spiral ganglion cells during aging (Watanabe & Bloch, 2013). But I would like to speculate further on epigenetic changes in some specific aspects of auditory pathology.

29.5.2.1 Aminoglycosides

Beginning with the early use of aminoglycosides, the notion has been spread among clinicians that patients who once received the drugs become more sensitive to the ototoxic effects of a second application, even months or years later. Drugs may persist in cochlear cells for a while but epigenetic changes can last a lifetime and might provide a better hypothesis for this observation. And if this is so, then we should be very concerned about babies in intensive care and infants receiving aminoglycosides. They might reap the negative rewards of drug treatment as aging adults and we better follow up on their late-life hearing.

29.5.2.2 Noise Exposure

We now know that youthful sins of exposing our ears to (seemingly) sub-damaging sounds will have dire consequences in old age (Kujawa & Liberman, 2006).

Again, I suggest that epigenetic changes induced by Ludwig van Beethoven (just think "Wellingtons Sieg"), *The Who* or your favorite dB racing team modulate our sensory organ's late-life performance.

29.5.2.3 Presbycusis

We do not have an established method to delay or ameliorate presbycusis despite all efforts and suggestive leads in animal experimentation. On the contrary and disconcerting for adherents of antioxidant supplementations, a recent prospective, placebo-controlled, double-blind, and randomized trial of antioxidant treatment in presbycusis (Polanski & Cruz, 2013) found no significant effect of any of the tested drug combinations. Not all is lost, though. Three studies on populations from three different continents have shown that older people who are moderate consumers of alcohol retain better hearing (Popelka et al., 2000; Fransen et al., 2008; Gopinath et al., 2010). These observations fit well with suggestions that light wine or alcohol intake is beneficial to health and might increase life expectancy (Streppel et al., 2009). I like to speculate that epigenetics is at work to save our ears because nutrition may modulate epigenetic events associated with disease states ((Hardy & Tollefsbol, 2011). Alcohol is one of the confirmed bioactive food ingredients that can affect DNA methylation or histone modifications, as are polyphenols such as resveratrol found in red wine (Vanden Berghe, 2012). A daily glass of wine as an epigenetic modifier and presbycusis antidote might appeal to many of us. Teetotalers will have to resort to a pill.

The way I see it, there is a good case to be made that acquired hearing loss is associated with and modulated by epigenetic changes. Once we elucidate those changes, the outlook to preserve hearing or ameliorate hearing loss seems promising because epigenetic changes can be reversed or modified not only by appropriate drugs but also by lifestyle (Alegria-Torres et al., 2011).

29.5.3 My Gut Feeling: It's the Microbiome

Our body harbors far more genetic material than is present in our own cells. Microbial cells outnumber our own by a factor of ten to one. The microbiome that developed with us during evolution is an integral part of our body and plays an almost invisible but important role in shaping our phenotype. The Human Microbiome Project has recognized its importance and announced a major milestone in June 2012 with a database on more than 10,000 commensal microbial species. We tend to tacitly accept the benefits of our gut microbiota in such daily tasks as digestion and the supply of some vitamins. When we become aware of our tenants it is mainly in the context of disease, although we might not even then clearly recognize its contributions. But we must accept the emerging reality that changes in the microbiota composition may be linked to altered immune responses, inflammation, liver injury, even to the determination of progression of obesity and cancer, and potentially of cardiovascular disease and rheumatoid arthritis (Cho & Blaser, 2012).

What do we know about a link between the microbiome and acquired hearing loss? Nothing. We should be suspicious, though, of its contributions and should pay more attention (and perhaps a little research money) to our boarders. In aging populations, microbiota composition correlates with frailty, comorbidities, and markers of inflammation. The interconnected processes of age-related physiological changes in the gastrointestinal tract and bacterial metabolism that contribute to the common symptom of chronic subclinical inflammation might also be detrimental to the preservation of a youthful hearing.

Animal experiments have long told us that nutritional status and general health can modulate the severity of noise trauma or antibiotic ototoxicity and we have attributed this phenomenon to external influences on internal cellular homeostasis. However, here it becomes interesting: just as in the case of epigenetics, microbial composition and function can be affected by diet. Rather than changing the physiology of our body's own cells with dietary supplements we might unknowingly be changing the composition and metabolism of our intestinal fauna. A recent clinical study (Queipo-Ortuño et al., 2012) showed the positive effect of red wine polyphenols on promoting a beneficial intestinal flora. So we are back to an intriguing circle of hearing loss and preservation, epigenetics, the microbiome, and red wine.

29.6 Afterthought

> A good traveler has no fixed plans
> and is not intent upon arriving.
> A good artist lets his intuition
> lead him wherever it wants.
> A good scientist has freed herself of concepts
> and keeps her mind open to what is.
>
> Lao-Tse (~6th century BCE): Tao Te Chin

It is inherent in scientific curiosity to speculate. However, although crystal balls might hold all the information on the future, we are limited in what we are seeing in them by our own imagination. So, after having filled these pages with suggestions and predictions, let me ask: if we can map out the directions of our research, will this really bring us forward? Or are we limiting ourselves to what we can envision? Let's hope that new and unexpected discoveries will meet us, those that were not planned in a grant application and instead arose from serendipity or were borne out of utter failures. Unforeseen breakthroughs and insights have advanced our knowledge in the past by leaps and bounds. Perhaps we should not think so much; just sit back, relax, and be ready when great ideas cross our way. With a glass of wine in our hands, of course. (Disclaimer: I do not own stock in wineries).

Acknowledgments I am grateful to my mentors, who tried their best to educate me and also showed me that the pursuit of science can be both intellectually rewarding and fun, most of all Joachim Knappe, my PhD advisor; Bernard W. Agranoff, my postdoctoral mentor; and Joseph Hawkins, my faculty colleague and friend. My thanks also go to all students, fellows, associates, and colleagues who worked with me over the last 40 years or so and helped to generate ideas and turn these ideas (or at least some of them) into results. My colleagues Rick Altschuler and Keith Duncan commented on this essay, and Debbie Stout and Andra E. Talaska provided editorial assistance. Finally, I am thankful to NIH for having supported my work over the decades, currently with grant DC003685 from the National Institutes of Deafness and Other Communication Disorders.

References

Adams, A. (1886). *The genuine works of Hippokrates translated from the Greek with a preliminary discourse and annotations*, Vol. 2. New York: William Wood & Company.

Alegria-Torres, J. A., Baccarelli, A., & Bollati, V. (2011). Epigenetics and lifestyle. *Epigenomics*, 3(3), 267–277.

Attias, J., Weisz, G., Almog, S., Shahar, A., Wiener, M., Joachims, Z., Netzer, A., Ising, H., Rebentisch, E., & Guenther, T. (1994). Oral magnesium intake reduces permanent hearing loss induced by noise exposure. *American Journal of Otolaryngology*, 15(1), 26–32.

Bielefeld, E.C., Coling, D., Chen, G.D., & Henderson, D. (2008). Multiple dosing strategies with acetyl L-carnitine (ALCAR) fail to alter age-related hearing loss in the Fischer 344/NHsd rat. *Journal of Negative Results in Biomedicine*, 7, 4.

Bouchoux, J., Beilstein, F., Pauquai, T., Guerrera, I.C., Chateau, D., Ly, N., Alqub, M., Klein, C., Chambaz, J., Rousset. M., Lacorte, J.M., Morel, E., & Demignot, S. (2011). The proteome of cytosolic lipid droplets isolated from differentiated Caco-2/TC7 enterocytes reveals cell-specific characteristics. *Biology of the Cell*, 103(11), 499–517.

Brand, Y., Setz, C., Levano, S., Listyo, A., Chavez, E., Pak, K., Sung, M., Radojevic, V., Ryan, A. F., & Bodmer, D. (2011). Simvastatin protects auditory hair cells from gentamicin-induced toxicity and activates Akt signaling in vitro. *BMC Neuroscience*, 12, 114.

Chen, F. Q., Schacht, J., & Sha, S. H. (2009). Aminoglycoside-induced histone deacetylation and hair cell death in the mouse cochlea. *Journal of Neurochemistry*, 108(5), 1226–1236.

Chen, F-Q., Hill, K., Guan, Y-J., Schacht, J., & Sha, S-H. (2011). Activation of apoptotic pathways in the absence of cell death in an inner-ear immortomouse cell line. *Hearing Research*, 284, 33–41.

Chiu, L. L., Cunningham, L. L., Raible, D. W., Rubel, E. W., & Ou, H. C. (2008). Using the zebrafish lateral line to screen for ototoxicity. *Journal of the Association for Research in Otolaryngology*, 9(2), 178–190.

Cho, I., & Blaser, M. J. (2012). The human microbiome: at the interface of health and disease. *Nature Reviews Genetics*, 13(4), 260–270.

Clerici, W. J., Hensley, K., DiMartino, D. L., & Butterfield, D. A. (1996). Direct detection of ototoxicant-induced reactive oxygen species generation in cochlear explants. *Hearing Research*, 98, 116–124.

Collins, B. C., Miller, C. A., Sposny, A., Hewitt, P., Wells, M., Gallagher, W. M., & Pennington, S. R. (2012). Development of a pharmaceutical hepatotoxicity biomarker panel using a discovery to targeted proteomics approach. *Molecular and Cellular Proteomics*, 11(8), 394–410.

Currall, B. B., Robertson, N. G., Lindgren, A. M., Talkowski, M. E., & Morton, C. C. (2012). Identifying genes associated with hearing loss through chromosomal translocations. Abstract 862 T. Presented at the *62nd Annual Meeting of The American Society of Human Genetics*, November 2012, San Francisco CA.

Dakshanamurthy, S., Issa, N. T., Assefnia, S., Seshasayee, A., Peters, O. J., Madhavan, S., Uren, A., Brown, M. L., & Byers, S. W. (2012). Predicting new indications for approved drugs using a proteochemometric method. *Journal of Medicinal Chemistry*, 55(15), 6832–6848.

dBDRA (2012) dB Drag Racing Association. World Records. http://www.termpro.com/asp/worldrecords.asp?ListMethod=1. Accessed 4 Nov. 2012.

Fischel-Ghodsian, N. (2005). Genetic factors in aminoglycoside toxicity. *Pharmacogenomics*, 6(1), 27–36.

Fransen, E., Topsakal, V., Hendrickx, J. J., Van Laer, L, Huyghe, J. R., Van Eyken, E., Lemkens, N., Hannula, S., Mäki-Torkko, E., Jensen, M., Demeester, K., Tropitzsch, A., Bonaconsa, A., Mazzoli, M., Espeso, A., Verbruggen, K., Huyghe, J., Huygen, P. L., Kunst, S., Manninen, M., Diaz-Lacava, A., Steffens, M., Wienker, T. F., Pyykkö, I., Cremers, C. W., Kremer, H., Dhooge, I., Stephens, D., Orzan, E., Pfister, M., Bille, M., Parving, A., Sorri, M., Van de Heyning, P., & Van Camp, G. (2008). Occupational noise, smoking, and a high body mass index are risk factors for age-related hearing impairment and moderate alcohol consumption is protective: A European population-based multicenter study. *Journal of the Association for Research in Otolaryngology*, 9(3), 264–276.

Garetz, S. L., Altschuler, R. A., & Schacht, J. (1994). Attenuation of gentamicin ototoxicity by glutathione in the guinea pig *in vivo*. *Hearing Research*, 77, 81–87.

Gopinath, B., Flood, V. M., McMahon, C. M., Burlutsky, G., Smith, W., & Mitchell, P. (2010). The effects of smoking and alcohol consumption on age-related hearing loss: The Blue Mountains Hearing Study. *Ear and Hearing*, 31(2), 277–282.

Goralczyk, R. (2009). Beta-carotene and lung cancer in smokers: Review of hypotheses and status of research. *Nutrition and Cancer*, 61(6), 767–774.

Grajetzki, W., & Quirke, S. (2002). Old age in Ancient Egyptian literary compositions. University College London. Retrieved from http://www.digitalegypt.ucl.ac.uk/age/ageliterary.html (Accessed October 22, 2012).

Greenberg, A. S., Coleman, R. A., Kraemer, F. B., McManaman, J. L., Obin, M. S., Puri, V., Yan, Q. W., Miyoshi, H., & Mashek, D. G. (2011). The role of lipid droplets in metabolic disease in rodents and humans. *Journal of Clinical Investigation*, 121(6), 2102–2110.

Hardy, T. M., & Tollefsbol, T. O. (2011). Epigenetic diet: Impact on the epigenome and cancer. *Epigenomics*, 3(4), 503–518.

Hawkins, E., Jr., & Schacht, J. (Eds.). (2008). *Sketches of Otohistory*. Basel: Karger.

Hodges, B. B., & Wu, C. C. (2010) Proteomic insights into an expanded cellular role for cytoplasmic lipid droplets. *Journal of Lipid Research*, 51(2), 262–273.

Hwang, J. H., Liu, K. S., Wu, C. C., & Liu, T. C. (2012). Association of cadherin23 single nucleotide polymorphism with age-related hearing impairment in Han Chinese. *Otolaryngology – Head and Neck Surgery*, 147(3), 531–534.

Jablonka, E. (2012). Epigenetic variations in heredity and evolution. *Clinical Pharmacology and Therapeutics*, 92(6), 683–688.

Jiang, H., Talaska, A. E., Schacht, J., & Sha, S-H. (2007). Oxidative imbalance in the aging inner ear. *Neurobiology of Aging*, 28,1605–1612.

Kalinec, F., Webster, P., Maricle, A., Guerrero, D., Chakravarti, D. N., Chakravarti, B., Gellibolian, R., & Kalinec, G. (2009). Glucocorticoid-stimulated, transcription-independent release of annexin A1 by cochlear Hensen cells. *British Journal of Pharmacology*, 158(7), 1820–1834.

Kong, W-J., Hao, J., & Schacht J. (2006). Sketches of otohistory. Part 12: The history of otology in traditional Chinese medicine. *Audiology and Neurotology*, 11, 145–150.

Kopke, R., Allen K. A., Henderson, D., Hoffer, M., Frenz, D., & Van de Water, T. (1999). A radical demise. Toxins and trauma share common pathways in hair cell death. *Annals of the New York Academy of Sciences*, 884, 171–191.

Kujawa, S. G., & Liberman, M. C. (2006). Acceleration of age-related hearing loss by early noise exposure: Evidence of a misspent youth. *Journal of Neuroscience*, 26(7), 2115–2123.

Lahne, M., & Gale, J. E. (2008). Damage-induced activation of ERK1/2 in cochlear supporting cells is a hair cell death-promoting signal that depends on extracellular ATP and calcium. *Journal of Neuroscience*, 28(19), 4918–4928.

Lautermann, J., McLaren, J., & Schacht, J. (1995). Glutathione protection against gentamicin ototoxicity depends on nutritional status. *Hearing Research*, 86(1–2), 15–24.

Lin, C. Y., Wu, J. L., Shih, T. S., Tsai, P. J., Sun, Y. M., Ma, M. C., & Guo, Y. L. (2010). N-Acetylcysteine against noise-induced temporary threshold shift in male workers. *Hearing Research*, 269(1–2), 42–47.

Masterson, E. A., Tak, S., Themann, C. L., Wall, D. K., Groenewold, M. R., Deddens, J. A., & Calvert, G. M. (2012). Prevalence of hearing loss in the United States by industry. *American Journal of Industrial Medicine*, doi: 10.1002/ajim.22082.

Matt, T., Ng, C. L., Lang, K., Sha, S. H., Akbergenov, R., Shcherbakov, D., Meyer, M., Duscha, S., Xie, J., Dubbaka, S. R., Perez-Fernandez, D., Vasella, A., Ramakrishnan, V., Schacht, J., & Böttger, E. C. (2012). Dissociation of antibacterial activity and aminoglycoside ototoxicity in the 4-monosubstituted 2-deoxystreptamine apramycin. *Proceedings of the National Academy of Sciences of the USA*, 109(27), 10984–10989.

Miller, R. A., Burke, D., & Nadon, N. (1999). Announcement: Four-way cross mouse stocks: A new, genetically heterogeneous resource for aging research. *Journals of Gerontology A. Biological Sciences and Medical Sciences*, 54A, B358.

Nudelman, I., Rebibo-Sabbah, A., Cherniavsky, M., Belakhov, V., Hainrichson, M., Chen, F., Schacht, J., Pilch, D. S., Ben-Yosef, T., & Baasov, T. (2009). Development of novel aminoglycoside (NB54) with reduced toxicity and enhanced suppression of disease-causing premature stop mutations. *Journal of Medicinal Chemistry*, 52, 2836–2845.

Oishi, N., & Schacht, J. (2011). Emerging treatments for noise-induced hearing loss. *Expert Opinion on Emerging Drugs*, 16, 235–245.

Polanski, J. F., & Cruz, O. L. (2013). Evaluation of antioxidant treatment in presbycusis: prospective, placebo-controlled, double-blind, randomized trial. *Journal of Laryngology and Otology*, January 15:1–8, Epub ahead of print.

Popelka, M. M., Cruickshanks, K. J., Wiley, T. L., Tweed, T. S., Klein, B. E., Klein, R., & Nondahl, D. M. (2000). Moderate alcohol consumption and hearing loss: A protective effect. *Journal of the American Geriatrics Society*, 48(10), 1273–1278.

Priuska, E., & Schacht, J. (1995). Formation of free radicals by gentamicin and iron and evidence for an iron/gentamicin complex. *Biochemical Pharmacology*, 50, 1749–1752.

Queipo-Ortuño, M. I., Boto-Ordonez, M., Murri, M., Gomez-Zumaquero, J. M., Clemente-Postigo, M., Estruch, R., Cardona Diaz, F., Andres-Lacueva, C., & Tinahones, F. J. (2012). Influence of red wine polyphenols and ethanol on the gut microbiota ecology and biochemical biomarkers. *American Journal of Clinical Nutrition*, 95(6), 1323–1334.

Schacht, J., Altschuler, R. A., Burke, D. T., Chen, S., Dolan, D., Galecki, A. T., Kohrman, D., & Miller, R. A. (2012). Alleles that modulate late life hearing in genetically heterogeneous mice. *Neurobiology of Aging*, 33, 1842.e15–1842.e29.

Seidman, M. D., Ahmad, N., & Bai, U. (2000). Molecular mechanisms of age-related hearing loss. *Ageing Research Reviews*, 1, 331–343.

Sha, S. H., Qiu, J. H., & Schacht, J. (2006). Aspirin to prevent gentamicin-induced hearing loss. *New England Journal of Medicine*, 354(17), 1856–1857.

Sha, S. H., Kanicki, A., Halsey, K., Wearne, K. A., & Schacht, J. (2012). Antioxidant-enriched diet does not delay the progression of age-related hearing loss. *Neurobiology of Aging*, 33(5), 1010 e1015–1016.

Slattery, E. L., Speck, J. D., & Warchol, M. E. (2009). Epigenetic influences on sensory regeneration: Histone deacetylases regulate supporting cell proliferation in the avian utricle. *Journal of the Association for Research in Otolaryngology*, 10(3), 341–353.

Streppel, M. T., Ocké, M. C., Boshuizen, H. C., Kok, F. J., & Kromhout, D. (2009). Long-term wine consumption is related to cardiovascular mortality and life expectancy independently of moderate alcohol intake: The Zutphen Study. *Journal of Epidemiology and Community Health*, 63(7), 534–540.

Vanden Berghe, W. (2012). Epigenetic impact of dietary polyphenols in cancer chemoprevention: Lifelong remodeling of our epigenomes. *Pharmacology Research*, 65(6), 565–567.

Watanabe, K. I., & Bloch, W. (2013). Histone methylation and acetylation indicates epigenetic change in the aged cochlea of mice. *European Archives of Otorhinolaryngology*, 270(6), 1823–1830.

WHO. (2012). Tuberculosis Fact Sheet No. 104, World Health Organization. Retrieved from http://www.who.int/mediacentre/factsheets/fs104/en/ (Accessed October 22, 2012).

Xie, J., Talaska, A. E., & Schacht, J. (2011). New developments in aminoglycoside therapy and ototoxicity. *Hearing Research*, 281, 28–37.

Yamane, H., Nakai, Y., Takayama, M., Iguchi, H., Nakagawa, T., & Kojima, A. (1995). Appearance of free radicals in the guinea pig inner ear after noise-induced acoustic trauma. *European Archives of Otorhinolaryngology*, 252, 504–508.

Yamashita, D., Jiang, H-Y., Schacht, J., & Miller, J.M. (2005). Post-exposure treatment attenuates noise-induced hearing loss. *Neuroscience*, 134, 633–642.

Yankaskas, K. (2013). Prelude: Noise-induced tinnitus and hearing loss in the military. *Hearing Research*, 295(1–2), 3–8.

Chapter 30
What's the Use of Genetics?

Karen P. Steel

Karen P. Steel

K.P. Steel (✉)
Wolfson Centre for Age-Related Diseases, King's College London,
Guy's Campus, London SE1 1UL, UK
e-mail: Karen.steel@kcl.ac.uk

A.N. Popper and R.R. Fay (eds.), *Perspectives on Auditory Research*, Springer Handbook of Auditory Research 50, DOI 10.1007/978-1-4614-9102-6_30,

30.1 Why Genetics?

Genetics might seem like an obscure obsession to people who are interested in hearing loss in the human population. After all, how can a mutation causing deafness in a single family in a remote part of the world be relevant to the large numbers of people who walk through the clinic door every day complaining that they cannot hear as well as they once did? Surely everyone knows that hearing loss is caused by hair cells degenerating and once they are gone, they're gone for good? And that young people listen to music that is far too loud and that is what is damaging their hearing? And that what we need are better hearing aids? Well, yes and no. In this chapter, I hope to explain why genetics is relevant to everyone affected by hearing impairment whatever the trigger, and how genetics can be used as a first step toward developing ways of curing deafness.

30.2 Some Background—Human Deafness

Deafness is very common in the human population. Early developmental defects lead to about 1 in 850 children being born with a significant, permanent hearing impairment, and this number doubles in the first decade of life (Fortnum et al., 2001). Thereafter, progressive hearing loss increases with each decade until more than half of the population in their 70s show a significant hearing loss of 25 dB or worse (Davis, 1995). Thus, despite the clear impact of early developmental defects of the ear on individuals and their families, the vast majority of people affected by deafness suffer progressive hearing loss, so this is the major problem to be solved. Hearing loss is profoundly isolating, both socially and economically, and has a major impact on the quality of life of those affected. The only remedial options commonly available are hearing aids and cochlear implants, prosthetic devices that provide some benefits but do not restore normal function. There is an unmet need for medical approaches to slow down or reverse progressive hearing loss.

What is the role of genetics in hearing loss? Deafness is a very heterogeneous disorder, with a wide range of causes, which makes it difficult to study directly in humans. Many different genes are known to be involved in deafness. For example, for nonsyndromic human deafness, more than 130 loci have been defined and 64 genes identified (http://hereditaryhearingloss.org/), and Online Mendelian Inheritance in Man (OMIM) lists more than 800 distinct syndromes including deafness as a feature (http://www.ncbi.nlm.nih.gov/omim). More than 200 different mouse mutants with some sort of auditory system defect have been described (Steel et al., 2002; http://hearingimpairment.jax.org/index.html). Our best estimates suggest there will be at least 500 and maybe as many as 1000 genes that can be involved in deafness, any one of which can be mutated and cause deafness in an individual. Minor variations in multiple different genes (genetic background) can also interact to make a person more or less likely to develop hearing loss as they get older, and twin, sib, and family studies have demonstrated a range of heritabilities of 0.5 or greater for age-related hearing loss indicating a significant contribution of genetics (Karlsson et al., 1997; Gates et al., 1999; Wolber et al., 2012). Noise, drugs, and infections can all contribute to hearing impairment. However, these insults do not act completely independently on hearing but instead will interact with the particular gene variants carried by an individual to influence the degree of damage. For example, the *A1555G* mutation of the human mitochondrial genome makes carriers highly susceptible to ototoxic drug-induced deafness (e.g., Estivill et al., 1998), and there are several mouse mutations that predispose the carriers to noise-induced hearing loss. Genetics (or more accurately, the assortment of genomic variants that each of us carry) is therefore an important factor in all forms of hearing impairment.

However, the vast majority of affected individuals have no molecular diagnosis. This is especially true in cases of later-onset, age-related progressive hearing loss, where we know very little about the molecular basis of the pathology. Until we have a molecular understanding of the processes underlying progressive hearing loss, it will be difficult to make progress toward developing treatments. Furthermore, improved understanding of the underlying causes of hearing loss will change the common perception of progressive hearing loss as an untreatable disorder, an inevitable part of growing older.

30.3 Genetics as a Tool

Genetics is the study of the inheritance of traits. It has two major uses for the study of deafness. First, it can be used to identify the mutations underlying hearing impairment in human families. For example, in the case of simple Mendelian inheritance in which a single gene mutation causes deafness, the trait can be linked to a chromosomal region by its co-inheritance with nearby DNA markers within the family. This will narrow down the number of genes that need to be examined for mutations that might cause the deafness. When a mutation is discovered to cause deafness in a family, this is useful in giving an explanation for the deafness and can be used to provide accurate genetic counseling to the family and better-informed prognosis of any likely future health developments, especially important for some syndromic forms of deafness. The importance to a family of knowing the cause of the deafness in their child should not be underestimated, even if there are no treatments available.

The second use of genetics is to identify molecules that are essential for normal hearing. Genetics plays a particularly important role in finding these molecules because a mutation can reveal an essential molecule even if there are only a few tens of copies in each cell (e.g., transduction channel components) and relatively few cells to analyze (e.g., inner hair cells). There are very few molecules that are present in the ear in such abundance that a biochemical or expression approach can be used for identifying them. The main examples where a biochemical approach led to discovery of an essential molecule are tectorin and prestin (Legan et al., 1997; Zheng et al., 2000), both abundant proteins in the cochlea. But these are rare exceptions, and almost all other critical molecules have been discovered using genetics as a tool. Assembling the molecular components required for normal hearing is an important step toward building the networks of molecules that operate in auditory development and function, and these networks will offer multiple possibilities for therapeutic targets for treatments.

30.4 The Mouse as a Model

Although many genes underlying deafness have been identified directly in human families with affected individuals, it is easier to identify the responsible genes in the mouse because we can establish very large families (e.g., 100–1000 offspring) for linkage analysis to localize the mutation to a small region of a specific chromosome. Furthermore, we can minimize the number of candidate mutations to be investigated by using inbred strains of mice with defined genome sequences. More importantly, in the mouse it is possible to analyze the development of auditory dysfunction and track it back to the initial stages, which is important for determining the mechanisms involved. In contrast, by the time a human ear becomes available for detailed histological study, the pathology often has progressed to an end-state with considerable degeneration, leaving few clues to the original cause of the deafness. There are many other advantages to using the mouse to study deafness. Detailed electrophysiological measurements can be recorded in mice but not in humans, for example, endocochlear potentials in the cochlear duct or transduction currents in single hair cells in the excised organ of Corti. The mouse can be genetically manipulated to create lesions in the genome, allowing us to answer questions about the role of specific genes in auditory function. With few exceptions, mice and humans with mutations in the equivalent (orthologous) gene show similar pathologies of the ear, making them good models. Finally, if we want to understand the full range of auditory deficits in humans, we need to study a mammal. Mammals have specific features not found in other vertebrates, such as the stria vascularis generating a high endocochlear potential that provides a strong driving force across the transduction channels of hair cells, and the outer hair cells, specialized hair cells that function as amplifiers of the stimulus delivered to inner hair cells. Progressive hearing loss is quite likely to involve outer hair cell function and/or deterioration of the homeostatic state of the cochlear duct, so these two structures will be important to study.

30.5 A Long History of Deaf Mice

The earliest accounts of deaf mice date from as early as 80 BC, when "dancing" mice were described in a Chinese manuscript from the Han dynasty (Keeler, 1931). Such mice were prized by collectors because of their hyperactive, circling movements, which, we now know, were almost certainly due to balance defects. Over the

Table 30.1 Genes involved in Usher syndrome

Human type	Gene	Mouse mutant
Usher 1B	*Myo7a*	Shaker1
Usher 1C	*Ush1c*	Deaf circler
Usher 1D	*Cdh23*	Waltzer
Usher 1F	*Pcdh15*	Ames waltzer
Usher 1G	*Sans*	Jackson shaker
Usher 1J	*Cib2*	None
Usher 2A	*Ush2a*	Knockout
Usher 2C	*Vlgr1*	Frings
Usher 2D	*Whrn*	Whirler
Usher 3A	*Clrn1*	Knockout

Seven of the ten known Usher syndrome genes were found to underlie hearing and balance defects in longstanding mouse mutants

centuries, similar dancing mice were reported, until in the early 20th century some were noticed by the scientific community and taken into laboratories to breed and be studied. For example, the *shaker1* mutant was first described by Lord and Gates in 1929, and others followed over the next few decades. It quickly became apparent that these mutant mice with balance problems were most often also deaf, not surprisingly because any fundamental abnormalities in sensory hair cells or malformations of the inner ear can lead to both hearing and balance defects. Mice with severe or profound deafness can be detected by the lack of a Preyer reflex (ear flick) in response to a sudden loud sound, but the first report of electrophysiological measurements of auditory responses of a mouse mutant was published in 1940, a study of the *shaker1* mutant (Grüneberg et al., 1940).

These mouse lines carried spontaneous mutations. A background level of spontaneous mutation continues, and with so many laboratories breeding mice any obvious phenotypes such as circling that occur are likely to be noticed and either studied or passed on to other scientists interested in the type of defect revealed. Once the mutations were identified, many of these spontaneous mutants turned out to have alterations of genes that underlie Usher syndrome in humans (Table 30.1). Usher syndrome often involves balance problems as well as early deafness and later retinitis pigmentosa. The first of these genes was identified as *Myo7a* in the *shaker1* mouse mutant in a collaboration between Steve Brown and myself (Gibson et al., 1995), followed rapidly by the discovery of mutations in human *MYO7A* in people with Usher syndrome by our collaborators Christine Petit in Paris and Bill Kimberling in Omaha, Nebraska (Weil et al., 1995).

However, not all humans with hearing impairment also have balance problems. It seemed likely that other genes might be involved in causing deafness alone in the mouse, but these mutants would not be noticed in a laboratory setting as easily as mice with balance defects. This thinking motivated Professor Malkiat S. Deol at University College London in the 1950s to screen his entire mouse collection for lack of a Preyer reflex in response to a sharp, high-pitched sound. He found two lines of previously unsuspected new deaf mutants segregating within his mouse colonies that were maintained for other purposes. These were the first two mouse mutant lines reported that showed deafness without any balance defects (Deol, 1956; Deol & Kocher, 1958). The first was called deaf (*df*, mutant line now extinct), and was an allele of the *waltzer* locus, and the second was called deafness (*dn*) and this one eventually led to the discovery of the *Tmc1* gene (Kurima et al., 2002), now thought to play a crucial role in the hair cell transduction channel. I studied for my PhD in Malkiat Deol's laboratory, and the deafness mutant was one of the first mouse mutants I worked on. During my first postdoc at the MRC Institute of Hearing Research in Nottingham, I collaborated with Greg Bock to measure cochlear function. We discovered that the deafness mutants showed no electrophysiological responses to sound stimuli at any stage (Steel & Bock, 1980), despite the presence of many intact hair cells, a subject I return to at the end of this chapter.

The mice showing hearing and balance abnormalities were initially characterized phenotypically, and this was useful in revealing a wide range of primary defects leading to deafness. However, the mutations causing these phenotypes were not discovered until much later. The identification of mutations leading to specific diseases was first reported from the mid 1980s onwards using a method called positional cloning (identifying a mutation on the basis of its position on a chromosome). One of the earliest mouse genes identified using this approach happened to be a gene involved in deafness; mutations were found in the *Kit* gene at the Dominant Spotting locus in different mouse lines with white spotting of the coat, mast cell deficiency, and deafness (Geissler et al., 1988).

In the years immediately following the emergence of the nuclear industry, research efforts were directed at studies of the biological effects of radiation and a number of new mouse mutants resulted, usually with DNA deletions, insertions, or other relatively large-scale genomic rearrangements. Several of these showed balance defects and deafness. Many of these mutations turned out to be difficult to analyze as the genomic changes were so complex, but nonetheless some have given useful insights into the role of key molecules in inner ear development. For example, we found that the Light coat and circling, *Lcc*, mutant showed local downregulation of *Sox2* and lack of sensory patch formation in the inner ear (Kiernan et al., 2005).

30.6 Chemical Mutagenesis—The Search for New Genes

As the available deaf mouse mutants were studied and the underlying genes involved identified by positional cloning, it became increasingly clear that there were far more genes involved in deafness (and other disorders) than there were mouse mutant lines. Each new mutant potentially can give access to a new molecule required for normal hearing, so we can use genetics as a tool to find the molecular components of the critical processes involved in auditory function. Therefore, efforts began to increase the number of mutant lines using chemical mutagenesis. Several programs were established using *n*-ethyl-*n*-nitrosourea (ENU), which creates point mutations (single DNA base changes) scattered around the genome. ENU is administered to males, mutagenizing their spermatogonia, and these are mated to wild-type females a few weeks later when the active sperm carry new mutations. The offspring can be screened directly to look for new phenotypes inherited in a dominant manner, or used to build pedigrees that reveal new recessively inherited phenotypes two generations later.

From 1997 to 2000 I led a European Commission (EC)-funded program that added a screen for deafness (lack of a Preyer reflex) and balance problems (head-bobbing or circling behavior) to a wider screen of first generation offspring from ENU-mutagenized males, looking for new dominant mutations. More than 50 founder offspring with hearing or balance phenotypes were recovered at the two mutagenesis centers led by Steve Brown in Harwell, UK and Martin Hrabé de Angelis in Munich, Germany, and the mice were characterized by my team in Nottingham, Karen Avraham in Tel Aviv, and Jean-Louis Guénet in Paris. Over the past 10 years more than half of these mutations have been identified by positional cloning by the five groups in the consortium. Some mutant lines had mutations in known deafness genes, and some loci showed multiple independent mutations possibly reflecting ascertainment bias due to a strong circling or head-bobbing phenotype (e.g., *Chd7* was mutated in multiple independent lines; Bosman et al., 2005), but altogether 10 new genes were identified as underlying hearing and balance defects through this program.

We learned some useful lessons from the hearing and balance screens as part of the ENU mutagenesis programs. First, there was a very wide range of phenotypes found, ranging from ossicle malformations through semicircular canal truncations to hair bundle orientation anomalies. Second, many of the phenotypes were not obvious without targeted screening, particularly hearing impairment, and several of the new mutations found were in genes that had already been knocked out and published but with no mention of any hearing or balance problem. Thus, as a general rule, we find only what we look for. Third, if we had started with the full list of 20,000 mouse genes, there was no prior evidence that would have led us to guess that these ten new genes would be required for normal hearing, suggesting that it is difficult to predict which genes are involved in deafness based on our current incomplete knowledge. Overall, the phenotype-driven approach (starting with a deaf mouse and finding the causative mutation) appeared to be a valuable way of identifying new genes involved in deafness and new pathological mechanisms.

The key to success for the large-scale ENU mutagenesis programs was close engagement between the screening teams and the groups analyzing the variant lines after initial discovery. Our program was a success because both screeners and experts in hearing and balance were funded by the same EC program facilitating close interaction. It was the experts in hearing and balance function who devised the screens and took on the resulting new mutant lines to study, while the experts in ENU mutagenesis produced the large numbers of mice with an optimum mutation density to screen. However, this success was not the case for all ENU screens, and researchers were sometimes disappointed at the difficulty in reproducing the initial phenotype in their own laboratory and were daunted by the prospect of having to identify the underlying mutations by positional cloning, which can be time-consuming with no guarantee of success. Furthermore, the benefits to be gained from ENU mutagenesis can be long term; for example, one of the last of the deaf mutants recovered from our 1997–2000 screen was published more than 10 years later, after exome sequencing became feasible (Hilton et al., 2011). High expectations of rapid progress were not fully realized and many ENU programs were funded for only a single grant cycle.

30.7 Targeted Mutagenesis

More recently, targeted disruption of specific genes has become the focus of attention. The advantage of this approach is that it should not be necessary to identify the gene by positional cloning, a significant shortcut. The first gene to be "knocked out" by genetic manipulation in embryonic stem (ES) cells that were then used to create a new mutant mouse line was a gene affecting inner ear development, *Fgf3*, previously known as *Int2* (Mansour et al., 1993). Since then, more than 6000 genes have been knocked out and reported (MGI; http://www.informatics.jax.org/). These have been generated in multiple laboratories, using different targeting strategies and various, often mixed, genetic backgrounds. Also, they are often not made available to other researchers.

Inspired by the success of targeting as a strategy and the drawbacks of the existing resource, an ongoing international effort (KOMP, EUCOMM) has resulted in more than 13,000 mouse genes being targeted in ES cells on a single inbred genetic background (C57BL/6N) and made available to all researchers from public repositories (Skarnes et al., 2011). When I moved from Nottingham to the Wellcome Trust Sanger Institute in 2003, I took advantage of the growing ES cell resource to establish a new screen, the Mouse Genetics Project, using these targeted ES cells to generate mice to screen for a wide range of phenotypes, including hearing impairment. Instead of using the Preyer reflex, which can detect only severe or profound deafness, we developed a rapid auditory brain stem response (ABR) protocol that takes only 15 minutes to perform and can detect mild and moderate hearing impairments too (Ingham et al., 2011). So far we have generated more than 800 new mutant lines and screened more than 600 of these by ABR. We have found 12 new genes associated with raised ABR thresholds, plus a number of further lines where we see normal thresholds but robust anomalies of the ABR waveform, suggesting that these mice may have a central auditory system processing defect.

Just as for the ENU screen, none of these genes was previously suspected of being involved in auditory function. The hearing impairment ranges widely, from mild or moderate threshold increases, to high-frequency hearing loss or severe deafness across all frequencies. Waveform anomalies are likewise varied, including some with small amplitudes of early waves but normal or near-normal later waves, and others with prolonged latencies. The underlying pathologies include middle ear inflammation, synaptic abnormalities and reduced endocochlear potential. Of these 12 new mutant lines with hearing impairment, only one would have been detected using the Preyer reflex, emphasising the benefit of using ABR to screen. The data from this screen can be viewed on the Sanger Institute website (http://www.sanger.ac.uk/mouseportal/) and all mutant lines are made available to the scientific community through public repositories (http://www.findmice.org/index.jsp).

30.8 What Have We Learned from Deaf Mouse Mutants?

Clearly we have learned a great deal about the role of many individual genes in auditory function over the past 20 years and more, but what general lessons have we learned from mice that are relevant to human deafness? Study of the mouse has led us to a number of key observations that really could not be deduced from study of human hearing alone, and I have listed a few below.

The fact that so many completely unexpected genes have been found by systematic screening for hearing impairment by ABR (12 with raised thresholds out of the first 600 mutant lines screened) suggests that there are many genes required for normal hearing waiting to be discovered and that there will probably be well over 500 genes associated with deafness. This suggestion is supported by the limited overlap in genes currently known to be involved in deafness in mice and humans (Fig. 30.1). Recently discovered genes lie mostly outside the region of overlap, because it takes time for a knockout mouse to be generated after a gene has been discovered in humans, and equally it takes time for a human family to be found with a mutation of a candidate gene identified in the mouse. Ultimately the gene sets will merge and then we will know we are close to finding all the genes associated with hearing impairment.

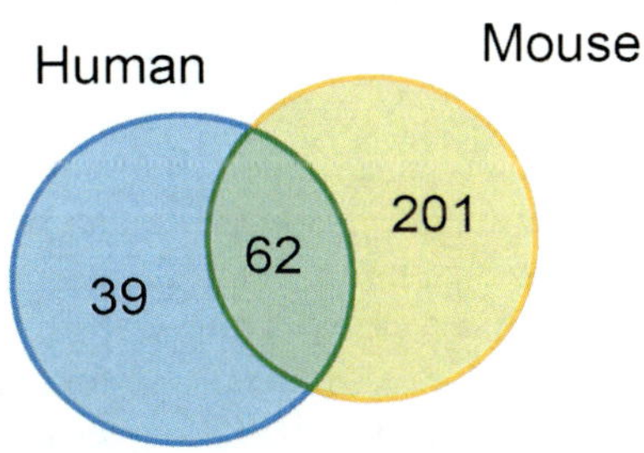

Fig. 30.1 Human and mouse deafness genes. The number of genes so far known to be associated with deafness in mice and humans. With very few exceptions, genes found associated with deafness in the mouse are eventually found in humans with deafness, and genes underlying human deafness are usually knocked out in the mouse, leading to deafness. The limited overlap reflects the rapid progress in recent years in identifying genes involved in deafness in both species, and when we approach the complete ascertainment of deafness genes the two circles are expected to merge. Genes included are those listed in the Hereditary Hearing Impairment in Mice website maintained by Ken Johnson at the Jackson Laboratory (http://hearingimpairment.jax.org/index.html) and the Hereditary Hearing Loss Homepage maintained by Guy Van Camp and Richard Smith (http://hereditaryhearingloss.org/)

Furthermore, more than 800 distinct human syndromes involving hearing impairment have been catalogued in OMIM, indicating that 500 may be an underestimate.

Access to tissues from normal mice at various developmental stages has been valuable for expression studies. The distribution of mRNA or protein derived from genes involved in deafness has demonstrated expression in many different locations in the auditory system, such as the stria vascularis and lateral cochlear wall, supporting cells, neurons, tectorial membrane, or middle ear epithelium, as well as in sensory hair cells, showing that the function of the entire auditory system is important for hearing.

Very few mouse mutants published so far have been shown to have deafness with a central auditory system origin and with a normally functioning cochlea, suggesting that most deafness is sensory rather than neural. However, as mentioned earlier, in our large-scale screen (the Mouse Genetics Project), we have observed a number of mutant lines with normal ABR thresholds but abnormal waveforms indicating a problem with central processing. Maybe we have not found many mice with central auditory defects because previously we focused on raised thresholds and so have not captured other anomalies.

We have found a very wide range of defects leading to hearing impairment in the mouse mutants we have studied. Although we can distinguish conductive from sensorineural hearing loss and sensory dysfunction from auditory neuropathy using standard audiological methods in humans, the wide range of pathologies found in the mouse indicates that better diagnostic methods will be critical to the choice of treatments when these are available in the future. For example, there will be no point in attempting to regenerate sensory hair cells that will not function due to an inherent defect in the hair cell itself or to dysfunction elsewhere in the cochlear duct.

Analysis of the time course of pathological events in the mouse indicates that although hair cells are often the earliest cell type to degenerate in the cochlea, in due course the surrounding supporting cells also degenerate. As these have key roles in hearing, any treatments involving stimulating regeneration will need to regenerate the whole sensory patch and not just hair cells.

Finally, hair cell degeneration is an extremely common feature in mouse mutants and humans with hearing impairment, and of course once a hair cell has degenerated it can no longer function. However, out of more than 100 different mouse mutant lines I have studied in my own laboratory, in no case was the hair cell loss the primary cause of the deafness. In every case, there was some form of damage or dysfunction of hair cells before they degenerated, suggesting that degeneration is a secondary effect following dysfunction, an epiphenomenon. There are many mutants in which for the first few weeks of life there is a complete set of hair cells present but no auditory responses can be obtained. It is not clear why dysfunction leads to hair cell death, but it seems to be a universal consequence. Detailed studies of noise-exposed cochleas also indicate that it is damage and not degeneration that corresponds most closely to threshold shifts (Liberman & Dodds, 1984). This observation is not clear from studying human pathology alone, because most inner ear samples from humans come from people who have suffered deafness for many years and the cochlea will be at an end-stage of the pathological process leaving few clues to the initial causes of hearing loss.

30.9 Goals for the Future

Assembling the molecular components supporting normal auditory function will be an important goal, and genetics will continue to be a key tool in identifying those components. We can think about hearing as a 1000-piece jigsaw puzzle; we are making some progress with putting together small sections of the puzzle by identifying clusters of interacting molecules but having the complete set of pieces will enable us to see the full picture. Much current research is focused on a well-known set of genes, but the picture will be complete only when light is shone on the total set of genes involved (Fig. 30.2; Edwards et al., 2011).

Generating and screening mouse mutations representing all 22,000 mouse genes is the long-term goal of the International Mouse Phenotyping Consortium (IMPC; Brown & Moore, 2012), and the success of the ABR screen within the Sanger Institute's Mouse Genetics Project has led to the adoption of ABR as a standard screen by the IMPC. This will be a major contribution to finding more genes required for normal hearing and candidates for human deafness.

The IMPC is currently using the EUCOMM/KOMP targeted ES cell resource as a source of the mutant lines to be screened. However, not all genes are targeted in

Fig. 30.2 Looking under the lamppost. We all spend time looking at our favorite genes that are well-studied and have good resources available, but it is important not to forget the dark matter outside the beam of light—the many genes that have not yet been identified as being involved in hearing

this library and the last 20 % may be particularly difficult to target for an assortment of technical reasons. At that point, it seems that a return to ENU as a mutagen will be useful, but using ENU in a gene-driven way rather than the phenotype-driven screens previously used. Several groups are building up libraries of DNA samples from thousands of male offspring of ENU-treated mice together with associated frozen sperm that can be used to recover living mice by in vitro fertilization techniques. The rapidly reducing cost of DNA sequencing has facilitated sequencing of the coding regions of the genome (the exome) from these samples, and mutations are detected, assessed, and displayed using bioinformatic tools. Thus it is a relatively simple matter to select a suitably damaging mutation of the gene of interest and order the resurrection of the line from the corresponding sperm sample. It is likely that this approach will enable the completion of the production of a set of mutant mouse lines representing all known mouse genes, and probably many nonprotein–coding elements like microRNAs and long noncoding RNAs as well. The use of these ENU resources will be valuable also for producing allelic series of different mutations of the same gene to confirm association of the phenotype with the gene and explore the role of specific sequence motifs in the function of the molecule.

A complete set of genes associated with deafness in the mouse will be a valuable aid to interpreting exome sequence in people with deafness. Every person carries a huge number of DNA variants (around 3 million) including potentially pathogenic mutations in many genes. Having a shortlist of likely candidate genes from the mouse will help reduce the number of variants that require further study to a more manageable level, even if that shortlist contains a thousand genes. The mouse can provide added confidence in the association between the phenotype and the sequence variant and hence support accurate diagnosis in humans. This will be particularly useful for cases of syndromic deafness.

However, it seems most unlikely that the cause of human nonsyndromic deafness could be diagnosed by genome analysis alone. Linkage analysis will help if the inheritance of deafness can be tracked in a family with linked DNA markers that can indicate which part of which chromosome contains the causative mutation, but often this will still result in a very large number of genes to consider. We need to know considerably more about the genes associated with deafness and the pathogenicity of specific types of mutation before sequence data in individuals could be used for diagnosis without other supporting evidence. For this reason, I envisage that further development of other diagnostic tools using audiological, psychoacoustic, electrophysiological, and imaging approaches would be a valuable complement to sequence analysis and point toward an underlying mechanism in each person before more sophisticated treatments can be applied. Well-characterized mouse mutants with different pathologies can be useful for linking known primary mechanisms with measurable features that can be transferred to clinical use. One key differential diagnosis required will be distinguishing a primary sensory hair cell defect from a problem in maintaining homeostasis of the cochlear fluids, as these types of deafness will need quite different approaches to treatments.

What about development of medical treatments for hearing loss? Despite the extreme heterogeneity of causes of deafness, there are several good reasons to believe that treatments are an achievable goal. First, a large proportion of people affected by deafness show progressive hearing loss, and it is much easier to imagine stopping or reversing the progressive deterioration of a system that once worked well than to find treatments for early developmental defects. Even slowing down the rate of progression of hearing loss would be useful. Second, the target population is considerable and getting larger as people live longer, making it more likely that large pharmaceutical companies will see development of treatments as a worthwhile investment. Third, it is likely that many different primary causes for hearing loss, such as mutations in different genes or responses to different environmental insults, may operate within a limited set of networks of molecular interactions. There may be common points within each network that could be targeted by small molecule or other interventions, meaning that people with different primary causes for deafness could be grouped together and benefit from a common treatment. Finally, there are already some forms of treatments that have been shown to be useful in animal models and as we understand more about new forms of deafness using the mouse, these opportunities are likely to increase. Some mechanisms we are studying in mice are known to be amenable to small molecule manipulation, such as systemic immune or cardiovascular diseases.

As we move closer to a complete catalogue of genes/molecules required for auditory function we will be able to explore the functional relationships between these molecules in pathways and networks, and then focus on those molecules/pathways/networks that are of greatest importance to the human population. The identification of these pathways through the use of genetics will be relevant to all causes of hearing impairment including those with a primary environmental trigger. Drawing up a preliminary network of gene interactions is a straightforward process using bioinformatic tools and available databases that utilize a wide range of sources of information. However, ensuring that each interaction (or edge) is relevant to the auditory system requires detailed follow-up to ask if the components are expressed in the relevant cell type (e.g., in the hair cell) and if the nature of the interaction (e.g., up- or downregulation) is supported by experimental evidence. This is not a simple task because such interactions between each pair of molecules may vary depending on the context—which cell type and which time of development is studied.

Despite the complications of building networks, these will prove to be an invaluable resource for supporting drug development. For example, some key molecules in the network may already have approved small molecules used for other disease indications, and repurposing is likely to be an important activity to ensure the maximum benefit is gained by both the patient and the organization that invested in development of each drug.

Building networks has another valuable purpose—identifying molecules that may play a critical role in hearing but also are essential for survival. These molecules would not be detected by a program that focuses on knocking out the function of the gene because there would be no offspring surviving to test for deafness. In our ABR screen of new mouse mutants described earlier, we screen heterozygotes in cases where the homozygote is lethal. The phenotypes we detect in heterozygotes suggest that knocking down the level of protein production has an impact on cellular function. Of course, mutations in humans are extremely variable and include gain-of-function as well as loss-of-function effects, so the system will be complex. Building networks that tap into the entire existing data set of molecular interactions can overcome the gaps that exist if only genes underlying deafness are considered. Thus, networks can generate hypotheses to be tested, for example, by using conditional knockouts affecting only the ear or relevant part of the auditory system.

Furthermore, construction of networks can point to genes that are redundant in auditory function. In these cases, knocking out the gene will not lead to deafness because an alternative gene can operate in its place. In normal circumstances, a level of redundancy leads to a more robust system. However, it may be that when the organism is put under stress, such as exposure to noise, the alternative gene alone is not as efficient at resisting the damaging effects as both genes together would be. Therefore, many of the genes that so far appear to have no obvious role might be required when the individual is exposed to damaging environments. Networks and pathways can reveal the redundant molecules operating between the nonredundant molecules known to be associated with deafness, and thus open up a broader range of targets for development of therapies.

Finally, I return to a question that has puzzled me since I first started working on deafness. Why do hair cells die? For part of my PhD, I studied three mutants (*deafness*, *jerker*, and *varitint-waddler*) plus mice treated with an antithyroid agent to produce hypothyroidism. All showed progressive degeneration of hair cells but had no responses to sound at a stage when most hair cells were still present, as mentioned previously in this chapter. The hair cells were clearly present and not functioning properly, but why did this lead to their death? Does the lack of normal function lead to a disruption in their cellular homoeostatic mechanisms, which must be highly adapted to manage a continuous flow of cations during transduction? Does the lack of normal synaptic activity lead to loss of a putative trophic role of auditory neurons? Or is there a loss of the normal function of supporting cells to support hair cell survival following abnormal hair cell activity? Although we now know the three mutations underlying deafness in these three mutants, I am still not sure we understand the reason for the hair cell death. There was a clue, however, in the pattern of hair cell loss common to all of them. The earliest signs of loss were a few scattered hair cells mostly in the basal half of the cochlear duct and over a few days this scattered loss extended towards the apical turn. Then, superimposed on the scattered pattern of loss there appeared patches where all hair cells as well as some of the supporting cells appeared to have degenerated. I wondered at the time if there was a tipping point where the loss of a single hair cell could be managed but if two or three hair cells close to each other died this led to a more widespread loss of homeostasis within the organ of Corti and a whole patch would degenerate rapidly. If this is the case, then are hair cells releasing a trophic agent that sustains the health of adjacent cells? The reason why hair cells die remains one of the key questions in auditory research, because interfering with that process might give us insights into how to preserve hair cells into old age.

Acknowledgments I thank Morag Lewis for drawing the figures and commenting on the manuscript and my many colleagues and collaborators who have helped me develop the thoughts presented in this essay. The work was supported by the Wellcome Trust, the MRC, and the EC.

References

Bosman, A. E., Penn, A. C., Ambrose, J. C., Kettleborough, R., Stemple, D. L., & Steel, K. P. (2005). Multiple mutations in mouse *Chd7* provide models for CHARGE syndrome. *Human Molecular Genetics*, 14, 3463–3476.

Brown, S. D., & Moore, M. W. (2012). The International Mouse Phenotyping Consortium: Past and future perspectives on mouse phenotyping. *Mammalian Genome*, 23, 632–640.

Davis, A. C. (1995). *Hearing in adults*. London: Whurr

Deol, M. S. (1956). A gene for uncomplicated deafness in the mouse. *Journal of Embryology & Experimental Morphology*, 4, 190–195.

Deol, M. S., & Kocher, W. (1958). A new gene for deafness in the mouse. *Heredity*, 12, 463–466.

Edwards, A. M., Isserlin, R., Bader, G. D., Frye, S. V., Willson, T. M., & Yu, F. H. (2011). Too many roads not taken. *Nature*, 470, 163–165.

Estivill, X., Govea, N., Barceló, E., Badenas, C., Romero, E., Moral, L., Scozzri, R., D'Urbano, L., Zeviani, M., & Torroni, A. (1998). Familial progressive sensorineural deafness is mainly due to the mtDNA A1555G mutation and is enhanced by treatment of aminoglycosides. *American Journal of Human Genetics*, 62, 27–35.

Fortnum, H. M., Summerfield, A. Q., Marshall, D. H., Davis, A. C., & Bamford, J. M. (2001). Prevalence of permanent childhood hearing impairment in the United Kingdom and implications for universal neonatal hearing screening: Questionnaire based ascertainment study. *British Medical Journal*, 323, 536–540.

Gates, G. A., Couropmitree, N. N., & Myers, R. H. (1999). Genetic associations in age-related hearing thresholds. *Archives of Otolaryngology, Head & Neck Surgery*, 125, 654–659.

Geissler, E. N., Ryan, M. A., & Housman, D. E. (1988). The dominant-white spotting (*W*) locus of the mouse encodes the c-kit proto-oncogene. *Cell*, 55, 185–192.

Gibson, F., Walsh, J., Mburu, P., Varela, A., Brown, K. A., Antonio, M., Beisel, K. W., Steel, K. P., & Brown, S. D. M. (1995). A type VII myosin encoded by the mouse deafness gene *shaker-1*. *Nature*, 374, 62–64.

Grüneberg, H., Hallpike, C. S., & Ledoux, A. (1940). Observations on the structure, development and electrical reactions of the internal ear of the shaker-1 mouse (*Mus musculus*). *Proceedings of the Royal Society of London B: Biological Sciences*, 129, 154–173.

Hilton, J. M., Lewis, M. A., Grati, M., Ingham, N., Pearson, S., Laskowski, R. A., Adams, D. J., & Steel, K. P. (2011). Exome sequencing identifies a missense mutation in *Isl1* associated with low penetrance otitis media in dearisch mice. *Genome Biology*, 12, R90.

Ingham, N. J., Pearson, S., & Steel, K. P. (2011). Using the auditory brainstem response (ABR) to determine sensitivity of hearing in mutant mice. *Current Protocols in Mouse Biology*, 1, 279–287.

Karlsson, K. K., Harris, J. R., & Svartengren, M. (1997). Description and primary results from an audiometric study of male twins. *Ear and Hearing*, 18, 114–120.

Keeler, C. (1931). *The laboratory mouse: Its origin, heredity and culture*. Cambridge, MA: Harvard University Press.

Kiernan, A. E., Pelling, A. L., Leung, K. K. H., Tang, A. S. P., Bell, D. M., Tease, C., Lovell-Badge, R., Steel, K. P., & Cheah, K. S. E. (2005). *Sox2* is required for sensory organ development in the mammalian inner ear. *Nature*, 434, 1031–1035.

Kurima, K., Peters, L. M., Yang, Y., Riazuddin, S., Ahmed, Z. M., Naz, S., Arnaud, D., Drury, S., Mo, J., Makishima, T., Ghosh, M., Menon, P. S., Deshmukh, D., Oddoux, C., Ostrer, H., Khan,

S., Riazuddin, S., Deininger, P. L., Hampton, L. L., Sullivan, S. L., Battey, J. F. Jr., Keats, B. J., Wilcox, E. R., Friedman, T. B., & Griffith, A. J. (2002). Dominant and recessive deafness caused by mutations of a novel gene, TMC1, required for cochlear hair-cell function. *Nature Genetics*, 30, 277–284.

Legan, P. K., Rau, A., Keen, J. N., & Richardson, G. P. (1997). The mouse tectorins: Modular matrix proteins of the inner ear homologous to components of the sperm-egg adhesion system. *Journal of Biological Chemistry*, 272, 8791–8801.

Liberman, M. C., & Dodds, L. W. (1984). Single-neuron labeling and chronic cochlear pathology. III. Stereocilia damage and alterations of threshold tuning curves. *Hearing Research*, 16, 55–74.

Lord, E. M., & Gates, W. H. (1929). Shaker, a new mutation of the house mouse. *The American Naturalist*, 63, 435–442.

Mansour, S. L., Goddard, J. M., & Capecchi, M. R. (1993). Mice homozygous for a targeted disruption of the proto-oncogene int-2 have developmental defects in the tail and inner ear. *Development*, 117, 13–28.

Skarnes, W. C., Rosen, B., West, A. P., Koutsourakis, M., Bushell, W., Iyer, V., Mujica, A. O., Thomas, M., Harrow, J., Cox, T., Jackson, D., Severin, J., Biggs, P., Fu, J., Nefedov, M., de Jong, P. J., Stewart, A. F., & Bradley, A. (2011). A conditional knockout resource for the genome-wide study of mouse gene function. *Nature*, 474, 337–342.

Steel, K. P., & Bock, G. R. (1980). The nature of inherited deafness in *deafness* mice. *Nature*, 288, 159–161.

Steel, K. P., Erven, A., & Kiernan, A. E. (2002). Mice as models for human hereditary deafness. In B. J. Keats, A. N. Popper, & R. R. Fay (Eds.), *Genetics and auditory disorders* (pp. 247–296). New York: Springer Science+Business Media.

Weil, D., Blanchard, S., Kaplan, J., Guilford, P., Gibson, F., Walsh, J., Mburu, P., Varela, A., Levilliers, J., Weston, M. D., Kelley, P. M., Kimberling, W. J., Wagenaar, M., Levi-Acobas, F., Larget-Piet, D., Munnich, A., Steel, K. P., Brown, S. D. M., & Petit, C. (1995). Defective myosin VIIA gene responsible for Usher syndrome type 1B. *Nature*, 374, 60–61.

Wolber, L. E., Steves, C. J., Spector, T. D., & Williams, F. M.(2012). Hearing ability with age in northern European women: A new web-based approach to genetic studies. *PLoS One*, 7, e35500.

Zheng, J., Shen, W., He, D. Z., Long, K. B., Madison, L. D., & Dallos, P. (2000). Prestin is the motor protein of cochlear outer hair cells. *Nature*, 405, 149–155.

Chapter 31
Advances in the Understanding of Binaural Information Processing: Consideration of the Stimulus as Processed

Constantine Trahiotis and Leslie R. Bernstein

Constantine Trahiotis and Leslie R. Bernstein

C. Trahiotis • L.R. Bernstein (✉)
Departments of Neuroscience and Surgery (Otolaryngology),
University of Connecticut Health Center, Farmington, CT 06030, USA
e-mail: tino@neuron.uchc.edu; Les@neuron.uchc.edu

A.N. Popper and R.R. Fay (eds.), *Perspectives on Auditory Research*, Springer Handbook of Auditory Research 50, DOI 10.1007/978-1-4614-9102-6_31,

31.1 Introduction

The goal of this chapter is to provide readers with an overview of selected aspects of the progress made over the last 20 years within the broad field of binaural information processing. The specific focus is restricted to empirical and theoretical contributions stemming from earphone-based studies concerning binaural detection, discrimination, and lateralization. Such knowledge has fostered important recent advances in the understanding of the mechanisms underlying such binaural capabilities. That knowledge, coupled with advances within other areas of auditory research, has yielded comprehensive, quantitative models of binaural processing that, in many cases, permit quite accurate accounts of human binaural performance.

At the outset, it is important to make clear the fundamental role played by the nature of the psychophysical procedures used to obtain behavioral data. The primary goal of binaural psychophysical experiments is to relate features of physical stimuli to performance-based measures of binaural information processing. In order to achieve that goal, one must be able to exercise precise control over the generation, measurement, and manipulation of the stimuli themselves. In addition, one must gather data that reflect the processing of those stimuli, per se, in a manner that minimizes or precludes the influence of potentially "confounding" variables. Those include such factors as: (1) stimulus uncertainty, (2) auditory memory, (3) the need to "label" percepts, and (4) learning effects. The objective is to maximize both validity (i.e., measuring what is intended to be measured) and reliability (i.e., both the within-listener and across-listener repeatability of the measures) while minimizing error of measurement.

Consistent with this, in order to measure binaural discrimination, the authors of this chapter often employ a four-interval, two-cue, two-alternative, forced choice procedure (e.g., Bernstein & Trahiotis, 1982). The advantage of this procedure is that the listener need only report which observation interval (the second or third) differed from the other three, regardless of the manner in which it differed. The first and fourth intervals provide cues because they never contain the binaural change of interest. In order to measure extents of perceived intracranial position (laterality), the authors have employed an "acoustic pointing task" (e.g., Feddersen et al., 1957; Domnitz & Colburn, 1977) in which the listeners adjust the intracranial position of one sound (the pointer) so that it "matches" the intracranial position of a second, experimenter-controlled sound that conveys the binaural cue of interest (the target). One advantage of both types of tasks is that listeners are not required to translate their percepts to another modality, for example, by translating their percepts to a linguistically or visually defined scale or dimension. Thus, the listener's responses are more directly attributable to experimenter-controlled changes in the physical stimulus.

31.2 Relating Binaural Detection, Discrimination, and Lateralization to the Stimuli "as Processed"

Historically, the primary focus of binaural psychophysics has been to relate behavioral measurements to experimenter-imposed changes in the physical, "external" stimuli while also considering other factors such as frequency resolution (i.e., critical bands) in order to provide qualitative and quantitative accounts of the results. Over the past 20 years, knowledge of binaural processing has expanded to reveal the import and explanatory power of the stimulus *as processed*. A number of important outcomes discussed below show that comprehensive explanations of binaural hearing, be they qualitative or quantitative, must incorporate transformations of the external auditory stimulus, especially those that occur as a consequence of peripheral auditory processing.

For example, consider that it was traditionally well-known and accepted that the ability to resolve changes in interaural temporal disparities (i.e., threshold ITDs) at high frequencies was typically an order of magnitude or more poorer than at low frequencies (e.g., Klumpp & Eady, 1956; Zwislocki & Feldman, 1956; McFadden & Pasanen, 1976). Likewise, binaural releases from masking (i.e., masking-level differences [MLDs]) were uniformly found to be smaller for high-frequency stimuli than for low-frequency stimuli (e.g., for a review, see Durlach & Colburn, 1978; Zurek & Durlach, 1987). Further, it was understood that, for a given ITD, intracranial images produced by high-frequency stimuli would be perceived to be (i.e., lateralized) much closer to the mid-line than would be their low-frequency counterparts (e.g., Blauert, 1983; Bernstein & Trahiotis, 1985).

It was not known whether these differences resulted from true differences in the *binaural* mechanisms at low and at high frequencies or, perhaps, from inherent differences in the *monaural* neural information serving as input to the binaural mechanisms, as suggested by Colburn and Esquissaud (1976). After all, the peripheral processing of low-frequency information would result in neural impulses synchronized to the waveform (fine-structure and envelope), while at high frequencies neural impulses would be synchronized only to the *envelope* of the waveform.

The key to understanding the differences in the behavioral data obtained at low and high frequencies was the use of "transposed stimuli" developed by van de Par and Kohlrausch (1997). They attempted to overcome temporal-coding limitations imposed by peripheral auditory processing at high frequencies by a process akin to "reverse engineering." Their idea was to construct a stimulus that, after peripheral auditory processing, would result in neural temporal information conveyed by high-frequency channels that effectively mimicked neural temporal information naturally conveyed by low-frequency channels. Figure 31.1 illustrates, schematically, van de Par and Kohlrausch's technique for transposing a low-frequency stimulus, in this case a sinusoid. The first step is to rectify and low-pass filter the tone. This mimics the basic properties of hair cell/neural transduction. The second step is to multiply the rectified/filtered tone by a high-frequency tonal "carrier." The resulting "transposed

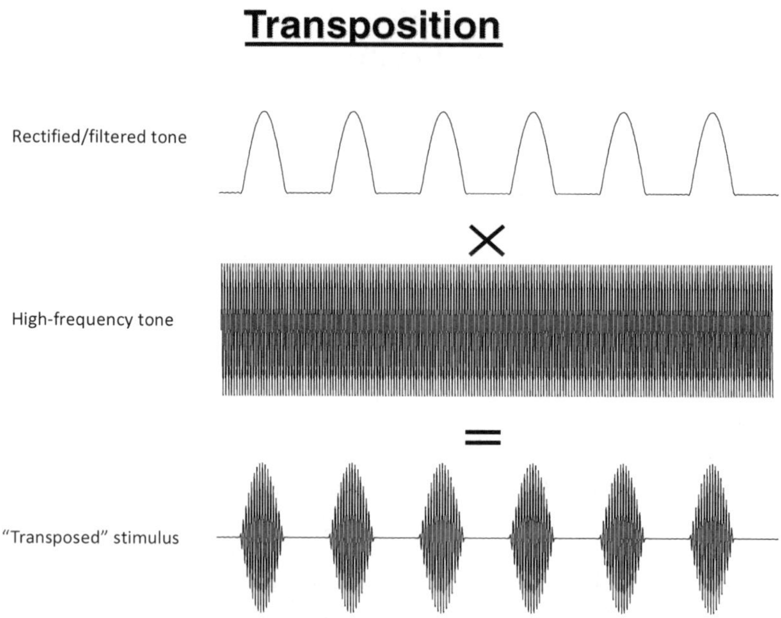

Fig. 31.1 A schematic representation of the technique of "transposition" introduced by van de Par and Kohlrausch (1997)

stimulus" is centered at the frequency of the carrier and has a temporal envelope that matches the rectified/filtered low-frequency waveform. Figure 31.2 provides a direct comparison among a low-frequency tone, its transposed counterpart, and a conventional high-frequency sinusoidally amplitude-modulated (SAM) tone. The left half of the figure depicts the three types of physical waveforms; the right half depicts the stimuli as they would be expected to be processed via hair cell–related rectification and low-pass filtering. Other factors notwithstanding, one would expect that the distributions of the neural impulses produced by the low-frequency sinusoid and its transposed counterpart would be the same. On the other hand, in accord with an argument made by Blauert (1983), the distributions of the neural impulses from the SAM tone would not be expected to be as temporally "sharp" because that stimulus, as processed, is sinusoidal and does not have the distinct "off-times" that characterize the other two stimuli as processed.

Transposed stimuli have been used to gain insight regarding the commonly found poorer binaural performance at high frequencies in three different experimental contexts. van de Par and Kohlrausch (1997) employed low-frequency signal-plus-noise and noise-alone waveforms transposed to 4 kHz in an experiment concerning binaural release from masking. They found that releases from masking using high-frequency transposed stimuli were quite similar in magnitude to those obtained with their low-frequency counterparts. In accord with those results, Bernstein and Trahiotis (2002) demonstrated that threshold ITDs obtained with low-frequency pure tones and transpositions of them to 4 kHz could yield threshold ITDs that were quite similar. Finally, Bernstein and Trahiotis (2003) measured extents of

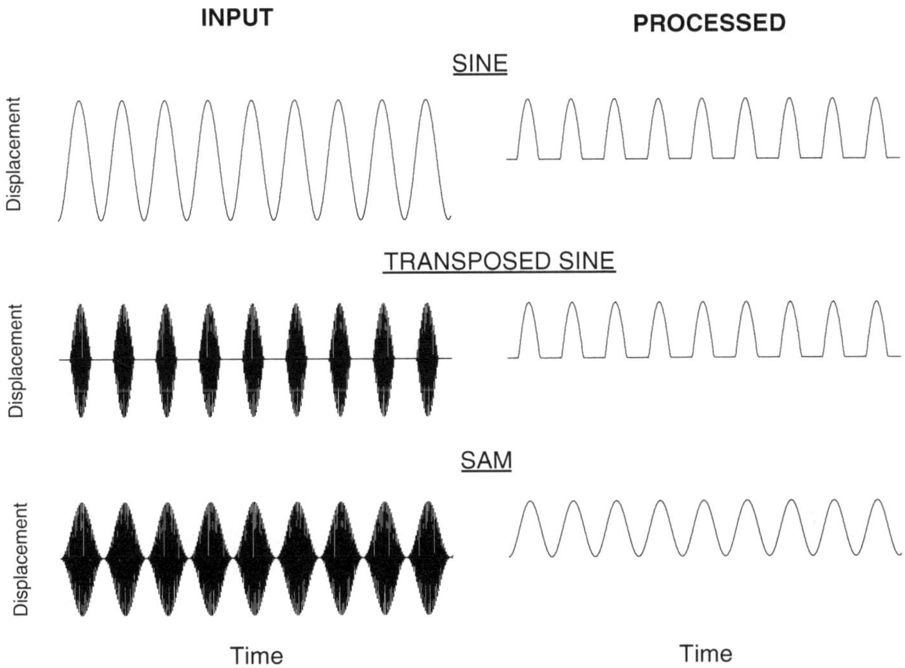

Fig. 31.2 A comparison among a low-frequency tone, its transposed counterpart, and a conventional high-frequency sinusoidally amplitude-modulated (SAM) tone. The left half of the figure depicts the three types of physical waveforms; the right half depicts the stimuli as they would be expected to be processed via hair-cell–related rectification and low-pass filtering

ITD-based laterality for (1) narrow bands of low-frequency Gaussian noise; (2) their counterparts transposed to 4 kHz; and (3) narrow bands of noise centered at 4 kHz. Some of their results are shown in Fig. 31.3, in which extent of laterality is indexed by the interaural intensitive difference (IID) of the acoustic pointer and is plotted as a function of the ongoing ITD imposed on the three different target stimuli. Note that extents of laterality for the low-frequency noises and their transposed counterparts were virtually identical and were substantially larger than the extents of laterality produced by conventional bands of high-frequency noise.

The results obtained in the three studies discussed in the preceding text (as well as others that followed) strongly suggest that the commonly-found differences in binaural performance found at high versus low frequencies are not attributable to any "deficit" in the central binaural comparator at high versus low frequencies. Rather, they are attributable, fundamentally, to differences in the nature of the neural inputs for binaural comparison produced at high versus low frequencies.

The processing of ITDs conveyed by the envelopes of high-frequency stimuli does, however, appear to be constrained by another "internal" factor that appears to

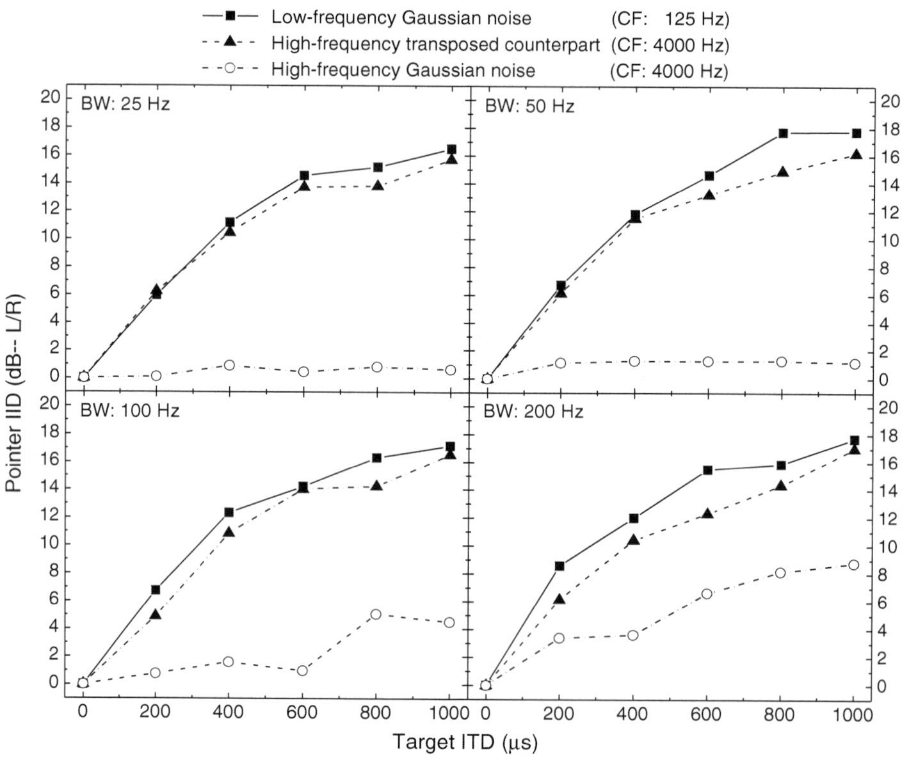

Fig. 31.3 The IID (in dB) of the acoustic pointer required to match the intracranial position of the target as a function of the ongoing ITD (left ear leading) of the target. Three different types of target stimuli were employed: a 125-Hz–centered Gaussian band of noise, its counterpart transposed to 4 kHz, and a Gaussian noise centered at 4 kHz. Each panel depicts the results for a given bandwidth of the 125-Hz–centered and 4-kHz–centered Gaussian noises. (Adapted from Fig. 31.2 of Bernstein & Trahiotis, 2003, with permission)

be unrelated to either peripheral auditory processing or central binaural comparison. Beginning with McFadden and Pasanen (1976), who employed two-tone complexes, and Nuetzel and Hafter (1981), who employed sinusoidally amplitude-modulated (SAM) tones, it became apparent that the efficiency of processing of changes in ITD is dependent on the rate of fluctuation of the envelopes. They found that listeners' ability to resolve ITDs was greatly degraded when the rate of fluctuation of the envelope exceeded approximately 250 Hz.

Specifically, there appears to be a process that functionally attenuates rapid fluctuations of the envelope, that is, those that occur at a rate above about 150 Hz. A variety of empirical outcomes and theoretical analyses have recently converged to demonstrate and to describe how this type of "envelope low-pass filtering" constrains the processing of ITDs conveyed by the envelopes of high-frequency stimuli.

Over the last 20 years, it became clear that the envelope "rate-limitation" did not stem from peripheral auditory processing. In binaural ITD-discrimination experiments,

Bernstein and Trahiotis (1994, 2002) showed that the envelope-rate-related increases in threshold ITD were, if anything, more severe at higher center frequencies (e.g., 8 kHz or 10 kHz) than at lower center frequencies (e.g., 4 kHz). This outcome was opposite to what one would expect were the limitation a manifestation of peripheral auditory filtering. This is so because auditory filters having higher center frequencies have broader bandwidths and thus would, for a given high rate of modulation, produce less attenuation of spectral "sidebands." This, in turn, would result in greater preservation of envelope information at higher center frequencies. Consequently, it appeared that some other factor was involved.

Three subsequent studies, employing either monaural or diotic stimuli (Kohlrausch et al., 2000; Ewert & Dau, 2000; Moore & Glasberg, 2001), showed that some type of envelope low-pass filtering also operates at high spectral frequencies when the tasks to be performed do not contain binaural cues. In those studies, temporal modulation transfer functions (TMTFs) were measured at various center frequencies using SAM tones. The data and their theoretical analyses led Kohlrausch et al. (2000) and Ewert and Dau (2000) to include in their modeling a low-pass filter that attenuates, independent of the center frequency, fluctuations of the envelope that are more rapid than 150 Hz.

The results of two recent neurophysiological investigations are consistent with the behavioral and theoretical results described in the preceding text. Both Rodríguez et al. (2010) and Middlebrooks and Snyder (2010) measured responses of neural units within the inferior colliculus (IC) of cats. The results of both studies are in agreement in that the rate of fluctuation beyond which envelope coding degrades for high-frequency stimuli systematically decreases with increases in the spectral frequency to which the unit is best tuned. In both studies, those rates of fluctuation were found range between about 100 and 250 Hz, values that are remarkably similar to the ones observed behaviorally.

These are clear examples of advances in understanding of fundamental aspects of binaural processing that can come about only when one suitably focuses on and gains an understanding how external stimuli are processed and transduced into neurally based cues. Another, related, example concerns research reported by Eddins and Barber (1998) and by Hall et al. (1998). They demonstrated that diotic narrow-band masking noise constructed to have minimal fluctuations in amplitude (so-called low-noise noise) produced 2 or 3 dB more masking of interaurally phase-reversed (Sπ), 500-Hz tones than did Gaussian noise of the same nominal bandwidth and power. Beyond their empirical import, the results were also important theoretically. This was so because, as Eddins and Barber demonstrated, the outcomes were directly opposite to what would be predicted from a modern, commonly successful, cross-correlation-based model of binaural detection.

Detailed consideration of how peripheral auditory processing could differentially affect Gaussian and low-noise noise was helpful. In this instance, it was the role played by basilar membrane-related compression that led to reconciliation between the data and the predictions of them via a cross-correlation-based model that contained stages of peripheral processing. Specifically, Bernstein et al. (1999) showed that incorporating into the model a physiologically valid stage of "envelope compression" *prior* to the stages of the model emulating hair-cell-related rectification and low-pass filtering provided excellent predictions of the data obtained by Eddins and Barber (1998) and by Hall et al. (1998). An important aspect of Bernstein et al.'s analysis was that the amount of compression was not arbitrary. Instead, the behavioral data were used to converge on the single estimate of the exponent of the compressive function which, when incorporated into the model, would yield accurate estimates of the binaural detection thresholds. That value, 0.23, was remarkably close to the value of 0.2 which was measured by Sellick et al. (1982) and Ruggero (1992) in their physiological studies of basilar membrane compression.

Yet another example of how detailed consideration of the stimulus as processed led to new insights and advances in the understanding binaural processing concerns the "precedence effect." In the context of this discussion, the effect refers to the fact that the perceived locations of external sources of sound are dominated by the interaural disparities conveyed by the first, or "direct" wavefronts reaching the two ears. Interaural disparities conveyed by later-arriving reflections of those sounds typically have far less influence on the perceived location. In the laboratory, the precedence effect has been studied by presenting, over earphones, successive pairs of binaural transients or clicks, each pair having an experimenter-determined value of ITD. The general finding is that ITDs conveyed by the first of the pair of binaural transients dominates the intracranial location of the unitary image produced by the composite stimulus. That outcome mirrors that found in the sound field with external sources (see Wallach et al., 1949).

An almost universally accepted explanation of the precedence effect in its many manifestations was that the later-arriving binaural cues are somehow inhibited or suppressed at some central level of binaural processing (Blauert, 1983; Lindemann, 1986a, b; Zurek, 1987; Litovsky et al., 1999). Hartung and Trahiotis (2001) sought to discover the extent to which the "earphone-based" precedence effect might be explained on the basis of the effects that monaural peripheral auditory processing would have on the binaural cues reaching the central binaural comparator. Using pairs of nonoverlapping binaural transients as inputs, they analyzed the outputs of left and right pairs of gammatone, auditory-like filters having various center frequencies and, concomitantly, different bandwidths. Hartung and Trahiotis found that the outputs of the filters within left and right "channels" often revealed temporal interactions and overlap between the filtered outputs of the first and second pairs of clicks. In addition, those interactions resulted, *across* right and left channels, in substantial departures from the *external* interaural cues. Specifically, very large and dynamically changing values of ITD and IID were shown to occur as a result of interactions within peripheral filters.

In order to evaluate the degree to which peripheral auditory processing affects or determines precedence, Hartung and Trahiotis (2001) constructed a model in which peripheral auditory processing was accomplished via a bank of gammatone filters followed by the "Meddis hair-cell model" (Meddis, 1986, 1988; Meddis et al., 1990). Cross-correlograms were then constructed and the predicted perceived position of the stimulus was taken to be the most central peak of the across-frequency averaged correlogram. This model accounted quantitatively for the patterning of precedence data obtained in the studies of Wallach et al. (1949) and Yost and Soderquist (1984), both of which employed binaural transients and the precedence data of Shinn-Cunningham et al. (1995), who employed low- and high-frequency bands of noise.

By considering, in detail, and incorporating peripheral auditory processing within a cross-correlation–based model, Hartung and Trahiotis (2001) were able to account, quantitatively, for classical results concerning binaural precedence without the need to resort to any type of central inhibition or suppression of binaural cues or the incorporation of any "top-down" cognitive or selective-attention mechanisms. Essentially the same conclusion was reached by Verhulst et al. (2012), who combined behavioral and physiological measures of the precedence effect. An interesting and recent extension of such ideas (Xia & Shinn-Cunningham, 2011) appears to be consistent with analyses and arguments put forward by Hartung and Trahiotis. On the other hand, and in accord with Hartung and Trahiotis' discussion, the results of these studies should not be taken to imply that central mechanisms, including inhibition, selective attention, etc., are not involved in any of the other outcomes of the myriad types of experiments falling under the rubric of the precedence effect.

The foregoing discussion provides primary examples of the import of considering the stimuli as processed in order to understand or explain binaural processing. There are many more deserving examples that are referenced and characterized briefly in what follows so that readers can gain further appreciation for progress that has been made along these lines. They are:

1. van der Heijden and Trahiotis (1999): This article shows how binaural detection thresholds measured with selective interaural delays of novel stimuli depend upon how the precision of *internal* interaural delay compensation varies as a function of the magnitude of the external delay. Historically, similar types of data could not be quantitatively accounted for because the models did not incorporate this aspect of internal processing.
2. Bernstein and Trahiotis (1996); van de Par et al. (2001): These studies make clear that, in order to account quantitatively for binaural detection and discrimination within a cross-correlation scheme, one must employ normalized indices of correlation that reflect underlying peripheral auditory processes. For example, in order to explain binaural detection at high center frequencies, one must employ an index of the interaural *correlation* of the envelope of the stimulus that preserves its DC or mean value, in accord with physiological results. Such a measure is very different from the commonly employed *coefficient* of correlation (or covariance) which overtly removes the mean value.
3. van de Par et al. (2000); Dreyer and Oxenham (2008); Bernstein and Trahiotis (2008); Bernstein and Trahiotis (2010): These studies reveal how taking into account listeners' potential use of information in "internal" "off-frequency" auditory filters can enhance the accuracy of quantitative predictions of threshold ITD, especially for low-modulation depths of the external stimuli that convey the ITD.

4. van der Heijden et al. (1997): This study showed that binaural detection of spectrally nonoverlapping tonal signals and noise maskers depended upon aural distortion products arising from peripheral auditory processing of the components within the masking noise. That is, the masking was dominated by spectral information at frequencies that were not present in the external masking noise.
5. Trahiotis et al. (2001): Extending and confirming the work of Stern and his colleagues (e.g., Stern & Colburn, 1978; Stern et al., 1988), who measured extent of laterality, threshold ITDs were measured using low-frequency bands of noise having parametric combinations of interaural time, interaural phase, bandwidth, and duration. The data and their quantitative analyses revealed that the resolution of ITD was explained via consideration of the complex patterning of the across-frequency cross-correlation function. Consistent with studies described earlier, accurate predictions depended upon computations made subsequent to stages representing peripheral auditory filtering, rectification, low-pass filtering, and compression.

Numerous types of advances in the understanding of auditory processing over the last two decades, be they monaural or binaural in nature or be they behavioral or physiological, have led, in a cumulative way, to the improvement, augmentation, and extension of comprehensive, quantitative models of binaural processing. A good illustration of the predictive power of such quantitative efforts comes from a recent study from the authors' laboratory (Bernstein & Trahiotis, 2012). Extents of laterality were measured using the acoustic pointing task for a variety of stimuli centered at 4 kHz. The basic stimuli were either classical SAM tones or specially modified, envelope-"sharpened" SAM tones termed "raised-sine" tones. In that study, the modification entailed raising the sinusoidal modulator to the power 8.0 before multiplying it with a 4-kHz carrier. The characteristics of the stimuli, which were varied parametrically, were frequency of modulation, depth of modulation, ITD, and IID. A subset of the data is shown in Fig. 31.4. Individual symbols represent the data averaged across three listeners. The lines represent predictions of the data obtained from a comprehensive cross-correlation–based model of the type described repeatedly in the preceding text.

Note that the predictions of the model are quite accurate. In fact, the model accounted for 94 % of the variance in the entire set of data, which comprised 960 stimulus conditions. Specific features of the data deserve mention. Note that extents of laterality obtained when the rate of modulation was 256 Hz and the depth of modulation was 1.0 did not increase linearly with ITD as they did when the rate of modulation was 128 Hz. The extents of laterality at the rate of modulation of 256 Hz were also generally smaller than those obtained when the rate was 128 Hz. The fact that the model accounts for these differences attests to the efficacy of predicting lateral position by incorporating the envelope low-pass filter that was useful in accounting for increases in threshold ITDs in the studies described in the preceding text.

Second, values of IID affect extent of laterality in a manner largely independent of the value of ITD. This property of the data is also captured well by the model. A more subtle success of the model is that it correctly predicts that a given external value of IID inserted into the 4-kHz-centered target requires the listener to use a

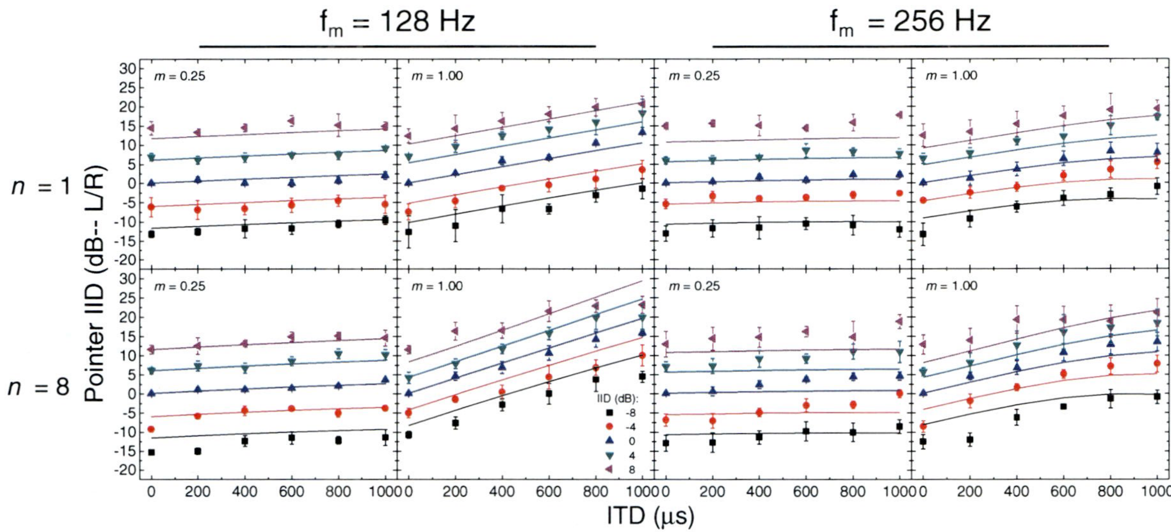

Fig. 31.4 The IID (in dB) of the acoustic pointer required to match the intracranial position of 4-kHz–centered raised-sine targets as a function of the ongoing ITD (left ear leading). The panels within the top and bottoms rows display data obtained when the exponent of the raised sine, *n*, was 1.0 and 8.0, respectively. The left and right halves of the figure depict the data obtained when the rate of modulation was 128 Hz and 256 Hz, respectively. Panels within the columns of the left and right halves contain data obtained for depths of modulation, *m*, of 0.25 and 1.00, respectively. Symbols represent the mean values computed across three listeners. The error bars represent ± 1 standard error of the mean. The parameter within each plot is the IID of the target. The lines represent predictions from a cross-correlation-based model. (Adapted from Fig. 1 of Bernstein & Trahiotis, 2012, with permission)

substantially larger value of the IID of the 500-Hz-centered pointer in order to match the intracranial position of the target. That is, the effective "internal" IID appears to be relatively magnified at 4 kHz as compared to the effective internal IID at 500 Hz. While the mechanistic explanation for this effect is beyond the scope of this chapter, suffice it to say that it stems from a complex interaction within the patterning of the cross-correlation. That is, the ability to predict the effects of IID "come free" within the model. In this case, it is a by-product of the general enterprise of constructing and augmenting the model over time in a manner that describes stimuli "as processed."

31.3 Future Directions

One enterprise that appears to hold particular promise for advancing understanding of binaural processing, both empirically and theoretically, would be a new, more integrative approach concerning behavioral and neurophysiological investigations. This would entail using, in both types of experiments, common sets of parametrically constructed stimuli that have been proven to be behaviorally diagnostic. The general idea would be to use those stimuli in order to gather large sets of neurophysiological data that would allow for the construction of suitable "population" responses. Specifically, one way for that to occur would be to measure, as a function of time, responses obtained from dozens and, perhaps, hundreds of single neural units, each being stimulated by exactly the same (i.e., reproducible) external stimulus. Then, taking into account the characteristic- or best-frequency and ITD-tuning of each unit, one could construct a "running-time" plot of neural activity that would be an analog of the across-frequency correlogram that has served to qualitatively and quantitatively account for a substantial set of behavioral data. Such a plot would, by its very nature, include all of the known (and any unknown) factors of the type discussed earlier concerning stimuli "as processed" which have proven to be essential to explain various binaural phenomena. The enterprise would not be expected to be unidirectional. That is, no doubt there would be instances wherein scrutiny of the neurophysiologically constructed correlograms would suggest new behavioral experiments and ways to augment, refine, and extend the best quantitative models of binaural hearing.

It would especially gratifying if such an approach would provide new levels of explanation of two important binaural phenomena. The first is what is termed "binaural sluggishness" and refers to the relative inability of the binaural system to allow one to "track," or "follow" closely, even what one could consider to be slowly varying values of ITD and IID (for a review, see Grantham, 1995). An interesting aspect of binaural sluggishness is that it is not manifest in the responses of single units stimulated by the rapid changes in the binaural cues (e.g., Yin & Kuwada, 1983; Joris et al., 2006). The second phenomenon is binaural "interference." This refers to the findings that the processing of binaural information within high-frequency "channels" can be highly degraded by the simultaneous presentation of low-frequency binaural information that is sufficiently spectrally remote so as to preclude masking and monaural interactions, in general, as explanations (e.g., Bernstein & Trahiotis, 1995; Heller & Trahiotis, 1996; Bernstein & Trahiotis, 2004; Best et al., 2007). There is currently no quantitative, mechanistically based account of binaural interference. What would help to foster explanations of both binaural sluggishness and binaural interference would be information within the type of population-based neural correlogram described earlier.

31.4 Summary

Progress made over the last 20 years within the broad field of binaural information processing has, in many ways, stemmed from a more sophisticated understanding of stimuli as processed as opposed to consideration of external, physical stimuli. The knowledge gained has permitted important advances in the understanding of specific, peripheral and/or central, mechanisms that underlie and constrain binaural capabilities. Many of the advances have stemmed from earphone-based studies concerning binaural detection, discrimination, and lateralization. The enterprise, taken together with advances in other areas of auditory research, has allowed for the development of more comprehensive and accurate, quantitative models of human binaural performance. At this juncture, it appears that progress in the future would be fostered by a more integrative approach in which common sets of parametrically constructed stimuli that have been proven to be theoretically diagnostic would be employed in "parallel" behavioral and neurophysiological investigations.

Acknowledgment This work was supported by research grant NIH DC-04147 from the National Institute on Deafness and Other Communication Disorders, National Institutes of Health.

References

Bernstein, L. R., & Trahiotis, C. (1982). Detection of interaural delay in high-frequency noise. *Journal of the Acoustical Society of America*, 71, 147–152.

Bernstein, L. R., & Trahiotis, C. (1985). Lateralization of sinusoidally-amplitude-modulated tones: Effects of spectral locus and temporal variation. *Journal of the Acoustical Society of America*, 78, 514–523.

Bernstein, L. R., & Trahiotis, C. (1994). Detection of interaural delay in high-frequency SAM tones, two-tone complexes, and bands of noise. *Journal of the Acoustical Society of America*, 95, 3561–3567.

Bernstein, L. R., & Trahiotis, C. (1995). Binaural interference effects measured with masking-level difference and with ITD- and IID-discrimination paradigms. *Journal of the Acoustical Society of America*, 98, 155–163.

Bernstein, L. R. & Trahiotis, C. (1996). Binaural beats at high frequencies: Listeners' use of envelope-based interaural temporal and intensitive disparities. *Journal of the Acoustical Society of America*, 99, 1670–1679.

Bernstein, L. R., & Trahiotis, C. (2002). Enhancing sensitivity to interaural delays at high frequencies by using "transposed stimuli. *Journal of the Acoustical Society of America*, 112, 1026–1036.

Bernstein, L. R., & Trahiotis, C. (2003). Enhancing interaural-delay-based extents of laterality at high frequencies by using 'transposed stimuli'. *Journal of the Acoustical Society of America*, 113, 3335–3347.

Bernstein, L. R., & Trahiotis, C. (2004). The apparent immunity of high-frequency "transposed" stimuli to low-frequency binaural interference. *Journal of the Acoustical Society of America*, 116, 3062–3069.

Bernstein, L. R., & Trahiotis, C. (2008). Discrimination of interaural temporal disparities conveyed by high-frequency sinusoidally amplitude-modulated tones and high-frequency transposed tones: Effects of spectrally flanking noises. *Journal of the Acoustical Society of America*, 124, 3088–3094.

Bernstein, L. R., & Trahiotis, C. (2010). Accounting quantitatively for sensitivity to envelope-based interaural temporal disparities at high frequencies. *Journal of the Acoustical Society of America*, 128, 1224–1234.

Bernstein, L. R., & Trahiotis, C. (2012). Lateralization produced by interaural temporal and intensitive disparities of high-frequency, raised-sine stimuli: Data and modeling. *Journal of the Acoustical Society of America*, 131, 409–415.

Bernstein, L. R., Par, Steven van de, & Trahiotis, C. (1999). The normalized correlation: Accounting for NoSπ thresholds obtained with Gaussian and "low-noise" masking noise. *Journal of the Acoustical Society of America*, 106, 870–876.

Best, V., Gallun, F. J., Carlile, S., & Shinn-Cunningham, B. (2007). Binaural interference and auditory grouping. *Journal of the Acoustical Society of America*, 121, 1070–1076.

Blauert, J. (1983). *Spatial hearing: The psychophysics of human Sound Localization*. Cambridge, MA: MIT Press.

Colburn, H. S., & Esquissaud, P. (1976). An auditory-nerve model for interaural time discrimination of high-frequency complex stimuli. *Journal of the Acoustical Society of America*, 59, S23.

Domnitz, R. H., & Colburn, H. S. (1977). Lateral position and interaural discrimination. *Journal of the Acoustical Society of America*, 61, 1586–1598.

Dreyer, A. A., & Oxenham, A. J. (2008). Effects of level and background noise on interaural time difference discrimination for transposed stimuli. *Journal of the Acoustical Society of America*, 123, EL1–EL7.

Durlach, N. I., & Colburn, H. S. (1978). Binaural phenomena. In E. C. Carterette & M. P. Friedman (Eds.), *Handbook of perception: Hearing* (pp. 365–466). New York: Academic Press.

Eddins, D. A., & Barber, L. E. (1998). The influence of stimulus envelope and fine structure on the binaural masking level difference. *Journal of the Acoustical Society of America*, 103, 2578–2589.

Ewert, S. D., & Dau, T. (2000). Characterizing frequency selectivity for envelope fluctuations. *Journal of the Acoustical Society of America*, 108, 1181–1196.

Feddersen, W. E., Sandel, T. T., Teas, D. C., & Jeffress, L. A. (1957). Localization of high-frequency tones. *Journal of the Acoustical Society of America*, 29, 988–991.

Grantham, D. W. (1995). Spatial hearing and related phenomena. In B. C. J. Moore (Ed.), *Handbook of perception and cognition: Hearing*. San Diego: Academic Press.

Hall, J. W. III, Grose, J. H., & Hartmann, W. M. (1998). The masking-level difference in low-noise noise. *Journal of the Acoustical Society of America*, 103, 2573–2577.

Hartung, K., & Trahiotis, C. (2001). Peripheral auditory processing and investigations of the 'precedence effect' which utilize successive transient stimuli. *Journal of the Acoustical Society of America*, 110, 1505–1513.

Heijden, M. van der, & Trahiotis, C. (1999). Masking with interaurally delayed stimuli: The use of 'internal' delays in binaural detection. *Journal of the Acoustical Society of America*, 105, 388–399.

Heijden, M. van der, Trahiotis, C., Kohlrausch, A., & Par, Steven van de (1997). Binaural detection with spectrally nonoverlapping signals and maskers: Evidence for masking by aural distortion products. *Journal of the Acoustical Society of America*, 102, 2966–2972.

Heller, L. M., & Trahiotis, C. (1996). Extents of laterality and binaural interference effects. *Journal of the Acoustical Society of America*, 99, 3632–3637.

Joris, P. X., van de Sande, B., Recio-Spinoso, A., & Heijden, M. van der (2006). Auditory midbrain and nerve responses to sinusoidal variations in interaural correlation. *Journal of Neuroscience*, 26, 279–289.

Klumpp, R. G., & Eady, H. R. (1956). Some measurements of interaural time difference thresholds. *Journal of the Acoustical Society of America*, 28, 859–860.

Kohlrausch, A., Fassel, R., & Dau, T. (2000). The influence of carrier level and frequency on modulation and beat-detection thresholds for sinusoidal carriers. *Journal of the Acoustical Society of America*, 108, 723–734.

Lindemann, W. (1986a). Extension of a binaural cross-correlation model by contralateral inhibition. I. Simulation of lateralization for stationary signals. *Journal of the Acoustical Society of America*, 80, 1608–1622.

Lindemann, W. (1986b). Extension of a binaural cross-correlation model by contralateral inhibition. II. The law of the first wave front. *Journal of the Acoustical Society of America*, 80, 1623–1630.

Litovsky, R. Y., Colburn, H. S., Yost, W. A., & Guzman, S. J. (1999). The precedence effect. *Journal of the Acoustical Society of America*, 106, 1633–1654.

McFadden, D., & Pasanen, E. G. (1976). Lateralization at high frequencies based on interaural time differences. *Journal of the Acoustical Society of America*, 59, 634–639.

Meddis, R. (1986). Simulation of mechanical to neural transduction in the auditory receptor. *Journal of the Acoustical Society of America*, 79, 702–711.

Meddis, R. (1988). Simulation of auditory-neural transduction: Further studies. *Journal of the Acoustical Society of America*, 83, 1056–1063.

Meddis, R., Hewitt M. J., & Shackleton, T. M. (1990). Implementation details of a computational model of the inner hair-cell/auditory-nerve synapse. *Journal of the Acoustical Society of America*, 87, 1013–1016.

Middlebrooks, J. C., & Snyder, R. L. (2010). Selective electrical stimulation of the auditory nerve activates a pathway specialized for high temporal acuity. *Journal of Neuroscience*, 30, 1937–1946.

Moore, B. C. J., & Glasberg, B. R. (2001). Temporal modulation transfer functions obtained using sinusoidal carriers with normally hearing and hearing-impaired listeners. *Journal of the Acoustical Society of America*, 110, 1067–1073.

Nuetzel, J. M., & Hafter, E. R. (1981). Discrimination of interaural delays in complex waveforms: Spectral effects. *Journal of the Acoustical Society of America*, 69, 1112–1118.

Par, S. van de, & Kohlrausch, A. (1997). A new approach to comparing binaural masking level differences at low and high frequencies. *Journal of the Acoustical Society of America*, 101, 1671–1680.

Par, S. van de, Trahiotis, C., & Bernstein, L. R. (2000). The use of off-frequency information in a high-frequency binaural discrimination task. *Acustica*, 86, 526–531.

Par, S. van de, Trahiotis, C., & Bernstein, L. R. (2001). A consideration of the normalization that is typically included in correlation-based models of binaural detection. *Journal of the Acoustical Society of America*, 109, 830–833.

Rodríguez, F. A., Read, H. L., & Escabí, M. A. (2010). Spectral and temporal modulation tradeoff in the inferior colliculus. *Journal of Neurophysiology*, 103, 887–903.

Ruggero, M. A. (1992). Responses to sound of the basilar membrane of the mammalian cochlea. *Current Opinion in Neurobiology*, 2, 449–456.

Sellick, P. M., Patuzzi, R., & Johnstone, B. M. (1982). Measurement of the basilar membrane motion in the guinea pig using the Mössbauer technique. *Journal of the Acoustical Society of America*, 72, 131–141.

Shinn-Cunningham, B. G., Zurek, P., Durlach, N. I., & Clifton, R. K. (1995). Cross-frequency interactions in the precedence effect. *Journal of the Acoustical Society of America*, 98,164–171.

Stern, R. M., & Colburn, H. S. (1978). Theory of binaural interaction based on auditory-nerve data. IV. A model for subjective lateral position. *Journal of the Acoustical Society of America*, 64, 127–140.

Stern, R. M., Zeiberg, A. S., & Trahiotis, C. (1988). Lateralization of complex binaural stimuli: A weighted image model. *Journal of the Acoustical Society of America*, 84, 156–165.

Trahiotis, C., Bernstein, L. R., & Akeroyd, M. A. (2001). Manipulating the "straightness" and "curvature" of patterns of interaural cross-correlation affects listeners' sensitivity to changes in interaural delay. *Journal of the Acoustical Society of America*, 109, 321–330.

Verhulst, S., Bianchi, F., & Dau, T. (2012). Cochlear contributions to the precedence effect. In B. C. J. Moore, R. D. Patterson, I. Winter, R. P. Carlyon, & H. E. Gockel (Eds.), *Basic aspects of hearing: Physiology and perception*, New York: Springer Science+Business Media.

Wallach, H., Newman, E. B. & Rosenzweig, M. R. (1949). The precedence effect in sound localization. *American Journal of Psychology*, 52, 315–336.

Xia, J. & Shinn-Cunningham, B. (2011). Isolating mechanisms that influence measures of the precedence effect: Theoretical predictions and behavioral tests. *Journal of the Acoustical Society of America*, 130, 866–882.

Yin, T. C., & Kuwada, S. (1983) Binaural interaction in low-frequency neurons in inferior colliculus of the cat. II. Effects of changing rate and direction of interaural phase. *Journal of Neurophysiology*, 50, 1000–1019.

Yost, W. A., & Soderquist, D. R. (1984). The precedence effect: Revisited. *Journal of the Acoustical Society of America*, 76, 1377–1383.

Zurek, P. M. (1987). The precedence effect. In W. A. Yost & G. Gourevitch (Eds.), *Directional hearing* (pp. 85–105). New York: Springer-Verlag.

Zurek, P. M., & Durlach, N. I. (1987). Masker-bandwidth dependence in homophasic and antiphasic tone detection. *Journal of the Acoustical Society of America*, 81, 459–464.

Zwislocki, J., & Feldman, R. S. (1956). Just noticeable differences in dichotic phase. *Journal of the Acoustical Society of America*, 28, 860–864.

Chapter 32
Temporal Processing: Observations on the Psychophysics and Modeling of Temporal Integration and Temporal Resolution

Neal F. Viemeister

Neal F. Viemeister

N.F. Viemeister (✉)
Department of Psychology, University of Minnesota,
75 East River Parkway, Minneapolis, MN 55455, USA
e-mail: nfv@umn.edu

A.N. Popper and R.R. Fay (eds.), *Perspectives on Auditory Research*, Springer Handbook of Auditory Research 50, DOI 10.1007/978-1-4614-9102-6_32,

32.1 Introduction

My interest in temporal processing began with reading Lloyd Jeffress' work on "leaky integrators" while a graduate student at Indianan University. This work, published in *Journal of the Acoustical Society of America* in the 1960s (e.g., Jeffress, 1968), modeled, somewhat successfully, detection of tones in noise with an electrical model using a bandpass filter followed by a half-wave rectification, and then a leaky integrator—a simple resistor-capacitor (RC) lowpass filter. This basic model, now digitally implemented, is widely used at present as a first approximation to single-channel auditory processing. It is essentially an envelope detector: The output of the model follows the slow-amplitude fluctuations present in the output of the relatively narrow filter. I was intrigued by Jeffress' approach and the notion that such a model might account for trial-by-trial decisions by humans in a simple tone-in-noise detection task. So, with the support of my PhD mentor James P. Egan (JP) and Air Force Office of Scientific Research (AFOSR), we bought an expensive Burr–Brown operational amplifier and a more expensive Tektronix storage oscilloscope. I learned more than I needed to know about op-amps and built a fancy Jeffress leaky integrator complete with multi-turn potentiometer to adjust the time constant precisely, a relay to reset the capacitor, rack mountable, etc. I experimented with it and was intrigued by its envelope detection aspects, especially the changes produced by adding a constant-amplitude signal. After many hours of watching the oscilloscope while listening to the input, I concluded that the Jeffress model did not adequately describe what I was hearing. In the model, decisions were based on the output of the leaky integrator at the end of the observation and did not capture the changes in the envelope produced by adding a signal. More importantly for me was that the model, with its long time constant, seemed to excessively smooth the envelope fluctuations in the narrow-band noise. I could hear the noise fluctuations clearly even though the model indicated they were negligible. I talked to JP about this and for various reasons he decided that a visit to Dr. J was appropriate. So, with my beautiful leaky integrator packed up, we drove from Bloomington to Austin in JP's hopped-up Pontiac. I met with Dr. J, showed him my work of art, and he seemed impressed. Then he showed me his electrical model. It was a breadboarded mess of wires, capacitors, and resistors seemingly randomly wired. But it worked, of course. He was a very gracious person and did not ask the obvious question of why I built such a fancy thing when relatively simple breadboarding did the job. My feeble response would have been that we did not do things that way in JP's elegant lab and would be an affront to Gordon Greenberg's beautiful work building the lab. Then over several meetings Dr. J and I talked good science and mulled over the problem of reconciling our own observations, the data on the perception of beats, and temporal integration such as captured by the long-time constant leaky integrator. This became later known as the resolution–integration paradox. With no resolution (pun intended), JP and I drove back with my precious leaky integrator that I put on a shelf in JP's lab and returned to my studies. But the general issue of temporal processing and developing plausible models was always in my thoughts.

This chapter offers my perspective on the psychophysics of temporal processing. This is a very broad topic, and much-needed thorough, current, and insightful reviews are not available. Such a review would be well beyond the scope of this chapter but some selected secondary references are listed (Plack, 2005; Viemeister & Plack, 1993) as general introductions. So with my apologies to those whose work on this topic I am about to ignore, distort, or slight, here is my perspective.

32.2 Temporal Integration

One of classic problems in perception is why detection and discrimination thresholds improve with increasing stimulus duration. This seems to be true of all sensory systems and has been studied extensively with an underlying goal of better understanding peripheral processing. For example, in the visual system, the decrease in the threshold for detecting a flash presented in darkness with increasing duration has been linked to photochemical processes (the Bunsen–Roscoe Law) occurring in the retina. In audition, it does not appear that a simple peripheral (cochlear) explanation for the integration phenomenon is tenable: responses of hair cells and of auditory nerve fibers are too fast and show no evidence of increasing response beyond the first few milliseconds. If true integration were occurring at this level the response would monotonically increase as the stimulus proceeded and the response would be greater at the end of the stimulus than at the beginning. Post-stimulus time histograms (PSTs) for auditory nerve fibers show constant or decreasing firing rates (adaptation). Because the behavioral data show threshold improvement out to hundreds of milliseconds, a strictly peripheral account of the integration process seems untenable. The question then is what more central processes account for this basic phenomenon and, ultimately, what does it tell us about auditory processing?

There is a wealth of psychophysical data on temporal integration. Temporal integration functions (threshold vs. duration) have been measured in many species using a variety of stimuli, typically for pure tones. The focus of most of the experimental work has been in humans and within that species there have been extensive, rigorous examinations of temporal integration in hearing-impaired (HI) listeners, cochlear implant (CI) users, young and old normal-hearing (NH) listeners, and persons with neurological disorders. To oversimplify, auditory temporal integration, at least for simple detection, is ubiquitous: Almost all studies show some type of improvement or increased response as stimulus duration is increased typically out to hundreds of milliseconds. To be sure, there are, across studies, possibly important differences in stimulus generation, experimental design, curve fitting, etc. but the basic data, at least for detection of tones in quiet and in noise, seem solid and deserving of theoretical attention. Less solid are the data on integration of suprathreshold stimuli. Whether loudness increases with duration, for example, still appears unclear and is perhaps of little relevance for understanding auditory processing.

Theoretical speculations about the basis for the phenomenon of temporal integration have a long history in auditory research. One that seems to been recently revived is "probability summation." This is a simple, high-threshold model, according to which the listener detects the signal if at least one event, for example, a spike, is elicited. Accordingly, as the duration of the signal is increased the probability of this event increases and thus the signal level necessary to elicit at least one event decreases. The major appeal of this notion is its simplicity. But it is, essentially, an obsolete idea in that it largely ignores the fact that sensory transmission is largely stochastic. For example, auditory nerve fibers that show highest sensitivity (in quiet) have a relatively high rate of spontaneous firing. A far more appealing classic idea is based on leaky integrator models according to which stimulus excitation is accumulated (integrated) over time but is dissipated at a rate that is proportional to the amount accumulated. For stimulus durations that are relatively brief, the accumulation is nearly perfect—there is not much to "leak." This conception is essentially a low-pass filter whose input is some transformation of the stimulus. The Jeffress-type envelope detector mentioned at the beginning of this chapter—half-wave rectification followed by lowpass filtering—is a classic example. Such leaky integrator models can provide a good quantitative description of the temporal integration data, at least for detection of tones in quiet and in noise. A variant of the leaky integration models is an energy detector in which stimulus power is perfectly integrated over some duration, usually the duration of the signal to be detected. To account for the behavioral data, integrations times/time constants of the order of hundreds of milliseconds are necessary. As discussed in the text that follows, such sluggishness seems incompatible with the data and everyday experience indicating that the auditory system is "fast." A solution to this paradox is that temporal integration reflects the "intelligent" combination of brief samples, "looks," from some short-term (2–3 ms) integration process and that these multiple looks are combined, perhaps nearly optimally, at some very central processor, such as the brain (Viemeister & Wakefield, 1991).

My perspective on temporal integration is that it has stimulated some fine, informative research but that the integration/combination process is far more complex than the hard-wired probability summation or leaky integrator accounts. It is a higher-level process and is not a fundamental property of audition. But it clearly is affected by more fundamental, for example, cochlear, processes and as such is a useful tool for understanding these processes and, perhaps more importantly, for understanding the peripheral processes that underlie temporal integration. For example, it appears that both HI and CI persons show less temporal integration than do NH individuals, that is, their thresholds do not decrease as rapidly (in dB per doubling of duration) as do those of NH listeners. This seems to be related to the limited dynamic range/recruitment/ growth of response seen in these persons. Although the old multiple looks notion clearly needs both experimental and theoretical refinement, it continues to offer a framework for understanding these processes.

32.3 Temporal Resolution

The auditory system is a fast system and shows excellent temporal resolution, at least compared to other sensory systems: We can detect interruptions in white noise up to several thousand interruptions per second versus less than a hundred interruptions per second for visual stimuli. Auditory communication signals, such as speech, take advantage of, or co-evolved with production, and utilize such rapid transitions, of the order of milliseconds, in frequency and amplitude as information-bearing elements.

There have been many approaches used to investigate various aspects temporal resolution and temporal processing in general. These include gap detection, duration discrimination, frequency sweep detection and discrimination, forward and backward masking, and temporal manipulations of speech such as voice onset time. These can be conceived as modulations in amplitude and in frequency. A fruitful approach, which began in the 1970s, has been to treat amplitude modulation (AM) and frequency modulation (FM) separately and try to analyze these aspects of temporal processing using well-known system-analysis approaches. This has been most successful in AM processing in which sinusoidally amplitude modulation (SAM) is used to estimate a temporal modulation transfer function (TMTF; Viemeister, 1979), which essentially relates sensitivity to modulation as a function of modulation frequency. The TMTF has been interpreted as sensitivity to temporal envelope fluctuations (vs. FMTFs for sinusoidal FM of pure tones, and SMTFs—spectral modulation of relatively broadband waveforms, e.g. spectral ripple detection). As with measuring the transfer function of a (linear) filter, the basic idea underlying the TMTF is that it may provide a general description of "the" envelope detector that underlies temporal resolution. Nearly impossible, but perhaps a useful approximation.

Psychophysical and physiological TMTFs have been measured extensively and it is impossible to simply summarize the literature on, for example, the effects of hearing impairment, on the single-unit physiological data recorded (unfortunately generally recorded under anesthesia) at various levels in the auditory pathway, on binaural processing, on TMTFs measured in CI patients, and on more theoretical aspects, concerning, for example, the role of adaptation. A noteworthy contribution that has received considerable attention is the notion of a modulation filterbank (Dau et al., 1997; Jepsen et al., 2008). This type of model postulates that there are channels (filters) that are selectively tuned to the modulation frequencies produced by AM. It is well known that we humans can detect differences in the modulation frequency of SAM and more directly that, analogous to pure tone masking in the spectral domain, modulation frequencies that are closer to that of a target SAM signal are more effective in elevating the threshold for detecting modulation of the target SAM (modulation masking). Such findings led to the development of models that incorporate filtering for AM analogous to critical band/auditory filters in the audio spectral domain. These modulation filterbank models provide a good account of AM frequency selectivity—not surprising considering that the models are based largely on data from experiments on modulation masking—but they do have predictive power such as accounting for differences between TMTFs measured using broadband noise and those measured using sinusoidal carriers. These models also have considerable heuristic value and have stimulated both experimental and theoretical work on modulation processing. For example, there have been attempts to generalize the modeling and experimental work beyond the AM domain and to describe across tonotopic channel effects such as modulation detection interference (MDI) and, more generally, to describe spectrotemporal processing.

Supporting the notion of modulation filters are psychophysical experiments and physiological experiments on selective adaptation to SAM. Selective adaptation has a long tradition in sensory psychology, especially vision, for identifying "channels" that are tuned to some stimulus feature. In audition, there was early work showing selective adaptation for SAM suggesting AM channels. Experiments that are more recent have avoided some of the controversial aspects of this early work and have demonstrated forward masking by SAM: Modulation thresholds for brief SAM signals are increased when preceded by a SAM masker whose modulation rate is similar to that of the signal (see Wojtczak et al., 2011). The tuning appears to be broad, and recovery times appear to be of the order of hundreds of milliseconds. An interpretation that is consistent with the notion of a modulation filterbank is that there are neural circuits tuned to modulation frequency. In response to SAM, these units adapt and recover to SAM with time constants of hundreds of milliseconds. Interestingly, single-unit recordings from the inferior colliculus (IC) of awake rabbits show no significant effects of a SAM forward masker (Wojtczak et al., 2011). This suggests that the masking seen in the human psychophysics, and presumably the tuning, occur central to the IC.

In general, however, physiologically measured responses to AM have provided little insight into the processes underlying the psychophysics, specifically whether there is tuning for AM rate as postulated by the modulation filterbank models. TMTFs recorded from the auditory nerve (AN) show a lowpass characteristic somewhat similar to that shown by the human psychophysical data obtained using continuous broadband noise carriers (Joris et al., 2004). The lowpass characteristic of the AN data is somewhat surprising considering the well-documented adaptation shown, for example, in the PSTs, which often show a relatively larger response at stimulus onset. Based on this, one might expect the TMTFs to emphasize higher AM rates and show a bandpass characteristic. Indeed, this seems to be occurring to some extent at more central locations in the auditory pathway, where bandpass characteristics are not atypical. However, such tuning for AM rate provides only weak evidence for AM filterbank models. There is little evidence of topicity for AM rate as suggested by such models and, perhaps more importantly, there is little evidence for selective responses to AM at modulation rates comparable to those that are detectable and discriminable by humans. There is the issue of the effects of anesthesia, which have been shown to affect the temporal response characteristics in the central nervous system (CNS; and possibly AN responses via cochlear efferent effects). Unfortunately, the physiological data seem to provide few constraints on models to describe AM perception. Are there channels that are systematically tuned to modulation frequency? What are the physiological correlates in the CNS of the demonstrated ability that we can detect, discriminate, and extract pitch from SAM noise at high modulation rates?

My concern is that models based on modulation filtering do not adequately capture some fundamental aspects of temporal processing because they minimize, or ignore, the temporal structure of the envelopes of complex stimuli. Time-reversed waveforms such as speech tokens and ramped vs. damped broadband noise can sound considerably different even though their amplitude spectra in both the AM domain and the audio frequency domain are identical. Modulation filters that ignore phase are missing a crucial aspect of temporal processing. However, temporally based models such as those based on cross correlation and template matching have not been able to provide a good quantitative account of AM frequency selectivity.

A set of general issues in AM processing is the role of peripheral filtering and frequency resolution. With sinusoidal carriers, there is the well-known problem of sideband detection—that the AM is not being detected based primarily on temporal fluctuations but on the presence/absence of "resolved" sidebands produced by AM. Experiments comparing the detectability of SAM versus phase-shifted SAM (e.g., quasi-frequency modulation [QFM]) perhaps have helped better understand this basic issue. A somewhat related issue is the possibility that AM at higher modulation rates is detected by using lower frequency distortion products such as the difference tone that reflects cochlear nonlinearities. The use of lowpass noise maskers minimizes this problem but, of course, there is the possibility that such maskers may affect temporal processing at the carrier frequency.

A related issue is the role of off-frequency listening, the idea that the information used to detect the AM is not being conveyed by the frequency channel tuned to the carrier frequency. One manifestation of this is the large decrease in modulation thresholds for SAM with increases in carrier level. Most likely this is attributable to the effects of spread of excitation that are seen with pure-tone intensity discrimination, the "near-miss" to Weber's Law: As level increases frequency channels are recruited that show increasingly less compression than that at the carrier frequency and thus changes in amplitude are, in effect, magnified and yield lower thresholds for detecting the change. A more subtle issue regarding off frequency listening is the possibility that by using frequency channels close to the carrier frequency the listener can maximize the effective envelope fluctuations of SAM. For example, by listening to a channel tuned below the carrier frequency the listener may able to effectively attenuate the carrier so as to better equate the levels of the carrier and lower sideband and thus increase the effective envelope fluctuations present in that channel. Another potential issue is the possibility that AM is detected and processed as FM: AM-to-FM conversion if you will. The instantaneous frequency of SAM with a pure-tone carrier is time invariant. A phase shift in, say, the carrier, as in typical QFM, will produce changes in instantaneous frequency that are related to the modulation frequency. Such phase shifts certainly occur in the auditory system and thus AM could be detected as FM, however, that is detected and processed. This notion is not being offered very seriously but may be a good exercise—why is it wrong?

Given the preceding discussion, one might ask whether we have learned anything useful using sinusoidal carriers. The answer is yes. This research has provided reasonably compelling evidence that a basic aspect of hearing, temporal resolution, is not strongly dependent on frequency. However, there are potentially serious problems that could undermine both experimental and theoretical development. As suggested earlier, these relate to auditory tonotopicity and the difficulty in distinguishing between time and frequency in processing by this diabolical system.

To avoid these spectral issues, SAM using broadband white noise carriers has been extensively used. The long-term spectrum of SAM noise is unaffected by the modulation and remains "flat." The rationale is that, unlike in SAM tones, there are no spectral cues and thus TMTFs based on SAM noise are a more pure measure of temporal resolution. The sacrifice, of course, is lack of frequency specificity—we do not know where in frequency the listener is extracting information about the presence of AM. This is not a trivial problem theoretically or practically. On the applied side, for example, the use of broadband noise carriers is somewhat problematical for assessing temporal processing in frequency regions of hearing loss. (Filtering the noise either before or after modulation raises its own problems.) Theoretically, there are several issues. One concerns the possibility that short-term spectral cues mediate detection and, more specifically, the pitch that is elicited by SAM noise. The question essentially is whether the modulation produces short-term increases in amplitude that differ across frequency and that could provide the basis for detecting SAM and extracting pitch. Based on experimental evidence this seems unlikely: Modulation thresholds and pitch extraction appear to be largely unaffected (aside from bandwidth effects) when the SAM noise is restricted to high-frequency regions where spectral resolution is poor.

A more pressing and continuing issue that arises from the use of broadband noise carriers is that TMTFs indicate that we can detect SAM noise up to modulation frequencies of several kilohertz. This seems incompatible with measures of audio frequency selectivity and indeed models such as the modulation filterbank models that attempt to incorporate realistic peripheral filtering fail to account for sensitivity to AM at high modulation rates. Thus, the modeling suggests that temporal resolution is considerably poorer than the psychophysical data indicate. More importantly, this suggests that somehow the auditory system can adjust its effective bandwidth depending on the task. For temporal resolution such as modulation detection, the system appears to be broadband whereas for frequency resolution, such as detection of tones in noise, the system appears to be sharply tuned. The auditory system somehow has solved, for its purposes, the time–frequency resolution tradeoff. How it does this is very uncertain. The usual theoretical approach would be to posit dual systems, one for temporal resolution and the other for frequency resolution. Indeed, physiological data indicate that in addition to sharply tuned neural responses in the CNS there are "broadband" units that respond over a wide range of frequencies. The problem, though, seems far more complex: How are the responses from narrowband units combined to result in a broadband response that permits some degree of envelope preservation at high modulation frequencies? This would seem to require preservation of phase across the narrowband channels so that a summation, for example, could reconstruct the envelope. The problem is that we can detect very high frequency SAM at carrier frequencies well above the usual limits of peripheral temporal synchronicity ("phase locking") and so simple combination of such narrowband channels would not preserve the envelope. Another possibility is that the estimates of frequency selectivity that are relevant for modeling temporal resolution are incorrect. The old notion that the critical band/auditory filter is broadly tuned at broad stimulus onset and increases its selectivity, an explanation for "overshoot," has received renewed attention with speculation that cochlear efferents may be involved in the sharpening. Broader peripheral filtering for onsets may account for our sensitivity to SAM noise at high modulation frequencies. This suggests the intriguing possibility auditory system may have solved the temporal versus frequency resolution tradeoff by dynamic filtering that is mediated by efferent influences.

The discussion so far has focused on temporal resolution that is primarily "within channel," that is, where the assumption is that there are no or limited spectral differences across the tonotopic frequency channels that are so basic to peripheral auditory processing. This focus has been dictated by both experimental and theoretical expediency—let's address and describe "one thing at a time"—and later tackle the much more difficult and relevant aspect of temporal processing that involves spectral changes that occur over time such as in virtually all auditory communication signals. An approach that has not proved to be especially informative has involved investigations of "across channel" temporal resolution. The experimental work includes, for example, psychophysical studies on detection and discrimination of onset asynchronies for tones that differ in frequency, envelope phase discrimination of SAM for carriers differing in frequency, comodulation masking release (CMR), modulation detection interference (MDI), discrimination of FM sweeps, discrimination of Huffman sequences, and numerous studies using speech tokens. An oversimplification of the findings from such studies is that such measures of temporal resolution suggest that across channel resolution is comparable to that for within-channel resolution—in the low-millisecond range. This similarity raises the question of whether these findings reflect within-channel effects. For example, in detecting onset asynchronies of tones there may be frequency channels in which there is an overlap of excitation produced by the tones. The responses from such channels will show differences in their envelopes depending on the onset disparity. In effect this make the nominally across-channel process a within-channel process involving detection of envelope changes. A possible solution is to force the system to make an across-channel comparison by, for example, using a masking noise to reduce information from overlapping channels or by presenting the stimuli to different ears. These manipulations introduce their own set of issues and are not entirely satisfactory as a means for assessing "pure" across-channel resolution. Compounding this is the deeper issue of what is meant by a channel. The idealization is that the early stages of auditory processing can be understood as a set of fixed filters tuned to different frequencies across the audio frequency range—the auditory filterbank. However, as previously mentioned in the discussion of SAM noise, the system in effect may adjust the bandwidth(s) of the filterbank depending on the acoustic context—sharply tuned channels for processing vowels and broadly tuned channels for processing the rapid transitions in speech, for example. If so, the fixed spectral filterbank model seems inappropriate as a basis for understanding both across- and within-channel temporal resolution.

32.4 Concluding Comments

Temporal processing is a fundamental aspect of audition. It underlies the basis for auditory communication in virtually all species, including humans. Some of the issues raised in this chapter are technical and in the broader scheme of things—understanding auditory perception—are relatively minor. I have attempted to raise some broader issues, such as the usefulness of the familiar fixed filterbank model or whether the system could be better described in terms of dynamic frequency selectivity. As has been said many times, "further research is necessary." My intent in this chapter was to suggest possible directions for this research and to be somewhat provocative.

References

Dau, T., Kollmeier, B., & Kohlrausch, A. (1997). Modeling auditory processing of amplitude modulation: II. Spectral and temporal integration. *Journal of the Acoustical Society of America*, 102, 2906–2919.

Jeffress, L. A. (1968). Mathematical and electrical models of auditory detection. *Journal of the Acoustical Society of America*, 44, 187–203.

Jepsen, M. L., Ewert, S., & Dau, T. (2008). A computational model of human auditory signal processing. *Journal of the Acoustical Society of America*, 124, 422–438.

Joris, P., Schreiner, C., & Rees, A. (2004). Neural processing of amplitude-modulated sounds. *Physiological Reviews*, 84 (2), 541–577.

Plack, C. J. (2005). *The sense of hearing*. Mahwah, NJ: Lawrence Erlbaum.

Viemeister, N. F. (1979). Temporal modulation transfer functions based upon modulation thresholds. *Journal of the Acoustical Society of America*, 66, 1364–1380.

Viemeister, N. F., & Wakefield, G. W. (1991). Temporal integration and multiple looks. *Journal of the Acoustical Society of America*, 90, 858–865.

Viemeister, N. F., & Plack, C. J. (1993). Time analysis. In W. Yost, A. Popper, & R. Fay (Eds.), *Human psychophysics* (pp. 116–154). New York: Springer-Verlag.

Wojtczak, M., Nelson, P. C., Viemeister, N. F., & Carney, L. H. (2011). Forward masking in the amplitude-modulation domain for tone carriers: Psychophysical and physiological correlates. *Journal of the Association for Research in Otolaryngology*, 12, 361–373.

Chapter 33
Psychoacoustics and Auditory Perception

William A. Yost

William A. Yost

W.A. Yost (✉)
Speech and Hearing Science, Arizona State University, Tempe, AZ 85287, USA
e-mail: william.yost@asu.edu

A.N. Popper and R.R. Fay (eds.), *Perspectives on Auditory Research*, Springer Handbook of Auditory Research 50, DOI 10.1007/978-1-4614-9102-6_33,

33.1 Introduction

At the Association for Research in Otolaryngology's (ARO) annual Award of Merit lecture in 1994, Peter Dallos, that year's Award of Merit winner, offered a provocative prediction. Peter suggested that in the not too distant future, psychoacoustics might be a very different field because of neural imaging. This comment reminded me of my time as a Program Officer at the National Science Foundation (NSF) in the mid-1980s. As Director of the NSF Program in Sensory Physiology and Perception I chaired the review panel for the Program. It was during this time that the 2-deoxyglucose (2DG) technique became a useful functional anatomical method, one of the early functional "imaging" methods (although used only in non-human animal models). Some of the 2DG images being produced from the visual cortex were "breathtaking" at that time. I recall several panel members (leaders in various fields of the sensory sciences) saying that this kind of technique was likely to replace single-unit electrophysiology. Some panel members conjectured that maybe even psychophysics would look very different in a few years. So far psychophysics and psychoacoustics in particular have not changed all that much since their beginnings in 1860, when Gustav Fechner published his two-volume book, *Elemente der Psychophysik*.

Why would informed scientists such as Peter Dallos and members of the NSF Sensory Physiology and Perception panel think that psychophysics would significantly change based on neural imaging technology? I believe it is because psychophysics from 1860 to today is the study of how sensations and perceptions are related to the physical properties produced by physical sources, objects, and events. These psychophysical relationships provided for many years the only integrated analysis of how peripheral sensory systems connected to the central nervous system processed information about the physical properties of sound, light, vibration, air-borne and fluid-borne molecules, etc. in order to produce sensations and perceptions informing an organism about what the sources of the physical properties are, where these sources are located, and what information or message the sources might provide. To many the promise of functional neural imaging is to provide these same types of relationships based on directly measuring neural activity, especially in human subjects. If direct neural measures can be made (via functional neural imaging) that relate to perception, perhaps psychophysics would lose its utility as a scientific approach.

The premise of this chapter is that although functional neural imaging has offered and will continue to offer new insights about auditory perception, psychoacoustics or, more generally, the behavioral study of the relationship between one's acoustic sensations and perceptions and the physical properties of sound and sound sources will continue to be required to advance knowledge about auditory perception.

When I think of the history of psychoacoustics and auditory perception, I tend to divide studies into several major time periods: The Dominance of Helmholtz (late 19th century into early 20th century), The Bell Lab Years (late 1920s to the late 1940s), The Influence of the Theory of Signal Detection (1950s through the 1970s), Appreciation of Complex Sounds (1970s to late 1980s), and Auditory Scene

Analysis (1990s to the present time). Clearly more was being investigated than these topics, but these are the major psychoacoustic topics that influenced and shaped my research program over the years. A caveat needs to be inserted before I proceed. My review will be very "English-centric." In the 21st century most that is published about psychoacoustics and auditory perception is published in English, and with the Internet, access to this literature is relatively easy. But when I started in the field and well before then, many important studies were published in non-English-language journals that were not easily accessible to an American. Early in my career many of my colleagues and I were often not sufficiently aware of important work being reported in non-English journals, so my work and theirs was usually not informed by this literature. As a result this chapter does not cover many important studies that were reported in non-English-language journals.

33.2 Psychoacoustics Prior to the 1990s

33.2.1 The Dominance of Helmholtz

Almost every graduate student in experimental psychology in my time read or was acquainted with E. G. Boring's (1942) *Sensation and Perception in the History of Experimental Psychology*. Larry Feth often referred to it as "Boring is." I still find it interesting to read. In reading Boring's book, is hard not to miss the huge influence Herman von Helmholtz had on all of science in the late 19th century and into the 20th century. Helmholtz's (1885) book, *On the Sensations of Tone as a Physiological Basis for the Theory of Music* had a major influence on understanding hearing and in the development of so-called "theories of hearing." Such theories of hearing were really theories of frequency processing, but given the importance Helmholtz attached to frequency processing, a theory of frequency perception was probably essentially a theory of hearing to Helmholtz's way of thinking.

33.2.2 The Bell Lab Years

Harvey Fletcher, former Director of Bell Labs, headed a formidable team of scientists who investigated a wide range of topics in psychoacoustics and speech perception. Most of the basic psychoacoustic measurements used today throughout the hearing sciences were originally made at Bell Labs in the 1920s through to the 1940s including thresholds of hearing, frequency and intensity discrimination thresholds, tone-in-noise masking, the critical band, loudness, speech recognition, articulation index, and much more. It has been said that Fletcher and his Bell Lab team had the most profound influence on the hearing sciences as any group that has ever existed. It is almost impossible to explore a topic in auditory perception without being able to find a direct connection to work done at Bell Labs from the 1920s to the 1940s.

33.2.3 The Influence of the Theory of Signal Detection

My introduction to hearing was via the study of the theory of signal detection (TSD) as a graduate student at Indiana University taking courses from Jim Egan and Don Robinson. TSD as it has been used in psychoacoustics was developed at the Electronic Defense Group at the University of Michigan in the early 1950s. David Green and John Swets were graduate students in the department of psychology working in the Electronic Defense Group with Ted Birdsall, Wilson (Spike) Tanner, and others. The book *Signal Detection Theory and Psychophysics* (Green & Swets, 1973), first published in 1966, is a landmark publication in the field of experimental psychology. TSD and its derivatives are probably the most widely used theories of decision making employed throughout the world across a wide range of topics.

Several interrelated aspects of TSD have influenced the hearing sciences: the decision process of separating bias from sensitivity, the ideal observer, an alternative view to threshold theory, new sensitive psychophysical measurement techniques, and the energy detection model for explaining detecting and discriminating signals in the background of noise. Some of these issues are not found as much in the literature now as in the 1960s and 70s, but most current psychophysical methods used to measure auditory sensations and perceptions are a direct result of the work on TSD. Threshold theory has mostly given way to the ideas of TSD of processing signals in uncertain conditions. The energy detection model still provides an excellent account of the detection and often the discrimination of well specified signals masked by well specified (in a statistical sense) maskers. Any psychoacoustician publishing credible research in the 21st century is fully aware of the distinctions between response bias/criteria and sensitivity. Many current models of auditory processing use concepts similar to those of the ideal observer used in the original development of TSD.

33.2.4 Appreciation of Complex Sounds

By the 1980s, based on the work surrounding TSD, there was a relatively good understanding of the detection and discrimination of simple, reasonably well-specified acoustic stimuli. To gain the control necessary to specify the stimuli adequately, most sounds were presented over headphones. There were good accounts and models of detection and discrimination of tones, clicks, and Gaussian noise. A lot was known about the processing of interaural time and level differences (cues used to locate a sound source in the horizontal or azimuth plane) presented over headphones, and about the detection of binaural signals with different interaural configurations than the interaural differences of binaural maskers (i.e., studies of the masking level difference, MLD). Loudness was relatively well described for simple sounds presented in simple contexts. There was a debate about the best way to characterize the pitch of sounds, especially a harmonic series of tones.

Some championed a "spectral" approach in which the sound's amplitude spectrum contained the cues responsible for pitch perception, while others argued for a "temporal" approach in which the timing (usually fine-structure timing) of sound pressure changes contained the appropriate pitch cues. And, there was an increased understanding of how the biomechanical and neural properties of the peripheral auditory system influenced many of the psychoacoustic results that had been obtained since the 1920s.

Chuck Watson had a strong influence on my view of psychoacoustics as we headed into the 1980s. He noted that simple tones, clicks, noises presented over headphones were a far cry from what we experience in the everyday world. Everyday sounds are complex (not simple tones, clicks, or noises), and we perceive them in complex and varying contexts (not occurring over headphones in prescribed psychophysical tasks). Chuck felt that we could study more complex sounds in more complex contexts using the rigorous methods and theories developed with the help of TSD. His attempt to do so was characterized by his many studies of the perception of "ten-tone patterns" (see Watson, 2005). Chuck's original goal was to construct complex acoustic patterns that were "speech-like" but had no language content, allowing one to study the psychoacoustics of acoustic patterns that contained similar acoustic information to that found in actual speech waveforms unconfounded by language. So a pattern of acoustics events (in simple cases, tones of different frequencies) presented sequentially with phoneme-like durations and timing and with about ten tones per pattern (close to the number of phonemes in a word) was used to start a pursuit of this goal. The first task was to determine listeners' acuity for discriminating a change in the frequency of any one of the 10 tones in the pattern. That is, what was the spectral resolution for processing changes in a word-like, 10-tone pattern? Chuck found that, with some practice, listeners were almost as accurate in discriminating a change in frequency for any tone presented in the context of the 10-tone pattern as they were when that tone (target tone) was presented by itself. This good frequency resolution occurred when the same 10-tone pattern was repeated throughout a block of trials. Chuck then reasoned that there are many changes in the spectral content of a speech utterance. Although the speech production mechanisms and language may constrain some of the possible variation, from an acoustic perspective these spectral changes are probably nearly random. So Chuck introduced random variability in the frequency content of the tones in the 10-tone patterns from pattern to pattern (there was a wide range of ways to vary the patterns). Using a same-different procedure, listeners were to determine on any one trial if the frequency of a target tone changed or did not change. Chuck found that random variation in the frequencies of the 10 tones increased frequency discrimination thresholds for the target tone, and the greater the amount of random variation the greater the threshold shifts. These threshold shifts were large, in some cases an order of magnitude or more larger than when there was no random variation. The acoustics of the target never changed as a function of presenting the 10-tone patterns with various amounts of spectral randomization, but the context in which the targets were presented did change. Context had a huge influence on performance even for well-practiced listeners. Chuck called the interference of frequency discrimination caused by context "informational masking," which he contrasted to "energetic masking" which led to thresholds in the absence of contextual variation. Informational masking was masking in addition to energetic masking.

Chuck Watson's work on 10-tone pattern processing is an example of changes that were taking place in psychoacoustics in the 1980s. More and more investigators began to explore complex sounds often presented in complex contents as a way to gain a better understanding of auditory processing in the real world. Chuck and I interacted on two projects during the 1980s that attempted to gain a better understanding of what we often referred to as auditory processing of complex sounds. John Tangney, a program officer for the Air Force Office of Scientific Research (AFOSR), contacted me when I was a program officer at the NSF about what I thought might be topic areas in hearing that the AFOSR could start supporting. I suggested that the AFOSR fund a study by the National Academies of Sciences/National Research Council that would identify key emerging areas of research in the hearing sciences. Such a study was conducted. Chuck chaired a workshop on the topic, and Terry Dolan and I co-edited a report of the workshop and following discussions that was published as a Supplement to the *Journal of the Acoustical Society of America* (Dolan & Yost, 1985). Several topics covered in the workshop and report dealt with complex sound perception. From that effort the AFOSR decided that it would develop a funding program in auditory processing of complex sounds. A few years later, Chuck and I received AFOSR funding for a workshop on auditory processing of complex sounds, which led to a book of the same title edited by Chuck and me (Yost & Watson, 1987). While Chuck and I were pleased at how well the workshop appeared to advance our understanding of complex sound perception, we had hoped for an even broader range of input as we recognized that many good ideas were emerging from fields outside of traditional psychoacoustics.

By the end of the 1980s studies on 10-tone patterns, profile analysis, stream segregation, comodulation masking release (CMR), modulation detection interference (MDI), informational masking, head-related transfer functions (HRTFs), and other topics had moved the field of psychoacoustics away from studies of simple tones, clicks, and noises to more complex sounds often presented in complex contexts. Although auditory processing of complex sounds was not an inappropriate description of these types of studies, there was not much of a theoretical view, nor was there really much of a theme that could be used to integrate work across various ways of investigating complex sound perception. That changed as we entered the 1990s.

33.3 Psychoacoustics Post-1990 (Auditory Scene Analysis)

In the very late 1980s and early 1990s interest in complex sound processing had created a zeitgeist in psychoacoustics/auditory perception. Within a short period of time around 1990 Bill Hartmann wrote a chapter about "auditory entities," Brian Moore added a topic on "auditory objects" to his popular textbook, I wrote articles on "auditory images" and "sound source determination," Steve McAdams wrote articles and chapters from a similar perspective, and others published related pieces. However, the publication that captured the attention of people in and outside the field of auditory perception was Al Bregman's book, *Auditory Scene Analysis: The Perceptual Organization of Sound* (1990).

Auditory Scene Analysis captured the essence of what all of us were attempting to discuss in our writings at this time. A great deal of psychoacoustics before the concept of auditory scene analysis came along dealt with the ways the physical parameters of sound affect a listener's ability to detect, discriminate, or recognize/identify one sound often in the presence of other sounds. Or psychoacoustic studies dealt with how the physical parameters of sound influence a listener's subjective judgments of pitch, loudness, complexity, timbre, etc. Auditory scene analysis and the ideas expressed by Hartmann, Moore, Yost, McAdams, Bregman, and many others since the early 1990s shifted the psychoacoustic question to how the physical parameters of sound enable the central nervous system to determine the sources that produced the sounds, especially when there are multiple sound sources. Real-world sources rarely generate the simple clicks, tones, and noises used so often in the previous years of psychoacoustic research. The more complex sounds used in studies of ten-tone patterns, profile analysis, informational masking, CMR, MDI, HRTFs, etc. were argued to be more like those originating from real-world sound sources. Many investigators argued that the results obtained in studying these phenomena suggested ways in which the auditory system might extract information from complex sounds to assist in determining the auditory scene. While studies in psychoacoustics and auditory perception since 1990 have been many and varied, interest in auditory scene analysis is a new view that occupies a lot of space in the journals, at scientific meetings, and in review chapters and books. The topic seems to be the dominant theme in auditory perception at this time.

Some have reflected that the auditory scene analysis idea really goes back at least to Collin Cherry's article in 1953 in which Cherry asked "on what logical basis could one design a machine ("filter") for carrying out such an operation?" The operation Cherry was addressing was the "cocktail party problem," by which he meant "how do we recognize what one person is saying when others are speaking at the same time?" The experiments that Cherry and others employed to study the "cocktail party problem" were selective attention-type experiments in which competing speech signals were often sent to the two ears over headphones and listeners were to attend to the message at one ear or the other. I believe that early on in the discussions of auditory scene analysis many associated the cocktail party idea with selective attention involving dichotic speech presentations. Today the cocktail party problem is viewed more broadly, as analogous to auditory scene analysis.

A major phenomenon studied by Bregman is auditory streaming. Streaming reflects both a description of one's perception and a technique. The simplest example of auditory streaming is a procedure in which two-tone bursts of different frequency are alternated in time. Under the proper conditions (e.g., the proper frequency difference between the two tones) listeners perceive an alternation in pitch from one tone to the other as if there are two sound sources each producing a different pitch. In other conditions (e.g., a different set of frequencies) listeners report a perception as if one sound source is producing a sound whose pitch is changing. "Fusion" is used to describe situations in which the percept is consistent with a single sound source, and "fission" pertains to a percept of two sound sources. Other physical differences can be used to differentiate one alternating sound burst from another (e.g., the same musical note played by two different instruments). The stimulus differences that lead to fission are those that are viewed as suggesting conditions that allow one sound source to be perceived as different (segregated) from another sound source. Whereas early in the discussion of auditory scene analysis distinctions were made between streaming and other ways to study complex sound processing, it is common today for streaming to be viewed as analogous to auditory scene analysis.

Auditory scene analysis, the cocktail party problem, and auditory streaming all focus on the problem of the perceptual segregation of the sources that produced the sounds. Thus, sound-source segregation is a key element of auditory scene analysis. This perhaps represents the most challenging aspect of the problem. That is, hearing involves the auditory system receiving a complex sound waveform made up of the physical interaction (sum) of the sounds from all of the sources producing sound at any one time. The auditory periphery provides a neural temporal-spectral code of this complex sound waveform for the central nervous system. There are no peripheral mechanisms that code for where or what the sources of the sounds are. The information about sound sources (what and where they are) is computed by the central nervous system using information in the peripheral temporal-spectral code. What are these computational neural mechanisms and how do they operate? This is the basic question of auditory scene analysis or this is the cocktail party problem.

Although a sound source can be fairly well defined in physical terms, the perception of a sound source or sources is more difficult to define in unambiguous ways. Do we perceive objects, images, events, etc. and how do these perceptions relate to the sound source itself? A recording of a piece of music allows one to determine the instruments (sound sources) playing the music essentially as well as listening to a live concert where the instruments are physically present. Thus, we form a perceptual representation of the sources of the sounds (e.g., the instruments) from the acoustic waveform that reaches our ear canals independent of the sound source that actually delivered the sound (the actual musical instruments or the loudspeaker playing a recording). That perceptual representation may not capture all aspects of the physical sound sources. For instance, if the recording and playback of a musical piece is monaural, the relative locations of the musical instruments could not be determined by our perceptual/neural systems, even if we could accurately "name" all of the instruments just from listening to the recording. Thus, it is sometimes necessary to have terms that refer to the physical sound sources and other terms to refer to perception. Trying to arrive at these perceptual terms has been debated over the past two or more decades with no clear agreement on the definitions. One consensus is that any term can probably be used as long as it is clearly defined and consistently used. My preference has been to describe the perceptual process as "sound source perception or processing" (e.g., as in the title of the SHAR book co-edited with Art and Dick; Yost et al., 2007). I also prefer the term "sound source perception" because it suggests that in addition to the issue of segregating two or more sound sources, it might be worthwhile to understand how a single sound source is perceived and how the physical properties of a single sound source might influence perception.

When a sound source produces sound, we might be able to identify or label the source (e.g., the source is a piano or it the source is to the right), but it isn't clear if labeling is a necessary and sufficient condition to determine that a sound source exists. Not much research has been devoted to understanding the extent to which one's ability to identify/label a sound source is important in segregating different sound sources. Many sounds we perceive are physically completed before perception has occurred; thus some form of auditory memory must play a role in perceiving real sounds in the real world. In the real world, we may have the perceptual ability to process more than we actually do, in that we depend on attention to determine what aspect of the acoustic world to process.

Sound source segregation, sound source identification, auditory memory, auditory attention all depend on the signals being peripheral processed, but these processes are all neural computations that take place in the central auditory nervous system, the auditory brain. I believe that the future of auditory perception will be based largely on what we learn about the auditory brain.

33.4 The Auditory Brain

Russell (Russ) DeValois, the well-known vision scientist who studied color and spatial vision, taught the sensory neurophysiology section of my graduate school proseminar at Indiana University. In his first lecture Russ said that most of his lectures would be about vision as this is what he studied, but he was always fascinated by the auditory system since the brain had to compute everything about sound to account for auditory perception and very little was known about these neural computations. The contrast between what is known about the auditory periphery and how these processes influence auditory perception and what is known about central neural processing and its influence on auditory perception is extremely large. Knowledge of the processing of simple sounds in simple contexts can often be explained based on knowledge of the biomechanical and neural processes of the auditory periphery. Understanding auditory perception of complex, everyday-like sounds, in everyday-like contexts cannot usually be accounted for by peripheral processes alone. Our knowledge of the auditory brain stem and cortex is too meager to provide theories, models, or accounts of the perception of complex sounds in complex contexts. I agree with Russ that the brain has to compute almost everything about sound in order for sound to provide useful information about the world. The auditory system is not a neural spatial system like vision or the somatosensory systems. Auditory information is not "mapped" and "remapped" from one neural center to another; the peripheral spectral-temporal code for sound must be analyzed and reanalyzed in the brain stem and cortex to allow an organism to determine what the sound source is, where it is located, and what message/information the sound source is providing. The analysis is neural computation. I believe that progress on understanding the functional role of the neural centers in the ascending and descending pathways in the auditory brain stem and cortex will be highly dependent on better understanding auditory perception of complex sounds in complex contexts.

I will use two areas that I have studied—sound source localization and complex pitch perception—to indicate what I believe to be some interesting problems of auditory perception that remain to be solved. I will indicate how answers will likely depend on additional work both in auditory psychoacoustics/perception and auditory neuroscience.

33.4.1 Sound Source Localization

In terms of describing the functional role of auditory pathways in the auditory brain, probably the most is known about the pathways responsible for processing interaural time differences (ITDs). This is the pathway (at least in mammals) from the cochlea to the cochlear nucleus through the medial superior olive (MSO) to the inferior colliculus (IC) often via the dorsal nucleus of the lateral lemniscus (DNLL) that computes interaural time differences. There is a wealth of psychoacoustical data, mainly obtained from headphone lateralization studies, on how changes in ITD affect perception. And it has been fairly well established that ITDs are the main cue used localizing the source of low-frequency sounds when the source of the sound is located in the free field in the azimuth plane. Not only do we know how many of the neurons in this circuit respond when ITD changes, but we have good estimates of the how some parts of the circuit compute information about ITD that accounts for many of the psychoacoustical data. This is especially true for the barn owl (*Tyto alba*). However, we are a long way from understanding what computations the auditory brain makes and how it makes these computations in terms of determining the location of a sound source in the real world. We are learning more about how interaural level differences (ILD) are processed, but the neural basis of ILD processing is not as well understood as that for ITD processing. We have little information about how the spectral information contained within HRTFs is neurally processed in order to determine sound source location in the vertical plane and along cones of confusion. And we know essentially nothing about what computations the nervous system makes that might explain the perceived distance of a sound source. One perceives the location of a sound source in three-dimensional space, not in just one plane or the other. What the computations are which allow for a neural map or a neural representation of a sound source in three-dimensional space is a complete mystery. Knowledge about how the nervous system of the barn owl (and to some extent in some bat species) makes these calculations is again a shining exception to these generalizations. I would argue that a big part of the reason we know less about ILD processing, how HRTF spectral information is processed, and the neural computations for distance perception than we do about ITD processing is because the psychophysical/perceptual data on ILD processing, use of HRTF cues, and distance perception are much more meager than those involving ITD and perception.

33.4.1.1 Distance Perception

Very little is known about mammals' ability to judge the distance of a sound source. There is a theory that the ratio of the direct to reverberant or reflective sound arriving at a listener provides a cue for judging the distance of a sound source, and there are a few data that are consistent with this theory. However, there are no theories or data (that I am aware of) that suggest how such a ratio would be computed in the auditory brain or that provide any details of how distance perception is related to the value of the ratio. Thus, I am not sure what a neural physiologist would look for in the auditory brain to help explain how the brain computes information about the perception of the distance of a sound source.

33.4.1.2 Interaural Level Difference

While the duplex theory of sound source localization has been around a very long time, our knowledge of how ILDs affect sound source location in the azimuth plane in the real world is sparse. The ILD varies both with frequency (due to the head shadow) and with azimuth location. In addition, because of the "bright spot" the largest ILD for any sinusoid is in the region of 45–66° of azimuth, not at 90° (at frequencies below about 750 Hz the bright spot is very small (<1 dB). This means there is a nonmonotonic relationship between ILD and azimuth caused by the "bright spot." Thus, a particular ILD value cannot indicate the position of a particular azimuthal sound source. Sound sources at different locations can produce the same ILD depending on the frequency of the sound and on the implications of the "bright spot." There are almost no psychophysical data that indicate how frequency, azimuth location, and the value of the ILD interact to provide an estimate of the location of a sound source (especially at high frequencies where azimuthal sound source location cannot be based on ongoing, fine-structure ITDs).

33.4.1.3 HRTF and Sound Source Localization

The role that the spectral transform provided by the HRTF plays in vertical sound source localization, sound source location along cones of confusion, and perceived externalization of sound sources has been at times a dominant topic in psychoacoustics (partially driven by the hope of finding a means to develop a commercial auditory virtual reality product). Although the psychoacoustic data are quite clear that something about the HRTF is important for sound source localization, it still isn't apparent what aspect or aspects provide the information used to locate sound sources. The data suggest that information in high-frequency regions of the HRTF are important, as listeners are poor at judging the vertical or cone of confusion source location of low-frequency sounds. But, what aspect or set of features has not been firmly established. There is still a debate as to whether or not the information necessary to locate a sound source in the vertical dimension or along cones of confusion is monaurally or binaurally mediated. A great deal of the lack of information about what aspect(s) of the HRTF is (are) used for sound source localization is probably a result of the very large individual differences in HRTFs. The HRTF (pinna shape) differences across individuals mean that an HRTF feature found for one person may not exist for another person. Perhaps there is not a universal set of features that are used by everyone; perhaps each of our auditory brains learns how the features of our own HRTFs relate to the location of sound sources (the basic idea behind the fact that sound source localization off the azimuth plane is dependent on individualized HRTFs, i.e., performance depends on using the HRTF for each particular person rather than on some average or generic HRTF). That is, the use of HRTF information to locate sound sources is a plastic neural process. How the auditory brain adapts to an individual's own HRTF is probably a crucial question that will require answers before there are any major discoveries of how the auditory brain calculates the location of sound sources in all three dimensions.

33.4.1.4 Multiple Sound Source Localization

In the everyday world a listener encounters many sound sources, several often producing sound at about the same time. The sounds from these various sources physically interact before the sounds reach the listener. Thus, the cues used for locating any particular single sound source when several sources produce sound at about the same time would be obscured by these physical interactions. Yet we appear to be able to locate more than one sound source when several sources produce sound (although there are surprisingly few data regarding multiple sound source localization). How humans do so is a mystery. It might be the case that understanding how the auditory brain calculates sound source locations could require more psychophysical information about multiple sound source localization.

33.4.1.5 Reverberation and the Effects of Precedence

Processing target sounds in reverberant environments is challenge for everyone, those with normal hearing and especially those with a hearing impairment. To the extent that being able to locate a sound source in reverberation is important for processing other aspects of a target sound in such environments, the "effects of precedence" (Litovsky et al., 1999) are probably important. Several studies have found physiological correlates in various auditory pathways of some of the "effects of precedence," especially fusion (perceiving one sound source when the source and reflected sound both occur). Several different psychoacoustic findings (especially the build-up and break-down of precedence) suggest that processing sound when a listener receives sound from the originating source along with that from reflective surfaces might be based on the auditory brain's prior experience with the particular acoustic space containing the source and reflected sounds.

The major paradigm that demonstrates buildup and breakdown of precedence involves presenting a brief sound from one loudspeaker (the lead loudspeaker, simulating the sound from the originating source) and a few milliseconds later the same sound is presented from a loudspeaker at a different location (the lag loudspeaker, simulating a reflected sound, which is a delayed version of the original sound coming from a different location). This lead-lag pair of sounds is repeated at a slow rate as if the sound from the originating source was repeating. For the first three or four presentations a listener often perceives a salient sound from or near the location of the lead sound source, and a less salient sound associated with the lag sound source. After three or four repetitions listeners with normal hearing then perceive a single sound (fusion) whose perceived spatial location is at or near that of the lead loudspeaker (localization dominance; it is as if the lag sound is completely suppressed, precedence has been fully established). If after many lead-lag repetitions, the lead and lag sounds are reversed, so that the lead sound is now the lag sound and the lag becomes the lead, two sounds are then perceived (breakdown of precedence) after the switch, with each sound perceived as coming from near each of the two loudspeakers. Two sound sources are perceived for three or four repetitions of the new lead-lag pairing until the buildup of precedence occurs, leading again to the perception of a single sound source but located at the new lead (old lag) loudspeaker location (i.e., precedence is re-established). Variations of this simple lead-lag paradigm have demonstrated similar build-up and break-down of precedence effects in more complex, real-world-like acoustic spaces.

Some models of the effects of precedence suggest that the information from a reflected sound is suppressed or combined with that from the original sound favoring information from the originating sound source (the first sound to reach the listener). However, the breakdown of precedence indicates that if information about reflected sound is suppressed, this information can essentially be fully recovered when certain changes in the acoustics occur. This result has led to the idea that in order to suppress a reflected sound, the auditory brain has to determine which sounds are reflections of which other sounds. Sounds from originating sound sources are not suppressed while reflections of these sounds are suppressed. Reflected sounds are attenuated (but early reflections may be attenuated by only a few decibels), correlated (due to the properties of absorption at a surface and the atmosphere, reflected sounds are low-passed versions of the original sounds), delayed copies of the original sound, and the reflection emanates from a different location than that of the originating sound source. The buildup of precedence suggests that the auditory brain quickly determines the properties of the source and reflected sound and uses these properties to suppress information about a reflection as long as the basic acoustic properties of the room does not change the sounds. These properties would depend on the reflected space. Breakdown of precedence suggests that if the information received by the auditory brain does not conform to prior information about how a room modifies a source and its reflection (e.g., because the room has changed and the originating and reflected sounds are coming from new locations), suppression is suspended until the buildup of precedence occurs for the new room arrangement. Nat Durlach has suggested that it is almost like the auditory brain quickly estimating the impulse response of a room (e.g., during the buildup stage) and the brain uses information gleaned from the impulse response to suppress information about reflections. If the impulse response changes, suppression is suspended until a new impulse response is fully obtained.

Although there are a growing number of studies of the effects of precedence (both psychophysical and physiological), understanding sound source localization in reverberant spaces is still poorly understood. Additional perceptual data, especially regarding how the auditory brain's past knowledge of source sounds and reflections effect sound source processing, will probably be required before a thorough description of the neural computations underlying the location of sound sources in reverberant spaces is obtained.

33.4.1.6 Moving Sound Sources and Moving Listeners

Our knowledge of sound source location is based almost entirely on data collected from stationary listeners. In the real world, listeners and sound sources each may or may not move. If either the sound source or the listener moves, the physical parameters arriving at a listener's ears (e.g., interaural differences) will change. If I remain stationary and a sound source moves, the ILDs and ITDs change, and I perceive a change in the location of the sound source (there are many studies of moving sound sources). If I move and the sound source is fixed in location, the ILDs and ITDs also change. But everyday experience suggests that I would not perceive a change in the location of the sound source (sound source localization would not be very useful if fixed sound sources were perceived to move when the listener moves). Why doesn't a fixed sound source appear to move when I do? The interaural cues that indicate a change in the location of a sound source changed. Although this issue has been studied for more than 100 years in vision and several neural circuits have been described to account for how fixed visual objects are perceived as fixed when an observer moves (e.g., collar discharge theory), there are only a handful of studies of this problem in the history of studying hearing and there are no known neural correlates (again studies of the barn owl and bats are exceptions). In vision when an observer moves and an object is fixed in location, there is a movement of the image across the retina similar to that occurring when an object moves and the observer is stationary. There are several circuits in the visual brain stem in which the retinal changes are "cancelled" when there are vestibular, proprioceptive, and/or eye control changes that occur when the observer moves. That is, if the retinal image indicates a retinal change of x degrees and a vestibular output indicates the observer has moved x degrees in the opposite direction, the retinal image is cancelled. The result is the perception of a stationary object when the object is fixed in space, but the observer moves. It is not unreasonable to imagine that similar "cancellation" circuits exist in the auditory brain. If so, then the "cancellation' could occur when the various cues for sound source localization (e.g., interaural differences) are being or have been computed. Or perhaps the cancellation occurs after all of the cues have been computed and the "read out" of the sound source's location in all three spatial dimensions has occurred. In the latter case, this "cancellation circuit" would probably be high in the auditory brain and the latency for such a cancellation would be expected to be long. A long latency might influence how quickly one perceives a fixed sound source to be stationary when the listener moves. In the former case, there might be cancellation circuits for each of the cues used to determine sound source location. In this case, there would need to be several well synchronized neural cancellation circuits. The problem of determining sound source location when listeners and sound sources move would appear to be important for understanding sound source localization in the everyday world. Again, such an understanding will be based on new psychoacoustic data and new neural studies of the calculations made by the auditory brain.

33.4.1.7 Sound Source Localization and Spatial Release from Masking

When a target sound source is at one location and interfering or masking sound sources are at different locations, the target can often be more easily detected, changes in the acoustics of the target can be more easily discriminated, and/or the target can be more easily recognized than when the target and interferers/masker sounds coexist at the same source location. The increase in performance based on spatial separation of target and masker sources is referred to as spatial release from masking (SRM). SRM has been studied almost exclusively when targets and masker sources are in the azimuth plane.

There are several reasons why SRM may occur (assuming a listener with normal hearing): head shadow, sound source localization, and binaural processing which enhances the target. If the target source is on one side of a listener's head and a masker source is on the opposite side, the signal-to-masker ratio is higher at the ear nearest the target source than at the opposite ear or for conditions in which the target and masker came from the same source. That is, the ear closest to the target receives an attenuated masker due to the head shadow. If one assumes that the auditory brain can process the sound arriving at just one ear when both ears receive sound, then the target would be easier to detect in the case in which target and masker sources are on opposite sides of the head (and as a consequence discrimination and recognition might also be better).

When the target and masker sources are at different locations, it is possible that each can be localized (but as mentioned previously not much is known about multiple sound source localization). The ability to localize the target source at a location different from the masker source may enable a listener to attend to the target and ignore the masker. As such the target may be more easily processed than when the target and masker sources are at the same location and listeners cannot as easily attend to one and ignore the other.

The interaural differences of time and level (ITDs and ILDs) are clearly the cues most listeners' use in localizing sound sources in the azimuth plane. An auditory brain processor may be able to enhance the target sound when it has a different set of interaural values than a masker sound. The Durlach Equalization-Cancellation (EC, Durlach; 1972) model does just that. As one example, suppose the target is directly in front of the listener producing zero ITDs and ILDs. We'll consider the case when the masker is co-located with the target in front and the case when the masker is located opposite one of the listener's ears. Let's keep things simple and imagine that the target is a 1000-Hz tone and the masker is a wideband noise and as such we will consider only ITDs (as ILDs for low-frequency sounds are very small). Let's assume that the auditory brain adds (an assumption of the EC model) the information arriving at both ears. If the target and masker both originated from the front loudspeaker, they would each be increased owing to the neural addition (ideally they would each be twice the amplitude of the original stimuli), but the target-to-masker ratio would be the same as that for the original signals. Now consider the case with the target source is in front and the masker source to the side. As in the case when the target and masker sources were co-located, the target would be enhanced because both ears receive the target sound without any changes. However, the masker arrives at one ear later than at the other ear (there is an ITD). Let's assume the ITD was 0.5 ms (a reasonable value for a sound source opposite one ear). At 1000-Hz this 0.5-ms delay is a 180° phase shift. When the noise maskers from each ear are added in the processor, the noise in the region of 1000 Hz (the spectral region containing the target) will be cancelled (adding two signals 180° out-of-phase is the same as subtraction). Thus, the processor has enhanced the target and decreased the masker in the spectral region containing the target yielding a higher target-to-masker ratio than when the target and masker sources were co-located, i.e., there would be SRM. The SRM in this case is based on ITDs, but not on sound source localization based on ITDs. Although this example was very simple, the same principles could lead to enhancement of target sounds under many conditions when the target and masker sources are at different spatial locations. In these conditions there is no need for the listener to be able to locate the sources of either the target or the masker in order for SRM to occur.

Thus, there are a variety of reasons why SRM may occur and the details of what the target and maskers are and where they are located may dictate what the cause of a SRM might be in any particular context. Although many psychophysical experiments have investigated and are investigating SRM, there is still much more to be done. We do not fully understand the extent to which listeners can use information arriving at only one ear while ignoring that arriving at the other ear (e.g., relative to issues of processing head-shadowed targets). We are still trying the figure out how attention and sound source localization interact. There is little clarity on the extent to which target enhancement occurs in the absence of sound source localization and what conditions would yield SRM for one type of processing or the other. And, although the neural circuits for processing ITDs and to a limited extent ILDs have been and are being studied, there is very little information about how these circuits or other circuits would perform the calculations that would yield SRM (e.g., is there something like an EC neural circuit?).

33.4.1.8 Sound Source Localization, SRM, and Hearing Impairment

Most, but not all, listeners with hearing impairment do not localize sound sources as well as listeners with normal hearing. Often listeners with hearing impairment do not gain the same SRM advantage as listeners with normal hearing. It is still not clear what aspects of hearing impairment cause a loss in sound source localization acuity or the amount of SRM. Reports of the success of hearing aids, cochlear implants (CIs), and combinations of hearing aids and cochlear implants in assisting patients in localizing sound sources and benefitting from SRM have been mixed. Issues of amplitude compression, envelope extraction, HRTF information, and low-frequency temporal fine-structure information are some of the variables that could affect how well a person fit with a hearing prosthetic device(s) might be able to localize sound sources or gain a SRM advantage. I am not sure real progress will be made in designing hearing prosthetic devices that help patients localize sound sources and gain a SRM advantage until we understand what aspects of biomechanical and/or neural processing lead to the poor performance by patients with hearing impairment when they attempt to localize sound sources or gain an SRM advantage. Although a fair amount is known about outer and inner hair cell function and neural response rate, much less is known about how damaged hair cells alter the temporal (e.g., fine-structure) information provided by the periphery. To the extent that ITD processing is important in sound source localization and obtaining an SRM advantage, understanding how a damaged auditory periphery affects temporal fine-structure processing seems crucial. Some of the variability seen in the performance of patients with hearing impairment in localizing sound sources and obtaining a SRM advantage might be due to some patient's ability to learn how to use impoverished cues to perform in such tasks. If some patients have been able to learn how to use cues better than others, then it might be very beneficial to figure out how to provide this ability to learn these cues to all patients. However, if such learning does occur, it will probably be necessary to understand what cues are impoverished and in what way and to understand what is being learned.

One topic that was not discussed previously in this chapter is the role envelope ITDs might play in localizing sound sources in the free field, especially for bilateral cochlear implant patients given that CIs extract the envelope of sounds across the sound's spectrum. There is a well-established literature based on headphone delivered stimuli that shows that high-frequency stimuli for which there are no usable ITD cues in the temporal fine structure of the high-frequency sounds can be lateralized based on ITDs imposed on an envelope applied to the high-frequency sound. For instance, varying the ITD of a 4000-Hz tone results in no perceptual changes. However, if the 4000-Hz tone is amplitude modulated, applying an ITD so the envelope is out-of-phase between the two ears leads to a change in perceived laterality and allows for measurable ITD difference thresholds. Such ITD difference thresholds are not as low as those obtained based on processing low-frequency temporal fine structure, but they are not far off. Thus, ITD processing does occur in high-frequency channels in the absence of any ILD cues when the ITD is an envelope ITD. What about in the free field? There are very few studies of ITD envelope processing in the free-field, that is, sound source localization based on envelope ITDs.

The challenge in performing free-field studies is that there will be naturally occurring ILDs that are large for high-frequency stimuli. Thus, in the free field the question becomes how does an envelope ITD cue interact with the always present ILD cue? Before one jumps to the conclusion that envelope ITD cues are useful for free-field sound source localization (e.g., in the case of bilateral CI patients), I believe we need to know about the interaction between ILD and envelope ITD cues for sound source localization.

33.4.2 Complex Pitch Perception

For more than 150 years hearing scientists have tried to determine what parameters of a complex sound or what parts of the peripheral code for a complex sound are used to determine pitch. The "debate" has almost always been between "temporal" versus "spectral" processing. The debate at the level of the stimulus can never be settled because time and frequency are inversely related, so temporal and spectral approaches are not independent. This is not necessarily true if one uses the peripheral code of sound to try to determine what aspects of this code might account for complex pitch. For instance, small spectral differences (e.g., 100-Hz spacing of the spectral components of a 100-Hz fundamental harmonic complex producing a 100-Hz perceived pitch) are not resolved for fibers tuned to high frequencies, but temporal aspects of the stimuli might be. Thus, if stimuli under high-pass conditions produce a pitch it might be more likely that temporal aspects as opposed to spectral aspects of the peripheral code are a cause of the perceived pitch.

At present it seems to me that the literature on complex pitch perception deals with three possible aspects of the peripheral code that could be used to account for pitch perception: spectral differences as exist in the excitation pattern of complex sounds, envelope cues as might exist in high-frequency fibers, and temporal regularity cues in the fine structure of complex sounds preserved within the phased-locked activity of low-frequency fibers. Only a spectral explanation seems capable of accounting for the pitch of high-frequency (>5000 Hz) tones because such tones have no envelope and fibers tuned at these frequencies do not phase lock to the temporal fine structure. Although such high-frequency sinusoids cannot support melody or other musical interval perceptions, such tones do produce a pitch, as pitch is typically defined. Only an envelope explanation can account for the weak pitch (including musical pitch) of amplitude modulated, wideband noise. The spectral content and the temporal fine structure of such modulated noise signals are random and, therefore, contain no information about the sound's pitch. Not everyone perceives a pitch for such modulated noise stimuli and only some musically trained people can judge melody and musical intervals for these modulated noises. Stimuli without spectrally resolved components can still produce pitch if the stimuli contain either envelope or phased-locked temporal fine structure cues. But such stimuli without resolved spectral structure usually produce weak pitches, often strong enough to support melody and musical interval judgments, but weaker than stimuli that have resolvable spectral components. The most salient pitches occur when complex sounds have both spectrally resolved components and temporal regularity in the phase locked activity associated with the sound's temporal fine structure. These are the sounds produced by musical instruments, including the voice.

Different models based on autocorrelation or autocorrelation-like processes can account for (qualitatively in some cases and quantitatively in many cases) the pitch of most sounds, except high-frequency tones. In most of these models the autocorrelation-like operations are probably revealing aspects of the temporal regularity of the stimuli (in the envelope or fine-structure) rather than spectral aspects of the stimuli (i.e., autocorrelation is the Fourier transform of the power spectrum, but this fact changes if autocorrelation is performed after stimuli are transformed by simulations of the actions of the inner ear and auditory nerve). Thus, currently autocorrelation-like models are the dominant way to model the pitch of complex sounds, but the details of how these computations are carried out are still in debate.

Roy Patterson and I have preferred to use the idea of temporal regularity rather than periodicity to describe the temporal characteristics of stimuli used to study complex pitch perception. Periodicity often implies a repeated constant temporal interval, that is, a constant first-order interval. Periodic signals like a pulse train are temporally regular, but so is a stimulus such as iterated ripple noise (IRN). The intervals in the fine structure of IRN stimuli are not periodic (they are often higher order intervals), but certain intervals (those associated with the delay used to generate an IRN stimulus) occur randomly but are more frequent than any other interval. The reciprocal of those regularly and often occurring intervals is the perceived pitch of IRN stimuli. The more regularly occurring intervals there are, the more salient or stronger the pitch of IRN stimuli becomes. Stimuli other than IRN stimuli that produce a perceived pitch also contain intervals that are regular, but not periodic.

As successful as autocorrelation-like approaches have been in accounting for many psychoacoustic data, there is no evidence that any process like autocorrelation exists in the auditory brain. If an autocorrelation-like device was conceived as some sort of coincidence detector (analogous to the way cross-correlation for ITD processing is modeled as a coincidence detector), the time delays in the coincidence network would have to be many milliseconds long, which seems neurally unlikely. While there have been some hints in the literature of other ways in which pitch might be calculated in the auditory brain, no substantial model (i.e., one that can account for most of the complex pitch psychoacoustic data) of actual neural processing of complex pitch has been proposed (at least that I know of).

What has been shown is that in humans cortical neural sites in the vicinity of Heschl's Gyrus respond robustly to a variety of stimuli that produce complex pitch. Positron emission tomography (PET), functional magnetic resonance imaging (fMRI), and magnetoencephalography (MEG) data all indicate that these cortical regions respond based on the pitch of the stimuli (as opposed to other stimulus or perceptual features), often based on the temporal regularity of the stimuli. Homologous regions in the monkey auditory cortex also seem "tuned" to the pitch of many different stimuli. These studies strongly implicate these cortical regions as part of a "pitch pathway." What such studies have not revealed is what neural processes determine the feature(s) that cause these neural sites to respond similarly when many different sounds all produce the same perceived pitch. How are these neural computations being made? Although there are still many psychoacoustic details to explore regarding complex pitch perception, I believe that major advances in understanding pitch perception are unlikely to occur until viable neural models of complex pitch processing are developed and tested.

33.5 Summary

Although neural imaging might allow "a look" at the human auditory brain, the technique has not and, in its present form, will not replace psychoacoustic and auditory perceptual approaches to better understand perception of the acoustic world. What I believe needs to occur to advance the understanding of auditory perception is a better idea of how the auditory brain makes the neural computations for determining what the sources of sounds are, where they are, and what information those sources provide. Functional neural imaging will be one of the techniques providing this information, but all of the other tools in the neuroscientist's tool box will also be required. Increasing our understanding of auditory perception will require important advances in psychoacoustics and behavioral measures of auditory perception.

Acknowledgment Work on this chapter was supported by a grant from the Air Force Office of Scientific Research (AFOSR).

References

Boring, E. G. (1942). *Sensation and perception in the history of experimental psychology*. New York: Appleton-Century-Crofts.

Bregman, A. S. (1990). *Auditory scene analysis: The perceptual organization of sound*. Cambridge, MA: MIT Press.

Cherry, C. (1953). Some experiments on the recognition of speech with one and with two ears. *Journal of the Acoustical Society of America*, 25, 975–981.

Dolan, T. R., & Yost, W. A. (1985). Basic research in hearing., *Journal of the Acoustical Society of America*, 78(2, Supplement), 295–388.

Durlach, N. I. (1972). Binaural signal detection: Equalization and cancellation theory. In J. V. Tobias (Ed.), *Foundations of modern auditory theory*, Vol. II. New York: Academic Press.

Green, D. M., & Swets, J. A. (1973). *Signal detection theory and psychophysics*. New York: Krieger.

Helmholtz, Hermann von (1885). *On the sensations of tone as a physiological basis for the theory of music*, 2nd English edition, translated by Alexander J. Ellis. London: Longmans, Green, and Co.

Litovsky, R., Colburn, S., Yost, W. A., & Guzman, S. (1999). The precedence effect. *Journal of the Acoustical Society of America*, 106, 1633–1654.
Watson, C. S. (2005). Some comments on informational masking. *Acustica*, 91(3), 502–512.
Yost, W. A., & Watson, C. S. (1987). *Auditory processing of complex sounds*. Hillsdale, NJ: Lawrence Erlbaum.
Yost, W. A., Fay, R. R., & Popper, A. N. (2007). *Auditory perception of sound sources*. New York: Springer Science+Business Media.

Appendix: The First 49 Volumes!

Volume 1: The Mammalian Auditory Pathway: Neuroanatomy (1992)

Editors: Douglas B. Webster, Arthur N. Popper, and Richard R. Fay

1. An Overview of Mammalian Auditory Pathways with an Emphasis on Humans
   Douglas B. Webster
2. The Auditory Nerve: Peripheral Innervation, Cell Body Morphology, and Central Projections
   David K. Ryugo
3. The Cochlear Nucleus: Neuronal Types and Their Synaptic Organization
   Nell Beatty Cant
4. The Superior Olivary Complex and Lateral Lemniscal Nuclei
   Ilsa R. Schwartz
5. Inferior and Superior Colliculi
   Douglas L. Oliver and Michael F. Huerta
6. The Functional Architecture of the Medical Geniculate Body and the Primary Auditory Cortex
   Jeffery A. Winer
7. Organization of Olivocochlear Efferent Systems in Mammals
   W. Bruce Warr

A.N. Popper and R.R. Fay (eds.), *Perspectives on Auditory Research*, Springer Handbook of Auditory Research 50, DOI 10.1007/978-1-4614-9102-6,

Volume 2: The Mammalian Auditory Pathway: Neurophysiology (1992)

Editors: Arthur N. Popper and Richard R. Fay

1. An Overview of Central Auditory Processing
   John F. Brugge
2. Physiology and Coding of Sound in the Auditory Nerve
   Mario A. Ruggero
3. Physiology of the Cochlear Nuclei
   William S. Rhode and Steven Greenberg
4. Physiology of the Auditory Brainstem
   Dexter R. F. Irvine
5. Physiology of Thalamus and Cortex
   Janine C. Clarey, Pascal Barone, and Thomas J. Imig
6. Electrophysiology of the Human Auditory System
   Nina Kraus and Therese McGee

Volume 3: Human Psychophysics (1993)

Editors: William A. Yost, Arthur N. Popper, and Richard R. Fay

1. Overview: Psychoacoustics
   William A. Yost
2. Auditory Intensity Discrimination
   David M. Green
3. Frequency Analysis and Pitch Perception
   Brian C. J. Moore
4. Time Analysis
   Neal F. Viemeister and Christopher J. Plack
5. Sound Localization
   Frederic L. Wightman and Doris J. Kistler
6. Auditory Perception
   William A. Yost and Stanley Sheft

Volume 4: Comparative Hearing: Mammals (1993)
Editors: Richard R. Fay and Arthur N. Popper

1. Comparative Auditory Research
   Richard R. Fay
2. Psychoacoustics
   Glenis R. Long
3. Sound Localization
   Charles H. Brown
4. How Monkeys Hear the World: Auditory Perception in Nonhuman Primates
   William C. Stebbins and David B. Moody
5. Structure of the Mammalian Cochlea
   Stephen M. Echteler, Richard R. Fay, and Arthur N. Popper
6. Outer and Middle Ears
   John J. Rosowski

Volume 5: Hearing by Bats (1995)
Editors: Arthur N. Popper and Richard R. Fay

1. Hearing in Bats: An Overview
   Alan D. Grinnell
2. Natural History and Biosonar Signals
   M. Brock Fenton
3. Behavioral Studies of Auditory Information Processing
   Cynthia F. Moss and Hans-Ulrich Schnitzler
4. Auditory Dimensions of Acoustic Images in Echolocation
   James A. Simmons, Michael J. Ferragamo, Prestor A. Saillant, Tim Haresign, Janine M. Wotton, Steven P. Dear, and David N. Lee
5. Cochlear Structure and Function in Bats
   Manfred Kössl and Marianne Vater
6. The Lower Brainstem Auditory Pathways
   Ellen Covey and John H. Casseday
7. The Inferior Colliculus
   George D. Pollak and Thomas J. Park
8. The Auditory Thalamus in Bats
   Jeffery J. Wenstrup

9. The Bat Auditory Cortex
William E. O'Neill

10. Perspectives on the Functional Organization of the Mammalian Auditory System: Why Bats Are Good Models
George D. Pollak, Jeffery A. Winer, and William E. O'Neill

Volume 6: Auditory Computation (1996)

Editors: Harold L. Hawkins, Teresa A. McMullen, Arthur N. Popper, and Richard R. Fay

1. Auditory Computation: An Overview
Harold L. Hawkins and Teresa A. McMullen

2. Models of External- and Middle-Ear Function
John J. Rosowski

3. Analysis and Synthesis of Cochlear Mechanical Function Using Models
Allyn E. Hubbard and David C. Mountain

4. Computational Analysis of Hair Cell and Auditory Nerve Processes
David C. Mountain and Allyn E. Hubbard

5. Physiological Models for Basic Auditory Percepts
Bertrand Delgutte

6. Auditory Representations of Timbre and Pitch
Richard Lyon and Shihab Shamma

7. Scene Analysis
David K. Mellinger and Bernard M. Mont-Reynaud

8. Computational Models of Binaural Processing
H. Steven Colburn

9. Auditory Computations for Biosonar Target Imaging in Bats
James A. Simmons, Prestor A. Saillant, Michael J. Ferragamo, Tim Haresign, Steven P. Dear, Jonathan Fritz, and Teresa A. McMullen

10. Further Computations Involving Time
Edwin R. Lewis

Volume 7: Clinical Aspects of Hearing (1996)
Editors: Thomas R. Van De Water, Arthur N. Popper, and Richard R. Fay

1. Introduction
   Robert J. Ruben
2. Approaches to Understanding the Molecular Genetics of Hearing and Deafness
   Karen P. Steel and William Kimberling
3. Regeneration of the Auditory Nerve: The Role of Neurotrophic Factors
   Thomas R. Van De Water, Hinrich Staecker, Stuart C. Apfel, and Philippe P. Lefebvre
4. Auditory Deprivation and Its Consequences: From Animal Models to Human
   Judith S. Gravel and Robert J. Ruben
5. Ototoxicity: Of Mice and Men
   Susan L. Garetz and Jochen Schacht
6. The Role of Viral Infection in the Development of Otopathology: Labyrinthitis and Autoimmune Disease
   Nigel K. Woolf
7. Otoacoustic Emissions: Animal Models and Clinical Observations
   Martin L. Whitehead, Brenda L. Longsbury-Martin, Glen K. Martin, and Marcy J. McCoy
8. Tinnitus: Psychophysical Observations in Humans and an Animal Model
   Merrilynn J. Penner and Pawel J. Jastreboff

Volume 8: The Cochlea (1996)
Editors: Peter Dallos, Arthur N. Popper, and Richard R. Fay

1. Overview: Cochlear Neurobiology
   Peter Dallos
2. Structure of the Mammalian Cochlea
   Norma B. Slepecky
3. Homeostatic Mechanisms in the Cochlea
   Philine Wangemann and Jochen Schacht
4. Cochlear Micromechanics and Macromechanics
   Robert Patuzzi
5. Mechanics of the Cochlea: Modeling Efforts
   Egbert de Boer

6. Physiology of Mammalian Cochlear Hair Cells
   Corné J. Kros
7. Outer Hair Cell Motility
   Matthew C. Holley
8. Physiology of Olivocochlear Efferents
   John J. Guinan Jr.
9. Neurotransmitters and Synaptic Transmission
   William F. Sewell

Volume 9: Development of the Auditory System (1998)

Editors: Edwin W Rubel, Arthur N. Popper, and Richard R. Fay

1. Overview: Personal Views on the Study of Auditory System Development
   Edwin W Rubel
2. Behavioral Studies of Hearing Development
   Lynne A. Werner and Lincoln Gray
3. Early Embryology of the Vertebrate Ear
   Bernd Fritzsch, Kate F. Barald, and Margaret I. Lomax
4. Development of Sensory and Neural Structures in the Mammalian Cochlea
   R. Pujol, M. Lavigne-Rebillard, and M. Lenoir
5. The Development of Cochlear Function
   Rudolph Rübsamen and William R. Lippe
6. The Development of Central Auditory Processing
   Dan H. Sanes and Edward J. Walsh
7. Structural Development of the Mammalian Central Auditory Pathways
   Nell Beatty Cant

Volume 10: Comparative Hearing: Insects (1998)

Editors: Ronald R. Hoy, Arthur N. Popper, and Richard R. Fay

1. Acute as a Bug's Ear: An Informal Discussion of Hearing in Insects
   Ronald R. Hoy
2. Biophysics of Sound Localization in Insects
   Axel Michelsen

3. The Sensory Ecology of Acoustic Communication in Insects
   Heiner Römer
4. Development of the Insect Auditory System
   George S. Boyan
5. Neural Processing of Acoustic Signals
   Gerald S. Pollack
6. The Evolutionary Innovation of Tympanal Hearing in Diptera
   Daniel Robert and Ronald R. Hoy
7. The Vibrational Sense of Spiders
   Friedrich G. Barth
8. The Sensory Coevolution of Moths and Bats
   James H. Fullard

Volume 11: Comparative Hearing: Fish and Amphibians (1998)

Editors: Richard R. Fay and Arthur N. Popper

1. Hearing in Fishes and Amphibians: An Introduction
   Richard R. Fay and Arthur N. Popper
2. Hearing in Two Worlds: Theoretical and Actual Adaptive Changes of the Aquatic and Terrestrial Ear for Sound Reception
   Bernd Fritzsch
3. The Auditory Periphery in Fishes
   Arthur N. Popper and Richard R. Fay
4. The Acoustic Periphery of Amphibians: Anatomy and Physiology
   Edwin R. Lewis and Peter M. Narins
5. Anatomy of the Central Auditory Pathways of Fish and Amphibians
   Catherine A. McCormick
6. Central Auditory Processing in Fishes and Amphibians
   Albert S. Feng and Nico A. M. Schellart
7. The Sense of Hearing in Fishes and Amphibians
   Richard R. Fay and Andrea Megela Simmons
8. The Enigmatic Lateral Line System
   Sheryl Coombs and John C. Montgomery
9. Acoustic Communications in Fishes and Frogs
   Randy Zelick, David A. Mann, and Arthur N. Popper

Volume 12: Hearing by Whales and Dolphins (2000)

Editors: Whitlow W. L. Au, Arthur N. Popper, and Richard R. Fay

1. Hearing in Whales and Dolphins: An Overview
   Whitlow W. L. Au
2. Cetacean Ears
   Darlene R. Ketten
3. In Search of Impulse Sound Sources in Odontocetes
   Ted W. Cranford
4. Communication and Acoustic Behavior of Dolphins and Wales
   Peter L. Tyack and Christopher W. Clark
5. Acoustics and Social Behavior of Wild Dolphins: Implications for a Sound Society
   Denise L. Herzing
6. The Auditory Central Nervous System of Dolphins
   Sam H. Ridgway
7. Electrophysiological Measures of Auditory Processing in Odontocetes
   William Ford Dolphin
8. Psychoacoustic Studies of Dolphin and Whale Hearing
   Paul E. Nachtigall, David W. Lemonds, and Herbert L. Roitblat
9. Echolocation in Dolphins
   Whitlow W. L. Au
10. Acoustic Models of Sound Production and Propagation
    James L. Aroyan, Mark A. McDonald, Spain C. Webb, John A. Hildebrand, David Clark, Jeffrey T. Laitman, and Joy S. Reidenberg

Volume 13: Comparative Hearing: Birds and Reptiles (2000)

Editors: Robert J. Dooling, Richard R. Fay, and Arthur N. Popper

1. Introduction
   Robert J. Dooling and Arthur N. Popper
2. The Middle Ear of Reptiles and Birds
   James C. Saunders, R. Keith Duncan, Daryl E. Doan, and Yehudah L. Werner
3. The Hearing Organs of Birds and *Crocodilia*
   Otto Gleich and Geoffrey A. Manley

4. The Hearing Organs of Lizards
   Geoffrey A. Manley
5. The Central Auditory System of Reptiles and Birds
   Catherine E. Carr and Rebecca A. Code
6. Sound Localization in Birds
   Georg Klump
7. Hearing in Birds and Reptiles
   Robert J. Dooling, Bernard Lohr, and Michael L. Dent

Volume 14: Genetics and Auditory Disorders (2002)

Editors: Bronya J. B. Keats, Arthur N. Popper, and Richard R. Fay

1. Introduction and Overview: Genetics in Auditory Science an Clinical Audiology
   Bronya J. B. Keats and Charles I. Berlin
2. Genes and Mutations in Hearing Impairment
   Karen B. Avraham and Tama Hasson
3. Mapping and Cloning of Genes for Inherited Hearing Impairment
   Robert F. Mueller, Guy Van Camp, and Nick J. Lench
4. Genetic Epidemiology of Deafness
   Walter E. Nance and Arti Pandya
5. Cytogenetics and Cochlear Expressed Sequence Tags (ESTs) for Identification of Genes Involved in Hearing and Deafness
   Anne B. Skvorak Giersch and Cynthia C. Morton
6. Autosomal and X-Linked Auditory Disorders
   Andrew J. Griffith and Thomas B. Friedman
7. Hearing Loss and Mitochondrial DNA Mutations: Clinical Implications and Biological Lessons
   Nathan Fischel-Ghodsian
8. Mice as Models for Human Hereditary Deafness
   Karen P. Steel, Alexandra Erven, and Amy E. Kiernan
9. Genetic Counseling for Deafness
   Kathy S. Arnos and M. Katherine Oelrich

Volume 15: Integrative Functions in the Mammalian Auditory Pathway (2002)

Editors: Donata Oertel, Richard R. Fay, and Arthur N. Popper

1. Introduction
   Donata Oertel

2. From the Cochlea to the Cortex and Back
   Philip H. Smith and George A. Spirou

3. Cellular Mechanisms for Information Coding in Auditory Brainstem Nuclei
   Laurence O. Trussell

4. Neural Mechanisms of Encoding Binaural Localization Cues in the Auditory Brainstem
   Tom C. T. Yin

5. Circuitry and Function of the Dorsal Cochlear Nucleus
   Eric D. Young and Kevin A. Davis

6. Ascending Pathways Through Ventral Nuclei of the Lateral Lemniscus and Their Possible Role in Pattern Recognition of Natural Sounds
   Donata Oertel and Robert E. Wickesberg

7. The Inferior Colliculus: A Hub for the Central Auditory System
   John H. Casseday, Thane Fremouw, and Ellen Covey

8. Location Signaling by Cortical Neurons
   John Middlebrooks, Li Xu, Shigeto Furukawa, and Brian J. Mickey

9. Feature Detection by the Auditory Cortex
   Israel Nelken

Volume 16: Acoustic Communication (2003)

Editors: Andrea Megela Simmons, Arthur N. Popper, and Richard R. Fay

1. Perspectives and Progress in Animal Acoustic Communication
   Andrea Megela Simmons

2. The Physical Acoustics of Underwater Sound Communication
   Andrew H. Bass and Christopher W. Clark

3. Unpacking "Honesty": Vertebrate Vocal Production and the Evolution of Acoustic Signals
   W. Tecumseh Fitch and Marc D. Hauser

4. Social Sounds: Vocal Learning and Development of Mammal and Bird Calls
   Janette Wenrick Boughman and Cynthia F. Moss

5. Selection on Long-Distance Acoustic Signals
   Michael J. Ryan and Nicole M. Kime
6. Hormonal Mechanisms in Acoustic Communication
   Ayako Yamaguchi and Darcy B. Kelley
7. The Neuroethology of Vocal Communication: Perception and Cognition
   Timothy Q. Gentner and Daniel Margoliash

Volume 17: Compression: From Cochlea to Cochlear Implants (2004)

Editors: Sid P. Bacon, Richard R. Fay, and Arthur N. Popper

1. Overview of Auditory Compression
   Sid P. Bacon
2. Compression in the Peripheral Auditory System
   Nigel P. Cooper
3. Psychophysical Manifestations of Compression: Normal-Hearing Listeners
   Andrew J. Oxenham and Sid P. Bacon
4. Psychophysical Manifestations of Compression: Hearing-Impaired Listeners
   Sid P. Bacon and Andrew J. Oxenham
5. Compression Amplification
   Harry Levitt
6. Compression and Cochlear Implants
   Fan-Gang Zeng

Volume 18: Speech Processing in the Auditory System

Editors: Steven Greenberg, William A. Ainsworth, Arthur N. Popper, and Richard R. Fay

1. Speech Processing in the Auditory System: An Overview
   Steven Greenberg and William A. Ainsworth
2. The Analysis and Representation of Speech
   Carlos Avendaño, Li Deng, Hynek Hermansky, and Ben Gold
3. Explaining the Structure of Feature and Phoneme Inventories: The Role of Auditory Distinctiveness
   Randy L. Diehl and Björn Lindblom

4. Physiological Representations of Speech
   Alan Palmer and Shihab Shamma
5. The Perception of Speech Under Adverse Acoustic Conditions
   Peter Assmann and Quentin Summerfield
6. Automatic Speech Recognition: An Auditory Perspective
   Nelson Morgan, Hervé Bourlard, and Hynek Hermansky
7. Hearing Aids and Hearing Impairment
   Brent Edwards
8. Cochlear Implants
   Graeme Clark

Volume 19: The Vestibular System (2004)

Editors: Stephen M. Highstein, Richard R. Fay, and Arthur N. Popper

1. Anatomy and Physiology of the Central and Peripheral Vestibular System: Overview
   Stephen M. Highstein
2. Molecular Genetics of Vestibular Organ Development
   Weise Chang, Laura Cole, Raquel Cantos, and Doris K. Wu
3. Morphophysiology of the Vestibular Periphery
   Anna Lysakowski and Jay M. Goldberg
4. Biomechanics of the Semicircular Canals and Otolith Organs
   Richard D. Rabbitt, Edward R. Damiano, and J. Wallace Grant
5. Sensory Processing and Ionic Currents in Vestibular Hair Cells
   Antoinette Steinacker
6. The Physiology of the Vestibuloocular Reflex (VOR)
   Bernard Cohen and Theodore Raphan
7. Vestibuloautonomic Interactions: A Teleologic Perspective
   C. D. Balaban and B. J. Yates
8. Vestibulocollic Reflexes
   Barry W. Peterson and Richard Boyle
9. Gain and Phase Control of Compensatory Eye Movements by the Flocculus of the Vestibulocerebellum
   Chris I. De Zeeuw, Sebastiaan K. E. Koekkoek, Arjan M. van Alphen, Chongde Luo, Freek Hoebeek, Johannes van der Steen, Maarten A. Frens, John Sun, Hieronymus H. L. M. Goossens, Dick Jaarsma, Michiel P. H. Coesmans, Matthew T. Schmolesky, Marcel T. G. De Jeu, and Niels Galjart

10. Localizing Sites for Plasticity in the Vestibular System
    M. Green, Y. Hirata, H. L. Galiana, and S. M. Highstein
11. Clinical Applications of Basic Vestibular Research
    G. Michael Halmagyi, Ian S. Curthoys, Swee T. Aw, and Joanna C. Jen

Volume 20: Cochlear Implants: Auditory Prostheses and Electric Hearing (2004)

Editors: Fan-Gang Zeng, Arthur N. Popper, and Richard R. Fay

1. Auditory Prostheses: Past, Present, and Future
   Fan-Gang Zeng
2. Engineering Design of Cochlear Implants
   Blake Wilson
3. Cochlear Implants: Clinical Applications
   John K. Niparko
4. Anatomical Considerations and Long-Term Effects of Electrical Stimulation
   Patricia A. Leake and Stephen J. Rebscher
5. Biophysics and Physiology
   Paul J. Abbas and Charles A. Miller
6. Central Responses to Electrical Stimulation
   Rainer Hartmann and Andrej Kral
7. Psychophysics and Electrical Stimulation
   Colette M. McKay
8. Speech Perception with Cochlear Implants
   Robert V. Shannon, Qian-Jie Fu, John Galvin, and Lendra Friesen
9. Learning, Memory, and Cognitive Processes in Deaf Children Following Cochlear Implantation
   David B. Pisoni and Miranda Cleary

Volume 21: Electroreception (2005)

Editors: Theodore H. Bullock, Carl D. Hopkins, Arthur N. Popper, and Richard R. Fay

1. Explaining Electroreception
   Theodore H. Bullock and Carl D. Hopkins
2. From Electrogenesis to Electroreception: An Overview
   Günther K. H. Zupanc and Theodore H. Bullock

Volume 22: Evolution of the Vertebrate Auditory System (2004)

Editors: Geoffrey A. Manley, Arthur N. Popper, and Richard R. Fay

Volume 23: Plasticity of the Auditory System (2004)

Editors: Thomas N. Parks, Edwin W Rubel, Richard R. Fay, and Arthur N. Popper

6. The Avian Song Control System: A Model for Understanding Changes in Neural Structure and Function
   Eliot A. Brenowitz and Sarah M. N. Woolley
7. Plasticity in the Auditory System of Insects
   Reinhard Lakes-Harlan

Volume 24: Pitch: Neural Coding and Perception (2005)

Editors: Christopher J. Plack, Andrew J. Oxenham, Richard R. Fay, and Arthur N. Popper

1. Overview: The Present and Future of Pitch
   Christopher J. Plack and Andew J. Oxenham
2. The Psychophysics of Pitch
   Christopher J. Plack and Andrew J. Oxenham
3. Comparative Aspects of Pitch Perception
   William P. Shofner
4. The Neurophysiology of Pitch
   Ian M. Winter
5. Functional Imaging of Pitch Processing
   Timothy D. Griffiths
6. Pitch Perception Models
   Alain de Cheveigné
7. Perception of Pitch by People with Cochlear Hearing Loss and by Cochlear Implant Users
   Brian C. J. Moore and Robert P. Carlyon
8. Pitch and Auditory Grouping
   Christopher J. Darwin
9. Effect of Context on the Perception of Pitch Structures
   Emmanuel Bigand and Barbara Tillman

Volume 25: Sound Source Localization (2005)

Editors: Arthur N. Popper and Richard R. Fay

1. Introduction to Sound Source Localization
   Richard R. Fay and Arthur N. Popper
2. Directional Hearing in Insects
   Daniel Robert

3. Sound Source Localization by Fishes
   Richard R. Fay
4. Directional Hearing in Nonmammalian Tetrapods
   Jakob Christensen-Dalsgaard
5. Comparative Mammalian Sound Localization
   Charles H. Brown and Bradford J. May
6. Development and the Auditory Centers Responsible for of Sound Localization
   M. Fabiana Kubke and Catherine E. Carr
7. Interaural Correlation as the Basis of a Working Model of Binaural Processing: An Introduction
   Constantine Trahiotis, Leslie R. Bernstein, Richard M. Stern, and Thomas N. Buell
8. Models of Sound Localization
   H. Steven Colburn and Abhijit Kulkarni

Volume 26: Development of the Inner Ear (2005)

Editors: Matthew W. Kelley, Doris K. Wu, Arthur N. Popper, and Richard R. Fay

1. Developmental Neurobiology of the Ear: Current Status and Future Directions
   Matthew W. Kelley and Doris K. Wu
2. The Induction of the Otic Placode
   Andrew K. Groves
3. Morphogenesis of the Inner Ear
   Suzanne L. Mansour and Gary C. Shoenwolf
4. Wiring the Ear to the Brain: The Molecular Basis of Neurosensory Development, Differentiation, and Survival
   Sarah Pauley, Veronica Matei, Kirk W. Beisel, and Bernd Fritzsch
5. Notch Signaling and Cell Fate Determination in the Vertebrate Inner Ear
   Pamela J. Lanford and Matthew W. Kelley
6. The Differentiation of Hair Cells
   Jane E. Bryant, Andrew Forge, and Guy P. Richardson
7. Developmental Genes Associated with Human Hearing Loss
   Ronna Hertzano and Karen B. Avraham

Volume 27: Vertebrate Hair Cells (2006)

Editors: Ruth Anne Eatock, Richard R. Fay, and Arthur N. Popper

1. Vertebrate Hair Cells: Modern and Historic Perspectives
   Ruth Anne Eatock
2. The Development of Hair Cells in the Inner Ear
   Richard J. Goodyear, Corné J. Kros, and Guy P. Richardson
3. The Structure and Composition of the Stereociliary Bundle of Vertebrate Hair Cells
   David N. Furness and Carole M. Hackney
4. Mechanoelectrical Transduction in Auditory Hair Cells
   Robert Fettiplace and Anthony J. Ricci
5. Contribution of Ionic Currents to Tuning in Auditory Hair Cells
   Jonathan J. Art and Robert Fettiplace
6. The Synaptic Physiology of Hair Cells
   Paul A. Fuchs and Thomas D. Parsons
7. The Piezoelectric Outer Hair Cell
   William E. Brownell
8. Mammalian Vestibular Hair Cells
   Ruth Anne Eatock and Anna Lysakowski

Volume 28: Hearing and Sound Communication in Amphibians (2007)

Editors: Peter M. Narins, Albert S. Feng, Richard R. Fay, and Arthur N. Popper

1. Hearing and Sound Communication in Amphibians: Prologue and Prognostication
   Peter M. Narins and Albert S. Feng
2. An Integrated Phylogeny of Amphibia
   David Cannatella
3. The Behavioral Ecology of Anuran Communication
   Kentwood D. Wells and Joshua J. Schwartz
4. Call Production and Neural Basis of Vocalization
   Wolfgang Walkowiak
5. Recognition and Localization of Acoustic Signals
   H. Carl Gerhardt and Mark A. Bee

6. Pathways for Sound Transmission to the Inner Ear in Amphibians
   Matthew J. Mason
7. Anatomy, Physiology, and Function of Auditory End Organs in the Frog Inner Ear
   Dwayne D. Simmons, Sebastiaan W. F. Meenderink, and Pantelis Vassilakis
8. Central Auditory Pathways in Anuran Amphibians: The Anatomical Basis of Hearing and Sound Communication
   Walter Wilczynski and Heike Endepols
9. Function of the Amphibian Central Auditory System
   Gary J. Rose and David M. Gooler
10. Plasticity in the Auditory System Across Metamorphosis
    Andrea Megela Simmons and Seth S. Horowitz
11. Sound Processing in Real-World Environments
    Albert S. Feng and Johannes Schul

Volume 29: Auditory Perception of Sound Sources (2007)

Editors: William A. Yost, Arthur N. Popper, and Richard R. Fay

1. Perceiving Sound Sources
   William A. Yost
2. Human Sound Source Identification
   Robert A. Lutfi
3. Size Information in the Production and Perception of Communication Sounds
   Roy D. Patterson, David R. R. Smith, Ralph van Dinther, and Thomas C. Walters
4. The Role of Memory in Auditory Perception
   Laurent Demany and Catherine Semal
5. Auditory Attention and Filters
   Ervin R. Hafter, Anastasios Sarampalis, and Psyche Loui
6. Informational Masking
   Gerald Kidd, Jr., Christine R. Mason, Virginia M. Richards, Frederick J. Gallun, and Nathaniel I. Durlach
7. Effects of Harmonicity and Regularity on the Perception of Sound Sources
   Robert P. Carlyon and Hedwig E. Gockel
8. Spatial Hearing and Perceiving Sources
   Christopher J. Darwin
9. Envelope Processing and Sound-Source Perception
   Stanley Sheft

10. Speech as a Sound Source
    Andrew J. Lotto and Sarah C. Sullivan
11. Sound Source Perception and Stream Segregation in Nonhuman Vertebrate Animals
    Richard R. Fay

Volume 30: Active Processes and Otoacoustic Emissions (2008)

Editors: Geoffrey A. Manley, Richard R. Fay, and Arthur N. Popper

1. Otoacoustic Emissions: Concepts and Origins
   David T. Kemp
2. Traveling Waves, Second Filters, and Physiological Vulnerability: A Short History of the Discovery of Active Processes in Hearing
   Nigel P. Cooper, James O. Pickles, and Geoffrey A. Manley
3. Critical Oscillators as Active Elements in Hearing
   Thomas A. J. Duke and Frank Jülicher
4. Active Hair-Bundle Motility of the Hair Cells of Vestibular and Auditory Organs
   Pascal Martin
5. The Morphological Specializations and Electromotility of the Mammalian Outer Hair Cell
   Richard Hallworth and Heather C. Jensen-Smith
6. Active Processes in Insect Hearing
   Martin C. Göpfert and Daniel Robert
7. Otoacoustic Emissions in Amphibians, Lepidosaurs, and Archosaurs
   Geoffrey A. Manley and Pim van Dijk
8. Otoacoustic Emissions: Basic Studies in Mammalian Models
   Brenda L. Lonsbury-Martin and Glen K. Martin
9. Mechanisms of Mammalian Otoacoustic Emission
   Christopher A. Shera and John J. Guinan, Jr.
10. Cellular and Molecular Mechanisms in the Efferent Control of Cochlear Nonlinearities
    Ian J. Russell and Andrei N. Lukashkin
11. Cochlear Models Incorporating Active Processes
    Stephen T. Neely and Duck O. Kim

12. Relationships Between Otoacoustic and Psychophysical Measures of Cochlear Function
    Tiffany A. Johnson, Michael P. Gorga, Stephen T. Neely, Andrew J. Oxenham, and Christopher A. Shera
13. Otoacoustic Emissions as a Diagnostic Tool in a Clinical Context
    Thomas Janssen and Jörg Müller
14. Future Directions in the Study of Active Processes and Otoacoustic Emissions
    Geoffrey A. Manley and William E. Brownell

Volume 31: Auditory Trauma, Protection, and Repair (2007)

Editors: Jochen Schacht, Arthur N. Popper, and Richard R. Fay

1. Auditory Pathology: When Hearing Is Out of Balance
   Jochen Schacht
2. Genetics of Hearing Loss
   Ella Shalit and Karen B. Avraham
3. Cochlear Homeostasis and Homeostatic Disorders
   Philine Wangemann
4. Tinnitus: Theories, Mechanisms, and Treatments
   Carol A. Bauer and Thomas J. Brozoski
5. Autoimmune Inner Ear Disease
   Quinton Gopen and Jeffrey P. Harris
6. Age-Related Hearing Loss and Its Cellular and Molecular Bases
   Kevin K. Ohlemiller and Robert D. Frisina
7. Patterns and Mechanisms of Noise-Induced Cochlear Pathology
   Donald Henderson, Bohua Hu, and Eric Bielefeld
8. Drug-Induced Hearing Loss
   Leonard P. Rybak, Andra E. Talaska, and Jochen Schacht
9. Central Consequences of Cochlear Trauma
   D. Kent Morest and Steven J. Potashner
10. Cell Death and Cochlear Protection
    Steven H. Green, Richard A. Altschuler, and Josef M. Miller
11. Emerging Strategies for Restoring the Cochlea
    Stefan Heller and Yehoash Raphael

Volume 32: Fish Bioacoustics (2008)

Editors: Jacqueline F. Webb, Arthur N. Popper, and Richard R. Fay

1. Introduction to Fish Bioacoustics
   Richard R. Fay, Arthur N. Popper, and Jacqueline F. Webb
2. Hearing and Acoustic Behavior: Basic and Applied Considerations
   Arthur N. Popper and Carl R. Schilt
3. Structures and Functions of the Auditory Nervous System of Fishes
   Richard R. Fay and Peggy L. Edds-Walton
4. Evolution of Peripheral Mechanisms for the Enhancement of Sound Reception
   Christopher B. Braun and Terry Grande
5. Bioacoustics and the Lateral Line System of Fishes
   Jacqueline F. Webb, John C. Montgomery, and Joachim Mogdans
6. Orientation to Auditory and Lateral Line Stimuli
   Olav Sand and Horst Bleckmann
7. Multipole Mechanisms for Directional Hearing in Fish
   Peter H. Rogers and David G. Zeddies
8. Vocal–Acoustic Communication: From Neurons to Behavior
   Andrew H. Bass and Friedrich Ladich
9. Active and Passive Acoustics to Locate and Study Fish
   David A. Mann, Anthony D. Hawkins, and J. Michael Jech

Volume 33: Hair Cell Regeneration, Repair, and Protection (2008)

Editors: Richard J. Salvi, Arthur N. Popper, and Richard R. Fay

1. Overview: Regeneration and Repair
   Richard J. Salvi
2. Morphological Correlates of Regeneration and Repair in the Inner Ear
   Jason R. Meyers and Jeffrey T. Corwin
3. Recovery of Function in the Avian Auditory System After Ototrauma
   James C. Saunders and Richard J. Salvi
4. Functional Recovery After Hair Cell Regeneration in Birds
   Robert J. Dooling, Michael L. Dent, Amanda M. Lauer, and Brenda M. Ryals

5. Hair Cell Regeneration: Mechanisms Guiding Cellular Proliferation and Differentiation
   Elizabeth C. Osterle and Jennifer S. Stone
6. Protection and Repair of Inner Ear Sensory Cells
   Andrew Forge and Thomas F. Van De Water
7. Gene Arrays, Cell Lines, Stem Cells, and Sensory Regeneration in Mammalian Ears
   Marcelo N. Rivolta and Matthew C. Holley

Volume 34: The Aging Auditory System

Editors: Sandra Gordon-Salant, Robert D. Frisina, Arthur N. Popper, and Richard R. Fay

1. Introduction and Overview
   Sandra Gordon-Salant and Robert D. Frisina
2. The Physiology of Cochlear Presbycusis
   Richard A. Schmiedt
3. Cell Biology and Physiology of the Aging Central Auditory Pathway
   Barbara Canlon, Robert Benjamin Illing, and Joseph Walton
4. Closing the Gap Between Neurobiology and Human Presbycusis: Behavioral and Evoked Potential Studies of Age-Related Hearing Loss in Animal Models and in Humans
   James R. Ison, Kelly L. Tremblay, and Paul D. Allen
5. Behavioral Studies with Aging Humans: Hearing Sensitivity and Psychoacoustics
   Peter J. Fitzgibbons and Sandra Gordon-Salant
6. Binaural Processing and Auditory Asymmetries
   David A. Eddins and Joseph W. Hall III
7. Effects of Senescent Changes in Audition and Cognition on Spoken Language Comprehension
   Bruce A. Schneider, Kathy Pichora-Fuller, and Meredyth Daneman
8. Factors Affecting Speech Understanding in Older Adults
   Larry E. Humes and Judy R. Dubno
9. Epidemiology of Age-Related Hearing Impairment
   Karen J. Cruickshanks, Weihai Zhan, and Wenjun Zhong
10. Interventions and Future Therapies: Lessons from Animal Models
    James F. Willott and Jochen Schacht

Volume 35: Computational Models of the Auditory System (2010)

Editors: Raymond Meddis, Enrique A. Lopez-Poveda, Arthur N. Popper, and Richard R. Fay

Volume 36: Music Perception (2010)

Editors: Mari Reiss Jones, Richard R. Fay, and Arthur N. Popper

4. Music Acquisition and Effects of Musical Experience
   Laurel L. Trainor and Kathleen A. Corrigall
5. Music and Emotion
   Patrick G. Hunter and E. Glenn Schellenberg
6. Tempo and Rhythm
   J. Devin McAuley
7. Neurodynamics of Music
   Edward W. Large
8. Memory for Melodies
   Andrea R. Halpern and James C. Bartlett

Volume 37: Loudness (2010)

Editors: Mary Florentine, Arthur N. Popper, and Richard R. Fay

1. Loudness
   Mary Florentine
2. Measurement of Loudness, Part I: Methods, Problems, and Pitfalls
   Lawrence E. Marks and Mary Florentine
3. Measurement of Loudness, Part II: Context Effects
   Yoav Arieh and Lawrence E. Marks
4. Correlates of Loudness
   Michael J. Epstein
5. Loudness in the Laboratory, Part I: Steady-State Sounds
   Walt Jesteadt and Lori J. Leibold
6. Loudness in the Laboratory, Part II: Non-Steady-State Sounds
   Sonoko Kuwano and Seiichiro Namba
7. Binaural Loudness
   Ville Pekka Sivonen and Wolfgang Ellermeier
8. Loudness in Daily Environments
   Hugo Fastl and Mary Florentine
9. Loudness and Hearing Loss
   Karolina Smeds and Arne Leijon
10. Models of Loudness
    Jeremy Marozeau

Volume 38: Auditory and Vestibular Efferents (2010)

Editors: David K. Ryugo, Richard R. Fay, and Arthur N. Popper

1. Introduction to Efferent Systems
   David K. Ryugo
2. Anatomy of Olivocochlear Neurons
   M. Christian Brown
3. Physiology of the Medial and Lateral Olivocochlear Systems
   John J. Guinan
4. Pharmacology and Neurochemistry of Olivocochlear Efferents
   William F. Sewell
5. Cholinergic Inhibition of Hair Cells
   Eleonora Katz, Ana Belén Elgoyhen, and Paul Albert Fuchs
6. The Efferent Vestibular System
   Joseph C. Holt, Anna Lysakowski, and Jay M. Goldberg
7. Development of the Inner Ear Efferent System
   Dwayne Simmons, Jeremy Duncan, Dominique Crapon de Caprona, and Bernd Fritzsch
8. Evolution of the Octavolateral Efferent System
   Christine Köppl
9. Central Descending Auditory Pathways
   Brett R. Schofield
10. Central Effects of Efferent Activation
    Donald Robertson and Wilhelmina H. A. M. Mulders
11. Corticofugal Modulation and Beyond for Auditory Signal Processing and Plasticity
    Nobuo Suga, Weiqing Ji, Xiaofeng Ma, Jie Tang, Zhongju Xiao, and Jun Yan

Volume 39: Auditory Prostheses: New Horizons (2011)

Editors: Fan-Gang Zeng, Arthur N. Popper, and Richard R. Fay

1. Advances in Auditory Prostheses
   Fan-Gang Zeng
2. Bilateral Cochlear Implants
   Richard van Hoesel

3. Combining Acoustic and Electric Hearing
   Christopher W. Turner and Bruce J. Gantz
4. Implantable Hearing Devices for Conductive and Sensorineural Hearing Impairment
   Ad Snik
5. Vestibular Implants
   Justin S. Golub, James O. Phillips, and Jay T. Rubinstein
6. Optical Stimulation of the Auditory Nerve
   Claus-Peter Richter and Angella Izzo Matic
7. A Penetrating Auditory Nerve Array for Auditory Prosthesis
   John C. Middlebrooks and Russell L. Snyder
8. Cochlear Nucleus Auditory Prostheses
   Douglas B. McCreery and Steven R. Otto
9. Midbrain Auditory Prostheses
   Hubert H. Lim, Minoo Lenarz, and Thomas Lenarz
10. Central Auditory System Development and Plasticity After Cochlear Implantation
    Anu Sharma and Michael Dorman
11. Auditory Training for Cochlear Implant Patients
    Qian-Jie Fu and John J. Galvin III
12. Spoken and Written Communication Development Following Pediatric Cochlear Implantation
    Sophie E. Ambrose, Dianne Hammes-Ganguly, and Laurie S. Eisenberg
13. Music Perception
    Hugh McDermott
14. Tonal Languages and Cochlear Implants
    Li Xu and Ning Zhou
15. Multisensory Processing in Cochlear Implant Listeners
    Pascal Barone and Olivier Deguine

Volume 40: Noise-Induced Hearing Loss: Scientific Advances (2012)

Editors: Colleen G. Le Prell, Donald Henderson, Richard R. Fay, and Arthur N. Popper

1. Perspectives on Noise-Induced Hearing Loss
   Colleen G. Le Prell and Donald Henderson
2. The Public Health Significance of Noise-Induced Hearing Loss
   Peter M. Rabinowitz

3. Noise-Induced Hearing Loss and Tinnitus: Challenges for the Military
   Marjorie A. M. Grantham
4. The Use of Kurtosis Measurement in the Assessment of Potential Noise Trauma
   Donald Henderson and Roger P. Hamernik
5. Noise-Induced Structural Damage to the Cochlea
   Bohua Hu
6. Neural Coding of Sound with Cochlear Damage
   Eric D. Young
7. Suprathreshold Auditory Processing in Noise-Induced Hearing Loss
   Mini N. Shrivastav
8. The Neurobiology of Noise-Induced Tinnitus
   James A. Kaltenbach and Ryan Manz
9. Genes That Influence Susceptibility to Noise-Induced Hearing Loss
   Tzy-Wen Gong and Margaret I. Lomax
10. Effects of Early Noise Exposure on Subsequent Age-Related Changes in Hearing
    Eric C. Bielefeld
11. Effects of Exposure to Chemicals on Noise-Induced Hearing Loss
    Thais C. Morata and Ann-Christin Johnson
12. Hearing Protection Devices: Regulation, Current Trends, and Emerging Technologies
    John G. Casali
13. Prevention of Noise-Induced Hearing Loss: Potential Therapeutic Agents
    Colleen G. Le Prell and Jianxin Bao
14. Frontiers in the Treatment of Hearing Loss
    Tatsuya Yamosoba, Josef M. Miller, Mats Ulfendahl, and Richard A. Altschuler

Volume 41: Synaptic Mechanisms in the Auditory System (2012)

Editors: Laurence O. Trussell, Arthur N. Popper, and Richard R. Fay

1. Sound and Synapse
   Laurence O. Trussell
2. Neuronal Response Properties and Voltage-Gated Ion Channels in the Auditory System
   Nace L. Golding

3. The Hair Cell Synapse
   Teresa Nicolson
4. The Endbulbs of Held
   Paul B. Manis, Ruili Xie, Yong Wang, Glen S. Marrs, and George A. Spirou
5. The Calyx of Held Synapse
   J. G. G. Borst and S. I. Rusu
6. Synaptic Mechanisms of Coincidence Detection
   Katrina M. MacLeod and Catherine E. Carr
7. Inhibitory Neurons in the Auditory Brainstem
   Laurence O. Trussell
8. Modulatory Mechanisms Controlling Auditory Processing
   Raju Metherate
9. Mechanisms of Memory and Learning in the Auditory System
   Thanos Tzounopoulos and Ricardo M. Leão

Volume 42: Human Auditory Development (2012)

Editors: Lynne A. Werner, Richard R. Fay, and Arthur N. Popper

1. Overview and Issues in Human Auditory Development
   Lynne A. Werner
2. Morphological and Functional Ear Development
   Caroline Abdala and Douglas H. Keefe
3. Morphological and Functional Development of the Auditory Nervous System
   Jos J. Eggermont and Jean K. Moore
4. Development of Auditory Coding as Reflected in Psychophysical Performance
   Emily Buss, Joseph W. Hall III, and John H. Grose
5. Development of Auditory Scene Analysis and Auditory Attention
   Lori J. Leibold
6. Development of Binaural and Spatial Hearing
   Ruth Y. Litovsky
7. Development of Speech Perception
   Robin Panneton and Rochelle Newman
8. Development of Pitch and Music Perception
   Laurel J. Trainor and Andrea Unrau
9. Atypical Auditory Development and Effects of Experience
   Laurie S. Eisenberg, Karen C. Johnson, Sophie E. Ambrose, and Amy S. Martinez

Volume 43: The Human Auditory Cortex (2012)

Editors: David Poeppel, Tobias Overath, Arthur N. Popper, and Richard R. Fay

1. Introduction: Why Human Auditory Cortex?
   David Poeppel and Tobias Overath
2. Architecture, Connectivity, and Transmitter Receptors of Human Auditory Cortex
   Stephanie Clarke and Patricia Morosan
3. Invasive Research Methods
   Matthew A. Howard III, Kirill V. Nourski, and John F. Brugge
4. Recording Event-Related Brain Potentials: Application to Study Auditory Perception
   Claude Alain and István Winkler
5. Magnetoencephalography
   Srikantan Nagarajan, Rodney A. Gabriel, and Alexander Herman
6. Hemodynamic Imaging: Functional Magnetic Resonance Imaging
   Thomas M. Talavage, Ingrid S. Johnsrude, and Javier Gonzalez-Castillo
7. Coding of Basic Acoustical and Perceptual Components of Sound in Human Auditory Cortex
   Deborah Hall and Daphne Barker
8. Auditory Object Analysis
   Timothy D. Griffiths, Christophe Micheyl, and Tobias Overath
9. Speech Perception from a Neurophysiological Perspective
   Anne-Lise Giraud and David Poeppel
10. Central Processing of Music
    Robert J. Zatorre and Jean Mary Zarate
11. Multisensory Role of Human Auditory Cortex
    Virginie van Wassenhove and Charles Schroeder
12. Redefining the Functional Organization of the Planum Temporale Region: Space, Objects, and Sensory–Motor Integration
    Gregory Hickok and Kourosh Saberi
13. Toward a Theory of Information Processing in Auditory Cortex
    Peter Cariani and Christophe Micheyl

Volume 44: Tinnitus (2012)

Editors: Jos J. Eggermont, Fan-Gang Zeng, Arthur N. Popper, and Richard R. Fay

1. Historical Issues on Current Reflections in Tinnitus
   Jos J. Eggermont and Fan-Gang Zeng
2. Behavioral Tests for Tinnitus in Animals
   Henry E. Heffner and Rickye S. Heffner
3. Molecular Mechanism of Tinnitus
   Marlies Knipper, Marcus Müller, and Ulrike Zimmermann
4. The Cochlea and the Auditory Nerve as a Primary Source of Tinnitus
   Régis Nouvian, Michel Eybalin, and Jean-Luc Puel
5. Dorsal Cochlear Nucleus: Somatosensory–Auditory Interactions in Tinnitus
   Susanne Dehmel, Seth D. Koehler, and Susan E. Shore
6. The Inferior Colliculus: Involvement in Hyperactivity and Tinnitus
   Donald Robertson and Wilhelmina Mulders
7. Cortex: Way Station or Locus of the Tinnitus Percept?
   Jos J. Eggermont
8. Human Brain Imaging of Tinnitus
   Jennifer R. Melcher
9. The Psychophysics of Tinnitus
   Brian C. J. Moore
10. Stimulating the Auditory System to Treat Tinnitus: From Alleviating the Symptoms to Addressing the Causes
    Arnaud J. Noreña
11. Treatment: Pharmacological, Repetitive Transcranial Magnetic Stimulation, Epidural Stimulation, and Deep Brain Stimulation
    Berthold Langguth, Dirk De Ridder, Tobias Kleinjung, and Ana Belén Elgoyhen

Volume 45: Neural Correlates of Auditory Cognition (2013)

Editors: Yale E. Cohen, Arthur N. Popper, and Richard R. Fay

1. Auditory Cognition: The Integration of Psychophysics with Neurophysiology
   Yale E. Cohen
2. Processing Streams in Auditory Cortex
   Joseph P. Rauschecker

Volume 46: The Middle Ear: Science, Otosurgery, and Technology (2013)

Editors: Sunil Puria, Richard R. Fay, and Arthur N. Popper

Volume 47: Deafness (2013)

Editors: Andrej Kral, Arthur N. Popper, and Richard R. Fay

Volume 48: The Lateral Line System (2013)

Editors: Sheryl Coombs, Horst Bleckmann, Richard R. Fay, and Arthur N. Popper

1. The Gems of the Past: A Brief History of Lateral Line Research in the Context of the Hearing Sciences
   Sheryl Coombs and Horst Bleckmann
2. Morphological Diversity, Development, and Evolution of the Mechanosensory Lateral Line System
   Jacqueline F. Webb
3. The Hydrodynamic of Flow Stimuli
   Matthew J. McHenry and James C. Liao
4. The Biophysics of the Fish Lateral Line
   Sietse M. van Netten and Matthew J. McHenry
5. Sensory Ecology and Neuroethology of the Lateral Line
   John C. Montgomery, Horst Bleckmann, and Sheryl Coombs
6. Information Encoding and Processing by the Peripheral Lateral Line System
   Boris Phillipe Chagnaud and Sheryl Coombs
7. The Central Nervous Organization of the Lateral Line System
   Mario F. Wullimann and Benedikt Grothe
8. Central Processing of Lateral Line Information
   Horst Bleckman and Joachim Mogdans
9. Functional Overlap and Nonoverlap Between Lateral Line and Auditory Systems
   Christopher B. Braun and Olav Sand
10. The Hearing Loss, Protection, and Regeneration in the Larval Zebrafish Lateral Line
    Allison B. Coffin, Heather Brignull, David W. Raible, and Edwin W Rubel

Volume 49: Insights from Comparative Hearing Research (2014)

Editors: Christine Köppl, Geoffrey A. Manley, Arthur N. Popper, and Richard R. Fay

Unique Contributions from Comparative Auditory Research
Christine Köppl and Geoffrey A. Manley

Transduction and Amplification in the Ear: Insights from Insects
Ryan G. Kavlie and Joerg T. Albert

Roles for Prestin in Harnessing the Basilar Membrane to the Organ of Corti
Ian Russell

Origin and Development of Hair Cell Orientation in the Inner Ear
Ulrike J. Sienknecht

The Remarkable Ears of Geckos and Pygopods
Geoffrey A. Manley, Christine Köppl, and Ulrike J. Sienknecht

Ultrasound Detection in Fishes and Frogs: Discovery and Mechanisms
Peter M. Narins, Maria Wilson, and David A. Mann

The Malleable Middle Ear: An Underappreciated Player in the Evolution of Hearing in Vertebrates
Jakob Christensen-Dalsgaard and Geoffrey Manley

Auditory Brain Stem Processing in Reptiles and Amphibians: Roles of Coupled Ears
Katie L. Willis, Jakob Christensen-Dalsgaard, and Catherine E. Carr

Modern Imaging Techniques as a Window to Prehistoric Auditory Worlds
Stig A. Walsh, Zhe-Xi Luo, and Paul M. Barrett

Emu and Kiwi: The Ear and Hearing in Paleognathous Birds
Jeremy R. Corfield, M. Fabiana Kubke, and Christine Köppl

Diversity in Hearing in Fishes: Ecoacoustical, Communicative, and Developmental Constraints
Friedrich Ladich

Reptile Auditory Neuroethology: What Do Reptiles Do with Their Hearing?
Bruce A. Young, Nicolas Mathevon, and Yezhong Tang

Advances in Understanding the Auditory Brain of Songbirds
Janie M. Ondracek and Richard H.R. Hahnloser

Printed by Publishers' Graphics LLC
KSO140327.15.19.4